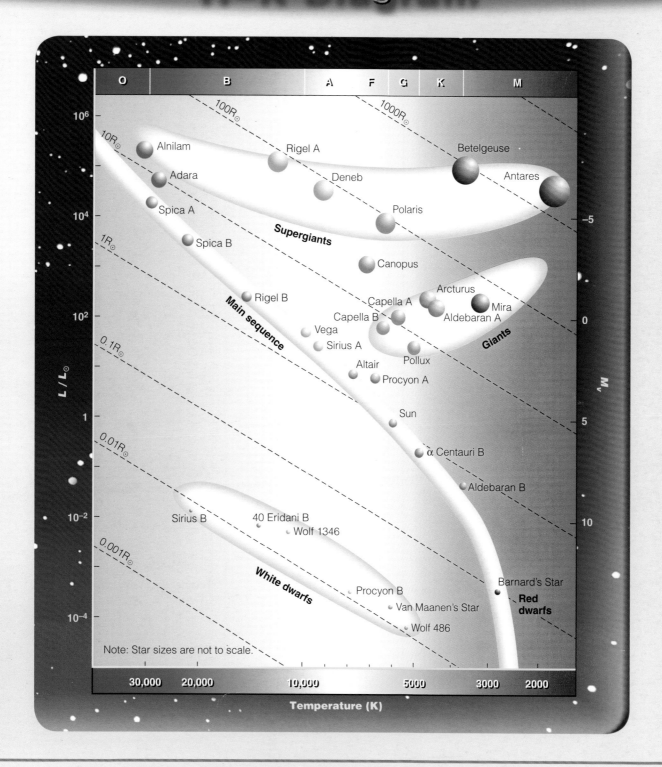

The H–R diagram is the key to understanding stars, their birth, their long lives, and their eventual deaths. Luminosity ($L / L_\odot$) refers to the total amount of energy that a star emits in terms of the Sun's luminosity, and the temperature refers to the temperature of its surface. Together, the temperature and luminosity of a star locate it on the H–R diagram and tell astronomers its radius, its family relationships with other stars, and a great deal about its history and fate.

The Terrestrial or Earthlike planets lie very close to the Sun, and their orbits are hardly visible in a diagram that includes the outer planets.

Mercury, Venus, Earth and its Moon, and Mars are small worlds made of rock and metal with little or no atmospheric gases.

The outer worlds of our Solar System orbit far from the Sun. Jupiter, Saturn, Uranus, and Neptune are Jovian or Jupiter-like planets much bigger than Earth. They contain large amounts of low-density gases.

Pluto is one of a number of small, icy worlds orbiting beyond Neptune. Astronomers have concluded that Pluto is not really a planet and now refer to it as a dwarf planet.

This book is designed to use arrows to alert you to important concepts in diagrams and graphs. Some arrows point things out, but others represent motion, force, or even the flow of light. Look at arrows in the book carefully and use this Flash Reference card to catch all of the arrow clues.

The Terrestrial Worlds

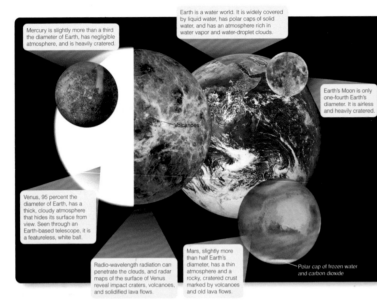

Mercury is slightly more than a third the diameter of Earth, has negligible atmosphere, and is heavily cratered.

Earth is a water world. It is widely covered by liquid water, has polar caps of solid water, and has an atmosphere rich in water vapor and water-droplet clouds.

Earth's Moon is only one-fourth Earth's diameter. It is airless and heavily cratered.

Volcanoes

Venus, 95 percent the diameter of Earth, has a thick, cloudy atmosphere that hides its surface from view. Seen through an Earth-based telescope, it is a featureless, white ball.

Radio-wavelength radiation can penetrate the clouds, and radar maps of the surface of Venus reveal impact craters, volcanoes, and solidified lava flows.

Mars, slightly more than half Earth's diameter, has a thin atmosphere and a rocky, cratered crust marked by volcanoes and old lava flows.

Polar cap of frozen water and carbon dioxide

Planetary Orbits

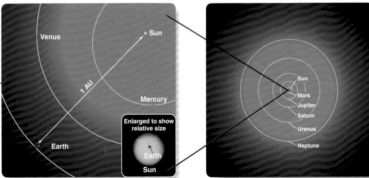

Venus

Sun

1 AU

Mercury

Earth

Enlarged to show relative size

Earth

Sun

Sun
Mars
Jupiter
Saturn
Uranus
Neptune

The Outer Worlds

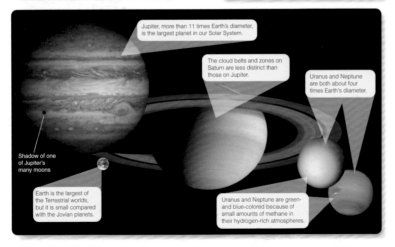

Jupiter, more than 11 times Earth's diameter, is the largest planet in our Solar System.

The cloud belts and zones on Saturn are less distinct than those on Jupiter.

Uranus and Neptune are both about four times Earth's diameter.

Shadow of one of Jupiter's many moons

Earth is the largest of the Terrestrial worlds, but it is small compared with the Jovian planets.

Uranus and Neptune are green- and blue-colored because of small amounts of methane in their hydrogen-rich atmospheres.

Point at things:

You are here

Force:

Process flow:

Measurement:

Direction:

Radio waves, infrared, photons:

Motion:

Rotation 2-D   *Rotation 3-D*   *Linear*

Light flow:

Updated arrow style

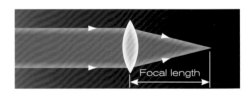

Focal length

• See pages 3 and 4 for the two orbital diagrams. See page 226 for the Terrestrial planets and page 298 for the Jovian planets.

The Sun is an ordinary star.

Nuclear reactions make energy.

Earth · · · · · Moon

The Last Inch

First hominids

One-inch line

Goal line

Ten thousand years ago, on the 0.0026 inch line, humans begin building cities and modern civilization begins.

50 40 30 20 10

Formation of the Sun and planets from a cloud of interstellar gas and dust

Life begins in Earth's oceans.

Cambrian explosion 540 million years ago: Life in Earth's oceans becomes complex.

Life first emerges onto the land.

Age of Dinosaurs

40 30 20 10

TODAY

Over billions of years, generation after generation of stars have lived and died, cooking the hydrogen and helium of the big bang into the atoms of which you are made. Study the last inch of the time line to see the rise of human ancestors and the origin of civilization. Only in the last flicker of a moment on the time line have humans begun to understand the story.

9

NINTH EDITION

The Solar System

Michael A. Seeds

Joseph R. Grundy Observatory
Franklin and Marshall College

Dana E. Backman

SOFIA (Stratospheric Observatory
for Infrared Astronomy)
SETI Institute & NASA Ames
Research Center

CENGAGE
Learning·

Australia • Brazil • Japan • Korea • Mexico • Singapore • Spain • United Kingdom • United States

The Solar System
Ninth Edition
Michael A. Seeds and Dana E. Backman

Product Director: Mary Finch

Senior Product Team Manager: Yolanda Cossio

Senior Product Manager: Aileen Berg

Content Developer: Margaret Pinette

Managing Developer, Life and Earth Sciences:
   Trudy Brown

Product Development Manager: Alex Brady

Associate Content Developer: Casey Lozier

Associate Content Developer: Kellie Petruzzelli

Product Assistant: Victor Luu

Media Developer: Stephanie Chase

Market Development Manager: Julie Schuster

Content Project Manager: Alison Eigel Zade

Senior Art Director: Cate Barr

Manufacturing Planner: Sandee Milewski

IP Analyst: Christine Myaskovsky

IP Project Manager: John Sarantakis

Production Service/Project Manager:
   Integra/Michelle Dellinger

Text and Cover Designer: Melissa Welch,
   Studio Montage

Cover Image Credit: NASA/JPL-Caltech/Space
   Science Institute

Compositor: Integra Software Services,
   Pvt. Ltd.

For product information and technology assistance, contact us at
Cengage Learning Customer & Sales Support, 1-800-354-9706

For permission to use material from this text or product,
submit all requests online at **www.cengage.com/permissions**.
Further permissions questions can be emailed to
permissionrequest@cengage.com.

Library of Congress Control Number: 2014953437

ISBN 13: 978-1-305-12076-1

Cengage Learning
20 Channel Center Street
Boston, MA 02210
USA

Cengage Learning is a leading provider of customized learning solutions with office locations around the globe, including Singapore, the United Kingdom, Australia, Mexico, Brazil and Japan. Locate your local office at **international.cengage.com/region**.

Cengage Learning products are represented in Canada by Nelson Education, Ltd.

For your course and learning solutions, visit **www.cengage.com**.

Purchase any of our products at your local college store or at our preferred online store **www.cengagebrain.com**.

Instructors: Please visit **login.cengage.com** and log in to access instructor-specific resources.

Printed in Canada
Printer Number: 01   Print Year: 2014

Dedication

In memory of Edward & Antonette Backman and Emery & Helen Seeds

Brief Contents

Contents

Part 2: The Stars

How Do We Know?

Concept Art Portfolios

Part 3: The Solar System

How Do We Know?

Concept Art Portfolios

Part 4: Life

How Do We Know?

Concept Art Portfolios

A Note to Students

From Dana and Mike

We are excited that you are taking an astronomy course and using our book. You are going to see and learn about some amazing things, from the icy rings of Saturn to monster black holes. We are proud to be your guides as you explore.

We have developed this book to help you expand your knowledge of astronomy, from recognizing the moon and a few stars in the evening sky, to a deeper understanding of the extent, power, and diversity of the universe. You will meet worlds where it rains methane, stars so dense their atoms are crushed, colliding galaxies that are ripping each other apart, and a universe that is expanding faster and faster.

Two Goals

This book is designed to help you answer two important questions:

- **What are we?**
- **How do we know?**

By the question *"What are we?"* we mean: How do we fit into the universe and its history? The atoms you are made of had their first birthday in the big bang when the universe began, but those atoms were cooked and remade inside stars, and now they are inside you. Where will they be in a billion years? Astronomy is the only course on campus that can tell you that story, and it is a story that everyone should know.

By the question *"How do we know?"* we mean: How does science work? What is the evidence, and how do we use it? For instance, how can anyone know there was a big bang? In today's world, you need to think carefully about the things so-called experts say. You should demand evidence, not just explanations. Scientists have a special way of knowing based on evidence that makes scientific knowledge much more powerful than just opinion, policy, marketing, or public relations. It is the human race's best understanding of nature. To comprehend the world around you, you need to understand how science works. Throughout this book, you will find boxes called *How Do We Know?* They will help you understand how scientists use the methods of science to know what the universe is like.

Expect to Be Astonished

One reason astronomy is exciting is that astronomers discover new things every day. Astronomers expect to be astonished. You can share in the excitement because we have worked hard to include new images, new discoveries, and new insights that will take you, in an introductory course, to the frontier of human knowledge. Huge telescopes on remote mountaintops and in space provide a daily dose of excitement that goes far beyond entertainment. These new discoveries in astronomy are exciting because they are about us. They tell us more and more about what we are.

As you read this book, notice that it is not organized as lists of facts for you to memorize. Rather, this book is organized to show you how scientists use evidence and theory to create logical arguments that explain how nature works. Look at the list of special features that follows this note. Those features were carefully designed to help you understand astronomy as evidence and theory. Once you see science as logical arguments, you hold the key to the universe.

Don't Be Humble

As teachers, our quest is simple. We want you to understand your place in the universe—your location not just in space but in the unfolding history of the physical universe. We want you not only to know where you are and what you are in the universe but also to understand how scientists know. By the end of this book, we want you to know that the universe is very big but that it is described and governed by a small set of rules and that we humans have found a way to figure out the rules—a method called science.

To appreciate your role in this beautiful universe, you need to learn more than just the facts of astronomy. You have to understand what we are and how we know. Every page of this book reflects that ideal.

Dana Backman
dbackman@sofia.usra.edu

Mike Seeds
mseeds@fandm.edu

Key Content and Pedagogical Changes for the Ninth Edition

- Every chapter has been revised and updated with new text and images regarding observatories, the heliopause, star-forming regions, supernova remnants, extrasolar planets, exploration of the surfaces of Mercury and Mars, large meteor impacts, asteroids and dwarf planets, comet nuclei, and extremophile habitats.
- Some chapters have been reorganized and rewritten to better present their topics, especially Chapter 10 ("Origin of the Solar System and Extrasolar Planets") and Chapter 13 ("Venus and Mars").
- Other chapters and sections with less substantial but still significant revisions are Chapter 7 ("Atoms and Spectra") and Chapter 15, Section 1 ("Uranus").
- The End-of-Chapter Review Questions, Discussion Questions, quantitative Problems, and Learning-to-Look questions have been substantially expanded, rewritten, and revised.
- All numerical values in the text and tables were checked and in some cases updated, figure credits were thoroughly checked and in many cases revised, and the style for figure wavelength labels was made uniform.
- The features known as *Scientific Arguments* in earlier editions were rewritten and renamed *Doing Science*.

Special Features

- *What Are We?* items are short summaries at the end of each chapter to help students see how they fit into the cosmos.
- *How Do We Know?* items are short boxes that help students understand how science works. For example, the *How Do We Know?* boxes discuss the difference between a hypothesis and a theory, the use of statistical evidence, the construction of scientific models, and so on.
- *Concept Art Portfolios* cover topics that are strongly graphic and provide an opportunity for students to create their own understanding and share in the satisfaction that scientists feel as they uncover the secrets of nature. Color and numerical keys in the introduction to the portfolios guide you to the main concepts.
- *Guideposts* on the opening page of each chapter help students see the organization of the book. The Guidepost connects the chapter with the preceding and following chapters and provides a short list of important questions as guides to the objectives of the chapter.

- *Doing Science* boxes at the end of most chapter sections begin with questions designed to put students into the role of scientists considering how best to proceed as they investigate the cosmos. These questions serve a second purpose as a further review of how we know what we know. Many of the *Doing Science* boxes end with a second question that points the student-as-scientist in a direction for investigation.
- *Celestial Profiles* of objects in our solar system directly compare and contrast planets with each other. This is the way planetary scientists understand the planets, not as isolated, unrelated bodies but as siblings with noticeable differences but many characteristics and a family history in common.
- End-of-chapter *Review Questions* are designed to help students review and test their understanding of the material.
- End-of-chapter *Discussion Questions* go beyond the text and invite students to think critically and creatively about scientific questions.
- End-of-chapter *Problems* promote quantitative understanding of the text contents.

Course Solutions That Fit Your Teaching Goals and Your Students' Learning Needs

MindTap Astronomy for Seeds/Backman, *Foundations of Astronomy,* can easily be adapted for use with Seeds/Backman, *Stars and Galaxies.* MindTap Astronomy is well beyond an eBook, a homework solution or digital supplement, a resource center website, a course delivery platform or a Learning Management System.

More than 70% of students surveyed said it was unlike anything they have seen before. MindTap is a new personal learning experience that combines all your digital assets—readings, multimedia, activities, and assessments—into a singular learning path to improve student outcomes.

MindTap Astronomy also features:

- CENGAGENOW CengageNOW™ is an integrated, online learning system that gives students 24/7 access to Study Tools and assignments. Working at their own pace, or within a schedule set up by their instructor, students can now do homework, read textbooks, take quizzes and exams, and track their grades in an easy-to-use, personalized online environment that they manage to best suit their needs.

Cognero

Cengage Learning Testing Powered by Cognero is a flexible, online system that allows you to:

- author, edit, and manage test bank content from multiple Cengage Learning solutions
- create multiple test versions in an instant
- deliver tests from your LMS, your classroom or wherever you want

Instructor Companion Site

Everything you need for your course in one place! This collection of book-specific lecture and class tools is available online via www.cengage.com/login. Access and download PowerPoint presentations, images, instructor's manual, videos, and more.

Acknowledgments

Over the years, we have had the guidance of a great many people who care about astronomy and teaching. We would like to thank all of the students and teachers who have contributed to this book. They helped shape the book through their comments and suggestions.

Many observatories, research institutes, laboratories, and individual astronomers have supplied figures and diagrams for this edition. They are listed on the credits page, and we would like to thank them specifically for their generosity.

We are happy to acknowledge the use of images and data from a number of important programs. In preparing materials for this book we used NASA's Sky View facility located at NASA Goddard Space Flight Center. We have used atlas images and mosaics obtained as part of the Two Micron All Sky Survey (2MASS), a joint project of the University of Massachusetts and the Infrared Processing and Analysis Center/California Institute of Technology, funded by the National Aeronautics and Space Administration and the National Science Foundation. A number of solar images are used courtesy of the *SOHO* consortium, a project of international cooperation between ESA and NASA. The SIMBAD database, operated at CDS, Strasbourg, France, was also used in preparation of this text.

It has been a great pleasure to work with our Cengage production team, Margaret Pinette, Victor Luu, Cate Barr, Alison Eigel Zade, and Aileen Berg, plus Michelle Dellinger of Integra and Sofia Priya Dharshini of PreMediaGlobal. Special thanks for help with this edition go to Professor Michele Montgomery of the University of Central Florida, who did a large fraction of the work needed to overhaul completely the end-of-chapter review questions and quantitative problems, and to Professor Gene Byrd of the University of Alabama, who worked effectively to review the book's page proofs.

Most of all, we would like to thank our families for putting up with "the books." They know all too well that textbooks are made of time.

Dana Backman
Mike Seeds

About the Authors

Dana Backman taught in the physics and astronomy department at Franklin and Marshall College in Lancaster, Pennsylvania, from 1991 until 2003. He invented and taught a course titled "Life in the Universe" in F&M's interdisciplinary Foundations program. Dana now teaches introductory solar system astronomy at Santa Clara University and introductory astronomy, astrobiology, and cosmology courses in Stanford University's Continuing Studies Program. His research interests focus on infrared observations of planet formation, models of debris disks around nearby stars, and evolution of the solar system's Kuiper belt. Dana is employed by the SETI Institute in Mountain View, California, as director of education and public outreach for SOFIA (the Stratospheric Observatory for Infrared Astronomy) at NASA's Ames Research Center. Dana is coauthor with Mike Seeds of *Horizons: Exploring the Universe,* 12th edition (2012); *Universe: Solar Systems, Stars, and Galaxies,* 7th edition (2012); *Stars and Galaxies,* 8th edition (2013); *The Solar System,* 8th edition (2013); and *ASTRO,* 2nd edition (2013), all published by Cengage.

Mike Seeds has been a professor of physics and astronomy at Franklin and Marshall College in Lancaster, Pennsylvania, from 1970 until his retirement in 2001. In 1989 he received F&M College's Lindback Award for Distinguished Teaching. Mike's love for the history of astronomy led him to create upper-level courses on archaeoastronomy and on the Copernican Revolution ("Changing Concepts of the Universe"). His research interests focus on variable stars and automation of astronomical telescopes. Mike is coauthor with Dana Backman of *Horizons: Exploring the Universe,* 12th edition (2012); *Universe: Solar Systems, Stars, and Galaxies,* 7th edition (2012); *Stars and Galaxies,* 8th edition (2013); *The Solar System,* 8th edition (2013); and *ASTRO,* 2nd edition (2013), all published by Cengage. He was senior consultant for creation of the 20-episode telecourse accompanying his book *Horizons: Exploring the Universe.*

Courtesy of March Dubroff

Courtesy of Kris Koenig

Here and Now 1

Guidepost As you study astronomy, you will learn about yourself. You are a planet-walker, and this chapter will give you a preview of what that means. The planet you live on whirls around a star that moves through a Universe filled with other stars and galaxies which are all results of billions of years of history and evolution. You owe it to yourself to know where you are in the Universe, and when you are in its history, because those are important steps toward knowing what you are.

In this chapter, you will consider three important questions about astronomy:

▶ **Where is Earth in the Universe?**

▶ **How does human history fit into the history of the Universe?**

▶ **Why study astronomy?**

This chapter is a jumping-off point for your exploration of deep space and deep time. The next chapter continues your journey by looking at the night sky as seen from Earth. As you study astronomy, you will see how science gives you a way to know how nature works. Later chapters will provide more specific insights into how scientists study and understand nature.

The longest journey begins with a single step.

—LAOZI

Visual

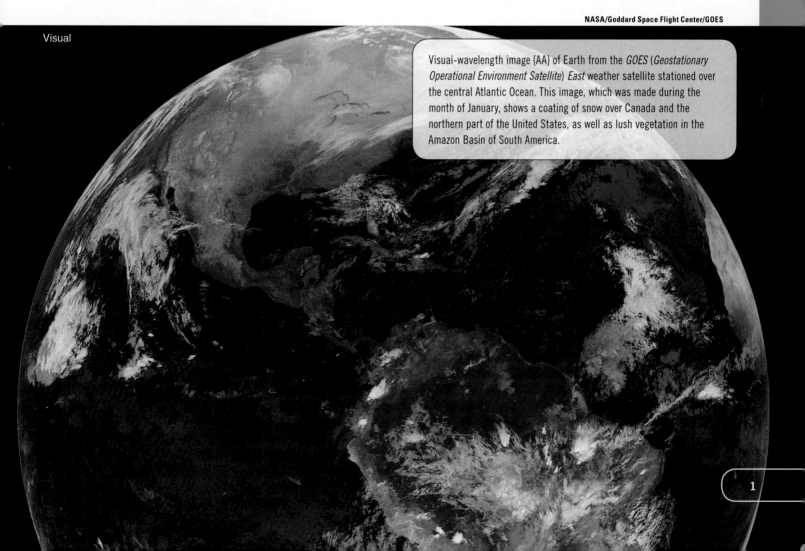

Visual-wavelength image {AA} of Earth from the *GOES* (*Geostationary Operational Environment Satellite*) *East* weather satellite stationed over the central Atlantic Ocean. This image, which was made during the month of January, shows a coating of snow over Canada and the northern part of the United States, as well as lush vegetation in the Amazon Basin of South America.

1-1 Where Are You?

To find your place among the stars, you can take a cosmic zoom—a ride out through the Universe to preview the kinds of objects you are about to study.

Begin with something familiar. **Figure 1-1** shows an area about 50 feet across on a college campus including a person, a sidewalk, and a few trees, which are all objects with sizes you can understand. Each successive picture in this "zoom" will show you a region of the Universe that is 100 times wider than the preceding picture. That is, each step will widen your **field of view**, which is the region you can see in the image, by a factor of 100.

Widening your field of view by a factor of 100 allows you to see an area 1 mile in diameter in the next image (**Figure 1-2**). People, trees, and sidewalks have become too small to discern, but now you can view an entire college campus plus surrounding streets and houses. The dimensions of houses and streets are familiar; this is still the world you know.

Before leaving this familiar territory, you need to change the units you use to measure sizes. All scientists, including astronomers, use the metric system of units because it is well understood worldwide and, more important, because it simplifies calculations. If you are not already familiar with the metric system, or if you need a review, study Appendix A (pages A-3–A-10) before reading on.

In metric units, the image in Figure 1-1 is about 16 meters across, and the 1-mile diameter of Figure 1-2 equals about 1.6 kilometers. You can see that a kilometer (abbreviated km) is a bit less than two-thirds of a mile—a short walk across a neighborhood. When you expand your field of view by another factor of 100, the neighborhood you saw in Figure 1-2 vanishes. Now your field of view is 160 km wide, and you see cities and towns as patches of gray (**Figure 1-3**). Wilmington, Delaware, is visible

▲ **Figure 1-2** This box ■ represents the relative size of the previous figure.

at the lower right. At this scale, you can see some of the natural features of Earth's surface. The Allegheny Mountains of southern Pennsylvania cross the image at the upper left, and the Susquehanna River flows southeast into Chesapeake Bay. What look like white bumps are a few puffs of cloud.

Figure 1-3 is an infrared photograph in which healthy green leaves and crops are shown as red. Human eyes are sensitive to only a narrow range of colors. As you explore the Universe, you will learn to use a wide range of other "colors," from X-rays to radio waves, to reveal sights invisible to unaided human eyes.

▲ **Figure 1-1**

▲ **Figure 1-3** This box ■ represents the relative size of the previous figure.

Visual

NASA

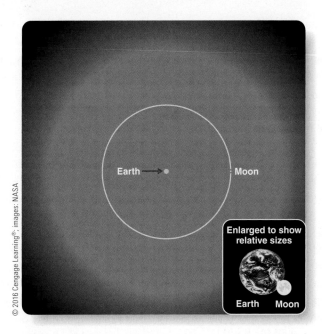

© 2016 Cengage Learning®; images: NASA

Enlarged to show
relative sizes

Earth   Moon

▲ **Figure 1-4** This box ■ represents the relative size of the previous figure.

▲ **Figure 1-5** This box ■ represents the relative size of the previous figure.

You will learn much more about infrared, X-ray, and radio energy in later chapters.

At the next step in your journey, you can see your entire planet, which is nearly 13,000 km in diameter (**Figure 1-4**). At any particular moment, half of Earth's surface is exposed to sunlight, and the other half is in darkness. As Earth rotates on its axis, it carries you through sunlight and then through darkness, producing the cycle of day and night. The blurriness at the right edge of the Earth image is the boundary between day and night—the sunset line. This is a good example of how a photo can give you visual clues to understanding a concept. Special questions called "Learning to Look" at the end of each chapter give you a chance to use your own imagination to connect images with explanations about astronomical objects.

Enlarge your field of view by another factor of 100, and you see a region 1,600,000 km wide (**Figure 1-5**). Earth is the small blue dot in the center, and the Moon, the diameter of which is only one-fourth of Earth's, is an even smaller dot along its orbit 380,000 km away. (The relative sizes of Earth and Moon are shown in the inset at the bottom right of Figure 1-5.)

The numbers in the preceding paragraph are so large that it is inconvenient to write them out. Soon you will be using numbers even larger than these to describe the Universe; rather than writing such astronomical numbers as they are in the previous paragraph, it is more convenient to write them in **scientific notation**. This is nothing more than a simple way to write very big or very small numbers without using lots of zeros. For example, in scientific notation 380,000 becomes 3.8×10^5. If you are not familiar with scientific notation, read the section on "Powers of 10 Notation" in Appendix A (pages A-4–A-5). The Universe is too big to describe without using scientific notation.

When you once again enlarge your field of view by a factor of 100 (**Figure 1-6**), Earth, the Moon, and the Moon's orbit that filled the previous figure all lie in the small red box at lower left of the new figure. Now you can see the Sun and two other planets that are part of our Solar System. Our **Solar System** consists of the Sun, its family of planets, and some smaller bodies such as moons, asteroids, and comets.

Earth, Venus, and Mercury are **planets**, which are spherical, nonluminous bodies that orbit a star and shine by reflected light. Venus is about the size of Earth, and Mercury has slightly

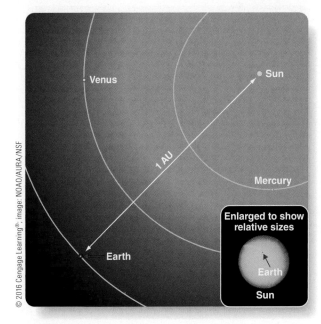

Venus

Sun

1 AU

Mercury

Earth

Enlarged to show
relative sizes

Earth

Sun

© 2016 Cengage Learning®; image: NOAO/AURA/NSF

▲ **Figure 1-6** The small red box around Earth at lower left contains the entire field of view of Figure 1-5.

more than one-third of Earth's diameter. On this diagram, they are both too small to be portrayed as anything but tiny dots. The Sun is a **star**, a self-luminous ball of hot gas. Even though the Sun is about 100 times larger in diameter than Earth (inset at bottom right of Figure 1-6), it, too, is no more than a dot in this diagram. Figure 1-6 represents an area with a diameter of 1.6×10^8 km.

Another way astronomers simplify descriptions and calculations that require large numbers is to define larger units of measurement. For example, the average distance from Earth to the Sun is a unit of distance called the **astronomical unit (AU)**; an AU is equal to 1.5×10^8 km. Using that term, you can express the average distance from Mercury to the Sun as about 0.39 AU and the average distance from Venus to the Sun as about 0.72 AU.

These distances are averages because the orbits of the planets are *not* perfect circles. This is especially apparent in the case of Mercury. Its orbit carries it as close to the Sun as 0.31 AU and as far away as 0.47 AU. You can see the variation in the distance from Mercury to the Sun in Figure 1-6. Earth's orbit is more circular than Mercury's; its distance from the Sun varies by only a few percent.

Enlarge your field of view again by a factor of 100, and you can see the entire planetary region of our Solar System (Figure 1-7). The Sun, Mercury, Venus, and Earth lie so closely together that you cannot see them separately at this scale, and they are lost in the red square at the center of this diagram that shows the size of the previous figure. You can see only the brighter, more widely separated objects such as Mars, the next planet outward. Mars is only 1.5 AU from the Sun, but Jupiter, Saturn, Uranus, and Neptune are farther from the Sun, and so they are easier to locate in this diagram. They are cold worlds that are far from the Sun's

▲ **Figure 1-8** The small red box at the center contains the entire field of view of Figure 1-7.

warmth. Light from the Sun reaches Earth in only 8 minutes, but it takes more than 4 hours to reach Neptune.

You can remember the order of the planets from the Sun outward by remembering a simple sentence such as: *My Very Educated Mother Just Served Us Noodles* (perhaps you can come up with a better one). The first letter of each word is the same as the first letter of a planet's name: Mercury, Venus, Earth, Mars, Jupiter, Saturn, Uranus, and Neptune. The list of planets once included Pluto, but in 2006, astronomers attending an international scientific congress made the decision that Pluto should be redefined as a **dwarf planet**. Although Pluto meets some of the criteria to be considered a planet, it is small and not alone in its orbit; Pluto is one of a group of small objects that have been discovered circling the Sun beyond Neptune.

When you again enlarge your field of view by a factor of 100, the Solar System vanishes (Figure 1-8). The Sun is only a point of light, and all the planets and their orbits are now crowded into the small red square at the center. The planets are too small and too faint to be visible so near the brilliance of the Sun.

Notice that no stars are visible in Figure 1-8 except for the Sun. The Sun is a fairly typical star, and it seems to be located in a fairly average neighborhood in the Universe. Although there are many billions of stars like the Sun, none is close enough to be visible in this diagram, which shows a region only 11,000 AU in diameter. Stars in the Sun's neighborhood are typically separated by distances about 30 times larger than that.

In **Figure 1-9**, your field of view has expanded again by a factor of 100 to a diameter of 1.1 million AU. The Sun is at the center, and at this scale you can see a few of the nearest stars. These stars are so distant that it is not convenient to give their distances in AU. To express distances so large, astronomers defined a new unit of distance, the light-year. One **light-year (ly)** is the distance that light travels in one year,

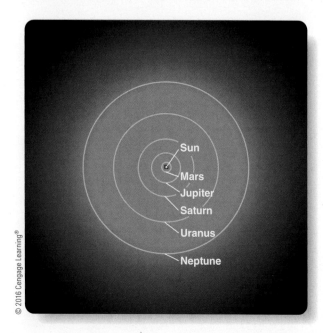

▲ **Figure 1-7** The small red box around the Sun at center contains the entire field of view of Figure 1-6.

© 2016 Cengage Learning®

▲ **Figure 1-9** This box ■ represents the relative size of the previous figure.

NOAO/AURA/NSF

Visual

▲ **Figure 1-10** This box ■ represents the relative size of the previous figure.

approximately 9.5×10^{12} km or 63,000 AU. It is a **Common Misconception** that a light-year is a unit of time, and you can sometimes hear the term misused in science fiction movies and TV shows. The next time you hear someone say, "It will take me light-years to finish my history paper," you could tell the person that a light-year is a distance, not a time (although perhaps that comment wouldn't be appreciated). The diameter of your field of view in Figure 1-9 is 17 ly.

Another **Common Misconception** is that stars look like disks when seen through a telescope. Although most stars are approximately the same size as the Sun, they are so far away that astronomers cannot see them as anything but points of light. Even the closest star to the Sun—Proxima Centauri, which is only 4.2 ly from Earth—looks like a point of light through the biggest telescopes on Earth. Figure 1-9 follows the common astronomical practice of making the sizes of the dots represent not the sizes of the stars but their brightness. This is how star images are recorded on photographs. Bright stars make larger spots on a photograph than faint stars, so the size of a star image in a photo tells you not how big the star is but rather how bright it is.

You might wonder whether other stars have families of planets orbiting around them as the Sun does. Such objects, termed **extrasolar planets**, are very difficult to see because they are generally small, faint, and too close to the glare of their respective parent stars. Nevertheless, astronomers have used indirect methods to find more than a thousand such objects, although only a handful have been photographed directly.

In **Figure 1-10**, you expand your field of view by another factor of 100, and the Sun and its neighboring stars vanish into the background of thousands of other stars. The field of view is now 1700 ly in diameter. Of course, no one has ever journeyed

thousands of light-years from Earth to look back and photograph our neighborhood, so this is a representative photograph of the sky. The Sun is a relatively faint star that would not be easily located in a photo at this scale.

If you again expand your field of view by a factor of 100, you see our galaxy, with a visible disk of stars about 80,000 ly in diameter (**Figure 1-11**). A **galaxy** is a great cloud of stars, gas, and dust held together by the combined gravity of all of its matter. Galaxies range from 1000 ly to more than 300,000 ly in diameter, and the biggest ones contain more than a trillion (10^{12}) stars. In the night sky, you can see our galaxy as a great, cloudy wheel

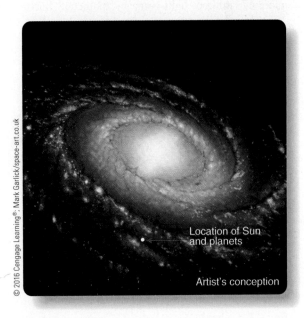

© 2016 Cengage Learning®; Mark Garlick/space-art.co.uk

Location of Sun and planets

Artist's conception

▲ **Figure 1-11** This box ■ represents the relative size of the previous figure.

of stars ringing the sky. This band of stars is known as the **Milky Way**, and our home galaxy is called the **Milky Way Galaxy**.

How does anyone know what the disk of the Milky Way Galaxy would look like from a vantage point tens of thousands of light years away? Astronomers use evidence to guide their explanations as they envision what our galaxy looks like. Artists can then use those scientific descriptions to create a painting. Many images in this book are artists' conceptions of objects and events that are too big or too dim to see clearly, emit energy your eyes cannot detect, or happen too slowly or too rapidly for humans to sense. These images are much better than guesses; they are scientifically based illustrations guided by the best information astronomers can gather. As you continue to explore, notice how astronomers use the methods of science to imagine, understand, and depict cosmic events.

The artist's conception of the Milky Way Galaxy reproduced in Figure 1-11 shows that our galaxy, like many others, has graceful **spiral arms** winding outward through its disk. In a later chapter, you will learn that the spiral arms are places where stars are formed from clouds of gas and dust. Our own Sun was born in one of these spiral arms, and, if you could see the Sun in this picture, it would be in the disk of the Galaxy about two-thirds of the way out from the center, at about the location of the marker dot indicated in the figure.

Ours is a fairly large galaxy. Only a century ago astronomers thought it was the entire Universe—an island cloud of stars in an otherwise empty vastness. Now they know that the Milky Way Galaxy is not unique; it is only one of many billions of galaxies scattered throughout the Universe.

You can see a few of these other galaxies when you expand your field of view by another factor of 100 (**Figure 1-12**). Our galaxy appears as a tiny luminous speck surrounded by other specks in a region 17 million light-years in diameter. Each speck

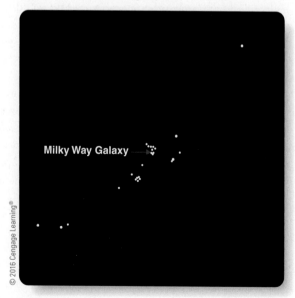

© 2016 Cengage Learning®

▲ **Figure 1-12** This box ■ represents the relative size of the previous figure.

Based on data from M. Seldner et al. 1977, *Astronomical Journal* 82, 249.

▲ **Figure 1-13** This box ■ represents the relative size of the previous figure.

represents a galaxy. Notice that our galaxy is part of a group of a few dozen galaxies. Galaxies are commonly grouped together in such clusters. Some galaxies have beautiful spiral patterns like our home, the Milky Way Galaxy, some are globes of stars without spirals, and some seem strangely distorted. In a later chapter, you will learn what produces these differences among the galaxies.

Now is a chance for you to spot another **Common Misconception**. People often say *Galaxy* when they mean *Solar System,* and they sometimes confuse both terms with *Universe*. Your cosmic zoom has shown you the difference. The Solar System is your local neighborhood, that is, the Sun and its planets, one planetary system. The Milky Way Galaxy contains our Solar System plus billions of other stars and whatever planets orbit around them—in other words, billions of planetary systems. The Universe includes everything: all of the galaxies, stars, and planets, including the Galaxy and, a very small part of that, our Solar System.

If you expand your field of view one more time, you can see that clusters of galaxies are connected in a vast network (**Figure 1-13**). Clusters are grouped into superclusters—clusters of clusters—and the superclusters are linked to form long filaments and walls outlining nearly empty voids. These filaments and walls appear to be the largest structures in the Universe. Were you to expand your field of view another time, you would probably see a uniform fog of filaments and walls. When you puzzle over the origin of these structures, you are at the frontier of human knowledge.

1-2 When Is Now?

Now that you have an idea where you are in space, you might also like to know where you are in time. The stars shone for billions of years before the first human looked up and wondered what they were. To get a sense of your place in time, all you need is a long ribbon.

Imagine stretching that ribbon from goal line to goal line down the center of a U.S. football field, a distance of 100 yards (about 91 meters), as shown on the inside front cover of this book. Imagine that one end of the ribbon represents *today*, and the other end represents the beginning of the Universe—the moment that astronomers call the *big bang*. Your ribbon represents 14 billion years, the entire history of the Universe.

Imagine beginning at the goal line labeled *BIG BANG* and replaying the entire history of the Universe as you walk along your ribbon toward the goal line labeled *TODAY*. Astronomers have evidence that the big bang initially filled the entire Universe with hot, glowing gas, but, as the gas cooled and dimmed, the Universe went dark. That all happened along the first half-inch of the ribbon. There was no light for the next 400 million years, until gravity was able to pull some of the gas together to form the first stars. That seems like a lot of years, but if you stick a little flag beside the ribbon to mark the birth of the first stars, it would be not quite 3 yards from the goal line where the Universe's history began.

You have to walk only about 4 or 5 yards along the ribbon before galaxies formed in large numbers. Our home galaxy would be one of those taking shape. By the time you cross the 50-yard line, the Universe is full of galaxies, but the Sun and Earth have not formed yet. You need to walk past the 50-yard line all the way to the other 33-yard line before you can finally stick a flag beside the ribbon to mark the formation of the Sun and planets—our Solar System—4.6 billion years ago and about 9 billion years after the big bang.

You can carry your flags a few yards further to about the 25-yard line, 3.4 billion years ago, to mark the earliest firm evidence for life on Earth—microscopic creatures in the oceans—and you have to walk all the way to the 3-yard line before you can mark the emergence of life on land only 0.4 billion (400 million) years ago. Your dinosaur flag goes inside the 2-yard line. Dinosaurs go extinct as you pass the one-half-yard line, 65 million years ago.

What about people? You can put a little flag for the first humanlike creatures, 4 million years ago, only about 1 inch (2.5 cm) from the goal line labeled *TODAY*. Civilization, the building of cities, began about 10,000 years ago, so you have to try to fit that flag in only 0.0026 inches from the goal line. That's less than the thickness of the page you are reading right now. Compare the history of human civilization with the history of the Universe. Every war you have ever heard of, the life of every person whose name is recorded, and the construction of every structure ever made from Stonehenge to the building you are in right now fits into that 0.0026 inches of the time ribbon.

Humanity is very new to the Universe. Our civilization on Earth has existed for only a flicker of an eyeblink in the history of the Universe. As you will discover in the chapters that follow, only in the last hundred years or so have astronomers begun to understand where we are in space and in time.

1-3 Why Study Astronomy?

Your exploration of the Universe will help you answer two fundamental questions:

> What are we?
> How do we know?

The question "What are we?" is the first organizing theme of this book. Astronomy is important to you because it will tell you what you are. Notice that the question is not "*Who* are we?" If you want to know who we are, you may want to talk to a paleontologist, sociologist, theologian, artist, or poet. "*What* are we?" is a fundamentally different question.

As you study astronomy, you will learn how you fit into the history of the Universe. You will learn that the atoms in your body had their birth in the big bang when the Universe began. Those atoms have been cooked and remade inside generations of stars, and now, after more than 10 billion years, they are inside you. Where will they be in another 10 billion years? This is a story everyone should know, and astronomy is the only course on campus that can tell you that story.

Every chapter in this book ends with a short segment titled "What Are We?" This summary shows how the astronomy in the chapter relates to your part in the story of the Universe.

The question "How do we know?" is the second organizing theme of this book. It is a question you should ask yourself whenever you encounter statements made by so-called experts in any field. Should you swallow a diet supplement recommended by a TV star? Should you vote for a candidate who warns of a climate crisis? To understand the world around you and to make wise decisions for yourself, for your family, and for your nation, you need to understand how science works.

You can use astronomy as a case study in science. In every chapter of this book, you will find short essays titled "How Do We Know?" They are designed to help you think not about *what* is known but about *how* it is known. To do that, these essays will explain different aspects of scientific thought processes and procedures to help you understand how scientists learn about the natural world.

Over the last four centuries, a way to understand nature has been developed that is called the **scientific method** (HOW DO WE KNOW? 1-1). You will see this process applied over and over as you read about exploding stars, colliding galaxies, and alien planets. The Universe is very big, but it is described by a small set of rules, and we humans have found a way to figure out the rules by using a method called science.

How Do We Know? 1-1

The Scientific Method

How do scientists learn about nature? You have probably heard several times during your education about the scientific method as the process by which scientists form hypotheses and test them against evidence gathered by experiments and observations. That is an oversimplification of the subtle and complex ways that scientists actually work. Scientists use the scientific method all the time, and it is critically important, but they rarely think of it while they are doing it, any more than you think about the details of what you are doing while you are riding a bicycle. It is such an ingrained way of thinking about and understanding nature that it is almost transparent to the people who use it most.

Scientists try to form hypotheses that explain how nature works. If a hypothesis is contradicted by evidence from experiments or observations, it must be revised or discarded. If a hypothesis is confirmed, it still must be tested further. In that very general way, the scientific method is a way of testing and refining ideas to better describe how nature works.

For example, Gregor Mendel (1822–1884) was an Austrian abbot who liked plants. He formed a hypothesis that offspring usually inherit traits from their parents not as a smooth blend, as most scientists of the time believed, but in discrete units according to strict mathematical rules. Mendel cultivated and tested more than 28,000 pea plants, noting which produced smooth peas and which produced wrinkled peas and how that trait was inherited by successive generations. His study of pea plants confirmed his hypothesis and allowed the development of a series of laws of inheritance. Although the importance of his work was not recognized in his lifetime, Mendel is now called the "father of modern genetics."

The scientific method is not a simple, mechanical way of grinding facts into understanding; a scientist needs insight and ingenuity both to form and to test good hypotheses. Scientists use the scientific method almost automatically, sometimes forming, testing, revising, and discarding hypotheses minute by minute as they discuss a new idea, other times spending years studying a single promising hypothesis.

The scientific method is, in fact, a combination of many ways of analyzing information, finding relationships, and creating new ideas, in order to know and understand nature. The "How Do We Know?" essays in the chapters that follow will introduce you to some of those techniques.

Whether peas are wrinkled or smooth is an inherited trait.

Inspirestock/Jupiter Images

What Are We? Participants

Astronomy will give you perspective on what it means to be here on Earth. This chapter has helped you locate yourself in space and time. Once you realize how vast our Universe is, Earth seems quite small. People on the other side of the world seem like neighbors. And, in the entire history of the Universe, the story of humanity is only the blink of an eye. This may seem humbling at first, but you can be proud of how much we humans have understood in such a short time.

Not only does astronomy locate you in space and time, it places you within the physical processes that govern the Universe. Gravity and atoms work together to make stars, generate energy, light the Universe, and create the chemical elements in your body. The chapters that follow will show how you fit into that cosmic process.

Although you are very small and your kind have existed in the Universe for only a short time, you are an important participant in something very large and beautiful.

Study and Review

Summary

▶ You surveyed the Universe by taking a cosmic zoom in which each **field of view (p. 2)** was 100 times wider than the previous field of view.

▶ Astronomers use the metric system because it simplifies calculations, and they use **scientific notation (p. 3)** for very large or very small numbers.

▶ You live on a **planet (p. 3)**, Earth, which orbits our **star (p. 4)**, the Sun, once per year. As Earth rotates once per day, you see the Sun rise and set.

▶ The Moon is approximately one-fourth the diameter of Earth, whereas the Sun is about 100 times larger in diameter than Earth—a typical size for a star.

▶ The **Solar System (p. 3)** includes the Sun at the center, all of the major planets that orbit around it—Mercury, Venus, Earth, Mars, Jupiter, Saturn, Uranus, and Neptune—plus the moons of the planets and other objects such as asteroids, comets, and **dwarf planets (p. 4)** like Pluto, bound to the Sun by its gravity.

▶ The **astronomical unit (AU) (p. 4)** is the average distance from Earth to the Sun. Mars, for example, orbits about 1.5 AU from the Sun. The **light-year (ly) (p. 4)** is the distance light can travel in one year. The nearest star is 4.2 ly from the Sun.

▶ Astronomers have found more than a thousand **extrasolar planets (p. 5)** orbiting stars other than our Sun, even though such distant and small bodies are very difficult to detect. So far only a few extrasolar planets are known to be Earth-like in size and temperature.

▶ The **Milky Way (p. 6)**, the hazy band of light that encircles the sky, is the **Milky Way Galaxy (p. 6)** seen from inside. The Sun is just one out of the billions of stars that fill the Milky Way Galaxy.

▶ **Galaxies (p. 5)** contain many billions of stars. The Milky Way Galaxy is about 80,000 ly in diameter and contains more than 100 billion stars.

▶ Some galaxies, including our own, have graceful **spiral arms (p. 6)** that are bright with stars. Many other galaxies are plain globes of stars without spiral arms, and a few galaxies have irregular shapes.

▶ Our galaxy is just one of billions of galaxies that fill the Universe in great clusters, clouds, filaments, and walls—the largest structures in the Universe.

▶ Astronomers have evidence that the Universe began about 14 billion years ago in an event called the big bang, which filled the Universe with hot gas.

▶ The hot gas cooled, the first galaxies began to form, and stars began to shine about 400 million years after the big bang.

▶ The Sun and planets of our Solar System formed about 4.6 billion years ago.

▶ Life began in Earth's oceans soon after Earth formed but did not emerge onto land until 400 million years ago, less than 1/30 of the age of the Universe. Dinosaurs evolved relatively soon after that and went extinct just 65 million years ago.

▶ Humanlike creatures developed on Earth only about 4 million years ago, less than 1/3000 of the age of the Universe, and human civilizations developed just 10,000 years ago.

▶ Although astronomy seems to be about stars and planets, it describes the Universe in which you live, so it is really about you. Astronomy helps you answer the question, "What are we?"

▶ As you study astronomy, you should ask, "How do we know?" and that will help you understand how science provides a way to understand nature.

▶ In its simplest outline, science follows the **scientific method (p. 7)**, in which scientists test hypotheses against evidence from experiments and observations. This method is a powerful way to learn about nature.

Review Questions

1. The field of view in Figure 1-2 is a factor of 100 larger than the field of view in Figure 1-1. What aspects of Figure 1-2 increased by a factor of 100 relative to Figure 1-1? Did the height increase by that amount? The diameter? The area?

2. What is the largest dimension of which you have personal sensory experience? Have you ever hiked 10 miles? Run a marathon? Driven across a continent? Flown to the opposite side of Earth?

3. What is the difference between the Solar System, the Galaxy, and the Universe?

4. What is the difference between the Moon and a moon?

5. Why do astronomers now label Pluto a "dwarf planet"?

6. Why are light-years more convenient than miles, kilometers, or AU for measuring certain distances?

7. Why is it difficult to detect extrasolar planets, that is, planets orbiting other stars?

8. What does the size of the star image in a photograph tell you?

9. What is the difference between the Milky Way and the Milky Way Galaxy?

10. When looking at the Milky Way in the night sky, are you seeing spiral arms of the Milky Way Galaxy? How do you know?

11. What are the largest known structures in the Universe?

12. Where are you in the Universe? If you had to give directions to your location in the Universe, what directions would you give?

13. What percentage is your life span compared to the age of the Solar System? Compared to the age of the Universe?

14. Why should you study astronomy? Do you anticipate needing to know astronomy 5 or 10 years from now? If so, where?

15. How does astronomy help answer the question, "What are we?"

16. **How do we know?** How does the scientific method give scientists a way to know about nature?

Discussion Questions

1. Do you think you have a responsibility to know the contents of this chapter? Are there ways this knowledge helps you enjoy a richer life and be a better citizen?

2. How is a statement in a political campaign speech different from a statement in a scientific discussion? Find examples in newspapers, magazines, and this book.

3. If *dwarf* means small, meaning dwarf planets are smaller than planets, should dwarf planets be considered planets, or not?

4. Is Earth an extrasolar planet to a planet that is orbiting around a star other than the Sun?

Problems

(Give your answers in scientific notation when appropriate.)

1. The equatorial diameter of Earth is 7928 miles. If a mile equals 1.609 km, what is Earth's diameter in kilometers? In centimeters?

2. The equatorial diameter of the Moon is 3476 kilometers. If a kilometer equals 0.6214 miles, what is the Moon's diameter in miles?

3. One astronomical unit (AU) is about 1.5×10^8 km. Explain why this is the same as 150×10^6 km.

4. A typical galaxy is shown on the first page of the Universe Bowl on the inside cover of this textbook. Express the number of stars in this typical galaxy in scientific notation.

5. The time of the Cambrian explosion is listed on the second page of the Universe Bowl on the inside cover of this textbook. Express that time in scientific notation.

6. Venus orbits 0.72 AU from the Sun. What is that distance in kilometers? (*Hint:* See Problem 3.)

7. Light from the Sun takes 8 minutes to reach Earth. How long does it take to reach Mars?

8. The Sun is almost 400 times farther from Earth than is the Moon. How long does light from the Moon take to reach Earth?

9. If the speed of light is 3.0×10^5 km/s, how many kilometers are in a light-year? How many meters? (*Hint:* How many seconds are in a year?)

10. Light from the star Betelgeuse takes 640 years to reach Earth. How far away is Betelgeuse in units of light-years? Name any historical event that was occurring on Earth at about the time the light left Betelgeuse. Is the distance to Betelgeuse unusual compared with other stars?

11. How long does it take light to cross the diameter of the Milky Way Galaxy?

12. The nearest galaxy to our home Galaxy is about 2.5 million light-years away. How many meters is that?

13. How many galaxies like our own would it take if they were placed edge-to-edge to reach the nearest galaxy? (*Hint:* See Problems 11 and 12.)

Learning to Look

1. Look at the center of Figure 1-4. Approximately what time of day is it at that location? Sunrise? Sunset? Noontime? Midnight? How do you know?

2. Look at Figure 1-6. How can you tell that Mercury does not follow a circular orbit?

3. Look at Figure 1-9. How many stars are within 5 ly of the Sun? Would that number be about the same or much different, if Earth orbited a different star than the Sun?

4. Look at Figure 1-12. Would you call the distribution of galaxies around the Milky Way Galaxy uniform? How do you know?

5. Of the objects listed here, which would be contained inside the object shown in the photograph at the right? Which would contain the object in the photo?

   star

   planet

   galaxy cluster

   supercluster filament

   spiral arm

Visual

Bill Schoening/NOAO/AURA/NSF

6. In the photograph shown here, which stars are brightest, and which are faintest? How can you tell? Why can't you tell which stars in this photograph are biggest or which have extrasolar planets?

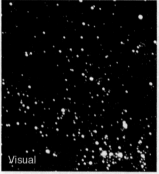

Visual

NOAO

A User's Guide to the Sky 2

Guidepost The previous chapter took you on a cosmic zoom through space and time. That quick preview prepared you for the journey to come. In this chapter you can begin your exploration by viewing the sky from Earth; as you do, consider five important questions:

▶ **How are stars and constellations named?**

▶ **How are the brightnesses of stars measured and compared?**

▶ **How does the sky appear to change and move in daily and annual cycles?**

▶ **What causes seasons?**

▶ **How do astronomical cycles affect Earth's climate?**

As you read about the sky and its motions, notice that the words often seem to imply that Earth is stationary at the center of the Universe. Remind yourself that Earth is really a planet spinning on an axis and moving in an orbit. The next chapter will introduce you to other impressive sky phenomena: phases of the Moon and eclipses.

The Southern Cross I saw every night abeam. The sun every morning came up astern; every evening it went down ahead. I wished for no other compass to guide me, for these were true.

CAPTAIN JOSHUA SLOCUM
SAILING ALONE AROUND THE WORLD

A long-exposure photograph of the Milky Way, the planet Jupiter (bright object at upper right), and the constellation Scorpius.

THE NIGHT SKY is the rest of the Universe as seen from Earth. When you look up at the stars, you are looking out through a layer of air only about 100 kilometers (60 miles) deep. Beyond that, space is nearly empty, with the planets of our Solar System several astronomical units away and the distant stars scattered many light-years apart.

As you read this chapter, you will learn about how Earth's motions affect what you can see from your planet, a moving platform:

▶ Because Earth rotates on its axis once a day, the sky appears to turn around you in a daily cycle. Not only does the Sun rise in the eastern part of the sky and set in the western part, but so do the stars and other celestial objects.

▶ Because Earth revolves around the Sun once a year, different stars are visible in the night sky in an annual cycle.

▶ The sequence of seasons you experience is caused by a combination of Earth's yearly motion plus the tilt of Earth's axis relative to its orbit.

2-1 Stars and Constellations

On a dark night far from city lights, you can see a few thousand stars. Long ago, humans tried to make sense of what they saw by naming stars and groups of stars. Some of those ancient names are still in use today.

Constellations

All around the world, native cultures celebrated heroes, gods, and mythical beasts by giving their names to groups of stars—**constellations** (Figure 2-1). You should not be surprised that the star patterns generally do not look like the creatures they are named after any more than Columbus, Ohio, looks like Christopher Columbus. The constellations named within Western culture originated in the civilizations of Assyria, Babylon, Egypt, and Greece more than 3000 years ago.

Different cultures grouped stars and named constellations differently. The constellation you call Orion was known in antiquity as Al Jabbār (the Giant) to the Arabs, as the White Tiger to the Chinese, and as Prajapati (a deity in the form of a stag) in India. The Pawnee Indians saw the constellation Scorpius as two groupings: The long tail of the scorpion was the Snake, and the bright stars at the tip of the scorpion's tail were the Two Swimming Ducks. In Hawai'i, the scorpion's tail was Maui's Fishhook that pulled the islands up from the bottom of the ocean.

On the other hand, many cultures, including the ancient Greeks, northern Asians, and Native Americans, all associated the stars in and around the Big Dipper with the figure of a bear. The concept of the celestial bear may have crossed the land bridge into North America with the first Americans more than 12,000 years ago. Hence, the names of some of the groups of stars you see in the sky may be among the oldest surviving traces of human culture.

Originally, constellations were simply loosely defined grouping of bright stars. Many of the fainter stars were not included in any constellation, and stars in the southern sky, not visible to

◀ **Figure 2-1** The constellations are an ancient heritage handed down for thousands of years as celebrations of mythical heroes and monsters. Here, Sagittarius and Scorpio appear above the southern horizon.

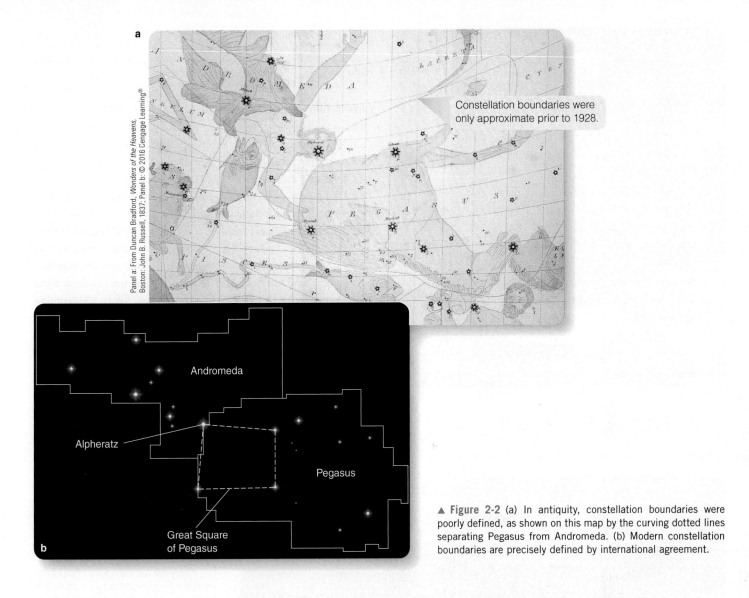

Panel a: From Duncan Bradford, *Wonders of the Heavens*, Boston: John B. Russell, 1837; Panel b: © 2016 Cengage Learning®

Constellation boundaries were only approximate prior to 1928.

▲ Figure 2-2 (a) In antiquity, constellation boundaries were poorly defined, as shown on this map by the curving dotted lines separating Pegasus from Andromeda. (b) Modern constellation boundaries are precisely defined by international agreement.

early astronomers observing from northern latitudes, were not included on their star maps. Constellation boundaries, when they were defined at all, were only approximate (Figure 2-2a), so a star like Alpheratz could be thought of as both part of Pegasus and part of Andromeda. To correct these gaps and ambiguities, modern astronomers invented more constellations, and in 1928 the **International Astronomical Union (IAU)** established 88 official constellations with carefully defined boundaries (Figure 2-2b) that together include every part of the sky. (The IAU is the same organization that redefined Pluto as a dwarf planet in 2006, as mentioned in Chapter 1.) Consequently, a constellation now represents not a group of stars, but a certain area of the sky, such that any star within the area belongs to just that one constellation. Now, Alpheratz is only in Andromeda.

In addition to the 88 official constellations, the sky contains a number of less formally defined groupings called **asterisms**. The Big Dipper, for example, is a well-known asterism that is part of the constellation Ursa Major (the Great Bear). Another asterism is the Great Square of Pegasus (Figure 2-2b) that

includes three stars from Pegasus plus Alpheratz from Andromeda. You can introduce yourself to the brighter constellations and asterisms using the star charts in Appendix B (pages A-11–A-13).

Although constellations and asterisms are groups of stars that appear close together in the sky, it is important to remember that most are made up of stars that are not physically associated with one another. Some stars may be many times farther away than others and moving through space in different directions. The only thing they have in common is that they happen to lie in approximately the same direction as seen from Earth (Figure 2-3).

Star Names

In addition to naming groups of stars, early astronomers gave individual names to the brightest stars. Modern astronomers still use many of those ancient names. Although the constellation names came from Greek translated into Latin—the languages of science until the 19th century—most individual star names come from Arabic and have been altered through the passing centuries. The name of Betelgeuse, the bright orange

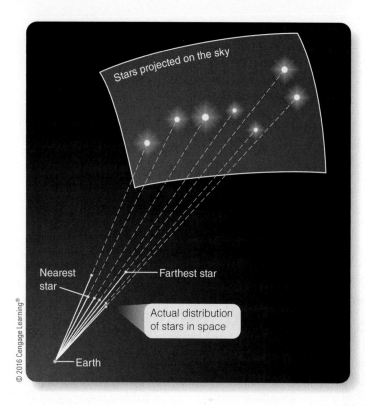

◀ **Figure 2-3** You see the Big Dipper in the sky because you are looking through a group of stars scattered through space at different distances from Earth. You view them as if they were projected on a screen, and they form the shape of the Dipper.

star in Orion, for example, comes from the Arabic *yad al-jawza*, meaning "Hand of Jawza [Gemini and Orion]." Names such as Sirius (Scorcher) and Aldebaran (the Follower [of the Pleiades]) are intriguing additions to the mythology of the sky.

Naming individual stars is not very helpful because you can see thousands of them. How many names could you possibly remember? Also, a simple name gives you little or no information about the star itself. A more useful way to identify stars is to assign letters to the bright stars in a constellation in approximate order of brightness. Astronomers use the Greek alphabet for this purpose. Thus, the brightest star in a constellation is usually designated Alpha, the second brightest Beta, and so on. Often the name of the Greek letter is spelled out, as in "Alpha," but sometimes the actual Greek letter is used, especially in charts. You can find the Greek alphabet in Appendix A (page A-9). For many constellations, the letters follow the order of brightness, but some constellations—by tradition, mistake, or the personal preferences of early chart makers—are exceptions, for example, Orion (Figure 2-4).

To identify a star by its Greek-letter designation, you would give the Greek letter followed by the genitive (possessive) form

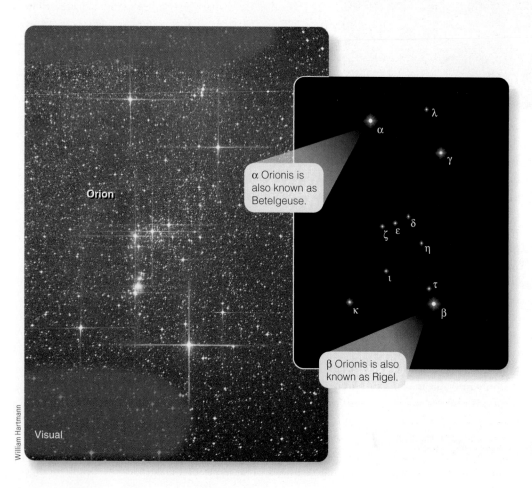

◀ **Figure 2-4** The stars in Orion do not quite follow the rule for assigning Greek letters in order of decreasing brightness. For example, β (Beta) is brighter than α (Alpha), and κ (Kappa) is brighter than η (Eta). Fainter stars do not have Greek letters or names, but if they are located inside the constellation boundaries, they are part of the constellation. The brighter stars in a constellation often also have individual names derived from Arabic. (The spikes on the star images in the photograph were produced by an optical effect in the telescope.)

of the constellation name; for example, the brightest star in the constellation Canis Major is Alpha Canis Majoris, which can also be written as α Canis Majoris. This name identifies the star and the constellation and gives a clue to the star's relative brightness. Compare this with the ancient "personal" name for this star, Sirius, which tells you nothing about its location or brightness.

Favorite Stars

It is fun to know the names of the brighter stars, but they are more than points of light in the sky. They are glowing spheres of gas resembling the Sun, each with its unique characteristics. Figure 2-5 identifies eight bright stars that can be adopted as Favorite Stars. As you study astronomy you will discover their colorful personalities and enjoy finding them in the evening sky. You will learn, for example, that Betelgeuse is not just an orange point of light but is an aging, cool star more than 500 times larger than the Sun. As you explore further in later chapters, you may want to add more Favorite Stars to your list.

You can use the star charts at the end of this book to help you locate these Favorite Stars. You can see Polaris throughout the year from the Northern Hemisphere, but Sirius, Betelgeuse, Rigel, and Aldebaran are only in the winter sky. Spica is a summer star, and Vega is visible in summer and fall evenings. Alpha Centauri, only 4.4 ly away, is the bright star that is nearest to us, but you have to travel to the latitude of southern Florida to glimpse it above the southern horizon.

Star Brightness

Astronomers describe the brightness of stars using the **magnitude scale**, a system that first appeared in the writings of the astronomer Claudius Ptolemaeus (pronounced TAHL-eh-MAY-us) about the year 140. The magnitude system probably originated even earlier; many historians attribute its invention to the Greek astronomer Hipparchus (about 190–120 BCE) who compiled the first known star catalog. Almost 300 years later, Ptolemy used the magnitude scale in his catalog, which was substantially based on Hipparchus's previous work, and successive generations of astronomers have continued to use their system.

Those early astronomers divided the stars into six classes. The brightest stars were called first-magnitude stars and the next brightest set, second-magnitude stars. The scale continued downward to sixth-magnitude stars, the faintest visible to the human eye. Thus, the larger the magnitude number, the fainter the star. This might make sense if you think of the brightest stars as first-class stars and the faintest visible stars as sixth-class stars.

Ancient astronomers could only estimate magnitudes by eye, but modern astronomers can use scientific instruments to measure the brightness of stars to high precision; so they have carefully redefined the magnitude scale. For example, instead of saying that the star known by the charming name Chort (Theta Leonis) is third magnitude, they can say its magnitude is 3.34. In the redefined scale, some stars are actually brighter than magnitude 1.0. For example, Favorite Star Vega (also known as Alpha Lyrae) is so bright that its magnitude, 0.03, is almost

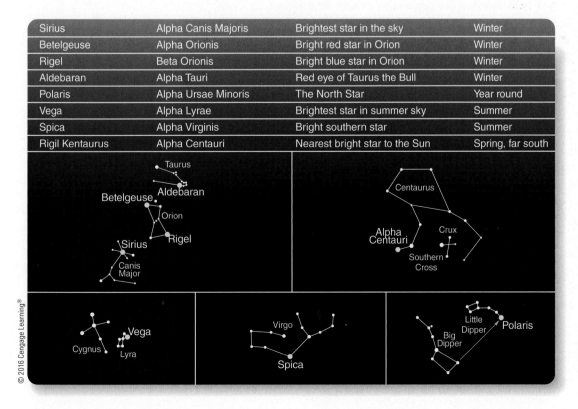

Sirius	Alpha Canis Majoris	Brightest star in the sky	Winter
Betelgeuse	Alpha Orionis	Bright red star in Orion	Winter
Rigel	Beta Orionis	Bright blue star in Orion	Winter
Aldebaran	Alpha Tauri	Red eye of Taurus the Bull	Winter
Polaris	Alpha Ursae Minoris	The North Star	Year round
Vega	Alpha Lyrae	Brightest star in summer sky	Summer
Spica	Alpha Virginis	Bright southern star	Summer
Rigil Kentaurus	Alpha Centauri	Nearest bright star to the Sun	Spring, far south

© 2016 Cengage Learning®

◀ **Figure 2-5** Favorite Stars: Locate these bright stars in the sky and learn about their characteristics. Refer to the star charts in Appendix B, pages A-11–A-13.

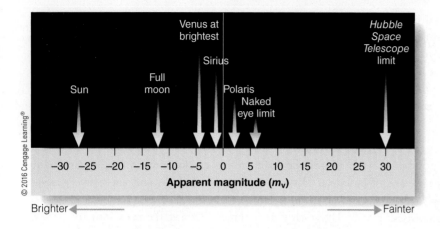

Brighter ◄——————————————————————————————► **Fainter**

◄ **Figure 2-6** The scale of apparent visual magnitudes extends into negative numbers to represent the brightest objects and to positive numbers larger than six to represent objects fainter than the unaided human eye can see.

zero. A few are so bright that the modern magnitude scale must extend into negative numbers (**Figure 2-6**). On this scale, Favorite Star Sirius—the brightest star in the sky—has a magnitude of −1.46. Modern astronomers have had to extend the faint end of the magnitude scale as well. The faintest stars you can see with your unaided eyes are about sixth magnitude, but if you use a telescope, you can detect stars much fainter than that. Magnitude numbers larger than 6 are needed to describe such faint stars.

These numbers are known as **apparent visual magnitudes** (m_V) because they describe how the stars look to human eyes observing from Earth. Although some stars emit relatively large amounts of infrared or ultraviolet light, human eyes can't see those types of radiation, and they are not included in the apparent visual magnitude. The subscript V stands for *visual* and reminds you that only visible light is included. Also, apparent visual magnitude does not take into account the distance to the stars. In other words, a star's apparent visual magnitude tells you only how bright the star looks as seen from Earth, not about its actual light output.

Magnitude and Flux

Your interpretation of brightness is quite subjective, depending on both the physiology of human eyes and the psychology of perception. As a careful investigator, you should refer to **flux**, which is a measure of the light energy from a star that hits a collecting area of one square meter in one second. Such measurements precisely and objectively define the brightness of starlight.

Astronomers use a simple formula to convert between magnitudes and flux. If two stars have fluxes F_A and F_B, then the ratio of their fluxes is F_A/F_B. To make today's measurements agree with ancient catalogs, astronomers have defined the modern magnitude scale so that two stars differing in brightness by five magnitudes have a flux ratio of exactly 100. Therefore, two stars that differ by 1 magnitude must have a flux ratio that equals the fifth root of 100, symbolized by $\sqrt[5]{100}$ or $100^{0.2}$, which is about 2.51; that is, the light from one star must be approximately 2.51 times brighter (has 2.51 times more flux arriving at Earth) than the other.

You can practice using this definition for other pairs of stars. For example, if two stars differ in brightness by 3 magnitudes they will have a flux ratio of approximately 2.51 × 2.51 × 2.51, which is 2.51^3 or about 15.8. **Table 2-1** shows the flux ratios corresponding to various magnitude differences. For example, suppose one star is third magnitude and another star is ninth magnitude. What is their flux ratio? In this case, the magnitude difference is six, and **Table 2-1** shows the equivalent flux ratio is about 251. Therefore, light from one star is about 251 times brighter than light from the other star.

A table is convenient, but for more precision you can use the relationship expressed as a simple formula. The flux ratio F_A/F_B is equal to 2.51 raised to the power of the magnitude difference $m_B - m_A$:

$$\frac{F_A}{F_B} = (2.51)^{(m_B - m_A)}$$

If, for example, the difference between the magnitudes of two stars is 6.32, then their flux ratio must be $2.51^{6.32}$. A calculator tells you the answer: 336. Star A is about 336 times brighter than Star B, meaning that the flux received on Earth from Star A is 336 times greater than that from Star B.

On the other hand, if you know the flux ratio of two stars and want to find their magnitude difference, it is convenient to rearrange the preceding formula and write it as:

$$m_B - m_A = 2.5 \log \left(\frac{F_A}{F_B} \right)$$

The expression *log* means logarithm to the base 10. For example, the light from Sirius is 24.2 times brighter than light from Polaris. Their magnitude difference is therefore 2.5 log (24.2). Your pocket calculator tells you the logarithm of 24.2 is 1.384, so the magnitude difference is 2.5 × 1.384 which equals 3.46 magnitudes. Thus, Sirius is 3.46 magnitudes brighter than Polaris.

TABLE 2-1	Magnitude Differences and Flux Ratios
Magnitude Difference	**Corresponding Flux Ratio**
0.00	1.00
1.00	2.51
2.00	6.31
3.00	15.8
4.00	39.8
5.00	100
6.00	251
7.00	631
8.00	1580
9.00	3980
10.0	10,000
:	:
:	:
15.0	1,000,000
20.0	100,000,000
25.0	10,000,000,000
:	:
:	:

© 2016 Cengage Learning®

The modern magnitude system, although seemingly complicated, has some advantages. It compresses a tremendous range of brightness into a small range of magnitudes, as you can see in Table 2-1. More important, it allows modern astronomers to measure and report the brightness of stars to high precision while remaining connected to observations of apparent visual magnitude that go back to the time of Hipparchus.

2-2 The Sky and Celestial Motions

The sky above seems like a great blue dome in the daytime and a sparkling ceiling at night. It was this domed ceiling that the first astronomers had in mind long ago as they tried to understand the Universe.

The Celestial Sphere

Ancient astronomers believed the sky was a great sphere surrounding Earth with the stars stuck on the inside like thumbtacks. Modern astronomers know that the stars are scattered through space at different distances, but it is still convenient to think of the sky as a great starry sphere enclosing Earth.

The Concept Art spread **The Sky Around You** on pages 18–19 takes you on an illustrated tour of the sky. Throughout this book, these Concept Art pages introduce new concepts and new terms through photos and diagrams, so be sure to examine them

carefully. **The Sky Around You** introduces you to three important principles and 16 new terms that will help you understand the sky:

1 The sky appears to rotate westward around Earth each day, but that is a consequence of the eastward rotation of Earth. That rotation produces day and night. Notice how reference points on the *celestial sphere* such as the *zenith, nadir, horizon, celestial equator, north celestial pole,* and *south celestial pole* define the four cardinal directions, *north point, south point, east point,* and *west point.*

2 Astronomers measure *angular distance* across the sky as angles and express them as degrees, *arc minutes,* and *arc seconds.* The same units are used to measure the *angular diameter* of an object.

3 What you can see of the sky depends on where you are on Earth. If you live in Australia, you can see many stars, constellations, and asterisms invisible from North America, but you would never see the Big Dipper. How many *circumpolar constellations* you see depends on where you are. Remember Favorite Star Alpha Centauri? It is in the southern sky and is not visible from most of the United States, but you can see it easily from Australia.

Pay special attention to the new terms on pages 18–19. You need to know these terms to describe the sky and celestial motions, but don't fall into the trap of just memorizing new terms. The goal of science is to understand nature, not to memorize definitions. Study the diagrams and see how the geometry of the celestial sphere and its apparent motions explain the changing appearance of the sky above you.

The celestial sphere is an example of a **scientific model**, a common feature of scientific thought (**How Do We Know? 2-1**). Notice that a scientific model does not have to be true to be useful. You will encounter many scientific models in the chapters that follow, and you will discover that some of the most useful models are highly simplified descriptions of the true facts.

This is a good time to consider a couple of **Common Misconceptions.** Many people, without thinking about it much, assume that the stars are not in the sky during the daytime. The stars are actually there day and night; they are just invisible during the day because the sky is lit by the Sun. Also, many people insist that Favorite Star Polaris is the brightest star in the sky. It is actually the 50th visually brightest star. Now you know that Polaris is important because of its position, not because of its brightness.

Precession

In addition to causing the obvious daily motion of the sky, Earth's rotation is connected with a very slow celestial motion that can be detected only over centuries. More than 2000 years ago, Hipparchus compared positions of some stars with their

The Sky Around You

1 The eastward rotation of Earth causes the Sun, Moon, planets, and stars to move westward in the sky as if the celestial sphere were rotating westward around Earth. From any location on Earth you see only half of the celestial sphere, the half above the **horizon**. The **zenith** marks the point of the celestial sphere directly above your head, and the **nadir** marks the point of the celestial sphere directly under your feet. The drawing at right shows the view for an observer in North America. An observer in South America would have a completely different horizon, zenith, and nadir.

The apparent pivot points are the **north celestial pole** and the **south celestial pole** located directly above Earth's north and south poles. Halfway between the celestial poles lies the **celestial equator.** Earth's rotation defines the directions you use every day: the **north point** and **south point** are the points on the horizon closest to the celestial poles, and the **east point** and the **west point** lie halfway between the north and south points. The celestial equator always meets the horizon at the east and west points.

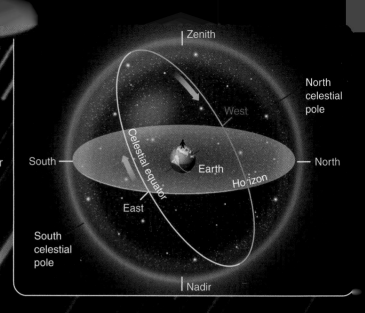

1a This time exposure of about 30 minutes shows stars as streaks, called star trails, rising behind an observatory dome lit from below by red night lights. The camera was facing northeast to take this photo. The motion you see in the sky depends on which direction you look, as shown at right. Looking north, you see Favorite Star Polaris (the North Star) located near the north celestial pole. As the sky appears to rotate westward, Polaris hardly moves, but other stars circle the celestial pole. Looking south from a location in North America you can see stars circling the south celestial pole, which is invisible below the southern horizon.

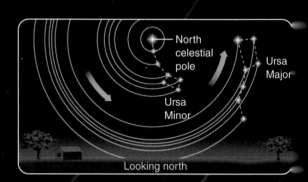

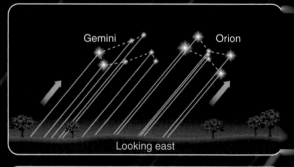

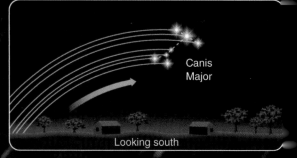

Astronomers measure distance across the sky as angles.

Angular distance

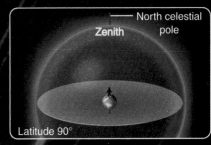

North celestial pole

Zenith

Latitude 90°

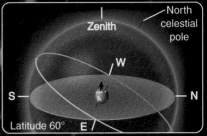

North celestial pole

Zenith

W

S — — N

E

Latitude 60°

2 Astronomers might say, "The star was two degrees from the Moon." Of course, the stars are much farther away than the Moon, but when you think of the celestial sphere, and pretend that all celestial objects are attached to it, you can measure distance on the sky as an angle. The **angular distance** between two objects is the angle between two lines extending from your eye to the two objects. Astronomers measure angles in degrees, **arc minutes** that are 1/60th of a degree, and **arc seconds** that are 1/60th of an arc minute. Using the term arc avoids confusion with minutes and seconds of time. The **angular diameter** of an object is the angular distance from one edge to the other. The Sun and Moon are each about half a degree in diameter, and the bowl of the Big Dipper is about 10 degrees wide.

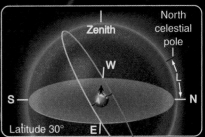

North celestial pole

Zenith

W

L

S — — N

E

Latitude 30°

3 What you see in the sky depends on your latitude, as shown at right. Imagine that you begin a journey in the ice and snow at Earth's North Pole with the north celestial pole directly overhead. As you walk southward, the celestial pole moves toward the horizon, and you can see further into the southern sky. The angular distance (L) from the horizon to the north celestial pole shown in the middle panel always equals your latitude—an important basis for celestial navigation. As you cross Earth's equator, the celestial equator would pass through your zenith, and the north celestial pole would sink below your northern horizon.

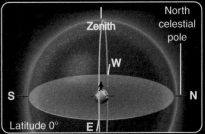

North celestial pole

Zenith

W

S — — N

E

Latitude 0°

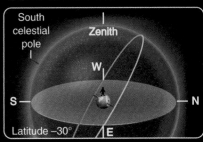

South celestial pole

Zenith

W

S — — N

E

Latitude −30°

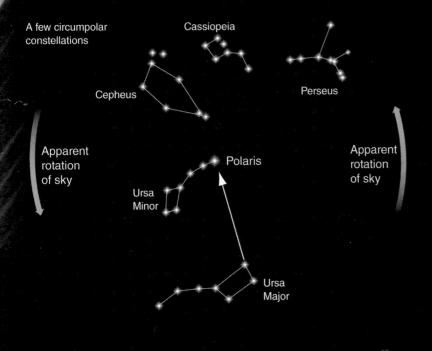

A few circumpolar constellations

Cassiopeia

Cepheus

Perseus

Apparent rotation of sky

Apparent rotation of sky

Polaris

Ursa Minor

Ursa Major

3a **Circumpolar constellations** are those that never rise or set. From mid-northern latitudes, as shown at left, you see a number of familiar constellations circling Polaris and never dipping below the horizon. As Earth turns and the sky appears to rotate, the pointer stars at the front of the Big Dipper always point approximately toward Polaris. Circumpolar constellations near the south celestial pole never rise as seen from mid-northern latitudes. From a high northern latitude location such as Norway (second panel from top), you would have more circumpolar constellations, and from Quito, Ecuador, located on Earth's equator (second panel from bottom), you would have no circumpolar constellations at all.

How Do We Know? 2-1

Scientific Models

How can a scientific model be useful if it is not entirely true? A scientific model is a carefully devised conception of how something works; that is, a framework that helps scientists think about some aspect of nature, just as the celestial sphere helps astronomers think about the motions of the sky.

Chemists, for example, use colored balls to represent atoms and sticks to represent the bonds between them, kind of like Tinkertoys. Using these molecular models, chemists can see the three-dimensional shape of molecules and understand how the atoms interconnect. The molecular model of DNA proposed by James D. Watson and Francis Crick in 1953 led to our modern understanding of the mechanisms of genetics. You have probably seen elaborate ball-and-stick models of DNA, but does the molecule really look like Tinkertoys? No, but the model is both simple enough and

accurate enough to help scientists think productively about the molecule.

A scientific model is not a statement of truth; it does not have to be precisely true to be useful. In an idealized model, some complex aspects of nature can be simplified or omitted. The ball-and-stick model of a molecule doesn't show the relative strength of the chemical bonds, for instance. A model gives scientists a way to think about some aspect of nature but need not be true in every detail.

When you use a scientific model, it is important to remember the limitations of that model. If you begin to think of a model as true, it can be misleading instead of helpful. The celestial sphere, for instance, can help you think about the sky, but you must remember that it is only a model. The Universe is much larger and much more interesting than this early scientific model of the heavens.

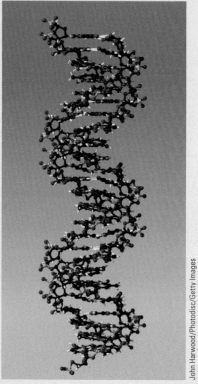

Balls represent atoms and rods represent chemical bonds in this model of a DNA molecule.

John Harwood/Photodisc/Getty Images

---

positions recorded nearly two centuries previously and realized that the celestial poles and equator were slowly moving across the sky. Later astronomers understood that this motion is caused by a toplike motion of Earth known as **precession**.

If you have ever played with a gyroscope or top, you have seen how the spinning mass resists any sudden change in the direction of its axis of rotation. The more massive the top and the more rapidly it spins, the more it resists your efforts to twist it out of position. You may recall that even the most rapidly spinning top slowly swings its axis around in a circle. The weight of the top tends to make it tip over, and this combines with its rapid rotation to make its axis sweep out the shape of a cone. That motion is precession (**Figure 2-7a**). In later chapters, you will learn that many celestial bodies precess.

Earth spins like a giant top, but it does not spin upright in its orbit; its axis is tipped 23.4 degrees from vertical. Earth's large mass and rapid rotation keep its axis of rotation pointed toward a spot near the star Polaris, and the axis would remain

pointed constantly in that direction except for the effect of precession.

Earth has a slight bulge around its middle because of its rotation. The gravity of the Sun and Moon pull on the bulge, tending to twist Earth's axis "upright" relative to its orbit. If Earth were a perfect sphere, it would not be subjected to this twisting force. Notice that the analogy to a spinning top is not perfect; gravity tends to make a top fall over, but it tends to twist Earth upright. In both cases, the twisting of the axis of rotation combined with the rotation of the object causes precession. The precession of Earth's axis takes about 26,000 years for one cycle (Figure 2-7b).

Because the locations of the celestial poles and equator are defined by Earth's rotation axis, precession slowly moves these reference marks. You would notice no change at all from night to night or even year to year, but precise measurements can reveal the slow precession of the celestial poles and the resulting change in orientation of the celestial equator.

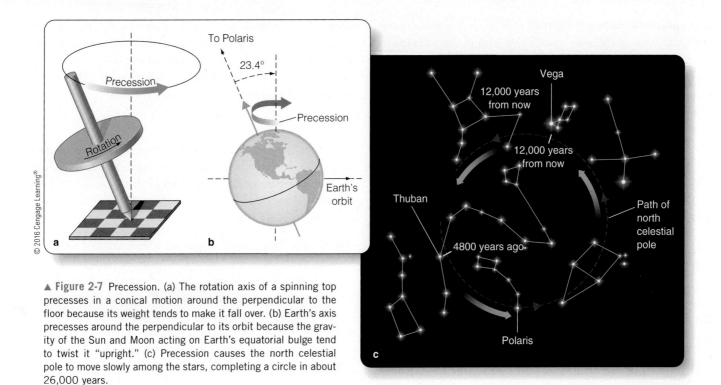

▲ **Figure 2-7** Precession. (a) The rotation axis of a spinning top precesses in a conical motion around the perpendicular to the floor because its weight tends to make it fall over. (b) Earth's axis precesses around the perpendicular to its orbit because the gravity of the Sun and Moon acting on Earth's equatorial bulge tend to twist it "upright." (c) Precession causes the north celestial pole to move slowly among the stars, completing a circle in about 26,000 years.

Over centuries, precession has significant effects. Egyptian records show that 4800 years ago, the north celestial pole was near the star Thuban (Alpha Draconis). The pole is now moving closer to Polaris and will be closest to it in about the year 2100. In about 12,000 years, the pole will have moved to within 5 degrees of Vega (Alpha Lyrae). Next time you glance at Favorite Star Vega, remind yourself that it will someday be an impressive north star. Figure 2-7c shows the path through the constellations followed by the north celestial pole during the 26,000-year precession cycle.

2-3 Sun and Planets

Earth's rotation on its axis causes the cycle of day and night, but its motion around the Sun in its orbit defines the year. Notice an important distinction. **Rotation** is the turning of a body on its axis, whereas **revolution** means the motion of a body around a point outside the body. Consequently, astronomers are careful to say Earth rotates once a day on its axis and revolves once a year around the Sun. (This may be difficult to keep straight because it is the opposite of the common English use of those words in relation to automobiles: People normally say that the tires revolve as a car moves, and every once in a while the tires have to be rotated. It would be astronomically correct to say the tires rotate as you drive, and you revolve your tires when they need it, but nobody except an astronomer would know what you meant.)

Because day and night are caused by the rotation of Earth, the time of day depends on your location on Earth. You can notice this if you watch live international news. It may be lunchtime where you are, but for a newscaster in the Middle East, it can already be dark. In **Figure 2-8**, you can see that four people in different places on Earth have different times of day.

Annual Motion of the Sun

The sky is filled with stars even in the daytime, but the glare of sunlight fills Earth's atmosphere with scattered light, and you can see only the brilliant Sun. If the Sun were fainter, you would be able to see it rise in the morning with certain stars in its background. During the day you would see the Sun and the

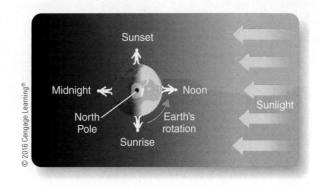

▲ **Figure 2-8** This view of Earth as if looking down from above the North Pole shows how the time of day or night depends on your location.

stars apparently moving westward, and the Sun would eventually set in front of the same stars it rose with in the morning. If you watched carefully as the day passed, you would notice that the Sun was creeping slowly eastward against the background of stars. It would move a distance roughly equal to its own diameter between sunrise and sunset. This motion is caused by the motion of Earth in its orbit around the Sun.

For example, from mid-December to mid-January, you would see the Sun in front of the constellation Sagittarius (Figure 2-9). As Earth moves along its orbit, the Sun appears to move eastward among the stars. By late February, you would see it in front of Aquarius.

Although people say the Sun is "in Sagittarius" or "in Aquarius," it isn't really correct to say the Sun is "in" any constellation. The Sun is only 1 AU away, and the stars visible in the sky are hundreds of thousands or millions of times more distant. Nevertheless, in March of each year, the Sun passes in front of the stars that make up Aquarius, and people conventionally use the expression, "The Sun is in Aquarius."

The apparent path of the Sun against the background of stars is called the **ecliptic**. If the sky were a great screen, the ecliptic would be the shadow cast by Earth's orbit. That is why the ecliptic is often called the *projection* of Earth's orbit on the sky.

Earth circles the Sun relative to the background stars in 365.26 days, and consequently the Sun appears to circle the sky, returning to the same position relative to the background stars, in the same period. That means the Sun, moving 360 degrees around the ecliptic in 365.26 days, travels approximately 1 degree eastward in 24 hours, about twice its angular diameter. You don't notice this apparent motion of the Sun because you can't see the stars in the daytime, but it does have an important consequence that you do notice—the seasons.

Seasons

Earth would not experience obvious seasons if it rotated upright in its orbit, but because its axis of rotation is tipped 23.4 degrees from the perpendicular to its orbit, it has seasons. Study **The Cycle of the Seasons** on pages 24–25 and notice two important principles plus six new terms:

1 Because Earth's axis of rotation is inclined 23.4 degrees, the Sun moves into the northern sky in the spring and into the southern sky in the fall. This is what causes the cycle of the seasons. Notice how the *vernal equinox,* the *summer solstice,* the *autumnal equinox,* and the *winter solstice* mark the beginnings of the seasons. Further, notice the very minor effects of Earth's slightly elliptical orbit as it travels from *perihelion* to *aphelion.*

2 Both of Earth's hemispheres go through cycles of seasons because of changes in the amount of solar energy they receive at different times of the year. Circulation patterns in Earth's atmosphere keep the Northern and Southern Hemispheres mostly isolated from each other, and they exchange little heat. When one hemisphere receives more solar energy than the other, it grows rapidly warmer.

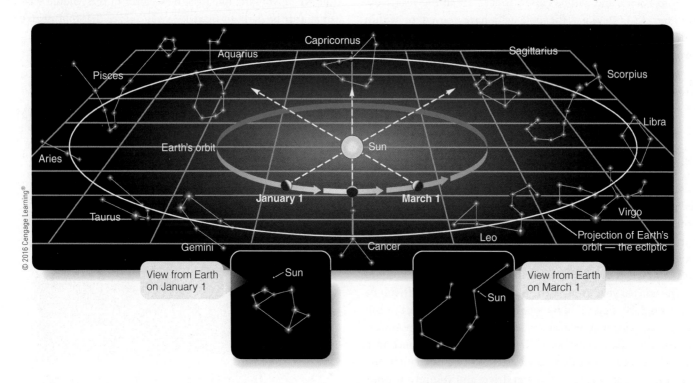

▲ Figure 2-9 Earth's orbit is a nearly perfect circle, but it is shown in an inclined view in this diagram and consequently looks oval. Earth's motion around the Sun makes the Sun appear to move against the background of the stars. Earth's orbit is thus projected on the sky as the path of the Sun, the ecliptic. If you could see the stars in the daytime, you would notice the Sun slowly crossing in front of the distant constellations as Earth moves along its orbit.

3 Notice that the seasons in Earth's Southern Hemisphere are reversed with respect to those in the Northern Hemisphere. Locations in the Southern Hemisphere experience winter from June 22 to September 22 and summer from December 21 to March 20.

Now you can set your friends straight if they mention two of the most **Common Misconceptions** about the seasons. First, the seasons are not caused by Earth moving closer to, or farther from, the Sun. If that were the cause, both of Earth's hemispheres would experience winter at the same time, when Earth is farthest from the Sun, and that's not what happens. Earth's orbit is nearly circular. Earth is actually closest to the Sun in January, but only 1.7 percent closer than the average, and 1.7 percent farther away than the average at its farthest, in July. That small variation isn't enough to cause noticeable seasons. Rather, the seasons arise because Earth's axis is not perpendicular to its orbit.

Here's a second **Common Misconception**: That it is easier to stand a raw egg on end on the day of the vernal equinox. Have you heard this one? Radio and TV personalities love to talk about it, but it just isn't true. It is one of the silliest misconceptions in science. You can stand a raw egg on end any day of the year if you have steady hands. (*Hint:* It helps to first shake the egg really hard to break the yoke inside so it can settle to the bottom.)

Throughout history, the cycle of the seasons, especially the dates of solstices and equinoxes, have been celebrated with rituals and festivals. Shakespeare's play *A Midsummer Night's Dream* describes the enchantment of the summer solstice night. (In many cultures, the equinoxes and solstices traditionally are taken to mark the midpoints, rather than the beginnings, of the seasons.) Many North American natives marked the summer solstice with ceremonies and dances. Early church officials placed Christmas day in late December to coincide with a previous celebration of the winter solstice.

Motions of the Planets

The planets of our Solar System produce no visible light of their own; they are visible only by reflected sunlight. Mercury, Venus, Mars, Jupiter, and Saturn are all easily visible to the unaided eye, but Uranus is usually too faint to be seen, and Neptune is never bright enough.

All of the planets of our Solar System, including Earth, move in nearly circular orbits around the Sun. If you were looking down on the Solar System from the north celestial pole, you would see the planets moving in the same counterclockwise direction around their orbits, with the planets farthest from the Sun moving the slowest. Seen from Earth, the outer planets move slowly eastward along the ecliptic. (You will learn about occasional exceptions to this eastward motion in Chapter 4.) In fact, the word *planet* comes from the Greek word meaning

"wanderer." Mars moves completely around the ecliptic in slightly less than 2 years, but Saturn, being farther from the Sun, takes nearly 30 years.

Mercury and Venus also stay near the ecliptic, but they move differently from the other planets. They have orbits inside Earth's orbit, which means they are never seen far from the Sun in the sky. Observed from Earth, they move eastward away from the Sun and then back toward the Sun, crossing the near part of their orbit. They continue moving westward away from the Sun and then move back, crossing the far part of their orbit before they move to the east of the Sun again. To find one of these planets, you need to look above the western horizon just after sunset or above the eastern horizon just before sunrise. Venus is easier to locate because it is brighter and because its larger orbit carries it higher above the horizon than does Mercury's (**Figure 2-10**). Mercury's orbit is smaller and it can never be farther than 28 degrees from the Sun. Consequently, Mercury is hard to see against the glare of the sky near the Sun, and also it is often hidden in the clouds and haze near the horizon.

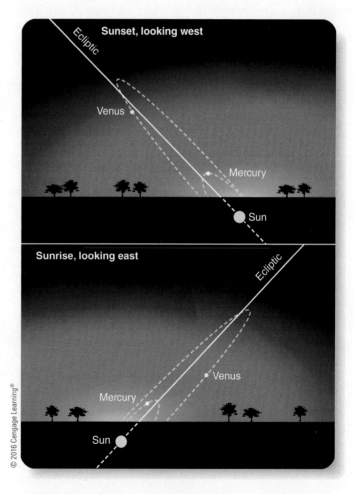

▲ **Figure 2-10** Mercury and Venus follow orbits that keep them near the Sun, and they are visible only soon after sunset or before sunrise when the brilliant Sun is hidden below the horizon. Venus takes 584 days to move from the morning sky to the evening sky and back again, but Mercury zips around in only 116 days.

The Cycle of the Seasons

1 You can use the celestial sphere to help you think about the seasons. The celestial equator is the projection of Earth's equator on the sky, and the ecliptic is the projection of Earth's orbit on the sky. Because Earth is tipped in its orbit, the ecliptic and equator are inclined to each other by 23.4 degrees, as shown at right. As the Sun moves eastward around the sky, it spends half the year in the southern half of the sky and half the year in the northern half. That causes the seasons.

The Sun crosses the celestial equator going northward at the point called the **vernal equinox**. The Sun is at its farthest north at the point called the **summer solstice**. It crosses the celestial equator going southward at the **autumnal equinox** and reaches its most southern point at the **winter solstice**.

North celestial pole

Celestial equator

Autumnal equinox

Winter solstice

Ecliptic

23.4°

Vernal equinox

Summer solstice

South celestial pole

1a The seasons are defined by the dates when the Sun crosses these four points, as shown in the table at the right. *Equinox* comes from the word for "equal"; the day of an equinox has equal amounts of daylight and darkness. *Solstice* comes from the words meaning "Sun" and "stationary." *Vernal* comes from the word for "green." The "green" equinox marks the beginning of spring in the Northern Hemisphere.

Event	Date*	N. Hemisphere
Vernal equinox	March 20	Spring begins
Summer solstice	June 22	Summer begins
Autumnal equinox	September 22	Autumn begins
Winter solstice	December 21	Winter begins

* Give or take a day due to leap year and other factors.

1b On the day of the summer solstice in late June, Earth's Northern Hemisphere is inclined toward the Sun, and sunlight shines almost straight down at northern latitudes. At southern latitudes, sunlight strikes the ground at an angle and spreads out. North America has warm weather, and South America has cool weather.

Earth's axis of rotation points toward Polaris, and, like a top, the spinning Earth holds its axis fixed as it orbits the Sun. On one side of the Sun, Earth's Northern Hemisphere leans toward the Sun; on the other side of its orbit, it leans away. The direction of the axis of rotation does not change during the year.

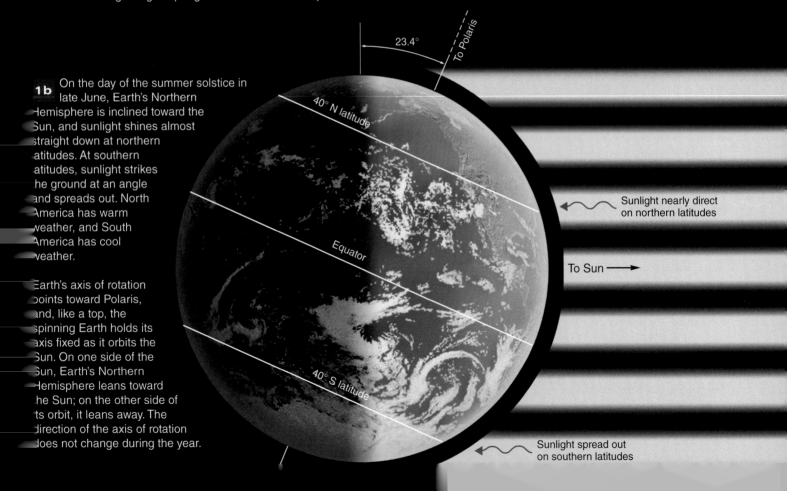

23.4°

To Polaris

40° N latitude

Equator

40° S latitude

Sunlight nearly direct on northern latitudes

To Sun →

Sunlight spread out on southern latitudes

Summer solstice light

1c Light striking the ground at a steep angle spreads out less than light striking the ground at a shallow angle. Light from the summer solstice Sun strikes northern latitudes from nearly overhead and is concentrated.

Winter solstice light

Light from the winter solstice Sun strikes northern latitudes at a much steeper angle and spreads out. The same amount of energy is spread over a larger area, so the ground receives less energy from the winter Sun.

2 The two causes of the seasons in the Northern Hemisphere are shown at right. First, the noon summer Sun is higher in the sky and the winter Sun is lower, as shown by the longer winter shadows. Thus, winter sunlight is more spread out. Second, the summer Sun rises in the northeast and sets in the northwest, spending more than 12 hours in the sky. The winter Sun rises in the southeast and sets in the southwest, spending less than 12 hours in the sky. Both of these effects mean that northern latitudes receive more energy from the summer Sun, and summer days are warmer than winter days.

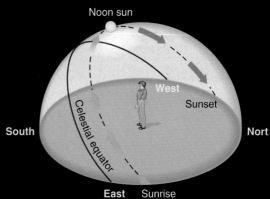

At summer solstice

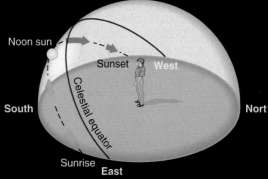

At winter solstice

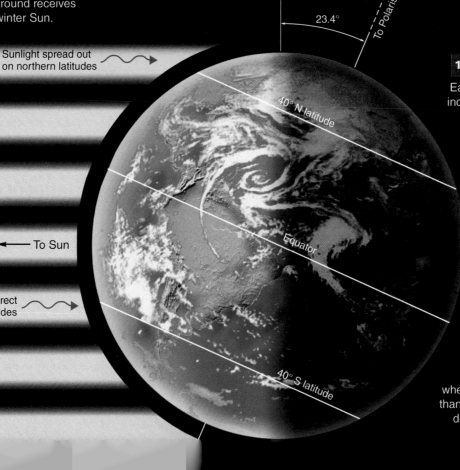

Sunlight spread out on northern latitudes

← To Sun

Sunlight nearly direct on southern latitudes

1d On the day of the winter solstice in late December, Earth's Northern Hemisphere is inclined away from the Sun, and sunlight strikes the ground at an angle and spreads out. At southern latitudes, sunlight shines almost straight down and does not spread out. North America has cool weather and South America has warm weather.

Earth's orbit is only very slightly elliptical. About January 3, Earth is at **perihelion**, its closest point to the Sun, when it is only 1.7 percent closer than average. About July 5, Earth is at **aphelion**, its most distant point from the Sun, when it is only 1.7 percent farther than average. This small variation does not significantly affect the seasons.

How Do We Know? 2-2

Pseudoscience

What is the difference between a science and a pseudoscience? Astronomers have a low opinion of beliefs such as astrology, mostly because they are groundless but also because they *pretend* to be a science. They are pseudosciences, from the Greek *pseudo*, meaning false.

A **pseudoscience** is a set of beliefs that appears to include scientific ideas but fails to follow the most basic rules of science. For example, in the 1970s a claim (in other words, a hypothesis) was made that pyramidal shapes focus cosmic forces on anything underneath and might even have healing properties. Supposedly, a pyramid made of paper, plastic, or other materials would preserve fruit, sharpen razor blades, and do other miraculous things. Many books promoted the idea of the special power of pyramids, and this idea led to a popular fad.

A key characteristic of science is that its claims can be tested and verified. In this case, simple experiments showed that any shape, not just a pyramid, protects a piece of fruit from airborne spores and allows it to dry without rotting. Likewise, any shape prevents

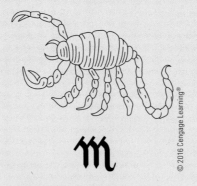

Astrology may be the oldest pseudoscience.

© 2016 Cengage Learning®

air flow and slows oxidation degradation of a razor blade's sharpness. Because experimental evidence contradicted the claim and because supporters of the hypothesis declined to abandon or revise their claims, you can recognize pyramid power as a pseudoscience. Disregard of contradictory evidence and alternate hypotheses is a sure sign of a pseudoscience.

Pseudoscientific claims can be self-fulfilling. For example, some believers in

pyramid power slept under pyramidal tents to improve their rest. There is no physical mechanism by which such a tent could affect a sleeper, but because people wanted and expected the claim to be true they reported that they slept more soundly. Vague claims based on personal testimony that cannot be tested are another sign of a pseudoscience.

Astrology is probably the best-known pseudoscience. It has been tested over and over for centuries, and it simply does not work: It has been proven beyond any reasonable doubt that there is no connection between the positions of the Sun, Moon, and planets with people's personalities or events in their lives. Nevertheless, many people believe in astrology despite contradictory evidence.

Pseudosciences appeal to our need to understand and control the world around us. Some such claims involve medical cures, ranging from using magnetic bracelets and crystals to focus mystical power, to astonishingly expensive, illegal, and dangerous treatments for cancer. Logic is a stranger to pseudoscience, but human fears and needs are not.

By tradition, any planet visible in the evening sky is called an **evening star**, even though planets are not stars. Similarly, any planet visible in the sky shortly before sunrise is called a **morning star**. Perhaps the most beautiful is Venus, which can become as bright as magnitude –4.7. As Venus moves around its orbit, it can dominate the western sky each evening for many weeks, but eventually its orbit carries it back toward the Sun, and it is lost in the haze near the horizon. In a few weeks, Venus reappears in the dawn sky, a brilliant morning star.

The cycles of the sky are so impressive that it is not surprising that people have strong feelings about them. Ancient peoples saw the motion of the Sun around the ecliptic as a powerful influence on their daily lives, and the motion of the planets along the ecliptic seemed similarly meaningful. The ancient superstition of astrology is based on the cycles of the Sun and planets around the sky. You have probably heard of the **zodiac**, a band around the sky extending about 9 degrees above and below the ecliptic in which the Sun, Moon, and planets are always found. The signs of the zodiac take their names from the 12 principal constellations along the ecliptic. A **horoscope** is just a diagram showing the location of the Sun, Moon, and planets around the ecliptic and their position above or below the horizon for a given date and

time. Centuries ago, astrology was an important part of astronomy, but the two are now almost exact opposites—astronomy is a science that depends on evidence, and astrology is a superstition that survives despite evidence (**How Do We Know? 2-2**). The signs of the zodiac are no longer important in astronomy.

2-4 Astronomical Influences on Earth's Climate

The seasons are produced by the annual motion of Earth around the Sun, but subtle changes in that motion can have dramatic effects on climate. You don't notice these changes during your lifetime, but, over thousands of years, they can bury continents under glaciers.

Earth has gone through ice ages when the worldwide climate was cooler and dryer and thick layers of ice covering the polar regions advanced multiple times partway to the equator and then retreated. The most recent ice age began about 3 million years ago and is still going on—you know that because there is ice at the poles. You are living during one of the periodic episodes in the middle of an ice age when Earth grows slightly warmer and the glaciers melt back closer to the poles. The current warm period began about 12,000 years ago. Between ice ages, Earth is warmer and there are no ice sheets even at the poles. Scientists have found evidence of at least four ice ages in Earth's past. One occurred 2.5 billion years ago, but the other three that have been identified most clearly have all occurred in the last billion years. There were probably others, but evidence of early ice ages is usually erased by more recent ice sheets. The lengths of ice ages range from tens of millions to hundreds of millions of years.

During ice ages, the advance and retreat of glaciers has a complicated pattern that involves cycles of about 40,000 years and 100,000 years. These cycles have no connection with the current global warming that has produced changes in Earth's climate over just a few decades. (Global warming is discussed in Chapter 11.) Evidence shows that these slow cycles of the ice ages have an astronomical origin.

Milankovitch Climate Cycles: Hypothesis

Sometimes a hypothesis is proposed long before scientists can find the critical evidence to test it. That happened in 1920 when engineer and mathematician Milutin Milankovitch proposed what became known as the **Milankovitch hypothesis**, which states that small changes in the shape of Earth's orbit and axis inclination, along with a subtle effect of precession, could combine to influence Earth's climate and cause ice ages. You should examine each of these motions separately.

First, Earth's orbit is slightly eccentric. As you have learned, Earth's distance from the Sun varies by ±1.7 percent from its average during each year's orbit. You also know that the primary cause of seasons is the tilt of Earth's axis. The variation in distance from the Sun each year has some effect on the Sun's heating of Earth, but that is minor compared with the effect of the axial tilt. However, astronomers know that because of gravitational interactions with other planets, the eccentricity of Earth's orbit varies between 0 and about 3 percent over a period of approximately 100,000 years. When the orbit has low eccentricity (is almost perfectly circular), summers in some locations can be not warm enough to melt all of the snow and ice from the previous winter, tending to make glaciers grow larger.

A second factor is the inclination of Earth's equator to its orbit, currently 23.4 degrees. Because of gravitational tugs of the Moon, Sun, and planets, this angle varies between 22 and almost 25 degrees with a period of 41,000 years. When the inclination is less, the summer/winter contrast is less and glaciers tend to grow.

A third factor is also involved. As you learned in the previous section, precession causes Earth's axis to sweep around a circle on the celestial sphere with a period of roughly 26,000 years. As a result, the points in Earth's orbit where a given hemisphere experiences each season gradually change. Northern Hemisphere summers now occur when Earth is farther from the Sun than average, making northern summers slightly cooler. (In this context, the Northern Hemisphere is the most important part of the globe: Most of the landmass where ice can accumulate is located in the north.) In 13,000 years, half a precession cycle from now, northern summers will occur on the other side of Earth's orbit where Earth is closer to the Sun. Northern summers will be slightly warmer, more able to melt all of the previous winter's snow and ice and prevent the growth of glaciers.

In 1920, Milankovitch proposed that these three cyclical factors combine with each other to produce complex periodic variations in Earth's climate that result in the advance and retreat of glaciers (**Figure 2-11a**). However, little evidence was available at the time to test the hypothesis, and scientists treated it with deep skepticism.

Milankovitch Climate Cycles: Evidence

By the mid-1970s, Earth scientists were able to collect the data that Milankovitch had lacked. Oceanographers drilled deep into the seafloor to collect long cores of sediment. In the laboratory, geologists could take samples from different depths in the cores and determine the age of the samples and the temperature of the oceans when they were deposited on the seafloor. From this, scientists constructed a history of ocean temperatures that convincingly matched the predictions of the Milankovitch hypothesis (Figure 2-11b).

The evidence seemed very strong, and by the 1980s the Milankovitch hypothesis was widely considered the leading hypothesis to explain climate cycles during the current ice age. But science follows a (mostly unstated) set of rules that holds that a hypothesis must be tested repeatedly against all available evidence (**How Do We Know? 2-3**). In 1988, scientists discovered some apparently contradictory evidence.

For 500,000 years rainwater has collected in a deep crack in a cave in Nevada called Devils Hole. That water has deposited the mineral calcite in layer on layer on the walls of the crack. It isn't easy to get to, and scientists had to dive with scuba gear to drill out samples of the calcite, but it was worth the effort. Back in the laboratory, they could determine the age of each layer in their core samples and the temperature of the rainwater that had formed the calcite in each layer. That gave them a history of temperatures at Devils Hole that spanned many thousands of years, and the results were a surprise. The new data seemed to show that Earth had begun warming up thousands of years too early for the last glacial retreat to have been caused by the Milankovitch cycles.

These contradictory findings were confusing because we humans naturally prefer certainty, but such circumstances are common in science. The disagreement between the ocean floor samples and the Devils Hole samples triggered a scramble to understand the problem. Were the age determinations of one or the other set of samples wrong? Were the calculations of prehistoric temperatures wrong? Or were scientists misunderstanding the significance of the evidence?

How Do We Know? 2-3

Evidence as the Foundation of Science

Why is evidence so important in science?
From colliding galaxies to the inner workings of atoms, scientists love to speculate and devise hypotheses, but all scientific knowledge is ultimately based on evidence from observations and experiments. Evidence is reality, and scientists constantly check their ideas against reality.

When you think of evidence, you probably think of criminal investigations in which detectives collect fingerprints and eyewitness accounts. In court, such evidence is used to try to understand the crime, but there is a key difference in how lawyers and scientists use evidence. A defense attorney can call a witness and intentionally fail to ask a question that would reveal evidence harmful to the defendant. In contrast, the scientist must be objective and not ignore any known evidence.

The attorney is presenting only one side of the case, but the scientist is searching for the truth. In a sense, the scientist must deal with the evidence as both the prosecution and the defense, and present all evidence—pro and con together—in the closing argument.

It is a characteristic of scientific knowledge that it is supported by evidence. A scientific statement is more than an opinion or a speculation because it has been tested objectively against reality.

As you read about any science, look for the evidence in the form of observations and experiments. Every theory or conclusion should have supporting evidence. If you can find and understand the evidence, the science will make sense. All scientists, from astronomers to zoologists, demand evidence. You should, too.

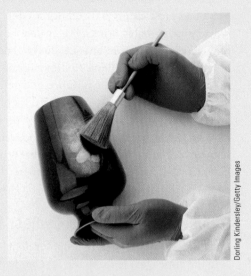

Fingerprints are evidence of past events.

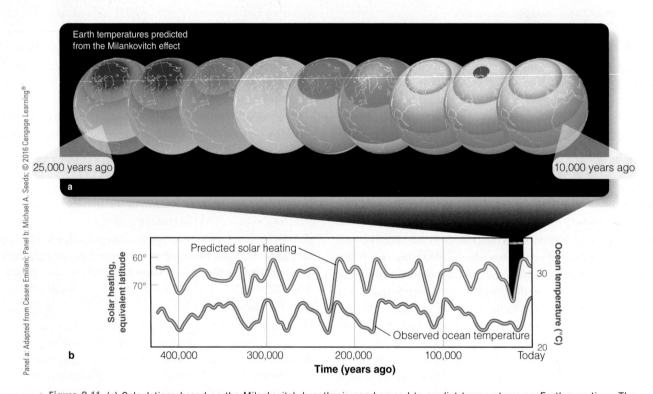

▲ **Figure 2-11** (a) Calculations based on the Milankovitch hypothesis can be used to predict temperatures on Earth over time. The warming illustrated by the Earth globes shown here took place from 25,000 to 10,000 years ago and ended the last glacial advance. Relatively cool temperatures are represented by violet and blue, warm temperatures by yellow and red. (b) Over the last 400,000 years, changes in ocean temperatures measured from fossils found in sediment layers on the seafloor approximately match calculated changes in solar heating. The globes in panel (a) illustrate events in only a short segment near the recent (right) end of the timelines.

How Do We Know? 2-4

Scientific Arguments

People in the legal profession and in science have a different meaning for the word *argument* than the sense of "verbal battle" that is often used in casual conversation. Lawyers and scientists use *argument* to mean a summary of evidence and principles leading to a conclusion.

How is a scientific argument different from a legal argument? A prosecuting attorney constructs an argument to persuade the judge or jury that the accused is guilty; a defense attorney in the same trial constructs an argument to persuade the same judge or jury toward the opposite conclusion. Neither prosecutor nor defender is obliged to consider anything that weakens their respective cases, and, in the United States, the defendant has a constitutional right not to say anything that might incriminate him or her. Lawyers rarely, if ever, include a statement in a closing argument about the possibility that that they might be wrong.

Scientists construct arguments because they want to test their own ideas and give an accurate explanation of some aspect of nature. For example, in the 1960s, biologist E. O. Wilson presented a scientific argument to show that ants communicate by smells.

The argument included a description of his careful observations and the ingenious experiments he had conducted to test his hypothesis. And, this is truly important: Wilson also considered other evidence and alternate explanations for ant communication. Scientists can include any evidence or hypothesis that supports their claim, but they must observe one fundamental rule of professional science: They must be as honest as possible; they must include all of the known evidence and all of the hypotheses previously proposed. Unlike lawyers, scientists must explicitly account for the possibility that they might be wrong.

Scientists publish their work in the form of scientific arguments, but they also think in scientific arguments. If, in thinking through his argument, Wilson had found a contradiction, he would have known he was on the wrong track. That is why scientific arguments must be complete and honest. Scientists who ignore inconvenient evidence or brush aside other explanations are only fooling themselves.

A good scientific argument gives you all the information you need to decide for yourself whether the argument is correct.

Wilson's study of ant communication is now widely understood and is being applied to other fields such as pest control and telecommunications networks.

Eye of Science/Science Source

Scientists have discovered that ants communicate with a large vocabulary of smells.

In 1997, a new study of the ages of the samples confirmed that those from the ocean floor were correctly dated. But the same study found that the ages of the Devils Hole samples were also correct. Evidently the temperatures at Devils Hole record local climate changes in the region that is now the southwestern United States. The ocean floor samples record global climate changes, and they fit well with the Milankovitch hypothesis. This has given scientists renewed confidence in the Milankovitch hypothesis in a general sense, but also made them aware that some regions of Earth do not exactly follow trends seen for the rest of the planet. Although it is widely accepted today, the Milaknovitch hypothesis is still being tested whenever scientists can find more evidence.

As you review this section, notice that it is a **scientific argument**, a careful presentation of hypothesis and evidence in a logical discussion. **How Do We Know? 2-4** expands on the ways scientists organize their ideas in logical arguments. Also, throughout this book, many chapter sections end with a short feature titled "Doing Science." These are case studies featuring one or two review questions that can be answered by imagining yourself as a scientist analyzing measurements, inventing

hypotheses, constructing scientific arguments, and so on. You can use the "Doing Science" features to review chapter material but also to practice thinking like a scientist.

DOING SCIENCE

Why was it critical, in testing the Milankovitch hypothesis about ice age climate change mechanisms, for scientists to determine the ages of ocean sediment? Ocean floors accumulate sediment in thin layers year after year. Scientists can drill into the ocean floor and collect cores of those sediment layers, and from chemical tests they can find the temperature of the seawater when each layer was deposited. Those determinations of past temperatures can be used as reality checks in the scientific argument regarding the Milankovitch hypothesis, but only if the ages of the sediment layers are determined correctly. When a conflict arose with evidence from Devils Hole in Nevada, the age determinations of the both the ocean floor and Devils Hole samples were carefully reexamined and found to be correct. After reviewing all of the evidence, scientists concluded that the ocean core samples did indeed support the Milankovitch hypothesis.

What Are We? Along for the Ride

Human civilization is spread over the surface of planet Earth like a thin coat of paint. Great cities of skyscrapers and tangles of superhighways may seem impressive, but if you use your astronomical perspective, you can see that we humans are confined to the surface of our world.

The rotation of Earth creates a cycle of day and night that controls everything from TV schedules to the chemical workings of our brains. We wake and sleep within that 24-hour cycle of light and dark. Furthermore, Earth's orbital motion around the Sun, combined with the inclination of its axis, creates a yearly cycle of seasons, and we humans, along with every other living thing on Earth, have evolved to thrive within those extremes of temperature. We protect ourselves from the largest extremes and have spread over most of Earth, hunting, gathering, and growing crops within the cycle of the seasons.

In recent times, we have begun to understand that conditions on Earth's surface are not entirely stable. Slow changes in Earth's motions and orientation of the planet produce irregular cycles of glaciation. All of recorded history, including the creation of all of the cities and roads that paint our globe, has occurred since the last glacial retreat ended only 12,000 years ago, so we humans have no recorded experience of Earth's coldest climate. We have never experienced our planet's icy personality. We are along for the ride and enjoying Earth's good times.

Study and Review

Summary

▶ Although the **constellations (p. 12)** currently used by astronomers originated in Middle Eastern and Greek mythology, the names are Latin. Even modern constellations, added to fill in the spaces between the ancient figures, have Latin names. Named groups of stars such as the Big Dipper or Orion's Belt that are not complete constellations are called **asterisms (p. 13)**.

▶ Astronomers now divide the sky into 88 constellations, defined in 1928 by the **International Astronomical Union (IAU) (p. 13)**.

▶ The names of individual stars usually come from old Arabic, though modern astronomers often refer to a bright star by its constellation plus a Greek letter assigned according to its brightness within the constellation.

▶ Astronomers describe the brightness of stars using the **magnitude scale (p. 15)**. First-magnitude stars are brighter than second-magnitude stars, which are brighter than third-magnitude stars, and so on. The magnitude describing what you see when you look at a star in the sky is its **apparent visual magnitude (m_V) (p. 16)**, which includes only types of light visible to the human eye and also does not take into account the star's distance from Earth.

▶ **Flux (p. 16)** is a measure of light energy striking one square meter per second. The magnitude of a star is related directly to the flux of light received on Earth from that star.

▶ The **celestial sphere (p. 18)** is a **scientific model (p. 17)** of the sky, to which the stars appear to be attached. Because Earth rotates eastward, the celestial sphere appears to rotate westward on its axis.

▶ The **north** and **south celestial poles (p. 18)** are the pivots on which the sky appears to rotate, and they define the four cardinal directions around the **horizon (p. 18)**: the **north**, **south**, **east**, and **west points (p. 18)**. The point directly overhead is the **zenith (p. 18)**, and the point on the sky directly underfoot is the **nadir (p. 18)**.

▶ The **celestial equator (p. 19)**, which is an imaginary line around the sky above Earth's equator, divides the sky into Northern and Southern Hemisphere.

▶ As the celestial sphere is curved, the distances between the stars "on" the sky are angular, not linear, distances. These **angular distances (p. 19)**, measured in degrees, **arc minutes (p. 19)**, and **arc seconds (p. 19)**, are not directly related to the true distance between the objects measured in units such as kilometers (km) or light-years (ly). The angular distance across an object is its **angular diameter (p. 19)**.

▶ What you see of the celestial sphere depends on your latitude. Much of the sky's Southern Hemisphere is not visible from northern latitudes. To see that part of the sky, you would have to travel southward over Earth's surface.

▶ **Circumpolar constellations (p. 19)** are those close enough to a celestial pole that they do not appear to rise from the east and set in the west.

▶ The angular distance from the horizon to the north celestial pole as measured from the north point always equals your latitude. This equality is an important basis for celestial navigation.

▶ **Precession (p. 20)** is caused by the gravitational forces of the Moon and Sun acting on the equatorial bulge of the spinning Earth, causing Earth's axis to sweep around in a conical motion like the motion of a wobbling top's axis. Earth's axis precesses with a period of 26,000 years. As a result, the positions of the celestial poles and celestial equator move slowly against the background of the stars.

▶ The **rotation (p. 21)** of Earth on its axis produces the daily cycle of day and night, and the **revolution (p. 21)** of Earth around the Sun produces the annual cycle of the seasons.

▶ Because Earth orbits the Sun, the Sun appears to move eastward along the **ecliptic (p. 22)**, through the constellations, completing a circuit of the sky in a year. Because the ecliptic is tipped 23.4 degrees to the celestial equator, the Sun spends half the year north of the celestial equator and half the year south of the celestial equator.

▶ In each hemisphere's summer, the Sun is above the horizon longer and shines more directly down on the ground. Both effects cause warmer weather in that hemisphere. In each hemisphere's winter, the Sun is above the sky fewer hours and also shines less directly than in summer, so the winter hemisphere has colder weather. When one hemisphere experiences summer, the opposite hemisphere experiences winter. When one hemisphere experiences spring, the opposite hemisphere experiences fall.

▶ The beginning of spring, summer, winter, and fall are marked by the **vernal equinox (p. 24)**, the **summer solstice (p. 24)**, the **autumnal equinox (p. 24)**, and the **winter solstice (p. 24)**.

▶ In its orbit around the Sun, Earth is slightly closer to the Sun at **perihelion (p. 25)** in January and slightly farther away from the Sun at **aphelion (p. 25)** in July. This change in distance to the Sun has almost no effect on Earth's seasons.

▶ The planets appear to move generally eastward along the ecliptic. They appear like bright, nontwinkling stars with the exception of Uranus and Neptune, which are too faint to be visible to the unaided eye. Mercury and Venus are never seen far from the Sun and are therefore seen either in the evening sky after sunset or in the dawn sky before sunrise.

▶ Planets visible in the sky at sunset are traditionally called **evening stars (p. 26)**, and planets visible in the dawn sky are called **morning stars (p. 26)** even though they are not actually stars.

▶ The locations of the Sun and planets along the **zodiac (p. 26)** are diagrammed in a **horoscope (p. 26)**, which is the basis for the ancient **pseudoscience (p. 26)** (or false science), known as astrology.

▶ According to the **Milankovitch hypothesis (p. 27)**, slow changes in the shape of Earth's orbit, the angle of axis tilt, and axis orientation can alter the planet's heat balance and cause the cycle of ice advances and retreats during an ice age. Evidence found in seafloor samples and other locations support the hypothesis, and the hypothesis is widely accepted today.

▶ Scientists routinely test their own ideas by organizing theory and evidence into a **scientific argument (p. 29)**.

Review Questions

1. Why are most of the constellations that were invented in modern times composed of faint stars or located in the southern sky?

2. How does the Greek letter designation of a star give you clues both to its location and its apparent brightness?

3. Which is the asterism and which is the constellation: Orion and Orion's belt? Name another asterism/constellation combination.

4. From your knowledge of star names and constellations, which of the following stars in each pair is probably brighter? Explain your answers.
   a. Alpha Ursae Majoris; Epsilon Ursae Majoris
   b. Epsilon Scorpii; Alpha Pegasi
   c. Alpha Telescopii; Alpha Orionis

5. How did the magnitude system originate in a classification of stars by apparent brightness?

6. What does the word *apparent* mean in *apparent visual magnitude*?

7. What does the word *visual* mean in *apparent visual magnitude*?

8. Does the apparent visual magnitude numbers of two stars take into account how far each of the stars is from Earth? Explain your answer.

9. Does the apparent visual magnitude numbers of two stars take into account the relative size of each celestial object? Explain your answer.

10. Does the apparent visual magnitude numbers of two stars tell us which star shines more light on the same area of Earth in the same time interval? Explain your answer.

11. Why doesn't a magnitude difference of one mean that the corresponding flux ratio is also one?

12. If Star B has $m_B = -2$ and Star A $m_A = 0$, from which star does Earth receive more flux? Which star emits more light at its surface? How do you know?

13. In what ways is the celestial sphere a scientific model?

14. Is the precessing top shown in Figure 2-7 an example of a scientific model? If so, which parts of the model are true and which parts are not necessarily true?

15. If Earth did not rotate, could you still define the celestial poles and celestial equator?

16. Where would you need to go on Earth if the celestial equator is to be seen very near your horizon?

17. Where would you go on Earth if you wanted to be able to see both the north celestial pole and the south celestial pole at the same time?

18. Your zenith is at your east point and your nadir is at your west point. Are you sitting, squatting, standing, lying prone on the ground, or lying supine on the ground? If your arms are outstretched and perpendicular to your body, in which direction(s) are your arms pointing?

19. Why does the number of circumpolar constellations depend on the latitude of the observer?

20. Explain two reasons why winter days are colder than summer days.

21. How does the date of the beginning of summer in Earth's Southern Hemisphere differ from the date in the Northern Hemisphere?

22. If it is the first day of spring in your hemisphere, what day is it in the opposite hemisphere?

23. It is the first day of summer. Will the days start getting longer tomorrow, start getting shorter tomorrow, or will the Sun stay up in the sky tomorrow for as long as it did today?

24. How much flux from the Sun does the Northern Hemisphere of Earth receive compared to the Southern Hemisphere on the first day of fall at noon?

25. Why does the eccentric shape of Earth's orbit make winter in Earth's Northern Hemisphere different from winter in Earth's Southern Hemisphere?

26. **How Do We Know?** How can a scientific model be useful if it is *not* a true description of nature?

27. **How Do We Know?** Why is astrology a pseudoscience?

28. **How Do We Know?** How is evidence a distinguishing characteristic of science?

29. **How Do We Know?** Why must a scientific argument dealing with some aspect of nature take all of the known evidence into account?

Discussion Questions

1. All cultures on Earth named constellations. Why do you suppose this was such a common practice?

2. If you were lost at sea in the Northern Hemisphere, you could find your approximate latitude by measuring the altitude of Polaris from the north point. Altitude is the angle above the horizon to the star. However, Polaris is not exactly at a celestial pole. What else would you need to know to be able to measure your latitude more precisely?

3. Do other planets in the Solar System, or extrasolar planets, have ecliptics? Could they have seasons? If so, name two possible sources to the seasons.

4. Alnitak, also called Zeta Orionis, is a blue supergiant in the belt of the constellation Orion. Suppose that we see this star become a supernova, and it now appears brighter than Venus in the night sky. Should we relabel the stars in Figure 2-4 to accommodate the new apparent brightness of Alnitak?

5. Why does Sirius have an $m_V = -1.46$ if ancient astronomers classed the brightest stars as first magnitude, the next brightest as second magnitude, and so on? Has Sirius changed apparent brightness since the time of the ancient astronomers?

Problems

1. Star A has a magnitude of 9.5; Star B, 5.5; and Star C, 3.5. Which is brightest? Which are visible to the unaided eye? Which pair of stars has a flux ratio of 16?

2. If one star is 7.3 times brighter than another star, how many magnitudes brighter is it?

3. If light from one star is 251 times brighter (has 251 times more flux) than light from another star, what is their difference in magnitudes?

4. If Earth receives twice as much light per unit area per unit time from Star A compared to Star B, what is the apparent visual magnitude difference between the stars? Which star is apparently brighter, Star A or Star B?

5. If Earth receives one-third as much light per unit area per unit time from Star A compared to Star B, what is the apparent visual magnitude difference between the stars? Which star is apparently brighter, Star A or Star B?

6. If in the previous problem Star A has an apparent visual magnitude of 5, what is the apparent visual magnitude of Star B? Which can you see on a clear night in your sky using your eyes: Star A, Star B, Star A and Star B, or neither Star A nor Star B?

7. If two stars differ by 8 magnitudes, what is their flux ratio?

8. If two stars differ by 5.6 magnitudes, what is their flux ratio?

9. If star A is magnitude 1.0 and star B is magnitude 9.6, which is brighter and by what factor?

10. By what factor is the full moon brighter than Venus at its brightest? (*Hint:* Refer to Figure 2-6.)

11. What is the angular distance from the north celestial pole to the point on the sky called the vernal equinox? To the summer solstice?

12. If you are at latitude 40 degrees north of Earth's equator, what is the angular distance from the northern horizon up to the north celestial pole? From the southern horizon down to the south celestial pole?

13. If you are at latitude 30 degrees north of Earth's equator, what is the angular distance from your zenith to the north celestial pole? From your nadir to the north celestial pole?

14. How many precession periods are in one "nodding" period of Earth's inclination axis and in one period of the changing shape of Earth's orbit? In the time span shown in Figure 2-11b, how many periods or fractions of periods did the Earth's axis precess, nod, and Earth's orbit change shape? Of the three periods, which is likely to have the most effect on the changes shown in Figure 2-11?

Learning to Look

1. Find the Big Dipper in the star trails photograph on the left side of **The Sky Around You** Concept Art spread.

2. Look at the five figures in **The Sky Around You**, item 3. Continue the series, drawing two more pictures. What latitudes are the next two pictures in the series? If you are at latitude −90 degrees, is your zenith the same as a person located at a latitude +90 degrees?

3. Look at **The Sky Around You**, item 2. What is the angular diameter of a typical star in the cartoon? (*Hint:* Compare the size of a star with that of the Moon in the cartoon.)

4. Look at **The Sky Around You**, item 1a. In the looking south illustration, is Canis Major a circumpolar constellation? Why or why not?

5. Look at the view from Earth on March 1 in Figure 2-9. Is the view from Earth's nighttime side or daytime side? How do you know? Which asterism or constellation is shown in this image?

6. Look at Figure 2-9. If you see Sagittarius high in your night sky on June 20 and today is your birthday, what is your zodiac constellation?

Moon Phases and Eclipses

3

Guidepost In the previous chapter, you learned about daily and yearly cycles of the Sun's appearance in Earth's sky. Now you can focus on phenomena of the next brightest object in the sky, the Moon. The Moon moves against the background of stars, changing its appearance in a monthly cycle and occasionally producing spectacular events called *eclipses*. This chapter will help you answer four important questions about Earth's natural satellite:

► **Why does the Moon go through phases?**

► **What causes a lunar eclipse?**

► **What causes a solar eclipse?**

► **How can eclipses be predicted?**

Understanding the phases of the Moon and eclipses will exercise your scientific imagination as well as help you enjoy the sight of the Moon crossing the sky.

Once you have an understanding of the sky as it appears from Earth, you will be ready to read the next chapter about how Renaissance astronomers analyzed motions of the Sun, Moon, and planets; used their imaginations; and came to a revolutionary conclusion—that Earth is a planet.

O, swear not by the moon, the fickle moon, the inconstant moon, that monthly changes in her circle orb, Lest that thy love prove likewise variable.

WILLIAM SHAKESPEARE, *ROMEO & JULIET*

Solar eclipses are dramatic. In June 2001, an automatic camera in southern Africa snapped pictures every 5 minutes as the afternoon Sun sank lower in the sky. From upper right to lower left, you can see the Moon crossing the disk of the Sun. A longer exposure was needed to record the total phase of the eclipse.

2001 F. Espanak, www.Mr.Eclipse.com

Visual images

SINCE ANCIENT TIMES, superstitious people have associated the Moon with insanity. The word *lunatic* comes from a time when even doctors thought that the insane were "moonstruck." Of course, the Moon does not cause madness, but it is bright and beautiful, moves relatively rapidly across the constellations, and is associated with eclipses and tides, so people might *expect* it to have a dramatic effect on them.

3-1 The Changeable Moon

Starting this evening, begin looking for the Moon in the sky. You might have to wait for almost a month before you find it appearing at a convenient time, but then, as you watch for the Moon on successive evenings, you will see it following its orbit around Earth and cycling through its phases as it has done for billions of years.

The Moon's Orbital Motion

Just as Earth would be seen to revolve counterclockwise around the Sun if viewed from the direction of the north celestial pole, the Moon revolves counterclockwise around Earth. Because the Moon's orbit is tipped about 5 degrees from the plane of Earth's orbit, the Moon's path is always near the ecliptic, sometimes slightly north of it and sometimes slightly south of it.

The Moon moves rapidly against the background of the constellations. If you watch the Moon for just an hour, you can see it move eastward by slightly more than its angular diameter. The Moon is about 0.5 degrees in angular diameter, and it moves eastward a bit more than 0.5 degrees per hour. In 24 hours, it moves 13 degrees. Thus, each night you see the Moon about 13 degrees eastward of its location the night before.

As the Moon orbits around Earth, its appearance changes from night to night in a monthly cycle.

The Cycle of Moon Phases

The changing appearance of the Moon as it revolves around Earth, called the **lunar phase** cycle, is one of the most easily observed phenomena in astronomy. Study **The Phases of the Moon** on pages 36–37 and notice three important points and two new terms:

1 The Moon always keeps the same side facing Earth. "The man in the moon" is produced by the familiar features on the Moon's near side, but you never see the far side of the Moon from Earth.

2 The changing shape of the Moon as it passes through its cycle of phases is produced by sunlight illuminating different portions of the side of the Moon you can see.

3 Notice the difference between the orbital period of the Moon around Earth relative to the background stars (the *sidereal period,* pronounced "si-DARE-ee-al," referring to stars) versus the length of the lunar phase cycle (the *synodic period*). That difference is a good illustration of how your view from Earth is produced by the combined motions of Earth and other celestial bodies such as the Sun and Moon.

The phases of the Moon are dramatic, and they have attracted lots of peculiar ideas. You have probably heard a number of **Common Misconceptions** about the Moon. Sometimes people are surprised to see the Moon in the daytime sky, and they think something has gone wrong! No, the gibbous (just pastfull) Moon is often visible in the daytime, although quarter moons and especially crescent moons can also be in the daytime sky but are harder to see when the Sun is above the horizon and the sky is bright. You may hear people mention "the dark side of the Moon," but you can assure them that this is a misconception; there is no permanently dark side. Any location on the Moon is sunlit for two weeks and is in darkness for two weeks as the Moon rotates. Finally, you have probably heard one of the strangest misconceptions about the Moon: that people tend to act up at full Moon. Careful statistical studies of records from schools, prisons, hospitals, and so on, show that it isn't true. There are always a few people who misbehave, and the Moon has nothing to do with it.

For billions of years, "the man in the moon" has looked down on Earth. People in the first civilizations saw the same monthly cycle of phases that you see (Figure 3-1), and even the dinosaurs may have noticed the changing phases of the Moon. Occasionally, however, the Moon displays more complicated moods during a lunar eclipse.

3-2 Lunar Eclipses

In cultures all around the world, the sky is a symbol of order and power, and the Moon is a regular counter of the passing days and months. It is not surprising that people are startled and sometimes worried when, once in a while, they see the full Moon become dark and coppery-red colored. Such events are neither mysterious nor frightening once you understand how they arise. To begin, you can think about Earth's shadow.

Earth's Shadow

As you just learned, the orbit of the Moon is tipped only a few degrees from the plane of Earth's orbit around the Sun. Earth's shadow points directly away from the Sun in the plane of Earth's orbit. A **lunar eclipse** can occur at full moon (and only full moon) if the Moon's path carries it through the shadow of Earth, sunlight is blocked, and the Moon grows dim temporarily. This

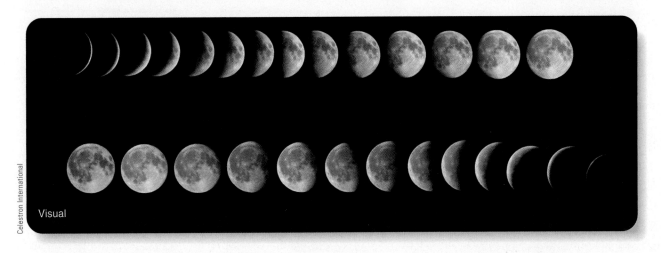

Visual

Celestron International

▲ **Figure 3-1** In this sequence of lunar phase snapshots taken at intervals of 1 day, the Moon cycles through its phases from crescent to full to crescent. From Earth you see the same face of the Moon, the same mountains, craters, and plains, at all times, but the changing direction of sunlight changes what part is illuminated and produces the lunar phases.

is somewhat unusual because most full moons pass north or south of ("above" or "below") Earth's shadow, and there is no eclipse (Figure 3-2). The conditions that allow eclipses to occur will be explained in detail later in this chapter.

A shadow consists of two parts (Figure 3-3). The **umbra** is the region of total shadow. If you were floating in a spacesuit in the umbra of Earth's shadow, the Sun would be completely hidden behind Earth, and you would not be able to see any part of the

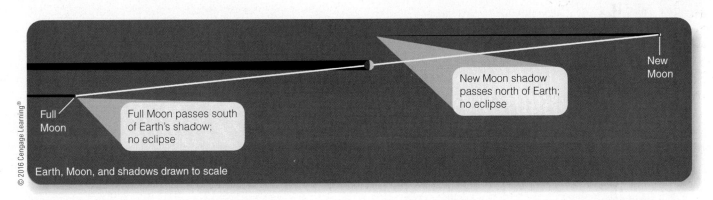

New Moon

New Moon shadow passes north of Earth; no eclipse

Full Moon

Full Moon passes south of Earth's shadow; no eclipse

Earth, Moon, and shadows drawn to scale

© 2016 Cengage Learning®

▲ **Figure 3-2** The Moon's orbit around Earth is tilted relative to Earth's orbit around the Sun, so the long, thin shadows of Earth and Moon usually miss each other. In most months there are no eclipses.

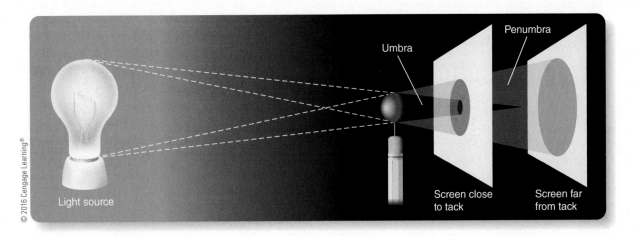

Penumbra

Umbra

Light source

Screen close to tack

Screen far from tack

© 2016 Cengage Learning®

▲ **Figure 3-3** The shadow cast by a map tack can be used to understand the shadows of Earth and the Moon. The umbra is the region of total shadow; the penumbra is the region of partial shadow.

The Phases of the Moon

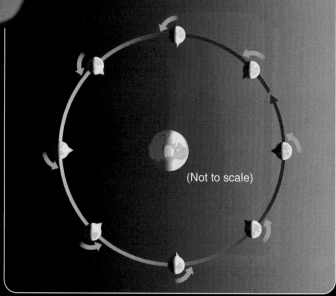

(Not to scale)

1 As the Moon orbits Earth, it rotates to keep the same side facing Earth, as shown at right. Consequently, you always see the same features on the Moon from Earth, and you never see the far side of the Moon. A mountain on the Moon that points at Earth will always point at Earth as the Moon rotates on its axis and revolves around Earth.

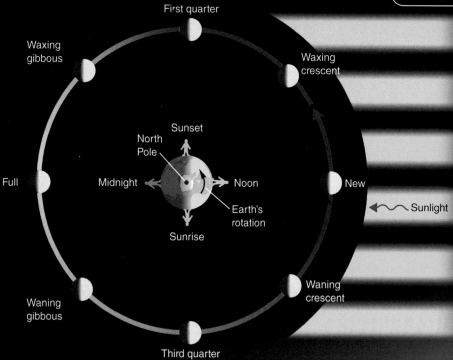

2 As seen at left, sunlight always illuminates half of the Moon. Because you see different amounts of this sunlit side, you see the Moon go through a cycle of phases. At the phase called "new Moon," sunlight illuminates the far side of the Moon, and the side you see is in darkness. In fact, at new Moon you cannot see the Moon at all in contrast to the bright daytime sky near the Sun. At full Moon, the side of the Moon you can see from Earth is fully lit, and the far side is in darkness. How much of the Moon you see illuminated depends on where the Moon is in its monthly orbit around Earth.

Notice that there is no such thing as the permanently "dark side of the Moon." All parts of the Moon experience day and night in a month-long cycle.

In the diagram at the left, you see that the new Moon is close to the Sun in the sky, and the full Moon is opposite the Sun. The observer's time of day depends on his or her location on Earth relative to the Sun.

2a The first two weeks of the Moon's monthly cycle are shown below by its position and phase as seen at sunset on 14 successive evenings. As the illuminated part of the Moon grows larger from new to full, it is said to "wax," an old-fashioned word meaning "to increase."

Gibbous comes from the Latin word for humpbacked.

The first-quarter Moon is 1 week through its 4-week cycle.

The full Moon is 2 weeks through its 4-week cycle.

Full Moon rises at sunset.

Waxing gibbous

Waxing crescent

New Moon is invisible near the Sun.

Moon Phase and Position at Sunset on the First 14 Nights of a Lunar Month

Days since new Moon

East   **South**   **West**

3 The Moon orbits eastward around Earth in 27.3 days, the Moon's **sidereal period**. This is how long the Moon takes to circle the sky once and return to the same position relative to the stars.

A complete cycle of lunar phases takes 29.5 days, the Moon's **synodic period**. (Synodic comes from the Greek words for "together" and "path.") To see why the synodic period is longer than the sidereal period, study the star charts at the right.

Although you think of the lunar cycle as being about 4 weeks long, it is actually 1.53 days longer than 4 weeks. The calendar divides the year into 30-day periods called months (literally "moonths") originating in recognition of the 29.5 day synodic cycle of the Moon.

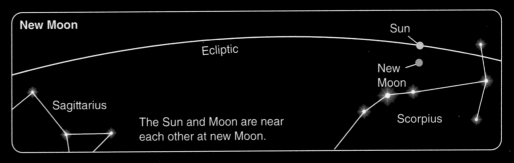

New Moon

Ecliptic

Sun

New Moon

Sagittarius

Scorpius

The Sun and Moon are near each other at new Moon.

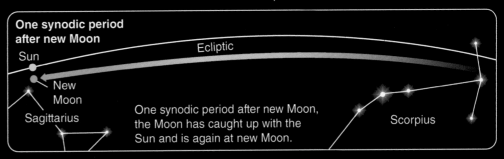

One sidereal period after new Moon

Ecliptic

Sun

Sagittarius

Moon

Scorpius

One sidereal period after new Moon, the Moon has returned to the same place among the stars, but the Sun has moved on along the ecliptic.

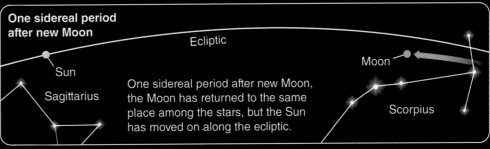

One synodic period after new Moon

Ecliptic

Sun

New Moon

Sagittarius

Scorpius

One synodic period after new Moon, the Moon has caught up with the Sun and is again at new Moon.

You can use the diagram on the opposite page to determine when the Moon rises and sets at different phases.

TIMES OF MOONRISE AND MOONSET

Phase	Moonrise	Moonset
New	Dawn	Sunset
First quarter	Noon	Midnight
Full	Sunset	Dawn
Third quarter	Midnight	Noon

2b The last 2 weeks of the Moon's monthly cycle are shown below by its position and phase as seen at sunrise on 14 successive mornings. As the illuminated portion of the Moon shrinks from the time of full Moon to new Moon, the Moon is said to "wane," an old-fashioned word for *decrease*.

The third-quarter Moon is 3 weeks through its 4-week cycle.

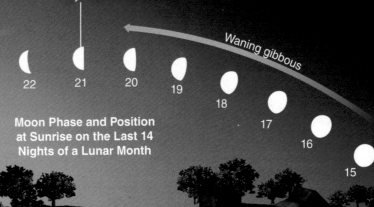

New Moon is invisible near the sun.

Waning crescent

Waning gibbous

27 26 25 24 23 22 21 20 19 18 17 16 15 14

Moon Phase and Position at Sunrise on the Last 14 Nights of a Lunar Month

Full Moon sets at sunrise.

East   **South**   **West**

Sun's bright disk. If you drifted into the **penumbra**, however, you would see part of the Sun peeking around the edge of Earth, so you would be in partial shadow. In the penumbra, the Sun is partly but not completely blocked.

The umbra of Earth's shadow is more than three times longer than the distance to the Moon and points directly away from the Sun. A giant screen placed in the shadow at the average distance of the Moon would reveal a dark umbral shadow about 2.5 times the diameter of the Moon. The faint outer edges of the penumbra would mark a circle about 4.6 times the diameter of the Moon. Consequently, when the Moon's orbit carries it through Earth's shadow, the shadow is plenty large enough for the Moon to become completely immersed.

Total Lunar Eclipses

Once or twice a year, the Moon's orbit carries it through the umbra of Earth's shadow, and if for some interval the Moon is completely within the umbra you see a **total lunar eclipse** (**Figure 3-4a**). As you watch the eclipse begin, the Moon first moves into the penumbra and dims slightly; the deeper it moves into the penumbra, the more it dims. Eventually, the Moon reaches the umbra, and you see the umbral shadow darken part, then all, of the Moon.

When the Moon is totally eclipsed, it does not disappear completely. While it is in the umbra it receives no direct sunlight, but the Moon is illuminated by some sunlight that is refracted (bent) through Earth's atmosphere. If you were on the Moon during **totality**, you would not see any part of the Sun because it would be entirely hidden behind Earth. However, you would see Earth's atmosphere lit from behind by the Sun. The red glow from this ring of sunsets and sunrises around the circumference of Earth shines into the umbra of Earth's shadow, making the umbra not completely dark. That glow illuminates the Moon during totality and makes it seem reddish in color, as shown in **Figure 3-4b** and in **Figure 3-5**.

How dim the totally eclipsed Moon becomes depends on a number of things. Total lunar eclipses tend to be darkest when the Moon's orbit carries it through the center of the umbra. Also, if Earth's atmosphere is especially cloudy in the regions that are bending light into the umbra, the Moon will be darker than during an average eclipse. An unusual amount of dust in Earth's atmosphere (for example, from volcanic eruptions) can also cause an exceptionally dark or especially red eclipse.

The exact timing of a lunar eclipse depends on where the Moon crosses Earth's shadow. If it crosses through the center of the umbra, the eclipse will have maximum length. For such an eclipse, the Moon spends about an hour crossing the penumbra and then another hour entering the darker umbra. Totality can last as long as 1 hour 45 minutes, followed by the emergence of the Moon into the penumbra, plus another hour as it emerges

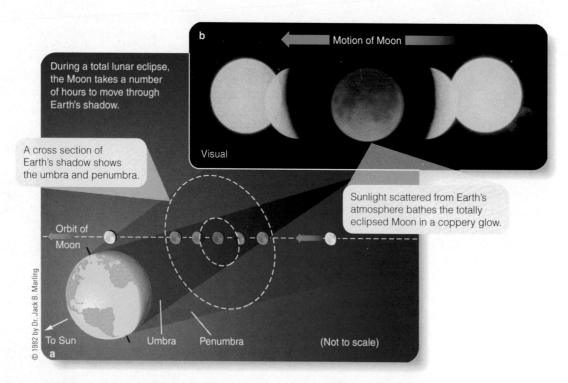

During a total lunar eclipse, the Moon takes a number of hours to move through Earth's shadow.

A cross section of Earth's shadow shows the umbra and penumbra.

Motion of Moon

Visual

Sunlight scattered from Earth's atmosphere bathes the totally eclipsed Moon in a coppery glow.

Orbit of Moon

To Sun   Umbra   Penumbra   (Not to scale)

© 1982 by Dr. Jack B. Marling

▲ **Figure 3-4** (a) In this diagram of a total lunar eclipse, the Moon passes from right to left through Earth's shadow. (b) A multiple-exposure photograph shows the Moon passing through the umbra of Earth's shadow. A longer exposure was used to record the Moon while it was totally eclipsed. The Moon's path appears curved in the photo because of photographic effects.

Celestron International

Visual

▲ **Figure 3-5** During a total lunar eclipse, the Moon turns a coppery-red color. In this photo, the Moon is darkest toward the lower right, the direction toward the center of the umbra. The edge of the Moon at upper left is brighter because it is near the edge of the umbra.

TABLE 3-1   Total and Partial Eclipses of the Moon, 2015 through 2024[a]

Date	Time of Mid-Eclipse (UTC)[b]	Length of Totality (Hr:Min)	Length of Eclipse[c] (Hr:Min)
2015 April 4	12:01	0:05	3:29
2015 September 28	02:48	1:12	3:20
2017 August 7	18:22	Partial	1:55
2018 January 31	13:31	1:16	3:23
2018 July 27	20:23	1:43	3:55
2019 January 21	05:13	1:02	3:17
2019 July 16	21:32	Partial	2:58
2021 May 26	11:20	0:15	3:07
2021 November 19	09:04	Partial	3:28
2022 May 16	04:13	1:25	3:27
2022 November 8	11:00	1:25	3:40
2023 October 28	20:15	Partial	1:17
2024 September 18	02:45	Partial	1:03

[a]There will be no total or partial lunar eclipses during 2016.

[b]Times are Universal Time. Subtract 5 hours for Eastern Standard Time, 6 hours for Central Standard Time, 7 hours for Mountain Standard Time, and 8 hours for Pacific Standard Time. For Daylight Savings Time (mid-March through early November), add 1 hour to Standard Time. Lunar eclipses that occur between sunset and sunrise in your time zone will be visible, and those at midnight will be best placed.

[c]Does not include penumbral phase.

Source: NASA Goddard Space Flight Center

into full sunlight. Thus, a total lunar eclipse can take nearly 6 hours from start to finish.

Partial and Penumbral Lunar Eclipses

Because the Moon's orbit is inclined by a bit more than 5 degrees to the plane of Earth's orbit around the Sun, the Moon does not always pass through the center of the umbra (look back to Figure 3-2). If the Moon passes a bit too far north or south, it may only partially enter the umbra, and we see a **partial lunar eclipse**. The part of the Moon that remains in the penumbra receives some direct sunlight, and the glare is usually great enough to prevent us from seeing the faint red glow of the part of the Moon in the umbra. For that reason, partial eclipses are not as beautiful as total lunar eclipses.

If the orbit of the Moon carries it far enough north or south of the umbra, the Moon may pass through only the penumbra and never reach the umbra. Such **penumbral lunar eclipses** are not dramatic at all. In the partial shadow of the penumbra, the Moon is only partially dimmed. Most people glancing at a penumbral eclipse would not notice any difference from a full Moon.

Total, partial, or penumbral lunar eclipses are interesting events in the night sky and are not difficult to observe. When the full Moon passes through Earth's shadow, the eclipse is visible from anywhere on Earth's dark side. Consult Table 3-1 to find the next lunar eclipse visible in your part of the world.

DOING SCIENCE

What would a total lunar eclipse look like if Earth had no atmosphere? As a way to test and improve their understanding, scientists often experiment with their ideas by imagining changing one part of a system and trying to figure out what would happen as a result. This is sometimes called a "thought experiment." In this example, the absence of an atmosphere around Earth would mean that no sunlight would be bent toward the eclipsed Moon, and it would not glow red. The Moon would be very dark in the sky during totality.

Now try a new thought experiment; imagine changing a different part of the Earth–Moon–Sun system and guess the result. ***What would a lunar eclipse look like if the Moon and Earth were the same diameter?***

3-3 Solar Eclipses

For millennia, cultures worldwide have understood that the Sun is the source of life, so you can imagine the panic people felt at the terrible sight of the Sun gradually disappearing in the middle of the day. Many imagined that the Sun was being devoured by a monster (Figure 3-6). Modern scientists must use their imaginations to visualize how nature works, but with a key difference: They test their ideas against reality (How Do We Know? 3-1).

How Do We Know? 3-1

Scientific Imagination

How do scientists produce hypotheses to test? Good scientists are invariably creative people with strong imaginations who can study raw data about some invisible aspect of nature such as an atom and construct mental pictures as diverse as a plum pudding or a solar system. These scientists share the same human impulse to understand nature that drove ancient cultures to imagine eclipses as serpents devouring the Sun.

As the 20th century began, physicists were busy trying to imagine what an atom was like. No one can see an atom, but English physicist J. J. Thomson used what he knew from his experiments and his powerful imagination to create an image of what an atom might be like. He suggested that an atom was a ball of positively charged material with negatively charged electrons distributed throughout like plums in a plum pudding.

The key difference between using a plum pudding to represent the atom and a hungry serpent to represent an eclipse is that the plum pudding model was based on experimental data and could be tested against new

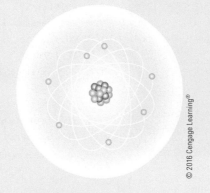

© 2016 Cengage Learning®

A model image of the atom as electrons orbiting a small nucleus has become the symbol for atomic energy.

evidence. As it turned out, Thomson's student, Ernest Rutherford, performed ingenious new experiments and showed that atoms can't be made like plum puddings. Rather, his data led him to imagine an atom as a tiny positively charged nucleus surrounded by negatively charged electrons, much like a tiny version of the Solar System with planets circling the Sun. Later experiments confirmed that Rutherford's description of atoms is closer to reality, and it has become a universally recognized symbol for atomic energy.

Ancient cultures pictured the Sun being devoured by a serpent. Thomson, Rutherford, and scientists like them used their scientific imaginations to visualize natural processes and then test and refine their ideas with new experiments and observations. The critical difference is that scientific imagination is continuously tested against reality and is revised when necessary.

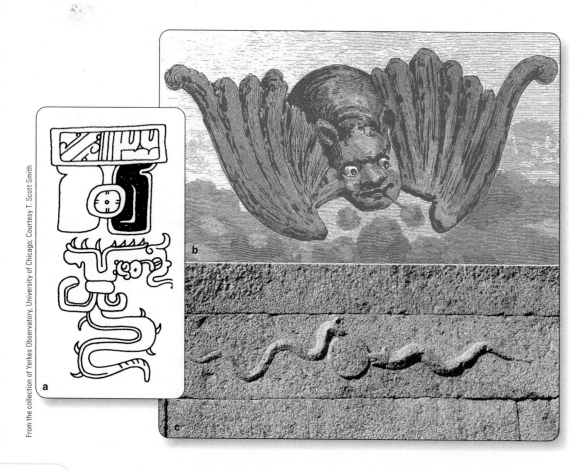

From the collection of Yerkes Observatory, University of Chicago; Courtesy T. Scott Smith

◀ **Figure 3-6** (a) A 12th-century Mayan symbol believed to represent a solar eclipse. The black-and-white Sun symbol hangs from a rectangular sky symbol, and a voracious serpent approaches from below. (b) The Chinese representation of a solar eclipse shows a monster, usually described as a dragon, flying in front of the Sun. (c) This wall carving from the ruins of a temple in Vijayanagaara in southern India symbolizes a solar eclipse as two snakes approach the disk of the Sun.

▲ **Figure 3-7** A total solar eclipse is really a lunar phenomenon. It occurs when the Moon crosses in front of the Sun and hides its brilliant surface. Then you can see the Sun's extended atmosphere.

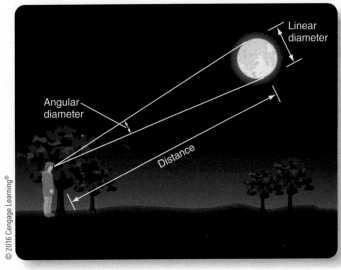

▲ **Figure 3-8** The angular diameter of an object is related to both its linear diameter and its distance.

A **solar eclipse** occurs when the Moon moves between Earth and the Sun. If the Moon covers the disk of the Sun completely, you see a spectacular **total solar eclipse** (Figure 3-7; also, look back to the image that opens this chapter, on page 33). If, from your location, the Moon covers only part of the Sun, you see a less dramatic **partial solar eclipse**. During a solar eclipse people in one place on Earth may see a total eclipse while people only a few hundred kilometers away see a partial eclipse.

The geometry of a solar eclipse is quite different from that of a lunar eclipse. You can begin by considering how big the Sun and Moon look in the sky.

The Angular Diameters of the Sun and Moon

Solar eclipses are spectacular because Earth's Moon happens to have nearly the same angular diameter as the Sun, so it can cover the Sun's disk almost exactly. You learned about angular diameter in Chapter 2; now you can consider how the size and distance of an object like the Moon to determine its angular diameter.

Linear diameter is simply the distance between an object's opposite sides. You use linear diameter when you order a 16-inch pizza—the pizza is 16 inches across. In contrast, the angular diameter of an object is the angle formed by lines extending toward you from opposite edges of the object and meeting at your eye (Figure 3-8). Clearly, the farther away an object is, the smaller its angular diameter.

To find the angular diameter of the Moon, you need to use the **small-angle formula**. That formula expresses the relationship of the linear (true) diameter, the angular (apparent) diameter, and the distance, of any object, whether it is a pizza or the Moon. If you know two of those quantities, you can find the third one by cross-multiplying. This formula is used very often in astronomy, and you will encounter its use many times in later chapters:

$$\frac{\text{angular diameter (in arc seconds)}}{2.06 \times 10^5} = \frac{\text{linear diameter}}{\text{distance}}$$

In the small-angle formula, you must always use the same units for distance and linear diameter. This version of the small-angle formula uses arc seconds as the unit of angular diameter. (The constant 2.06×10^5 in the formula is the number of arc seconds in one radian.)

You can now find the angular diameter of the Moon using its linear diameter, 3480 km (2160 mi), and its average distance from Earth, 384,000 km (both values have been rounded to a precision of 3 digits). Because the Moon's linear diameter and distance are both given in the same units, kilometers, you can put them directly into the small-angle formula:

$$\frac{\text{angular diameter}}{2.06 \times 10^5} = \frac{3480 \text{ km}}{384,000 \text{ km}}$$

When you do the calculation you will find that the angular diameter of the Moon is 1870 arc seconds (rounded to three digits of precision). If you divide by 60, you get 31 arc minutes; dividing by 60 again, you get about 0.5 degrees. The Moon's orbit is slightly elliptical, so the Moon can sometimes look a bit larger or smaller, but its angular diameter is always close to 0.5 degrees. It is a **Common Misconception** that the Moon is larger when it is on the horizon. Certainly the rising full moon looks big when you see it on the horizon, but that is an optical illusion. In reality, the Moon is the same size on the horizon as when it is high overhead.

Now, do another small-angle calculation to find the angular diameter of the Sun. The Sun has a linear diameter of 1.39×10^6 km and its average distance from Earth is 1.50×10^8 km. If you put these numbers into the small-angle formula, you will find that the Sun has an angular diameter of 1910 arc seconds, which is about 32 arc minutes, or 0.5 degrees.

By fantastic good luck, you live on a planet with a moon that is almost exactly the same angular diameter as its sun.

Thanks to that coincidence, when the Moon passes in front of the Sun, it is almost exactly the right size to cover the Sun's brilliant surface but leave the Sun's atmosphere visible.

The Moon's Shadow

To see a solar eclipse, you have to be in the Moon's shadow. Like Earth's shadow, the Moon's shadow consists of a central umbra of total shadow and a penumbra of partial shadow. The Moon's umbral shadow produces a spot of darkness no more than 270 km (170 mi) in diameter on Earth's surface. (The exact size of the umbral shadow depends on the location of the Moon in its elliptical orbit and the angle at which the shadow strikes Earth.) The combination of Earth's rotation with the Moon's orbital motion causes the shadow to rush across Earth at speeds of at least 1700 km/h (1060 mph), sweeping out a **path of totality** (Figure 3-9). People lucky enough to be in the path of totality will see a total eclipse of the Sun while the umbral spot

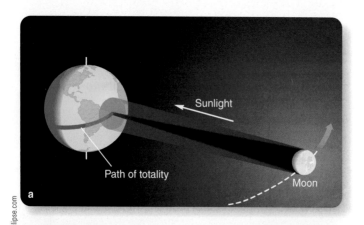

Visual image

Panel a: © 2016 Cengage Learning®. Panel b: NASA *GOES* images courtesy of MrEclipse.com

▲ **Figure 3-9** (a) The umbra of the Moon's shadow sweeps from west to east across Earth, and observers in the path of totality see a total solar eclipse. Those outside the umbra but inside the penumbra see a partial eclipse. (b) Eight photos made by a weather satellite have been combined to show the Moon's shadow moving across the eastern Pacific, Mexico, Central America, and Brazil during an eclipse in 1991.

sweeps over them. Observers just outside the path of totality will see a partial solar eclipse as the penumbral shadow sweeps over their location. Those living even farther from the path of totality will see no eclipse.

The orbit of the Moon is slightly elliptical, and its distance from Earth varies. When it is at **apogee**, its farthest point from Earth, the Moon's angular diameter is 5.5 percent smaller than average, and when it is at **perigee**, its closest point to Earth, its angular diameter is 5.5 percent larger than average. Another factor is Earth's slightly elliptical orbit around the Sun. When Earth is closest to the Sun in January, the Sun looks 1.7 percent larger in angular diameter; and when Earth is farthest from the Sun in July, the Sun looks 1.7 percent smaller. As a result of those effects, sometimes the disk of the Moon is not big enough to cover the Sun as seen from Earth's surface. Or, to put it another way, sometimes the Moon's umbral shadow is not long enough to reach Earth. If the Moon crosses in front of the Sun when the Moon's disk is smaller in angular diameter than the Sun's, it produces an **annular eclipse**, a solar eclipse in which a ring (or annulus) of light is visible around the disk of the Moon (Figure 3-10).

Total solar eclipses are rare if you are not willing to leave home to see one. If you stay in one location, you will see a total solar eclipse on average about once every 360 years. On the other hand, some people are eclipse chasers: They plan years in advance and travel halfway around the world to place themselves in the path of totality. Table 3-2 shows the date and location of solar eclipses over the next few years.

Features of Solar Eclipses

A solar eclipse begins when you first see the edge of the Moon encroaching on the Sun. This is the moment when the edge of the *penumbra* sweeps over your location.

During the partial phases of a solar eclipse, the Moon gradually covers the bright disk of the Sun (Figure 3-11). Totality begins as the last sliver of the Sun's bright surface disappears behind the Moon. This is the moment when the edge of the *umbra* sweeps over your location. So long as any of the Sun is visible, the countryside remains bright, but, as the last of the Sun disappears, darkness falls in a few seconds. Automatic streetlights come on, drivers switch on their headlights, and birds go to roost. The darkness of totality depends on a number of factors, including the weather at the observing site, but it is usually dark enough to make it difficult to read the settings on cameras.

The totally eclipsed Sun is a spectacular sight. With the Moon covering the bright surface of the Sun, called the **photosphere**, you can see the Sun's faint outer atmosphere, called the **corona**, glowing with a pale white light faint enough that you can safely look at it directly. The corona is made of hot, low **density** gas that is given a wispy appearance by the solar magnetic field, as shown in the bottom frame of Figure 3-11 and even more so in Figure 3-7. Also visible just above the photosphere is a thin layer of bright gas called the **chromosphere**. The

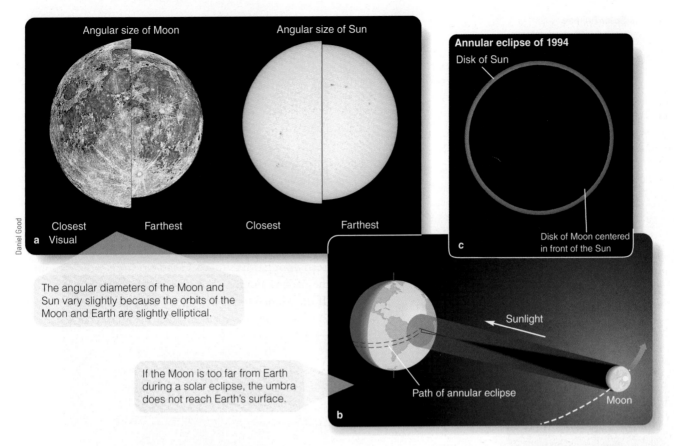

The angular diameters of the Moon and Sun vary slightly because the orbits of the Moon and Earth are slightly elliptical.

If the Moon is too far from Earth during a solar eclipse, the umbra does not reach Earth's surface.

Angular size of Moon

Closest · Farthest

a Visual

Angular size of Sun

Closest · Farthest

Annular eclipse of 1994
Disk of Sun

Disk of Moon centered in front of the Sun

c

Sunlight

Path of annular eclipse

Moon

b

▲ Figure 3-10 Because the angular diameter of the Moon and the Sun vary slightly, the disk of the Moon is sometimes too small to cover the disk of the Sun. That means the umbra of the Moon does not reach Earth, and the eclipse is annular, meaning a ring ("annulus") of the Sun's disk can be seen around the Moon. In this photograph of an annular eclipse in 1994, the dark disk of the Moon is almost exactly centered on the bright disk of the Sun.

TABLE 3-2 Total and Annular Eclipses of the Sun, 2015 through 2024[a]

Date	Total/Annular (T/A)	Time of Mid-Eclipse (UTC)[b]	Maximum Length of Total or Annular Phase (Min:Sec)	Area of Visibility
2015 March 20	T	09:47	2:47	North Atlantic, Arctic
2016 March 9	T	01:58	4:09	Borneo, Pacific
2016 September 1	A	09:08	3:06	Atlantic, Africa, Indian Ocean
2017 February 26	A	14:55	0:44	South America, Atlantic, Africa, Antarctica
2017 August 21[c]	T	18:27	2:40	Pacific, United States, Atlantic
2019 July 2	T	19:24	4:33	Pacific, South America
2019 December 26	A	05:19	3:39	Southeast Asia, Pacific
2021 June 10	A	10:43	3:51	North America, Arctic
2021 December 4	T	07:35	1:54	Antarctica, South Atlantic
2023 April 20	A/T[d]	04:18	1:16	Southeast Asia, Philippines, Indonesia, Australia
2023 October 14	A	18:00	5:17	United States, Central America, South America
2024 April 8	T	18:18	4:28	North America, Central America
2024 October 2	A	18:46	7:25	Pacific, South America

[a]There will be no total or partial solar eclipses in 2018.

[b]Times are Universal Time. Subtract 5 hours for Eastern Standard Time, 6 hours for Central Standard Time, 7 hours for Mountain Standard Time, and 8 hours for Pacific Standard Time. For Daylight Savings Time (mid-March through early November), add 1 hour to Standard Time.

[c]The next major total solar eclipse visible from the United States will occur on August 21, 2017, when the path of totality will cross the United States from Oregon to South Carolina.

[d]Hybrid eclipse: begins as annular, becomes total, ends as annular.

Source: NASA Goddard Space Flight Center

The Moon moving from the right just begins to cross in front of the Sun.

The disk of the Moon gradually covers the disk of the Sun.

Sunlight begins to dim as more of the Sun's disk is covered.

During totality, pink prominences are often visible.

A longer-exposure photograph during totality shows the fainter corona.

Daniel Good

Visual images

▲ **Figure 3-11** This sequence of photos shows the first half of a total solar eclipse.

chromosphere is often marked by eruptions on the solar surface called **prominences** (Figure 3-12a) that glow with a clear pink color because of the high temperature of the gases involved. A large prominence can be wider than three times the diameter of Earth. (The nature of the photosphere, chromosphere, corona, and prominences as components of the Sun's atmosphere will be described in detail in Chapter 8.)

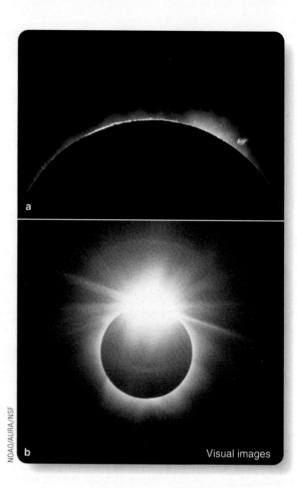

NOAO/AURA/NSF

a

b

Visual images

▲ **Figure 3-12** (a) During a total solar eclipse, the Moon covers the photosphere, and the ruby-colored chromosphere and prominences are visible. Only the lower corona is visible in this image. (b) The diamond ring effect can sometimes occur momentarily at the beginning or end of totality if a small segment of the photosphere peeks out through a valley at the edge of the lunar disk.

Totality during a solar eclipse cannot last longer than 7.5 minutes under any circumstances, and the average is only 2 to 3 minutes. Totality ends when the Sun's bright surface reappears at the trailing edge of the Moon. Daylight returns quickly, and the corona and chromosphere vanish. This corresponds to the moment when the trailing edge of the Moon's umbra sweeps over the observer.

Just as totality begins or ends, a small part of the photosphere can peek through a valley at the edge of the lunar disk. Although it is intensely bright, such a small part of the photosphere does not completely drown out the fainter corona, which forms a silvery ring of light with the brilliant spot of photosphere gleaming like a diamond (Figure 3-12b). This **diamond ring effect** is one of the most spectacular of astronomical sights, but it is not visible during every solar eclipse. Its occurrence depends on the exact orientation and motion of the Moon.

Observing an Eclipse

Not too many years ago, astronomers traveled great distances to exotic places to get their instruments into the path of totality and study the faint outer corona that is visible only during the few minutes of a total solar eclipse. Now, many of those observations can be made every day by solar telescopes in space, but eclipse enthusiasts still journey to remote corners of the world for the thrill of seeing a total solar eclipse.

No matter how thrilling a solar eclipse is, you must be cautious when viewing it. During the partial phase, part of the brilliant photosphere remains visible, so it is hazardous to look at the eclipse without protection. Dense filters and exposed film do not necessarily provide protection because some filters do not block the invisible infrared (heat) radiation that can burn the retina of your eyes. Dangers like these have led officials to warn the public not to look at solar eclipses at all and have even frightened some people into locking themselves and their children into windowless rooms. It is a **Common Misconception** that sunlight is somehow more dangerous during an eclipse. In fact, it is always dangerous to look at the Sun. The danger posed by an eclipse is that people are tempted to ignore common sense and look at the Sun directly, which can burn their eyes even when the Sun is almost totally eclipsed.

The safest and simplest way to observe the partial phases of a solar eclipse is to use pinhole projection. Poke a small pinhole in a sheet of cardboard. Hold the sheet with the hole in sunlight and allow light to pass through the hole and onto a second sheet of cardboard (Figure 3-13). On a day when there is no eclipse, the result is a small, round spot of light that is an image of the Sun. During the partial phases of a solar eclipse, the image will show the dark silhouette of the Moon obscuring part of the Sun. Pinhole images of the partially eclipsed Sun can also be seen in the shadows of trees as sunlight peeks through the tiny openings between the

leaves and branches. This can produce an eerie effect just before totality as the remaining sliver of Sun produces thin crescents of light on the ground under trees. Once totality begins, it is safe to look directly. The totally eclipsed Sun is fainter than a full moon.

3-4 Predicting Eclipses

A Chinese legend tells of two astronomers, Hsi and Ho, who were too drunk to predict the solar eclipse of October 22, 2137 BCE. Or perhaps they failed to conduct the proper ceremonies to scare away the dragon that, according to Chinese tradition, was snacking on the Sun's disk. When the emperor recovered from the terror of the eclipse, he had the two astronomers beheaded.

Making exact eclipse predictions requires a computer and proper software, but astronomers in early civilizations could make educated guesses as to which full moons and which new moons might result in eclipses. There are three good reasons to review their methods. First, it is an important chapter in the history of science. Second, it will illustrate how apparently complex phenomena can be analyzed in terms of cycles. Third, eclipse prediction will exercise your scientific imagination and help you visualize Earth, the Moon, and the Sun as objects moving through space.

Conditions for an Eclipse

You can predict eclipses by thinking about the motion of the Sun and Moon in the sky. Imagine that you can look up into the sky from your home on Earth and see the Sun appearing to

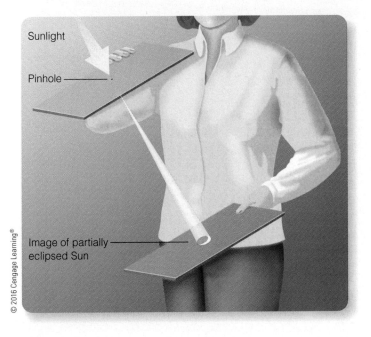

Sunlight

Pinhole

Image of partially eclipsed Sun

© 2016 Cengage Learning®

▲ **Figure 3-13** A safe way to view the partial phases of a solar eclipse. Use a pinhole in a card to project an image of the Sun on a second card. The greater the distance between the cards, the larger (and fainter) the image will be.

move along the ecliptic and the Moon moving along its orbit. Because the orbit of the Moon is tipped slightly more than 5 degrees to the plane of Earth's orbit, you see the Moon follow a path tipped by the same angle to the ecliptic. Each month, the Moon crosses the ecliptic at two points called **nodes**. It crosses at one node going southward, and about two weeks later it crosses at the other node going northward.

Eclipses can occur only when, viewed from Earth, the Sun is near one of the nodes of the Moon's orbit. Only then can the new moon cross in front of the Sun and produce a solar eclipse, as shown in Figure 3-14a, and only then can the full moon enter Earth's shadow and be eclipsed. Most new moons pass too far north or too far south of the ecliptic to cause an eclipse (look again at Figure 3-2, page 35). (Note that this requirement for eclipses is the reason the Sun's apparent path through the sky is called the *ecliptic*.) Also, when the Sun is near one node, Earth's shadow points near the other node, and a lunar eclipse is possible. A lunar eclipse doesn't happen at every full moon because most full moons pass too far north or too far south of the ecliptic and miss the umbra of Earth's shadow. Some months you might see a partial lunar eclipse, as illustrated in Figure 3-14b.

Thus, there are two conditions for an eclipse: The Sun must be near one of the two nodes of the Moon's orbit, and the Moon must pass near either the same node (solar eclipse) or the other node (lunar eclipse). This means, of course, that solar eclipses can occur only when the Moon is new, and lunar eclipses can occur only when the Moon is full.

Now you can understand the ancient secret of predicting eclipses. An eclipse can occur only in a period called an **eclipse season**, during which the Sun is close to a node in the Moon's orbit. For solar eclipses, an eclipse season is about 32 days long. Any new moon during this period will produce a solar eclipse. For lunar eclipses, the eclipse season is a bit shorter, about 22 days. Any full moon in this period will encounter Earth's shadow and be eclipsed.

This makes eclipse prediction easy. All you have to do is keep track of where the Moon crosses the ecliptic (where the nodes of its orbit are). Then, when the Sun approaches either of the nodes you can warn everyone that eclipses are possible. This system works fairly well, and astronomers in early civilizations such as the Maya may have used such a system. You could have been a very successful Mayan astronomer with what you know about eclipse seasons, but you can do even better if you change your point of view.

The View from Space

Change your point of view and imagine that you are looking at the orbits of Earth and the Moon from a point far away in space. Recall that the Moon's orbit is tipped at an angle to Earth's orbit. The shadows of Earth and Moon are long and thin, as shown in Figure 3-2. That is why it is so easy for them to miss their mark at new moon or full moon and usually fail to produce an eclipse. As Earth orbits the Sun, the Moon's orbit remains approximately fixed in orientation. The nodes of the Moon's orbit are the points where it passes through the plane of Earth's orbit; an eclipse season occurs

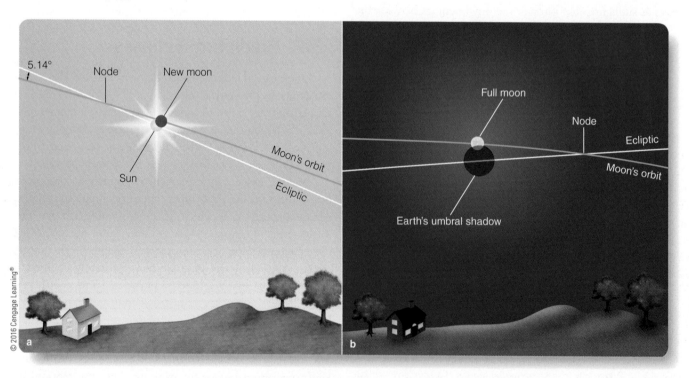

▲ **Figure 3-14** Eclipses can occur only when the Sun appears from Earth to be near one of the nodes of the Moon's orbit. (a) A solar eclipse occurs when the Moon meets the Sun near a node. (b) A lunar eclipse occurs when the Sun and Moon are near opposite nodes. Partial eclipses are shown here for clarity.

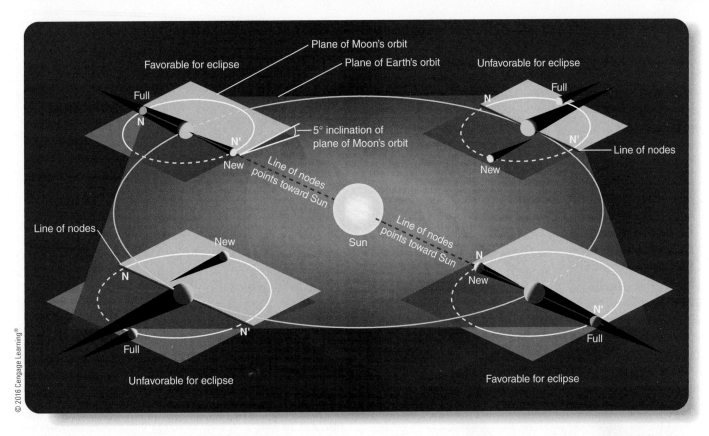

▲ **Figure 3-15** The Moon's orbit is tipped a bit more than 5 degrees to Earth's orbit. The nodes N and N' are the points where the Moon passes through the plane of Earth's orbit. At those parts of Earth's orbit where the line of nodes points toward the Sun, eclipses are possible at new moon and full moon.

each time the line connecting these nodes, the **line of nodes**, points directly toward the Sun, allowing the shadows of Earth and Moon then to hit their marks. Look at **Figure 3-15** and notice that the line of nodes does not point at the Sun in the example at lower left, and no eclipses are possible at that time of the year; the shadows miss. At lower right, during an eclipse season, the line of nodes points toward the Sun, and the shadows produce eclipses.

If you watched for years from your point of view in space, you would see the orbit of the Moon precess like a hubcap spinning on the ground. This precession is caused mostly by the gravitational influence of the Sun, and it makes the line of nodes seem to rotate around the sky, as viewed from Earth, once every 18.6 years. As a result, the nodes slip westward along the ecliptic at a rate of 19.4 degrees per year. Consequently, the Sun does not need a full year to go from a node all the way around the ecliptic and appear back at that same node. Because the node is moving westward to meet the Sun, the Sun will cross the node after only 346.6 days (an **eclipse year**). This means that eclipse seasons begin about 19 days earlier every year (**Figure 3-16**). If you see an eclipse in late December one year, you can see eclipses in early December the next year, and so on.

Eclipses follow a pattern, and if you were an astronomer in an early civilization who understood the pattern, you could predict eclipses without ever knowing what the Moon is or how an orbit

works. Once you have observed a few eclipses from a given location, you know when the eclipse seasons are occurring, and you can predict next year's eclipse seasons by subtracting 19 days. New moons and full moons near those dates are candidates for eclipses.

The Saros Cycle

Astronomers of antiquity could predict eclipses in an approximate way using eclipse seasons, but they could have been much more accurate if they had recognized that eclipses occur following certain patterns. The most important of these is the **Saros cycle** (sometimes referred to simply as the Saros). After one Saros cycle of 18 years 11⅓ days, the pattern of eclipses repeats. In fact, *Saros* comes from a Greek word that means "repetition."

One Saros cycle contains 6585.321 days, which is equal to 223 lunar synodic months. Therefore, after one Saros, the Moon is back to the same phase it had when the cycle began. But, one Saros is also equal to exactly 19 eclipse years. After one Saros cycle, the Sun has returned to the same place it occupied with respect to the nodes of the Moon's orbit when the cycle began. If an eclipse occurs on a given day, then 18 years 11⅓ days later, the Sun, the Moon, and the nodes of the Moon's orbit return to nearly the same relationship, and an eclipse with almost exactly the same geometry (length of totality, general direction of motion of the Moon's shadow, and so on) occurs again.

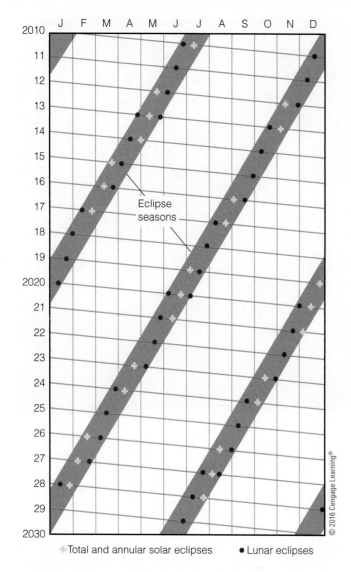

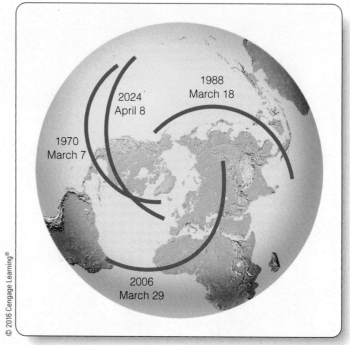

▲ **Figure 3-17** The Saros cycle at work: An eclipse with a track having nearly the same shape as that of the total solar eclipse of March 7, 1970, recurred 18 years 11⅓ days later over the Pacific Ocean. After another interval of 18 years 11⅓ days, an eclipse with a similar track was visible from Asia and Africa. After another 18 years 11⅓ days, in the year 2024, an eclipse like the 1970 eclipse will again be visible from the United States.

✚ Total and annular solar eclipses   ● Lunar eclipses

▲ **Figure 3-16** A calendar of eclipse seasons: Each year the eclipse seasons begin about 19 days earlier than in the previous year. Any new moon or full moon that occurs during an eclipse season results in an eclipse. Only total and annular eclipses are shown here.

mid-afternoon so alarmed the two armies—thinking they must have offended the gods somehow—that they concluded a truce.

Although there are historical reasons to doubt that Thales actually predicted the eclipse, the important point is that he could have done it. If he had records of past eclipses of the Sun visible from the area, he could have discovered that they tended to recur with a period of 54 years plus 34 days (three Saros cycles). Indeed, he could have predicted the eclipse without having any understanding of the cause of the Saros cycle.

Although the eclipse geometry repeats almost exactly, it is not visible from the same place on Earth. The Saros cycle is ⅓ of a day longer than 18 years 11 days. When the eclipse happens again, Earth will have rotated ⅓ of a turn farther east, and the eclipse will occur ⅓ of the way westward around Earth (**Figure 3-17**). That means that after three Saros cycles—a period of 54 years plus 34 days—the same eclipse occurs in about the same part of Earth.

One of the most famous predictors of eclipses was the Greek philosopher Thales of Miletus (about 640–546 BCE), who supposedly learned of the Saros cycle from Babylonian astronomers. No one knows for certain which eclipse Thales predicted, but some scholars suspect it was the eclipse of May 28, 585 BCE. In any case, the eclipse occurred at the height of a battle between the Lydians and the Medes, and the mysterious darkness in

---

DOING SCIENCE

Why can't two successive full moons be totally eclipsed? Most people suppose that eclipses occur at random or in some pattern so complex you need a big computer to make predictions. In fact, like many natural events, eclipses occur in a cycle that can be observed, and predicting eclipses can be reduced to a series of simple steps.

A lunar eclipse can happen only when the Sun is near one node and the Moon crosses Earth's shadow at the other node. A lunar eclipse season is only 22 days long, and any full moon in that time will be eclipsed. However, the Moon takes 29.5 days to go from one full moon to the next. If one full moon is totally eclipsed, the next full moon 29.5 days later will occur long after the end of the eclipse season, and there can't be a second eclipse.

Now use your knowledge of the cycles of the Sun and Moon to consider a similar question: ***How can the Sun be eclipsed by two successive new moons?***

The Moon is a companion in our daily lives, in our history, and in our mythology. It makes a dramatic sight as it moves through the sky, cycling through a sequence of phases that has repeated for billions of years. The Moon has been humanity's timekeeper. Moses, Jesus, and Muhammad saw the same Moon that you see. The Moon is part of our human heritage, and famous paintings, poems, plays, and music celebrate the beauty of the Moon.

Eclipses of the Sun and Moon have frightened and fascinated for millennia, and some of humanity's earliest efforts to understand nature were focused on counting the phases of the Moon and predicting eclipses. Some astronomers have found evidence that Stonehenge could have been used for eclipse prediction, and the ancient Maya in Central America left behind elaborate tables that allowed them to predict eclipses.

Our lives are ruled by the Moon as it divides our year into months, and its cycle from new to first quarter to full to third quarter and back to full divides the month into four weeks. In a Native American story, Coyote gambles with the Sun to see if the Sun will return after the winter solstice to warm Earth. The Moon keeps score. The Moon is a symbol of regularity, reliability, and dependability. It is the scorekeeper counting out our weeks and months.

Study and Review

Summary

▶ The Moon orbits eastward around Earth once a month and rotates on its axis so as to keep the same side facing Earth throughout its orbit.

▶ Because you see the Moon by reflected sunlight, its appearance changes as it orbits Earth and sunlight illuminates different amounts of the side facing Earth. This repeating pattern of Moon shapes as viewed from Earth is called the **lunar phase (p. 34)** cycle.

▶ The lunar phases are said to "wax" from new moon to first quarter to full moon (meaning, the portion of the Moon seen to be illuminated increases) and "wane" from full moon to third quarter to new moon (illuminated portion decreases).

▶ A complete cycle of lunar phases takes 29.5 days, which is known as the Moon's **synodic period (p. 37)**. The **sidereal period (p. 37)** of the Moon—its orbital period with respect to the stars—is shorter, about 27.3 days.

▶ If a full moon passes through Earth's shadow, sunlight is cut off, and the Moon darkens in a **lunar eclipse (p. 34)**. If the Moon fully enters the dark **umbra (p. 35)** of Earth's shadow, the eclipse is a **total lunar eclipse (p. 38)**; but if it only grazes the umbra, the eclipse is a **partial lunar eclipse (p. 39)**. If the Moon enters the **penumbra (p. 38)** but not the umbra, the eclipse is a **penumbral lunar eclipse (p. 39)**.

▶ During **totality (p. 38)**, the eclipsed Moon looks copper red because sunlight refracts through Earth's atmosphere and bounces off the Moon to the night side of Earth.

▶ The **small-angle formula (p. 41)** allows you to calculate an object's angular diameter from its linear diameter and distance. The angular diameter of the Sun and Moon is about 0.5 degrees.

▶ A **solar eclipse (p. 41)** occurs if a new moon passes between the Sun and Earth and the Moon's shadow sweeps over Earth's surface along the **path of totality (p. 42)**. Observers inside the path of totality see a **total solar eclipse (p. 41)**, and those just outside the path of totality but inside the penumbra see a **partial solar eclipse (p. 41)**.

▶ When the Moon is near **perigee (p. 42)**—the closest point in its orbit—its angular diameter is large enough to cover the Sun's photosphere and produce a total eclipse. But if the Moon is near **apogee (p. 42)**—the farthest point in its orbit—it looks too small and can't entirely cover the photosphere. A solar eclipse occurring then would be an **annular eclipse (p. 42)**.

▶ During a total eclipse of the Sun, the bright **photosphere (p. 42)** of the Sun is covered, and the faint low-density **corona (p. 42)**, the **chromosphere (p. 42)**, and **prominences (p. 44)** become visible.

▶ Sometimes at the beginning or end of the totality phase of a total solar eclipse, a small piece of the Sun's photosphere can peek out through a valley at the edge of the Moon and produce a **diamond ring effect (p. 44)**.

▶ Looking at the Sun is dangerous and can burn the retinas of your unprotected eyes. The safest way to observe the partial phases of a solar eclipse is by pinhole projection. Only during totality, when the photosphere is completely hidden, is it safe to look at the Sun directly.

▶ Solar eclipses must occur at new moon, and lunar eclipses must occur at full moon. Because the Moon's orbit is tipped a few degrees from the plane of Earth's orbit, most new moons cross north or south of the Sun, and there are no solar eclipses in those months. Similarly, most full moons cross north or south of Earth's shadow, and there are no lunar eclipses in those months.

▶ The Moon's orbit crosses the ecliptic at two locations called **nodes (p. 46)**, and eclipses can occur only when the Sun and Moon are simultaneously near a node. During these periods, called **eclipse seasons (p. 46)**, a new moon will cause a solar eclipse, and a full moon can have a lunar eclipse. An eclipse season occurs each time the **line of nodes (p. 47)** points toward the Sun. Knowing when the eclipse seasons occur would allow you to guess which new moons and full moons could cause eclipses.

▶ Because the orbit of the Moon precesses in the retrograde direction, the nodes slip westward along the ecliptic, and it takes the Sun only about 347 days to go from a node around the ecliptic and back to the same node. This length of time is called an **eclipse year (p. 47)**. For that reason, eclipse seasons begin about 19 days earlier each year.

▶ Eclipses follow a pattern called the **Saros cycle (p. 47)**. After one Saros of 18 years $11^1/_3$ days, the pattern of eclipses repeats. After three Saros, which is 54 years and 34 days, the pattern of eclipses will repeat in approximately the same parts of Earth. Some ancient astronomers knew of the Saros cycle and used it to predict eclipses.

Review Questions

1. Tonight you see the Moon at midnight at its highest point in your sky. What is the phase of the Moon? Three hours later, what phase will the Moon be in?

2. You are located in Syracuse, NY, United States, the time is 6 PM, and you see the full moon in your clear winter night sky. You call your aunt in San Diego, CA, United States, and ask her to go outside to view the Moon with you. In which direction are you looking to see the full moon: north, south, east, or west? Your aunt tells you that the weather in San Diego is clear as she is stepping outside. What does your aunt report about the moon's phase and location in her sky?

3. You are located in Knoxville, TN, United States. Your friend is located in Lima, Peru. You see a waning gibbous in your clear night sky. What phase, if any, will your friend see if the night sky in Lima is also clear?

4. Which lunar phases would be visible in the sky at dawn? At midnight?

5. Tonight you see a waxing crescent moon. Seven days from now, which phase will you see if the night sky is clear?

6. You look along the easterly horizon and see a crescent moon rising in the clear night sky. What time is it? Which side—right or left—of the Moon's near side is illuminated?

7. If you looked back at Earth from the Moon, what phase would Earth have when the Moon was full? New? At first quarter? A waxing crescent?

8. The phase of the Moon is a waning crescent as viewed from Earth. You are located in the dark side of the Moon's near side, near the Moon's equatorial region. Which side of the Earth from your vantage point on the Moon—the left, the right, the top, or the bottom side—is illuminated by the Sun?

9. If a planet has a moon, must that moon go through the same phases that Earth's Moon displays?

10. Could a solar powered spacecraft generate any electricity while passing through Earth's umbral shadow? Through Earth's penumbral shadow?

11. If a lunar eclipse occurred at midnight, where in the sky would you look to see it?

12. If Earth had no atmosphere, what color would the Moon appear in the sky?

13. If the Moon orbited the Earth from North Pole to South Pole instead of near the ecliptic, would lunar and solar eclipses still occur? Would the moon phase still have to be full or new?

14. Why do solar eclipses happen only at new moon? Why not every new moon?

15. Why isn't the corona visible during partial or annular solar eclipses?

16. Which has the larger angular diameter in the sky—the Sun or Moon—during an annular eclipse? If you wanted to be in the umbra, where would you have to physically be located to see this annular eclipse as a total solar eclipse?

17. What is the angular diameter of the Moon if in the third-quarter phase? What is the shortest/longest angular distance from the horizon to the Moon if in the third-quarter phase and the time is midnight or noon?

18. Why can't the Moon be eclipsed when it is halfway between the nodes of its orbit?

19. Why are solar eclipses separated by one Saros cycle not visible from the same location on Earth?

20. How could Thales of Miletus have predicted the date of a solar eclipse without observing the location of the Moon in the sky?

21. Will an eclipse occur in February 2015? In July 2028? If so, what kind?

22. **How Do We Know?** Some people think science is like a grinder that cranks data into hypotheses. What would you tell them about the need for scientists to be creative and imaginative?

Discussion Questions

1. Can you see the dark side of the Moon from Earth?

2. What would the correct response be to someone who refers to "the dark side of the Moon" as if there were a side of the Moon that is always dark?

3. How would eclipses be different if the Moon's orbit were not tipped with respect to the plane of Earth's orbit?

4. Is it possible for a planet to have a moon but never to have "lunar" (moon) eclipses? To never have total solar eclipses? Why or why not?

5. If nodes occur when the Moon's orbit and the ecliptic cross, what do you suppose antinodes are? Along a line of antinodes, are conditions favorable or unfavorable for an eclipse?

Problems

1. Pretend the Moon's orbit around Earth is a perfect circle. How long does it take in units of days for the Moon to move 90 degrees relative to the stars? Is this number tracking with the synodic period or the sidereal period?

2. Identify the phases of the Moon if on March 20 the Moon is located at the point on the ecliptic called (a) the vernal equinox, (b) the autumnal equinox, (c) the summer solstice, (d) the winter solstice.

3. Identify the phases of the Moon if at sunset in the Northern Hemisphere the Moon is (a) near the eastern horizon, (b) high in the southern sky, (c) in the southeastern sky, (d) in the southwestern sky.

4. What fraction of the Moon's surface area is the far side? Of the near side of a third-quarter moon, what fraction is dark? What fraction of the far side is in the dark that cannot be seen by an observer from Earth viewing the Moon in its third-quarter phase?

5. About how many days must elapse between first-quarter moon and third-quarter moon in the same cycle?

6. Tonight you see a waning crescent in the night sky. A few days later, the night is once again clear and you see a waning crescent. How many degrees did the Moon advance in its orbit during this time frame?

7. If on March 1 the Moon is full and is near Favorite Star Spica, when will the Moon next be near Spica? When will it next be full? Are these values the same? Why or why not?

8. How many times larger than the Moon is the diameter of Earth's umbral shadow at the Moon's distance? (*Hint:* See the photo in Figure 3-4.)

9. Use the small-angle formula to calculate the angular diameter of Earth as seen from the Moon.

10. Use the small-angle formula to calculate the angular diameter of the Sun as seen from Earth if the Sun were at the location of the Moon. Show your answer in units of degrees. (*Hint:* The diameter of the Sun is given in this chapter.)

11. At perigee, the Moon is closer than average by 21,100 km. At apogee, the Moon is further than average by 21,100 km. Is the angular diameter more or less at perigee than at apogee? What is the angular diameter of the Moon at perigee? At apogee? By how much greater a percentage is the angular diameter larger or smaller at perigee than at the average distance? At apogee? (*Hint:* The Moon's average distance from Earth is given in this chapter.)

12. Examine the list of upcoming lunar eclipses in Table 3-1. What fraction of years have two eclipses?

13. During solar eclipses, large prominences are often seen extending as much as 5 arc minutes from the edge (limb) of the Sun's disk. How far is that in kilometers? In Earth diameters? If you used protective glasses, do you think you could see prominences around the Sun's limb? Why or why not? (*Hints:* Use the small-angle formula. The Sun's average distance is given in this chapter, and Earth's radius [half its diameter] can be found in Appendix Table A-10.)

14. If a solar eclipse occurs on October 3: (a) Why can't there be a lunar eclipse on October 13 of that same year? (b) Why can't there be a solar eclipse on December 28 of that same year?

15. A total eclipse of the Sun was visible from Canada on July 10, 1972. When did an eclipse occur next with the same Earth–Moon–Sun geometry? From what part of Earth was it total?

16. When will the eclipse described in Problem 15 next be total as seen from Canada?

17. When will the eclipse seasons occur during the current year? How many total of all types will occur? Which type of eclipse(s) will occur?

18. Examine Figure 3-16. List the letter *S* for each total or annular solar eclipse that occurs from July 2019 through July 2028 in chronological order. When only a lunar eclipse occurs, put *N* for the word *none*. Do you see a pattern? If so, identify it and predict the next two letters.

Learning to Look

1. Look at **The Sky Around You** concept art spread in Chapter 2. What phase of the Moon is the woman viewing?

2. To take the photos that are combined on the opening page of this chapter, was the photographer located on the day, or night, side of Earth? Was the photographer in the Moon's umbra, or penumbra, or both? How do you know?

3. Look at **The Phases of the Moon** concept art spread in this chapter. Find the person looking at the third-quarter phase of the Moon at sunrise. What percentage of the near side of the Moon is illuminated? Likewise, what percentage is in the dark? Repeat the exercise for the new phase of the Moon.

4. Look at **The Phases of the Moon** concept art spread in this chapter. Find the waxing gibbous phase of the Moon. If this phase could be seen at its highest point in your winter sky, is it daytime or nighttime? Approximately what time is it? At approximately what time did that phase rise over the eastern horizon? At approximately what time will that phase set over your western horizon? Repeat the exercise for the waning crescent phase.

5. Use the photos in Figure 3-1 as evidence to show that the Moon always keeps the same side facing Earth.

6. Draw the umbral and penumbral shadows onto the diagram in the middle of page 36. Use the diagram to explain why lunar eclipses can occur only at full moon and solar eclipses can occur only at new moon.

7. Look at Figure 3-4. What phase of the Moon is being viewed? Are you looking at the near side or far side? Are you on the daytime or nighttime side of Earth? Are you in the umbra or penumbra, and which celestial object is casting the shadow?

8. What is odd about Figure 3-6 with regard to the Sun in the total solar eclipse and the red color in the picture? In which general direction—north, south, east, or west—is the picture of the tree taken? Approximately what time of day was the picture of the total solar eclipse taken?

9. Do you think the color of the chromosphere in Figure 3-11a is falsely colored ruby red? Why or why not?

10. Can you see evidence of the Saros cycle in Figure 3-15?

11. The accompanying cartoon shows a crescent moon. Explain why the Moon could never look this way at night.

"Fixing a leak—and you?"

12. The photo at right shows the annular eclipse of May 30, 1984. How is it different from the annular eclipse shown in Figure 3-9?

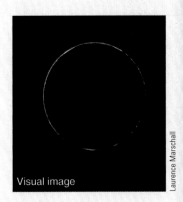

Visual image

4 Origins of Modern Astronomy

Guidepost The preceding three chapters gave you a modern view of the ways in which Earth, the Moon, and the Sun move through space, and how those motions produce the sights you see in the sky. But how did humanity first realize that we live on a planet moving through space? That required the overthrow of ancient and honored ideas of Earth's place in the Universe.

By the 16th century, many astronomers were uncomfortable with the long-standing model that an unmoving Earth sits at the center of a spherical universe. In this chapter, you will discover how an astronomer and Church official named Nicolaus Copernicus created a new model, how a mathematician and schoolteacher named Johannes Kepler discovered the laws of planetary motion, and how a physicist named Galileo Galilei changed the way we understand nature. Here you will find answers to four important questions about the transition from ancient to modern views of the Universe:

- ► **How did classical philosophers describe Earth's place in the Universe?**
- ► **How did Copernicus revise those ancient ideas?**
- ► **How did Kepler discover the laws of planetary motion?**
- ► **How did Galileo's observations support the Copernican model?**

This chapter is not just about the history of astronomy. As the astronomers of the Renaissance struggled to understand Earth, the Solar System, and the Universe, they invented a new way of understanding nature—a way of thinking that is now called *science*.

(a) On the night of January 7, 1610, Galileo saw three small "stars" near the bright disk of Jupiter and sketched them in his notebook. On subsequent nights (except January 9, which was cloudy), he saw that the stars were actually four moons orbiting Jupiter. (b) This photo taken through a modern telescope shows the overexposed disk of Jupiter and three of the four moons discovered by Galileo.

Michael A. Seeds; Grundy Observatory/Franklin & Marshall College

Jan. 7, 1610

Jan. 8, 1610

Jan. 9, 1610

Jan. 10, 1610

Jan. 11, 1610

Jan. 12, 1610

Jan. 13, 1610

a

b

FOUR CENTURIES AGO, Galileo was condemned by the Inquisition for his part in a huge controversy about the nature of the Universe, a controversy that focused on two problems. The place of Earth was the most acrimonious issue: Was it the center of the Universe, or was the Sun at the center? A related issue was the nature of planetary motion. Ancient astronomers could see the Sun, Moon, and planets moving along the ecliptic, but they could not describe or predict those motions precisely. To understand the place of Earth in the Universe, astronomers first had to understand planetary motion.

4-1 Roots of Astronomy

Astronomy has its origin in a noble human trait: curiosity. Just as modern children ask their parents what the stars are and why the Moon has phases, early humans asked themselves those same questions. Their answers, which were often couched in mythical or religious terms, reveal great reverence for the order of the heavens.

Archaeoastronomy

Most of the history of astronomy is lost forever. You can't go to a library or search the Internet to find out what the first astronomers thought about the world because they left no written records. The study of the astronomy of ancient peoples, called **archaeoastronomy** (a combination of "archaeology" and "astronomy"), yields abundant evidence that seeking to understand the heavens is part of human nature.

Perhaps the best-known object investigated by archaeoastronomy is also a major tourist attraction. Stonehenge, standing on Salisbury Plain in southern England, was built in stages from about 3100 BCE to about 1600 BCE, a period extending from the late Stone Age into the Bronze Age. Though the public is most familiar with the monument's massive stones, they were added late in its history. During its first stages, Stonehenge consisted of a circular ditch slightly larger in diameter than the length of a U.S. football field, with a concentric bank just inside the ditch and a long avenue leading away toward the northeast. A massive stone, the Heelstone, stood then, as it does now, outside the ditch in the opening of the avenue.

As early as 1740, the English scholar William Stukely suggested that the avenue pointed toward the rising Sun at the summer solstice, but few historians accepted that it was intentional. Nevertheless, seen from the center of the monument, the summer solstice sun does rise behind the Heelstone. More recently, astronomers have recognized other significant astronomical alignments at Stonehenge. For example, there are sight lines that point toward the most northerly and most southerly horizon rising points of the Moon (**Figure 4-1**).

The significance of these alignments has been debated. Some have claimed that the Stone Age people who built Stonehenge were using it as a device to predict lunar eclipses. After studying eclipse prediction in the previous chapter, you understand that predicting eclipses is easier than most people realize, so perhaps it was used in that way, but the truth may never be known. The builders of Stonehenge had no written language and left no records of their intentions. Nevertheless, the presence of solar and lunar alignments at Stonehenge and at many other Stone Age monuments dotting England and continental Europe shows that so-called primitive peoples were paying close attention to the sky. Building astronomical alignments into structures gives the structures special, even holy, meaning by connecting them with the heavens. The roots of astronomy lie not in sophisticated science and mathematics but in human curiosity and awe.

Astronomical alignments in sacred structures are common all around the world. For example, many tombs are oriented toward the rising Sun, and Newgrange, a 5000-year-old passage grave in Ireland (**Figure 4-2**), faces southeast so that, at dawn on the day of the winter solstice, light from the rising Sun shines into its long passageway and illuminates the central chamber. No one today knows what the alignment meant to the builders of Newgrange. Whatever its original purpose, Newgrange is clearly a sacred site linked by its alignment to the order and power of the sky.

Some alignments may have served purposes related to keeping an annual calendar. The 2000-year-old Temple of Isis in Dendera, Egypt, was built to align with the rising point of the bright star Sirius. Each year, the first appearance of this star in the dawn twilight marked the flooding of the Nile, so it was an important date indicator. The link between Sirius and the Nile was described in Egyptian mythology; the goddess Isis was associated with the star Sirius, and her husband, Osiris, was linked to the constellation now called Orion, and also to the Nile, the source of Egypt's agricultural fertility.

An intriguing American site in New Mexico, known as the Sun Dagger, unfortunately has no surviving mythology to tell its story. At noon on the day of the summer solstice, a narrow dagger of sunlight shines across the center of a spiral carved on a cliff face high above the desert floor (**Figure 4-3**). The purpose of the Sun Dagger is open to debate, but similar examples have been found throughout the U.S. Southwest. It may have had more of a symbolic and ceremonial purpose than a precise calendar function. In any case, it is just one of the many astronomical alignments that ancient people built into their structures to link themselves with the sky.

Sunrise on the morning
of the summer solstice

◄ **Figure 4-1** (a) The central horseshoe of upright stones is only the most obvious part of Stonehenge. (b) The best-known astronomical alignment at Stonehenge is the summer solstice sun rising over the Heelstone. (c) Although a number of astronomical alignments, indicated on the diagram, have been found at Stonehenge, experts debate their significance.

Highway A344 (modern)

N

E

Most northerly moonset
Most northerly moonset
Summer solstice sunrise
Most northerly moonrise
Summer solstice sunrise
Most southerly moonrise at winter solstice
Winter solstice sunset
Summer solstice sunrise
Winter solstice sunrise
Most southerly moonrise
Winter solstice sunset
Most southerly moonrise

Ditch
Bank
● Stone
○ Aubrey Hole

◄ **Figure 4-2** Newgrange was built on a small hill in Ireland about 3200 BCE. A long passageway extends from the entryway back to the center of the mound, and sunlight shines down the passageway into the central chamber at dawn on the day of the winter solstice. Other passage graves have similar alignments, but their purpose is unknown.

High on Fajada Butte, the Sun Dagger is off limits to visitors.

The spiral pattern is the size of a dinner plate.

▶ Figure 4-3 (a) In the ancient Native American settlement known as Chaco Canyon, New Mexico, sunlight shines between two slabs of stone high on the side of 440-foot-high Fajada Butte to form a dagger of light on the cliff face. (b) About noon on the day of the summer solstice, the dagger of light slices through the center of a spiral pecked into the sandstone.

Some archaeoastronomers study small artifacts made thousands of years ago rather than large structures. Scratches on certain bone and stone implements follow a pattern that may record the phases of the Moon (Figure 4-4). Although controversial, such finds suggest that some of the first human written records were of astronomical phenomena.

Archaeoastronomy research uncovers the earliest roots of astronomy and, in so doing, reveals some of the first human efforts at systematic inquiry. The most important lesson of archaeoastronomy is that humans don't have to be technologically advanced to have a sophisticated understanding of celestial cycles.

Although the methods of archaeoastronomy can show how ancient people observed the sky, their thoughts about the Universe are, in many cases, unknown. Many cultures had no written language. In other cases, the written record has been lost or even intentionally destroyed. For instance, dozens, possibly hundreds, of beautiful Mayan manuscripts were burned by Spanish missionaries who thought these were the work of the Devil. Only four of the Mayan books survived, and all four include astronomical references. One contains complicated tables that allowed the Maya to predict the motion of Venus and eclipses of the Moon. However, no one will ever know the extent of what was lost.

The fate of the Mayan books illustrates one reason why histories of astronomy usually begin with the Greeks. Our culture and language is partly descended from theirs, and some of their writing has survived. From that, you can discover what the Greeks thought about the geometry and motions of the heavens.

The Astronomy of Classical Greece

Greek astronomy was derived from Babylon and Egypt, but the Greek philosophers took a new approach. Rather than relying on religion and astrology, the Greeks proposed a rational universe whose secrets could be understood through logic and reason.

As you study early Greek astronomy, you should keep in mind that the Greeks knew much less about the Universe than you do. As you learned or were reminded of in Chapter 1, the stars are other suns scattered through space within the galaxy called the Milky Way. You also realize that out to the greatest distances the largest telescopes can see, the Universe is filled with other galaxies. Early astronomers knew none of this. They saw the Sun, Moon, and planets moving in regular patterns against a background of stars, and they imagined that the entire Universe was hardly more than the bright objects in the Solar

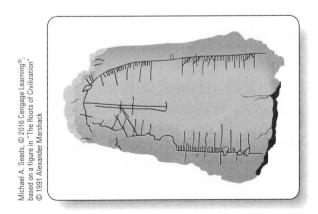

▲ Figure 4-4 A fragment of a mammoth tusk found in Ukraine with an age of at least 15,000 years contains scribe marks on its edge, simplified in this drawing. These markings have been interpreted as a record of four cycles of lunar phases.

System, extending just a little way beyond the planets to an enclosing sphere carrying the stars. In most cases, they believed we inhabit a **geocentric universe**, with Earth at the center. But you will see in this chapter that a small number of astronomers proposed a **heliocentric universe** with the Sun at the center.

When classical Greek astronomers analyzed the heavens using logic and reason, they took the first step toward modern science, which was made possible especially by two early Greek philosophers. Thales of Miletus (c. 624–c. 546 BCE) lived and worked in what is now Turkey. He taught that the Universe is rational and that the human mind can understand why the Universe works the way it does. This view contrasts sharply with that of earlier cultures, which believed that the ultimate causes of things are mysteries beyond human understanding. To Thales and his followers, the mysteries of the Universe were mysteries only because they were unknown, not because they were unknowable.

The second philosopher who made the new scientific attitude possible was Pythagoras (c. 570–c. 495 BCE). He and his students noticed that many things in nature seem to be governed by geometrical or mathematical relations. Musical pitch, for example, is related in a regular way to the lengths of plucked strings. This led Pythagoras to propose that all nature was underlain by musical principles, by which he meant mathematics. One result of this philosophy was the later belief that the harmony of celestial movements produced actual music that was called "the music of the spheres." At a deeper level, the teachings of Pythagoras made Greek astronomers look at the Universe in a new way. Thales said that the Universe could be understood, and Pythagoras said that the underlying rules were mathematical.

In trying to understand the Universe, Greek astronomers did something that Babylonian astronomers had never done: They tried to describe the Universe using geometrical forms. Philolaus (5th century BCE) argued that Earth moved in a circular path around a central fire (not the Sun), which was always hidden behind a counter-Earth located between the fire and Earth. This was the earliest known example of a hypothesis that Earth is in motion.

The great philosopher Plato (c. 424–347 BCE) was not an astronomer, but his teachings influenced astronomy for 2000 years. Plato argued that the reality humans see is only a distorted shadow of a perfect, ideal form. If human observations are distorted, then observation can be misleading, and the best path to truth, said Plato, is through pure thought on the ideal forms that underlie nature.

Plato agreed with other philosophers on a principle known as the *perfection of the heavens*. They saw the beauty of the night sky and the regular motion of the heavenly bodies, and they concluded that the heavens represented perfection. Plato argued that the most perfect geometrical form was the sphere; therefore, the perfect heavens must be made up of spheres rotating at constant rates and carrying objects around in circles. Consequently, later astronomers tried to describe the motions of the heavens by imagining multiple rotating spheres. This became known as the principle of **uniform circular motion**.

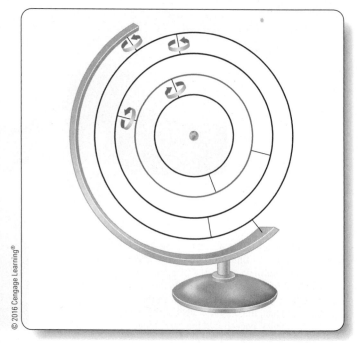

▲ Figure 4-5 The spheres of Eudoxus explain the motions in the heavens by means of nested spheres rotating about various axes at different rates. Earth is located at the center. In this illustration, only 4 of the 27 spheres are shown.

Eudoxus of Cnidus (408–355 BCE), a student of Plato, applied this principle when he devised a system of 27 nested spheres that rotated at different rates about different axes to produce a mathematical description of the motions of the Universe (**Figure 4-5**).

At the time of the Greek philosophers, it was common to refer to systems such as that of Eudoxus as descriptions of the world, where the world included not only Earth but all of the heavenly spheres; today, you would say they were attempting to describe the Universe. The reality of these spheres was open to debate. Some thought of the spheres as nothing more than mathematical ideas that described motion in the world model, whereas others began to think of the spheres as real objects made of perfect celestial material. Aristotle, for example, seems to have thought of the spheres as real.

Aristotle and the Nature of Earth

Aristotle (384–322 BCE), another of Plato's students, made his own unique contributions to philosophy, history, politics, ethics, poetry, drama, and other subjects (**Figure 4-6**). Because of his far-ranging and penetrating insights, he became the greatest authority of antiquity, and later philosophers referred to him as "The Philosopher." His astronomical model was accepted with only minor alterations for almost 2000 years.

Much of what Aristotle wrote about scientific subjects was wrong, but that is not surprising. The modern scientific method, with its insistence on evidence and hypothesis, had not yet been

▲ **Figure 4-6** Aristotle wrote on such a wide variety of subjects and with such deep insight that he became the great authority on all matters of learning. His opinions on the nature of Earth and the sky were widely accepted for almost two millennia.

invented. Aristotle, like other philosophers of his time, attempted to understand his world by reasoning logically and carefully from **first principles**. A first principle is something that is held to be obviously true and needs no further examination. The perfection of the heavens, for instance, was, for Aristotle, a first principle. Once a principle is recognized as true, whatever can be logically derived from it must also be true.

Aristotle believed that the Universe was divided into two parts: Earth, imperfect and changeable, and the heavens, perfect and unchanging. Like most of his predecessors, he believed that Earth was the center of the Universe, so his model is a geocentric universe. The heavens surrounded Earth, and he added parts to the model proposed by Eudoxus to bring the total to 55 crystalline spheres turning at different rates and at different angles to carry the Sun, Moon, and planets across the sky with their observed motions. The lowest sphere, that of the Moon, marked the boundary between the changeable imperfect region containing Earth and the unchanging perfection of the celestial realm beyond the Moon.

Because he believed Earth to be unmoving, Aristotle had to assume that the entire nest of spheres whirls westward around Earth every 24 hours to produce day and night. Different spheres had to move at different rates to produce the motions of the Sun, Moon, and planets against the background of the stars. Because his model was geocentric, he taught that Earth could be the only center of motion, meaning that all of the whirling spheres had to be centered on Earth.

About a century after Aristotle, the Alexandrian philosopher Aristarchus proposed that Earth rotates on its axis and revolves around the Sun. These ideas are, of course, generally correct, but most of the writings of Aristarchus were lost, and his theory was not well known to subsequent scholars. In fact, virtually all later astronomers rejected any suggestion that Earth could move not only because they could feel no motion but also because it conflicted with the teachings of the great philosopher Aristotle.

Aristotle taught that Earth must be a sphere because it always casts a round shadow during lunar eclipses, but he could only roughly estimate its size. About 200 BCE, Eratosthenes (c. 276–c. 195 BCE), working in the great library in the Egyptian city of Alexandria, found a way to measure Earth's radius. He learned from travelers that the city of Syene (Aswan) in southern Egypt contained a well into which sunlight shone vertically on the day of the summer solstice. This told him that the Sun was at the zenith at Syene, but on that same day in Alexandria, he noted that the Sun was 1/50 of the circumference of the sky (about 7 degrees) south of the zenith.

Because sunlight comes from such a great distance, its rays arrive at Earth traveling almost parallel. That allowed Eratosthenes to use simple geometry to conclude that the distance from Alexandria to Syene is 1/50 of Earth's circumference (**Figure 4-7**).

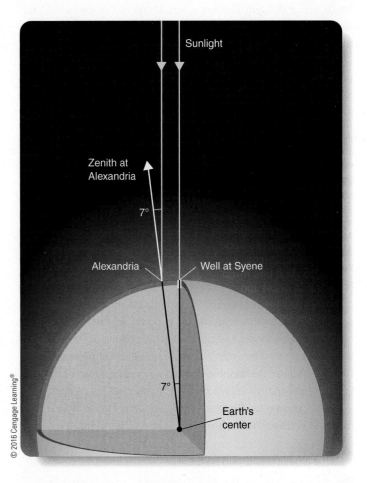

▲ **Figure 4-7** On the day of the summer solstice, sunlight fell to the bottom of a well at Syene, but on the same day the Sun was about 1/50 of a circle (7 degrees) south of the zenith at Alexandria. From this, Eratosthenes was able to calculate Earth's radius.

To find Earth's circumference, Eratosthenes had to learn the distance from Alexandria to Syene. Travelers told him it took 50 days to cover the distance, and he knew that a camel can travel about 100 stadia per day. This meant the total distance was about 5000 stadia. If 5000 stadia is 1/50 of Earth's circumference, then Earth must be 250,000 stadia in circumference, and, dividing by 2π, Eratosthenes found Earth's radius to be about 40,000 stadia.

How accurate was Eratosthenes' estimate? The stadium (singular of stadia) had different lengths in ancient times. If Eratosthenes used the Olympic stadium, his result was 14 percent bigger than the true value—not bad at all. This was a much better measurement of Earth's radius than Aristotle's estimate, which was much smaller, about 40 percent of the true radius.

You might think this is just a disagreement between two ancient philosophers, but it is related to a **Common Misconception.** Christopher Columbus did not have to convince Queen Isabella that the world was round. At the time of Columbus, all educated people (a small fraction of the population) knew that the world was round and not flat, but they weren't sure how big it was. Columbus, like many others, adopted Aristotle's diameter for Earth, so he thought Earth was small enough that he could sail west and reach Japan and the Spice Islands of the East Indies in a couple of months. If he had accepted Eratosthenes' diameter, Columbus would never have risked the voyage. He and his crew were lucky that America was in the way because if there had been open ocean all the way to Japan, they would have starved to death long before they reached land.

Aristotle, Aristarchus, and Eratosthenes were philosophers, but the next person you will meet was a real astronomer who observed the sky in detail. Little is known about Hipparchus, who lived about two centuries after Aristotle, during the 2nd century BCE. He is usually credited with the invention of trigonometry, the creation of the first star catalog, and the discovery of precession (look back to Chapter 2, pages 17 and 20–21). Hipparchus also described the motion of the Sun, Moon, and planets as following circular paths with Earth near—but not at—their centers. These off-center circles are now known as **eccentrics.** Hipparchus recognized that he could reproduce this motion in a model where each celestial body traveled around a small circle that followed a larger circle around Earth. The compounded circular motion that he devised became the key element in the masterpiece of the last great astronomer of classical times, Claudius Ptolemaeus, who is better known as Ptolemy (pronounced *TAHL-eh-mee;* the initial *P* is silent) and whom you met in Chapter 2.

The Ptolemaic Universe

Ptolemy was one of the great astronomer–mathematicians of antiquity. His nationality and birth year are unknown, but around the year 140 he lived and worked in the Greek settlement at Alexandria in what is now Egypt. He ensured the continued acceptance of Aristotle's universe by transforming it into a sophisticated mathematical model.

When you read **An Ancient Model of the Universe** on pages 60–61, notice three important ideas and five new terms that show how first principles influenced early descriptions of the Universe and its motions:

1 Ancient philosophers and astronomers accepted as first principles that the heavens were geocentric with Earth located at the center and the Sun, Moon, and planets moving in uniform circular motion. It seemed clear to them that Earth was not moving because they saw no *parallax* in the positions of the stars.

2 The observed motion of the planets did not fit the theory very well. The *retrograde motion* of the planets was difficult to explain with a model of an unmoving Earth at the center of the Universe and uniform circular motion of celestial objects.

3 In his book that later came to be known as the *Almagest,* Ptolemy attempted to explain the motion of the planets by devising a small circle, an *epicycle,* that rotated along the edge of a larger circle, the *deferent,* which enclosed a slightly off-center Earth. An *equant* was a point from which the center of an epicycle appeared to move at a constant rate. That meant the speed of the planets would vary slightly as viewed from Earth.

Ptolemy lived about five centuries after Aristotle, and although Ptolemy based his work on the Aristotelian universe, he was interested in a specific problem—the motions of the planets. Ptolemy was a brilliant mathematician, and he was mainly interested in creating a mathematical description of the motions he saw in the heavens. For him, first principles took second place to mathematical precision.

Aristotle's universe, as embodied in the mathematics of Ptolemy, dominated ancient astronomy, but it was wrong. The planets do not follow circles at uniform speeds. At first, the Ptolemaic system predicted the positions of the planets well, but as centuries passed, errors accumulated. If your watch gains only one second each day, it will keep time pretty well for months, but the error will gradually become noticeable. So, too, the errors in the Ptolemaic system gradually accumulated as the centuries passed, but because of the deep respect people had for the writings of Aristotle, the Ptolemaic system was not abandoned.

Islamic and later European astronomers tried to update the system, computing new constants and adjusting epicycles. In the middle of the 13th century a team of astronomers supported by King Alfonso X of Castile studied the *Almagest* for ten years. Although they did not revise the theory very much, they simplified the calculation of the positions of the planets using the Ptolemaic system and published the result as *The Alfonsine*

Tables, named in honor of the king, the last great attempt to make the Ptolemaic system of practical use.

4-2 The Copernican Revolution

You would not have expected Nicolaus Copernicus (1473–1543) to trigger a revolution in astronomy and science. He was born to a prosperous and politically connected merchant family in Poland. Orphaned at the age of 10, he was raised by his uncle, an important bishop, who sent him to the University of Krakow and then to the best universities in Italy. Copernicus studied law and medicine before pursuing a lifelong career as a Church administrator. Nevertheless, he had a passion for astronomy (Figure 4-8).

Copernicus and the Heliocentric Hypothesis

If you could go back in time and sit beside Copernicus in his astronomy classes, you would study the Ptolemaic model, a detailed version of Aristotle's universe. The central location of Earth was widely accepted, and everyone knew that the heavens moved in uniform circular motion. For most scholars, questioning these principles was not an option because, over the course of centuries, Aristotle's universe had become linked with Christian theology. According to the Aristotelian view, the most perfect region was in the heavens and the most imperfect at Earth's center. That classical geocentric universe matched the commonly held Christian geometry of heaven and hell, so anyone who criticized the Ptolemaic model was not only questioning Aristotle's universe but also indirectly challenging belief in heaven and hell.

For this reason, Copernicus probably found it difficult at first to consider alternatives to the Ptolemaic universe. Throughout his life, he was associated with the Catholic Church, which had adopted many of Aristotle's ideas. Through the influence of his uncle the bishop, Copernicus was appointed a canon (a type of Church official) of the cathedral in Frauenberg at the unusually young age of 24. This gave Copernicus an income, although he continued his studies at the universities in Italy. When he finally left Italy, he joined his uncle and served as his secretary and personal physician until the uncle's death in 1512. At that point, Copernicus moved into quarters adjoining the cathedral in Frauenberg, where he lived and worked for the rest of his life.

His close connection with the Church notwithstanding, Copernicus began to consider an alternative to the Ptolemaic model, probably while he was still at university. Sometime before 1514, he wrote an essay proposing a heliocentric universe model in which the Sun, not Earth, is the center of the Universe. To explain the daily and annual cycles of the sky, Copernicus proposed that Earth rotates on its axis and revolves around the Sun. He distributed this commentary in handwritten form, without a title, and in some cases anonymously, to friends and astronomical correspondents. He may have been cautious out of modesty, out of respect for the Church, or out of fear that his unorthodox ideas would be attacked unfairly. Although his early essay discussed every major aspect of his later work, it did not include observations and calculations. His ideas needed supporting evidence, so Copernicus began gathering observations and making detailed calculations to be published as a book that would demonstrate the truth of his revolutionary idea.

◀ **Figure 4-8** Nicolaus Copernicus (Latinized version of his birth name, Mikolaj Kopernik) pursued a lifetime career in the Church, but he was also a talented mathematician and astronomer. His work triggered a revolution in human thought.

An Ancient Model of the Universe

1 For 2000 years, the minds of astronomers were shackled by a pair of ideas. The Greek philosopher Plato argued that the heavens were perfect. Because he considered the only perfect geometrical shape to be a sphere, and a turning sphere carries a point on its surface around in a circle, and because the only perfect motion is uniform motion, Plato concluded that all motion in the heavens must be made up of combinations of circles turning at uniform rates. This idea was called *uniform circular motion.*

Plato's student Aristotle argued that Earth was imperfect and lay at the center of the Universe. Such a model is known as a *geocentric universe.* His model contained 55 spheres turning at different rates and at different angles to carry across the sky the seven celestial objects called planets at the time (the Moon, Mercury, Venus, the Sun, Mars, Jupiter, and Saturn).

Aristotle was known as the greatest philosopher in the ancient world, and for 2000 years his authority limited the imaginations of astronomers to uniform circular motion and geocentrism. See the model at right.

From *Cosmographica* by Peter Apian (1539).

Seen by left eye   Seen by right eye

1a Early astronomers believed that Earth did not move because they saw no **parallax**, the apparent motion of an object because of the motion of the observer. (To demonstrate parallax, close one eye and cover a distant object with your thumb held at arm's length. Switch eyes, and your thumb seems to shift position as shown at left.) If Earth moves, they reasoned, you should see the sky from different locations at different times of the year, and you should see parallax distorting the shapes of the constellations. They saw no parallax, so they concluded Earth does not move. Actually, the parallax of the stars is too small to see with the unaided eye.

Every 2.14 years, Mars passes through a retrograde loop. Two successive loops are shown here. Each loop occurs farther east along the ecliptic and has its own shape. Simple uniform circular motion centered on Earth could not explain retrograde motion, so ancient astronomers combined uniformly rotating circles much like gears in a machine to try to reproduce the motion of the planets.

2 Planetary motion was a big problem for ancient astronomers. In fact, the word planet comes from the Greek word for "wanderer," referring to the eastward motion of the "planets" (including the Moon and the Sun) against the background of the fixed stars and constellations. The planets do not, however, move at a constant rate, and they could occasionally stop and move westward for a few months before resuming their eastward motion. This backward motion is called *retrograde motion*.

Leo

West

April 16, 2016

Regulus

June 30, 2016

Virgo

Ecliptic

June 27, 2018

2a Simple uniform circular motion centered on Earth could not explain retrograde motion, so ancient astronomers combined uniformly rotating circles much like gears in a machine to try to reproduce the motion of the planets.

3 Uniformly rotating circles were key elements of ancient astronomy. Ptolemy created a mathematical model of the Aristotelian universe in which the planet followed a small circle called an **epicycle** that slid around a larger circle called a **deferent**. By adjusting the size and rate of rotation of the circles, he could approximate the retrograde motion of a planet. See illustration at right.

To adjust the speed of the planet, Ptolemy supposed that Earth was slightly off-center, and that the center of the epicycle moved such that it appeared to move at a constant rate as seen from a point called the **equant**.

To further adjust his model, Ptolemy added small epicycles (not shown here) riding on top of larger epicycles, producing a highly complex model.

3a Ptolemy's great book *Mathematike Syntaxis* (published around the year 140) contained the details of his model. Islamic astronomers preserved and studied the book through the Middle Ages; they called it *Al Magisti (The Greatest)*. When the book was translated from Arabic to Latin in the 12th century, it became known as the *Almagest*.

3b The Ptolemaic model of the Universe shown below was geocentric and based on uniform circular motion. Note that Mercury and Venus were treated differently from the rest of the planets. The centers of the epicycles of Mercury and Venus had to remain on the Earth–Sun line as the Sun circled Earth through the year.

Equants and smaller epicycles are not shown here. Some versions of the model contained nearly 100 epicycles added by generations of astronomers trying to fine-tune the model to better reproduce the observed motion of the planets.

Notice that this modern illustration shows rings around Saturn and sunlight illuminating the globes of the planets, features that would not have been known before the invention of the telescope.

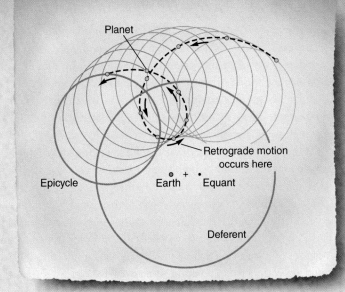

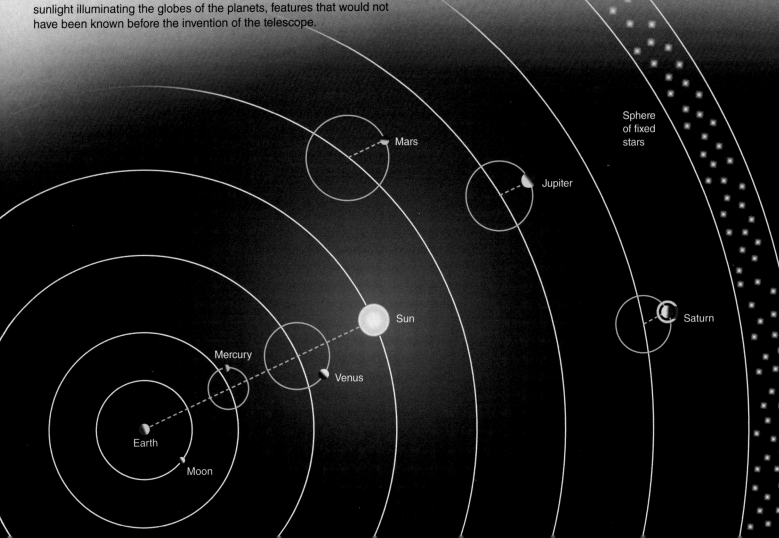

Copernicus's Book: *De Revolutionibus*

Copernicus worked on his book *De Revolutionibus Orbium Coelestium (On the Revolutions of the Celestial Spheres)* over a period of many years and was essentially finished by about 1529. Even some Church officials, concerned about the reform of the calendar, sought his advice and looked forward to the book's publication. Nevertheless, Copernicus hesitated to publish it although some astronomers already knew of his work.

One reason he hesitated was that he knew the idea of a heliocentric universe would be highly controversial. This was a time of rebellion in the Church; Martin Luther was criticizing many Church teachings, and others—both scholars and scoundrels—were questioning the Church's authority. As you have learned, Earth's place in the Universe was linked in people's minds to the locations of heaven and hell, so moving Earth from its central place could be considered a heretical idea. Even matters as abstract as astronomy had a potential to stir possibly dangerous controversy.

Another reason Copernicus may have hesitated to publish his work was that it was incomplete. His model could not accurately predict planetary positions, so he continued to refine it. Finally in 1540 he allowed the visiting astronomer Joachim Rheticus to publish an account of the Copernican universe in Rheticus's book *Narratio Prima (First Narrative)*. In late 1542, Copernicus sent the manuscript of *De Revolutionibus* off to be printed. He died in the spring of 1543 before the printing was completed.

The most important idea in the book was placing the Sun at the center of the Universe. That single innovation had an astonishing consequence: The retrograde motion of the planets was immediately explained in a straightforward way without the many epicycles used by Ptolemy.

In the Copernican system, Earth moves faster along its orbit than the planets that lie farther from the Sun. Consequently, Earth periodically overtakes and passes these planets. To visualize this, imagine that you are in a race car, driving rapidly along the inside lane of a circular racetrack. As you pass slower cars driving in the outer lanes, they fall behind, and if you did not realize you were moving, it would look as if the cars in the outer lanes occasionally slowed to a stop and then backed up for a short interval as you lapped them. Figure 4-9 shows how the same thing happens as Earth passes a planet such as Mars. Although Mars moves steadily along its orbit, as seen from Earth it appears to slow to a stop and move westward (backward, retrograde) as Earth passes it. This happens to any planet whose orbit lies outside Earth's orbit, so Mars, Jupiter, and Saturn are all seen occasionally to move retrograde along the ecliptic. Because the planetary orbits do not lie precisely in the ecliptic, a planet does not resume its eastward, forward motion in precisely the same path it followed before the retrograde motion began. Consequently, the planet's path describes a loop with a shape that depends on the angle between the planet's orbital plane and the ecliptic and on the planet's location along the ecliptic.

Copernicus could explain retrograde motion without epicycles, and that was impressive. The Copernican system was

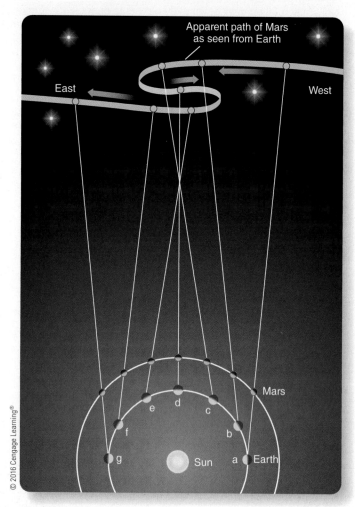

© 2016 Cengage Learning®

▲ **Figure 4-9** The Copernican explanation of retrograde motion. As Earth overtakes Mars (positions a through c), Mars appears to slow its forward motion. As Earth passes Mars (d), Mars appears to move backward. As Earth draws ahead of Mars (e–g), Mars resumes its forward motion against the background stars. The positions of Earth and Mars are shown at equal intervals of one month.

elegant and simple compared with the multiple whirling epicycles and off-center equants of the Ptolemaic model. You can see Copernicus's own diagram for his heliocentric model in Figure 4-10a. However, *De Revolutionibus* failed in one critical way: The Copernican model could not predict the positions of the planets any more accurately than the Ptolemaic model could. To understand why it failed, you need to understand Copernicus's mind-set.

Copernicus proposed an astonishingly bold idea when he made the Universe (meaning, the Solar System) heliocentric. Nevertheless, Copernicus was a classically trained astronomer with tremendous respect for the old concept of uniform circular motion. In fact, Copernicus objected strongly to Ptolemy's use of the equant. It seemed arbitrary to Copernicus, an obvious violation of the elegance of Aristotle's philosophy of the heavens. Copernicus called equants "monstrous" because they undermined both geocentrism and uniform circular motion. In devising his

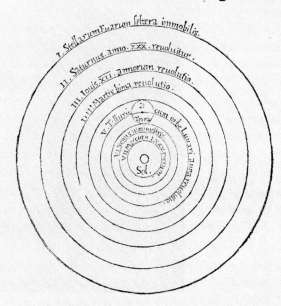

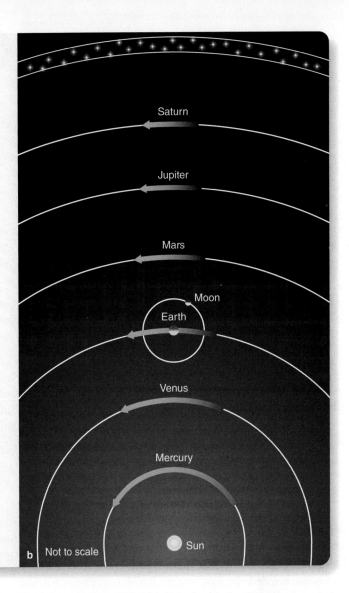

▲ **Figure 4-10** (a) The Copernican universe as drawn by Copernicus in his book *De Revolutionibus*. Earth and all the known planets revolve in separate circular orbits about the Sun (Sol) at the center. The outermost sphere carries the immobile stars of the celestial sphere. Notice the orbit of the Moon around Earth (Terra). (b) The model was elegant not only in its arrangement of the planets but also in their motions. Orbital speed (blue arrows) decreased from Mercury, the fastest, to Saturn, the slowest. Compare the elegance of this model with the complexity of the Ptolemaic model on page 61.

model, Copernicus managed to abandon geocentrism but not his strong belief in uniform circular motion.

Although he did not need epicycles to explain retrograde motion, Copernicus quickly discovered that the Sun, Moon, and planets showed other small variations in their motions that he could not explain using uniform circular motion centered on the Sun. Today astronomers recognize those variations as the result of planets following orbits that are not quite circular; the orbits actually are very slightly elliptical. Because Copernicus held firmly to uniform circular motion, he had to include his own new, small epicycles to try to reproduce these minor variations observed in the motions of the objects in the Solar System.

Because Copernicus retained uniform circular motion in his model, it could not accurately predict the motions of the planets. Only 9 years after the publication of the Copernican theory, a set of planetary tables called the *Prutenic Tables* (named after Prussia) was calculated and published by the astronomer Erasums Reinhold. Although these tables were based on the new Copernican model, they were not significantly more accurate than the 300-year-old *Alfonsine Tables* that were based on Ptolemy's model. Both could be in error by as much as 2 degrees, which is four times the angular diameter of the full moon.

At this point you should consider an important distinction: The Copernican *model* is inaccurate. It includes uniform circular motion and consequently does not precisely describe the motions of the planets. But the Copernican *hypothesis*—that the Solar System is heliocentric—is correct. The planets in fact circle the Sun, not Earth.

Although astronomers throughout Europe read and admired *De Revolutionibus,* they did not immediately accept the

How Do We Know? (4-1)

Scientific Revolutions

How do scientific revolutions occur? You might think from what you know of the scientific method that science grinds forward steadily as new hypotheses are tested against evidence and accepted or rejected. In fact, science sometimes leaps forward in scientific revolutions. The Copernican Revolution is often cited as the perfect (and first) example. In the course of a few decades, astronomers rejected the 2000-year-old geocentric model and adopted the heliocentric model. Why does that happen? It's all because scientists are human.

The philosopher of science Thomas Kuhn has referred to a commonly accepted set of scientific ideas and assumptions as a scientific **paradigm**. The pre-Copernican astronomers shared a geocentric paradigm that included uniform circular motion, geocentrism, and the perfection of the heavens. Although they were intelligent, they were prisoners of that paradigm. A scientific paradigm is powerful because it shapes your perceptions. It determines what you judge to be important questions and what you judge to be significant evidence. Consequently, the ancient astronomers could not recognize how their geocentric paradigm limited their understanding.

The works of Copernicus, Galileo, and Kepler overthrew the geocentric paradigm. Scientific revolutions occur when the deficiencies of the old paradigm build up until finally someone has the insight to think "outside the box." Pointing out the failings of the old ideas and proposing a new paradigm with supporting evidence is like poking a hole in a dam; suddenly the pressure is released, and the old paradigm is swept away.

Scientific revolutions are exciting because they give you a dramatic new understanding of nature, but they are also times of conflict as new insights sweep away old ideas.

Visual

People originally believed stars are attached to the surface of a sphere that enclosed the Universe.

NSF/AURA/NOAO/N. Sharp

Copernican hypothesis. The mathematics were elegant, and the astronomical observations and calculations were of tremendous value, but few astronomers believed, at first, that the Sun actually was the center of the planetary system and that Earth moved. How the Copernican hypothesis was gradually recognized as correct has been called the Copernican Revolution because it was not just the adoption of a new idea but a total change in the way astronomers, and, truly, all of humanity, thought about the place of Earth. In fact, our modern use of the words *revolution* and *revolutionary* to describe philosophical, political, and social upheavals comes from the title of Copernicus's book, *On Revolutions* (**How Do We Know? 4-1**).

There are several reasons why the Copernican hypothesis gradually won support, including the revolutionary (!) temper of the times, but the most important factor may have been the elegance and simplicity of the idea. Placing the Sun at the center of the Universe produced a symmetry among the motions of the planets that is pleasing to the eye as well as to the intellect. In the Ptolemaic model, Mercury and Venus had to be treated differently from the rest of the planets; their epicycles had to remain centered on the Earth–Sun line because they are never seen far from the Sun. In contrast, in the Copernican model, all of the planets were treated the same. They all followed orbits that circled the Sun at the center. Furthermore, their speed depended in an orderly way on their distance from the Sun, with those closest moving fastest (Figure 4-10b).

The most astonishing consequence of the Copernican hypothesis was not what it said about the Sun but what it said about Earth. By placing the Sun at the center, Copernicus made Earth into a planet moving along an orbit like the other planets. By making Earth a planet, Copernicus completely changed humanity's view of its place in the Universe and triggered a controversy that would eventually bring the astronomer Galileo

DOING SCIENCE

Why would you say the Copernican hypothesis was correct but the Copernican model was inaccurate? Distinguishing between hypotheses and models is an important part of doing science.

The Copernican hypothesis was that the Sun and not Earth is the center of the Universe. Given the limited knowledge of the Renaissance astronomers about distant stars and galaxies, that hypothesis was correct.

The Copernican model, however, included not only the heliocentric hypothesis but also an assumption of uniform circular motion. The model is inaccurate because the planets don't really follow circular orbits, nor travel at constant rates, and the small epicycles that Copernicus added to his model never quite reproduced the motions of the planets.

Now, look back to Chapter 2. The celestial sphere is a model. **How is it accurate—how does it allow precise calculations of astronomical phenomena? If, instead, the celestial sphere were to be considered a hypothesis, is it correct?**

Galilei before the Inquisition. This controversy over the apparent conflict between scientific knowledge and philosophical and theological ideas continues even today.

4-3 Tycho, Kepler, and Planetary Motion

The Copernican hypothesis solved the problem of the place of Earth, but it did not completely explain the details of planetary motion. If planets do not move in uniform circular motion, how do they move? The puzzle of planetary motion was solved during the century following the death of Copernicus almost entirely through the work of two scientists. One compiled the observations, and the other did the analysis.

Tycho Brahe

Tycho Brahe (1546–1601), known today simply as Tycho (pronounced *TEE-co*), was not a churchman like Copernicus but rather a nobleman from an important family, but like Copernicus he was educated at the finest universities. He was infamous for his vanity and his lordly, haughty manners. Tycho's disposition perhaps was not improved by a dueling injury from his university days. His nose was badly disfigured, and for the rest of his life he wore false noses made of gold and silver, stuck on with wax (Figure 4-11a).

Although Tycho officially studied law at the university, his real passions, much to his family's disappointment, were mathematics and astronomy, and early in his university days he began measuring the positions of the planets in the sky. In 1563, Jupiter and Saturn passed very near each other in the sky, nearly merging into a single point on the night of August 24. Tycho found that the *Alfonsine Tables,* based on the Ptolemaic model, were a full month in error in their prediction of that event, and the *Prutenic Tables,* based on the Copernican model and calculated just a few years earlier, were in error by several days.

In 1572, a "new star" (now called Tycho's supernova) appeared in the sky, shining more brightly than Venus, and Tycho carefully measured its position. According to classical astronomy, the new star represented a change in the heavens and therefore had to lie below (closer than) the sphere of the Moon. To Tycho, who at this time still believed in a geocentric universe, this meant that the new star should show parallax, meaning that it would appear slightly too far east as it rose and slightly too far west as it set (Figure 4-12). But Tycho saw no parallax in the position of the new star, so he concluded that it must lie above the sphere of the Moon and was probably on the starry sphere itself. This contradicted Aristotle's conception of the starry sphere as perfect and unchanging.

No one before Tycho could have made this discovery because no one had ever measured the positions of celestial

▲ Figure 4-11 Tycho Brahe was, during his lifetime, the most famous astronomer in the world. Tycho's model of the Universe retained the first principles of classical astronomy; it was geocentric with the Sun and Moon revolving around Earth, but the planets revolved around the Sun. All motion was along circular paths.

objects as accurately as he did. Tycho had great confidence in the precision of his measurements, and he had studied astronomy thoroughly, so when he failed to detect parallax for the new star, he knew it was important evidence against the Ptolemaic theory. He announced his discovery in a small book, *De Stella Nova (On the New Star),* published in 1573.

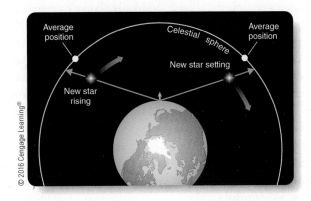

▲ Figure 4-12 According to Aristotle, the new star of 1572 should have been located below the sphere of the Moon; consequently, reasoned Tycho, it should display parallax and be seen east of its average position as it was rising and west of its average position when it was setting. Because he did not detect this daily parallax, Tycho concluded that the new star of 1572 had to lie on the celestial sphere, far beyond the Moon.

The book attracted the attention of astronomers throughout Europe, and soon Tycho's family, concerned about his professional future, introduced him to the court of the Danish King Frederick II, where he was offered funds to build an observatory on the island of Hveen just off the Danish coast. To support his observatory, Tycho was given a steady income as lord of a coastal district from which he collected rents. (It is said that he was not a popular landlord.) On Hveen, Tycho constructed a luxurious home with six towers especially equipped for astronomy and populated it with servants, assistants, and a dwarf to act as jester. Soon Hveen was an international center of astronomical study.

Tycho's Legacy

Tycho made no direct contribution to astronomical theory. Because he could measure no parallax for the stars, he concluded that Earth had to be stationary, thus rejecting the Copernican hypothesis. However, he also rejected the Ptolemaic model because of its inaccuracy. Instead, he devised a complex model in which Earth was the unmoving center of the Universe around which the Sun and the Moon moved while the other planets circled the Sun (Figure 4-11b). The model thus incorporated part of the Copernican model, but Tycho preserved the central immobile Earth. Although Tycho's model was popular at first, the Copernican model replaced it within a century.

The true value of Tycho's work was the quality of his observational data. Because he was able to devise new and better instruments, he was able to make highly accurate observations of the position of the stars, Sun, Moon, and planets. Tycho had no telescopes—they were not invented until the following century—so his observations were made by naked eye, peering along devices much like gun sights. He and his assistants made precise observations for 20 years at Hveen.

Unhappily for Tycho, King Frederick II died in 1588, and his young son took the throne. Suddenly, Tycho's temper, vanity, and noble presumptions threw him out of favor. In 1596, taking most of his instruments and books of observations, he went to Prague, the capital of Bohemia, and became imperial mathematician to the Holy Roman Emperor Rudolph II. His goal was to revise *The Alfonsine Tables* and publish the result as a monument to his new patron. He promised to call it *The Rudolphine Tables*.

Tycho did not intend to base *The Rudolphine Tables* on the Ptolemaic system but rather on his own Tychonic system, proving once and for all the validity of his hypothesis. To assist him, he hired a few mathematicians and astronomers, including one Johannes Kepler. Then, in November 1601, Tycho collapsed while visiting a nobleman's home in Prague. Before he died 11 days later, he asked Rudolph II to make Kepler imperial mathematician. The newcomer, not a nobleman at all but a commoner, became Tycho's replacement (at one-sixth Tycho's salary).

Johannes Kepler

No one could have been more different from Tycho Brahe than Johannes Kepler (Figure 4-13a). Kepler was born in 1571 to a poor family in a region that is now part of southwest Germany. His father was unreliable and shiftless, principally employed as a mercenary soldier fighting for whomever paid enough. He was often absent for long periods and finally failed to return from a military expedition. Kepler's mother was apparently an unpleasant and unpopular woman. She was accused of witchcraft, and Kepler had to defend her in a trial that dragged on for three years. She was finally acquitted but died the following year.

Despite family disadvantages and chronic poor health, Kepler did well in school, winning promotion to a Latin school and eventually a scholarship to the university at Tübingen, where he studied to become a Lutheran pastor. During his last year of study, Kepler accepted a job in Graz teaching mathematics and astronomy. His superiors put him to work teaching a few introductory courses and preparing an annual almanac that contained astronomical, astrological, and weather predictions. Evidently he was not a good teacher; he had few students his first year and none at all his second. Fortunately, and of course only by pure luck, some of his predictions were considered to be fulfilled, and he gained a reputation as an astrologer and seer. Even in later life he earned money from his almanacs.

While still a college student, Kepler had become a believer in the Copernican hypothesis, and at Graz he used his extensive spare time to study astronomy. By 1596, the same year Tycho arrived in Prague, Kepler was sure he had solved the mystery of the Universe. That year he published a book in Latin with a 28-word title that is usually abbreviated as *Mysterium Cosmographicum (Mystery of the Universe)*.

By modern standards, *Mysterium Cosmographicum* contains almost nothing of value. It begins with a long appreciation of Copernicanism and then goes on to speculate on the reasons for the spacing of the planetary orbits. Kepler assumed that the heavens could be described by only the most perfect of shapes. Therefore, he felt that he had found the underlying architecture of the Universe in the sphere plus the five regular solids. In Kepler's model, the five regular solids became spacers for the orbits of the six planets, which were represented by nested spheres (Figure 4-13b). In fact, Kepler went so far as to conclude that there must be only six planets (Mercury, Venus, Earth, Mars, Jupiter, and Saturn) because there were only five regular solids to act as spacers between their spheres. He provided astrological, numerological, and even musical arguments for his theory.

The second half of the book is no better than the first, but it has one virtue: As Kepler tried to fit the five solids to the planetary orbits, he demonstrated that he was a talented mathematician and that he was well versed in astronomy. He sent copies of his book to well-known astronomers, including Tycho and Galileo.

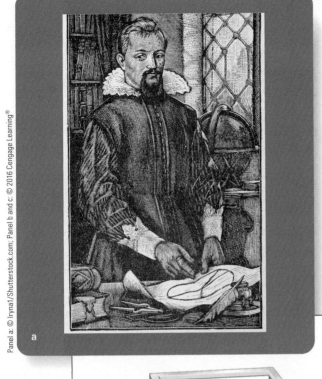

◀ **Figure 4-13** (a) Johannes Kepler was Tycho Brahe's successor. (b) This diagram, based on one drawn by Kepler, shows how he believed the sizes of the celestial spheres carrying the outer three planets—Saturn, Jupiter, and Mars—are determined by spacers (blue) consisting of two of the five regular solids. Inside the sphere of Mars, the remaining regular solids separated the spheres of Earth, Venus, and Mercury (not shown in this drawing). The Sun lay at the very center of this Copernican universe based on geometrical spacers. (c) The five regular solids are the tetrahedron, cube, octahedron, dodecahedron, and icosahedron, the only shapes with all faces of equal sizes and all equal angles between the faces.

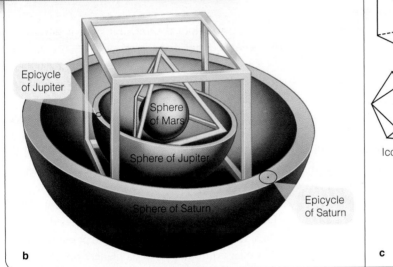

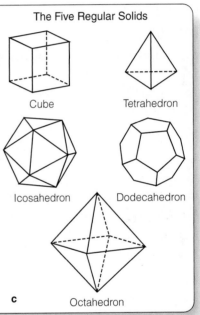

Working with Tycho

Life became unsettled for Kepler because of recurring persecution of Protestants in the region where he lived, so when Tycho Brahe invited him to Prague in 1600, Kepler went readily, eager to work with a famous astronomer. Tycho's sudden death in 1601 left Kepler, the new imperial mathematician, in a position to use Tycho's data to analyze the motions of the planets and complete *The Rudolphine Tables.* Tycho's family, recognizing that Kepler was a Copernican and guessing that he would not follow the Tychonic system in completing *The Rudolphine Tables,* sued to recover the instruments and books of observations. The legal wrangle went on for years. Tycho's family did get back the instruments Tycho had brought to Prague, perhaps because Kepler,

with his poor eyesight, couldn't use them, but Kepler had the books of observational data, and he kept them.

Whether Kepler had any legal right to Tycho's records is debatable, but he put them to good use. He began by studying the motion of Mars, trying to deduce from the observations how the planet moved. By 1606, he had solved the mystery. The orbit of Mars is not a circle but an ellipse. With that, Kepler gave up the 2000-year-old belief in the circular motion of the planets. But even this insight was not enough to explain the observations. The planets do not move at uniform speeds along their elliptical orbits. Kepler's analysis showed that they move faster when close to the Sun and slower when farther away. With those two brilliant discoveries, Kepler abandoned both uniform motion and circular motion, and thereby finally solved the puzzle of planetary motion. He published his results in 1609 in a book called *Astronomia Nova (New Astronomy)*.

Despite the abdication of Rudolph II in 1611, Kepler continued his astronomical work. He wrote about a supernova that appeared in 1604 (now known as Kepler's supernova) and about comets, and he produced a textbook about Copernican astronomy. In 1619, he published *Harmonices Mundi (The Harmony of the World)*, in which he returned to the cosmic mysteries of *Mysterium Cosmographicum*. The only thing of note in *Harmonices Mundi* is Kepler's discovery that the radii of the planetary orbits are related to the planets' orbital periods. That and his two previous discoveries are so important that they have become known as the three fundamental laws of orbital motion.

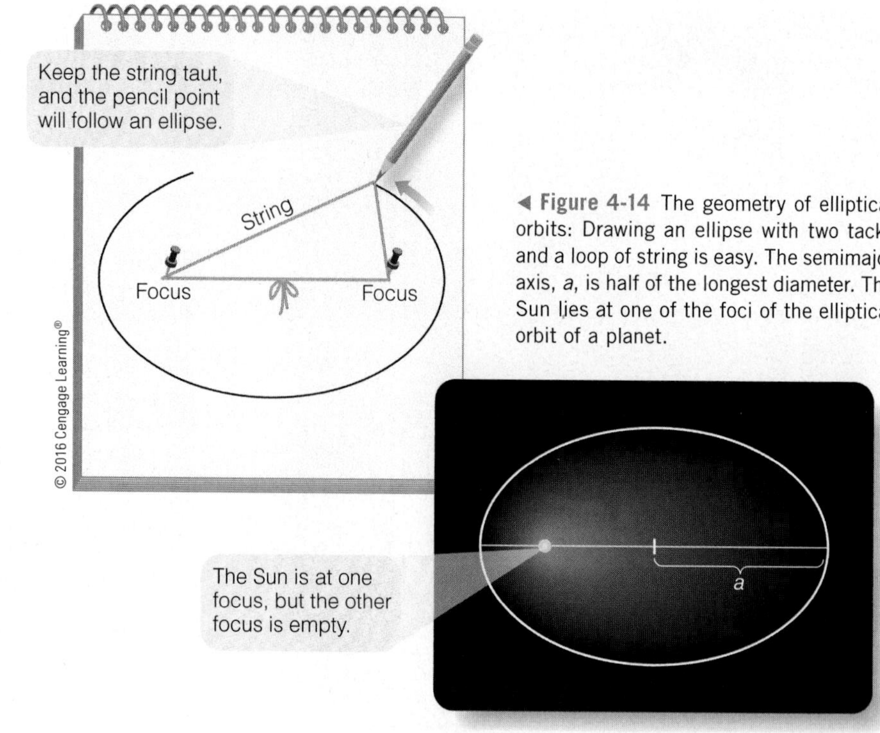

Keep the string taut, and the pencil point will follow an ellipse.

String

Focus   Focus

© 2016 Cengage Learning®

◀ **Figure 4-14** The geometry of elliptical orbits: Drawing an ellipse with two tacks and a loop of string is easy. The semimajor axis, *a*, is half of the longest diameter. The Sun lies at one of the foci of the elliptical orbit of a planet.

The Sun is at one focus, but the other focus is empty.

a

Kepler's Three Laws of Planetary Motion

Although Kepler dabbled in the philosophical arguments of his day, he was at heart a mathematician, and his triumph was his mathematical explanation of the motion of the planets. The key to his solution was the ellipse.

An **ellipse** is a figure that can be drawn around two points, called the foci, in such a way that the distance from one focus to any point on the ellipse and back to the other focus equals a constant. This makes it easy to draw ellipses using two thumbtacks and a loop of string. Press the thumbtacks into a board, loop the string about the tacks, and place a pencil in the loop. If you keep the string taut as you move the pencil, it traces out an ellipse (**Figure 4-14**).

The geometry of an ellipse is described by two simple numbers: (1) The **semimajor axis, *a*,** is half of the longest diameter, as you can see in (**Figure 4-14**). (2) The **eccentricity, *e*,** of an ellipse is half the distance between the foci divided by the semimajor axis. The eccentricity of an ellipse tells you its shape; if *e* is nearly equal to one, the ellipse is very elongated. If *e* is close to zero, the ellipse is more circular. To draw a circle with the string and tacks shown in Figure 4-14, you would have to move the two thumbtacks together because a circle is really just an ellipse with eccentricity equal to zero.

Ellipses are an essential part of Kepler's three fundamental rules of planetary motion (**Table 4-1**). Those rules have been tested and confirmed so many times that astronomers now refer to them as natural laws (**How Do We Know? 4-2**).

Kepler's first law says that the orbits of the planets around the Sun are ellipses with the Sun at one focus. Thanks to the precision of Tycho's observations and the sophistication of Kepler's mathematics, Kepler was able to recognize the elliptical shape of the orbits even though they are nearly circular. Mercury

has the most elliptical orbit, but even it deviates only slightly from a circle (**Figure 4-15**).

Kepler's second law says that an imaginary line drawn from the planet to the Sun always sweeps over equal areas in equal intervals of time. This means that when the planet is closer to the Sun and the line connecting it to the Sun is shorter, the planet must move more rapidly so that the line sweeps over the same area per time interval that it sweeps over when the planet is farther from the Sun. For example, the hypothetical planet in Figure 4-15 that follows a highly elongated ellipse moves from point *A* to point *B* in one month, sweeping over the area shown. When the planet is farther from the Sun, one month's motion along the orbit would be less, from *A'* to *B'*, but the area swept out would be the same.

Kepler's third law relates a planet's orbital period to its average distance from the Sun. The orbital period, *P*, is the time a planet takes to travel around the Sun once. The average distance of a planet from the Sun around its elliptical path turns out simply to equal the semimajor axis of its orbit, *a*. Kepler's third law says that a planet's orbital period squared is proportional to the

TABLE 4-1   Kepler's Laws of Planetary Motion

I. The orbits of the planets are ellipses with the Sun at one focus.

II. A line from a planet to the Sun sweeps over equal areas in equal intervals of time.

III. A planet's orbital period squared is proportional to its average distance from the Sun cubed:

$$P_{yr}^2 = a_{AU}^3$$

Hypotheses, Theories, and Laws

Why is a theory much more than just a guess? Scientists study nature by devising and testing new hypotheses and then developing the successful ideas into theories and laws that describe how nature works. A good example is the connection between sour milk and the spread of disease.

A scientist's first step in solving a natural mystery is to propose a reasonable explanation based on what is known so far. This proposal, called a **hypothesis**, is a single assertion or statement that can be tested through observations and experiments. If the explanation is not testable somehow, it is not really a scientific hypothesis.

From the time of Aristotle, philosophers believed that food spoils as a result of the spontaneous generation of life—for example, mold growing out of drying bread. French chemist Louis Pasteur hypothesized instead that microorganisms were not spontaneously generated but were carried through the air. To test his hypothesis, he sealed an uncontaminated nutrient broth in glass, completely protecting it from the spores, microorganisms, and dust particles in the air. No mold grew, effectively disproving spontaneous generation. Although others had argued against spontaneous generation before Pasteur, it was Pasteur's meticulous testing of his hypothesis through experimentation that finally convinced the scientific community.

A **theory** generalizes the specific results of well-confirmed hypotheses to give a broad description of nature that can be applied to a wide variety of circumstances. For instance, Pasteur's specific hypothesis about mold growing in broth contributed to a broader theory that disease is caused by microorganisms transmitted from sick people to healthy people. This theory, called the *germ theory of disease,* is a cornerstone of modern medicine. It is a **Common Misconception** that the

A fossil of a 500-million-year-old trilobite: Darwin's theory of evolution has been tested many times and is universally accepted in the life sciences as a natural law, but by custom it is called Darwin's theory and not Darwin's law.

Collection of John Coolidge III/Franklin & Marshall College

word *theory* means a tentative idea, a guess. Actually, "hypothesis" is the word that scientists use to describe what a layperson would call a guess. Scientists generally use the word *theory* to mean an idea that is widely applicable, has been tested in many ways, and confirmed by abundant evidence—solid and trustworthy, much better than a guess.

Sometimes, when a theory has been refined, tested, and confirmed so often that scientists have great confidence in it, and is so basic that it is applicable everywhere and every time, it is called a **natural law**. Natural laws are the most fundamental principles of scientific knowledge. Kepler's laws of planetary motion are good examples.

Confidence is the key criterion. In general, scientists have more confidence in a theory than in a hypothesis and the most confidence in a natural law. However, there is no precise distinction among a hypothesis, a theory, and a law, and use of these terms is sometimes a matter of tradition. For instance, some textbooks refer to the Copernican "theory" of heliocentrism, but it had not been well tested when Copernicus proposed it, and it is more rightly called the Copernican hypothesis. At the other extreme, Darwin's "theory" of evolution, containing many hypotheses that have been tested and confirmed over and over for nearly 150 years, might more correctly be called a natural law.

---

semimajor axis of its orbit cubed. Measuring P in years and a in astronomical units, you can summarize the third law as:

$$P_{yr}^2 = a_{AU}^3$$

For example, Jupiter's average distance from the Sun is 5.2 astronomical units (AU). That value cubed is about 140, so the period must be the square root of 140, which equals slightly less than 12 years.

Notice that Kepler's three laws are **empirical**. That is, they describe a phenomenon without explaining why it occurs. Kepler derived the laws from Tycho's extensive observations, not from any first principle, fundamental assumption, or theory. In fact, Kepler never knew what held the planets in their orbits or why they continued to move around the Sun.

Kepler's Final Book: *The Rudolphine Tables*

Kepler continued his mathematical work on *The Rudolphine Tables,* and at last, in 1627, it was ready. He financed the printing himself, dedicating the book to the memory of Tycho Brahe. In fact, Tycho's name appears in larger type on the title page than Kepler's own. This is surprising because the tables were not based on the Tychonic system but on the heliocentric model of Copernicus and the elliptical orbits of Kepler. The reason for Kepler's evident deference was Tycho's family, still powerful and still intent on protecting Tycho's reputation. They even demanded a share of the profits and the right to censor the book before publication, though they changed nothing but a few words on the title page and added an elaborate dedication to the emperor.

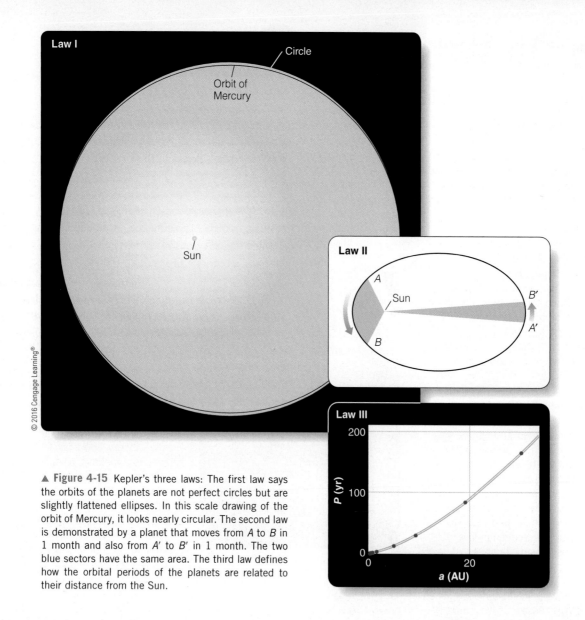

© 2016 Cengage Learning®

▲ **Figure 4-15** Kepler's three laws: The first law says the orbits of the planets are not perfect circles but are slightly flattened ellipses. In this scale drawing of the orbit of Mercury, it looks nearly circular. The second law is demonstrated by a planet that moves from *A* to *B* in 1 month and also from *A'* to *B'* in 1 month. The two blue sectors have the same area. The third law defines how the orbital periods of the planets are related to their distance from the Sun.

The Rudolphine Tables was Kepler's masterpiece. The tables could predict the positions of the planets 10 to 100 times more accurately than previous tables. Kepler's tables were the precise model of planetary motion that Copernicus had sought but failed to find because he could not give up the idea of perfectly circular motions. The accuracy of *The Rudolphine Tables* was strong evidence that both Kepler's laws of planetary motion and the Copernican hypothesis for the place of Earth were correct. Copernicus would have been pleased.

Kepler died in 1630. He had solved the problem of planetary motion, and his *Rudolphine Tables* demonstrated his solution. Although he did not understand why the planets moved or why they followed ellipses—insights that had to wait half a century for Isaac Newton—Kepler's three laws worked. In science the only test of a hypothesis is, "Does it describe reality?" (In other words, does it match what is observed?) Kepler's laws have been used for almost four centuries as a true description of orbital motion.

DOING SCIENCE

How were Kepler's three laws of planetary motion based on evidence? Kepler derived his three laws of planetary motion from the observations made by Tycho Brahe during 20 years on Hveen.

The observations were the evidence, and they gave Kepler a reality check each time he tried a new calculation. He chose ellipses because they were the best fit to the data, not because he had some reason before he started his work to expect that the answer must be ellipses. Trying not to have a load of preconceptions about the final results, like striving to identify and question first principles, is an important part of the modern way of doing science.

The Copernican model was a poor predictor of planetary motion, but Kepler's *Rudolphine Tables* were much more accurate. ***Why might you expect that, given how the Rudolphine Tables were produced?***

4-4 Galileo Finds Conclusive Evidence

Most people think they know two "facts" about Galileo, but both facts are wrong They are **Common Misconceptions**, so you have probably heard them. The truth is, Galileo did not invent the telescope, nor was he condemned by the Inquisition for believing Earth moves around the Sun. Then why is Galileo so famous? Why did the Vatican reopen his case in 1979, almost 400 years after his trial? As you learn about Galileo, you will discover that his legacy concerns not only advancing understanding of the place of Earth and the motion of the planets but also helping establish a new and powerful method of understanding nature, a method called *science.*

Telescope Observations

Galileo Galilei (1564–1642) (**Figure 4-16a**) was born in Pisa, a city in what is now Italy, and he studied medicine at the university there. His true love, however, was mathematics, and, although he had to leave school early for financial reasons, he returned only four years later as a professor of mathematics. Three years after that he became professor of mathematics at the university in Padua, where he remained for 18 years.

During this time, Galileo seems to have adopted the Copernican model, although he admitted in a 1597 letter to Kepler that he did not support Copernicanism publicly. At that time, the Copernican hypothesis was not officially considered heretical, but it was hotly debated among astronomers, and Galileo, living in a region controlled by the Catholic Church, cautiously avoided trouble. It was the telescope that finally drove Galileo to publicly defend the heliocentric model.

The telescope seems to have been invented around 1608 by lens makers in Holland. Galileo, hearing descriptions in the fall of 1609, was able to build telescopes in his workshop (Figure 4-16b). In fact, Galileo was not even the first person to look at the sky through a telescope, but he was the first person to write down what he discovered and apply telescopic observations to the theoretical problem of the day: the place of Earth.

What Galileo saw through his telescopes was so amazing that he rushed a small book into print. *Sidereus Nuncius (The Sidereal Messenger)* reported three major discoveries. First, the Moon was not perfect. It had mountains and valleys on its surface, and Galileo even used some of the mountains' shadows to calculate

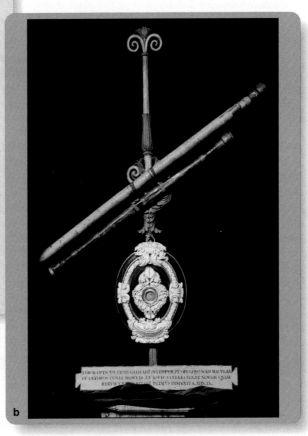

Panel a: Irynai/Shutterstock.com; Panel b: Leemage/UIG/Getty Images

▲ **Figure 4-16** (a) Galileo is remembered as the great defender of Copernicanism. (b) Two of Galileo's telescopes, on display in a museum in Florence. Although he did not invent telescopes, Galileo will always be associated with them because they were the source of much of the observational evidence he used to try to understand the Universe.

their height. Aristotle's philosophy held that the Moon was perfect; the "Man in the Moon" markings were believed to be reflections of continents on Earth. Galileo showed that the Moon is not only imperfect but is a world with features like Earth's.

The second discovery reported in the book was that the Milky Way was made up of myriad stars too faint to see with the unaided eye. Although intriguing, this could not match Galileo's third discovery. Galileo's telescope revealed four new "planets" circling Jupiter, objects known today as the **Galilean moons** of Jupiter (look back to the figure that opens this chapter on page 52).

The moons of Jupiter were strong evidence for the Copernican model. Critics of Copernicus had said Earth could not move because the Moon would be left behind, but Galileo's discovery showed that Jupiter, which everyone agreed was moving, was able to keep its satellites. This indicated that Earth, too, could move yet still hang on to the Moon. Aristotle's philosophy also included the belief that all heavenly motion was centered on Earth. Galileo's observations showed that Jupiter's moons revolve around Jupiter, which proved there could be other centers of motion besides Earth.

Sometime after *Sidereus Nuncius* was published, Galileo noticed something else that made Jupiter's moons even stronger evidence for the Copernican model. When he measured the orbital periods of the four moons, he found that the innermost moon had the shortest period and that the moons farther from Jupiter had proportionally longer periods. Jupiter's moons made up a harmonious system ruled by Jupiter, just as the planets in the Copernican universe were a harmonious system ruled by the Sun (look back to Figure 4-10b.) The similarity is not proof, but Galileo saw it as an argument that the Solar System could be Sun-centered rather than Earth-centered.

In the years following publication of *Sidereus Nuncius,* Galileo made two additional discoveries. When he observed the

Sun, he discovered sunspots, raising the suspicion that the Sun, like the Moon, is imperfect. Further, by noting the movement of the spots, he concluded that the Sun is a sphere and that it rotates on its axis.

His most dramatic discovery came when he observed Venus. Galileo saw that it was going through phases like those of the Moon. In the Ptolemaic model, Venus moves around an epicycle centered on a line between Earth and the Sun. That means it would always be seen as a crescent (Figure 4-17a). But Galileo saw Venus go through a complete set of phases, which proved that it did indeed revolve all the way around the Sun (Figure 4-17b). There is no way the Ptolemaic model could produce the observed phases. This was the strongest evidence in support of the Copernican model that came from Galileo's telescope. Nevertheless, when controversy erupted, Galileo's critics focused more on claims about the imperfection of the Sun and Moon and the motion of the satellites of Jupiter.

Sidereus Nuncius was popular and made Galileo famous. He became chief mathematician and philosopher to the Grand Duke of Tuscany in Florence. In 1611, Galileo visited Rome and was treated with great respect. He had long, friendly discussions with the powerful Cardinal Barberini, but he also made enemies. Galileo was outspoken, forceful, and sometimes tactless. He enjoyed argument, but most of all he enjoyed being right. In lectures, debates, and letters he offended important people who questioned his telescope discoveries.

By 1616, Galileo was the center of a storm of controversy. Some critics said he was wrong, and others said he was lying. Some refused to look through a telescope lest it mislead them, and others looked and claimed to see nothing, which is hardly surprising, given the awkwardness of those first telescopes (Figure 4-18). Pope Paul V decided to end the disruption, so when Galileo visited Rome in 1616, Cardinal Bellarmine interviewed

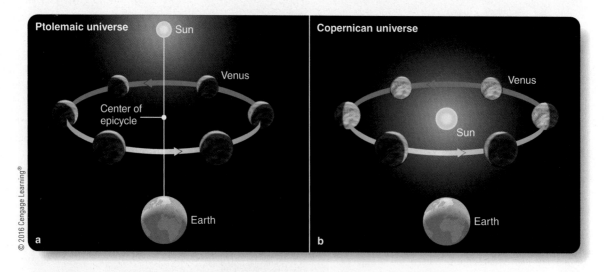

© 2016 Cengage Learning®

▲ Figure 4-17 (a) If Venus moved in an epicycle centered on the Earth–Sun line (see page 61), it would always appear as a crescent. (b) Galileo observed through his telescope that Venus goes through a full set of phases, proving that it must orbit the Sun.

▲ **Figure 4-18** Galileo's telescope revealed such things as craters on the Moon, phases of Venus, and the existence of four moons orbiting Jupiter. He demonstrated his telescope and discussing his observations with powerful people, explained how observational evidence could be used to test the prevailing Earth-centered model of the Universe. Some of the viewers thought the telescope was the work of the devil and would deceive anyone who looked. Galileo's discoveries produced intense and, in some cases, angry debate; he was condemned by the Inquisition in 1633.

him privately and ordered him to cease debate. There is some controversy today about the nature of Galileo's instructions, but he did not pursue astronomy for some years after the interview. Books relevant to Copernicanism were banned in all Catholic lands, although *De Revolutionibus,* recognized as an important and useful book in astronomy, was only suspended pending revision. Everyone who owned a copy of the book was required to cross out certain statements and add handwritten corrections stating that Earth's motion and the central location of the Sun were only theories and not facts. (You might recognize a similarity to current controversies about biological evolution.)

Dialogo and Trial

In 1621 Pope Paul V died, and his successor, Pope Gregory XV, died in 1623. The next pope was Galileo's friend Cardinal Barberini, who took the reign name of Urban VIII. Galileo rushed to Rome hoping to have the prohibition of 1616 lifted, and although the new pope did not revoke the orders, he did apparently encourage Galileo. Soon after returning home, Galileo began to write his great defense of Copernicanism, finally completing it at the end of 1629. After some delay, the book was approved by both the local censor in Florence and the head censor of the Vatican in Rome. It was printed in February 1632.

Called *Dialogo Sopra i Due Massimi Sistemi del Mondo (Dialogue Concerning the Two Chief World Systems),* it confronts the Aristotelian/Ptolemaic model with the Copernican model, using telescopic observations as evidence. Galileo wrote the book in the form of a debate among three friends: Salviati, a swift-tongued defender of Copernicus, dominates the book; Sagredo, intelligent but largely uninformed; Simplicio, dismal defender of Ptolemy, who makes all the old arguments and sometimes doesn't seem very bright.

The publication of *Dialogo* created a storm of controversy, and it was sold out by August 1632, when the Inquisition ordered sales stopped. The book was a clear and strong defense of Copernicus, and, perhaps unintentionally, Galileo exposed papal authority to ridicule. Urban VIII was fond of arguing that, given that God is omnipotent, He could construct the Universe in any form while making it appear to humans to have a different form, and thus its true nature cannot be deduced by mere observation. Galileo placed Urban VIII's argument in the mouth of Simplicio, and Galileo's enemies showed the passage to the pope as an example of Galileo's disrespect. The pope thereupon ordered Galileo to face the Inquisition.

Galileo was interrogated by the Inquisition four times and was threatened with torture. He must have thought often of Giordano Bruno, a philosopher, poet, and Dominican monk, who was tried, condemned, and burned at the stake in Rome in 1600. One of Bruno's offenses had been advocacy of Copernicanism. However, Galileo's trial did not center on his belief in Copernicanism; *Dialogo* had been approved by two censors. Rather, the trial centered on the instructions given Galileo in 1616. From his file in the Vatican, Galileo's accusers produced a record of the meeting between Galileo and Cardinal Bellarmine that included the statement that Galileo was "not to hold, teach, or defend in any way" the principles of Copernicus. Some historians believe that this document, which was signed neither by Galileo nor by Bellarmine nor by a legal secretary, was a forgery. Others suspect it may be a draft that was never used. It is quite possible that Galileo's actual instructions were much less restrictive, but, in any case, Bellarmine was dead and could not testify at Galileo's trial.

The Inquisition condemned Galileo not for heresy but for disobeying the orders given to him in 1616. On June 22, 1633, at the age of 69, kneeling before the Inquisition, Galileo read a recantation admitting his errors. Tradition has it that as he rose he whispered *"E pur si muove"* ("Still it moves"), referring to Earth.

Although he was sentenced to life imprisonment, he was, possibly through the intervention of the pope, confined at his villa for the next 10 years. He died there on January 8, 1642, 99 years after the death of Copernicus.

Galileo was not condemned for heresy, nor was the Inquisition concerned when he tried to defend Copernicanism. He was tried and condemned on a charge you might consider a technicality. Nevertheless, in his recantation he was forced to abandon all belief in heliocentrism. His trial has been held up as an example of the suppression of free speech and free inquiry and as a famous attempt to deny reality. Some of the world's greatest authors, including Bertolt Brecht, have written about Galileo's trial. That is why Pope John Paul II created a commission in 1979 to reexamine the case against Galileo.

To understand the trial, you must recognize that it was the result of a conflict between two ways of understanding the

Universe. Since the Middle Ages, biblical scholars had taught that the only path to true understanding was through religious faith. St. Augustine (354–430) wrote *"Credo ut intelligam,"* which can be translated as "Believe in order to understand." Galileo and other scientists of the Renaissance, however, used their own observations as evidence to try to understand nature. When their observations contradicted Scripture, they assumed that their observations represented reality. Galileo paraphrased Cardinal Baronius in saying, "The Bible tells us how to go to heaven, not how the heavens go." The trial of Galileo was not really about the place of Earth in the Universe. It was not about Copernicanism. It wasn't even about the instructions Galileo received in 1616. It was, in a larger sense, about the birth of modern science as a rational way to understand the physical Universe (**Figure 4-19**).

The commission appointed by Pope John Paul II in 1979, reporting its conclusions in October 1992, said of Galileo's inquisitors, "This subjective error of judgment, so clear to us today, led them to a disciplinary measure from which Galileo 'had much to suffer.'" Galileo was not found innocent in 1992 so much as the clerical judges were deemed, 400 years afterward, of having failed to fully grapple with the real questions involved.

DOING SCIENCE

How were Galileo's observations of the moons of Jupiter evidence against the Ptolemaic model? In this case, the distinction scientists usually make between model and hypothesis is blurred.

The Ptolemaic model was considered to be more than just a method for calculating planetary motions. That model was understood to represent the Aristotelian universe, believed to be true by virtually all astronomers and philosophers. The Ptolemaic model implied predictions about what the first astronomer with a telescope should see. Making predictions that can be used to check and compare hypotheses, as well as making observations to check predictions of hypotheses, are both central to doing science.

Galileo presented his scientific reasoning in the form of evidence and conclusions; the moons of Jupiter were some of his key evidence. Moons circling Jupiter did not fit the implicit Aristotelian belief that all motion is centered on Earth. Obviously, there can be other centers of motion.

The phases of Venus provide even better evidence against the Ptolemaic model and for the Copernican model. Both models imply specific, and very different, predictions about how the phases of Venus should appear. In the Ptolemaic model, Venus will always be seen as a crescent from Earth. In the Copernican model, Venus will go through a Moon-like sequence of phases, from crescent to full and back to crescent, and the planet should have the smallest angular size when it is at full phase. What Galileo saw was exactly what the Copernican model would predict.

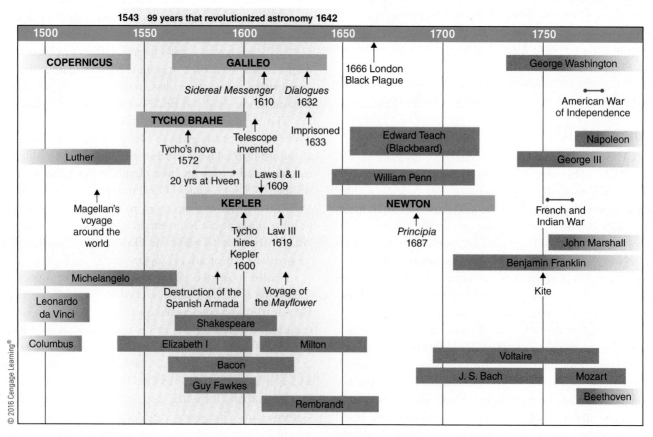

▲ **Figure 4-19** The 99 years between the death of Copernicus in 1543 and the death of Galileo in 1642 marked the transition from the ancient astronomy of Ptolemy and Aristotle to the revolutionary theory of Copernicus, and, simultaneously, the invention of science as a way of understanding nature.

4-5 Ninety-Nine Years That Revolutionized Astronomy

The transition from ancient to modern astronomy began during the 99 years between the publication of *De Revolutionibus* (1543) and Galileo's death (1642). That transition began with the replacement of the Ptolemaic model of the Universe by the Copernican model, with a closely related controversy over the place of Earth in the Universe. That same period also saw an evolution in the methods of science in general, illustrated by the solution of the puzzle of planetary motion. That puzzle was not solved by philosophical arguments about the perfection of the heavens or by debate over the meanings of scripture. It was solved by precise observations and careful computations, techniques that are the foundations of modern science.

The discoveries made by Kepler and Galileo found acceptance in the 1600s because the world was in transition. Astronomy was not the only thing changing during this period. The Renaissance is commonly taken to be the period between about the years 1300 and 1600; these 99 years of astronomical history thus lie at the culmination of a reawakening of learning in all fields (Figure 4-16). Ships were sailing to new lands and encountering new cultures. The world was open to new ideas and new observations. Martin Luther started a reformation of the Christian religion, and other philosophers and scholars rethought their respective areas of human knowledge. Had Copernicus not published his hypothesis, someone else would have suggested that the Solar System is heliocentric. History was ready to shed Aristotle's universe and the Ptolemaic model.

Beginning with Copernicus, scientists such as Tycho, Kepler, and Galileo depended more and more on evidence, observation, and measurement rather than on first principles. That change was connected with the Renaissance via advances in metalworking and lens making. At the time of Copernicus, no astronomer had looked through a telescope because one could not be made. By 1642, not only telescopes but also other sensitive measuring instruments had transformed science into something new and precise. As you can imagine, scientists were excited by these discoveries, and they founded scientific societies that increased the exchange of observations and hypotheses, and stimulated more and better work. The most important advance, however, was the application of mathematics to scientific questions. Kepler's work demonstrated the power of mathematical analysis, and as the quality of these numerical techniques improved, the progress of science accelerated. The story of the birth of modern astronomy is actually the story of the birth of modern science as well.

What Are We? Thinkers

The scientific revolution began when Copernicus made humanity part of the Universe. Before Copernicus, people thought of Earth as a special place different from any of the objects in the sky, but in trying to explain the motions in the sky, Copernicus made Earth one of the planets. Galileo and those who brought him to trial understood the significance of making Earth a planet. It made Earth and humanity part of nature, part of the Universe.

Kepler showed that the planets move according to simple rules. We are not in a special place ruled by mysterious planetary forces. Earth, the Sun, and all of humanity are part of a Universe in which motions can be described by a few fundamental laws. If simple laws describe the motions of the planets, then the Universe is not ruled by mysterious influences as in astrology or the whims of the gods atop Mount Olympus. And if the Universe can be described by simple rules, then it is open to scientific study.

Before Copernicus, people felt they were special because they thought they were at the center of the Universe. Copernicus, Kepler, and Galileo showed that we are not at the center but nevertheless are part of an elegant and complex Universe. Astronomy tells us that we are special because we can study the Universe and eventually understand what we are. It also tells us that we are not only part of nature; we are the part of nature that can think about nature.

Study and Review

Summary

- **Archaeoastronomy (p. 53)** is the study of the astronomical knowledge of ancient peoples. Many cultures around the world observed the sky and marked important alignments. Structures such as Stonehenge, Newgrange, and other human-made phenomena such as the Sun Dagger involve astronomical alignments.

- In most cases, ancient cultures, having no written language, left no detailed records of their astronomical beliefs. Greek astronomy, derived in part from Babylon and Egypt, is better known because written documents have survived.

- A **first principle (p. 57)** is an idea considered so obviously true that the idea does not need to be questioned. Classical philosophers accepted as a first principle that Earth was the unmoving center of the Universe. Another first principle was that the heavens were perfect, so philosophers such as Plato argued that, because the sphere was the most perfect geometrical form, the heavens must be made up of spheres in uniform rotation. This led to the belief in **uniform circular motion (p. 56)**.

- Many astronomers argued that Earth could not be moving because they could see no **parallax (p. 60)** in the positions of the stars.

- Aristotle's estimate for the size of Earth was only about 40 percent of its true size. Eratosthenes used sunlight shining to the bottom of a well at the southern Egyptian city of Syene to measure the diameter of Earth and produced an accurate estimate.

- The **geocentric universe (p. 56)** became part of the teachings of the great philosopher Aristotle, who argued that the Sun, Moon, and stars were carried around Earth on rotating crystalline spheres.

- Hipparchus, who lived about two centuries after Aristotle, devised a model in which the Sun, Moon, and planets revolved in circles called **eccentrics (p. 58)** with Earth near, but not precisely at, their centers.

- **Retrograde motion (p. 60)**, the occasional westward (backward) motion of the planets, was difficult for astronomers to explain.

- About the year 140, Aristotle's model was given mathematical form in Claudius Ptolemy's book *Almagest*. Ptolemy preserved the principles of geocentrism and uniform circular motion, but he added **epicycles (p. 61)**, **deferents (p. 61)**, and **equants (p. 61)**. Ptolemy's epicycles could approximate retrograde motion, but the Ptolemaic model was not accurate, and it had to be revised a number of times as centuries passed.

- Copernicus devised a **heliocentric universe (p. 56)**. He preserved the principle of uniform circular motion, but he argued that Earth rotates on its axis and revolves around the Sun once a year. His theory was controversial because it contradicted Church teaching about Earth's place in the Universe. Copernicus published his theory in his book *De Revolutionibus* in 1543, the same year he died.

- A **hypothesis (p. 69)** is a specific statement about nature that needs further testing, but a **theory (p. 69)** is usually a general description of some aspect of nature that is well understood, has been thoroughly tested, and is widely accepted. A **natural law (p. 69)** is a fundamental and universal principle in which scientists have great confidence.

- Because Copernicus kept uniform circular motion, his model did not offer improved predictions of planetary motions, but it did offer a simple explanation of retrograde motion without using big epicycles.

- One reason the Copernican model won converts was that it was more elegant than the Ptolemaic model. Venus and Mercury were treated the same as all the other planets, and the speed of each planet was related to its distance from the Sun.

- The shift from the geocentric **paradigm (p. 64)** to the heliocentric paradigm is an example of a scientific revolution.

- Although Tycho Brahe developed his own model in which the Sun and Moon circled Earth and the planets circled the Sun, his great contribution was to compile detailed, precise observations of the positions of the Sun, Moon, and planets over a period of 20 years, observations that were later used by Johannes Kepler.

- Kepler inherited Tycho's books of observations in 1601 and used them to discover three laws of planetary motion. He found that the planets follow orbits that are **ellipses (p. 68)** with the Sun at one focus, that they move faster when near the Sun, and that a planet's orbital period squared is proportional to the **semimajor axis, *a* (p. 68)**, of its orbit cubed. These laws are all **empirical (p. 69)**, describing the phenomena of planetary motion without supplying explanations.

- The **eccentricity, *e* (p. 68)**, of an orbit is a measure of its departure from a perfect circle. A circle is an ellipse with an eccentricity of zero.

- Kepler's final book, *The Rudolphine Tables* (1627), combined heliocentrism with elliptical orbits and predicted the positions of the planets 10 to 100 times more accurately than any previous effort.

- Galileo used the newly invented telescope to observe the heavens, and he recognized the significance of what he saw there. His discoveries of the phases of Venus, the satellites of Jupiter now known as the **Galilean moons (p. 72)**, the mountains of Earth's Moon, and other phenomena helped undermine the Ptolemaic universe.

- Galileo based his analysis on observational evidence. In 1633, he was condemned by the Inquisition for disobeying instructions not to hold, teach, or defend Copernicanism.

- Historians of science view Galileo's trial as a conflict between two ways of knowing about nature, reasoning from first principles versus depending on evidence.

- The 99 years from the death of Copernicus to the death of Galileo marked the birth of modern science. From that time on, science depended on evidence to test theories and relied on the mathematical analytic methods first demonstrated by Kepler.

Review Questions

1. What evidence suggests that early human cultures observed astronomical phenomena?
2. Why did early human cultures observe astronomical phenomena? Was it for scientific research?
3. Early cultures believed that ultimate causes of things are mysteries beyond human understanding. Is this a first principle? Why or why not?
4. Name one example each of a famous politician, mathematician, philosopher, observer, and theoretician in this chapter.
5. Why did Plato propose that all heavenly motion was uniform and circular?

6. On what did Plato base his knowledge? Was it opinions, policies, marketing, public relations, myths, evidence, hypotheses, beliefs, laws, principles, theories? On what do modern astronomers base their knowledge?

7. Which two-dimensional (2D) and three-dimensional (3D) shapes did Plato and Aristotle consider perfect? Give an example of a 2-D and 3-D nonperfect geometrical shape.

8. Are the spheres of Eudoxus a scientific model? If so, is it entirely true?

9. In Ptolemy's model, how do the epicycles of Mercury and Venus differ from those of Mars, Jupiter, and Saturn?

10. In Ptolemy's model, did all planets travel at constant speeds as viewed from Earth, and did each planet have its own rotating sphere around Earth?

11. In Ptolemy's model, which of the following—epicycle, equant, or deferent—travels in uniform circular motion as viewed from a particular point? Name and describe that point. Are these uniform circular motions at the same speeds and in the same directions?

12. Why did Copernicus have to keep small epicycles in his model? Which planet has the longest duration of retrograde motion as viewed from Earth? The shortest?

13. Was the belief held by ancient astronomers that the Moon and Sun are unblemished a paradigm, a scientific revolution, or neither?

14. When Tycho observed the new star of 1572, he could detect no parallax. Why did that undermine belief in the Ptolemaic system? In the perfect heavens idea of Aristotle?

15. Does the Moon have parallax? Assume the night is clear and the Moon's phase is full so you can see it all night long.

16. Does Tycho's model of the Universe explain the phases of Venus that Galileo observed? Why or why not?

17. Name an empirical law. Why is it considered empirical?

18. How does Kepler's first law of planetary motion overthrow one of the basic beliefs of classical astronomy? How about Kepler's second law?

19. When Mercury is at aphelion (farthest from the Sun) in Figure 4-15, compare *a*, the semimajor axis, to *r*, the distance from the Sun to Mercury. Is *a* greater than *r*, or less than *r*, or equal to *r?*

20. In the *Law II* panel of Figure 4-15, did the planet travel faster from points *A* to *B*, travel faster from points *A'* to *B'*, travel the same speed between *A* to *B* and *A'* to *B'*, or can you not determine speeds of a planet's orbit based on the information given? How do you know?

21. What is *P* for Earth? What is *a* for Earth? Do these values support or disprove Kepler's third law?

22. Based on the Law III panel of Figure 4-15, do planets with larger *a* take longer, shorter, or the same time to orbit the Sun?

23. How did *The Alfonsine Tables, The Prutenic Tables,* and *The Rudolphine Tables* differ?

24. Explain how each of Galileo's telescopic discoveries contradicted the Ptolemaic theory.

25. How did discovery of the Galilean moons disprove Plato's and Aristotle's perfect heavens first principle(s)?

26. **How Do We Know?** What is a paradigm, and how it is related to a scientific revolution? Give an example of a paradigm and a scientific revolution.

27. **How Do We Know?** Describe the differences between a hypothesis, a theory, and a law. Give an example of each.

Discussion Questions

1. Pythagoras proposed that all nature was underlain by musical, or mathematical, principles. For example, Western music is based on factors of 2 (that is, in 2, 4, 8, 16, 32, 64, ... counts) whereas money, lengths, and temperature are in base 10 (for example, the U.S. dollar has 100 pennies, a decimeter is 100 millimeters, the difference between boiling and freezing water on the Celsius scale is 100 degrees). Why are money, lengths, and temperature not in base 2?

2. Historian of science Thomas Kuhn has said that *De Revolutionibus* was a revolution-making book but not a revolutionary book. How was it an old-fashioned, classical book?

3. Why might Tycho Brahe have hesitated to hire Kepler? Why do you suppose he appointed Kepler his scientific heir? What is limited about Kepler's third law $P^2 = a^3$, where P is the time in units of years a planet takes to orbit the Sun and a is the planet's average distance from the Sun in units of AU? (*Hint:* Look at the units.) What does this tell you about Kepler and his laws?

4. Galileo was condemned, but Kepler, also a Copernican, was not. Why not?

5. How does the modern controversy over creationism and evolution reflect two ways of knowing about the physical world?

Problems

1. Draw and label a diagram of the western horizon from northwest to southwest and label the setting points of the Sun at the solstices and equinoxes for a person in the Northern Hemisphere. (*Hint:* See pages 19 and 25 and Figure 4-1.)

2. If you lived on Mars, which planets would exhibit retrograde motion like that observed for Mars from Earth? Which would never be visible as crescent phases?

3. How long does it take for one retrograde cycle of Mars as viewed from Earth, and in which direction is the retrograde motion? What fraction of Mars's orbit around the Sun is the duration of retrograde motion as viewed from Earth?

4. If a planet has an average distance from the Sun of 2.0 AU, what is its orbital period?

5. If a space probe is sent into an orbit around the Sun that brings it as close as 0.4 AU and as far away as 5.4 AU, what will be its orbital period? Is the orbit a circle or an ellipse?

6. Uranus orbits the Sun with a period of 84.0 years. What is its average distance from the Sun?

7. A celestial object takes 5.2 years to orbit the Sun. How far is the celestial object from the Sun?

8. One planet is three times further from the Sun than another. Will the further planet take more, less, or the same amount of time to orbit the Sun? Will the closer planet orbit slower, faster, or the same speed? How much longer will the further planet take to orbit than the closer planet? If the closer planet is located at 10 AU, how far is the further planet, and what are the two planet's names?

9. Galileo's telescope showed him that Venus has a large angular diameter (61 arc seconds) when it is a crescent and a small angular diameter (10 arc seconds) when it is nearly full. Use the small-angle formula to find the ratio of its maximum to minimum distance from Earth. Is this ratio compatible with the Ptolemaic universe shown on page 61?

10. Which is the phase of Venus when it is closest? Which when furthest? How do you know?

11. Galileo's telescopes were not of high quality by modern standards. He was able to see the moons of Jupiter, but he never reported seeing features on Mars. Use the small-angle formula to find the angular diameter of Mars when it is closest to Earth. How does that compare with the maximum angular diameter of Jupiter?

Learning to Look

1. With an outstretched arm, hover your thumb about ½ inch above the page and over the thumb shown in the "Seen by right eye" image in part 1a of **An Ancient Model of the Universe.** Close one eye and rapidly blink closed one eye then the other and repeat several more times. Why does the building in the picture not disappear as shown in the "Seen by left eye" picture? Is this an example of seeing parallax? What do you need to do to see a very large parallax?

2. Study Figures 4-11 and 4-17 and describe the phases that Venus would have displayed to Galileo's telescope if the Tychonic universe had been correct.

3. What three astronomical objects are represented here? What are the two rings?

4. Use the figure below to explain how the Ptolemaic model treated some planets differently from the rest. How did the Copernican model treat all of the planets the same?

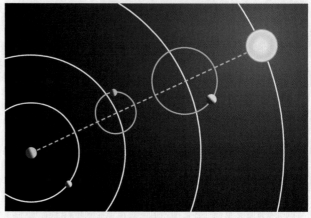

Gravity 5

Guidepost If Renaissance astronomers had understood gravity, they would have had much less trouble describing the motions of the planets, but real insight about gravity didn't come until three decades after Galileo's trial. Isaac Newton started from Galileo's work and devised a way to explain motion and gravity that allowed astronomers to understand orbits and tides with great precision. Later, in the early 20th century, Albert Einstein found an even better way to explain motion and gravity that included and broadened Newton's previous explanations.

This chapter is about gravity, the master of the Universe. Here you will find answers to five important questions:

► **What were Galileo's insights about motion and gravity?**

► **What were Newton's insights about motion and gravity?**

► **How does gravity explain orbital motion?**

► **How does gravity explain tides?**

► **What were Einstein's insights about motion and gravity?**

Gravity rules. From the Moon orbiting Earth, to matter falling into black holes, to the formation of galaxy clusters, as you study the Universe you will see gravity in action.

Nature and Nature's laws lay hid in night: God said, "Let Newton be!" and all was light.

ALEXANDER POPE

NASA

NASA astronaut Group 15 (nicknamed "The Flying Escargot") training in the KC-135 zero-gravity simulator aircraft.

OESN'T IT SEEM strange that Isaac Newton(1642–1727) is said to have "discovered" gravity in the late 17th century—as if people didn't have gravity before that, as if they floated around holding onto tree branches? Of course, everyone experienced gravity while taking it for granted. Newton's insight was to see that the force of gravity that makes apples fall to Earth also keeps moons and planets in their orbits. That realization changed the way people thought about nature (**Figure 5-1**).

5-1 Galileo's and Newton's Two New Sciences

Newton was born in Woolsthorpe, England, on December 25, 1642, and on January 4, 1643. This was not a biological anomaly but a quirk of the calendar. Most of Europe, following the lead of the Catholic countries, had adopted the new Gregorian calendar, but Protestant England continued to use the older Julian calendar. So December 25 in England was January 4 in Europe. If you use the English date, then Newton was born in the same year that Galileo died.

Newton became one of the greatest scientists in history, but even he admitted the debt he owed to those who had studied nature before him. Newton said, "If I have seen farther it is by standing on the shoulders of giants." Surely one of those giants was Galileo. In the previous chapter you learned that Galileo was the great defender of Copernicanism who made the first recorded use of an astronomical telescope, but he was also the first scientist who carefully studied the motions of falling bodies. Galileo provided the key information that helped Newton understand gravity.

Galileo's Observations of Motion

Galileo (look back to Figure 4-16a) was performing experiments to study forces and motion years before he built his first telescope in 1609. After the Inquisition condemned and imprisoned him in 1633, he continued his study of motion. He seems to have realized that he needed to understand motion before he could truly comprehend the Copernican system.

In addition to writing about a geocentric universe, Aristotle also wrote about the nature of motion, and those ideas still held sway in Galileo's time. Aristotle said that the world is made up of

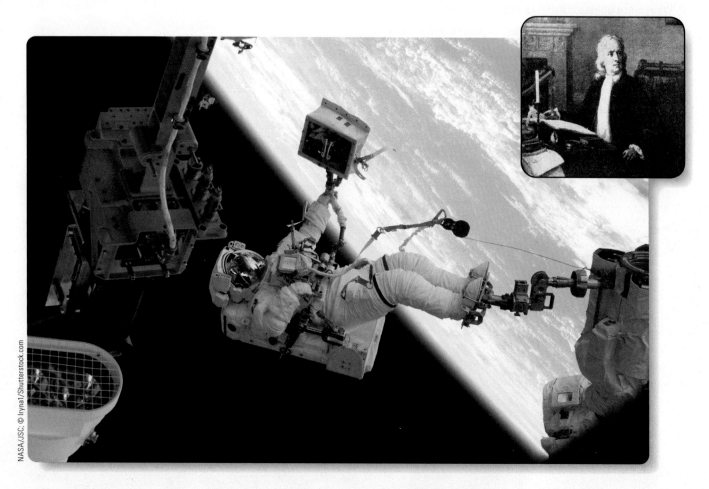

▲ **Figure 5-1** Space stations and astronauts, as well as planets, moons, stars, and galaxies, follow paths called *orbits* that are described by three simple laws of motion and a theory of gravity first proposed by Isaac Newton.

four elements: earth, water, air, and fire, with each located in its proper place. The proper place for earth (meaning soil and rock) is the center of the Universe, and the proper place of the water element is just above the earth element. Air and then fire form higher layers, and above them is the realm of the planets and stars, made of a celestial substance unknown on Earth. (The Earthly elements are shown as layers in the center of the diagram at the top of page 60.)

Aristotle wrote, and so everyone believed for almost 2000 years afterward, that objects have a natural tendency to move toward their proper places in the cosmos. Things made up mostly of air or fire—smoke, for instance—tend to move upward. Things composed mostly of earth and water—wood, rock, flesh, bone, and so on—tend to move downward. Therefore, according to Aristotle, objects that fall downward do so because they are moving toward their proper place. That is one reason why Aristotle's universe had to be geocentric. His explanation of gravity—why things fall down—works only if the center of Earth is also the center of the Universe. (Ironically, it is accurate to think of living creatures especially as being really composed of a combination of earth = soil, water, air, and fire = energy.)

Aristotle called these motions **natural motions;** this was to distinguish them from **violent motions** that are produced when, for instance, you push on an object and force it to move other than toward its proper place. According to Aristotle, such motions stop as soon as the force is removed. To explain how an arrow could continue to move upward even after it had left the bowstring, he said that currents in the air around the arrow carried it forward even though the bowstring was no longer pushing it.

In Galileo's time, as well as during the two preceding millennia, scholars had tended to resolve problems by referring to authority. To analyze the flight of a cannonball, for instance, they would turn to the writings of Aristotle and other classical philosophers and try to deduce what those philosophers would have said on the subject. This generated a great deal of discussion but little real progress. Galileo broke with this tradition when he conducted his own experiments and, furthermore, believed the results were more informative than ancient authority.

Galileo began by studying the motions of falling bodies, but he quickly discovered that the velocities were so great and the times so short that he could not measure them accurately. To solve that problem, Galileo began using polished metal balls rolling down gently sloping inclines so the velocities were lower and the times longer. Using an ingenious water clock he invented, Galileo was able to measure the amount of time the balls took to roll given distances down the incline and, most important, correctly recognized that those times are proportional to the times he would have measured for freely falling bodies.

Galileo found that falling bodies do not fall at constant rates, as Aristotle had said, but are accelerated. That is, they move faster with each passing second. Near Earth's surface, a falling object will have a velocity of 9.8 m/s (32 ft/s) at the end of 1 second, 19.6 m/s after 2 seconds, 29.4 m/s after 3 seconds, and so on. Each passing second adds 9.8 m/s (32 ft/s) to the object's velocity (**Figure 5-2**). In modern terms, this steady increase in the velocity of a falling body by 9.8 m/s each second (usually written 9.8 m/s^2, read as "9.8 meters per second squared") is called the **acceleration of gravity** at Earth's surface.

© 2016 Cengage Learning®

▲ **Figure 5-2** Galileo found that a falling object is accelerated downward. Each second its velocity increases by 9.8 m/s (32 ft/s).

Air resistance would have slowed the wooden ball more and ruined Galileo's demonstration.

On the airless Moon, there is no air resistance to slow the feather.

Hammer Feather

Panel a: © 2016 Cengage Learning®; Panel b: NASA

▲ **Figure 5-3** (a) According to a traditional story, Galileo demonstrated that the acceleration of a falling body is independent of its weight by dropping balls of iron and wood from the Leaning Tower of Pisa. In fact, air resistance would have confused the result. (b) In a historic television broadcast from the Moon on August 2, 1971, Apollo 15 Commander David Scott dropped a hammer and a feather at the same instant. They fell with the same acceleration and hit the surface together.

Galileo also discovered that the acceleration does not depend on the weight of the object. This, too, is contrary to the teachings of Aristotle, who believed that heavy objects, containing more earth and water, fell with higher velocity. Galileo found that the acceleration of a falling body is the same whether it is heavy or light. According to some accounts, he demonstrated this by dropping balls of iron and wood from the top of the Leaning Tower of Pisa to show that they would fall together and hit the ground at the same time (**Figure 5-3a**). In fact, he probably did not perform this experiment. It would not have been conclusive anyway because of the effect of air resistance. More than 300 years later, Apollo 15 astronaut David Scott, standing on the airless Moon, demonstrated the truth of Galileo's discovery by simultaneously dropping a feather and a steel geologist's hammer. They fell at the same rate and hit the lunar surface at the same time (Figure 5-3b).

Having described natural motion, Galileo turned his attention to what Aristotle called "violent" motion—that is, motion directed other than toward an object's proper place in the cosmos. Aristotle said that such motion must be sustained by a cause. Today we would say "sustained by a force." Galileo pointed out that an object rolling down an incline is accelerated and that an object rolling up the same incline is decelerated. If the incline were perfectly horizontal and frictionless, he reasoned, there could be no acceleration or deceleration to change the object's velocity, and in the absence of friction, the object would continue to move forever. In Galileo's own words, "Any velocity once imparted to a moving body will be rigidly maintained as long as the external causes of acceleration or retardation are removed." In other words, motion does not need to be sustained by a force, said Galileo, disagreeing with Aristotle. Once begun, motion continues until something changes it. This property of matter is called **inertia**. In fact, Galileo's description of what is essentially a law of inertia is a perfectly valid summary of the principle that became known as Newton's first law of motion.

Galileo published his work on motion in 1638, two years after he had become entirely blind and only four years before his death. The book was named *Discourses and Mathematical Demonstrations Concerning Two New Sciences, Relating to Mechanics and to Local Motion* (in Italian, *Discorsi e Dimostrazioni Matematiche, intorno à due nuove Scienze, Attenenti alla Mecanica & i Movimenti Locali*). It is known today as *Two New Sciences*.

The book is a brilliant achievement for a number of reasons. To understand motion, Galileo had to abandon ancient authority, devise his own experiments, and draw his own conclusions. In a sense Galileo's work was the first example of modern experimental science. Also notice that Galileo was able to make valid general conclusions about how nature works based on his limited experiments. Though his apparatus was finite and his results affected by friction, he was able to imagine an infinite, frictionless plane on which a body moves at constant velocity. In his workshop, the law of inertia was obscure, but in his imagination it was clear and precise.

Newton's Laws of Motion

From the work of Galileo, Kepler, and other early scientists, Newton was able to deduce three laws of motion (Table 5-1) that describe any moving object, from an automobile driving along a highway to galaxies colliding with each other. Those laws of motion led Newton further, to an understanding of gravity.

Newton's first law of motion is really a restatement of Galileo's law of inertia: An object continues at rest or in uniform motion in a straight line unless acted on by some force. For example, an astronaut drifting in space will travel at a constant speed in a straight line forever if no forces act on her (Figure 5-4a).

Newton's first law also explains why a projectile continues to move after all forces have been removed—for instance, how an arrow continues to move after leaving the bowstring. The object continues to move because it has **momentum**.

An object's momentum is a measure of its amount of motion, equal to its velocity multiplied by its mass. A paper clip tossed across a room has low velocity and therefore little momentum, and you could easily catch it in your hand. But the same paper clip fired at the speed of a rifle bullet would have tremendous momentum, and you would not dare try to catch it. Momentum also depends on the mass of an object (**Focus on Fundamentals 1**). Now imagine that, instead of tossing a paper clip, someone tosses you a bowling ball. A bowling ball contains much more mass than a paper clip and therefore has much greater momentum, even though it is moving with the same velocity.

Newton's second law of motion is about forces. Where Galileo spoke only of accelerations, Newton saw that acceleration is the result of force acting on a mass (Figure 5-4b). Newton's second law is commonly written as:

$$F = ma$$

As always, you need to define terms carefully when you look at an equation. **Acceleration** is a change in velocity, and **velocity** is speed with a specific direction. Most people use the words *speed* and *velocity* interchangeably, but they mean two different things. Speed is a rate of motion and does not have any direction implied, but velocity does. If you drive a car in a circle at 55 mph, your speed is constant, but your velocity is changing because your direction of motion is changing. An object experiences acceleration if its speed changes or *if its direction of motion changes*. Every automobile has three accelerators—the gas pedal, the brake pedal, and the steering wheel. All three change the car's velocity, in the technical sense of the word.

In a way, the second law is just common sense; you experience its consequences every day. The acceleration of a body is proportional to the force applied to it. If you push gently against a grocery cart, you expect a small acceleration. The second law of motion also says that the acceleration depends on the mass of the body. If your grocery cart is filled with bricks and you push it gently, you expect little result. If it is full of inflated balloons, however, it begins accelerating easily in response to a gentle push. Finally, the second law says that the resulting acceleration is in the direction of the force. This is also what you would expect. If you push on a cart that is not moving, you expect it to begin moving in the direction you push.

The second law of motion is important because it establishes a precise relationship between cause and effect (**How Do We Know? 5-1**). Objects do not just move; they accelerate as a result of the action of a force. Moving objects do not just stop; they decelerate as a result of a force. And moving objects don't just change direction for no reason. A change in direction is a

TABLE 5-1 Newton's Three Laws of Motion

I. A body continues at rest or in uniform motion in a straight line unless acted on by an external force.

II. The acceleration of a body is inversely proportional to its mass, directly proportional to the force, and in the same direction as the force.

III. To every action, there is an equal and opposite reaction.

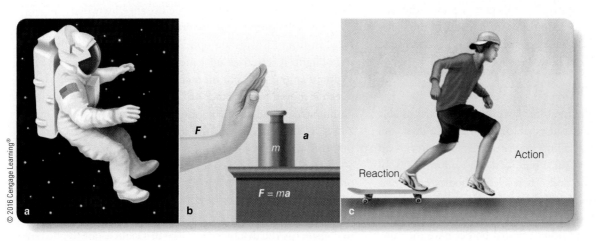

F

m

a

F = ma

Reaction

Action

◀ **Figure 5-4** Newton's three laws of motion.

© 2016 Cengage Learning®

Mass

One of the most fundamental parameters in science is **mass**, which is a measure of the amount of matter in an object. A bowling ball, for example, contains a large amount of matter and so is more massive than a child's rubber ball of the same size.

Mass is not the same as weight. Your weight is the force that Earth's gravity exerts on the mass of your body. Because gravity pulls you downward, you press against the bathroom scale, and you can measure your weight. Floating in space, you would have no weight at all; a bathroom scale would be useless. But your body would still contain the same amount of matter, so you would still have the same mass you do on Earth.

Sports analogies illustrate the importance of mass in dramatic ways. A bowling ball, for example, must be massive to have a large effect on the pins it strikes. Imagine trying to knock down all the pins with a balloon instead of a bowling ball. In space, where the bowling ball would be weightless, a bowling ball would still have more effect on the pins than a balloon would. On the other hand, runners want track shoes that have low mass so that they are easy to move. Imagine trying to run a 100-meter dash wearing track shoes that were as massive as bowling balls. It would be difficult to accelerate away from the starting blocks. Finally, think of the shot put. It takes muscle because the shot is massive, not because it is heavy. Imagine throwing the shot in space

where it would have no weight. It would still be massive, and it would take great effort to start it moving.

Mass is a unique measure of the amount of material in an object. Mass is expressed in kilograms in the metric system.

Mass is not the same as weight.

| MASS | ENERGY | TEMPERATURE AND HEAT | DENSITY | PRESSURE |

change in velocity and requires the presence of a force. Aristotle said that objects move because they have a tendency to move. Newton said that objects move because of a specific cause, a force. And Newton's second law is in the form of an equation, so, for example, you can calculate an object's precise numerical amount of acceleration if you know the numerical amounts of its mass and the force acting on it.

Newton's third law of motion specifies that for every action there is an equal and opposite reaction. In other words, forces must occur in pairs directed in opposite directions. For example, if you stand on a skateboard and jump forward, the skateboard will shoot away backward. As you jump, your feet must exert a force against the skateboard, which accelerates it toward the rear. But, as Newton realized, forces must occur in pairs, so the skateboard must exert an equal but opposite force on your feet, and that is what accelerates your body forward (Figure 5-4c).

Mutual Gravitation

Once Newton understood the three laws of motion, he was able to consider the force that causes objects to fall. The first and second laws tell you that falling bodies accelerating downward means there must be some force pulling downward on them. In Aristotle's view, the Moon and other celestial bodies move perpetually in circles because that is the nature of whatever heavenly (un-Earthly) substance they are composed. Newton assumed instead that the Moon and other celestial bodies have the same

nature and follow the same rules as objects on Earth, which meant that some force must act on the Moon to keep it in orbit. The Moon follows a curved path around Earth, and motion along a curved path is accelerated motion. The second law of motion says that acceleration requires a force, so a force must be making the Moon follow that curved path.

Newton wondered if the force that holds the Moon in its orbit could be the same force that causes apples to fall—gravity. He was aware that gravity extends at least as high as the tops of mountains, but he did not know if it could extend all the way to the Moon. He believed that it could, but he thought it would be weaker at greater distances, and he assumed that its strength would decrease as the square of the distance increased.

This relationship, the **inverse square law**, was familiar to Newton from his work on optics, where it applied to the intensity of light. A screen set up 1 meter from a candle flame receives a certain amount of light on each square meter. However, if that screen is moved to a distance of 2 meters, the light that originally illuminated 1 square meter must now cover 4 square meters (Figure 5-5). Consequently, the intensity of the light is inversely proportional to the square of the distance to the screen.

Newton made a second assumption that enabled him to predict the strength of Earth's gravity at the distance of the Moon. He assumed not only that the strength of gravity follows the inverse square law but also that the important distance to consider is the distance from Earth's center, not the distance

How Do We Know? 5-1

Cause and Effect

Why is the principle of cause and effect so important to scientists? One of the most often used, and least often stated, principles of science is cause and effect. Modern scientists all believe that events have causes, but ancient philosophers such as Aristotle argued that objects moved because of tendencies. They said that earth and water, and objects made mostly of earth and water, had a natural tendency to move toward the center of the Universe. This natural motion had no cause but was inherent in the nature of the objects. Newton's second law of motion ($F = ma$) was the first clear statement of the principle of cause and effect. If an object (of mass m) changes its motion by a certain amount (a in the equation), then it must be acted on by a force of a certain size (F in the equation). That effect (a) must be the result of a cause (F).

The principle of cause and effect goes far beyond motion. Scientists have confidence that every effect has a cause. The struggle against disease is an example. Cholera is a horrible disease that can kill its victims in hours. Long ago it was blamed on such things as bad magic or the will of the gods, and only two centuries ago it was blamed on "bad air." When an epidemic of cholera struck England in 1854, Dr. John Snow carefully mapped cases in London showing that the victims had drunk water from a small number of wells contaminated by sewage. In 1876, the German Dr. Robert Koch traced cholera to an even more specific cause when he identified the microscopic bacillus that causes the disease. Step by step, scientists tracked down the cause of cholera.

If the Universe did not depend on cause and effect, then you could never expect to understand how nature works. Newton's second law of motion was arguably the first clear statement that the behavior of the Universe depends on causes.

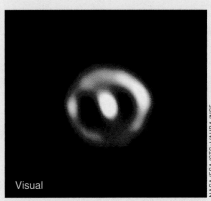

Visual

NASA/ESA/STScI/AURA/NSF

Cause and effect: Why did this star explode in 1992? There must have been a cause.

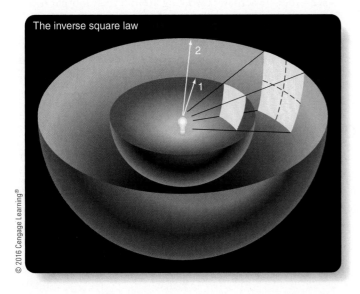

© 2016 Cengage Learning®

The inverse square law

▲ **Figure 5-5** As light radiates away from a source, it spreads out and becomes less intense. Here the amount of light falling on 1 square meter on the inner sphere must cover 4 square meters on a sphere twice as big. This shows how the intensity of light is inversely proportional to the square of the distance.

from Earth's surface. (By the way, Newton invented calculus to verify that assumption.) Because the Moon is about 60 Earth radii away, Earth's gravity at the distance of the Moon should be about 60^2 (that is, 3600) times weaker than at Earth's surface. The acceleration as a result of Earth's gravity is 9.8 m/s² at Earth's surface, so Newton estimated it should be about 0.0027 m/s² at the distance of the Moon.

Now, Newton wondered, was this enough acceleration to keep the Moon in orbit? He knew the Moon's distance and its orbital period, so he could calculate the actual acceleration needed to keep it in its curved path. The answer turned out to be 0.0027 m/s², as his inverse-square-law calculations predicted. Thus, Newton became certain that the Moon is held in its orbit by gravity, and gravity obeys the inverse square law.

Newton's third law says that forces always occur in pairs, so if Earth pulls on the Moon, then the Moon must pull on Earth. This is called mutual gravitation and is a general property of the Universe. The Sun, the planets, and all their moons must also attract each other by mutual gravitation. In fact, every particle with mass in the Universe must attract every other particle, which is why Newtonian gravity is often called universal mutual gravitation.

Clearly the force of gravity depends on mass. Your body is made of matter, and you have your own personal gravitational field, but your gravity is weak and does not cause personal satellites to orbit around you. Larger masses have stronger gravity. From an analysis of the third law of motion, Newton realized that the mass that resists acceleration in the first law must be identical to the mass causing gravity. Newton performed precise experiments with pendulums and confirmed this equivalence between the mass that resists acceleration and the mass that produces gravity.

From this, combined with the inverse square law, he was able to write the famous formula for the gravitational force between two masses, M and m:

$$F = -\frac{GMm}{r^2}$$

The constant G is the gravitational constant that connects units of mass to units of gravitational force. In the equation, r is the distance between the masses. The negative sign means that the force is attractive, pulling the masses together and making r decrease. In plain language, Newton's law of gravitation states: The force of gravitational attraction between two masses, M and m, is proportional to the product of the masses and inversely proportional to the square of the distance between them.

Newton's description of gravity was a difficult idea for scientists of his time to accept because it involved the puzzling notion of action at a distance. In other words, Earth and Moon somehow exert forces on each other even though there is no physical connection between them. Modern scientists conceptualize this by referring to gravity as a **field**. Earth's mass produces a gravitational field throughout space that is directed toward Earth's center. The strength of the field decreases according to the inverse square law. Any particle with mass in that field experiences a force that depends on the mass of the particle and the strength of the field at the particle's location. The force is directed toward the center of the field.

A field is an elegant way to describe gravity, but it still does not say what gravity is and why there is a field. Later in this chapter, when you learn about Einstein's theory of curved space-time, you may get a better idea of what gravity really is.

DOING SCIENCE

What do the words universal and mutual mean in the phrase universal mutual gravitation? Scientists often work by making step-by-step logical arguments involving careful definitions of words, and this is a good example.

Newton argued that the force that makes an apple accelerate downward is the same as the force that accelerates the Moon and holds it in its orbit. The third law of motion says that forces always occur in pairs, so if Earth attracts the Moon, then the Moon must attract Earth. That is, gravitation is mutual between any two objects.

Furthermore, if Earth's gravity attracts the apple and the Moon, then it must attract the Sun, and the third law says that the Sun must attract Earth. But if the Sun attracts Earth, then it must also attract the other planets and even distant stars, which, in turn, must attract the Sun and each other. Step by step, Newton's third law of motion leads logically to the conclusion that gravitation must apply to all masses in the Universe. That is, gravitation must be universal.

5-2 Orbital Motion and Tides

Orbital motion and tides are two different kinds of gravitational phenomena. As you think about the orbital motion of the Moon and planets, you are considering how gravity pulls on an entire object. When you think about tides, you are considering how gravity pulls on different parts of an object. Analyzing these two phenomena will give you a deeper insight into how gravity works.

Orbits

Newton was the first person to realize that objects in orbit are falling. You can explore Newton's insight by analyzing the motion of objects orbiting Earth. Carefully read **Orbits** on pages 88–89 and notice three important concepts and six new terms:

1 An object orbiting Earth is actually falling (being accelerated) toward Earth's center. The object continuously misses colliding with Earth because of its lateral ("sideways") orbital velocity. To follow a circular orbit, the object must move at *circular velocity*. Placed in a circular orbit at the right distance from Earth, it could be an especially useful *geosynchronous satellite*.

2 Notice that it is more accurate to say that objects orbiting each other are actually revolving around their mutual *center of mass*.

3 Finally, notice the difference between *closed orbits* and *open orbits*. If you want to leave Earth forever, you need to accelerate your spaceship at least until it is moving at *escape velocity*, (V_e), so it will follow an *open orbit* and never return.

Orbital Velocity

To successfully ride a rocket into orbit, you first need to answer a critical question: "How fast must I go to stay in orbit?" An object's **circular velocity** is the lateral velocity it must have to remain in a circular orbit. If you assume that the mass of your spaceship is small compared with the mass of Earth, then the circular velocity is:

$$V_c = \sqrt{\frac{GM}{r}}$$

In this formula, M is the mass of the central body (Earth in this case) in kilograms, r is the radius of the orbit in meters, and G is the gravitational constant, $6.67 \times 10^{-11} \ \text{m}^3/\text{s}^2/\text{kg}$. This simple formula is all you need to calculate how fast an object must travel to stay in a circular orbit.

For example, how fast does the Moon travel in its orbit? (Assume that the Moon's orbit is perfectly circular even though it is actually slightly elliptical.) Earth's mass is $5.97 \times 10^{24} \ \text{kg}$, and the radius of the Moon's orbit is $3.84 \times 10^8 \ \text{m}$. Therefore, the Moon's orbital velocity is:

$$V_c = \sqrt{\frac{6.67 \times 10^{-11} \times 5.97 \times 10^{24}}{3.84 \times 10^8}} = \sqrt{\frac{39.8 \times 10^{13}}{3.84 \times 10^8}}$$

$$= \sqrt{1.04 \times 10^6} = 1020 \text{ m/s} = 1.02 \text{ km/s}$$

This calculation shows that the Moon travels 1.02 km along its orbit each second. That is the circular velocity at the average distance of the Moon.

A satellite just above Earth's atmosphere is only about 200 km above Earth's surface, or 6570 km from Earth's center; so Earth's gravity is much stronger than at the Moon's position and the satellite must travel much faster than the Moon to stay in a circular orbit. You can use the preceding formula to find that the circular velocity for a low orbit 200 km above Earth's surface—just above the atmosphere—is about 7790 m/s, or 7.9 km/s. This is about 17,400 miles per hour, which shows why putting satellites into Earth orbit takes such large rockets. Not only must the rocket lift the satellite above Earth's atmosphere, but the rocket's trajectory must also then curve over and accelerate the satellite horizontally to reach this circular velocity.

A **Common Misconception** holds that there is no gravity in space. You can see that space is filled with gravitational forces from Earth, the Sun, and all other objects in the Universe. An astronaut who appears weightless in space is actually falling along a path at the urging of the combined gravitational fields of the rest of the Universe. Just above Earth's atmosphere, the orbital motion of the astronaut is almost completely due to Earth's gravity.

Calculating Escape Velocity

If you launch a rocket upward, it will consume its fuel in a few moments and reach its maximum speed. From that point on, it will coast upward. How fast must a rocket travel to coast away from Earth and escape? Of course, no matter how far it travels, it can never escape from Earth's gravity. The effects of Earth's gravity (and the gravity of all other objects) extend to infinity. It is possible, however, for a rocket to travel so fast initially that gravity can never slow it to a stop. Then the rocket could leave Earth permanently.

Escape velocity is the velocity required to escape an astronomical body. Here you are interested in escaping from the surface of Earth; in later chapters you will consider the escape velocity from other planets, the Sun, stars, galaxies, and even black holes.

Escape velocity, V_e, is given by a simple formula:

$$V_e = \sqrt{\frac{2GM}{r}}$$

Again, G is the gravitational constant, 6.67×10^{-11} m³/s²/kg, M is the mass of the central body in kilograms, and r is its radius in meters. (Notice that this formula is similar to the formula for

circular velocity; in fact, the escape velocity is $\sqrt{2}$ times the circular velocity.

You can find the escape velocity from Earth by again using its mass, 5.97×10^{24} kg, and the value of Earth's average radius, 6.37×10^6 m. The escape velocity from Earth's surface is:

$$V_e = \sqrt{\frac{2 \times 6.67 \times 10^{-11} \times 5.97 \times 10^{24}}{6.37 \times 10^6}} = \sqrt{\frac{7.96 \times 10^{14}}{6.37 \times 10^6}}$$

$$= \sqrt{1.25 \times 10^8} = 11,200 \text{ m/s} = 11.2 \text{ km/s}$$

This is equal to 11.2 km/s, or about 25,100 mph.

Notice from the formula that the escape velocity from a body depends on both its mass and radius. A massive body might have a low escape velocity if it has a large radius. You will meet such objects when you consider giant stars. On the other hand, a rather low-mass body could have a large escape velocity if it had a small radius, a condition you will encounter when you study black holes.

Once Newton understood gravity and motion, he could do what Kepler had not done; he could explain why the planets obey Kepler's laws of planetary motion.

Kepler's Laws Revisited

Now that you understand Newton's laws, gravity, and orbital motion, you can look at Kepler's laws of planetary motion in a new and more sophisticated way.

Kepler's first law states that the orbits of the planets are ellipses with the Sun at one focus. In one of his most famous mathematical proofs, Newton showed that if a planet moves in a closed orbit under the influence of an attractive force that follows the inverse square law, then the orbit must be an ellipse. In other words, orbits of planets are ellipses because gravity follows the inverse square law.

Even though Kepler correctly identified the shape of the planets' orbits, he still wondered why the planets keep moving along these orbits, and now you know the answer. They move because there is nothing to slow them down. Newton's first law says that a body in motion stays in motion unless acted on by some force. In the absence of friction, the planets must continue to move.

Kepler's second law states that a planet moves faster when it is near the Sun and slower when it is farther away. Once again, Newton's discoveries explain why. Imagine you are in an elliptical orbit around the Sun. As you move around the most distant part of the ellipse, aphelion, you begin to move back closer to the Sun, and the Sun's gravity pulls you slightly forward in your orbit. You pick up speed as you fall toward the Sun, so, of course, you go faster as you approach the Sun. As you move around the closest point to the Sun, perihelion, you begin to move away from the Sun, and the Sun's gravity pulls slightly backward on you, slowing you down as you recede from the Sun. If you were in a circular orbit, the Sun's gravity would always pull perpendicular to your motion, and you would not

Orbits

1   You can understand orbital motion by thinking of a cannonball moving around Earth in a circular path. Imagine a cannon on a high mountain aimed horizontally as shown at right. A little gunpowder gives the cannonball a low velocity, and it doesn't travel very far before falling to Earth. More gunpowder gives the cannonball a higher velocity, and it travels farther. With enough gunpowder, the cannonball travels so fast it never strikes the ground. Earth's gravity pulls it toward Earth's center, but Earth's surface curves away from it at the same rate it falls. It is in orbit. The velocity needed to stay in a circular orbit is called **circular velocity**. Just above Earth's atmosphere, at an altitude of 200 km, circular velocity is 7790 m/s or about 17,400 miles per hour, and the orbital period is about 90 minutes.

A satellite above Earth's atmosphere feels no friction and will fall around Earth indefinitely.

North Pole

Earth satellites eventually fall back to Earth if they orbit too low and experience friction with the upper atmosphere.

1a   A **geosynchronous satellite** orbits eastward with the rotation of Earth and remains above a fixed spot on the equator, which is ideal for communications and weather satellites.

A Geosynchronous Satellite

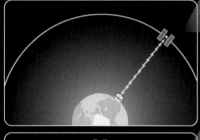

At a distance of 42,230 km (26,240 miles) from Earth's center, a satellite orbits with a period of 24 hours.

The satellite orbits eastward, and Earth rotates eastward under the moving satellite.

The satellite remains fixed above a spot on Earth's equator.

1b   According to Newton's first law of motion, the Moon should follow a straight line and leave Earth forever. Because it follows a curve, Newton knew that some force must continuously accelerate it toward Earth, gravity. Every second the Moon moves 1020 m (3350 ft) eastward and falls about 1.4 mm (1/18 in.) toward Earth. The combination of these motions produces the Moon's curved orbit. The Moon is falling *all* the time.

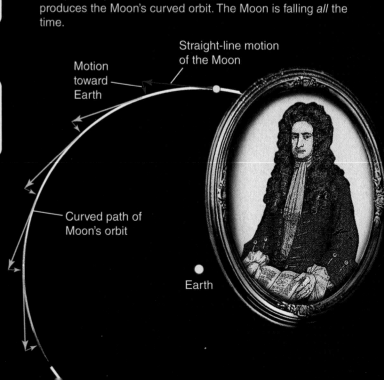

Straight-line motion of the Moon

Motion toward Earth

Curved path of Moon's orbit

Earth

1c Astronauts in orbit around Earth feel weightless, but they are not—to use a term from old science fiction movies—"beyond Earth's gravity." Like the Moon, the astronauts are accelerated toward Earth by Earth's gravity, but they travel fast enough along their orbits that they continually "miss the Earth." They are literally falling around Earth. Inside or outside a spacecraft, astronauts feel weightless because they and their spacecraft are falling at the same rate. Rather than saying they are weightless, you should more accurately say they are in free fall.

2 To be accurate you should not say that an object orbits Earth. Rather, the two objects orbit each other. Gravitation is mutual, and if Earth pulls on the Moon, the Moon pulls on Earth. The two bodies revolve around their common **center of mass**, the balance point of the system.

2a Two bodies of different mass balance at their center of mass, which is located closer to the more massive object. As the two objects orbit each other, they revolve around their common center of mass as shown at right. The center of mass of the Earth–Moon system lies only 4670 km (2900 mi) from the center of Earth—inside Earth. As the Moon orbits the center of mass on one side, Earth swings around the center of mass on the opposite side.

Center of mass

3 Closed orbits are repeating cycles. The Moon and artificial satellites orbit Earth in closed orbits. Below, the cannonball could follow an elliptical or a circular **closed orbit**. If the cannonball travels at the velocity needed to leave Earth permanently, called **escape velocity**, it will enter an **open orbit**. An open orbit does not return the cannonball to Earth; it will escape.

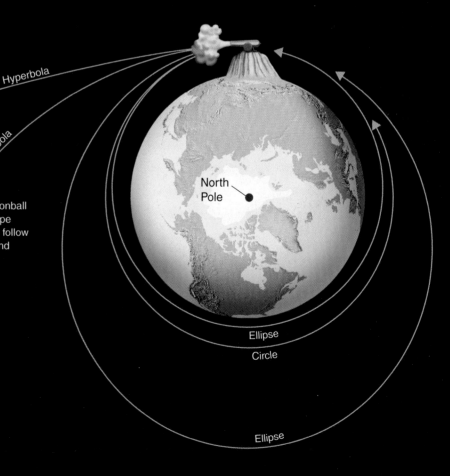

Hyperbola

A cannonball with a velocity greater than escape velocity will follow a hyperbola and escape from Earth.

Parabola

A cannonball with escape velocity will follow a parabola and escape.

North Pole

Ellipse

Circle

Ellipse

3a As described by Kepler's second law, an object in an elliptical orbit has its lowest velocity when it is farthest from Earth (apogee), and its highest velocity when it is closest to Earth (perigee). Perigee must be above Earth's atmosphere, or friction will rob the satellite of energy and it will quickly fall back to Earth.

speed up or slow down. Kepler's second law makes sense when you analyze it in terms of forces and motions.

There is a more elegant and profound way to think about Kepler's second law. Previously you learned about Galileo's insight that a body moving on a frictionless surface continues to move in a straight line until it is acted on by some force; that is, the object has momentum. In a similar way, an object rotating on a frictionless surface will continue rotating until something acts to speed up or slow down its rotation. Such an object has **angular momentum,** a combination of the object's mass with its speed of rotation or revolution. A planet circling the Sun in an orbit has a given amount of angular momentum, and, with no outside influences to alter its motion, its angular momentum must remain constant; physicists say that angular momentum is "conserved." Mathematically, a planet's angular momentum around the Sun is the product of its mass, velocity, and distance from the Sun. This provides another way to understand why a planet must speed up as it comes closer to the Sun along an elliptical orbit. Because its angular momentum is conserved, as its distance from the Sun decreases its velocity must increase, and as its distance from the Sun increases its velocity must decrease.

The conservation of angular momentum is actually a common human experience. Skaters spinning slowly can draw their arms and legs closer to their axis of rotation and, through conservation of angular momentum, spin faster (Figure 5-6). To slow their rotation, they can extend their arms again. Similarly, divers can spin rapidly in the tuck position and then slow their rotation by stretching into the extended position.

Kepler's third law states that a planet's orbital period depends on its distance from the Sun. That law is also explained by a certain measure of the planet's motion remaining constant. In this case, it is the law of conservation of energy (**Focus on Fundamentals 2**). A planet orbiting the Sun has a specific amount of energy that depends only on its average distance from the Sun. That energy is the sum of the energy of motion plus energy involved in the gravitational attraction between the planet and the Sun. The energy of motion depends on how fast the planet moves, and the gravitational attraction energy depends on the size of its orbit. The relation between those two kinds of energy underlies Newton's laws. That means there has to be a fixed relationship between the rate at which a planet moves around its orbit and the size of the orbit—between its orbital period, P, and the orbit's semimajor axis, a. You can even derive Kepler's third law from Newton's laws of motion, as shown in the next section.

Newton's Version of Kepler's Third Law

The equation for circular velocity is actually a version of Kepler's third law, as you can prove with three lines of simple algebra. The result is one of the most useful formulas in astronomy.

The equation for circular velocity, as you have seen, is:

$$V_c = \sqrt{\frac{GM}{r}}$$

The circular orbital velocity of a planet is simply the circumference of its orbit divided by the orbital period:

$$V = \frac{2\pi r}{P}$$

If you substitute this for V in the first equation and solve for P^2, you get:

$$P^2 = \left(\frac{4\pi^2}{GM}\right)r^3$$

Here M is the total mass of the two-body system in kilograms. For a planet orbiting the Sun, you can use just the mass of the Sun as a good approximation for M because the mass of the planet is negligible compared to the mass of the Sun, so adding in the planet's mass makes only a tiny difference. (In a later chapter, you will apply this formula to two stars orbiting each other, and then the mass M will be the sum of the two masses.) For a circular orbit, the semimajor axis, a, equals the radius of the circle. As you can see, this formula is a general version of Kepler's third law, $P^2 = a^3$, and you can use it for any object orbiting any other object. In Kepler's version, you used astronomical units (AU) for distance and years for time, but in Newton's version of the formula, you need to use units of meters, seconds, and kilograms. G is the gravitational constant, defined previously.

This is a powerful formula. It is important to realize that there is no other way to find masses of objects in the Universe than by measuring their effects on other objects. If, for example, you observe a moon orbiting a planet and you can measure the size of that moon's orbit, r, and its orbital period, P, you can use this formula to solve for M, the total mass of the planet plus the moon. It is important for you to realize that *there is no other way*

© 2016 Cengage Learning®

▲ **Figure 5-6** Skaters demonstrate conservation of angular momentum when they spin faster by drawing their arms and legs closer to their axis of rotation.

Energy

Physicists define **energy** as "the ability to do work," but you might paraphrase that definition in less technical vocabulary as "the ability to produce a change." A moving body has energy called **kinetic energy**. A planet moving along its orbit, a cement truck rolling down the highway, and a golf ball sailing down the fairway all have the ability to produce a change. Imagine colliding with any of these objects!

Energy need not be represented by motion. Sunlight falling on a green plant, on photographic film, or on unprotected skin can produce chemical changes, and thus light is a form of energy. Batteries and gasoline are examples of chemical energy, and uranium fuel rods contain nuclear energy. A tank of hot water contains thermal energy.

Potential energy is the energy an object has because of its position; for example, its position in a gravitational field. A bowling ball on a shelf above your desk has potential energy. It is only potential, however, and does not produce any changes until the bowling ball descends onto your desk. The higher the shelf, the more potential energy the ball has.

Energy constantly flows through nature and produces changes. Sunlight (energy) is absorbed by plants and stored as sugars and starches (energy). When the plant dies, it and other organic remains are buried and become oil (energy), which gets pumped to the surface and burned in automobile engines to produce motion (energy).

Aristotle believed that all change originated in the motion of the starry sphere and flowed down to Earth. Modern science has found a more sophisticated description of the continuous change you see around you. In a way, science is simply the study of the many ways and forms in which energy flows through the world and produces change. Energy is the heartbeat of the natural world.

Energy is expressed in **joules (J)** in the metric system. One joule is about as much energy as that released when an apple falls from a table to the floor.

© 2016 Cengage Learning®

Energy is the ability to cause change.

MASS | ENERGY | TEMPERATURE AND HEAT | DENSITY | PRESSURE

---

to precisely measure masses of objects in the Universe than by using this formula. In later chapters, you will see this formula used over and over to find the masses of stars, galaxies, and planets.

This discussion is a good illustration of the power of Newton's work. By carefully defining motion and gravity and by giving them mathematical expression, Newton was able to derive new truths, among them his version of Kepler's third law. His work finished the transformation of what were once considered the mysterious wanderings of the planets into understandable motions that follow simple rules. In fact, his discovery of gravity explained something else that had mystified philosophers for millennia: the ebb and flow of ocean tides.

Tides and Tidal Forces

Newton understood that gravity is mutual—Earth attracts the Moon, and the Moon attracts Earth—and that means the Moon's gravity can explain the ocean tides.

Tides are caused by small differences in gravitational forces. For example, Earth's gravity attracts your body downward with a force equal to your weight. The Moon is less massive and more distant, so it attracts your body with a force that is a tiny percentage of your weight. You don't notice that little force, but Earth's oceans respond visibly.

The side of Earth that faces the Moon is about 6400 km (4000 mi) closer to the Moon than is the center of Earth. Consequently, the Moon's gravity, small though it is at the distance of Earth, is just a bit stronger on the near side of Earth than on the center. It pulls on the oceans on the near side of Earth a bit more strongly than on Earth's center, and the oceans respond by flowing to make a bulge of water on the side of Earth facing the Moon. There is also a bulge on the side of Earth that faces away from the Moon because the Moon pulls more strongly on Earth's center than on its far side. Thus, on the far side of Earth the Moon pulls Earth away from the oceans, which flow into a second bulge, this one pointing away from the Moon, as shown in **Figure 5-7a**. The ocean tides are caused by the accelerations Earth and its oceans feel as they orbit around the Earth–Moon **center of mass.**

A **Common Misconception** holds that the Moon's effect on tides means that the Moon has an affinity for water—including the water in your body—and, according to some people, that's how the Moon affects you. That's not true. If the Moon's gravity affected only water, then there would be only one tidal bulge, the one facing the Moon. As you know, the Moon's gravity acts on all of Earth, the rock as well as the water, and that produces the tidal bulge in the oceans on the far side of Earth. In fact, small tidal bulges occur in the rocky bulk of Earth because it is deformed by the Moon's gravity. Although you do not notice it, as Earth rotates the landscape rises and falls by a few centimeters with the tides. The Moon has no special affinity for water, and,

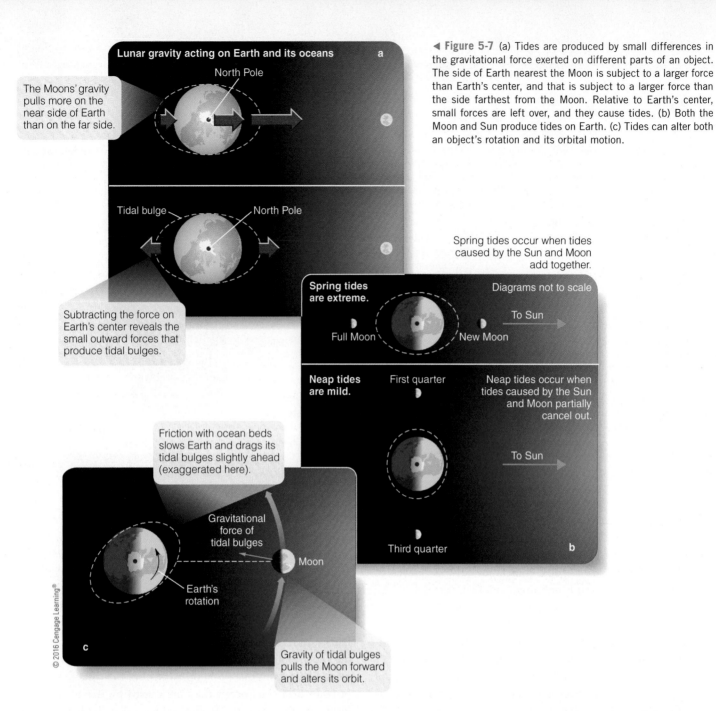

Lunar gravity acting on Earth and its oceans *a*

The Moons' gravity pulls more on the near side of Earth than on the far side.

North Pole

Tidal bulge · North Pole

Subtracting the force on Earth's center reveals the small outward forces that produce tidal bulges.

Friction with ocean beds slows Earth and drags its tidal bulges slightly ahead (exaggerated here).

Gravitational force of tidal bulges

Moon

Earth's rotation

Gravity of tidal bulges pulls the Moon forward and alters its orbit.

c

© 2016 Cengage Learning®

◀ **Figure 5-7** (a) Tides are produced by small differences in the gravitational force exerted on different parts of an object. The side of Earth nearest the Moon is subject to a larger force than Earth's center, and that is subject to a larger force than the side farthest from the Moon. Relative to Earth's center, small forces are left over, and they cause tides. (b) Both the Moon and Sun produce tides on Earth. (c) Tides can alter both an object's rotation and its orbital motion.

Spring tides occur when tides caused by the Sun and Moon add together.

Spring tides are extreme.

Diagrams not to scale

Full Moon · New Moon · To Sun

Neap tides are mild. First quarter

Neap tides occur when tides caused by the Sun and Moon partially cancel out.

To Sun

Third quarter *b*

because your body is so much smaller than Earth, any tides the Moon raises in your body are immeasurably small. Ocean tides are large because oceans are large.

You can see obvious evidence of tides if you watch the ocean shore for a few hours. The tidal bulges remain fixed in position with respect to the Moon as Earth rotates. The turning Earth carries you and your beach into a tidal bulge, the ocean water deepens, and you see the tide crawling up the sand. The tide does not really "come in"; it's more accurate to say you are carried into the tidal bulge. Later, when Earth's rotation carries you out of the bulge, the ocean becomes shallower, and the tide falls. The tides rise and fall twice a day on a normal coastline because there are two bulges on opposite sides of Earth.

The tidal cycle at any given location can be quite complex because it is affected by the latitude of the site, shape of the shoreline, wind strength, and so on. For example, tides in the Bay of Fundy (New Brunswick, Canada) occur twice a day and can exceed 40 feet (12 m). In contrast, the northern coast of the Gulf of Mexico has only one tidal cycle a day of roughly 1 foot (30 cm).

Gravity is universal, so the Sun also produces tides on Earth. The Sun is 27 million times more massive than the Moon, but it lies almost 400 times farther from Earth. Tides on Earth caused by the Sun are less than half as high as those caused by the Moon. Twice a month, at new moon and at full moon, the Moon and Sun produce tidal bulges that add together and produce extreme tidal changes: At those moon phases, high tides are exceptionally high, and low tides are exceptionally low. Such tides are called **spring tides**. Here the word *spring* does not refer to the season of the year but to the rising up of water. At first- and third-quarter moons, the Sun and Moon pull at right angles to each other, and the tides caused by the Sun partly cancel out the tides caused by the Moon. These less extreme tides are called **neap tides**. The word *neap* comes from an Old English word, *nep,* that meant something like *weak*. Spring tides and neap tides are illustrated in Figure 5-7b.

Galileo tried to understand tides, but it was not until Newton described gravity that astronomers could analyze tidal forces and recognize their surprising effects. For example, the moving water in tidal bulges experiences friction with the ocean beds and resistance as it rises onto continents. That friction slows Earth's rotation and makes the length of a day grow by 0.0023 second per century. Thin layers of silt laid down millions of years ago where rivers emptied into oceans contain a record of tidal cycles as well as daily, monthly, and annual cycles. Those data confirm that only 620 million years ago Earth's day was less than 22 hours long.

Tidal forces can also affect orbital motion. Earth rotates eastward, and friction with the ocean beds drags the tidal bulges slightly eastward out of a direct Earth–Moon line. These tidal bulges are massive, and their gravitational field pulls the Moon forward in its orbit, as shown in Figure 5-7c. As a result, the Moon's orbit is growing larger by about 3.8 cm a year, an effect that astronomers can measure by bouncing laser beams off reflectors left on the lunar surface by the Apollo astronauts.

Earth's gravitation exerts tidal forces on the Moon, and, although there are no bodies of water on the Moon, friction within the flexing rock has slowed the Moon's rotation to the point that it now keeps the same face toward Earth.

Tides are much more than just the cause of oceans rising and falling in daily and monthly rhythms. In later chapters, you will see how tides can pull gas away from stars to feed black holes, rip galaxies apart, and melt the interiors of small moons orbiting massive planets. Tidal forces produce some of the most surprising and impressive processes in the Universe.

Astronomy After Newton

Newton published his work in 1687 in a book titled, in Latin, *Philosophiae Naturalis Principia Mathematica (Mathematical Principles of Natural Philosophy)*, now known simply as the *Principia* (pronounced *prin-KIP-ee-uh;* Figure 5-8a). It is one of the most important books ever written. The *Principia* changed astronomy, changed science, and changed the way people think about nature.

The *Principia* changed astronomy by ushering in a new age. No longer did people have to appeal to whims of the gods to explain things in the heavens. No longer did they speculate on why the planets wander across the sky. After the *Principia* was published, physicists and astronomers understood that the motions of celestial bodies are governed by simple, universal rules that describe the motions of everything from orbiting planets to falling apples. Suddenly the Universe was understandable in simple terms, and astronomers could accurately predict future planetary motions (**How Do We Know? 5-2**).

The *Principia* also changed science in general. The works of Copernicus and Kepler had been mathematical, but no book before the *Principia* had so clearly demonstrated the power of mathematics as a language of precision. Newton's arguments in his book were such powerful illustrations of the quantitative study of nature that scientists around the world adopted mathematics as their most powerful tool.

Also, the *Principia* changed the way people thought about nature. Newton showed that the rules that govern the Universe are simple. Particles move according to just three laws of motion and attract each other with a force called *gravity*. These motions

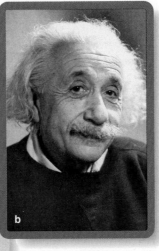

Panel a: © Iryna1/Shutterstock.com; Panel b: Fred Stein Archive/Archive Photos/Getty Images

◄ **Figure 5-8** (a) Newton, working from the discoveries of Galileo and Kepler, derived three laws of motion and the law of mutual gravitation. Understanding Newtonian physics is necessary, and sufficient, for solving problems ranging from sending astronauts to the Moon to analyzing the rotation of the largest galaxies. (b) Einstein has become a symbol of the brilliant scientist. His fame began when he was a young man and thought deeply about the nature of motion. That led him to revolutionary insights into the meaning of space and time and a new understanding of gravity.

How Do We Know? 5-2

Testing a Hypothesis by Prediction

How are the predictions of a hypothesis useful in science? Scientific hypotheses face in two directions. They look back into the past and explain phenomena previously observed. For example, Newton's laws of motion and gravity explained observations of the movements of the planets made over many centuries. But hypotheses also look forward, making predictions about what you should find as you explore further. For example, Newton's laws allowed astronomers to calculate the orbits of comets, predict their return, and eventually understand their origin.

Scientific predictions are important in two ways. First, if a prediction of a hypothesis is confirmed, scientists gain confidence that the hypothesis is a true description of nature. But predictions are important for a second reason. They can point the way to unexplored avenues of knowledge.

Particle physics is a field in which predictions have played a key role in directing research. In the early 1970s, physicists proposed a hypothesis, later named the *Standard Model*, regarding particles inside atoms and the forces between them. This hypothesis explained what scientists had already observed in experiments, but it also predicted the existence of particles that had not yet been observed. To test the hypothesis, scientists focused their efforts on building more and more powerful particle accelerators in the hopes of detecting the predicted particles.

A number of these particles have since been discovered, and they do match the characteristics predicted by the Standard Model, further confirming the hypothesis. Existence of the Higgs boson, a fundamental particle predicted by the Standard Model, was confirmed in 2012. There are still some predictions of the Standard Model remaining to be checked, and this scientific detective story is ongoing.

You learned in the previous chapter that a hypothesis that has passed many tests and has wide predictive value can "graduate" to being considered a theory, and if it is considered fundamental enough, it may be called a "law." Newton's laws of motion are at the end of the journey toward powerful reliability from their beginning as tentative hypotheses. The Standard Model is pretty far along in its version of the same journey. As you read about any scientific hypothesis, think about both what it can explain that has been observed already and what it can predict that can be observed in future.

Physicists build huge accelerators to search for subatomic particles predicted by their hypotheses.

Brookhaven National Laboratory

are predictable, and that makes the Universe seem like a vast machine, but one with operations based on a few simple rules. The Universe is complex only in that it contains a vast number of particles. In Newton's view, if he knew the location and motion of every particle in the Universe, he could—in principle—derive the past and future of the Universe in every detail. This idea of mechanical determinism has been modified by modern quantum mechanics (laws that govern behavior of particles inside atoms), but it dominated science for more than two centuries. During those years, scientists thought of nature primarily as a beautiful clockwork that would be perfectly predictable if they knew how all the gears meshed.

Most of all, Newton's work broke the last bonds between science and formal philosophy. Newton did not speculate on the good or evil of gravity. Not more than a hundred years before, scientists would have argued over the "meaning" of gravity. Newton didn't care for these debates. He wrote, "It is enough that gravity exists and suffices to explain the phenomena of the heavens."

Newton's laws were foundations of astronomy and physics for two centuries. Then, early in the 20th century, a physicist named Albert Einstein proposed a new way to describe gravity. The new theory did not replace Newton's laws but rather showed that they were only approximately correct and could be seriously in error under certain special circumstances. Einstein's theories further extended the scientific understanding of the nature of gravity. Just as Newton stood on the shoulders of Galileo, Einstein stood on the shoulders of Newton.

DOING SCIENCE

How do Newton's laws of motion and gravity explain the orbital motion of the Moon? What scientific argument—chain of evidence and logical statements—can you use, as scientists have done many times since Newton's day, to verify that the orbit of the Moon shows the operation of Newton's laws of motion and gravity?

If Earth and the Moon did not attract each other, the Moon would move in a straight line in accord with Newton's first law of motion and vanish into deep space. Instead, gravity pulls the Moon toward Earth's center, and the Moon accelerates toward Earth. This acceleration is just enough to pull the Moon away from its straight-line motion and cause it to follow a curve around Earth.

In fact, it is correct to say that the Moon is falling, but because of its lateral motion it continuously misses Earth. Every orbiting object is falling toward the center of its orbit but is also moving laterally fast enough to compensate for the inward motion, and it follows a curved orbit.

5-3 Einstein and Relativity

In the early years of the last century, Albert Einstein (1879–1955; Figure 5-8b) began thinking about how motion and gravity are related. He soon gained international fame by showing that Newton's laws of motion and gravity were only partially correct. The revised theory became known as the theory of relativity. As you will see, there are really two theories of relativity.

Special Relativity

Einstein began by thinking about how moving observers see events around them. His analysis led him to the first postulate of relativity, also known as the principle of relativity:

> **First postulate** (the principle of relativity): Observers can never detect their *uniform* motion except relative to other objects.

You may have had experiences that illuminate the first postulate while sitting on a train in a station. You suddenly notice that the train on the next track has begun to creep out of the station. However, after several moments you realize that it is your own train that is moving and that the other train is still motionless on its track. You can't tell which train is moving until you look at external objects such as the station platform.

Consider a second example. Suppose you are floating in a spaceship in interstellar space, and another spaceship comes coasting by (Figure 5-9a). You might conclude that it is moving and you are not, but someone in the other ship might be equally sure that you are moving and it is not. Of course, you could just look out a window and compare the motion of your spaceship with a nearby star, but that just expands the problem. Which is moving, your spaceship or the star? The principle of relativity says that there is no experiment you can perform inside your ship to decide which ship is moving and which is not. This means that all motion is relative.

Because no internal experiment can detect either spaceship's absolute motion through space, the laws of physics must have the same form inside both ships. Otherwise, experiments would produce different results in the two ships, and you could decide

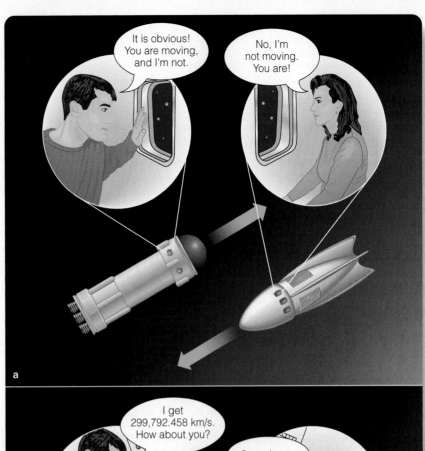
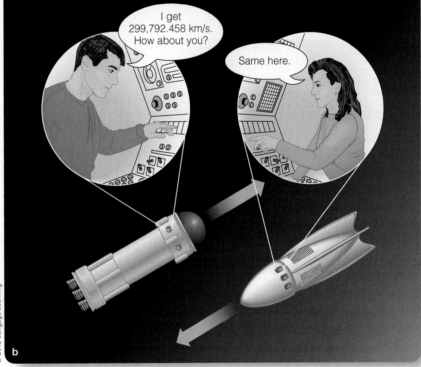

© 2016 Cengage Learning®

▲ **Figure 5-9** (a) The principle of relativity says that observers can never detect their uniform motion, except relative to other objects. Neither of these travelers can decide who is moving and who is not. (b) If the speed of light depended on the motion of the observer through space, then these travelers could perform measurements inside their spaceships to discover who was moving. If the principle of relativity is correct, then the speed of light must be a constant when measured by any observer.

who was moving. Thus, a more general way of stating Einstein's first postulate refers to the laws of physics:

First postulate (more sophisticated version): The laws of physics are the same for all observers, no matter what their motion, so long as they are not *accelerated*.

The word *accelerated* is important. If either spaceship were to fire its rockets, then its velocity would change. The crew of that ship would know it because they would feel the acceleration pressing them into their couches. Accelerated motion, therefore, is different; the pilots of the spaceships can always tell which ship is accelerating and which is not. The postulates of relativity discussed here apply only to the special case of observers in *uniform* motion, which means *unaccelerated* motion. That is why the theory is called the **special theory of relativity**.

The first postulate led Einstein to the conclusion that the speed of light must be constant for all observers. No matter how you are moving, your measurement of the speed of light has to give the same result (Figure 5-9b). This became the second postulate of special relativity:

Second postulate: The speed of light in a vacuum is constant and will have the same value for all observers independent of their motion relative to the light source.

You can see that this is required by the first postulate; if the speed of light were not constant, then the pilots of the spaceships could measure the speed of light inside their spaceships and decide who was moving. (Note the phrase "in a vacuum." Light slows down when it passes through a medium; the speed of light in water is one-third less than the speed of light in a vacuum. It is the speed of light in a vacuum, not affected by passing through any medium, to which the first postulate refers.)

Once Einstein had thought through the basic postulates of relativity (Table 5-2), he was led to some startling discoveries. Newton's laws of motion and gravity work well as long as distances are small and velocities were low. But when Einstein began to think about very large distances or very high velocities, he realized that Newton's laws were no longer always adequate to describe what happens. Instead, the postulates led Einstein to derive a more accurate description of nature that is now known as the special theory of relativity. It

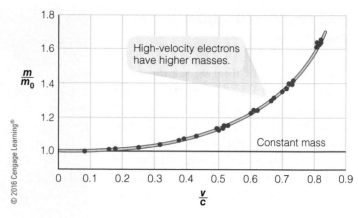

© 2016 Cengage Learning®

▲ **Figure 5-10** The observed mass of moving electrons depends on their velocity. As the ratio of their velocity to the velocity of light, *v/c*, increases, the mass of the electrons relative to their mass at rest, m/m_0, increases. Such relativistic effects are quite evident in particle accelerators, which accelerate atomic particles to very high velocities.

predicts some peculiar effects. For example, special relativity predicts that the observed mass of a moving particle depends on its velocity. The higher the velocity, the greater will be the mass of the particle. This effect is not significant at low velocities, but it becomes important as the velocity approaches the speed of light. As strange as that may seem, such increases in mass are reliably observed whenever physicists accelerate particles to high velocities (Figure 5-10).

This discovery led to yet another insight. The relativistic equations that describe the energy of a moving particle predict that the energy of a motionless particle is not zero. Rather, its energy at rest is m_0c^2. This is, of course, the famous equation:

$$E = m_0c^2$$

The constant *c* is the speed of light, and m_0 is the mass of the particle when it is at rest. This simple formula shows that mass and energy are related, and you will see in later chapters how nature can convert one into the other inside stars.

For example, suppose that you convert 1 kg of matter into energy. The speed of light is 3×10^8 m/s, so your result is 9×10^{16} joules (J), approximately equal to the energy released by a 20-megaton nuclear bomb. Recall that a joule is a unit of energy roughly equivalent to the energy given up when an apple falls from a table to the floor. This simple calculation shows that the energy equivalent of even a small mass is very large.

Other relativistic effects include the slowing of moving clocks and the shrinkage of lengths measured in the direction of motion. A detailed discussion of the major consequences of the special theory of relativity is beyond the scope of this book, but you can be confident that these strange effects have been confirmed many times in experiments. Einstein's work is called the special *theory* of relativity because it meets the scientific definition of a *theory*: It is well understood, has been checked many times in many ways, and is widely applicable (look back to How Do We Know? 4-2, page 69).

TABLE 5-2 Postulates of Relativity
I. (Relativity principle) Observers can never detect their *uniform* motion except relative to other objects.
II. The speed of light is constant and will be the same for all observers independent of their motion relative to the light source.
III. (Equivalence principle) Observers cannot distinguish between inertial forces due to acceleration and uniform gravitational forces.

The General Theory of Relativity

In 1916, Einstein published a more general version of the theory of relativity that dealt with accelerated as well as uniform motion. This **general theory of relativity** contained a new description of gravity.

Einstein began by thinking about observers in accelerated motion. Imagine an observer sitting in a windowless spaceship. Such an observer cannot distinguish between the force of gravity and the inertial forces produced by the acceleration of the spaceship (**Figure 5-11**). This led Einstein to conclude that gravity and acceleration are related, which is a conclusion now known as the equivalence principle:

> **Equivalence principle:** Observers cannot distinguish locally between inertial forces due to acceleration and uniform gravitational forces due to the presence of a massive body.

This should not surprise you. Previously in this chapter, you read that Newton concluded that the mass that resists acceleration is the same as the mass that exerts gravitational forces and then performed experiments to confirm that principle.

The importance of the general theory of relativity lies in its description of gravity. Einstein concluded that gravity, inertia, and acceleration are all associated with the way space and time are connected as a single entity referred to as space-time. This relation is often referred to as curvature, and a one-line description of general relativity is that it explains a gravitational field as a curved region of space-time:

> **Gravity according to general relativity:** Mass tells space-time how to curve, and the curvature of space-time (gravity) tells mass how to accelerate.

Therefore, you feel gravity because Earth's mass causes a curvature of space-time. The mass of your body responds to that curvature by accelerating toward Earth's center (**Figure 5-12**), and that presses you downward in your chair. According to general relativity, all masses cause curvature of the space around them, and the larger the mass, the more severe the curvature. That's gravity.

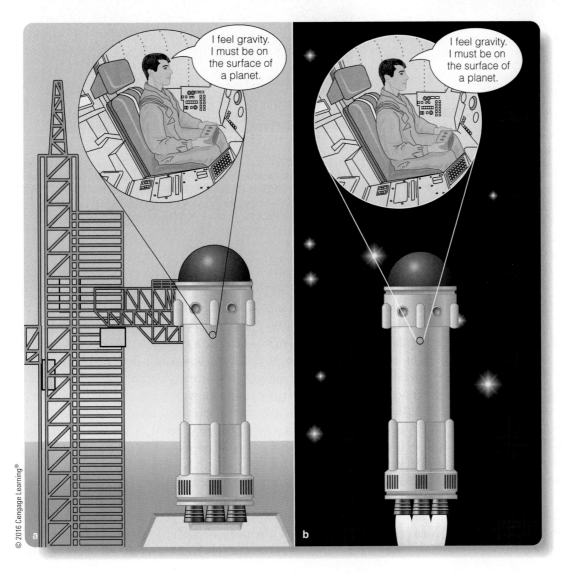

◀ **Figure 5-11** (a) An observer in a closed spaceship on the surface of a planet feels gravity. (b) In space, with the rockets smoothly firing and accelerating the spaceship, the observer feels inertial forces that are equivalent to gravitational forces.

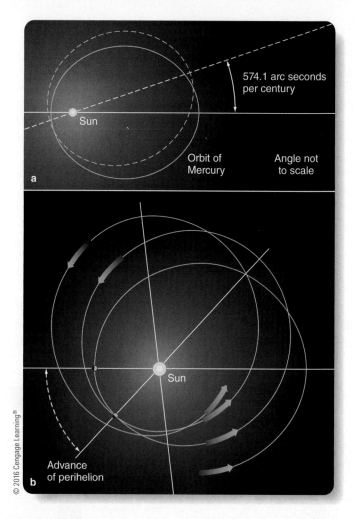

▲ **Figure 5-12** These Acapulco cliff divers are navigating rapidly through curved space-time.

Confirmation of the Curvature of Space-Time

Einstein's general theory of relativity has been confirmed by a number of experiments, but two are worth mentioning here because they were among the first tests of the theory, and required astronomical observations. One involves Mercury's orbit, and the other involves eclipses of the Sun.

Kepler understood that the orbit of Mercury is elliptical. Later, astronomers discovered that the long axis of Mercury's orbit sweeps around the Sun in a motion that is an example of precession (look back to Chapter 2, pages 17 and 20–21). The total observed precession is almost 600 arc seconds per century (**Figure 5-13**). Most of this precession is caused by the gravitation of Venus, Earth, and the other planets. However, when astronomers take all known effects into account and use Newton's description of gravity to account for the gravitational influence of all of the planets, they are left with a small excess. Mercury's orbit is precessing 43 arc seconds per century faster than Newton's laws predict.

This is a tiny effect. Each time Mercury returns to perihelion—its closest point to the Sun—it is about 29 km (18 mi) past the position predicted by Newton's laws. This is such a small distance compared with the planet's diameter of 4880 km that it could never have been detected had it not been cumulative. Each orbit, Mercury gains only 29 km, but after a century it's ahead by more than 12,000 km—more than twice its own diameter. This tiny effect, called the "advance" of Mercury's orbital perihelion, accumulated from the time of Newton to the

▲ **Figure 5-13** (a) Mercury's orbit precesses 574.1 arc seconds per century, 43.0 arc seconds more than predicted by Newton's laws. (b) Even when you ignore the influences of the other planets, Mercury's orbit is not a perfect ellipse. Curved space-time near the Sun distorts the orbit from an ellipse into a rosette. The advance of Mercury's perihelion is exaggerated by a factor of about one million in this figure.

time of Einstein into a serious discrepancy in the Newtonian description of the Universe.

The advance of perihelion of Mercury's orbit was one of the first problems to which Einstein applied the principles of general relativity. First he calculated how much the Sun's mass curves space-time in the region of Mercury's orbit, and then he calculated how Mercury moves through the space-time. The theory predicted that the curved space-time should cause Mercury's orbit to advance by 43.03 arc seconds per century, exactly the same as the observed excess to within the measurement uncertainty (Figure 5-13b).

When his theory matched observations, Einstein was so excited he could not return to work for three days. He would be even happier with modern studies that have shown that Venus, Earth, and even Icarus, an asteroid that comes close to the Sun, also have orbits observed to be slipping forward as a result of the curvature of space-time near the Sun. This same effect has been detected in pairs of stars that orbit each other.

A second test of general relativity was related to the motion of light through the curved space-time near the Sun. Because light has a limited speed, Newton's laws predict that the gravity of an object should slightly bend the paths of light beams passing nearby. The equations of general relativity indicated that light should have an extra deflection caused by traveling through curved space-time, just as a rolling golf ball is deflected by undulations in a putting green. Einstein predicted that starlight grazing the Sun's surface would be deflected by 1.75 arc seconds, twice the deflection that Newton's law of gravity would predict (**Figure 5-14**). Starlight passing near the Sun is normally lost in the Sun's glare, but during a total solar eclipse, stars beyond the Sun can be seen. As soon as Einstein published his theory, astronomers rushed to observe such stars and test the curvature of space-time.

The first total solar eclipse following Einstein's publication of his theory of general relativity in 1916 occurred during June 1918. It was cloudy at some observing sites, and results from other sites were inconclusive. The next occurred in May 1919, only months after the end of World War I, and was visible from Africa and South America. British teams went to both Brazil and Príncipe, an island off the coast of Africa. Months before the eclipse, they photographed the part of the sky where the Sun would be located during the eclipse and measured the positions of the stars on the photographic plates. Then, during the eclipse, they photographed the same star field with the eclipsed Sun located in the middle. After measuring the plates, they found slight changes in the positions of the stars. Stars seen near the edge of the solar disk were observed to be shifted outward, away from the Sun, by about 1.8 arc seconds, close to the theory's prediction.

Because the angles are so small, this is a delicate observation, and it has been repeated at many total solar eclipses since 1919, with similar results (**Figure 5-15**). The most accurate results were

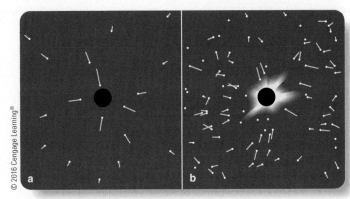

▲ **Figure 5-15** (a) Schematic drawing of the deflection of starlight by the Sun's gravity. Dots show the true positions of the stars as photographed months before the eclipse. Lines point toward the positions of the stars during the eclipse. (b) Actual data from an eclipse in 1922. Random uncertainties of observation cause some scatter in the data, but on average the stars appear to move away from the Sun by 1.77 arc seconds at the edge of the Sun's disk. The deflection of stars is magnified by a factor of 2300 in both (a) and (b).

obtained in 1973 when a combined University of Texas and Princeton team measured a deflection of 1.66 ± 0.18 arc seconds, which is in very good agreement with the prediction of Einstein's theory.

The general theory of relativity is critically important in modern astronomy. You will encounter the theory again in the discussions of black holes, distant galaxies, and the big bang Universe. Einstein revolutionized modern physics by providing an explanation of gravity based on the geometry of curved space-time. Galileo's inertia and Newton's mutual gravitation are shown to be not just descriptive rules but fundamental properties of space and time.

DOING SCIENCE

What does the equivalence principle tell you? Einstein began his work, as scientists sometimes do, by thinking carefully about common things such as what you feel when you are moving uniformly versus accelerating. This led him to deep insights now called *postulates*, one of which is known as the *equivalence principle*.

The equivalence principle states that there is no observation you can make inside a closed spaceship to distinguish between uniform acceleration and gravitation. Of course, you could open a window and look outside, but then you would no longer be in a closed spaceship. As long as you make no outside observations, you can't tell whether your spaceship is firing its rockets and accelerating through space or resting on the surface of a planet where gravity gives you weight.

Einstein took the equivalence principle to mean that gravity and acceleration through space-time are somehow related. The general theory of relativity gives that relationship a mathematical form and shows that gravity is really a distortion in space-time that physicists refer to as *curvature*. Consequently, it is said that "mass tells space-time how to curve, and curved space-time tells mass how to move." The equivalence principle led Einstein to an explanation for gravity.

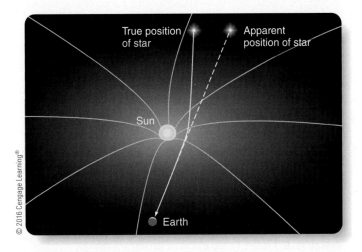

▲ **Figure 5-14** Like a depression in a putting green, the curved space-time near the Sun deflects light from distant stars and makes them appear to lie slightly farther from the Sun than their true positions.

Everything in the Universe is falling. The Moon is falling around Earth. Earth is falling along its orbit around the Sun, and the Sun and every other star in our galaxy are falling along their orbits around the galactic center. Stars in other galaxies are falling around the center of those galaxies, and every galaxy in the Universe is falling as it feels the gravitational tugs of every bit of matter that exists.

Newton's explanation of gravity as a force between two unconnected masses was action at a distance, and it offended many of the scientists of his time. They thought Newton's gravity seemed like magic. In the 20th century, Einstein explained that gravity is a curvature of space-time and that every mass accelerates according to the curvature it feels around it. That's not action at a distance, and it gives new insight into how the Universe works.

The mass of every atom in the Universe contributes to the curvature, creating a universe filled with three-dimensional hills and valleys of curved space-time. You and the Earth, the Sun, our galaxy, and every other object in the Universe are falling through space guided by the curvature of space-time.

Study and Review

Summary

▶ Aristotle argued that the universe was composed of four elements, each with its proper place: earth (rock and soil) at the center, with water, air, and fire in layers above. **Natural motion (p. 81)** occurred when a displaced object returned to its proper place. **Violent motion (p. 81)** was motion other than natural motion and had to be sustained by a force. Aristotle's ideas were considered authoritative until the time of Copernicus, Kepler, and Galileo.

▶ Galileo found that a falling object is accelerated; that is, it falls faster and faster with each passing second. The rate at which a falling object changes its speed, termed the **acceleration of gravity (p. 81)**, is 9.8 m/s^2 (32 ft/s^2) at Earth's surface and does not depend on the weight of the object, contrary to what Aristotle said. Supposedly, Galileo dramatically demonstrated this lack of dependency on weight by dropping balls of iron and wood from the Leaning Tower of Pisa to show that the balls would fall together. Air resistance would have complicated that experiment, but a feather and a hammer dropped in a vacuum do fall together.

▶ Galileo reasoned that, in the absence of friction, a moving body on a horizontal plane will continue moving forever. The first of Newton's three laws of motion, "A body continues at rest or in uniform motion in a straight line unless it is acted on by some force," was based on Galileo's insight. Resistance of matter to changes in motion is called **inertia (p. 82)**.

▶ Newton's second law says that an **acceleration (p. 83)**, which is defined as a change in velocity, must be caused by a force and vice versa. A **velocity (p. 83)** is a speed with a specific direction, so a change in speed *or* direction is an acceleration.

▶ **Mass (p. 84)** is the amount of matter in a body. **Momentum (p. 83)** is a measure of a body's amount of motion, which is a combination of its velocity and mass.

▶ Newton's third law says that forces occur in pairs acting in opposite directions.

▶ Newton realized that the curved path of the Moon meant that the Moon was being accelerated toward Earth and away from a straight-line path. That required the presence of a force—the gravitational attraction between two bodies.

▶ From his mathematical analysis, Newton was able to show that the force of gravity between two masses is proportional to the product of their masses and depends on distance with an **inverse square law (p. 84)**. That is, the force of gravity is inversely proportional to the square of the distance between the two masses.

▶ To explain how gravity can act at a distance, scientists now describe it as a **field (p. 86)**, which means that there is a strength and direction of gravitational force associated with every point in space.

▶ An object in space near Earth would move along a straight line and quickly leave Earth were it not for Earth's gravity accelerating the object toward Earth's center and forcing it to follow a curved path, an orbit. Objects in orbit around Earth are falling (being accelerated) toward Earth's center. If there is no friction, the object will fall around its orbit forever.

▶ An object in a **closed orbit (p. 89)** follows an elliptical path. A circle is simply a special case of an ellipse with zero eccentricity. To follow a circular orbit, an object must orbit with **circular velocity (p. 89)**. At a certain distance from Earth, a **geosynchronous satellite (p. 88)** can stay above a spot on Earth's equator as Earth rotates.

▶ If a body's velocity equals or exceeds the **escape velocity, V_e (p. 89)**, it will follow a parabola or hyperbola. These orbits are termed **open orbits (p. 89)** because the object never returns to its starting place.

▶ Two objects in orbit around each other actually orbit their common **center of mass (p. 89)**.

▶ Newton's laws of motion and of gravity explain Kepler's three laws of planetary motion. The planets follow elliptical orbits because gravity obeys the inverse square law. The planets move faster when closer to the Sun and slower when farther away because they conserve **angular momentum (p. 90)**. A planet's orbital period squared is proportional to its orbital radius cubed because the moving planet conserves energy.

▶ **Energy (p. 91)** refers to the ability to produce a change. **Kinetic energy (p. 91)** is an object's energy of motion, and **potential energy (p. 91)** is the energy an object has because of its position. The unit of energy is the **joule (J) (p. 91)**.

- Tides are caused by differences in the force of gravity acting on different parts of a body. Tides on Earth occur because the Moon's gravity pulls more strongly on the near side of Earth than on the center of Earth and also more strongly on the center of Earth than on the far side of Earth. As a result, there are two tidal bulges on Earth caused by the Moon's gravity, one toward the Moon on Earth's near side and one away from the Moon on Earth's far side.

- Tides produced by the Moon combine with tides produced by the Sun to cause extreme tides, called **spring tides (p. 93)**, at new and full moons. The Moon and Sun work against each other to produce the smallest tides, called **neap tides (p. 93)**, at quarter moons.

- Friction from tides can slow the rotation of a rotating object, and the gravitational pull of tidal bulges can make orbits change slowly.

- Einstein published two theories that extended Newton's laws of motion and gravity, the special theory of relativity and the general theory of relativity.

- The first postulate of **special theory of relativity (p. 96)** is that observers cannot detect their uniform motion through space by internal tests, only by observation of outside objects. In other words, uniform (unaccelerated) motion is relative. This leads to the second relativity postulate: The speed of light measured in a vacuum is a constant for all observers.

- A consequence of special relativity is that mass and energy are related. That relationship is expressed by the famous equation, $E = m_0c^2$.

- The **general theory of relativity (p. 97)** says that a gravitational field is a curvature of space-time caused by the presence of a mass. For example, Earth's mass curves space-time, and the mass of your body responds to that curvature by accelerating toward Earth's center.

- General relativity's prediction of the curvature of space-time was confirmed by observations of the slow advance in perihelion (precession) of the orbit of Mercury and by the deflection of starlight passing near the Sun observed during a 1919 total solar eclipse. Further observations confirming Einstein's general theory of relativity continue to be made up to the present day.

Review Questions

1. According to Aristotle, if earth and water were displaced then they would return naturally to their proper place. Today, what do we call this Aristotelean natural motion?

2. Today, what do we call the Aristotelean violent motion?

3. Which of Kepler's or Newton's laws best describes Aristotelean violent motions?

4. Why would Aristotle's explanation of gravity not work if Earth is not the center of the universe?

5. According to the principles of Aristotle, what part of the motion of a baseball pitched across home plate is natural motion? What part is violent motion?

6. If you drop a feather and a steel hammer at the same moment, they should hit the ground at the same instant. Why doesn't this work on Earth, and why does it work on the Moon? Will it work on Phobos, a moon of Mars?

7. What is the difference between mass and weight?

8. When a person says he gained weight, does he mean that he gained in mass, gravity, or both mass and gravity?

9. An astronaut working in space near the International Space Station says she feels weightless. What does she mean? Does the astronaut not have weight?

10. What is the difference between speed and velocity?

11. A car is on a circular off ramp of an interstate and is traveling at exactly 25 mph around the curve. Does the car have velocity? Does the car have acceleration? Is the car decelerating?

12. How many accelerators does a car have?

13. You put your astronomy textbook and your No. 2 pencil on a ceramic tile floor, and you blow on each. Which has more inertia—the pencil, the textbook, or neither? Why? Which has more momentum? Why?

14. An astronaut is in space with a baseball and a bowling ball. The astronaut gives both objects an equal push in the same direction. Does the baseball have the same inertia as the bowling ball? Why? Does the baseball have the same acceleration as the bowling ball from the push? Why? If both balls are traveling at the same speed, does the baseball have the same momentum as the bowling ball?

15. You are at a red light in your car. The red light turns green, and you put your car in first gear and step on the gas pedal. The speedometer changes from 0 to 10 to 20 mph in 3 seconds. In this span of time, did the car accelerate or decelerate, and in which direction? Did the car increase in speed, decrease in speed, or travel at the same speed, and in which direction?

16. You weigh 100 pounds, your friend weighs 200 pounds, and you are in an arm wrestling contest with each other. Neither person is winning, but each of you is struggling to push the other's forearm over to the tabletop. Which of Kepler's or Newton's laws applies to this scenario and why? Now in one swift motion you plant your friend's forearm on the table, winning the contest. Which of Kepler's or Newton's laws applies to this motion and why?

17. Why did Newton conclude that some force had to pull the Moon toward Earth?

18. Why did Newton conclude that gravity has to be mutual and universal?

19. You have the same mass as a person sitting next to you. Are you gravitationally attracted to him? If so, why don't you instantly zoom over and stick to him? Is the attraction mutual? If not, why not?

20. You are sitting next to a person who has twice as much weight. Are you gravitationally attracted to her? If so, is it twice as much as her attraction to you? If not, why not?

21. You are sitting next to a person who has twice as much weight. You get up and move one seat over, doubling your distance from him. Did the gravitational force between you increase, decrease, or stay the same?

22. You are sitting next to a person who has twice as much weight as you do. A friend comes by and gives you a marshmallow, and you eat it. Did your gravitational force to your neighbor increase, decrease, or stay the same? Why?

23. How does the concept of a field explain action at a distance? Name another kind of field also associated with action at a distance.

24. Why can't a spacecraft go "beyond Earth's gravity"?

25. Where is the center of mass of your body?

26. Balance a pencil lengthwise on the side of your finger. Where is the center of mass? Balance a pencil widthwise (for example, on the eraser side) on your finger. Where is the center of mass? Is the center of mass a plane, sphere, circle, point, or a line?

27. What is the center of mass of two bodies? Where is the center of mass of the Earth–Moon system?

28. Why can't you leave Earth's gravitational field when jumping vertically?

29. According to Kepler's first law, planets move in elliptical orbits. Why is that considered accelerated motion? According to Newton, what is the force causing that acceleration?

30. How do planets orbiting the Sun and skaters doing a spin both conserve angular momentum?

31. If a planet were to migrate inward toward the Sun, would its orbital speed increase, decrease, or stay the same? Would its angular momentum change? Which of Kepler's laws or Newton's version of Kepler's laws does this scenario describe?

32. If you hold this textbook out at shoulder height and let go, at the instant you let go, does the book have potential energy? Kinetic energy?

33. Today at the beach you see the highest of all high tides in the last month. You see the Moon in the daytime sky. What is the most likely Moon phase?

34. Why is the period of an open orbit undefined?

35. In what conditions do Newton's laws of motion and gravity need to be modified?

36. How does the first postulate of special relativity imply the second postulate?

37. When you ride a fast elevator upward, you feel slightly heavier as the trip begins and slightly lighter as the trip ends. How is this phenomenon related to the equivalence principle?

38. From your knowledge of general relativity, would you expect radio waves from distant galaxies to be deflected as they pass near the Sun? Why or why not?

39. How is gravity related to acceleration? Are all accelerations the result of gravity?

40. Near a massive planet, is gravitational acceleration large or small? Is space strongly curved, or not? What about near a small marble?

41. **How Do We Know?** Why would science be impossible if some natural events happened without causes?

42. **How Do We Know?** Why is it important that a theory make testable predictions?

Discussion Questions

1. How did Galileo idealize his inclines to conclude that an object in motion stays in motion until it is acted on by some force?

2. Give an example from everyday life to illustrate each of Newton's laws.

3. Where in the Universe can you be weightless?

4. People who lived before Newton may not have believed in cause and effect as strongly as you do. How do you suppose that affected how they saw their daily lives?

5. Is everything gravitationally attracted to other things in the Universe, and thus is everything in a state of falling?

6. Give an example from everyday life of kinetic energy and gravitational potential energy.

7. If Newton modified Kepler's laws and Einstein modified Newton's laws, is it possible Einstein's laws (postulates) might be modified?

Problems

1. This astronomy textbook is to be dropped from a tall building on Earth. One second after dropped, what are the textbook's speed, velocity, and acceleration? After 2 seconds? After 3 seconds? The book hits the ground; what are the book's speed, velocity, and acceleration?

2. Compared to the strength of Earth's gravity at its surface $r = R_E$ where R_E is the radius of Earth, how much weaker is gravity at a distance of $r = 10\ R_E$? At $r = 20\ R_E$?

3. Compare the force of gravity on a 1 kg mass on the Moon's surface with the force that mass on Earth's surface. Which force is greater, why, and by how much?

4. A satellite is in orbit at a distance r from the center of Earth. If the orbit radius is halved so that the satellite is orbiting closer to Earth's surface, will the field strength increase, decrease, or stay the same and by how much?

5. The International Space Station is in orbit around the Earth at a distance r from the center of Earth. A recent addition increased the Station's mass by a factor of 3. Did Earth's gravitational force on the Station increase, decrease, or stay the same and by how much?

6. If a small lead ball falls from a high tower on Earth, what will be its velocity after 2 seconds? After 4 seconds?

7. What is the circular velocity of an Earth satellite 1000 km above Earth's surface? (*Note:* Earth's average radius is 6371 km. *Hint:* Convert all quantities to m, kg, s.)

8. What is the circular velocity of an Earth satellite 36,000 km above Earth's surface? What is its orbital period? (*Note:* Earth's average radius is 6371 km. *Hint:* Convert all quantities to m, kg, s.)

9. What is the orbital speed at Earth's surface? Ignore atmospheric friction. (*Note:* Earth's average radius is 6371 km. *Hint:* Convert all quantities to m, kg, s.)

10. What is the orbital speed at Earth's surface? Ignore atmospheric friction. (*Note:* Earth's average radius is 6371 km. *Hint:* Convert all quantities to m, kg, s.)

11. Repeat the previous problem for Mercury, Venus, the Moon, and Mars. (*Note:* You can find the mass and radius of each of these objects in the Appendix A tables).

12. Describe the orbit followed by the slowest cannonball on page 88 pretending that the cannonball could pass freely through Earth. (Newton got this problem wrong the first time he tried to solve it.)

13. If you visited a spherical asteroid 30 km in radius with a mass of 4.0×10^{17} kg, what would be the circular velocity at its surface? A major league fastball travels about 90 mph. Could a good pitcher throw a baseball into orbit around the asteroid? (*Note:* 90 mph is 40 m/s.)

14. What is the orbital period of a satellite orbiting just above the surface of the asteroid in Problem 13?

15. What is the escape velocity from you if your mass is 60 kg and your radius is 1 m? Is it easy or difficult for a fly to leave your gravitational pull?

16. What would be the escape velocity at the surface of the asteroid in Problem 13? Could a major league pitcher throw a baseball off the asteroid so that it never came back?

17. A moon of Jupiter takes 1.8 days to orbit at a distance of 4.2×10^5 km from the center of the planet. What is the mass of Jupiter plus its moon? Which moon is it? (*Note:* One day is 86,400 seconds. *Hint:* See Appendix Table A-11.)

Learning to Look

1. Why can the object shown at the right be bolted in place and used 24 hours a day without adjustment?

Larry Mulvehill/The Image Works

2. What is the flux at position 2 compared to position 1 in Figure 5-5? How does the distance from the center to position 2 compare with the distance to position 1?

3. Why is it a little bit misleading to say that this astronaut is weightless?

NASA/JSC

Light and Telescopes 6

Guidepost In previous chapters of this book, you viewed the sky the way the first astronomers did, with the unaided eye. Then, in Chapter 4, you got a glimpse through Galileo's small telescope that revealed amazing things about the Moon, Jupiter, and Venus. Now you can consider the telescopes, instruments, and techniques of modern astronomers.

Telescopes gather and focus light, so you need to study what light is, and how it behaves, on your way to understanding how telescopes work. You will learn about telescopes that capture invisible types of light such as radio waves and X-rays. These enable astronomers to reach a more complete understanding of the Universe. This chapter will help you answer these five important questions:

► **What is light?**

► **How do telescopes work?**

► **What are the powers and limitations of telescopes?**

► **What kind of instruments do astronomers use to record and analyze light gathered by telescopes?**

► **Why are some telescopes located in space?**

Science is based on observations. Astronomers cannot visit distant stars and galaxies, so they must study them using telescopes. Eleven chapters of exploration remain, and everyone will present information gained by astronomers using telescopes.

The strongest thing that's given us to see with's
A telescope. Someone in every town
Seems to me owes it to the town to keep one.

ROBERT FROST, "THE STAR-SPLITTER"
Reprinted by permission of Henry Holt & Company

ESO/B. Tafreshi (twanight.org)

A portion of the ALMA millimeter/submillimeter telescope array, at work on a high plateau in the Chilean Andes. Each dish is 12 meters (40 ft) in diameter. The photograph's long exposure shows star trails caused by Earth's rotation. In the Southern Hemisphere, stars appear to circle around the south celestial pole that lies in the faint constellation of Octans (the Octant), not marked by any bright star.

LIGHT FROM THE SKY is a treasure that links you to the rest of the Universe. Astronomers strive to study light from the Sun, planets, moons, asteroids, comets, stars, nebulae, and galaxies, extracting information about their natures. Most celestial objects are very faint sources of light, so large telescopes are built to collect the greatest amount of light possible.

Some types of telescopes, for example radio telescopes like the ones featured on the previous page, gather light that is invisible to the human eye, but all telescopes work by the same basic principles. Some telescopes are used on Earth's surface, but others must go high in Earth's atmosphere, or even above the atmosphere into space, to work properly.

There is more to astronomy than amazing technology and brilliant scientific analysis. Astronomy helps us understand what we are. In the quotation that opens this chapter, the poet Robert Frost suggests that someone in every town should own a telescope to help us look upward and outward.

6-1 Radiation: Information from Space

Astronomers no longer spend their time mapping constellations or charting phases of the Moon. Modern astronomers analyze light using sophisticated instruments and techniques to investigate the temperatures, compositions, and motions of celestial objects to be able to make inferences about their internal processes and evolution. To understand how astronomers gain such detailed information about distant objects, you first need to learn about the nature of light.

Light as Waves and Particles

When you admire the colors of a rainbow, you are seeing an effect of light acting as a wave (Figure 6-1a). When you use a digital camera to take a picture of the same rainbow, the light acts

like particles as it hits the camera's detectors (Figure 6-1b). Light has both wave-like and particle-like properties, and how it behaves depends partly on how you treat it.

Light is referred to as **electromagnetic radiation** because it is made up of both electric and magnetic fields. (You encountered the concept of a field in the previous chapter in the context of gravitational fields.) The word *light* is commonly used to refer to electromagnetic radiation that humans can see, but visible light is only one among many types of electromagnetic radiation that include X-rays and radio waves.

Some people are wary of the word *radiation*, but that involves a **Common Misconception**. *Radiation* refers to anything that radiates away from a source. Dangerous high-energy particles emitted from radioactive atoms are also called radiation, and you have learned to be concerned when you hear that word. But light, like all electromagnetic radiation, spreads outward from its origin, so you can correctly refer to light as a form of radiation.

Electromagnetic radiation travels through space at a speed of 3.00×10^8 m/s (186,000 mi/s). This is commonly referred to as the speed of light, symbolized by the letter c, but it is in fact the speed of all types of electromagnetic radiation.

Electromagnetic radiation can act as a wave phenomenon—that is, it is associated with a periodically repeating disturbance—a wave—that carries energy. You are familiar with waves in water: If you disturb a pool of water, waves spread across the surface. Imagine placing a ruler parallel to the travel direction of the wave. The distance between peaks of the wave is called the **wavelength**, usually represented by the Greek lowercase letter lambda (λ) (Figure 6-2).

Wavelength is related to **frequency**, the number of waves that pass a stationary point in 1 second. Frequency is often represented by the Greek lowercase letter *nu* (ν). The relationship among the wavelength, frequency, and speed of a wave can be expressed by:

$$\lambda \nu = c$$

If your favorite FM station is on the dial at 89.5, that means the station's radio waves have a frequency $\nu = 89.5$ megahertz. In other words, 89.5 million radio wave peaks pass by you each second. You already know that the

Panel a: © Gail Johnson/Shutterstock.com; Panel b: Roy McMahon/Cardinal/Corbis

◄ Figure 6-1 The wavelike properties of light produce a rainbow, whereas the particle-like properties are involved in the operation of a digital camera.

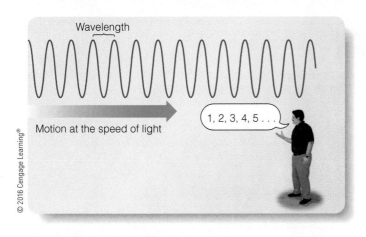

Wavelength

Motion at the speed of light

1, 2, 3, 4, 5 . . .

▲ **Figure 6-2** All electromagnetic waves travel at the speed of light. The wavelength is the distance between successive peaks. The frequency of the wave is the number of peaks that pass you in 1 second.

radio waves are traveling at the speed of light $c = 3.00 \times 10^8$ m/s. Using the formula at the bottom of page 104, you can calculate that your favorite station is radiating radio waves with a wavelength of $\lambda = 3.35$ m. Note that wavelength and frequency have an inverse relationship: The higher the frequency, the shorter the wavelength.

Sound is another example of a wave—in this case, a periodically repeating pressure disturbance that moves from source to ear. Sound requires a medium, meaning a substance such as air, water, or rock to travel through. In contrast, light is made up of electric and magnetic fields that do not require a medium and can travel through empty space. For example, on the Moon, where there is no air, there can be no sound, but there is plenty of light. This brings up a **Common Misconception** that radio waves are related to sound. Actually, radio waves are a type of light (electromagnetic radiation) that your radio receiver transforms into sound so you can listen. Radio communication works just fine between astronauts standing on the airless Moon; radio signals travel through the vacuum, and then the spacesuit radios convert the radio signals to sound that is heard in the air inside their helmets.

Although electromagnetic radiation can behave as a wave, it can also behave as a stream of particles. A particle of electromagnetic radiation is called a **photon**. You can think of a photon as a packet of waves. The amount of energy a photon carries is inversely proportional to its wavelength. The following simple formula describes that relationship:

$$E = \frac{hc}{\lambda}$$

Here h is Planck's constant (6.63×10^{-34} joule s), c is the speed of light in meters per second, and λ is the wavelength in meters.

This equation expresses the important point that there is a relationship between the energy E of a photon, a particle property of light, and the wavelength λ, a wave property. The inverse proportion means that as λ gets smaller E gets larger: Shorter-wavelength photons carry more energy, and longer-wavelength photons carry less energy. You can see that the relationship between wavelength and frequency means there must also be a simple relationship between photon energy and frequency. That is, short wavelength, high frequency, and large photon energy go together; long wavelength, low frequency, and small photon energy go together.

The Electromagnetic Spectrum

A **spectrum** is an array of electromagnetic radiation displayed in order of wavelength. You are most familiar with the spectrum of visible light that you see in rainbows. The colors of the rainbow differ in wavelength, with red having the longest wavelength and violet the shortest. The visible spectrum is shown in **Figure 6-3a**.

The average wavelength of visible light is about 0.0005 mm. This means that roughly 50 light waves would fit end to end across the thickness of a sheet of household plastic wrap. It is awkward to describe such short distances in millimeters, so scientists usually give the wavelength of light using **nanometer (nm)** units, equal to one-billionth of a meter (10^{-9} m). Another unit that astronomers commonly use is called the **angstrom (Å)**, named after the Swedish astronomer Anders Jonas Ångström. One angstrom is 10^{-10} m, that is, one-tenth of a nanometer. The wavelength of visible light ranges from about 400 to 700 nm (4000 to 7000 Å).

Just as you sense the wavelength of sound as pitch, you sense the wavelength of light as color. Light with wavelengths at the short-wavelength end of the visible spectrum ($\lambda =$ about 400 nm) appears violet to your eyes, and light with wavelengths at the long-wavelength end ($\lambda =$ about 700 nm) appears red.

Figure 6-3a shows that the visible spectrum makes up only a small part of the entire electromagnetic spectrum. Beyond the red end of the visible spectrum lies **infrared (IR)** radiation, with wavelengths ranging from 700 nm to about 1 mm (1 million nm). Your eyes do not detect infrared, but your skin senses it as heat. A heat lamp warms you by giving off infrared radiation. Infrared radiation was discovered in the year 1800, the first known example of "invisible light" (**Figure 6-4**).

Beyond the infrared part of the electromagnetic spectrum lie **microwaves** and **radio waves**. Microwaves, used for cooking food in a microwave oven, as well as for radar and some long-distance telephone communications, have wavelengths from a few millimeters to a few centimeters. The radio waves used for FM, television, military, government, and cell phone

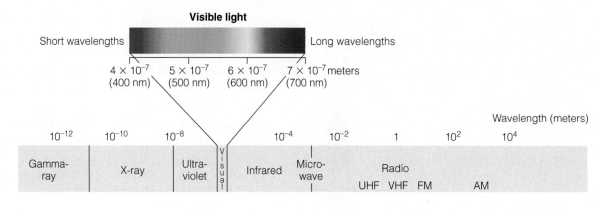

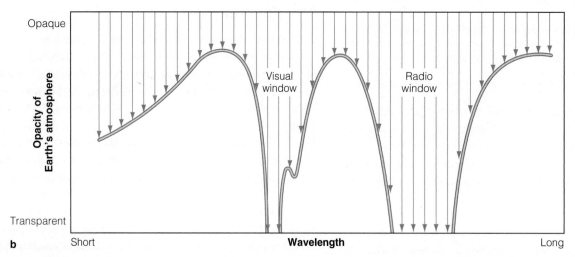

▲ **Figure 6-3** (a) The spectrum of visible light, extending from red to violet, is only part of the electromagnetic spectrum. (b) Most forms of light (electromagnetic radiation) are absorbed in Earth's atmosphere. Light can reach Earth's surface only through the visual and radio "windows."

radio transmissions have wavelengths of a few centimeters to a few meters, whereas AM and other types of radio transmissions have wavelengths of a few hundred meters to a few kilometers.

Now look at the other end of the electromagnetic spectrum in Figure 6-3a and notice that electromagnetic waves shorter than violet are called **ultraviolet (UV)**. Electromagnetic waves that are even shorter are called **X-rays**, and the shortest are **gamma-rays**.

Recall the formula for the energy of a photon. Extremely short-wavelength, high-frequency photons, such as X-rays and gamma-rays, have high energies and can be dangerous. Even ultraviolet photons have enough energy to harm you. Small amounts of ultraviolet radiation produce a suntan, and larger doses cause sunburn and skin cancers. Contrast this to the lower-energy infrared photons. Individually they have too little energy to affect skin pigment, a fact that explains why you can't get a tan from a heat lamp. Only by concentrating many low-energy photons in a small area, as in a microwave oven, can you transfer significant amounts of energy.

The boundaries between these wavelength ranges are defined only by conventional usage, not by natural divisions. There is no real distinction between short-wavelength ultraviolet light and long-wavelength X-rays. Similarly, long-wavelength infrared radiation is indistinguishable from short-wavelength microwaves.

Astronomers collect and study electromagnetic radiation from space because it carries almost the only clues available about the nature of stars, planets, and other celestial objects. Earth's atmosphere is opaque to most electromagnetic radiation, as shown in the graph in Figure 6-3b. Gamma-rays and X-rays are absorbed high in Earth's atmosphere, and a layer of ozone (O_3) at altitudes of about 15 to 30 km (10 to 20 mi) absorbs most ultraviolet radiation. Water vapor in the lower atmosphere absorbs most long-wavelength infrared radiation and microwaves. Only visible light, some short-wavelength infrared radiation, and some radio waves reach Earth's surface through wavelength bands called **atmospheric windows**. Obviously, if you wish to study the Universe from Earth's surface, you have to "look through" one of those windows.

▲ Figure 6-4 Depiction of Sir William Herschel discovering that sunlight contains radiation detectable by thermometers but not by human eyes. He named that invisible light "infrared," meaning "below red."

DOING SCIENCE

What would you see if your eyes were sensitive only to radio wavelengths? An important part of doing science is being able to observe and measure things that cannot be detected with unaided human senses.

The world is much richer and more complicated than the aspects we can see, hear, taste, smell, and feel. If you had radio vision, you would probably be able to see through walls because ordinary walls are transparent to most radio wavelengths. But remember that your eyes don't give off light; they only detect light that already exists. What you would see through the walls would be the many strong radio wave sources on Earth—radio and TV stations, cell phones, power lines, and even electric motors. Your radio eyes would see many bright "lights" nearby, but they would all be artificial.

As you have learned, Earth's atmosphere is mostly transparent to radio waves. If, after looking around the surface of Earth, you looked up at the sky, you would see the Sun and Jupiter, which are both strong natural radio sources, but probably nothing else in the Solar System. You would also see numerous radio stars arranged in unfamiliar constellations because few if any of the stars that are bright at visual wavelengths are also strong radio sources.

Now imagine a slightly different situation. ***Would you be in the dark if your eyes were sensitive only to X-ray wavelengths?***

6-2 Telescopes

Astronomers build telescopes to collect light from distant, faint objects for analysis. That requires very large telescopes built by careful optical and mechanical engineering work. You can understand these ideas more completely by learning about the two types of telescopes and their relative advantages and disadvantages.

Two Ways to Do It: Refracting and Reflecting Telescopes

Light can be focused into an image in one of two ways (Figure 6-5). Either (1) a lens refracts ("bends") light passing through it, or (2) a mirror reflects ("bounces") light from its surface.

These two ways to manipulate light correspond to two astronomical telescope designs. **Refracting telescopes** use a lens to gather and focus light, whereas **reflecting telescopes** use a mirror (Figure 6-6). You learned in Chapter 4 that Galileo was the first person to systematically record observations of celestial objects using a telescope, beginning a little more than 400 years ago in 1610. Galileo's telescope was a refractor. In Chapter 5 you learned about the amazing range of Isaac Newton's scientific work; among his many accomplishments was the invention of the reflecting telescope.

The main lens in a refracting telescope is called the **primary lens**, and the main mirror in a reflecting telescope is called the **primary mirror**. The distance from a lens or mirror to the image it forms of a distant light source such as a star is called the **focal length**. Both refracting and reflecting telescopes form an image that is small, inverted, and difficult to observe directly, so a lens called the **eyepiece** normally is used to magnify the image and make it convenient to view.

Manufacturing a lens or mirror to the proper shape and necessary smoothness is a delicate, time-consuming, and expensive process. Short focal-length lenses and mirrors must be made with more curvature than ones with long focal lengths. The surfaces of lenses and mirrors then must be polished to eliminate irregularities larger than the wavelengths of light. Creating the optics for a large telescope can take months or years; involve huge, precision machinery; and employ several expert optical engineers and scientists.

Refracting telescopes suffer from a serious optical distortion that limits their use. When light is refracted through glass, shorter-wavelength light bends more than longer wavelengths; so, for example, blue light comes to a focus closer to the lens than does red light (Figure 6-7a). That means if you focus the eyepiece on the blue image, the other colors are out of focus, and you see a colored blur around the image. If you focus instead on the red image, all the colors except red are blurred, and so on. This color separation is called **chromatic aberration**. Telescope designers can grind a telescope lens with two components made of different kinds of glass and thereby bring two different

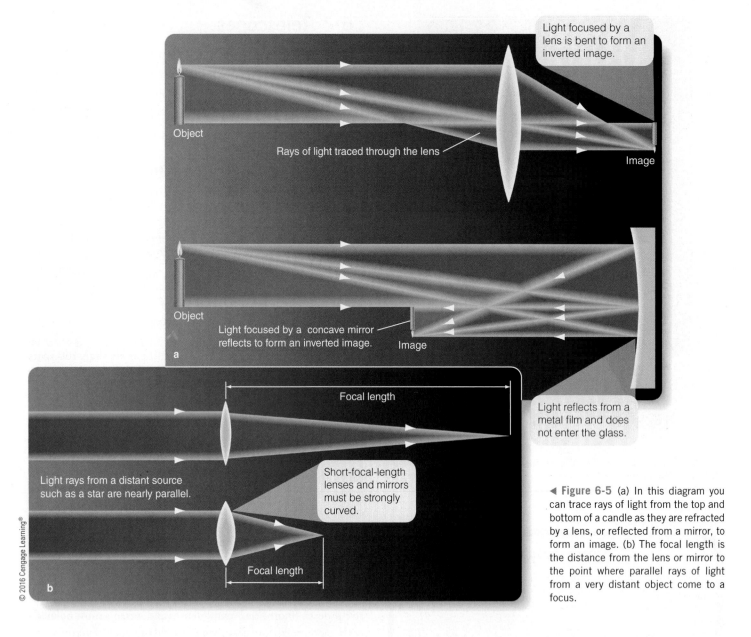

Light focused by a lens is bent to form an inverted image.

Object

Rays of light traced through the lens

Image

Object

Light focused by a concave mirror reflects to form an inverted image.

Image

a

Light reflects from a metal film and does not enter the glass.

Focal length

Light rays from a distant source such as a star are nearly parallel.

Short-focal-length lenses and mirrors must be strongly curved.

Focal length

b

© 2016 Cengage Learning®

◀ **Figure 6-5** (a) In this diagram you can trace rays of light from the top and bottom of a candle as they are refracted by a lens, or reflected from a mirror, to form an image. (b) The focal length is the distance from the lens or mirror to the point where parallel rays of light from a very distant object come to a focus.

wavelengths to the same focus (Figure 6-7b). That improves the image, but these so-called **achromatic lenses** are not totally free of chromatic aberration. Even though two colors have been brought together, the others are still out of focus.

A refracting telescope's primary lens is much more difficult to manufacture than a mirror of the same size. The interior of the glass must be pure and flawless because the light passes through it. Also, if the lens is achromatic, it must be made of two different kinds of glass requiring four precisely ground surfaces. The largest refracting telescope in the world was completed in 1897 at Yerkes Observatory in Wisconsin. Its achromatic primary lens has a diameter of 1 m (40 in.) and weighs half a ton. Refracting telescopes larger than that would be prohibitively expensive.

The primary mirrors of reflecting telescopes are much less expensive than lenses because the light reflects off the front

surface of the mirror. This means that only the front surface needs to be made with a precise shape and that surface is coated with a highly reflective surface of aluminum or silver. Consequently, the glass of the mirror does not need to be transparent, and the mirror can be supported across its back surface to reduce sagging caused by its own weight. Most important, reflecting telescopes do not suffer from chromatic aberration because the light does not pass through the glass, so reflection does not depend on wavelength. For these reasons, all large astronomical telescopes built since the start of the 20th century have been reflecting telescopes.

Telescopes intended for the study of visible light are called **optical telescopes** (Figure 6-8a). As you learned previously, radio waves as well as visible light from celestial objects can penetrate Earth's atmosphere and reach the ground. Astronomers gather radio waves using **radio telescopes** such as the one in

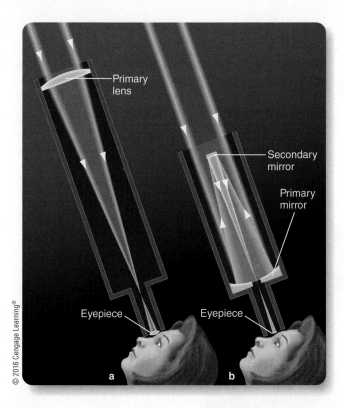

▲ **Figure 6-6** (a) A refracting telescope uses a primary lens to focus starlight into an image that is magnified by another lens called an eyepiece. The primary lens has a long focal length, and the eyepiece has a short focal length. (b) A reflecting telescope uses a primary mirror to focus the light by reflection. In this particular reflector design, called a Cassegrain telescope, a small secondary mirror reflects the starlight back down through a hole in the middle of the primary mirror to the eyepiece lens.

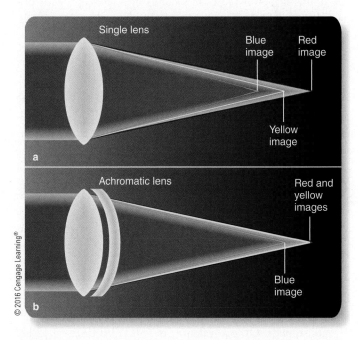

▲ **Figure 6-7** (a) An ordinary lens suffers from chromatic aberration because short wavelengths bend more than long wavelengths. (b) An achromatic lens, with two components made of two different kinds of glass, can bring any two colors to the same focus, but other colors remain slightly out of focus.

Figure 6-8b that resemble giant TV satellite dishes. It is technically extremely difficult to make a lens that can focus radio waves, so all radio telescopes, including small ones, are reflecting telescopes; the dish is the primary mirror.

The Powers and Limitations of Telescopes

A telescope's capabilities are described in three important ways that are called the three powers of a telescope. The two most important of these powers depend on the diameter of the telescope.

Light-Gathering Power: Nearly all of the interesting objects in the sky are faint sources of light, so astronomers need telescopes that can collect large amounts of light to be able to study those objects. **Light-gathering power** refers to the ability of a telescope to collect light. Catching light in a telescope is like catching rain in a bucket—the bigger the bucket, the more rain it can catch (**Figure 6-9**).

Light-gathering power is proportional to the *area* of the telescope primary lens or mirror; a lens or mirror with a large area gathers a large amount of light. The area of a circular lens or mirror written in terms of its diameter D is $\pi D^2/4$. To compare the relative light-gathering powers (LGP) of two telescopes A and B, you can calculate the ratio of the areas of their primaries, which equals the ratio of the primaries' diameters squared:

$$\frac{(LGP_A)}{(LGP_B)} = \left(\frac{D_A}{D_B}\right)^2$$

Suppose you compare telescope A, which is 24 cm in diameter, with telescope B, which is 4 cm in diameter. The ratio of their diameters is 24/4, or 6, but the light-gathering power increases as the ratio of their diameters *squared,* so telescope A gathers 36 times more light than telescope B. Because the diameter ratio is squared, even a small increase in diameter produces a relatively large increase in light-gathering power and allows astronomers to study significantly fainter objects. This principle holds not just at visual wavelengths but also for telescopes collecting any kind of radiation.

Resolving Power: The second power of a telescope, called **resolving power**, refers to the ability of the telescope to reveal fine detail. One consequence of the wavelike nature of light is that there is an unavoidable blurring called **diffraction fringes** around every point of light in an image, and you cannot see any detail smaller than the fringes (**Figure 6-10**).

Astronomers can't eliminate diffraction fringes, but the size of the diffraction fringes is inversely proportional to the diameter of the telescope. This means that the larger the telescope, the better its resolving power. However, the size of diffraction fringes is also proportional to the wavelength of light being focused. In other words, an infrared or radio telescope has less resolving power than an optical telescope of the same size.

You can imagine testing the resolving power of a telescope by measuring the angular distance between two stars that are

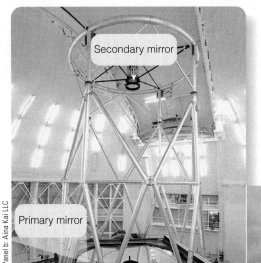

◀ **Figure 6-8** (a) The Gemini-North optical telescope on Mauna Kea in Hawai'i stands more than 19 m (62 ft) high when pointed straight up. The primary mirror (at bottom) is 8.1 m (26.5 ft) in diameter—larger than some classrooms. The sides of the telescope dome can be opened, allowing quick equalization of inside and outside temperatures at sunset, reducing air turbulence and improving seeing. (b) The largest fully steerable radio telescope in the world is at the National Radio Astronomy Observatory in Green Bank, West Virginia. The telescope stands higher than the Statue of Liberty and has a reflecting surface 100 × 110 m (330 × 360 ft) in diameter, more than big enough to hold an entire football field. Its surface consists of 2004 computer-controlled panels that adjust to maintain the shape of the reflecting surface.

Secondary mirror

Primary mirror

a

Receiver

Primary mirror

Telescope engineer in white hard hat.

b

▲ **Figure 6-9** Gathering light is like catching rain in a bucket. A large-diameter telescope gathers more light and produces a brighter image than a smaller telescope of the same focal length.

just barely distinguishable as separate objects (Figure 6-10b). The resolving power α in arc seconds of a telescope with primary diameter D that is collecting light of wavelength λ equals:

$$\alpha \left(\text{arc seconds}\right) = 2.06 \times 10^5 \left(\frac{\lambda}{D}\right)$$

To use the formula correctly, the units of D and λ need to be the same, for example, meters and meters, or centimeters and centimeters. The multiplication factor of 2.06×10^5 is the conversion between radians and arc seconds that you first saw in the small-angle formula (Chapter 3, page 41). If the wavelength of light being studied is assumed to be 550 nm, in the middle of the visual band, then the preceding formula simplifies to:

$$\alpha \left(\text{arc seconds}\right) = \frac{0.113}{D}$$

For example, the resolving power of a telescope with a diameter of 0.100 m (about 4 in.) observing at visual wavelengths is about $\alpha = (2.06 \times 10^5) \times (550 \times 10^{-9})/(0.100) = 1.13$ arc seconds. Or, equivalently, $\alpha = 0.113/0.100 = 1.13$ arc seconds. In other words, using a telescope with a diameter of 4 in., you should be able to distinguish as separate points of light any pair of stars farther apart than about 1.1 arc seconds if the optics are of good quality and if the atmosphere is not too

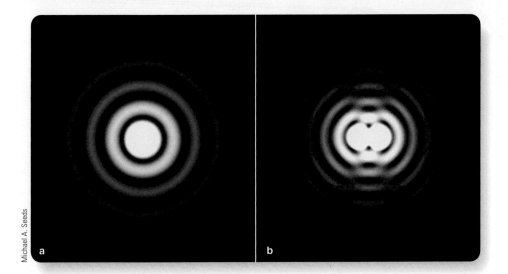

a b

◀ **Figure 6-10** (a) Stars are so far away that their images are points, but the wavelike characteristic of light causes each star image to be surrounded with diffraction fringes, much magnified in this computer model. (b) Two stars close to each other have overlapping diffraction fringes and become impossible to detect separately.

turbulent. Stars any closer together than that will be blurred together into a single image by the diffraction fringes.

Aside from diffraction, two other factors—optical quality and atmospheric conditions—limit resolving power. A telescope must have high-quality optics to achieve its full potential resolving power. Even a large telescope reveals little detail if its optical surfaces are marred by imperfections. Also, when you look through a telescope, you are looking up through miles of turbulent air in Earth's atmosphere, inevitably making images wiggle and blur to some extent. Astronomers use the term **seeing** to refer to the amount of image wiggling and blurring as a result of atmospheric conditions. A related phenomenon is the twinkling of stars. Star twinkles are caused by turbulence in Earth's atmosphere, and a star near the horizon, where you look through more air, will twinkle and blur more than a star overhead.

On a night when the atmosphere is unsteady, images are badly blurred, and astronomers say that "the seeing is bad" (**Figure 6-11a**). Generally, even under relatively good seeing conditions, the detail visible through a large telescope is limited not by its diffraction fringes but by the turbulence of the air through which the telescope must look. An optical telescope performs better on a high mountaintop where the air is thin and steady. But even in that situation, Earth's atmosphere spreads star images at visual wavelengths into blobs about 0.5 to 1.0 arc second in diameter. Radio telescopes are also affected by atmospheric seeing, but less than optical telescopes, so they do not benefit much in this respect by being located on mountains. You will learn later in this chapter about special techniques that improve seeing from ground-based telescopes and also about telescopes that orbit above Earth's atmosphere and are not limited by seeing.

Seeing and diffraction both limit the precision of any measurement that can be made using that image, and that limits the amount of information in the image. All measurements have some built-in uncertainty (**How Do We Know? 6-1**), and scientists

must learn to work within those limitations. Have you ever tried to magnify a newspaper photo to distinguish some detail? Newspaper photos are composed of tiny dots of ink, and no detail smaller than a single dot will be visible no matter how much you magnify the photo. In an astronomical image, the resolution is limited by seeing, or diffraction, or both. You can't see any detail in the image that is smaller than the telescope's resolution. That's why stars look like fuzzy points of light no matter how big the telescope.

a b

Visual

▲ **Figure 6-11** (a) The left half of this photograph of a galaxy is from an image recorded on a night of poor seeing. Small details are blurred. (b) The right half of the photo is from an image recorded on a night when Earth's atmosphere above the telescope was steady and the seeing was better. Much more detail is visible under good seeing conditions.

How Do We Know? 6-1

Resolution and Precision

What limits the precision of an observation?
As an example, think about observations that are in the form of images. All images have limited resolution. You can see on your computer screen that images there are made up of picture elements, pixels. If your screen has low resolution, it has large pixels, and you can't see much detail. In an astronomical image, the practical size of a pixel is set by the resolution limit, a combination of atmospheric seeing, telescope optical quality, and telescope diffraction. You can't see details smaller than the resolution limit. This limitation on the level of detail viewable in an image is one example of the limited precision, and therefore unavoidable uncertainty, of all scientific measurements.

Now imagine a zoologist trying to measure the length of a live snake by holding it along a meter stick. Meter sticks are usually not marked with resolution smaller than millimeters. Also, the wriggling snake is hard to hold, so it is difficult to measure

accurately. Both factors—the meter stick's resolution and the snake's wriggling—together limit the precision of the measurement. If the zoologist said the snake is 432.8932 mm long, you might wonder if that is really true. The best resolution possible in that zoologist's situation does not justify the precision implied by all those digits. Images made with even the largest and best telescopes do not show surface details on stars because of limits on precision (resolution) set by diffraction and atmospheric seeing (a stellar equivalent of the snake wriggling).

If you are a scientist, one question you must ask yourself routinely is: How precise are the measurements you and other investigators have made? Precision of measurements is limited by the resolution of the measurement technique such as the size of the pixels in a photograph or the finest markings on a meter stick as much as by variability in what is being observed such as

the snake wriggling or atmospheric turbulence. And, because precision is always limited, uncertainty is always present.

Visual

NASA/ESA/STScI

A high-resolution visual-wavelength image of Mars made by the Hubble Space Telescope reveals details such as mountains, craters, and the south polar cap.

Magnifying Power: It is a **Common Misconception** that the purpose of an astronomical telescope is to magnify images. In fact, the **magnifying power** of a telescope—its ability to make images bigger—is the least important of the three powers. Because the amount of detail that a telescope can discern is limited generally either by its resolving power or the seeing conditions, very high magnification does not necessarily show more detail. The magnifying power of a telescope equals the focal length of the primary mirror or lens divided by the focal length of the eyepiece.

$$M = \left(\frac{F_p}{F_e}\right)$$

For example, if a telescope has a primary with a focal length $F_p = 80$ cm and you use an eyepiece with a focal length $F_e = 0.5$ cm, the magnification is 80/0.5, or 160. Radio telescopes, of course, don't have eyepieces, but they do have instruments that examine the radio waves focused by the telescope, and each such instrument would, in effect, have its own magnifying power.

As was mentioned previously, the two most important powers of the telescope—light-gathering power and resolving power—depend on the diameter of the telescope that is essentially impossible to change. In contrast, you can change the magnification of a telescope simply by changing the eyepiece.

This explains why astronomers describe telescopes by diameter and not by magnification. Astronomers will refer to a telescope as a 4-meter telescope or a 10-meter telescope, but they would never identify a research telescope as being, say, a 1000-power telescope.

6-3 Observatories on Earth: Optical and Radio

The quest for light gathering power and good resolution explains why nearly all the world's major observatories are located far from big cities and, especially in the case of optical telescopes, usually on top of mountains. Astronomers avoid cities because **light pollution**, which is the brightening of the night sky by light scattered from artificial outdoor lighting, can make it impossible to see faint objects (**Figure 6-12**). In fact, many residents of cities are unfamiliar with the beauty of the night sky because they can see only the brightest stars. Even far from cities, the Moon, nature's own light pollution, is sometimes so bright it drowns out fainter objects, and astronomers are unable to perform certain types of observations during nights near full moon. On such nights, faint objects cannot be detected even with large telescopes at good locations.

Astronomers no longer build large observatories in populous areas.

A number of major observatories are located on mountaintops in the Southwest.

Visual

NOAA

▲ **Figure 6-12** This satellite view of the continental United States at night shows the light pollution and energy waste resulting from outdoor lighting. Observatories are best located far from large cities.

Radio astronomers face a problem of radio interference comparable to visible light pollution. Weak radio waves from the cosmos are easily drowned out by human-made radio noise—everything from automobiles with faulty spark plugs to poorly designed communication systems. A few narrow radio bands are reserved for astronomy research, but even those are often contaminated by stray signals. To avoid that noise and have the radio equivalent of a dark sky, astronomers locate radio telescopes as far from civilization as possible. Hidden in mountain valleys or in remote deserts, they are able to study the Universe protected from humanity's radio output.

As you have already learned, astronomers prefer to put optical telescopes on high mountains for several reasons. To find sites with the best seeing, astronomers carefully select mountains where the airflow is measured to be smooth and not turbulent. Also, the air at high altitude is thin, dry, and more transparent, which is important not only for optical telescopes but also for other types of telescopes. Building an observatory on top of a remote high mountain is difficult and expensive, as you can imagine from **Figure 6-13**, but the dark sky, good seeing, and transparent atmosphere make it worth the effort.

Modern Optical Telescopes

For most of the 20th century, astronomers faced a serious limitation on the size of astronomical telescopes. Telescope mirrors were made thick to avoid bending that would distort the reflecting surface, but those thick mirrors were heavy. The 5-m (200-in.) mirror on Mount Palomar weighs 14.5 tons. Those old-fashioned telescopes were massive and expensive. Today's

astronomers have solved these problems in a number of ways. Look at **Modern Optical Telescopes** on pages 114–115 and notice three important points about telescope design and ten new terms that describe optical telescopes and their operation:

❶ Conventional-design reflecting telescopes use large, solid, heavy mirrors to focus starlight to a *prime focus,* or by using a *secondary mirror,* to a *Cassegrain focus* (pronounced *KASS-uh-grain*). Other telescopes have a *Newtonian focus* or a *Schmidt-Cassegrain focus.*

❷ Telescopes must have a *sidereal drive* to follow the stars. An *equatorial mount* with motion around a *polar axis* is the conventional way to provide that motion. Today, astronomers can build simpler, lighter-weight telescopes on *alt-azimuth mounts* that depend on computers to move the telescope so that it follows the apparent motion of stars as Earth rotates without having an equatorial mount and polar axis.

❸ *Active optics,* computer control of the shape of a telescope's main mirrors, allows the use of thin, lightweight mirrors—either "floppy" mirrors or segmented mirrors. Reducing the weight of the mirror reduces the weight of the rest of the telescope, making it stronger and less expensive. Also, thin mirrors cool and reach a stable shape faster at nightfall, producing better images during most of the night.

Richard Wainscorr, Aina Kai LLC

▲ **Figure 6-13** Aerial view of the optical, infrared, and radio telescopes on Mauna Kea in Hawai'i, 4200 m (nearly 14,000 ft) above sea level. The high altitude, low atmospheric moisture, lack of nearby large cities, and location near the equator make this mountain one of the best places on Earth to build an observatory.

Modern Optical Telescopes

1 Reflecting telescopes with standard designs depicted on this page have capabilities limited by complexity, weight, and turbulence in Earth's atmosphere. Modern design solutions are shown on the opposite page.

The primary mirror makes light converge to a **prime focus** position high in the telescope tube, as shown at the right. Although the prime focus is a good place to image faint objects, it is inconvenient for large instruments. A **secondary mirror** can reflect the light through a hole in the primary mirror to a **Cassegrain focus**. This focal arrangement is the most common one for large telescopes.

Secondary mirror

With large enough telescopes, astronomers can actually ride inside a prime-focus "cage", although observations are usually made using instruments connected to computers in a separate control room.

Conventional primary mirrors are thick to prevent the optical surface from sagging and distorting the image as the telescope is moved around the sky. But, large mirrors can weigh many tons, are difficult to support, and are expensive to make. Also, large mirrors take a long time to cool slowly after nightfall. Changes in shape as the mirror cools down make the telescope difficult to focus and cause image distortions.

The Cassegrain focus is convenient to access and has room for large, heavy instruments.

1a Smaller telescopes are often built with a **Newtonian focus**, the arrangement that Isaac Newton used in his first reflecting telescope. The Newtonian focus is inconvenient for large telescopes, as shown at right.

Newtonian focus

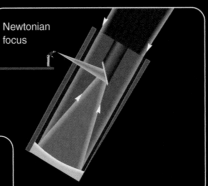

Thin correcting lens

Schmidt-Cassegrain telescope

1b Many small telescopes such as the one on the left use a **Schmidt-Cassegrain** focus. A thin correcting plate improves the image but is not curved enough to introduce serious chromatic aberration.

1c Shown below, observations using the 4-m Mayall Telescope at Kitt Peak National Observatory in Arizona can be made at either the prime focus or the Cassegrain focus. Note the human figure at lower right.

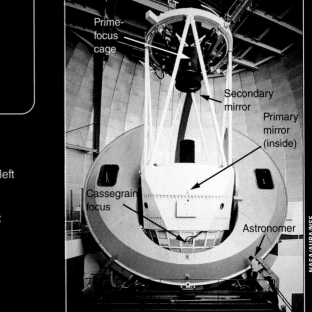

Prime-focus cage

Secondary mirror

Primary mirror (inside)

Cassegrain focus

Astronomer

NASA/AURA/NSF

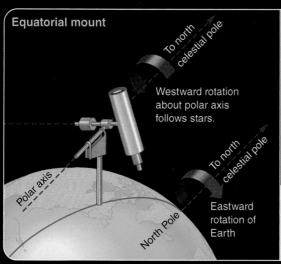

Equatorial mount

To north celestial pole

Westward rotation about polar axis follows stars.

Polar axis

To north celestial pole

Eastward rotation of Earth

North Pole

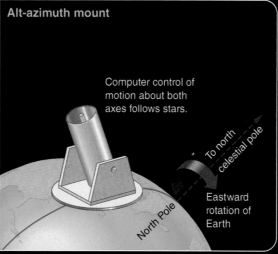

Alt-azimuth mount

Computer control of motion about both axes follows stars.

To north celestial pole

Eastward rotation of Earth

North Pole

2 Telescope mountings must contain a **sidereal drive** to move the telescope smoothly westward, countering the eastward rotation of Earth. The earlier **equatorial mount** (*far left*) has a **polar axis** parallel to Earth's axis, but the mount used for the largest modern telescopes is **alt-azimuth mount** (altitude-azimuth; near left) moves like a cannon—up and down, left and right. Alt-azimuth mountings are simpler to build than equatorial mountings but require computer control to follow the stars.

3 Unlike traditional thick mirrors, thin mirrors, sometimes called "floppy" mirrors as shown at right, weigh less and require less massive support structures. Also, they cool rapidly at nightfall and there is less distortion from uneven expansion and contraction.

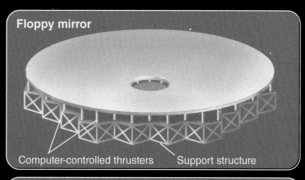

Floppy mirror

Computer-controlled thrusters   Support structure

3a Grinding a large mirror may remove tons of glass and take months, but new techniques speed the process. Some large mirrors are cast in a rotating oven that causes the molten glass to flow to form a concave upper surface. Grinding and polishing such a preformed mirror is much less time consuming.

3b Mirrors made of segments are economical because the segments can be made separately. The resulting mirror weighs less and cools rapidly.

Segmented mirror

Computer-controlled thrusters   Support structure

3c Both floppy mirrors and segmented mirrors sag under their own weight. Their optical shapes must be controlled by computer-driven thrusters behind the mirrors, a technique called **active optics**.

Keck I telescope mirror segments

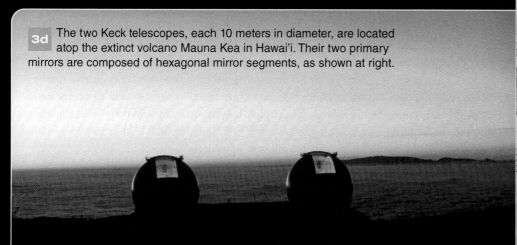

3d The two Keck telescopes, each 10 meters in diameter, are located atop the extinct volcano Mauna Kea in Hawai'i. Their two primary mirrors are composed of hexagonal mirror segments, as shown at right.

W.M. Keck Observatory

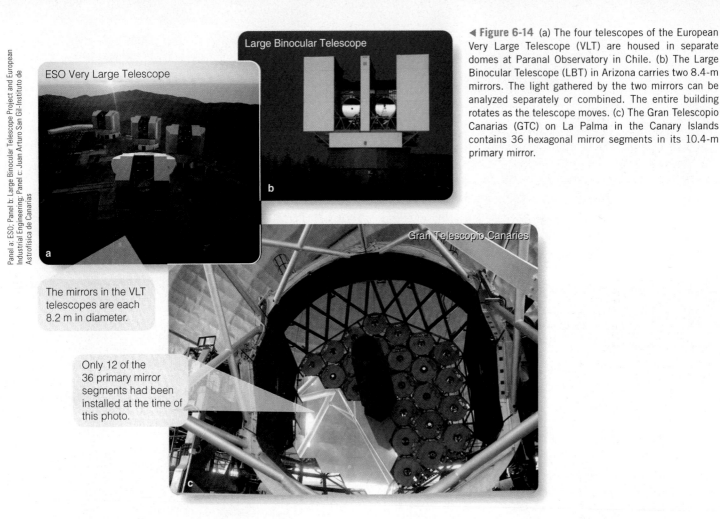

Panel a: ESO; Panel b: Large Binocular Telescope Project and European Industrial Engineering; Panel c: Juan Arturo San Gil-Instituto de Astrofísica de Canarias

ESO Very Large Telescope

Large Binocular Telescope

Gran Telescopio Canaries

The mirrors in the VLT telescopes are each 8.2 m in diameter.

Only 12 of the 36 primary mirror segments had been installed at the time of this photo.

◀ **Figure 6-14** (a) The four telescopes of the European Very Large Telescope (VLT) are housed in separate domes at Paranal Observatory in Chile. (b) The Large Binocular Telescope (LBT) in Arizona carries two 8.4-m mirrors. The light gathered by the two mirrors can be analyzed separately or combined. The entire building rotates as the telescope moves. (c) The Gran Telescopio Canarias (GTC) on La Palma in the Canary Islands contains 36 hexagonal mirror segments in its 10.4-m primary mirror.

Modern engineering techniques and high-speed computers have allowed astronomers to build and use new, giant telescopes with unique designs. A few are shown in **Figure 6-14**. The European Southern Observatory built the Very Large Telescope (VLT) in the foothills of the Andes Mountains in northern Chile. The VLT actually consists of four telescopes, each with a mirror 8.2 m (323 in., about 27 ft) in diameter and only 17.5 cm (6.9 in.) thick. U.S. and Italian astronomers have built the Large Binocular Telescope (LBT) on Mount Graham in Arizona. The LBT carries a pair of 8.4-m (331-in.) mirrors on a single mount. The twin Keck telescopes on Mauna Kea in Hawai'i have primary mirrors 10 m (400 in.) in diameter that are each made of 36 individually controlled hexagonal segments. The Gran Telescopio Canarias (GTC), located atop a volcanic peak in the Canary Islands, carries a segmented mirror 10.4 m (410 in., over 34 ft) in diameter and is, at the time of this writing, the largest single telescope in the world.

Other giant telescopes are being planned for completion in the 2020s, all with segmented or multiple mirrors (**Figure 6-15**). The Giant Magellan Telescope (GMT) will carry seven asymmetrically curved thin mirrors, each 8.4 m in diameter, on a single mounting. It will be located in Chile and have the light-gathering power of a single 24.5-m telescope. The Thirty Meter Telescope (TMT), now under development by a consortium of countries, including the United States, Canada, Japan, China, and India, is planned to have a mirror up to 30 m (100 ft) in diameter comprised of 492 hexagonal segments and will be placed on Mauna Kea in Hawai'i. An international team is designing the European Extremely Large Telescope (E-ELT) to carry 798 segments, making up a mirror 39 m (nearly 130 ft) in diameter. The E-ELT will be built on Cerro Armazones, a mountain in Chile's Atacama Desert.

A ground-based telescope is normally operated by astronomers and technicians working in a control room in the same building, but some telescopes are now used by astronomers many miles, even thousands of miles, from the observatory. Other telescopes are fully automated and operate without direct human supervision. That, plus continuous improvement in computer speed and storage capacity, has made possible huge surveys of the sky in which millions of objects have been observed or are planned for observation. For example, the Sloan Digital Sky Survey (SDSS) mapped the entire Northern Hemisphere sky, measuring the position and brightness of 100 million stars and galaxies at five ultraviolet, optical, and infrared wavelengths.

The data from SDSS are available for you to examine and manipulate at this website: skyserver.sdss.org. Some of the

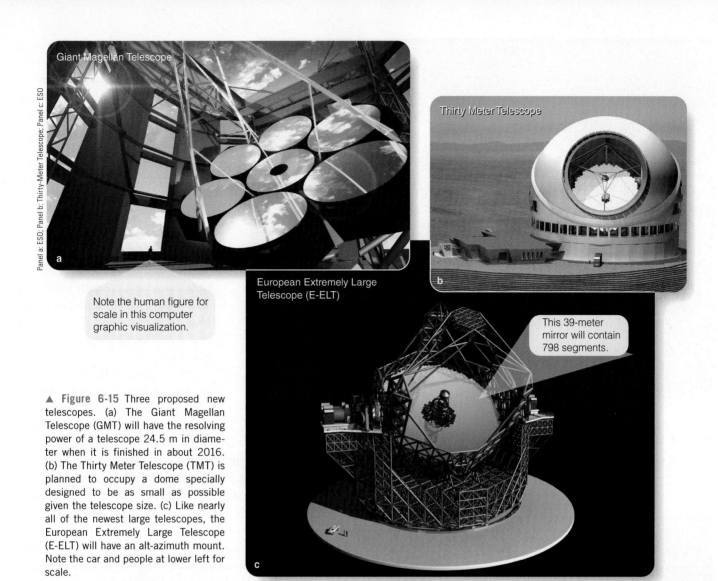

Giant Magellan Telescope

Thirty Meter Telescope

Note the human figure for scale in this computer graphic visualization.

European Extremely Large Telescope (E-ELT)

This 39-meter mirror will contain 798 segments.

▲ **Figure 6-15** Three proposed new telescopes. (a) The Giant Magellan Telescope (GMT) will have the resolving power of a telescope 24.5 m in diameter when it is finished in about 2016. (b) The Thirty Meter Telescope (TMT) is planned to occupy a dome specially designed to be as small as possible given the telescope size. (c) Like nearly all of the newest large telescopes, the European Extremely Large Telescope (E-ELT) will have an alt-azimuth mount. Note the car and people at lower left for scale.

SDSS data are also at Microsoft's World Wide Telescope (www .worldwidetelescope.org) and Google Sky (www.google.com/ sky/). The "citizen science" Galaxy Zoo site (www.galaxyzoo .org) allows volunteers (this could mean you) to classify galaxies based on their appearances in SDSS images.

The future Large Synoptic Survey Telescope (LSST) has an 8.4-m primary mirror already completed; construction of facilities on Cerro Pachón in Chile began in 2014 (**Figure 6-16**). Using a 3.2-billion-pixel charge-coupled device (CCD) camera, LSST will be able to record the brightness at selected ultraviolet, visual, and infrared wavelengths of every object in one hemisphere of the sky brighter than magnitude 24.5 every three nights. Astronomers and private citizens will be studying those data for decades to come.

Modern Radio Telescopes

The dish reflector of a radio telescope, like the mirror of a reflecting telescope, collects and focuses radiation. Although a radio telescope's dish may be tens or hundreds of meters in diameter, the receiver antenna may be as small as your hand. Its function is to absorb the radio energy collected by the dish. Because radio wavelengths are in the range of a few millimeters to a few tens of meters, the dish only needs to be shaped to that level of accuracy, much less smooth than a good optical mirror. In fact, wire mesh works well as a mirror for all but the shortest-wavelength radio waves.

The largest single radio dish in the world at the time of this writing (mid-2014) is 305 m (1000 ft) in diameter. Such a large dish can't be supported easily, so it is built into a mountain valley in Arecibo, Puerto Rico. The primary mirror is a thin metallic surface supported above the valley floor by cables attached near the rim, and the antenna platform hangs above the dish on cables from towers built on three mountain peaks around the valley's rim (**Figure 6-17**). By moving the antenna above the dish, radio astronomers can point the telescope at any object that passes within 20 degrees of the zenith as Earth rotates. Since completion in 1963, the Arecibo telescope has been an international center of radio astronomy research. A 500-m (1650-ft) diameter radio telescope named FAST is being built by the Chinese

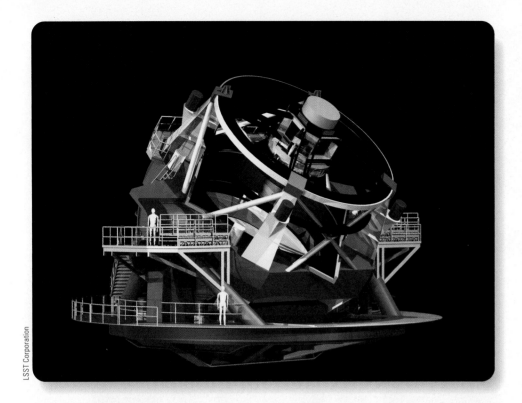

▲ **Figure 6-16** The 8.4-m Large Synoptic Survey Telescope (LSST) will use a special three-mirror design to create an exceptionally wide field of view, with the ability to survey the entire southern sky every three nights.

▲ **Figure 6-17** The 305-m (1000-ft) radio telescope in Arecibo, Puerto Rico, is nestled in a naturally bowl-shaped valley. The receiver platform is suspended over the dish. A consortium led by SRI International and Universities Space Research Association (USRA) manages Arecibo Observatory for the National Science Foundation (NSF).

government in a mountain-ringed bowl-shaped valley like the one in Puerto Rico. FAST will have more than 2.5 times the collecting area of Arecibo's dish.

A radio astronomer works under two disadvantages relative to optical astronomers: poor resolution and low signal intensity. Recall that the resolving power of a telescope depends on the diameter of the primary lens or mirror but also on the wavelength of the radiation. At very long wavelengths like those of radio waves, the diffraction fringes are quite large. This means that images or maps from individual radio telescopes generally don't show such fine details as are seen in optical images.

The second handicap radio astronomers face is the low intensity of the radio signals. You learned previously that the energy of a photon depends on its wavelength. Photons of radio energy have such long wavelengths that their individual energies are quite low. The cosmic radio signals arriving on Earth are astonishingly weak—as little as one-billionth the strength of the signal from a commercial radio station. To get detectable signals focused on the antenna, radio astronomers must build large collecting areas either as single large dishes or by combining arrays of smaller dishes. Even then, because the radio energy from celestial objects is so weak, it must be strongly amplified before it can be measured and recorded.

DOING SCIENCE

Why do astronomers build optical observatories at the tops of mountains? Precise and accurate measurements are so fundamental to doing science that scientists often take extreme steps to get them.

It is certainly not easy to build a large, complicated, and fragile optical telescope at the top of a high mountain, but it is worth the effort. A telescope on top of a high mountain is above the densest part of Earth's atmosphere, so there is less air to dim the incoming light. Even more important, the turbulence of thin air on a mountaintop is less able to disturb light waves than that of thick air, so the seeing is better. The resolving power of a large optical telescope on Earth's surface is set by atmospheric seeing rather than by the telescope's diffraction. It really is worth the trouble to build telescopes on the peaks of high mountains.

Astronomers also put radio observatories in special locations. *What considerations might astronomers make in choosing the location for a new radio telescope?*

6-4 Airborne and Space Observatories

Ground-based telescope performance is limited by Earth's atmospheric turbulence and transparency. There are sophisticated techniques that partly compensate for atmospheric seeing, but a telescope in space has no such problem, and its resolution is defined only by diffraction.

Also, as you learned earlier in this chapter, a telescope on the ground must look through one of the open atmospheric "windows" (wavelength ranges). Most types of electromagnetic radiation arriving here from the Universe—gamma-rays, X-rays, ultraviolet, and much of the infrared—do not reach Earth's surface because they are partly or completely absorbed by Earth's atmosphere. To gather light with those blocked wavelengths, telescopes must go to high altitudes or into space. As you will learn in the next few chapters, objects that are cooler than stars, such as stars that are forming, produce lots of infrared and microwave radiation but relatively little visible or ultraviolet light. In contrast, cosmic catastrophes such as exploding stars make mostly gamma-rays and X-rays. Combining information from as wide a variety of wavelengths as possible allows astronomers to gain a more comprehensive understanding of the Universe.

Airborne Telescopes

In addition to the atmospheric windows at visual and radio wavelengths you have already learned about, there are also a few narrow windows at short infrared wavelengths accessible from the ground, especially from high mountains such as Mauna Kea

(Figure 6-13). However, most infrared wavelengths are blocked, especially by water vapor absorption. Also, Earth's atmosphere itself produces a strong infrared "glow." Observations at very long infrared wavelengths can only be made using telescopes carried to high altitudes by aircraft or balloons or launched entirely out of the atmosphere onboard spacecraft. (Notice that the reasons to put an infrared telescope above the atmosphere are not the same as the reasons to send an optical telescope into space.)

Starting in the 1960s, NASA developed a series of infrared observatories with telescopes carried above Earth's atmospheric water vapor by jet aircraft. Such airborne observatories are also able to fly to remote parts of Earth to monitor astronomical events not observable by any other telescope. The modern successor to those earlier flying observatories is the *Stratospheric Observatory for Infrared Astronomy,* or *SOFIA* (Figure 6-18). *SOFIA* consists of a 2.5-m (100-in.) telescope looking out an opening with a rollback door in the left side of a modified Boeing 747SP aircraft.

Space Telescopes

The most successful observatory in history, the *Hubble Space Telescope* (Figure 6-19a), is named after Edwin Hubble, the astronomer who discovered the expansion of the Universe. The *Hubble* telescope, also known as *HST,* was launched in 1990 and contains a 2.4-m (95-in.) mirror plus three instruments with which it can observe visible light plus some ultraviolet and infrared wavelengths. Its greatest advantage is the lack of seeing distortion, located completely above Earth's atmosphere. *Hubble* therefore can detect fine detail, and because it concentrates light into sharp images, it can detect extremely faint objects. It is

▲ **Figure 6-18** (a) The *Stratospheric Observatory for Infrared Astronomy (SOFIA),* a joint project of NASA and the German Aerospace Center (DLR), flies at altitudes up to 14 km (45,000 ft) where it can collect infrared radiation with wavelengths that are unobservable even from high mountaintops. (b) A visual-wavelength image of the planet Jupiter (*left*) compared with a composite infrared image (*right*) using images at wavelengths of 5.4, 24, and 37 microns made during *SOFIA*'s "First Light" flight in 2010. The white stripe in the infrared image is a region of relatively transparent clouds through which the warm interior of the planet can be seen.

Panel a: NASA; Panels b & c: NASA/SAO/CXC

◄ **Figure 6-19** (a) The *Hubble Space Telescope (HST)* orbits Earth at an average altitude of 570 km (355 mi) above the surface. In this image, the telescope is viewing toward the upper left. (b) Artist's conception of *HST*'s eventual successor, the *James Webb Space Telescope (JWST)*. *JWST* will be located in solar orbit almost 1 million miles from Earth, four times as far away as the Moon. It will not have an enclosing tube, thus resembling a radio dish more than a conventional optical telescope. *JWST* will observe the Universe from behind a multi-layered sunscreen larger than a tennis court. (c) Artist's conception of the *Herschel* infrared space telescope that carried a 3-m mirror and instruments cooled almost to absolute zero.

controlled from a research center on Earth and observes almost continuously. Nevertheless, the telescope has time to complete only a fraction of the many projects proposed by astronomers from around the world.

Hubble has been visited a number of times by the space shuttle so that astronauts could service its components and install new cameras and other instruments. Thanks to the work of the space shuttle crew who visited in 2009 and accomplished another refurbishment of the telescope's instruments, batteries, and gyroscopes, *Hubble* will almost certainly last until it can be replaced by the *James Webb Space Telescope (JWST)*, which is expected to be ready in about the year 2018. *JWST* telescope will be launched into a solar orbit to avoid interference from Earth's strong infrared glow. Its primary mirror is a cluster of beryllium mirror segments that will open in space to form a 6.5-m (256-in.) mirror (Figure 6-19b).

Telescopes carrying long-wavelength infrared detectors must carry coolant such as liquid helium to chill their optics to near absolute zero temperature ($-273°C$ or $-460°F$) so that heat radiation from the insides of the telescope and instruments does

not blind the detectors. Such observatories have limited lifetimes because the coolant eventually runs out. The European Space Agency's *Herschel* 3-meter infrared space telescope (Figure 6-19c), named after the scientist who discovered infrared radiation (Figure 6-4), was launched into solar orbit in 2009 together with the smaller *Planck* space observatory that studied millimeter-wavelength radiation. *Herschel* and *Planck* made important discoveries concerning distant galaxies, star formation, planets orbiting other stars, and the origin of the Universe during their 4-year lifetimes.

High-Energy Astronomy

Like infrared-emitting objects, gamma-ray, X-ray, and ultraviolet sources in the Universe are difficult to observe because the telescopes must be located high in Earth's atmosphere or in space. Also, high-energy photons are difficult to bring to a focus.

The first high-energy astronomy satellite, *Ariel 1*, was launched by the United Kingdom in 1962 and made solar observations in the ultraviolet and X-ray segments of the electromagnetic spectrum. Since then, many more space telescopes have followed *Ariel's* lead. Some high-energy astronomy satellites such as *XMM-Newton,* an X-ray observatory developed by a consortium of European and British astronomers, have been general-purpose telescopes that observe many different kinds of objects. In contrast, some space telescopes are designed to study a single question or a single object. For example, the Japanese satellite *Hinode* (pronounced, *hee-no-day*) studies the Sun continuously at visual, ultraviolet, and X-ray wavelengths, and the *Kepler* space observatory operated for 4 years detecting planets orbiting stars other than the Sun.

The largest X-ray telescope to date is the *Chandra X-ray Observatory (CXO)*. *Chandra* operates in an orbit that extends a third of the way to the Moon so that it spends 85 percent of the time above the belts of charged particles surrounding Earth that would produce electronic noise in its detectors. (*Chandra* is named for the late Indian American Nobel laureate Subrahmanyan Chandrasekhar, who was a pioneer in many branches of theoretical astronomy.) Focusing X-rays is difficult because they penetrate into most mirrors, so astronomers devised cylindrical mirrors in which the X-rays reflect at shallow angles from the polished inside of the cylinders to form images on X-ray detectors, as shown in Figure 6-20. The *Chandra* observatory has made important discoveries about everything from star formation to monster black holes in distant galaxies that will be described in later chapters.

The first large gamma-ray space telescope was the *Compton Gamma Ray Observatory,* launched in 1991. It mapped the entire sky at gamma-ray wavelengths. The European-built *INTernational Gamma-Ray Astrophysics Laboratory (INTEGRAL)* satellite was launched in 2002 and has been very productive in the study of violent eruptions of stars and black holes. The *Fermi Gamma-ray Space Telescope,* launched in 2008 and operated by a consortium of nations led by the United States, is capable of making highly sensitive gamma-ray maps of large areas of the sky.

NASA/SAO/CXC

▲ **Figure 6-20** X-rays that hit a mirror at grazing angles are reflected like a pebble skipping across a pond. Thus, X-ray telescope mirrors like the ones in *Chandra* are shaped like barrels rather than dishes.

Modern astronomy has come to depend on observations that cover the entire electromagnetic spectrum. More orbiting space telescopes are planned that will be even more versatile and sensitive than the ones operating now.

6-5 Astronomical Instruments and Techniques

Just looking through a telescope doesn't tell you much. A star looks like a point of light. A planet looks like a little disk. A galaxy looks like a hazy patch. To use a research telescope to learn about the Universe, you need to carefully analyze the light the telescope gathers. Special instruments attached to the telescope make that possible.

Cameras and Photometers

The **photographic plate** was the first device used by astronomers to record images of celestial objects. Photographic plates can detect faint objects in long time exposures and can be stored for later analysis. Brightness of objects imaged on a photographic plate can be measured with a lot of hard work that yields only moderate precision. Astronomers also build **photometers**, sensitive light meters used to measure the brightness of individual objects very precisely.

Present-day astronomers use **charge-coupled devices (CCDs)** as both image-recording devices and photometers. A CCD is a specialized computer chip containing millions of microscopic light detectors arranged in an array as small as a postage stamp. CCD chips have replaced photographic plates because they have some important advantages. CCDs are much more sensitive than photographic plates and can detect both bright and faint objects in a single exposure. Also, CCD images are **digitized**, meaning converted to numerical data, and thus can be stored in a computer's memory for later analysis. Although astronomy research-grade CCDs are extremely sensitive and therefore expensive, less sophisticated CCDs are now part of everyday life. You are familiar with them in digital cameras (both still and video) as well as in cell phone cameras.

Infrared astronomers use **array detectors** that are similar in operation to optical CCDs. At other wavelengths, photometers are still used for measuring brightness of celestial objects. Array detectors and photometers generally must be cooled to operate properly (Figure 6-21).

The digital data representing an image from a CCD or other array detector are easy to manipulate to bring out details that would not otherwise be visible. For example, astronomical images are often reproduced as negatives, with the sky white and the stars dark. That makes the faint parts of the image easier to see (Figure 6-22). Astronomers also can manipulate images to produce **false-color** (or **representational-color**) **images** in which the colors represent different aspects of the object such as

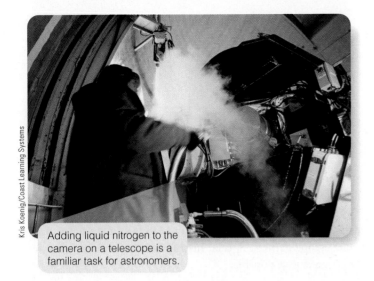

Adding liquid nitrogen to the camera on a telescope is a familiar task for astronomers.

▲ **Figure 6-21** Astronomical cameras with CCD and other types of array detectors must be cooled to low temperatures to operate properly, and that is especially true for infrared cameras.

intensity, rather than visual color. For example, because humans can't see radio waves, astronomers must convert radio data into something perceptible. One way is to measure the strength of the radio signal at various places in the sky and produce a representational-color map in which each color marks areas of similar radio intensity. You can compare such a map to a weather map in which the different colors mark areas forecast to have different types and amounts of precipitation (Figure 6-23a). Representational-color images and maps are very commonly used in nonoptical astronomy (Figures 6-23b and 6-23c).

Spectrographs

To analyze light in detail, astronomers spread out the light according to wavelength (color), a function performed by a **spectrograph**. You can understand how this instrument works if you imagine repeating an experiment performed by Isaac Newton in 1666. Newton bored a small hole in the window shutter of his room to admit a thin beam of sunlight. When he placed a prism in the beam, it spread the light into a beautiful spectrum that splashed across his wall. From that and related experiments, Newton concluded that white light is made of a mixture of all the colors.

As you learned previously in regard to chromatic aberration of refracting telescopes, light passing from one medium such as air into another medium such as glass has its path bent at an angle that depends on its wavelength. For example, blue (short-wavelength) light passing through a prism bends the most, and red (long-wavelength) light bends least. Thus, the white light entering the prism is spread into a spectrum exiting the prism (Figure 6-24). You can build a simple spectrograph by using a narrow opening to define the incoming light beam, a prism to spread the light into its component colors, and a lens to guide the light into a camera.

Almost all modern spectrographs use a **grating** rather than a prism. A grating is a piece of glass or metal with thousands of parallel microscopic grooves scribed onto its surface. Different wavelengths of light reflect from or pass through the grating at slightly different angles, so white light encountering the grating is spread into a spectrum. You have probably noticed this effect when you look at the closely spaced lines etched onto a CD or DVD: As you tip the disk, different colors flash across its surface. A modern spectrograph can be built using a high-quality grating

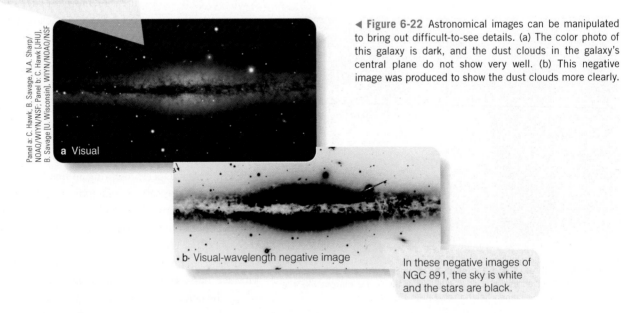

Galaxy NGC 891 in true color. It is edge-on and contains thick dust clouds.

a Visual

b Visual-wavelength negative image

◀ **Figure 6-22** Astronomical images can be manipulated to bring out difficult-to-see details. (a) The color photo of this galaxy is dark, and the dust clouds in the galaxy's central plane do not show very well. (b) This negative image was produced to show the dust clouds more clearly.

In these negative images of NGC 891, the sky is white and the stars are black.

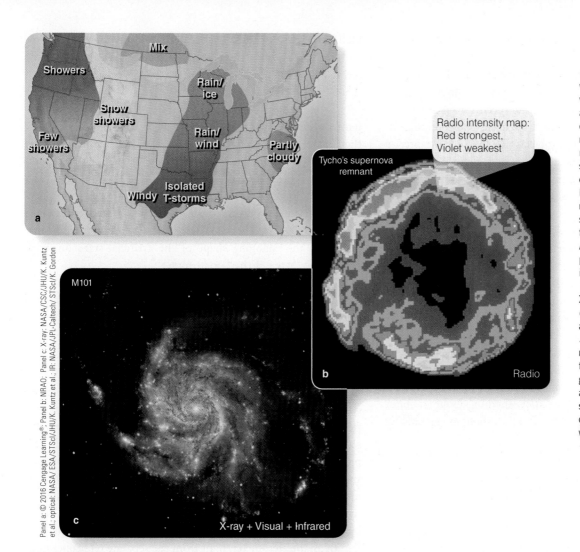

Radio intensity map:
Red strongest,
Violet weakest

Tycho's supernova remnant

b Radio

M101

c X-ray + Visual + Infrared

◀ **Figure 6-23** (a) A typical weather map uses contours with added color to show which areas are likely to receive precipitation, and what type. (b) A radio image of Tycho's supernova remnant, the expanding shell of gas produced by the explosion of a star first seen on Earth in 1572. This image's representational-color code shows intensity of radio radiation at just one wavelength. (c) A picture of galaxy M101 composed of a visual-wavelength image from the *Hubble Space Telescope* combined with an X-ray image made by the *Chandra X-ray Observatory* plus an infrared image from the *Spitzer Space Telescope.* In this representational-color image, the blue shows X-rays from hot gas heated by exploding stars and black holes, whereas red shows infrared emission from cool, dusty clouds of gas in which stars are being born.

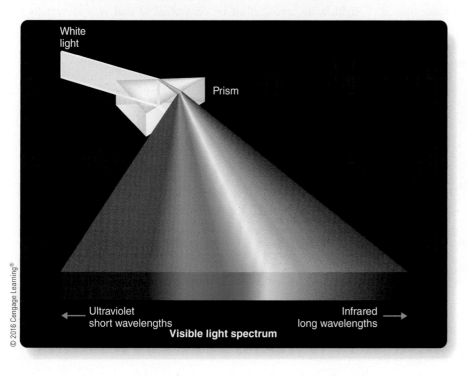

White light

Prism

Ultraviolet
short wavelengths

Infrared
long wavelengths

Visible light spectrum

◀ **Figure 6-24** A prism bends light by an angle that depends on the wavelength of the light. Short wavelengths bend most and long wavelengths least. Thus, white light passing through a prism is spread into a spectrum.

to separate light by wavelength, plus a CCD detector to record the resulting spectrum.

You will learn in the next chapter that spectra of astronomical objects such as stars and planets usually contain **spectral lines**—dark or bright lines that occur in spectra at specific wavelengths. Spectral lines are produced by atoms and molecules in the atmospheric gases of those objects. To measure the precise wavelengths of individual spectral lines and identify the atoms that produced them, astronomers use a **comparison spectrum**. Special light bulbs in the spectrograph produce bright spectral lines, or cells of gas in the spectrograph add dark lines, that are recorded next to the unknown spectrum. The wavelengths of the comparison spectral lines have been measured to high precision in laboratories, so astronomers can use comparison spectra as standards to measure wavelengths and identify spectral lines in the spectra of stars, galaxies, or planets.

Because scientists understand the details of how light interacts with matter, a spectrum carries a tremendous amount of information. That makes a spectrograph the astronomer's most powerful instrument. In the next chapter, you will learn more about the information astronomers can extract from a spectrum. Some astronomers say, "We don't know anything about an object until we get a spectrum," and that is only a slight exaggeration.

Adaptive Optics

You have already learned about active optics, which is a technique to adjust the shape of telescope optics slowly, compensating for effects of changing temperature as well as gravity bending the mirror when the telescope points at different locations in the sky. **Adaptive optics** is a more sophisticated technique that uses high-speed computers to monitor the distortion produced by turbulence in Earth's atmosphere and rapidly alter some optical components to correct the telescope image, sharpening a fuzzy blob into a crisp picture. The resolution of the image is still limited by diffraction in the telescope, but removing much of the seeing distortion produces a dramatic improvement in the detail that is visible (**Figure 6-25a**).

To monitor the distortion in an image, adaptive optics systems must look at a fairly bright star in the field of view, but there is not always such a star conveniently located near a target object such as a faint galaxy. In that case, astronomers can point a laser in a direction very close to that of their target object (Figure 6-25b). The laser causes gas in Earth's upper atmosphere to glow, producing an artificial star called a **laser guide star** in the field of view. The adaptive optics system can use information from the changing shape of the artificial star's image to correct the image of the fainter target.

You have read about huge existing and planned optical telescopes 10 or more meters in diameter composed of segmented mirrors. Those telescopes would be much less useful without the addition of adaptive and active optics.

Interferometry

One of the reasons astronomers build big telescopes is to increase resolving power. Astronomers have been able to achieve very high resolution by connecting multiple telescopes together to work, in a sense, as if they comprised a single, very large telescope. This method of synthesizing a large "virtual" telescope from two or more smaller telescopes is known as **interferometry** (**Figure 6-26**). The images from such an interferometric telescope are not limited by the diffraction fringes of the individual small telescopes but rather by the diffraction fringes of the much larger virtual telescope.

In an interferometer, light from the separate telescopes must be brought together and combined carefully. The path that each light beam travels must be controlled so that it is known to a precision of a small fraction of the light's wavelength. Turbulence

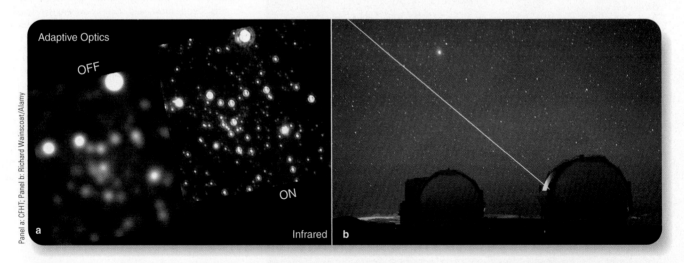

▲ **Figure 6-25** (a) In these images of the center of our galaxy, the adaptive optics system was turned "Off" for the left image and "On" for the right image. In the "On" image, the images of stars are sharper because the light is focused into smaller images; fainter stars are visible. (b) The laser beam shown leaving one of the Keck telescopes produces an artificial star in the field of view, and the adaptive optics system uses that laser guide star as a reference to reduce seeing distortion in the entire image.

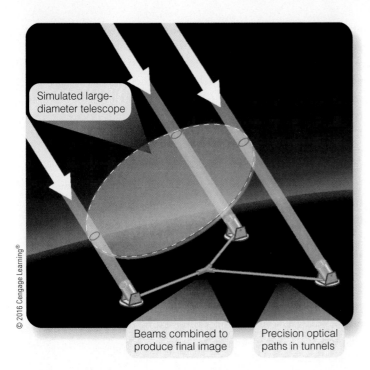

© 2016 Cengage Learning®

Simulated large-diameter telescope

Beams combined to produce final image

Precision optical paths in tunnels

▲ **Figure 6-26** In an astronomical interferometer, smaller telescopes can combine their light to simulate a larger telescope with a resolution set by the separation of the smaller telescopes.

in Earth's atmosphere constantly distorts incoming light, so high-speed computers must continuously adjust the light paths.

As you already know, the resolving power of a radio telescope is relatively low. A dish 30 m in diameter receiving radiation with a wavelength of 21 cm has a resolution of only about 0.5 degrees. In other words, a radio telescope 100 ft across is unable to detect any details in the sky smaller than the apparent size of the Moon.

But, because long-wavelength radio waves are relatively easy to manipulate, radio astronomers were the first to learn how to combine two or more telescopes to form an interferometer capable of much higher resolution than a single telescope.

Radio interferometers must be quite large. The Very Large Array (VLA) consists of 27 dish antennas spread across the New Mexico desert (**Figure 6-27a**). In combination, they have the resolving power of a radio telescope up to 36 km (22 mi) in diameter. The VLA can resolve details smaller than 1 arc second, rivaling the performance of a large optical telescope at a good site. The Very Long Baseline Array (VLBA) that includes the VLA consists of matched radio dishes spread from Hawai'i to the Virgin Islands, with an effective diameter almost as large as Earth.

The Atacama Large Millimeter/submillimeter Array (ALMA) is an interferometric facility located on the Chajnantor plateau in northern Chile at an altitude of 5050 m (16,600 ft; see image on the first page of this chapter). It is described as the most powerful telescope ever built because of its combination of total mirror collecting area and high spatial resolution. ALMA began supporting research observations in 2011 when the planned array of 66 high-precision dish antennas was about half complete. Astronomers from the entire world will be able to use ALMA without having to travel to Chile; because of the extreme altitude of the facility, observations and data analyses will all be done over the Internet.

Radio astronomers are now planning the Square Kilometer Array (SKA) that will contain thousands of radio receivers with total collecting area of a square kilometer (1 million m^2, 15 times larger than the Arecibo dish) spread over a distance of 6500 km (4000 mi; Figure 6-27b). These giant radio interferometers depend on state-of-the-art computers to combine signals properly and create radio maps.

Panel a: NRAO/AUI; Panel b: SKA Organisation

a

b

▲ **Figure 6-27** (a) The Very Large Array (VLA) radio dishes in New Mexico can be moved to different positions along a Y-shaped set of tracks. They are shown here in their most compact arrangement. Signals from the dishes are combined to create very high-resolution radio maps of celestial objects. (b) Artist's conception of the proposed Square Kilometer Array (SKA) that will have concentrations of radio receivers in two clusters, one in South Africa and one in Australia, separated by 6500 km (4000 mi).

Recall that the wavelength of light is very short, roughly 0.0005 mm, so building optical interferometers is one of the most difficult technical problems that astronomers face, but the challenge has been met in several instances. The European VLT (Figure 6-14a) consists of four 8.2-m telescopes that can operate separately, but the light they collect, along with light from three 1.8-m telescopes on the same mountaintop, can be brought together through underground tunnels. The resulting optical interferometer, known as the VLTI, can provide the resolution (but, of course, not the light-gathering power) of a telescope 200 m in diameter (660 ft, bigger than two football fields).

Astronomers using the VLTI in 2009 made an image of the red giant star T Leporis with a resolution of 0.004 arc second, equivalent to being able to discern a two-story house on the Moon. The Center for High Angular Resolution Astronomy (CHARA) telescope array on Mt. Wilson in Southern California combines six 1-m telescopes to create resolving power equivalent to a telescope 300 m (one-fifth of a mile) in diameter. Other facilities such as the two Keck 10-m telescopes in Hawai'i and the Large Binocular Telescope in Arizona also are capable of operating as interferometers. Although turbulence in Earth's atmosphere can be partially averaged out in an interferometer, astronomers are considering the possibility of putting interferometers in space to avoid atmospheric turbulence altogether.

6-6 Non-Electromagnetic Astronomy

This chapter is focused on how to collect and analyze electromagnetic radiation from space. Other types of energy also arrive here bearing information from the rest of the Universe and deserve at least a brief mention.

Particle Astronomy

Cosmic rays are subatomic particles traveling through space at tremendous velocities. Almost no cosmic rays reach the ground, but some of them smash into gas atoms in Earth's upper atmosphere, and fragments of those atoms shower down to the ground. Those secondary cosmic rays are passing through you as you read this sentence and will continue to do so throughout your life. Other types of particles from space interact weakly and seldom with Earth atoms, so huge detectors must be built to catch and count them. Detectors for some kinds of cosmic rays have been carried on balloons or launched into orbit, whereas others have been built deep underground where layers of rock filter out all but the most penetrating particles.

Astronomers are not yet sure what produces cosmic rays. Incoming particles that have electric charges have been deflected by electromagnetic forces as they traveled through our galaxy, which means astronomers can't easily tell where their original sources are located. Some lower-energy particles of various types are known to come from the Sun, and there are indications that at least a few high-energy cosmic rays are produced by the violent explosions of dying stars or supermassive black holes at the centers of galaxies. You will meet these exotic objects again in later chapters.

Gravity Wave Astronomy

Gravity waves are predicted by Einstein's general theory of relativity, which you read about in Chapter 5. Gravity waves should be produced by any mass that accelerates. The greater the mass and the more abrupt the acceleration, the stronger the gravity waves that will be produced. Nevertheless, even the strongest gravity waves are expected to be extremely weak and difficult to detect; the existence of gravity waves has been inferred, but so far they have never been observed directly. The Laser Interferometer Gravitational Wave Observatory (LIGO) is a ground-based facility intended to be sensitive enough to detect cosmic gravity waves after an advanced version begins operating around 2014. The *Laser Interferometry Space Antenna (LISA)* is its planned highly sensitive space-based counterpart, a collaboration among U.S. and European space agencies.

What Are We? Curious

Telescopes are creations of curiosity. You look through a telescope to see more and to understand more. The unaided eye is a detector with limited sensitivity, and the history of astronomy is the history of bigger and better telescopes gathering more and more light to search for fainter and more distant objects.

The old saying "Curiosity killed the cat" is an insult to the cat and to curiosity. We humans are curious, and curiosity is a noble trait—the mark of an active, inquiring mind. At the base of human curiosity lies the fundamental question, "What are we?"

Telescopes extend and amplify our senses, but they also allow us to extend and amplify our curiosity about our place in the Universe.

When people find out how something works, they say their curiosity is satisfied. Curiosity is an appetite like hunger or thirst, but it is an appetite for understanding. As astronomy expands our horizons and we learn about how distant stars and galaxies form and evolve, we feel satisfaction partly because we are learning about ourselves and about how we fit in the Universe. We are beginning to understand what we are.

Summary

▶ Visual light is the visible form of **electromagnetic radiation (p. 104)**, which is an electric and magnetic disturbance that transports energy at the speed of light c. The **wavelength (λ) (p.104)** of light, or the distance between the peaks of a wave, is usually measured in **nanometers (nm) (p. 105)** (10^{-9} m) or **angstroms (Å) (p. 105)** (10^{-10} m). The wavelength band of visual light is from 400 nm to 700 nm (4000 to 7000 Å).

▶ **Frequency (ν) (p. 104)** is the number of waves that pass a stationary point in 1 second. The frequency ν of an electromagnetic wave equals the speed of light c divided by the wave's wavelength λ.

▶ A **photon (p. 105)** is a packet of light waves that can act as a particle or as a wave. The energy carried by a photon is proportional to its frequency and inversely proportional to its wavelength.

▶ A **spectrum (p. 105)** is a display of light that is viewed or recorded after being sorted in order of wavelength or frequency. The complete electromagnetic spectrum includes **gamma-rays (p. 106)**, **X-rays (p. 106)**, **ultraviolet (UV) (p. 106)** radiation, visible light, **infrared (IR) (p. 105)** radiation, **microwaves (p. 105)**, and **radio waves (p. 105)**.

▶ Gamma-rays, X-rays, and ultraviolet radiation have shorter wavelengths and higher frequencies and carry more energy per photon than visible light. Infrared, microwave, and radio waves have longer wavelengths and lower frequencies and carry less energy per photon than visible light.

▶ Earth's atmosphere is transparent in some **atmospheric windows (p. 106)**: visible light, shorter-wavelength infrared, and short-wavelength radio.

▶ **Refracting telescopes (p. 107)** use a **primary lens (p. 107)** to bend and focus the light into an image. **Reflecting telescopes (p. 107)** use a **primary mirror (p. 107)** to focus the light. The image produced by the telescope's primary lens or mirror can be magnified by an **eyepiece (p. 107)**. Lenses and mirrors with short **focal lengths (p. 107)** must be strongly curved and are more expensive to grind to an accurate shape.

▶ Because of **chromatic aberration (p. 107)**, refracting telescopes cannot bring all colors to the same focus, resulting in color fringes around the images. An **achromatic lens (p. 108)** partially corrects for this, but such lenses are expensive and cannot be made much larger than about 1 m (40 in.) in diameter.

▶ Reflecting telescopes are easier to build and less expensive than refracting telescopes of the same diameter. Also, reflecting telescopes do not suffer from chromatic aberration. Most large **optical telescopes (p. 108)** and all **radio telescopes (p. 108)** are reflecting telescopes.

▶ **Light-gathering power (p. 109)** refers to the ability of a telescope to collect light. **Resolving power (p. 109)** refers to the ability of a telescope to reveal fine detail. **Diffraction fringes (p. 109)** in an image, caused by the interaction of light waves with the telescope's apertures, limit the amount of detail that can be seen. **Magnifying power (p. 112)** is the ability of a telescope to make an object look bigger. This power is less important because it is not a property of the telescope itself; this power can be altered simply by changing the eyepiece.

▶ Astronomers build optical observatories on remote, high mountains for two reasons: (1) Turbulence in Earth's atmosphere blurs the image of an astronomical object, a phenomenon that astronomers refer to as **seeing (p. 111)**. The air on top of a mountain is relatively steady, and the seeing is better. (2) Observatories are located far from cities to avoid **light pollution (p. 112)**. Astronomers also build radio telescopes remotely but more for the reason of avoiding interference from human-produced radio noise.

▶ In a reflecting telescope, light first comes to a focus at the **prime focus (p. 114)**, but a **secondary mirror (p. 114)** can direct light to other locations such as the **Cassegrain focus (p. 114)**. The **Newtonian focus (p. 114)** and **Schmidt-Cassegrain focus (p. 114)** are other focus locations used in some smaller telescopes.

▶ Because Earth rotates, telescopes must have a **sidereal drive (p. 115)** to remain pointed at celestial objects. An **equatorial mount (p. 115)** with a **polar axis (p. 115)** is the simplest way to accomplish this. An **alt-azimuth mount (p. 115)** can support a more massive telescope but requires computer control to compensate for Earth's rotation.

▶ Very large telescopes can be built with **active optics (p. 115)** to control the mirror's optical shape. These telescopes usually have either one large, thin, flexible mirror or a mirror broken into many small segments. Advantages include mirrors that weigh less, are easier to support, and cool faster at nightfall. A major disadvantage is that the optical shape needs to be adjusted gradually and continuously to maintain a good focus.

▶ The turbulence in Earth's atmosphere distorts and blurs images. Telescopes in orbit are above this seeing distortion and are limited only by diffraction in their optics. Earth's atmosphere absorbs gamma-ray, X-ray, ultraviolet, far-infrared, and microwave light. To observe at these wavelengths, telescopes must be located at high altitudes or in space.

▶ Astronomers in the past used **photographic plates (p. 121)** to record images at the telescope and **photometers (p. 121)** to precisely measure the brightness of celestial objects. Modern electronic systems such as **charge-coupled devices (CCDs) (p. 121)** and other types of **array detectors (p. 121)** have replaced both photographic plates and photometers in most applications.

▶ Electronic detectors have the advantage that data from them are automatically **digitized (p. 121)** in numerical format and can be easily recorded and manipulated. Astronomical images in digital form can be computer-enhanced to produce **false-color images (p. 121)**, also called **representational-color images (p. 121)**, which bring out subtle details.

▶ **Spectrographs (p. 122)** using prisms or a **grating (p. 122)** spread light out according to wavelength to form a spectrum, revealing hundreds of **spectral lines (p. 124)** produced by atoms and molecules in the object being studied. A **comparison spectrum (p. 124)** that contains lines of known wavelengths allows astronomers to measure the precise wavelengths of individual spectral lines produced by an astronomical object.

- **Adaptive optics (p. 124)** techniques involve measuring seeing distortions caused by turbulence in Earth's atmosphere, then partially canceling out those distortions by rapidly altering some of the telescope's optical components. In some facilities a powerful laser beam is used to produce an artificial **laser guide star (p. 124)** high in Earth's atmosphere that can be monitored by an adaptive optics system.

- **Interferometry (p. 124)** refers to the technique of connecting two or more separate telescopes to act as a single large telescope that has a resolution equivalent to that of a single telescope with a diameter that is as large as the separation between the individual telescopes. The first working interferometers were composed of multiple radio telescopes.

- **Cosmic rays (p. 126)** are not electromagnetic radiation; they are subatomic particles such as electrons and protons traveling at nearly the speed of light, arriving from mostly unknown cosmic sources.

Review Questions

1. Does light include radio waves?
2. Why would you not include sound waves in the electromagnetic spectrum? (*Hint:* Look at Figure 6-2.)
3. If the frequency of an electromagnetic wave increases, does the number of waves passing by you increase, decrease, or stay the same? Does the wavelength increase, decrease, or stay the same? Does the energy of the photon increase, decrease, or stay the same?
4. Compared to infrared waves, do UV rays have longer or shorter wavelengths? Do UV rays have higher or lower energy?
5. Does red light have a higher or lower energy than blue light? Does red light have higher or lower frequency than blue light? Does red light have a longer or shorter wavelength than blue light?
6. If you had limited funds to build a large telescope, which type would you choose, a refractor or a reflector? Why?
7. Why do nocturnal animals usually have large pupils in their eyes? How is that related to the way astronomical telescopes work?
8. If you were in a deep, dark cave alone and you turned off all the lights, could you see? If so, what would you see?
9. Why do optical astronomers often put their telescopes at the tops of mountains, whereas radio astronomers sometimes put their telescopes in deep valleys?
10. What advantage, if any, would radio astronomers have by building their telescopes at the tops of mountains?
11. What are the advantages of making a telescope mirror thin? What problems result?
12. Small telescopes are often advertised as "200 power" or "magnifies 200 times." How would you change these advertisements to market to astronomers?
13. Why do single-dish radio telescopes have poor resolving power compared to optical telescopes of the same diameter?
14. The Moon has no sustained atmosphere. What advantages would you have if you built an observatory on the lunar surface?
15. Why must telescopes observing at long infrared (that is, far-infrared) wavelengths be cooled to low temperatures?
16. What purpose do the colors in a representational-color (or false-color) image or map serve?
17. What might you detect with an X-ray telescope that you could not detect with an infrared telescope?
18. If you were looking for exploding stars, which wavelength band would you likely like to observe?
19. How is the phenomenon of chromatic aberration related to how a prism spectrograph works?
20. What are prisms and gratings compared to spectrographs?
21. How is active optics different from adaptive optics?
22. Why would radio astronomers build identical radio telescopes in many different places around the world?
23. Are cosmic rays waves?
24. Give an example of a cosmic ray.
25. **How Do We Know?** How is the resolution of an astronomical image related to the precision of a measurement?

Discussion Questions

1. Why does the wavelength response of the human eye match the visual window of Earth's atmosphere so well?
2. Most people like beautiful sunsets with brightly glowing clouds, bright moonlit nights, and twinkling stars. Astronomers don't. Why?
3. You would like to compare and contrast the large features on the near side of the Moon such as the mountains, craters, and plains shown in Figure 3-1. Which telescope and instruments should you apply for time to use? That is, which band of the electromagnetic spectrum? Would you use a refracting or reflecting telescope? Would you use a ground-based, airborne, or space-based telescope? What size telescope do you need? Do you need active optics, adaptive optics, a laser guide star, a spectrograph, or interferometry? Do you have to worry about cosmic rays? Why or why not?
4. Is the left panel of Figure 6-25a an example of good seeing? How do you know? If that is your first image from tonight's observing, should you continue taking data, move to another target, or shut down the telescope and call it a night? Why?

Problems

1. Plastic bags have a thickness about 0.001 mm. How many wavelengths of red light is that?
2. What is the wavelength of radio waves transmitted by a radio station with a frequency of 100 million cycles per second?
3. What is the frequency and wavelength of an FM radio station on your radio dial at 102.2?
4. Does a 700-nm wavelength photon have more or less energy than a 400-nm wavelength photon? What colors are associated with these wavelengths? Which color is associated with higher energy? How much more or less?
5. Compare the light-gathering power of a 10-m Keck telescope with that of a 0.6-m telescope.
6. How does the light-gathering power of one of the 10-m Keck telescopes compare with that of the human eye? Assume that the pupil of your eye can open to a diameter of about 0.8 cm in dark conditions.
7. Telescope A has a 60-in. diameter whereas telescope B has a 4-cm diameter. Which telescope gathers more light and how much more?
8. What is the resolving power of Telescope A and Telescope B in the previous problem in the visual band? Explain which telescope has the better resolving power. How do you know?
9. In general, does a telescope resolve a close double star, such as in Figure 6-10, better at blue wavelengths or red? How do you know?
10. What is the resolving power of a 25-cm (10-in.) telescope at a wavelength of 550 nm? What do two stars 1.5 arc seconds apart look like through this telescope?

11. Most of Galileo's telescopes were only about 2 cm in diameter. Should he have been able to resolve the two stars mentioned in Problem 10?

12. How does the resolving power of the Mount Palomar 5-m telescope compare with that of the 2.4-m *Hubble Space Telescope?* Why does *HST* generally still outperform the ground-based 5-m telescope?

13. If you build a telescope with a focal length of 1.3 m, what eyepiece focal length is needed for a magnification of 100 times?

14. Astronauts observing from a space station need a telescope with a light-gathering power 15,000 times that of the dark-adapted human eye (*Note:* See Problem 6), capable of resolving detail as small as 0.1 arc second at a wavelength of 550 nm, and a magnifying power of 250. Design a telescope to meet their needs. Could you test your design by using your telescope to observe stars from the surface of Earth?

15. A spy satellite orbiting 400 km above Earth is supposedly capable of counting individual people in a crowd in visual-wavelength images. Assume that the middle of the visual wavelength band is at 550 nm. Assume an average person has a size of 0.7 m as seen from above. Estimate the minimum telescope diameter that the satellite must carry. (*Hint:* Use the small-angle formula [Chapter 3] to convert linear size to angular size.)

Learning to Look

1. What is the wavelength of the wave shown in Figure 6-2 in units of mm?

2. How many atmospheric windows are shown in Figure 6-3, and which bands of the electromagnetic spectrum are they in?

3. Locate the primary optical element in Figure 6-17. Is this a refracting telescope or a reflecting telescope? How can you tell?

4. Did the magnification, resolving, or light-gathering power change from the left image to the right image in Figure 6-25a? How do you know?

5. Explain what is meant by "intensity" in the single-wavelength false-color representation of Figure 6-23b. Would you have selected this false-color code pattern, or would you have selected red to represent a wavelength with low intensity?

6. The two images at right show a star before and after an adaptive optics system attached to the telescope was switched on. What causes the distortion in the first image, and how do adaptive optics improve the image?

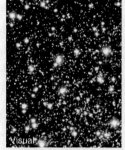

ESO

7. The star images in the photo at the right are tiny disks, but the diameters of these disks are not related to the diameter of the stars. Explain why the telescope can't resolve the diameter of the stars. What causes the apparent diameters of the stars?

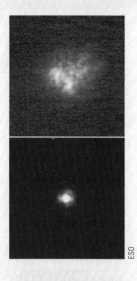

Visual

NASA, ESA, and G. Meylan (Ecole Polythenique Federal de Lausanne)/ STSci-PRC06-33

8. The X-ray image at right shows the remains of an exploded star. Explain why images recorded by telescopes in space are often displayed in representational ("false") color rather than in the "colors" (that is, wavelengths) received by the telescope. What color would we see this image if the image was not falsely colored?

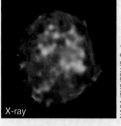

X-ray

NASA/CXC/PSU/S. Park

7 Atoms and Spectra

Guidepost In the previous chapter, you read how telescopes gather light, cameras record images, and spectrographs spread light into spectra. Now you can consider why astronomers make such efforts. Here you will find answers to three important questions:

▶ **How do atoms interact with light to produce spectra?**

▶ **What are the types of spectra that can be observed?**

▶ **What can be learned from spectra of celestial objects?**

Up to this point, you have been considering what you can see with your eyes alone or aided by telescopes and astronomical instruments such as spectrographs. This chapter marks a change in the way you study nature: You will begin learning about the modern field of **astrophysics** that links physics experiments and theory to astronomical observations, and realize why the spectrum of an object can be so informative. In the chapters that follow, you will learn about the rich information derived from spectra of planets, stars, and galaxies that reveals the secrets of their internal structures and histories.

Awake! for Morning in the Bowl of Night
Has flung the Stone that puts the Stars to Flight:
And Lo! the Hunter of the East has caught
The Sultan's Turret in a Noose of Light.

THE RUBÁIYÁT OF OMAR KHAYYÁM,
TRANSLATION BY EDWARD FITZGERALD

NASA/ESA/STScI/AURA/NSF

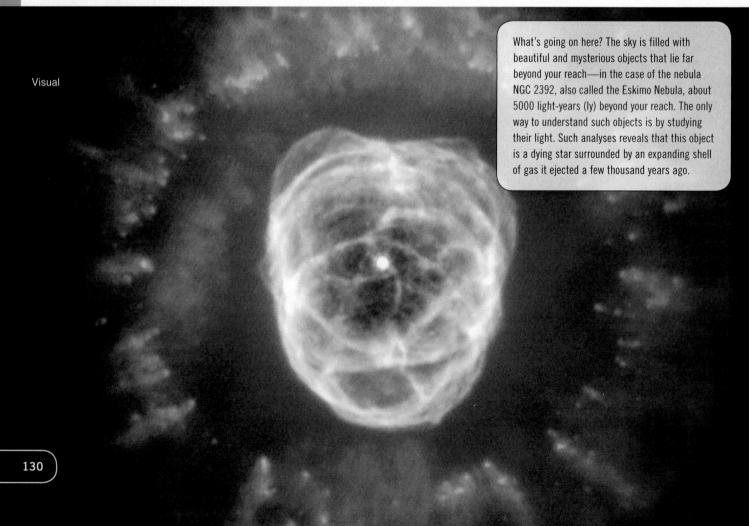

Visual

What's going on here? The sky is filled with beautiful and mysterious objects that lie far beyond your reach—in the case of the nebula NGC 2392, also called the Eskimo Nebula, about 5000 light-years (ly) beyond your reach. The only way to understand such objects is by studying their light. Such analyses reveals that this object is a dying star surrounded by an expanding shell of gas it ejected a few thousand years ago.

THE UNIVERSE IS POPULATED with brilliant stars illuminating exotic planets and fabulously beautiful clouds of glowing gas. But other than the objects in our tiny local Solar System, they are all out of reach for the foreseeable future. No human space probe has visited another star, and no telescope can directly examine the insides of any celestial object. The information obtained about most of the Universe is contained in the light reaching Earth across space.

Earthbound humans knew almost nothing about the composition of celestial objects until the early 19th century. First, the German optician Joseph von Fraunhofer studied the spectrum of the Sun and discovered that it is interrupted by more than 600 narrow dark lines which are colors missing from the sunlight that Earth receives. Then other scientists performed laboratory experiments showing that those spectral lines are related to the presence of various atoms in the Sun's atmosphere. Finally, astronomers observed that the spectra of other stars have similar patterns of lines, opening a window to real understanding of how the Sun and stars are related.

In this chapter, you will see how the Sun and other stars produce light and how atoms in the atmospheres of stars, planets, and gas clouds in space interact with light to cause spectral lines (Figure 7-1). Once you understand that, you will know how astronomers determine the chemical composition of distant objects, as well as measure motions of gas in and around them.

7-1 Atoms

Atoms in stars and planets leave their fingerprints on the light we receive from them. By first reviewing what atoms are and then learning how they interact with light, you can understand how the spectra of objects in space are decoded.

A Model Atom

To think about how atoms interact with light, you need a working model of an atom. In Chapter 2, you used a model of the sky, the celestial sphere. In this chapter, you will begin your study of atoms by using a mental model of an atom. Remember that such a model can have practical value without being true. The stars are not actually attached to a sphere surrounding Earth, but to navigate a ship or point a telescope, it is convenient and practical to pretend they are. The electrons in an atom are not actually little beads orbiting the nucleus the way planets orbit the Sun, but for some purposes it is useful to picture them as such.

A single atom is not a massive object. A hydrogen atom, for example, has a mass of only 1.7×10^{-27} kg, about a trillionth of a trillionth of a gram. Positively charged **protons** and uncharged **neutrons** have masses almost 2000 times greater than that of negatively charged **electrons**, so most of the mass of an atom lies in the **nucleus**. Normally the number of electrons equals the number of protons so the positive and negative charges balance to produce a neutral atom. In this atomic model, the electrons can be pictured as being in a cloud that completely surrounds the nucleus.

An atom is mostly empty space. To see this, imagine constructing a simple scale model of a hydrogen atom. Its nucleus is a single proton with a diameter of approximately 0.0000017 nm, or 1.7×10^{-15} m. If you multiply that by 1 trillion (10^{12}), you can represent the nucleus of your model atom with something about 1.7 mm in diameter—a grape seed would do. The region of a hydrogen atom that normally contains the electron has an effective diameter of about 0.24 nm, or 2.4×10^{-10} m. This is two times a hydrogen atom's Van der Waals radius, which characterizes the distance over which it interacts with other atoms. Multiplying that by a trillion increases the diameter to about 240 m, or almost

◄ **Figure 7-1** A display of aurora borealis, also known as "northern lights." Gas in Earth's upper atmosphere is excited by electrical currents caused by interactions of charged particles in the solar wind with Earth's magnetic field. This particular auroral glow shows spectral emission lines of ionized oxygen (*green*) and nitrogen (*red*).

Anna Henly/Photolibrary/Getty Images

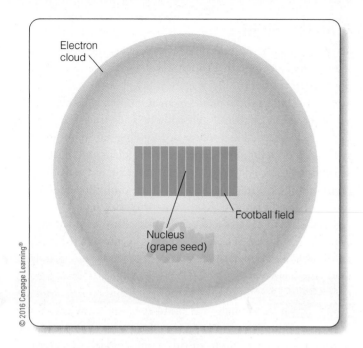

▲ **Figure 7-2** Magnifying a hydrogen atom by 10^{12} makes the nucleus the size of a grape seed and the diameter of the electron cloud more than 2.6 times larger than the length of a U.S. football field.

three U.S. football fields laid end to end (**Figure 7-2**). When you imagine a grape seed in the middle of a sphere nearly three football fields in diameter, you can see that an atom is mostly empty space.

Now you can consider a **Common Misconception.** Most people, without thinking about it much, imagine that matter is solid, but you have seen that atoms are mostly empty space. The chair you sit on, the floor you walk on, are mostly not there. When you study the deaths of stars in the next chapter, you will see what happens to a star when the empty space gets squeezed out of its atoms.

Different Kinds of Atoms

There are more than a hundred chemical elements. The number of protons in the nucleus of an atom determines which element it is. For example, a carbon atom has six protons in its nucleus. An atom with one more proton than that is nitrogen, and an atom with one fewer proton is boron.

Although an atom of a given element always has the same number of protons in its nucleus, the number of neutrons is less restricted. For instance, if a neutron is added to a carbon nucleus, it would still be carbon, but it would be slightly heavier. Atoms that have the same number of protons but a different number of neutrons are **isotopes.** Carbon has two stable isotopes. One contains six protons and six neutrons for a total of 12 particles and is thus called carbon-12. Carbon-13 has six protons and seven neutrons in its nucleus.

The number of electrons in an atom of a given element can vary. Protons and neutrons are bound tightly into the nucleus,

but the electrons are held loosely in the electron cloud. Running a comb through your hair creates a static charge by removing a few electrons from their atoms. An atom that has lost or gained one or more electrons is called an **ion.** A neutral carbon atom has six electrons that balance the positive charge of the six protons in its nucleus. If you **ionize** the atom by removing one or more electrons, the atom is left with a net positive charge. Under other circumstances, an atom may capture one or more extra electrons, giving it more negative charges than positive. Such a negatively charged atom is also considered an ion.

Atoms that collide may form bonds with each other by exchanging or sharing electrons. As you already know, two or more atoms bonded together form a **molecule.** Atoms do collide in stars, but the high temperatures cause violent collisions that are unfavorable for chemical bonding. Only in the coolest stars are the collisions gentle enough to permit the formation of chemical bonds. The presence of molecules such as titanium oxide (TiO) detected in some stars is one clue that those stars are very cool compared with other stars. In later chapters, you will see that molecules also can form in cool gas clouds in space and in the atmospheres of planets.

Electron Orbits

So far you have been considering the cloud of electrons in atoms only in a general way. Now it is necessary to be more specific about how electrons behave within the cloud on the way to understanding how light interacts with atoms.

Electrons are bound to the atom by the attraction between their negative charge and the positive charge on the nucleus. This attraction is known as the **Coulomb force**, after the physicist Charles-Augustin de Coulomb. To ionize an atom, you need a certain amount of energy to pull an electron completely away from the nucleus. This energy is the electron's **binding energy**, the energy that holds it to the atom.

The size of an electron's orbit is related to the energy that binds it to the atom. If an electron orbits close to the nucleus, it is tightly bound, and a large amount of energy is needed to pull it away. In other words, its binding energy is large. An electron orbiting farther from the nucleus is held more loosely, and less energy is needed to pull it away. That means it has small binding energy.

Nature permits atoms only certain amounts (quanta) of binding energy. The laws that describe how atoms behave are called the laws of **quantum mechanics** (How Do We Know? 7-1). Much of this discussion of atoms is based on the laws of quantum mechanics that were discovered by physicists early in the 20th century.

Because atoms can have only certain amounts of binding energy, electrons can have orbits only of certain sizes, called **permitted orbits.** These are like steps in a staircase: You can stand on the number one step or the number two step, but not

How Do We Know? 7-1

Quantum Mechanics

How can you understand nature if it depends on the atomic world you cannot see? You can see objects such as stars, planets, aircraft carriers, and hummingbirds, but you can't see individual atoms. As scientists apply the principle of cause and effect, they study the natural effects they can see and work backward to find the causes. Invariably, that quest for causes in the physical world leads back to the invisible world of atoms.

Quantum mechanics is the set of rules that describe how atoms and subatomic particles behave. On the atomic scale, particles behave in ways that seem unfamiliar and difficult to comprehend. One of the principles of quantum mechanics specifies that you cannot know simultaneously the exact location and exact motion of a particle. This is why, in practice, physicists go beyond the simple atomic model you may have learned in high school that imagines electrons as particles following orbits, and instead describe the electrons in an atom as if they are each clouds of negative charge surrounding the nucleus. That's a much better model, although it's still only a model.

This raises some serious questions about reality. Is an electron really a particle at all? If you can't know simultaneously the position and motion of a specific particle, how can you know how it will react to a collision with a photon or another particle? The surprising and puzzling answer is that you can't know certainly and completely. That seems to violate the principle of cause and effect.

Many of the phenomena you can see depend on the behavior of huge numbers of atoms, so quantum mechanical uncertainties average out. Nevertheless, the ultimate causes that scientists seek lie at the level of atoms, and modern physicists are trying to understand the nature of the particles that make up atoms. That is one of the most exciting frontiers of science.

The world you see, including these neon signs, is animated by the properties of atoms and subatomic particles.

© Antonio V. Oquias/Shutterstock.com

---

on the number one and one-half step because there isn't one. The electron can occupy any permitted orbit, but there are no orbits in between.

The arrangement of permitted orbits depends primarily on the charge of the nucleus, which in turn depends on the number of protons. Consequently, each chemical element—each type of atom—has its own pattern of permitted orbits (Figure 7-3). Isotopes of the same elements have nearly the same pattern because they have the same number of protons in their nuclei but slightly different masses. Ionized atoms, with altered electrical charges, have orbital patterns that differ greatly from their unionized forms.

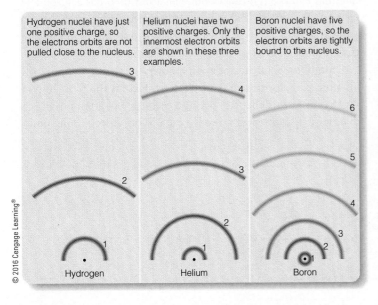

Hydrogen nuclei have just one positive charge, so the electrons orbits are not pulled close to the nucleus.

Helium nuclei have two positive charges. Only the innermost electron orbits are shown in these three examples.

Boron nuclei have five positive charges, so the electron orbits are tightly bound to the nucleus.

© 2016 Cengage Learning®

▲ **Figure 7-3** An electron in an atom may occupy only certain permitted orbits. Because each element has a different number of protons and therefore a different electrical charge in the nucleus attracting the electrons, each element has a different, unique pattern of permitted orbits.

DOING SCIENCE

How many hydrogen atoms would it take to cross the head of a pin? By answering this question, you will discover how small atoms really are and also see how important mathematics are in doing science and understanding nature.

To begin, assume that the head of a pin is about 1 mm in diameter—that is, 0.001 m. The size of a hydrogen atom is represented by the diameter of the electron cloud, roughly 0.24 nm. Because 1 nm equals 10^{-9} m, you can multiply and discover that 0.24 nm equals 2.4×10^{-10} m. To find out how many atoms would stretch 0.001 m, you can divide the diameter of the pinhead by the diameter of an atom. That is, divide 0.001 m by 2.4×10^{-10} m, and you get 4.2×10^6. That means it would take 4.2 million hydrogen atoms lined up side by side to cross the head of a pin.

Now you can see how tiny an atom is and also how powerful a bit of simple math can be. It reveals a view of nature beyond the capability of your eyes. Now do some more science with another bit of math. ***How many hydrogen atoms would you need to add up to the mass of a paper clip (1 g)?***

7-2 Interactions of Light and Matter

If light and matter did not interact, you would not be able to see these words. In fact, you would not exist because, among other problems, photosynthesis would be impossible, so there would be no grass, wheat, bread, beef, yogurt, or any other kind of food. The interaction of light and matter makes life possible, and it also makes it possible for you to understand the Universe.

You have already been considering a model hydrogen atom. Now you can use that model as you begin your study of light and matter by thinking about hydrogen. It is both simple and common: Roughly 90 percent of all atoms in the Universe are hydrogen.

The Excitation of Atoms

Because each electron orbit in an atom represents a specific amount of binding energy, physicists commonly refer to the orbits as **energy levels**. Using this terminology, you can say that an electron in its smallest and most tightly bound orbit is in its lowest permitted energy level, which is called the atom's **ground state**. You could move the electron from one energy level to another by supplying enough energy to make up the difference between the two energy levels. It would be like moving a package from a low shelf to a high shelf; the greater the distance between the shelves, the more energy you would need to raise the package. The amount of energy needed to move the electron is the difference in the binding energy between the two levels. Giving the package a different amount of energy would put it between the shelves, where it would not be able to stay.

If you move an electron from a low energy level to a higher energy level, the atom becomes an **excited atom**. That is, you have added energy to the atom by moving its electron outward from the nucleus. One way an atom can become excited is by collision. If two atoms collide, one or both may have electrons knocked into a higher energy level. This happens very commonly in hot gas, where atoms move rapidly and collide often.

Another way an atom can become excited is to absorb a photon. As you learned in the previous chapter, a photon is a bundle of electromagnetic waves with a specific energy. Only a photon with exactly the right amount of energy can move the electron from one level to another. If the photon has too much or too little energy, that atom cannot absorb it. Because the energy of a photon depends on its wavelength, only photons of certain wavelengths (colors) can be absorbed by a given kind of atom.

Figure 7-4 shows the lowest four energy levels of the hydrogen atom, along with three photons the atom is capable of absorbing. The photon with the longest wavelength has only enough energy to excite (move) the electron up to the second energy level, but the photons with shorter

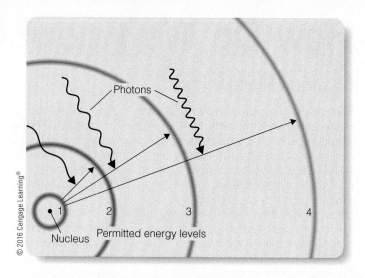

▲ **Figure 7-4** A hydrogen atom can absorb only those photons that have the right energy to move the atom's electron to one of the higher-energy orbits. Here three photons with different wavelengths are shown along with the changes they each would produce in the electron's orbit if they were absorbed.

wavelengths have more energy and can excite the electron to higher levels. An actual hydrogen atom has many more energy levels than shown in Figure 7-4, and it can absorb photons of many different wavelengths.

Atoms, like humans, cannot exist in an excited state forever. An excited atom is unstable and must eventually (usually within 10^{-9} to 10^{-6} seconds) give up the energy it has absorbed and return its electron to a lower energy level. Thus, the electron in an excited atom tends to tumble down to its lowest energy level, which is its ground state. When an electron drops from a higher to a lower energy level, it moves from a loosely bound level to one that is more tightly bound. The atom then has a surplus of energy—the energy difference between the levels—that it can emit as a photon of light with a wavelength corresponding to that amount of energy (look back to Chapter 6).

Study the sequence of events shown in **Figure 7-5** to see how an atom can absorb and emit photons. Here is the most important point in this chapter: Because each type of atom or ion has a unique set of energy levels, each type absorbs and emits

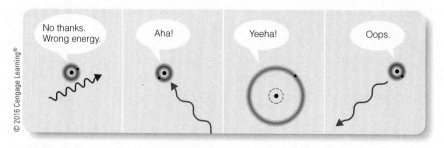

▲ **Figure 7-5** An atom can absorb a photon only if the photon has the correct amount of energy. The excited atom is unstable and within a fraction of a second returns to a lower energy level, radiating new photons in random directions relative to the original photon's direction.

photons with a unique set of wavelengths. As a result, you can identify the elements in a gas by studying the characteristic wavelengths of light that are absorbed or emitted.

The processes of atomic excitation and photon emission are common sights in urban areas at night. A neon sign glows when atoms of neon gas in a glass tube are excited by electricity flowing through the tube. As the electrons in the electric current flow through the gas, they collide with the neon atoms and excite them. Almost immediately after a neon atom is excited, its electron drops back to a lower energy level, emitting the surplus energy as a photon of a certain wavelength. The photons emitted by excited neon blend to produce a reddish-orange glow. Signs of other colors, generically called *neon signs,* contain other gases or mixtures of gases. Whenever you look at a "neon" sign, you are seeing atoms emitting energy in the form of photons with specific colors determined by the structure of electron orbits in those atoms.

Neon signs are simple, but stars are complex. Stars have colors, but those colors are not determined by the composition of the gases they contain. In the next section, you will discover why some stars are red and some are blue, and that will give you further insight into how light interacts with matter.

Radiation from a Heated Object

When you view the stars in the constellation Orion, you notice that they are not all the same color (look back to Figure 2-4a). One of the Favorite Stars, Betelgeuse, in the upper left corner of Orion, is quite red; another Favorite Star, Rigel, in Orion's lower right corner, is blue. These differences in color arise from differences in temperature.

The starlight that you see comes from gases that make up the visible surface of the star, its photosphere. (Recall that you first learned about the photosphere of the Sun in Chapter 3, in the context of solar eclipses.) Layers of gas deeper inside the star also emit light, but that light is reabsorbed before it can reach the surface. The gas above the photosphere is too thin to emit much light. The photosphere is the visible surface of a star because it is dense enough to emit lots of light but transparent enough to allow that light to escape.

Stars produce their light for the same reason heated horseshoes glow in a blacksmith's forge—because they are hot. If a horseshoe is moderately hot, it glows ruddy red, but as it heats up further it grows brighter and yellower. Yellow hot is hotter than red hot, but not as hot as blue hot.

The light from stars and from glowing horseshoes is produced by the acceleration of charged particles. Usually the accelerated particles are electrons because they are the least massive charged particles, and they are on the outsides of atoms, so they are the easiest to get moving. An electron produces a surrounding electric field, and if an electron is accelerated, the change in its electric field spreads outward at the speed of light as electromagnetic radiation. You learned in Chapter 5 that *acceleration* means any change in motion—not only increasing

speed, as that word means in everyday language, but also decreasing speed, and keeping constant speed while changing direction. Whenever the motion of any charged particle is changed, electromagnetic waves are generated. If you run a comb through your hair, you disturb electrons in both hair and comb, producing static electricity. That produces electromagnetic radiation, which you can hear as snaps and crackles if you are standing near an AM radio. Stars are hot, and they are made up of ionized gases, so there are plenty of electrons zipping around and being accelerated.

The molecules and atoms in any object are in constant motion, and in a hot object they are more agitated than in a cool object. This agitation is called **thermal energy**. If you touch an object that contains lots of thermal energy, it will feel hot as the thermal energy flows into your fingers. The flow of thermal energy is called **heat**. In contrast, **temperature** refers to the average speed or intensity of agitation of the particles in an object. A metal speck from a fireworks sparkler can be much hotter than a hot metal clothes iron, but the iron contains much more thermal energy and therefore can burn your hand much more badly (**Focus on Fundamentals 3**).

When astronomers refer to the temperature of a star, they are talking about the temperature of the gases in the photosphere, and they express those temperatures with the **Kelvin temperature scale**. On this scale, zero degrees Kelvin (written 0 K) is **absolute zero** ($-273.2°C$ or $-459.7°F$), the temperature at which an object contains no thermal energy that can be extracted. Water at sea level atmospheric pressure freezes at 273 K and boils at 373 K. The Kelvin temperature scale is used in astronomy and physics because it is based on absolute zero and therefore is related most directly to the motions of particles in an object.

Now you can understand why a hot object glows or, to put it another way, why a hot object emits photons (bundles of electromagnetic energy). The hotter an object is, the more motion there is among its particles. The agitated particles, including electrons, collide with each other, and when electrons change their motion—accelerate—part of the energy of motion is carried away as electromagnetic radiation. The typical spectrum of an opaque heated object, meaning the amount and color distribution of radiation it emits, is called **blackbody radiation**. That name is translated from a German term referring to an object that is a perfectly efficient absorber and emitter of radiation. At room temperature, such a perfect absorber and emitter would look black, but at higher temperatures it would glow at wavelengths visible to a human eye. That explains why in astronomy and physics you will see the term *blackbody* referring to objects that are actually glowing brightly.

Blackbody radiation is quite common. In fact, it describes the light emitted by an incandescent lightbulb. Electricity flowing through the opaque filament of the bulb heats it to high temperature, and it glows with a blackbody spectrum. You can also recognize the light emitted by hot lava as blackbody radiation. Many objects in the sky, including the Sun and other stars,

Temperature, Heat, and Thermal Energy

One of the most **Common Misconceptions** in science involves temperature. People often say "temperature" when they really mean "heat," and sometimes they say "heat" when they mean something entirely different. These are fundamental ideas, and it is important for you to understand the difference.

Even in an object that is solid, the atoms and molecules are continuously jiggling around and bumping into each other. When something is hot, the particles are moving rapidly. Temperature is a measure of the average motion of the particles. (Mathematically, temperature is proportional to the square of the average velocity.) If you have your temperature taken, it will probably be 37.0°C (98.6°F), which is an indication that the atoms and molecules in your body are moving at a normal pace. If you measure the temperature of a baby, the thermometer should register the same temperature, showing that the atoms and molecules in the baby's body

are moving at the same average velocity as the atoms and molecules in your body.

The total energy of all of the moving particles in a body is called *thermal energy*. People often confuse temperature and thermal energy, so be careful to distinguish between them. You have much more mass than the baby, so you must contain more thermal energy even though you have the same temperature. The thermal energy in your body and in the baby's body have the same intensity (temperature) but different total amounts.

Many people say "heat" when they should say "thermal energy." Heat is the thermal energy moving from a hot object to a cool object. If two objects have the same temperature—you and the infant for example—when they touch there is no transfer of thermal energy and therefore no heat.

You may have burned yourself on cheese pizza, but you probably haven't burned yourself on green beans. Cheese is denser than green beans, so at the same temperature,

cheese holds more thermal energy than green beans. It isn't the temperature that burns your tongue, but the flow of thermal energy, and that's heat. When you hear someone say "heat," consider whether that person really means thermal energy.

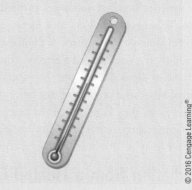

What's the difference between temperature and heat?

© 2016 Cengage Learning®

MASS | ENERGY | TEMPERATURE AND HEAT | DENSITY | PRESSURE

---

emit radiation approximately as blackbodies because they are mostly opaque.

Hot objects emit blackbody radiation, but so do cold objects. Ice cubes are cold, but their temperature is higher than absolute zero, so they contain some thermal energy and must emit some blackbody radiation. The coldest gas drifting in space has a temperature only a few degrees above absolute zero, but it also emits a blackbody spectrum.

Two Blackbody Radiation Laws

Two features of blackbody radiation are important. First, the hotter an object is, the more radiation it emits. Hot objects emit more radiation because their agitated particles collide more often and more violently with each other. That's why a glowing coal from a fire emits more total energy than an ice cube of the same size.

Figure 7-6 shows the intensity of radiation versus wavelength for three objects of different temperatures. The total area under each curve is proportional to the total energy emitted. You can see that the hottest object emits more total energy than the two cooler objects. This rule is known as the **Stefan-Boltzmann law**, named after Jozef Stefan and Ludwig Boltzmann, the physicists who discovered it.

The Stefan-Boltzmann law mathematically relates the temperature of a blackbody to the total radiated energy. Recall from Chapter 5 that energy is expressed in units of joules (symbolized by capital J; the energy gained by an apple falling from a table onto the floor is approximately 1 joule). The total radiation in units of joules per second given off by 1 square meter of the surface of an object equals a constant number called the Stefan-Boltzmann constant, represented by the Greek lowercase letter sigma, σ, multiplied by the temperature in degrees Kelvin raised to the fourth power:

$$E = \sigma T^4 \ (\text{J/s/m}^2)$$

(For the sake of completeness, note that the constant σ equals 5.67×10^{-8} J/s/m^2/K^4, units of joules per second per square meter per degree Kelvin to the fourth power).

The second feature of blackbody radiation is the relationship between the temperature of the object and the wavelengths of the photons it emits. The wavelength of the photon emitted when electrons collide with other particles depends on the violence of the collision; a violent collision can produce a short wavelength (high energy) photon. The electrons in an object have a distribution of speeds; a few move very rapidly, and a few move very slowly, but most travel at intermediate speeds. Because electrons

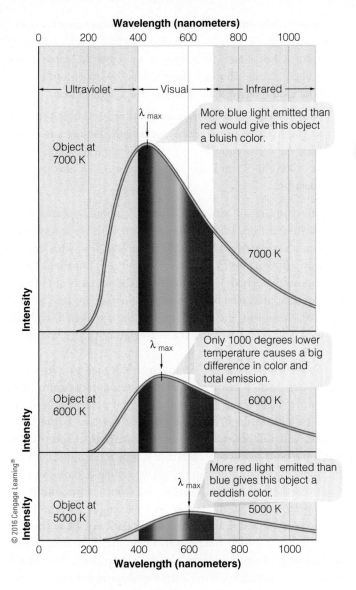

Wavelength (nanometers)

Ultraviolet — Visual — Infrared

λ_{max}

More blue light emitted than red would give this object a bluish color.

Object at 7000 K

7000 K

Only 1000 degrees lower temperature causes a big difference in color and total emission.

λ_{max}

Object at 6000 K

6000 K

More red light emitted than blue gives this object a reddish color.

λ_{max}

Object at 5000 K

5000 K

Wavelength (nanometers)

Intensity

© 2016 Cengage Learning®

▲ **Figure 7-6** Graphs of blackbody radiation intensity versus wavelength for three objects at temperatures of 7000 K, 6000 K, and 5000 K, respectively (*top to bottom*). Comparison of the graphs demonstrates that a hot object radiates more total energy per unit area than a cooler object (Stefan-Boltzmann law) and that the wavelength of maximum intensity is shorter for hotter objects than for cooler objects (Wien's law). The hotter object here would look blue to your eyes, whereas the cooler object would look red.

with speeds much greater or much less than the average speed are rare, extremely violent collisions and extremely gentle collisions don't occur very often. Consequently, blackbody radiation is composed of a distribution of photons in which very short-wavelength and very long-wavelength photons are rare; photons with intermediate wavelengths are most common.

Look again at Figure 7-6, showing the intensity of blackbody radiation emitted versus wavelength for three objects of different temperatures. The curves are high in the middle and low at either end because the objects emit radiation most intensely at intermediate wavelengths. The **wavelength of maximum intensity (λ_{max})** is the wavelength at which the object emits the most intense radiation.

(Note that λ_{max} refers not to the maximum wavelength but to the wavelength of the maximum.) You can see by comparing the three curves in Figure 7-6 that the wavelength of maximum intensity depends on temperature: The hottest object has the shortest wavelength of maximum emitted intensity. In other words, the hotter object emits more blue light than red and thus looks blue, and the cooler object emits more red than blue and consequently looks red. This rule is known as **Wien's law**, named after Wilhelm Wien.

Wien's law quantitatively expresses the relation between a blackbody's temperature and the wavelength of maximum intensity (λ_{max}) of its emitted spectrum. Written for conventional intensity units, the law is:

$$\lambda_{max} = 2.90 \times 10^6/T$$

That is, the wavelength, in nanometers, of the radiation with maximum intensity emitted by a blackbody equals 2.9 million divided by the blackbody's temperature on the Kelvin scale.

You can see both Wien's law and the Stefan-Boltzmann law in operation if you look down into a toaster after you start the toast. First, you see a faint deep red glow that, as the coils get hotter, becomes brighter (Stefan-Boltzmann law) and more orange-yellow in color (Wien's law). It's important to note that objects too cool to glow at visible wavelengths still produce blackbody radiation. For example, the human body has a temperature of 310 K and emits blackbody radiation mostly in the infrared part of the spectrum. Infrared security cameras can detect intruders by the radiation they emit, and mosquitoes can track you down in total darkness by homing in on your infrared radiation. Although you emit lots of infrared radiation, you almost never emit a gamma-ray or radio photon. The wavelength of maximum intensity of your glow lies in the infrared part of the spectrum.

How do the Stefan-Boltzmann law and Wien's law help you understand stars and other celestial objects? Suppose a star the same size as the Sun has a surface temperature twice as hot as the Sun's surface. Then, according to the Stefan-Boltzmann law, each square meter of that star radiates $2^4 = 16$ times as much energy as a square meter of the Sun's surface. You can see that a small difference in temperature between two stars can produce a large difference in the amount of energy emitted from their surfaces. The Stefan-Boltzmann law relates temperature, surface area, and total energy emitted by stars and other blackbodies. If you know two of those three quantities, you can determine the other one.

Using Wien's law, you can measure the temperatures of distant objects without having to travel there and stick in a thermometer. A cool star with a temperature of 2900 K will emit most intensely at a wavelength of 1000 nm, which is infrared light. In comparison, a very hot star with a temperature of 29,000 K radiates most intensely at a wavelength of 100 nm, which is ultraviolet light. Now you can understand why the two Favorite Stars in Orion mentioned previously, Betelgeuse and Rigel, have such different colors. Betelgeuse is relatively cool and therefore looks red, but Rigel is hot and looks blue.

7-3 Understanding Spectra

Science is a way of understanding nature, and the spectrum of a star can tell you a great deal about the star's temperature, motion, and composition. In later chapters, you will use spectra to study other astronomical objects such as galaxies and planets, but you can begin by looking at the spectra of stars, including that of the Sun.

The spectrum of a star is formed as light passes outward through the gases near its surface. Read **Atomic Spectra** on pages 140–141 and notice that it describes important properties of spectra and defines 12 new terms that will help you understand astronomical spectra:

1 There are three types of spectra: (i) *continuous spectra;* (ii) absorption or *dark-line spectra,* which contain *absorption lines;* and (iii) *emission* or *bright-line spectra,* which contain *emission lines.* These types of spectra are described by *Kirchhoff's laws.* When you see one of these types of spectra, you can recognize the arrangement of matter that emitted the light.

2 Photons are emitted or absorbed when an electron in an atom makes a *transition* from one energy level to another. The wavelengths of the photons depend on the energy difference between the two levels, so (note, this is especially important) each spectral line represents not one energy level but rather an electron transition between two energy levels. Hydrogen atoms can produce many spectral lines that are grouped in series such as the *Lyman, Balmer,* and *Paschen*

series. Only three hydrogen lines—all in the Balmer series—are visible to human eyes. The emitted photons coming from a hot cloud of hydrogen gas have the same wavelengths as the photons absorbed by hydrogen atoms in a cool cloud between an observer and a light source.

3 Most modern astronomy books and articles display spectra as graphs of intensity versus wavelength. Be sure you recognize the connection between dark absorption lines, bright emission lines, and the dips and peaks in the graphed spectrum.

Imagine you are an astronaut with a handheld spectrograph, approaching a fresh lava flow on a moon with no atmosphere. You aim your spectrograph straight at the lava flow; what kind of spectrum do Kirchhoff's laws say you will see? You should see a continuous (blackbody) spectrum, with all the colors of the rainbow present, produced by the opaque, glowing hot lava. And as a bonus, measuring the wavelength of the strongest blackbody emission lets you determine the temperature of the lava using Wien's law.

Suddenly, the lava flow begins to bubble, and gas trapped in the molten rock is released, making a temporary, warm, thin atmosphere right above the lava flow. If you point your spectrograph to look through that gas with the hot lava as the background, you will observe an absorption (dark-line) spectrum in which the lava's blackbody spectrum is now interrupted by missing colors caused by atoms in the gas absorbing photons on their way from the lava to you. Finally, you crouch down and point your spectrograph at the gas at such an angle that the background is not hot lava but cold, empty, dark sky. Now you will see an emission (bright-line) spectrum, produced by atoms in the gas releasing photons as their electrons drop down toward the ground state, with lines at the same wavelengths as in the absorption spectrum of the same gas.

Chemical Composition

Identifying the elements that are present in a star, planet, or gas cloud by identifying the lines in that object's spectrum is a relatively straightforward procedure. For example, two dark absorption lines appear in the yellow region of the Sun's spectrum at the wavelengths 589.0 nm and 589.6 nm. The only atom that can produce this pair of lines is sodium, so the Sun must contain sodium. More than 90 elements in the Sun have been identified this way (**Figure 7-7a**).

However, just because the spectral lines that identify an element are missing, you cannot conclude that the element itself is absent. For example, the spectral lines in the hydrogen Balmer series are weak in the Sun's spectrum, even though 90 percent of the atoms in the Sun are hydrogen. The next chapter will explain how it was discovered that this occurs because the Sun is too cool to produce strong hydrogen Balmer lines. Astronomers must consider that an element's spectral

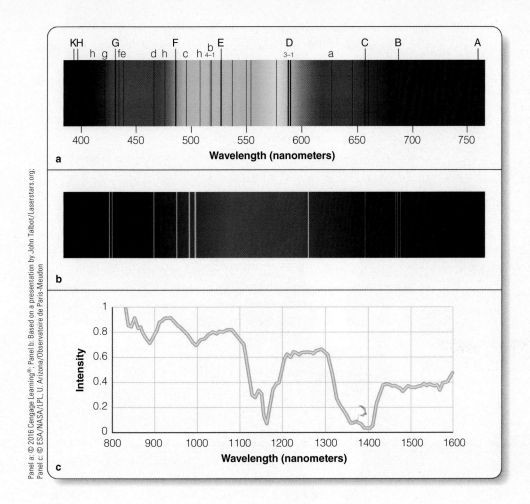

Panel a: © 2016 Cengage Learning®. Panel b: Based on a presentation by John Talbot/Laserstars.org;
Panel c: © ESA/NASA/LPL, U. Arizona/Observatoire de Paris-Meudon

◀ **Figure 7-7** (a) The Sun's spectrum at visual wavelengths. The bright colored background shows the continuous spectrum of blackbody emission from the Sun's photosphere. The dark spectral absorption lines represent precise colors (photons of exact energies) removed from the Sun's radiation by atoms in its transparent atmosphere. (b) A model of the visual-wavelength emission (*bright-line*) spectrum of NGC 2392, the nebula in the image that opens this chapter. Emission lines from ionized atoms of hydrogen (*red*), nitrogen (*red*), and oxygen (*green*), among others, are seen. (c) Graph of the near-infrared spectrum of the atmosphere and surface of Saturn's moon Titan measured by the *Huygens* probe at an altitude of 20 meters (about 65 feet) using a light source on the bottom of the probe.

lines may be absent from an object's spectrum not because that element is missing but because that object has the wrong temperature to excite those atoms to the energy levels that produce detectable spectral lines.

To derive accurate chemical abundances, astronomers must use the laws that describe the interaction of light and matter to analyze a spectrum, take into account the object's temperature, and calculate the amounts of the elements present there. Such results show that nearly all stars, and most of the visible matter in the Universe, have a chemical composition similar to the Sun's—about 91 percent of the atoms are hydrogen, and 8.9 percent are helium, with small traces of heavier elements. You will use these results in later chapters when you study the life stories of the stars, the history of our galaxy, and the origin of the Universe.

Measuring Velocities—The Doppler Effect

Surprisingly, one of the pieces of information hidden in a spectrum is the velocity of the light source. Astronomers can measure the wavelengths of the lines in a star's spectrum and find the velocity of the star. The **Doppler effect** is the apparent change in the wavelength of radiation from a source caused by the relative motion of the source and observer.

When astronomers talk about the Doppler effect, they are talking about a shift in the wavelength of electromagnetic radiation. But the Doppler shift can occur in any type of wave phenomena, including sound waves. You probably hear the Doppler effect several times every day without noticing. Every time a car or truck passes you and the pitch of its engine noise seems to drop, that's the Doppler effect. The pitch of a sound is determined by its wavelength; sounds with long wavelengths have low pitches, and sounds with short wavelengths have higher pitches. The vehicle's sound is shifted to shorter wavelengths and higher pitches while it is approaching to longer wavelengths and lower pitches after it passes as a result of the Doppler effect.

To see why the sound waves are shifted in wavelength, consider a fire truck approaching you with its siren blaring (**Figure 7-8a**). The sound coming from the siren will be a wave that can be depicted as a series of "peaks" and "valleys" representing compressions and decompressions. If the truck and siren are moving toward an observer, the peaks and valleys of the siren's sound wave will arrive closer together—at a higher frequency—than if the truck were not moving, and the observer will hear the siren at a higher pitch than the same siren when it is stationary. If the truck and siren are moving away,

Atomic Spectra

1 To understand how to analyze a spectrum, begin with a simple incandescent lightbulb. The hot filament emits blackbody radiation, which forms a **continuous spectrum**.

An **absorption spectrum** results when radiation passes through a cool gas. In this case you can imagine that the lightbulb is surrounded by a cool cloud of gas. Atoms in the gas absorb photons of certain wavelengths that are then missing from the observed spectrum, and you see dark **absorption lines** at those wavelengths. Such absorption spectra are also called **dark-line spectra**.

An **emission spectrum** is produced by photons emitted by an excited gas. You could see **emission lines** by turning your telescope aside so that photons from the bright bulb do not enter the telescope and the excited gas has a dark background. The photons you would see would be those emitted by the excited atoms near the bulb, and the observed spectrum is mostly dark with a few bright emission lines. Such spectra are also called **bright-line spectra**.

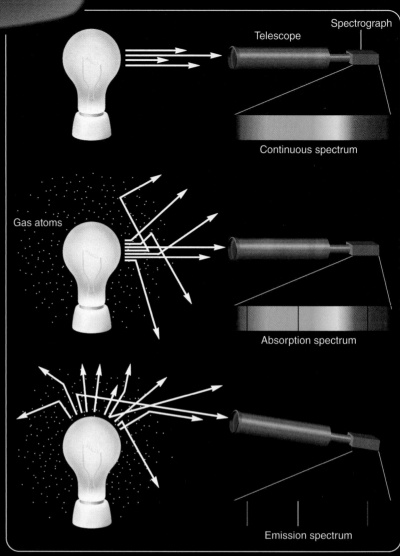

Telescope · Spectrograph

Continuous spectrum

Gas atoms

Absorption spectrum

Emission spectrum

1a The spectrum of a star is an absorption spectrum. The denser layers of the photosphere emit blackbody radiation. Gases in the atmosphere of the star absorb their specific wavelengths and form dark absorption lines in the spectrum.

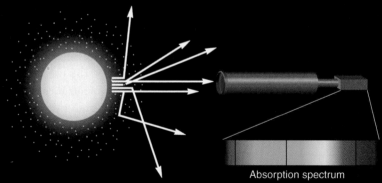

Absorption spectrum

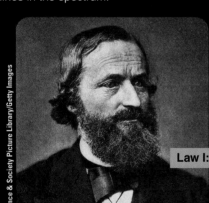

1b In 1859, long before scientists understood atoms and electron energy levels, the German scientist Gustav Kirchhoff formulated three rules—now known as **Kirchhoff's laws**—describing the three types of spectra.

KIRCHHOFF'S LAWS

Law I: The Continuous Spectrum

A solid, liquid, or dense gas excited to emit light will radiate at all wavelengths and thus produce a continuous spectrum.

Law II: The Emission Spectrum

A low-density gas excited to emit light will do so at specific wavelengths and thus produce an emission spectrum.

Law III: The Absorption Spectrum

If light comprising a continuous spectrum passes through a cool, low-density gas, the result will be an absorption spectrum.

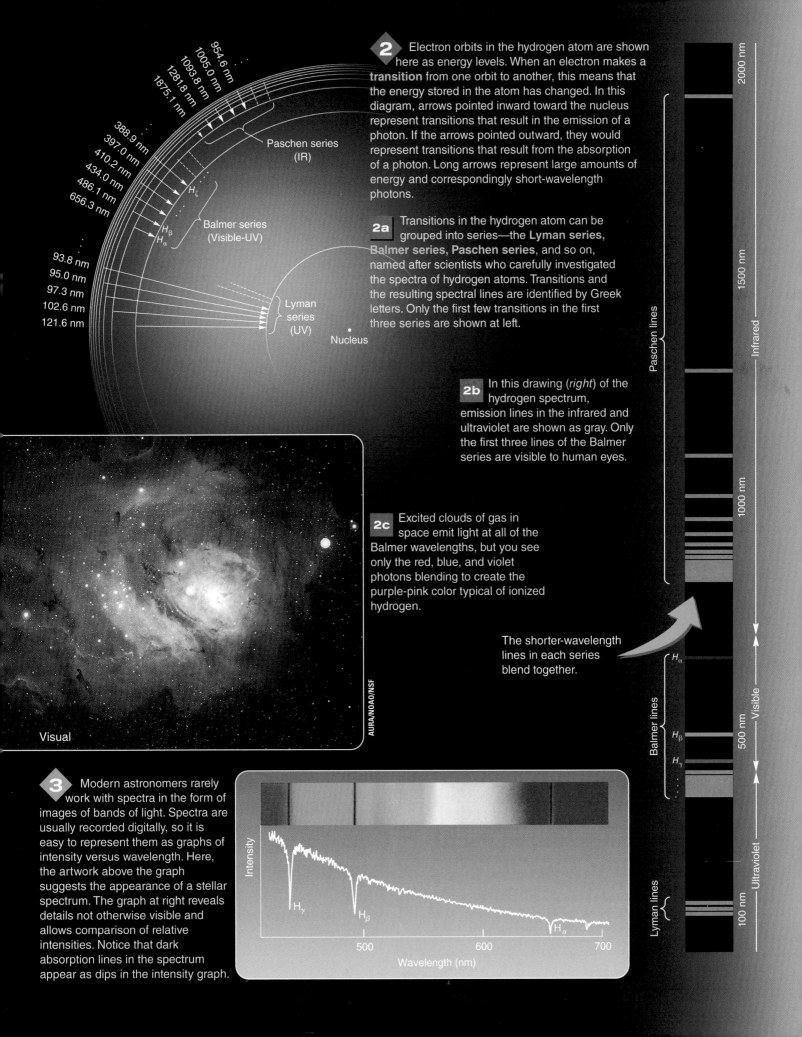

2 Electron orbits in the hydrogen atom are shown here as energy levels. When an electron makes a **transition** from one orbit to another, this means that the energy stored in the atom has changed. In this diagram, arrows pointed inward toward the nucleus represent transitions that result in the emission of a photon. If the arrows pointed outward, they would represent transitions that result from the absorption of a photon. Long arrows represent large amounts of energy and correspondingly short-wavelength photons.

2a Transitions in the hydrogen atom can be grouped into series—the **Lyman series**, **Balmer series**, **Paschen series**, and so on, named after scientists who carefully investigated the spectra of hydrogen atoms. Transitions and the resulting spectral lines are identified by Greek letters. Only the first few transitions in the first three series are shown at left.

2b In this drawing (*right*) of the hydrogen spectrum, emission lines in the infrared and ultraviolet are shown as gray. Only the first three lines of the Balmer series are visible to human eyes.

2c Excited clouds of gas in space emit light at all of the Balmer wavelengths, but you see only the red, blue, and violet photons blending to create the purple-pink color typical of ionized hydrogen.

The shorter-wavelength lines in each series blend together.

954.6 nm
1005.0 nm
1093.8 nm
1281.8 nm
1875.1 nm

Paschen series (IR)

388.9 nm
397.0 nm
410.2 nm
434.0 nm
486.1 nm
656.3 nm

H_ζ

Balmer series (Visible-UV)

H_β
H_α

93.8 nm
95.0 nm
97.3 nm
102.6 nm
121.6 nm

Lyman series (UV)

Nucleus

Visual

AURA/NOAO/NSF

3 Modern astronomers rarely work with spectra in the form of images of bands of light. Spectra are usually recorded digitally, so it is easy to represent them as graphs of intensity versus wavelength. Here, the artwork above the graph suggests the appearance of a stellar spectrum. The graph at right reveals details not otherwise visible and allows comparison of relative intensities. Notice that dark absorption lines in the spectrum appear as dips in the intensity graph.

Intensity

H_γ H_β H_α

500   600   700

Wavelength (nm)

2000 nm
1500 nm
Infrared
Paschen lines
1000 nm

H_α
Balmer lines
Visible
500 nm
H_β
H_γ

100 nm
Ultraviolet
Lyman lines

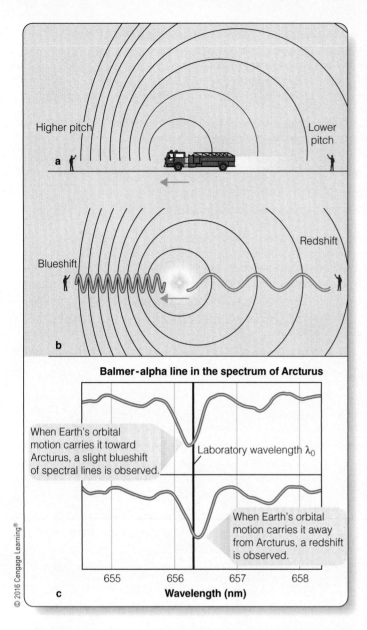

Balmer-alpha line in the spectrum of Arcturus

When Earth's orbital motion carries it toward Arcturus, a slight blueshift of spectral lines is observed.

Laboratory wavelength λ_0

When Earth's orbital motion carries it away from Arcturus, a redshift is observed.

655   656   657   658

Wavelength (nm)

c

© 2016 Cengage Learning®

▲ **Figure 7-8** The Doppler effect. (a) The sound waves (*black circles*) emitted from a siren on an approaching truck will be received more often, and thus be heard with a higher pitch, than the sound waves from a stationary truck. The siren will have a lower pitch if it is going away from the observer. (b) A moving source of light emits waves that move outward (*black circles*). An observer toward whom the light source is moving observes a shorter wavelength (*a blueshift*); an observer for whom the light source is moving away observes a longer wavelength (*a redshift*). (c) Absorption lines in the spectrum of the bright star Arcturus are blueshifted in winter, when Earth's orbital motion carries it toward the star, and redshifted in summer when Earth moves away from the star.

the sound waves will arrive farther apart—at a lower frequency, a lower pitch.

Now, substitute a source of light for the siren (Figure 7-8b). Imagine the light source emitting waves continuously as it approaches you. Each time the source emits the peak of a wave (meaning, the strongest electric and magnetic fields in the

wave), it will be slightly closer to you than when it emitted the peak of the previous wave. From your vantage point, the successive peaks of the light wave will arrive closer together in the same way that the successive peaks of the siren's sound wave seemed closer together. The light will appear to have a shorter wavelength. Because shorter wavelengths are toward the blue, this is called a **blueshift**. After the light source has passed you and is moving away, the peaks of successive waves arrive farther apart, so the light has a longer wavelength and is redder. This is a **redshift**.

The terms *redshift* and *blueshift* are used to refer to any range of wavelengths. The light does not actually have to be red or blue visible light; the terms apply just as well to wavelengths in other parts of the electromagnetic spectrum such as X-rays and radio waves. *Red* and *blue* refer to the direction of the shift, not to actual color.

The amount of change in wavelength, and thus the size of the Doppler shift, depends on the velocity of the source. A moving car has a smaller Doppler shift than a jet plane, and a slow-moving star has a smaller Doppler shift than one that is moving more quickly. You can measure the velocity of a star by measuring the size of its Doppler shift. If a star is moving toward Earth, it has a blueshift, which means that each of its spectral lines is shifted toward shorter wavelengths. If it is receding from Earth, it has a redshift. The shifts are normally much too small to change the overall color of a star noticeably, but they are easily detected in spectra. In the next section, you will learn how astronomers can convert Doppler shifts into velocities.

When you think about the Doppler effect in relation to celestial objects, it is important to understand two things. Earth itself moves, so measurement of a Doppler shift really measures the relative motion between Earth and the object. Figure 7-8c shows the Doppler effect in two spectra of the star Arcturus. Lines in the top spectrum are slightly blueshifted because the spectrum was recorded when Earth, in the course of its orbit, was moving toward Arcturus. Lines in the bottom spectrum are redshifted because it was recorded six months later, when Earth was moving away from Arcturus. To find the true motion of Arcturus through space, astronomers must first account for the motion of Earth.

The second point to remember is that the Doppler shift is sensitive only to the part of the velocity directed away from you or toward you—the **radial velocity** (V_r). You cannot use the Doppler effect to detect any part of the velocity that is perpendicular to your line of sight. That is why police using radar guns park right next to the highway (**Figure 7-9a**). They want to measure your full velocity as you drive toward them, not just part of your velocity. For the same reason, a star moving only across your field of view would have no blueshift or redshift because its distance from Earth would not be decreasing or increasing.

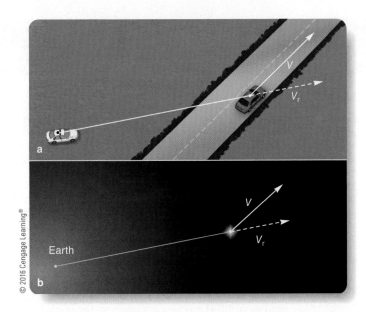

▲ Figure 7-9 (a) Police radar can measure only the radial part of your velocity (V_r) as you drive down the highway, not your true velocity along the pavement (V). That is why police using radar should never park far from the highway. This police car is actually poorly placed to make a good measurement. (b) From Earth, astronomers can use the Doppler effect to measure the radial velocity (V_r) of a star, but that is less than its true total velocity, V, through space.

Calculating Doppler Velocities

It is easy to calculate the radial velocity of an object from its Doppler shift. The formula is a simple proportion relating the radial velocity V_r divided by the speed of light c, to the change in wavelength $\Delta\lambda$ of a line divided by the "laboratory" wavelength of the line, λ_0. (This simple version of the formula is used if the velocity is much less than the speed of light.) The laboratory wavelength λ_0 of a spectral line (subscript 0) is the wavelength it will have if the source of the light is not moving relative to the spectrograph. In the actual spectrum of a star, this spectral line's wavelength is observed shifted by some small amount, $\Delta\lambda$ (pronounced *delta-lambda;* delta conventionally symbolizes a small change). If the wavelength is increased (a redshift), $\Delta\lambda$ is positive; if the wavelength is decreased (a blueshift), $\Delta\lambda$ is negative. The radial velocity, V_r, of the star is given by the Doppler formula:

$$\frac{V_r}{c} = \frac{\Delta\lambda}{\lambda_0}$$

For example, suppose the laboratory wavelength λ_0 of a certain spectral line is 600.00 nm, but the line is observed in a star's spectrum at a wavelength $\lambda = 600.10$ nm. What is the star's radial velocity? First note that the change in wavelength $\Delta\lambda$ is +0.10 nm so that:

$$\frac{V_r}{c} = \frac{0.1}{600} = 0.000167$$

The radial velocity is 0.10/600 multiplied by the speed of light. In astronomy, velocities are almost always given in kilometers per second, so the speed of light c is expressed in those units, as 3.00×10^5 km/s. Therefore, the radial velocity of this star is $(0.000167) \times 3.00 \times 10^5$ km/s), which equals 50 km/s. Because $\Delta\lambda$ is positive, you know the star is receding from you.

Armed with your new understanding of light and spectra, you are ready to focus on your first astrophysical object, the star that supports life on Earth—the Sun—which is the subject of the next chapter.

What Are We? Stargazers

Do you suppose chickens ever look at the sky and wonder what the stars are? Probably not. Chickens are very good at the chicken business, but they are not known for big brains and deep thought. Humans, in contrast, have highly evolved, sophisticated brains and are extremely curious. In fact, curiosity may be the most reliable characteristic of intelligence, and curiosity about the stars could have been the start of our ongoing attempts to understand the world around us.

For astronomers up to the time of Copernicus and Kepler, the stars were just points of light. There seemed to be no way to learn anything about them. Galileo's telescope revealed surprising details about the planets, but, even viewed through a large telescope, the stars are just points of light. Even when later astronomers began to realize that the stars were other suns, the stars seemed forever beyond human knowledge.

As you have seen, the key to understanding the Universe is knowledge about how light interacts with matter. In the past 150 years or so, scientists have discovered how atoms and light interact to produce the spectra we observe, and astronomers have applied those discoveries to the ultimate object of human curiosity—the stars.

Chickens may never wonder what the stars are, or even wonder what chickens are, but humans are curious animals, and we do wonder about the stars and about ourselves. Our yearning to understand the stars is just part of our quest to understand what we are.

Study and Review

Summary

- Modern astronomy is more properly called **astrophysics (p. 130)**, a field of study that interprets astronomical observations in terms of physics theory and laboratory experiments to understand the compositions, internal processes, and histories of celestial objects.

- An atom consists of a **nucleus (p. 131)** surrounded by a cloud of **electrons (p. 131)**. The nucleus is made up of one or more positively charged **protons (p. 131)** and, except for hydrogen, uncharged **neutrons (p. 131)**.

- The number of protons in an atom determines which element it is. Atoms of the same element (that is, having the same number of protons) with different numbers of neutrons are called **isotopes (p. 132)**.

- A neutral atom is surrounded by a number of negatively charged electrons equal to the number of protons in the nucleus. An atom that has lost or gained an electron is said to be **ionized (p. 132)** and is called an **ion (p. 132)**.

- Two or more atoms joined together form a **molecule (p. 132)**.

- Electrons in an atom are attracted to the nucleus by the **Coulomb force (p. 132)**. As described by **quantum mechanics (p. 132)**, the **binding energy (p. 132)** that holds electrons in an atom is limited to certain energies, and thus electrons may occupy only certain **permitted orbits (p. 132)**.

- The size of an electron's orbit depends on its energy, so the orbits can be thought of as **energy levels (p. 134)**, with the lowest possible energy level known as the **ground state (p. 134)**.

- An **excited atom (p. 134)** is one in which an electron is raised to a higher orbit by a collision between atoms or the absorption of a photon having the proper energy.

- **Temperature (p. 135)** refers to the average intensity of the agitation among the atoms and molecules of an object that can be expressed on the **Kelvin temperature scale (p. 135)**, which gives temperature above **absolute zero (p. 135)**. The sum of the agitation of the particles in an object is called **thermal energy (p. 135)**, and the flow of thermal energy is **heat (p. 135)**.

- Collisions among the particles in a hot, dense object accelerate electrons and cause the emission of **blackbody radiation (p. 135)**. The hotter an object, the more total energy the blackbody radiates (this principle is known as the **Stefan-Boltzmann law, (p. 136)** and the shorter is the blackbody's **wavelength of maximum intensity, λ_{max} (p. 137)**. This principle is known as **Wien's law (p. 137)**, and it allows astronomers to estimate the surface temperatures of stars from their colors.

- **Kirchhoff's laws (p. 140)** summarize how (1) a hot solid, liquid, or dense gas emits electromagnetic radiation at all wavelengths and produces a **continuous spectrum (p. 140)**; (2) an excited, low-density gas produces an **emission (bright-line) spectrum (p. 140)** containing **emission lines (p. 140)**; and (3) a light source viewed through a low-density, cool gas produces an **absorption (dark-line) spectrum (p. 140)** containing **absorption lines (p. 140)**.

- An atom can emit or absorb a photon when an electron makes a **transition (p. 141)** between orbits.

- Because orbits of only certain energy differences are permitted in an atom, photons of only certain wavelengths can be absorbed or emitted. Each kind of atom has its own characteristic set of spectral lines. The hydrogen atom has the **Lyman series (p. 141)** of lines in the ultraviolet, the **Balmer series (p. 141)** partially in the visible, and the **Paschen series (p. 141)** (plus others) in the infrared.

- An emission or absorption spectrum can tell you the chemical composition of the stars. The presence of spectral lines of a certain element is evidence that element is present in the star, but you need to proceed with care. The strengths of the spectral lines of the chemical elements are not directly related to the elements' respective abundances. Lines of a certain element may be weak or absent in the observed spectra if the star is too hot or too cool, even if that element is present in the star's atmosphere.

- The **Doppler effect (p. 139)** can provide clues to the motions of the stars. When a star is approaching, you observe a chemical element with slightly shorter wavelengths than the pattern produced by that same chemical element in a lab, a **blueshift (p. 142)**. When a star is receding, you observe a chemical element with slightly longer wavelengths than the pattern produced by that same chemical element in the lab, a **redshift (p. 142)**. This Doppler effect reveals a star's **radial velocity, V_r (p. 142)**, the part of its velocity directed toward or away from Earth.

Review Questions

1. Why might you say that an atom is mostly composed of empty space?

2. How many protons, neutrons, and electrons are in a neutral hydrogen atom? In a neutral helium atom? How many times heavier is the He atom compared to the H atom?

3. How is an isotope different from an ion?

4. Deuterium has a proton and a neutron in the nucleus surrounded by an electron. Is deuterium an element, atom, ion, isotope, and/or molecule? Is it neutral or ionized? How do you know?

5. He-3 (helium-3) contains two protons and one neutron in the nucleus. If neutral, how many electrons orbit a He-3 atom? Is He-3 an element, atom, ion, isotope, and/or molecule? How do you know?

6. Name a molecule. What atoms make up that molecule?

7. Why is the binding energy of an electron related to the size of the electron's orbit?

8. Explain why ionized calcium can form absorption lines, but ionized hydrogen cannot.

9. Describe two ways an atom can become excited.

10. An electron in a boron atom makes a transition from the fifth excited state to the third excited state. Did the atom become ionized as a result? Did the atom become excited as a result? Was a blackbody spectrum produced? How do you know?

11. Why do different atoms have different lines in their spectra?

12. Why does the amount of blackbody radiation emitted depend on the temperature of the object?

13. What is the wavelength of maximum intensity and the total energy emitted by a celestial object at absolute zero?

14. Why do hot stars look bluer than cool stars?

15. Why does a fireplace poker appear black when not in the fire and red, yellow, or white when left in the fire?

16. Celestial object A has a temperature of 60 K, and celestial object B has a temperature of 600 K. Which object emits the shorter wavelength of maximum intensity? Which objects has the least total energy emitted?

17. How is heat different from temperature?

18. What kind of spectrum does a neon sign produce? What colors are associated with a neon sign?

19. How can the Doppler effect explain wavelength shifts in both light and sound?

20. The emission spectra you obtained from a star shows a hydrogen spectrum that is shifted toward the blue end of the electromagnetic spectrum compared to the hydrogen spectrum you obtained from the lab. Is the star moving toward Earth, away from Earth, or is not enough information provided to determine its motion?

21. Which kind of spectrum is produced by a white household incandescent lightbulb?

22. Why does the Doppler effect detect only radial velocity?

23. Could an object be orbiting another object and we only detect the radial motion via the Doppler effect?

24. If the Doppler effect of light is a shift of spectral lines toward the blue or red end of the electromagnetic spectrum, what is the Doppler effect of sound?

25. **How Do We Know?** How is the macroscopic world you see around you determined by a microscopic world you cannot see?

Discussion Questions

1. In what ways is the model of an atom a scientific model? In what ways is it incorrect?

2. A perfect blackbody is a dense, hot body that absorbs and subsequently emits all incident electromagnetic radiation. An imperfect blackbody is still hot and dense but does not absorb all, or reemit all, incident electromagnetic radiation. Are you a perfect blackbody, an imperfect blackbody, or neither?

3. If all the lights are turned off in a room and there is no ambient light in the room, do you or your neighbor standing nearby emit any light? If so, can you see that light with your eyes? If not, why not?

4. Before Fraunhofer and others worked to observe and interpret spectra, most people were of the opinion that we would never know the composition of the Sun and stars. Can you think of any scientific question today that most people believe will probably never be answered?

5. List the "from" and "to" orbital numbers needed to generate the nebula in part 2c of **Atomic Spectra**.

Problems

1. Human body temperature is about 310 K (3.10×10^2 K). At what wavelength do humans radiate the most energy? In which part of the electromagnetic spectrum (gamma-ray, X-ray, UV, visible light, IR, microwave, or radio) do we emit?

2. A celestial body has a temperature of 50 K. What is the wavelength of maximum intensity? In which part of the electromagnetic spectrum (gamma-ray, X-ray, UV, visible light, IR, microwave, or radio) does this peak wavelength lie? Give an example of an object that might have this temperature. Another celestial body has a temperature of 500 K. What is the wavelength of maximum intensity? In which part of the electromagnetic spectrum is this? Give an example of an object that might have this temperature. A third celestial body has a temperature of 5000 K. What is the wavelength of maximum intensity? In which part of the electromagnetic spectrum is this? Give an example of an object that might have this temperature. A fourth celestial body has a temperature of 50,000 K. What is the wavelength of maximum intensity? In which part of the electromagnetic spectrum is this? Give an example of an object that might have this temperature. Draw a conclusion about temperature and wavelength of maximum intensity trends.

3. If a star has a surface temperature of 20,000 K (2.00×10^4 K), at what wavelength will it radiate the most energy? Is this a cool or hot star?

4. Infrared observations of a star show that the star is most intense at a wavelength of 2000 nm (2.00×10^3 nm). What is the temperature of the star's surface?

5. If you double the temperature of a blackbody, by what factor will the total energy radiated per second per square meter increase?

6. If one star has a temperature of 6000 K and another star has a temperature of 7000 K, how much more energy per second will the hotter star radiate from each square meter of its surface?

7. What is the wavelength of maximum intensity and the total energy emitted by a celestial object at 2 K above absolute zero? Which part of the EM spectrum does the wavelength of maximum intensity lie?

8. Electron orbital transition A produces light with a wavelength of 500 nm. Transition B involves twice the energy of transition A. What wavelength is the light it produces?

9. Photon energy is related to wavelength by the Planck equation, $E = hc/\lambda$, where h is Planck's constant, 6.63×10^{-34} J/s, and c is the speed of light, 3.00×10^8 m/s. If the energy released by an electron making a transition from one hydrogen atom orbit to another is 1.64×10^{-18} J, what is the wavelength of the photon? Which part of the electromagnetic spectrum is that in? In which orbit did the electron start, and in which orbit did it finish? (*Hint:* examine **Atomic Spectra.**)

10. In a laboratory, the Balmer-beta spectral line of hydrogen has a wavelength of 486.1 nm. If the line appears in a star's spectrum at 486.3 nm, what is the star's radial velocity? Is it approaching or receding? Is this a blueshift or a redshift?

11. An astronomer observes the Balmer-beta line in a celestial object's spectrum at a wavelength 996.5 nm. Is the object approaching or receding? If you can find the object's radial velocity, what is it? (*Note:* The laboratory wavelength of Balmer-beta is given in Problem 10.)

12. The highest-velocity stars an astronomer might observe in the Milky Way Galaxy have radial velocities of about 400 km/s (4.00×10^2 km/s). What change in wavelength would this cause in the Balmer-beta line? (*Note:* The laboratory wavelength of Balmer-beta is given in Problem 10.)

Learning to Look

1. Consider Figure 7-3. Does an electron in the fifth excited state of a boron atom have more or less binding energy than the third excited state of helium? What about the electron in the first excited state of hydrogen compared to the second excited state of helium? Why?

2. Consider Figure 7-3. When an electron in a hydrogen atom moves from the third orbit to the second orbit, the atom emits a Balmer-alpha photon in the red part of the spectrum. In what part of the spectrum would you look to find the photon emitted when an electron in a helium atom makes the same transition?

3. Did ionized hydrogen produce the spectrum shown in part 3 of **Atomic Spectra?**

4. What colors are the 589.0 nm and 589.6 nm sodium lines in the Sun's absorption spectrum, Figure 7-7a? Do these sodium lines shown in Figure 7-7a originate in the Sun's photosphere?

5. Where should the police car in Figure 7-9a have parked to make a good measurement?

6. The nebula shown at right contains mostly hydrogen excited to emit photons. What kind of spectrum would you expect this nebula to produce?

7. If the nebula shown below crosses in front of the star, and the nebula and star have different radial velocities, what might the spectrum of the star look like?

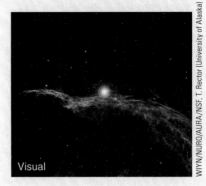

Visual

WIYN/NURO/AURA/NSF, T. Rector (University of Alaska)

The Sun 8

Guidepost The Sun is the source of light and warmth in our Solar System, so it has always been a primary object of human curiosity and awe. It is also the star that is most easily visible from Earth. Understanding the interactions of light and matter that you studied in Chapter 7 can help reveal the secrets of the Sun and introduce you to the stars.

In this chapter, you will discover how analysis of the solar spectrum paints a detailed picture of the Sun's atmosphere and how basic physics has solved the mystery of what goes on in the Sun's core. Here you will find answers to four important questions:

► **What can be learned about the Sun by observing its surface and atmosphere?**

► **What are the dark sunspots?**

► **Why does the Sun go through 11- and 22-year cycles of activity?**

► **What is the source of the Sun's energy?**

Although this chapter considers only the star at the center of our Solar System, introducing you first to one star in detail lets you continue onward and outward in later chapters among the other stars that fill the Universe.

All cannot live on the piazza,
but everyone may enjoy the sun.

ITALIAN PROVERB

Venus transiting the setting Sun's disk in 2004, photographed from Flagler Beach Pier in Florida.

SCIENTISTS JOKE THAT we would know a lot more about the Sun if it were farther away. The Sun is so close that Earth's astronomers can see swirling currents of gas and arched bridges of magnetic force with a level of detail that seems overwhelming. But the Sun is just a normal star. In a sense, it is a simple object. The Sun is made up almost entirely of hydrogen and helium gas confined by its own gravity in a sphere 109 times Earth's diameter (■ Celestial Profile 1). The gases of the Sun's photosphere (surface) and atmosphere are hot, radiating the light and heat that are visible and make life possible on Earth. That part of the Sun is where you can begin your exploration.

8-1 The Solar Photosphere and Atmosphere

The part of the Sun you can see directly from Earth is made up of three layers. The visible surface is the **photosphere**, and above that are the **chromosphere** and the **corona**. (Note that astronomers normally speak of the interior of the Sun as being "below" or "under" the photosphere, and the Sun's atmosphere as being "above" or "over" the photosphere.) You first learned about these components of the Sun in the context of observing them during solar eclipses (look back to Chapter 3).

When you view the Sun, you see the photosphere as a hot, glowing surface with a temperature of about 5800 K. That temperature is determined by precisely measuring the spectrum of sunlight, then making a calculation using Wien's law (look back to Chapter 7). At that temperature, every square millimeter of the Sun's surface is radiating more energy than a 60-watt lightbulb (Stefan-Boltzmann law, Chapter 7). With all that energy radiating into space, the Sun's surface would cool rapidly if new energy did not arrive to keep the surface hot, so simple logic tells you that there must be heat flowing outward from the Sun's interior. Not until the 1930s did astronomers understand that the Sun creates energy by nuclear reactions at its center. Those nuclear reactions are described in detail at the end of this chapter.

For now, you can consider the Sun's atmosphere in its relatively quiet, average state. Later you can add details regarding the types of activity produced by heat flow that makes the Sun's outer layers churn like a pot of boiling water.

The Photosphere

The visible surface of the Sun seems to be a distinct surface, but it is not solid. In fact, the Sun is gaseous from its outer atmosphere right down to its center. The photosphere is a thin layer of gas from which Earth receives most of the Sun's light. It is less than 500 km (300 mi) deep. In a model of the Sun the size of a

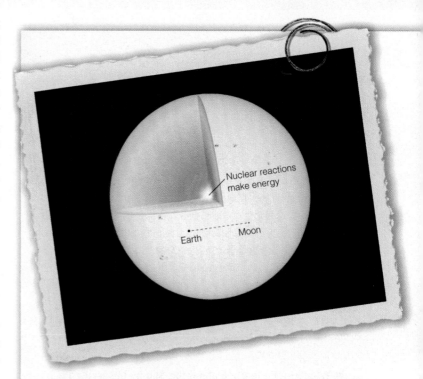

This visible image of the Sun shows a few sunspots and is cut away to show the location of energy generation at the Sun's center. The Earth, Moon, and the distance between them are shown for scale.

Celestial Profile 1 The Sun

From Earth:

Average distance from Earth	1.000 AU (1.496 × 10⁸ km)
Maximum distance from Earth	1.017 AU (1.521 × 10⁸ km)
Minimum distance from Earth	0.983 AU (1.471 × 10⁸ km)
Average apparent diameter	0.533° (1920 arc seconds)
Period of rotation (sidereal)	24.5 days at equator
Apparent visual magnitude	−26.74

Physical Characteristics:

Radius	6.96 × 10⁵ km
Mass	1.99 × 10³⁰ kg
Average density	1.41 g/cm³
Escape velocity at surface	618 km/s
Luminosity (electromagnetic)	3.84 × 10²⁶ J/s
Surface temperature	5780 K
Central temperature	15.7 × 10⁶ K

Personality Profile:

In Greek mythology, the Sun was carried across the sky in a golden chariot pulled by powerful horses and guided by the Sun god, Helios. When Phaeton, the son of Helios, drove the chariot one day, he lost control of the horses, and Earth was nearly set ablaze before Zeus smote Phaeton from the sky. Even in classical times, people understood that life on Earth depends critically on the Sun.

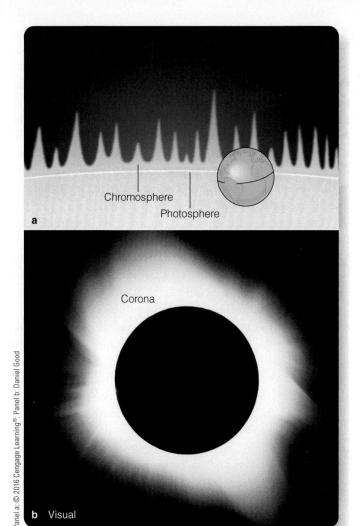

a

Chromosphere

Photosphere

b Visual

Corona

▲ **Figure 8-1** (a) A cross section at the edge of the Sun shows the relative thickness of the photosphere and chromosphere. Earth is shown for scale. On this scale, the disk of the Sun would be more than 1.5 m (5 ft) in diameter. The corona extends from the top of the chromosphere to a great distance above the photosphere. (b) This photograph, made during a total solar eclipse, shows only the inner part of the corona.

bowling ball, the photosphere would be no thicker than a layer of tissue paper wrapped around the ball (**Figure 8-1**).

The photosphere is the layer in the Sun's atmosphere that is dense enough to emit plenty of light but not so dense that light can't escape. Below the photosphere, the gas is denser and hotter and therefore radiates plenty of light, but that light cannot escape because it is blocked by the outer layers of gas. Therefore, Earth does not receive light from deeper layers under the photosphere. In contrast, the gas above the photosphere is less dense, so although Earth does receive that light, there is not much of it.

The photosphere appears to be substantial, but it is really a very low-density gas. Even in its deepest and densest layers, the photosphere is less than 1/3000 as dense as the air you breathe.

To find gases as dense as the air at Earth's surface, you would have to descend about 15,000 km (10,000 mi) below the photosphere. With fantastically efficient insulation, you could fly a spaceship right through the photosphere.

The spectrum of the Sun is an absorption spectrum, and that can tell you a great deal about the photosphere. You know from Kirchhoff's third law (Chapter 7) that an absorption spectrum is produced when the source of a continuous (blackbody) spectrum is viewed through a transparent gas. The deeper layers of the photosphere are dense enough to produce a continuous spectrum. Atoms in higher, transparent layers of the photosphere and in the Sun's atmosphere absorb photons with unique energies (wavelengths) corresponding to the jumps between each type of atom's electron orbits, producing absorption lines that allow you to identify hydrogen, helium, and other elements.

In high-resolution photographs, the photosphere has a mottled appearance because it is made up of dark-edged regions called *granules*. The overall pattern is called **granulation** (**Figure 8-2a**). Granules can be several thousand kilometers across but last for only 10 to 20 minutes each before fading, shrinking, and being replaced by new granules. Detailed observations of the granules show that their centers emit more blackbody radiation and are slightly bluer than the edges. From this information, plus the Wien and Stefan-Boltzmann laws, astronomers can calculate that the granule centers are a few hundred degrees hotter than the edges (Figure 8-2b). Doppler shifts of spectral lines (Chapter 7) reveal that the granule centers are rising and the edges are sinking at speeds of about 0.4 km/s (900 mph).

From this evidence, astronomers recognize granulation as the surface effects of **convection** currents just below the photosphere. Convection occurs when hot material rises and cool material sinks, as when, for example, a current of hot gas rises above a candle flame. You can watch convection in a liquid by adding a bit of cool creamer to an unstirred cup of hot coffee. The cool creamer sinks, gets warmer, expands, rises, cools, contracts, sinks again, and so on, creating small regions on the surface of the coffee that mark the tops of convection currents. Viewed from above, these coffee regions look something like solar granules. The presence of granulation is clear evidence that energy is flowing upward through the photosphere. You will learn more about the Sun's convection currents and internal structure later in this chapter.

Spectroscopic studies of the solar surface have revealed another larger but less obvious kind of granulation. **Supergranules** are regions a little over twice Earth's diameter that include an average of about 300 granules each. These supergranules are regions of very slowly rising currents that last a day or two. They appear to be produced by larger gas currents that begin deeper under the photosphere than the ones that produce the granules.

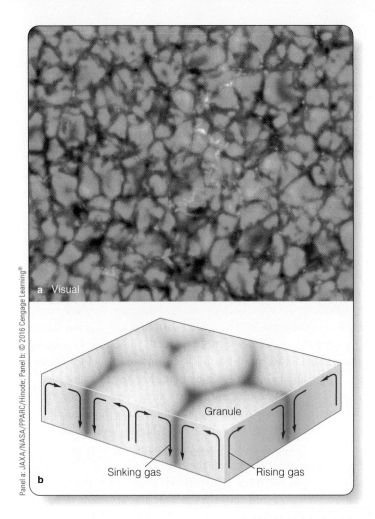

Panel a. JAXA/NASA/PPARC/Hinode; Panel b: © 2016 Cengage Learning®

a Visual

Granule

Sinking gas   Rising gas

b

▲ **Figure 8-2** (a) This ultra-high-resolution image of the photosphere shows granulation. The largest granules here are about the size of Texas. (b) This model explains granulation as the tops of rising convection currents just below the photosphere. Heat flows upward as rising currents of hot gas and downward as sinking currents of cool gas. The rising currents heat the solar surface in small regions seen from Earth as granules.

The edge, or **limb**, of the solar disk is dimmer than the center (see the figure in **Celestial Profile 1**, page 148). This **limb darkening** is caused by the absorption of light in the photosphere. When you look at the center of the solar disk, you are looking directly down into the Sun, and you see deeper, hotter, brighter layers in the photosphere. In contrast, when you look near the limb of the solar disk, you are looking at a steep angle to the surface and cannot see as deeply. The photons you see come from shallower, cooler, dimmer layers in the photosphere. Limb darkening proves that the temperature in the photosphere increases with depth, another confirmation that energy is flowing up from below.

The Chromosphere

Above the photosphere lies the chromosphere. Solar astronomers define the lower edge of the chromosphere as lying just above the visible surface of the Sun, with its upper regions blending gradually with the corona. The chromosphere is a layer with an irregular thickness on average less than Earth's diameter (Figure 8-1a). Because the chromosphere is roughly 1000 times fainter than the photosphere, you can see it with your unaided eyes only during a total solar eclipse when the Moon covers the brilliant photosphere. Then, the chromosphere flashes into view as a thin pink layer just above the photosphere. The word *chromosphere* comes from the Greek word *chroma,* meaning "color." The pink color is produced by the combined light of three bright emission lines—the red, blue, and violet Balmer lines of hydrogen (Chapter 7).

The chromosphere produces an emission spectrum, and Kirchhoff's second law tells you it therefore must be an excited, low-density gas. The chromosphere's density ranges from 10,000 times less dense than the air you breathe at the bottom of the chromosphere (near the photosphere) to 100 billion times less dense at the top (near the corona). Further analysis of solar spectra reveals that atoms in the lower chromosphere are ionized, and atoms in the higher layers of the chromosphere are even more highly ionized, having lost most or all of their electrons. Astronomers can find the temperature in different parts of the chromosphere from the amount of ionization. Just above the photosphere, the temperature falls to a minimum of about 4500 K and then rises rapidly (**Figure 8-3**) to the extremely high temperatures of the corona. The upper chromosphere is hot enough to emit X-rays and can be studied by X-ray telescopes in space (look back to Chapter 6).

Solar astronomers can take advantage of the way spectral lines form to map the chromosphere. The gases of the chromosphere are transparent to nearly all wavelengths of visible light, but atoms in that gas are very good at absorbing photons of a few specific wavelengths. This produces some exceptionally strong dark absorption lines in the solar spectrum. A photon at one of those wavelengths is unlikely to escape from deeper layers to be received at Earth and can come to you only from higher in the Sun's atmosphere. A **filtergram** is an image of the Sun made using light only at the wavelength of one of those strong absorption lines such as the hydrogen Balmer series (Chapter 7, page 141) to reveal detail in the upper regions of the chromosphere. Another way to study these layers of gas high in the Sun's atmosphere is to record solar images in the far-ultraviolet or in the X-ray part of the spectrum because those layers are very hot and emit most of their light at short wavelengths.

Figure 8-4 shows a filtergram made at the wavelength of the Balmer H-alpha line. This image shows complex structure in the chromosphere. **Spicules** are flamelike jets of gas extending upward into the chromosphere and lasting about 5 to 15 minutes. Seen at the limb of the Sun's disk, these spicules blend together and look like flames covering a burning prairie (Figure 8-4c), but they are more like the opposite of flames; spectra show that spicules are cooler gas from the lower chromosphere extending upward into hotter regions. Images at the center of the solar disk show that

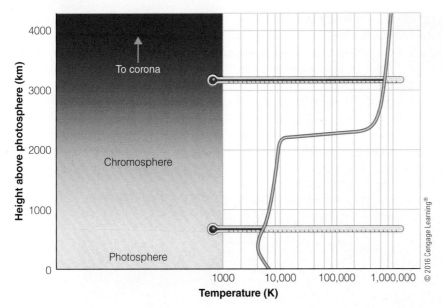

© 2016 Cengage Learning®

◄ **Figure 8-3** A plot showing the chromosphere's temperature profile. If you could place thermometers in the Sun's atmosphere, you would discover that the temperature increases from about 5800 K at the photosphere to 1 million K at the top of the chromosphere.

spicules spring up around the edge of supergranules like weeds around paving stones (Figure 8-4b).

The Corona

The outermost part of the Sun's atmosphere is called the corona, after the Greek word for crown. The corona is so dim that, like the chromosphere, it is not visible in Earth's daytime sky because of the glare of scattered light from the Sun's brilliant photosphere. During a total solar eclipse, the innermost parts of the corona are visible to the unaided eye, as shown in Figure 8-1b (also, Chapter 3). Observations made with specialized telescopes called **coronagraphs** can block the light of the photosphere and record the corona out beyond 20 solar radii, almost 10 percent of the way to Earth. Such images reveal streamers in

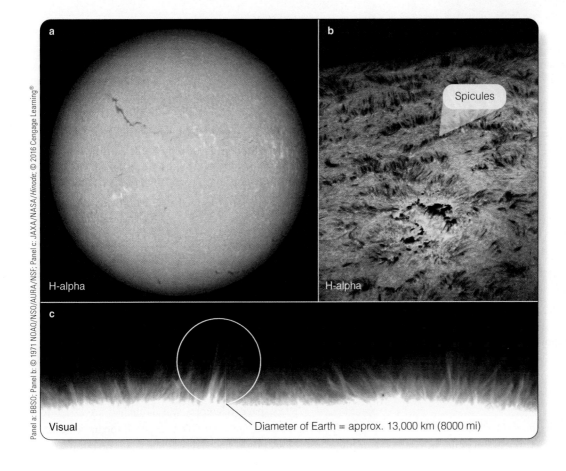

Panel a: BBSO; Panel b: © 1971 NOAO/NSO/AURA/NSF; Panel c: JAXA/NASA/Hinode; © 2016 Cengage Learning®

◄ **Figure 8-4** (a) H-alpha filtergram of the Sun's disk. (b) H-alpha filtergrams reveal complex structures in the chromosphere that cannot be seen in ordinary visual-wavelength images, including spicules springing from the edges of supergranules. (c) Seen at the edge of the solar disk, spicules look like a burning prairie, but they are not at all related to burning. The white circle shows the size of Earth to scale. Compare with Figure 8-1a.

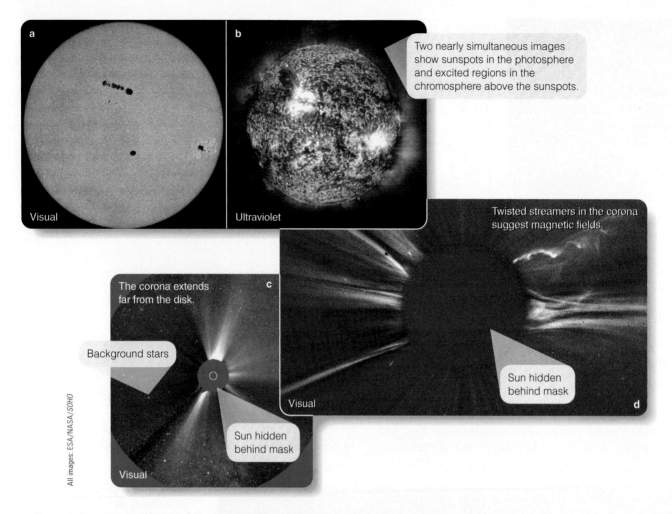

Two nearly simultaneous images show sunspots in the photosphere and excited regions in the chromosphere above the sunspots.

a

Visual

b

Ultraviolet

Twisted streamers in the corona suggest magnetic fields.

The corona extends far from the disk.

c

Background stars

Sun hidden behind mask

Visual

Sun hidden behind mask

Visual

d

All images: ESA/NASA/SOHO

▲ Figure 8-5 Images of the photosphere, chromosphere, and corona show relationships among the layers of the Sun's atmosphere. The visual-wavelength image in panel (a) was taken through a dense filter that produced the orange tint.

the corona that follow magnetic lines of force in the Sun's magnetic field (Figure 8-5). Later in this chapter you will learn more about how features and activity in the Sun's atmosphere are controlled by magnetic fields.

The corona's spectrum, like that of the upper chromosphere, includes emission lines of highly ionized gases. In the lower corona, the atoms are not as highly ionized as they are at higher altitudes, and this tells you that the temperature of the corona rises with altitude. Just above the chromosphere the temperature is about 500,000 K, and in the outer corona the temperature can be 2 million K or more. The corona is hot enough to emit X-rays, but the coronal gas is not very luminous because its density is very low, with only 10^6 atoms/cm^3 in its lower regions. That is about 1000 trillion times less dense than the air you breathe. In its outer regions the corona contains only 1 to 10 atoms/cm^3, which is fewer than in the best vacuum in laboratories on Earth.

Astronomers continue to wonder how the corona and chromosphere can be so hot. Heat flows from hot regions to cool regions, never from cool to hot. So, how can the heat from the photosphere, with a temperature of only 5800 K, flow out into the much hotter chromosphere and corona? Observations made by the *Solar and Heliospheric Observatory (SOHO)* satellite have mapped a **magnetic carpet** of looped magnetic fields extending up through the photosphere (Figure 8-6). Because the gas in the chromosphere and the corona is ionized and has very low densities, it can't resist being accelerated by movements of the magnetic fields. Turbulence below the photosphere seems to flick the magnetic loops back and forth and whip the gas about, heating it. Furthermore, observations by the *Hinode* spacecraft reveal magnetic waves generated by turbulence below the photosphere traveling up into the chromosphere and corona and heating the gas. In both cases, energy appears to flow out from the interior of the Sun to the chromosphere and corona not by radiation but by agitation of magnetic fields. The *Interface Region Imaging Spectrograph (IRIS)* space telescope was launched in 2013 to make rapid, high-resolution ultraviolet images of the upper chromosphere and lower corona. Those data will help advance

Extreme-UV image plus artist's conception

▲ Figure 8-6 Flying through the magnetic carpet. This computer model shows an extreme-ultraviolet image of a section of the Sun's lower corona (*green*) with black and white areas marking regions of opposite magnetic polarity. The model includes lines to show how the areas are linked by loops of magnetic force. The largest loops could encircle Earth.

investigations of the Sun's outer atmosphere and its puzzlingly high temperatures.

Ionized, low-density gas cannot cross magnetic fields, so in places where the Sun's field loops back toward the surface, the corona's gas is trapped in the vicinity of the Sun. However, some of the magnetic field lines are "open" and lead outward into space. At those locations the gas flows away from the Sun in the **solar wind** that can be considered an extension of the corona. The low-density gases of the solar wind blow past Earth at 300 to 800 km/s with gusts as high as 1000 km/s (more than 2 million mph).

Earth is bathed in the corona's hot breeze, but that breeze blows all the way to the outskirts of the Solar System. The *Voyager* spacecraft, launched in the 1970s to explore Jupiter and other outer planets, are now traveling through and investigating the region known as the **heliopause** where the solar wind collides with material in interstellar space. The *Interstellar Boundary Explorer (IBEX)* spacecraft, on the other hand, is able to analyze particles emitted from the heliopause without leaving Earth orbit.

Because of the solar wind, the Sun loses about 1 million tons per second, but that is only 10^{-14} of its total mass per year. Later in life, the Sun, like many other stars, will lose mass rapidly in a more powerful wind. You will see in future chapters how rapid outflowing winds affect the evolution of stars.

Do other stars have chromospheres, coronae (plural of corona), and stellar winds like the Sun? Stars are so far away they appear only as points of light even in the largest telescopes, but ultraviolet and X-ray observations suggest that the answer is yes, other stars have atmospheric features analogous to the Sun's. The spectra of many stars contain emission lines at far-ultraviolet wavelengths that could have formed only in the low-density, high-temperature gases of a chromosphere and corona. Also, many stars are sources of X-rays that seem to be produced by high-temperature gas in their chromospheres and coronae. This observational evidence gives astronomers good reason to consider the Sun to be a typical star, despite all its complexity that can seen from our nearby viewpoint.

Composition of the Sun

It seems as though it should be easy to learn the composition of the Sun just by studying its spectrum, but this is actually a difficult problem that wasn't well understood until the 1920s. The solution to that problem is part of the story of an important American astronomer who waited decades to get proper credit for her work.

In her 1925 PhD thesis, Cecilia Payne invented the modern methods of interpreting spectra of the Sun and stars. For example, sodium lines are observed in the Sun's spectrum, so you can be sure that the Sun's atmosphere contains some sodium atoms. Payne came up with a mathematical procedure to determine just how many sodium atoms are there. She also proved that if spectral lines of a certain element are not detected in the Sun's spectrum, that element might still be present, but the gas is too hot or too cool, or the wrong density, for that type of atom to have electrons in the right energy levels to produce visible lines.

Payne's first calculations showed that more than 90 percent of the atoms in the Sun must be hydrogen and most of the rest are helium. In contrast, atoms like calcium, sodium, and iron that have strong lines in the Sun's spectrum are actually not very abundant. Rather, at the temperature of the Sun's atmosphere, those atoms are especially efficient at absorbing photons with wavelengths of visible light. At the time Payne did her original work, astronomers found it hard to believe her calculated abundances of hydrogen, helium, and other elements in the Sun. They especially found such a high abundance of helium unacceptable because helium lines are nearly invisible in the Sun's spectrum. Eminent astronomers dismissed Payne's results as obviously wrong; it was only several decades later that the scientific community realized the value of her work. Now we know that she was correct.

Abundances of elements in the Sun are presented in Table 8-1. Some of the abundances in the table, particularly of carbon, nitrogen, and oxygen (CNO), are subjects of ongoing controversy because of revised calculations of solar atmospheric conditions. Those calculations indicate that the CNO abundances might need to be lowered somewhat, but even so, the modern values are close to the ones determined in the 1920s by Payne.

Payne's work on the composition of the Sun illustrates the importance of fully understanding the interaction between light and matter in order to investigate objects in the Universe.

The layers of the solar atmosphere are all that astronomers can observe directly, but there are phenomena in those layers that reveal what it's like inside the Sun, your next destination.

TABLE 8-1 The Most Abundant Elements in the Sun

Element	Percentage by Number of Atoms	Percentage by Mass
Hydrogen	91.0	70.6
Helium	8.9	27.5
Carbon	0.03	0.3
Nitrogen	0.01	0.1
Oxygen	0.05	0.6
Neon	0.01	0.2
Magnesium	0.003	0.07
Silicon	0.003	0.07
Sulfur	0.002	0.04
Iron	0.003	0.1

© 2016 Cengage Learning®

Below the Photosphere

Almost no light emerges from below the photosphere, so you can't see into the solar interior. However, solar astronomers using a technique called **helioseismology** can analyze naturally occurring vibrations in the Sun to explore its depths. Convective movements of gas in the Sun constantly produce vibrations—rumbles that would be much too low to hear with human ears even if your ears could survive a visit to the Sun's atmosphere. Some of these vibrations resonate in the Sun like sound waves in organ pipes. A vibration with a period of 5 minutes is strongest, but other vibrations have periods ranging from 3 to 20 minutes. These are very, very low-pitched sounds!

Astronomers can detect these vibrations by observing Doppler shifts in the solar surface. As a sound wave travels down into the Sun's interior, the changing density and temperature of the gas it moves through curves its path, and it returns to the surface. At the surface, it makes the photosphere heave up and down by small amounts—roughly plus or minus 15 km (10 mi). Multiple vibrations occurring simultaneously cover the surface of the Sun with a pattern of rising and falling regions that can be mapped using the Doppler effect (Figure 8-7). For example, the *SOHO* space telescope can observe solar oscillations continuously and is able to detect motions as slow as 1 mm/s (0.002 mph). Short-wavelength waves penetrate less deeply and travel shorter distances than longer-wavelength waves, so the vibrations of different wavelengths explore different layers in the Sun. Just as geologists can study Earth's interior by analyzing seismic waves from earthquakes, so solar astronomers can use helioseismology to explore the Sun's interior.

You can better understand how helioseismology works if you think of a duck pond. If you stood at the shore of a duck

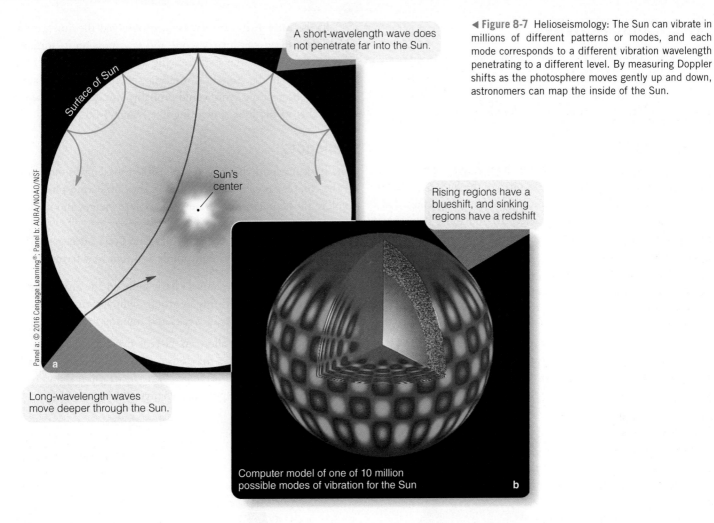

Panel a: © 2016 Cengage Learning®.; Panel b: AURA/NOAO/NSF

A short-wavelength wave does not penetrate far into the Sun.

Surface of Sun

Sun's center

Long-wavelength waves move deeper through the Sun.

a

◄ **Figure 8-7** Helioseismology: The Sun can vibrate in millions of different patterns or modes, and each mode corresponds to a different vibration wavelength penetrating to a different level. By measuring Doppler shifts as the photosphere moves gently up and down, astronomers can map the inside of the Sun.

Rising regions have a blueshift, and sinking regions have a redshift

Computer model of one of 10 million possible modes of vibration for the Sun

b

pond and looked down at the water, you would see ripples arriving from all parts of the pond. Because every duck on the pond contributes to the ripples, you could, in principle, study the ripples near the shore and draw a map showing the position and velocity of every duck on the pond. Of course, it would be difficult to untangle the different ripples. Nevertheless, all of the information would be there, lapping at the shore.

Helioseismology requires huge amounts of data, so astronomers have used a network of telescopes around the world operated by the Global Oscillation Network Group (GONG). The network can observe the Sun continuously for weeks at a time as Earth rotates; in other words, the Sun never sets on GONG. Solar astronomers can then use supercomputers to separate the different vibration patterns on the solar surface and determine the strength of the waves at many different wavelengths.

Helioseismology has allowed astronomers to map the temperature, density, and rate of rotation in the interior of the Sun, as well as find the positions and speeds of great currents of gas flowing below the photosphere. For example, the depth of the region of convection is now known to be exactly 29 percent of

the radius of the Sun. That detailed information confirms a model developed to understand the cycles of solar activity that you will learn about in the next section.

DOING SCIENCE

What evidence leads astronomers to conclude that temperature increases with height in the chromosphere and corona? In astronomy, as in any science, evidence is crucial, and gathering evidence means making observations and measurements.

Solar astronomers can observe the spectrum of the chromosphere, and they find that atoms there are more highly ionized (have lost more electrons) than atoms in the photosphere. Atoms in the corona are even more highly ionized. That must mean the chromosphere and corona are hotter than the photosphere.

A central part of doing science is gathering, evaluating, and understanding evidence. Now, continue investigating the Sun by comparing it with other stars. ***What evidence leads astronomers to conclude that some stars have chromospheres and coronae like those of the Sun?***

8-2 Solar Activity

The Sun is not quiet. It has storms larger than Earth that last for weeks, and unimaginably vast eruptions. All of these seemingly different forms of solar activity have one thing in common—magnetic fields. The weather on the Sun is magnetic.

Observing the Sun

Solar activity is often visible with even a small telescope, but you should be very careful if you try to observe the Sun. Sunlight is intense, and the infrared radiation in sunlight is especially dangerous because your eyes can't detect it. You don't sense how intense the infrared is, but it is converted to thermal energy in your eyes and can burn and scar your retinas.

It is not safe to look directly at the Sun, and it is even more dangerous to look at the Sun through any optical instrument such as a telescope, binoculars, or even the viewfinder of a camera. The light-gathering power of such an optical system concentrates the sunlight and can cause severe injury. Never look at the Sun with any optical instrument unless you are certain it is safe. Figure 8-8a shows a safe way to observe the Sun with a small telescope.

In the early 17th century, Galileo observed the Sun with a thick, dark filter over his telescope and saw spots on its surface. Day by day, he saw the spots moving across the Sun's disk and concluded that the Sun is a rotating sphere. If you repeated Galileo's observations, you would probably also detect **sunspots**, a view that would look something like Figure 8-8b.

Sunspots

The dark sunspots that you see at visible wavelengths only hint at the complex processes that go on in the Sun's atmosphere. To explore those processes, you need to analyze images and spectra at a wide range of wavelengths.

Study **Sunspots and the Solar Magnetic Cycle** on pages 158–159 and notice five important points and four new terms:

1 Sunspots are cool, relatively dark spots on the Sun's photosphere, usually appearing in groups, which form and disappear over time scales of weeks and months.

2 Sunspot numbers follow an 11-year cycle, becoming more numerous, reaching a maximum, and then becoming much less numerous. The *Maunder butterfly diagram* shows how the location of sunspots also changes during a cycle.

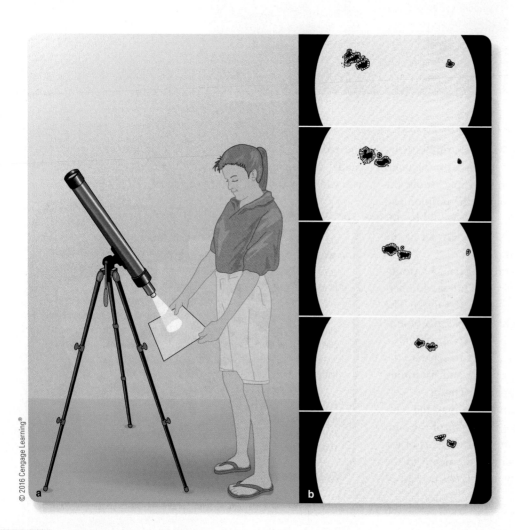

◀ **Figure 8-8** (a) Looking through a telescope at the Sun is dangerous, but you can always view the Sun safely with a small telescope by projecting its image on a white screen. (b) If you sketch the location and structure of sunspots on successive days, you will see the rotation of the Sun and gradual changes in the size and structure of sunspots, just as Galileo did in 1612.

3 The *Zeeman effect* gives astronomers a way to measure the strength of magnetic fields on the Sun and provides evidence that sunspots contain, and are caused by, strong local magnetic fields. When the magnetic properties of sunspots are considered, the 11-year cycle is understood to be really a 22-year cycle.

4 The intensity of the sunspot cycle can vary from cycle to cycle and appears to have almost faded away during the *Maunder minimum* in the late 17th century. Some scientists hypothesize that this solar activity minimum was somehow connected with a significant cooling of Earth's climate that lasted for several centuries.

5 The evidence is clear that sunspots are parts of *active regions* dominated by magnetic fields that involve all layers of the Sun's atmosphere.

Sunspot groups are merely the visible traces of magnetically active regions. But what causes this magnetic activity? The answer is linked to a growth and decay cycle of the Sun's overall magnetic field.

The Sun's Magnetic Cycle

You are familiar with magnetic fields from classroom demonstrations with magnets and iron filings and from seeing the effect of Earth's magnetic field on a compass needle. The Sun's magnetic field is powered by the energy flowing outward through the moving currents of gas. The gas is highly ionized, so it is a very good conductor of electricity. When that electrically conducting gas rotates or is stirred by convection, some of the energy in the gas motion can be converted into magnetic field energy. This process is called the **dynamo**

effect; it is understood to operate in Earth's core and produce Earth's magnetic field. Helioseismologists have found evidence that the dynamo effect generates the Sun's magnetic field at the bottom of the convection currents, deep below the photosphere.

Another important connection between solar gas motions and magnetic field lies in details of the Sun's rotation: The Sun does not rotate as a rigid body; this is possible because the Sun is entirely gas. For example, the equatorial region of the photosphere has a shorter rotation period than regions at higher latitudes (Figure 8-9a). At the equator, the photosphere rotates once every 24.5 days, but at latitude 45 degrees one rotation takes 27.8 days. This phenomenon is called **differential rotation**. Helioseismology maps of rotation in the Sun's interior (Figure 8-9b) reveal that the gas at different levels also rotates with different periods, another type of differential rotation. Both types of differential rotation, latitude-dependent and depth-dependent, seem to be involved in the Sun's magnetic cycle.

Although the magnetic cycle is not fully understood, the **Babcock model**, invented by astronomer Horace Babcock, explains the magnetic cycle as repeated tangling and untangling of the solar magnetic field. You have learned that an ionized gas is a very good conductor of electricity. This means that if the gas moves, embedded electrical currents and resulting magnetic fields must move with it. As a result, differential rotation drags the magnetic field along and wraps it around the Sun like a long string caught on a turning wheel. Rising and sinking convection currents then twist and concentrate the field into ropelike tubes. The Babcock model predicts that pairs of sunspots should occur where these tubes of

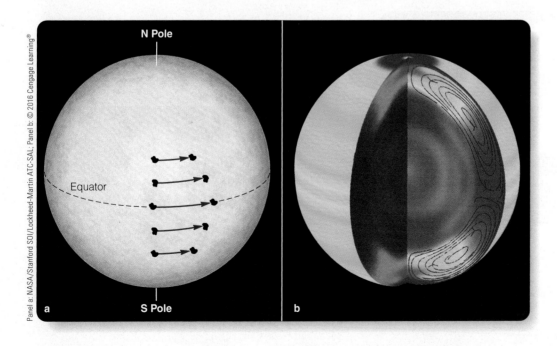

Panel a: NASA/Stanford SOI/Lockheed-Martin ATC-SAL; Panel b: © 2016 Cengage Learning®

◄ **Figure 8-9** (a) The photosphere of the Sun rotates faster at the equator than at higher latitudes. If you started five sunspots in a row along a north-south line, they would not stay lined up as the Sun rotates. (b) Detailed analysis of the Sun's rotation from helioseismology reveals that the interior of the Sun rotates differentially, with regions of relatively slow rotation (*blue*) and rapid rotation (*red*). Currents of gas are also detected moving from the equator toward the poles and back toward the equator.

Sunspots and the Solar Magnetic Cycle

1 The dark spots that appear on the Sun are only the visible traces of complex regions of activity. Evidence gathered over many years and at a wide range of wavelengths shows that sunspots are clearly linked to the Sun's magnetic field.

Spectra show that sunspots are cooler than the photosphere with a temperature of about 4200 K. The photosphere has an average temperature of about 5800 K. Because the total amount of energy radiated by a surface depends on its temperature raised to the fourth power, sunspots look dark in comparison with the photosphere. Actually, a sunspot emits quite a bit of radiation. If the Sun were removed and only an average-size sunspot were left behind, it would be brighter than a full moon.

Visual

Umbra

Penumbra

A typical sunspot is about twice the size of Earth, but there is a wide range of sizes. Sunspots appear, last for a few weeks to a few months, and then shrink away. Usually, sunspots occur in pairs or complex groups.

Earth to scale

NASA

JAXA/NASA/*Hinode*

Sunspots are not shadows, but astronomers refer to the dark core of a sunspot as its *umbra* and the outer, lighter region as the *penumbra*.

Streamers above a sunspot suggest a magnetic field.

JAXA/NASA/*Hinode*

Number of sunspots

Sunspot minimum

Sunspot maximum

2 The number of spots visible on the Sun varies in a cycle with a period of 11 years. At maximum, there are often more than 100 spots visible. At minimum, there are very few or zero.

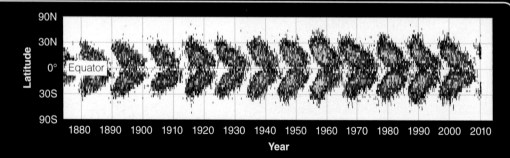

Latitude

Equator

2a Early in a cycle, spots appear at high latitudes north and south of the Sun's equator. Later in the cycle, new spots appear closer to the Sun's equator. If you plot the latitude of sunspots versus time, the graph looks like butterfly wings, as shown in this **Maunder butterfly diagram**, named after E. Walter Maunder of Greenwich Observatory.

NASA MSFC/D. Hathaway

3 Astronomers can measure magnetic fields on the Sun using the **Zeeman effect**, as shown below. When an atom is in a magnetic field, the electron energy levels are altered, and the atom is able to absorb photons with a greater variety of wavelengths than the same atom not in a magnetic field. In this spectrum you see single spectral lines split into multiple components, with the separation between the components proportional to the strength of the magnetic field.

Slit allows light from sunspot to enter spectrograph.

Visual

Spectral line split by Zeeman effect

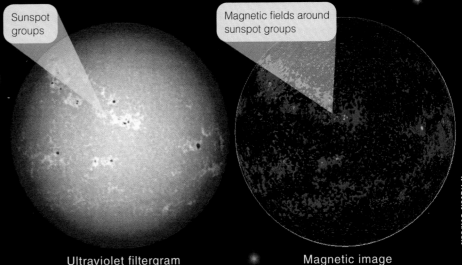

Sunspot groups

Magnetic fields around sunspot groups

Ultraviolet filtergram

Magnetic image

Simultaneous images

3a Images of the Sun above show that sunspots contain magnetic fields a few thousand times stronger than Earth's. Such strong fields inhibit motions of ionized gas below the photosphere; consequently, convection is reduced below the sunspot, less energy is transported from the interior, and the Sun's surface at the position of the spot is cooler. Heat that is prevented from emerging at the sunspot's position is deflected and emerges around the sunspot, making the surrounding area hotter than the average photosphere. The deflected heat can be detected in ultraviolet and infrared images; the result is that the entire active region, including the sunspots, is actually emitting more energy than the same area of normal photosphere.

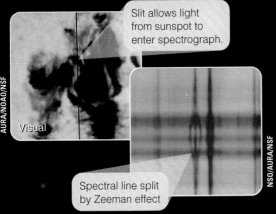

Maunder minimum few spots colder winters

Winter severity in London and Paris

Warmer winters

↑ Warm
↓ Cold

Number of sunspots

Year

4 Historical records show that there were very few sunspots from about 1645 to 1715, a phenomenon known as the **Maunder minimum**. This coincides with the middle of a period called the "Little Ice Age," a time of unusually cool weather in Europe and North America from about 1500 to about 1850, as shown in the graph at left. Other such periods of cooler climate are known. Evidence suggests that there is a link between solar activity and the amount of solar energy Earth receives. This link has been confirmed by measurements made by spacecraft above Earth's atmosphere.

Magnetic fields can reveal themselves by their shape. For example, iron filings sprinkled over a bar magnet reveal an arched shape.

The complexity of an active region becomes visible at short wavelengths.

Far-UV

5 Observations at nonvisual wavelengths reveal that the chromosphere and corona above sunspots are violently disturbed in what astronomers call **active regions**. Spectrographic observations show that active regions contain powerful magnetic fields. If all wavelengths are included, Earth receives more radiation from the spotted, active Sun than at times of low activity.

Arched structures above an active region are evidence of gas trapped in magnetic fields.

Visual

Simultaneous

Images

Far-UV

The Solar Magnetic Cycle

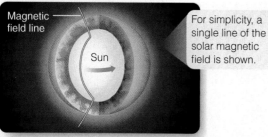

Magnetic field line

Sun

For simplicity, a single line of the solar magnetic field is shown.

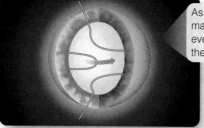

Differential rotation drags the equatorial part of the magnetic field ahead.

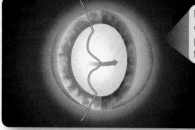

As the Sun rotates, the magnetic field is eventually dragged all the way around.

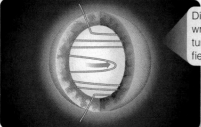

Differential rotation wraps the Sun in many turns of its magnetic field.

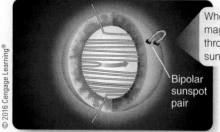

Where loops of tangled magnetic field rise through the surface, sunspots occur.

Bipolar sunspot pair

© 2016 Cengage Learning®

▲ **Figure 8-10** The Babcock model of the solar magnetic cycle explains the sunspot cycle as primarily a consequence of the Sun's differential rotation gradually winding up and tangling the magnetic field near the base of the Sun's outer, convective layer.

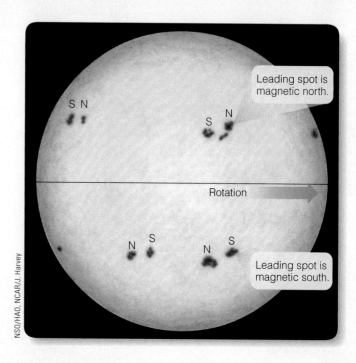

▲ **Figure 8-11** In sunspot groups, here simplified into pairs of major spots, the leading spot and the trailing spot have opposite magnetic polarity. Spot pairs in the Southern Hemisphere have reversed polarity from those in the Northern Hemisphere.

magnet, with one end being magnetic north and the other end magnetic south. That is just what is expected if magnetic tubes, produced by convection and differential rotation according to the Babcock model, emerge from the Sun's surface through one sunspot in a pair and reenter through the other. At any one time, sunspot pairs south of the Sun's equator have reversed **polarity** (orientation of their magnetic poles) relative to those north of the Sun's equator. **Figure 8-11** illustrates this by showing sunspot pairs south of the Sun's equator moving with magnetic south poles leading, and sunspots north of the Sun's equator moving with magnetic north poles leading. At the end of an 11-year sunspot cycle, spots appear in the next cycle with reversed magnetic polarities relative to the spots in the previous cycle.

The Babcock model accounts for the reversal of the Sun's magnetic field from cycle to cycle. As the magnetic field becomes more and more tangled, adjacent regions of the Sun are dominated by magnetic fields that point in different directions. After years of tangling, the field becomes very complicated. Regions of weak north or south polarity "flip" into alignment with neighboring regions of stronger polarity. The entire field then quickly rearranges itself into a simpler pattern, the number of sunspots drops nearly to zero, and the cycle ends. Then, differential rotation and convection begin winding up the magnetic field to start a new cycle. The newly organized field is reversed relative to its predecessor, and the new sunspot cycle begins with the magnetic north end of sunspot groups replaced by magnetic south. Thus, although the solar activity cycle is

concentrated magnetic energy burst through the Sun's surface (**Figure 8-10**).

Sunspots do tend to occur in groups or pairs, and the magnetic field around the pair resembles that around a bar

How Do We Know? (8-1)

Confirmation and Consolidation

What do scientists do all day? The scientific method is sometimes portrayed as a kind of assembly line where scientists crank out new hypotheses and then test them through observation. In reality, scientists don't often generate entirely new hypotheses. And it is rare that an astronomer makes an observation that disproves a long-held theory and triggers a revolution in science. Then what is the daily grind of science really about?

Many observations and experiments confirm already-tested hypotheses. The biologist knows that all worker bees in a hive are sisters because they are all female, and they all had the same mother, the queen bee. A biologist can study the DNA from many workers and confirm that hypothesis. By repeatedly checking and thereby *confirming* a hypothesis, scientists build confidence in the hypothesis and may be able to extend it. Do all of the workers in a hive have the same father, or did the queen mate with more than one male drone?

Another aspect of routine science is *consolidation,* the linking of a hypothesis to other well-studied phenomena. A biologist can study yellow jacket wasps from a single nest and discover that the wasps, too, are sisters. There must be a queen wasp who lays all of the eggs in a nest. But, in a few nests, the scientist may find two sets of unrelated sister workers. Those nests evidently contain two queens sharing the nest for convenience and protection. From his study of wasps, the biologist consolidates what he knows about bees with what others have learned about wasps and reveals something new: that bees and wasps have evolved in similar ways for similar reasons.

Confirmation and consolidation allow scientists to build confidence in their understanding and extend it to explain more about nature.

A yellow jacket is a wasp from a nest containing a queen wasp.

Michael Durham/Minden Pictures//Getty Images

---

11 years long if you count numbers of sunspots, it is 22 years long if you consider sunspot magnetic field directions.

This magnetic cycle also explains the Maunder butterfly diagram. The Babcock model predicts that, as a sunspot cycle begins, the twisted tubes of magnetic force should first produce sunspot pairs at high latitudes on the Sun, exactly as is observed. In other words, the first sunspots in a new cycle appear far from the Sun's equator. Later in the cycle, when the field is more tightly wound, the tubes of magnetic force arch up through the surface at lower latitudes. As a result, sunspot pairs later in a cycle appear closer to the equator.

A refinement of the Babcock model includes the **meridional flow**, which involves slow movements of gas from the Sun's equator to each pole and back, tens of thousands of kilometers below the photosphere, that are detected by helioseismology measurements (Figure 8-9b). The meridional flow carries magnetic field bundles toward the poles from active regions at lower latitudes during each sunspot cycle, thereby establishing the foundation of the next cycle's magnetic field.

Notice the power of a scientific model. Even though the model of the sky in Chapter 2 and the model of atoms in Chapter 7 are only partially correct, they serve as organizing themes to guide further exploration. Similarly, many of the details of the solar magnetic cycle are not yet understood. The Babcock model may be partly incorrect or incomplete. For example, the start of the solar cycle that should have begun around mid-2008 was delayed by 15 months to late 2009. The subsequent cycle maximum in 2013 was the weakest in 100 years in terms of the number of sunspots and amount of solar magnetic activity. The simple Babcock model does not easily account for such an anomaly. Nevertheless, the model provides a framework around which to organize descriptions and investigations of complex solar activity (**How Do We Know? 8-1**).

Chromospheric and Coronal Activity

The solar magnetic fields extend high into the chromosphere and corona, where they produce beautiful and powerful phenomena. Study **Solar Activity and the Sun–Earth Connection** on pages 162–163 and notice three important points and seven new terms:

1 All solar activity is magnetic. The arched shapes of *prominences* are produced by magnetic fields, and *filaments* are prominences seen from above.

2 Tremendous amounts of energy can be stored in arches of magnetic fields, and when two arches encounter each other, a *reconnection event* can cause powerful eruptions called *flares.* Although these eruptions occur far from Earth, they can affect us in dramatic ways. For example, *coronal mass ejections (CMEs)* can trigger communications blackouts and *auroras.*

3 In some regions of the solar surface, the magnetic field does not loop back. High-energy gas from these *coronal holes* flows outward and produces much of the solar wind.

Solar Activity and the Sun–Earth Connection

1 Magnetic phenomena in the chromosphere and corona, like magnetic weather, result as constantly changing magnetic fields in the Sun's atmosphere trap ionized gas to produce beautiful arches and powerful outbursts. Some of this solar activity can affect Earth's magnetic field and atmosphere.

This ultraviolet image of the solar surface was made by the NASA *TRACE* spacecraft. It shows hot gas trapped in magnetic arches extending above active regions. At visual wavelengths, you would see sunspot groups in these active regions.

Sacramento Peak Observatory/NSO/AURA/NSF

H-alpha

1a A **prominence** is composed of ionized gas trapped in a magnetic arch rising up through the photosphere and chromosphere into the lower corona. Seen during total solar eclipses at the edge of the solar disk, prominences look pink because of emission in the three hydrogen Balmer lines, H-alpha, H-beta, and H-gamma. The image above shows the arch shape suggestive of magnetic fields. Seen from above against the Sun's bright surface, prominences form dark **filaments**.

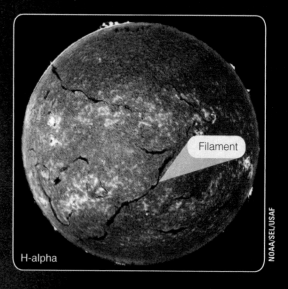

Filament

H-alpha

NOAA/SEL/USAF

1b Quiescent prominences may hang in the lower corona for many days, whereas eruptive prominences burst upward in hours. The eruptive prominence below is many Earth diameters in length.

The gas in prominences may be 60,000 to 80,000 K, quite cold compared with the low-density gas in the corona, which may be as hot as a million Kelvin.

TRACE/NASA

Far-UV

Earth shown for size comparison

ESA/NASA/SOHO EIT

2 Solar flares rise to maximum in minutes and decay in an hour. They occur in active regions where oppositely directed magnetic fields meet and cancel each other in what are called **reconnection events**. Energy stored in the magnetic fields is released as short-wavelength photons plus high-energy protons and electrons. X-ray and ultraviolet photons reach Earth in 8 minutes and increase ionization in our upper atmosphere, which can interfere with radio communications. Particles from flares reach Earth hours or days later as gusts in the solar wind, which can distort Earth's magnetic field and disrupt navigation systems. Solar flares can also cause surges in electrical power lines and damage to Earth satellites.

This multiwavelength image shows a sunspot interacting with a neighboring magnetic field to produce a solar flare.

JAXA/NASA/Hinode

2a At right, waves rush outward at 50 km/s from the site of a solar flare 40,000 times stronger than the 1906 San Francisco earthquake. The biggest solar flares can be a billion times more powerful than a hydrogen bomb.

Helioseismology image

ESA/NASA/SOHO MDI

2b The solar wind, enhanced by eruptions on the Sun, interacts with Earth's magnetic field and can create electrical currents with up to a million megawatts of power. Those currents flowing down into a ring around Earth's magnetic poles excite atoms in Earth's upper atmosphere to emit photons, as shown below. The emission results in glowing clouds and curtains of **auroras**.

Auroras occur about 130 km above the Earth's surface.

Coronal Mass Ejection

Ring of aurora around the north magnetic pole

ESA/NASA/SOHO

NSSDC, Holzworth and Meng

2c Reconnection events can release enough energy to blow large amounts of ionized gas outward from the corona in **coronal mass ejections (CMEs)**. If a CME strikes Earth, it can produce especially violent disturbances in Earth's magnetic field.

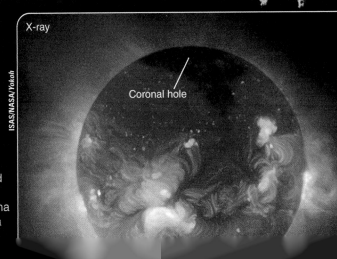

X-ray

Coronal hole

ISAS/NASA/Yokoh

3 Much of the solar wind comes from **coronal holes** where the magnetic field does not loop back into the Sun. These open magnetic fields allow ionized gas in the corona to flow away as the solar wind. The dark area in the X-ray image at right is a coronal hole.

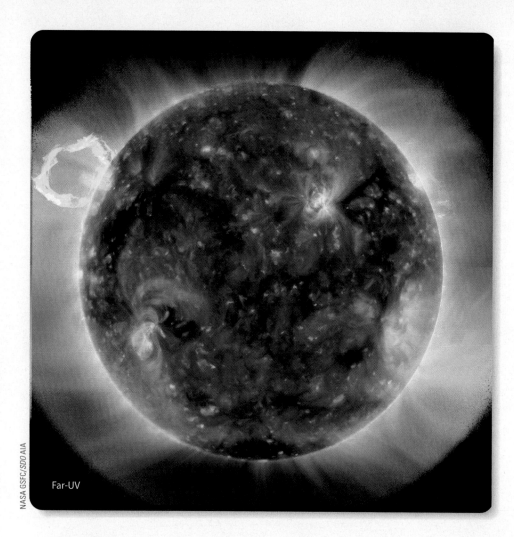

Far-UV

◄ **Figure 8-12** A false-color far-ultraviolet image of the Sun taken by the *Solar Dynamics Observatory (SDO)* spacecraft in 2010. Colors show different gas temperatures: Reds and yellows are relatively cool material (about 60,000 to 100,000 K); blues and greens are hotter (1 million K or more). Note the large eruptive loop prominence at upper left.

Images of the active Sun often show eruptive prominences, enormous arches that are shaped by magnetic fields and have sizes that dwarf Earth, standing above active regions on the solar limb (Figure 8-12).

Auroras are sometimes called the "northern lights," but they can be viewed often from high latitudes in both the Northern and Southern Hemisphere. Now, if you ever have an opportunity to watch a beautiful aurora display, you will know that you are actually seeing spectral emission lines from gases in Earth's upper atmosphere excited to glow by a complicated interaction with the solar wind and Earth's magnetic field (look back to Figure 6-1).

A series of solar eruptions in August–September 1859 produced electromagnetic disturbances at Earth's surface so severe that telegraph equipment caught on fire and operators received painful electrical shocks. A 2010 study by the Metatech Corporation (funded by NASA) indicated that if a solar eruption as large as the ones in 1859 occurred today, it would produce electrical blackouts affecting 40 percent of U.S. households that could last for months, waiting for destroyed electrical transmission and generation equipment to be replaced. There was a solar eruption of that magnitude in 2010, but the active

region was pointed mostly away from Earth. Humanity became fully aware of the size of that event only because of data from a combination of the *Solar Dynamics Observatory (SDO)* and the twin *Solar TErrestrial RElations Observatory (STEREO)* spacecraft monitoring the Sun from different positions in the Solar System. You can imagine that scientists would like to identify early warning signs that such an eruption is imminent, as well as strategies to protect the electrical and communications equipment on which human civilization is now so dependent.

The Solar Constant

Even a small change in the Sun's energy output could produce dramatic changes in Earth's climate, but humanity knows very little about long-term variations in the Sun's energy output. The energy production of the Sun can be monitored by measuring the amount of solar radiation reaching Earth. Of course, you should include all wavelengths of electromagnetic radiation from X-rays to radio waves, so you need to correct for absorption by Earth's atmosphere or make the measurement from space. The result, called the **solar constant**, amounts to about 1370 W/m². (The conventional units for this measurement are joules per square meter per second, but 1 joule per second is defined as 1 watt [W], so those units are equivalent to watts per square meter.) But is the Sun really constant?

Measurements by the *Solar Maximum Mission* satellite showed variations in the energy received from the Sun by about 0.1 percent that last for days, weeks, or years, including one pattern of variation that appears to be timed with the magnetic activity cycle. Superimposed on both the random and cyclical variations is a very slight long-term decrease of about 0.018 percent per year that has been confirmed by observations made with other instruments. This long-term decrease may be related to a cycle of activity on the Sun with a period longer than the 22-year magnetic cycle. Thus, careful measurements show that the solar constant is not really constant.

As you saw on page 159, the "Little Ice Age" was a period of unusually cool weather in Europe and America that lasted from about 1500 to 1850. The average temperature worldwide was about 1°C cooler than it is now. This period of cool weather corresponded very roughly to the Maunder minimum, a period of reduced solar activity—few sunspots and auroral displays and little or no corona visible during solar eclipses. Scientists do not yet completely understand how those changes

in the Sun's surface activity would connect to changes in Earth's average temperature. The measured changes in the modern solar "constant" seemingly would cause Earth to become slightly cooler, yet our planet is clearly observed to be warming. In a later chapter, you will learn more about the complex interaction between solar input, human activity, and changes in Earth's climate.

Spots and Magnetic Cycles of Other Stars

The Sun seems to be a representative star, so you should expect other stars to have cycles of "starspots" and magnetic activity similar to the Sun's. This is difficult to demonstrate observationally because, with few exceptions, the stars are too small or too far away to allow detection of surface detail. Some stars, however, vary in brightness in ways that suggest that they are covered by dark spots. As these stars rotate, their total brightness changes slightly, depending on the number of spots on the side facing Earth. High-precision spectroscopic analysis has even allowed astronomers to map the locations of spots on the surfaces of certain stars (**Figure 8-13a**). Such results confirm that the sunspots seen on our Sun are not unusual.

Panel a: NOAO/AURA/NSF/K. Strassmeier, University of Vienna; Panel b: Ca II emission adapted from data by Baliunas and Saar

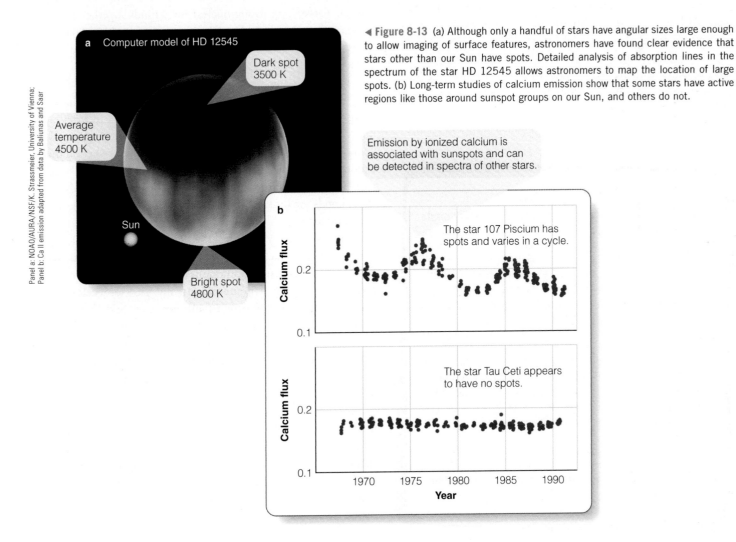

◄ **Figure 8-13** (a) Although only a handful of stars have angular sizes large enough to allow imaging of surface features, astronomers have found clear evidence that stars other than our Sun have spots. Detailed analysis of absorption lines in the spectrum of the star HD 12545 allows astronomers to map the location of large spots. (b) Long-term studies of calcium emission show that some stars have active regions like those around sunspot groups on our Sun, and others do not.

Certain features found in stellar spectra might be associated with magnetic fields by analogy with the Sun. Regions of strong magnetic fields on the solar surface emit strongly at the central wavelengths of the two strongest lines of ionized calcium. This calcium emission appears in the spectra of other Sun-like stars and suggests that these stars, too, have strong magnetic fields at some locations on their surfaces. In some cases, the strength of this calcium emission varies over periods of days or weeks and indicates that the stars have active regions and are rotating with periods similar to those of the Sun. These stars presumably have "starspots" as well.

In 1966, astronomers at Mt. Wilson observatory began a long-term project that monitored the strengths of these calcium emission features in the spectra of 91 stars with photosphere temperatures ranging from 1000 K hotter than the Sun to 3000 K cooler that were considered most likely to have Sun-like magnetic activity on their surfaces. The observations show that the strength of the calcium emission varies over periods of years. The calcium emission averaged over the Sun's disk varies with the sunspot cycle, and similar periodic variations can be seen in the spectra of some of the stars studied (Figure 8-13b). The star 107 Piscium, for example, appears to have a starspot cycle lasting nine years. At least one star, tau Boötis, has been observed to reverse its magnetic field. This kind of evidence shows that stars like the Sun have similar magnetic cycles, and that the Sun is normal in this respect. It is interesting to note that 15 percent of the Sun-like stars in the Mt. Wilson study were found to have very low activity levels; some astronomers have speculated that those stars are in phases equivalent to the Sun's Maunder minimum.

DOING SCIENCE

What kind of activity would the Sun have if it didn't rotate differentially? Imagining a physical system with one factor changed is something scientists do to help them understand a concept.

Consider the Babcock model for solar magnetic activity cycles. If the Sun didn't rotate differentially, with its equator turning in a shorter period than its higher latitudes, then the magnetic field would not get so tangled. As a result, there might not be a solar cycle because twisted tubes of magnetic field might not form and rise through the photosphere to produce sunspots and active regions with prominences and flares. On the other hand, convection might still tangle the magnetic field and produce some activity. Is the magnetic activity that causes sunspots and heats the chromosphere and corona, driven mostly by differential rotation, or by convection? Astronomers are not sure, but it seems likely that without differential rotation the Sun would not have a strong magnetic field and resulting high-temperature gas above its photosphere.

This is very speculative, but speculation can be revealing. For example, consider a complementary scenario to the one discussed. ***How do you think the Sun's appearance would differ if it had no convection inside?***

8-3 Nuclear Fusion in the Sun

Like soap bubbles, stars are structures balanced between opposing forces that, if unbalanced, can destroy them. The Sun is a ball of hot gas held together by its own gravity. If it were not for the Sun's gravity, the hot, high-pressure gas in the Sun's interior would explode outward. Likewise, if the Sun were not so hot, its gravity would compress it into a small, dense body.

In this section, you will discover that the Sun is powered by nuclear reactions occurring near its center. The energy released by those reactions keeps the interior hot and the gas totally ionized (meaning, all electrons moving unattached to nuclei). How exactly can the nucleus of an atom yield energy? The answer lies in the force that holds the particles in nuclei together.

Nuclear Binding Energy

The Sun generates its energy by breaking and reconnecting the bonds between the particles *inside* atomic nuclei. There are only four different ways in which matter affects other matter. These are called the four forces of nature: gravity, the electromagnetic force, the **weak nuclear force**, and the **strong nuclear force**. The strong nuclear force binds together atomic nuclei, and the weak nuclear force is involved in the radioactive decay and other interactions of certain kinds of nuclear particles.

The strong and weak nuclear forces are short-range forces that are effective only within the nuclei of atoms. Nuclear energy originates from the strong force, as nuclear reactions break and re-form the bonds that hold atomic nuclei together. In contrast, the process of burning wood is a chemical reaction that extracts energy by breaking and rearranging chemical bonds among atoms in the wood. The chemical energy released when those bonds are broken and rearranged comes from the electromagnetic force.

There are two types of reactions by which atomic nuclei can release energy. Nuclear power plants on Earth use **nuclear fission** reactions that split uranium nuclei into less massive fragments. The isotope of uranium normally used for nuclear fuel contains a total of 235 protons and neutrons. Splitting such a nucleus produces a range of possible fragment nuclei, each containing roughly half as many particles. Because the fragment nuclei are more tightly bound (have lower total potential energy) than the original uranium nucleus, binding energy is released during uranium fission.

Stars make energy by another type of nuclear reaction—**nuclear fusion**—that combines small nuclei into larger, more massive nuclei. The most common reaction inside stars, the one that occurs in the Sun, fuses hydrogen nuclei (single protons) to produce helium nuclei, which contain two protons and two neutrons. Just as with fission, because the nuclei produced by fusion are more tightly bound than the original nuclei, net energy is released.

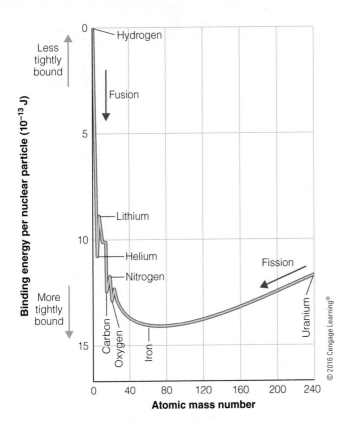

▲ **Figure 8-14** The orange curve in this graph shows the binding energy per particle, the energy that holds particles inside atomic nuclei. The horizontal axis gives the atomic mass number of each element, the number of protons and neutrons in the nucleus. Both fission and fusion nuclear reactions "move" downward in the diagram (*arrows*), meaning the nuclei produced by a reaction are more tightly bound than the nuclei that went into the reaction, and the reaction resulted in a net release of energy. Iron has the most tightly bound nucleus, so no nuclear reactions can use iron and release energy.

The curve plotted in **Figure 8-14** shows the nuclear binding energy that holds various atomic nuclei together. If the data point for a given type of nucleus is low in the diagram, the particles in that nucleus are held together tightly. Notice that both fusion and fission reactions involve moving downward in the diagram from less tightly bound toward more tightly bound nuclei. Both types of nuclear reaction produce energy by releasing binding energy of atomic nuclei.

Hydrogen Fusion

The fusion reaction in the Sun combines four hydrogen nuclei to make one helium nucleus. Because one helium nucleus has 0.7 percent less mass than four hydrogen nuclei, it seems that some mass vanishes in the process. To see this, subtract the mass of a helium nucleus from the mass of four hydrogen nuclei:

$$4 \text{ hydrogen nuclei} = 6.690 \times 10^{-27} \text{ kg}$$
$$\underline{-1 \text{ helium nucleus} = 6.646 \times 10^{-27} \text{ kg}}$$
$$\text{Difference in mass} = 0.044 \times 10^{-27} \text{ kg}$$

That mass difference, 0.044×10^{-27} kg, does not actually disappear but is converted to energy according to Einstein's famous equation (look back to Chapter 5):

$$E = m_0 c^2$$
$$= (0.044 \times 10^{-27} \text{ kg}) \times (3.0 \times 10^8 \text{ m/s})^2$$
$$= 4.0 \times 10^{-12} \text{ J}$$

You can symbolize the fusion reactions in the Sun with a simple equation:

$$4 \text{ }^1\text{H} \rightarrow {}^4\text{He} + \text{energy}$$

In that equation, the superscripts indicate the number of nucleons (protons plus neutrons) in each of the nuclei. ^{1}H represents a proton—the nucleus of a hydrogen atom—and ^{4}He represents the nucleus of a helium atom.

The actual steps in the process are more complicated than this convenient summary suggests. Instead of waiting for four hydrogen nuclei to collide simultaneously, a highly unlikely event, the process normally proceeds step by step in a series of reactions called the **proton–proton chain** (**Figure 8-15**). The proton–proton chain consists of three nuclear reactions that build a helium nucleus by adding protons one at a time. Those three reactions are:

$$^1\text{H} + {}^1\text{H} \rightarrow {}^2\text{H} + e^+ + \nu$$
$$^2\text{H} + {}^1\text{H} \rightarrow {}^3\text{He} + \gamma$$
$$^3\text{He} + {}^3\text{He} \rightarrow {}^4\text{He} + {}^1\text{H} + {}^1\text{H}$$

In the first reaction, two protons (two hydrogen nuclei) combine. The strong nuclear force binds the protons together, whereas the weak nuclear force causes one of them to transform into a neutron and emit two particles: a **positron**, which is the positively charged version of an electron (e^+); and a neutrino (μ), which is a subatomic particle having an extremely low mass and a velocity nearly equal to the velocity of light. The combination of a proton with a neutron forms a heavy hydrogen nucleus called **deuterium**.

In the second reaction, a deuterium nucleus absorbs another proton and, with the emission of a gamma-ray photon (γ) becomes a lightweight helium nucleus. Finally, two lightweight helium nuclei combine to form a nucleus of normal helium plus two hydrogen nuclei. Because the last reaction needs two^3 He nuclei, the first and second reactions must occur twice. The net result of this sequence of reactions is the transformation of four hydrogen nuclei into one helium nucleus plus energy.

The energy released in the proton–proton chain appears in the form of gamma-rays, positrons, neutrinos, and the energy of motion of all the particles. The gamma-rays are photons that are absorbed by the surrounding gas before they can travel more than a fraction of a millimeter. That heats the gas. The positrons produced in the first reaction combine with free electrons, and both particles vanish, converting their mass into gamma-rays, which are also absorbed and help keep the gas hot. In addition, when fusion produces new nuclei, they fly apart at high speed

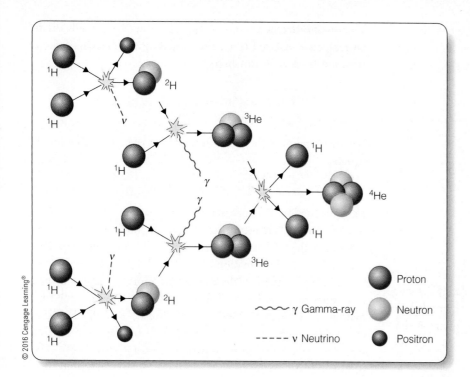

© 2016 Cengage Learning®

◄ **Figure 8-15** The proton–proton chain combines four protons (*at far left*) to produce one helium nucleus (*at right.*) Energy is produced mostly as gamma-rays (γ) and as positrons (e⁺), which combine with electrons and convert their mass into more gamma-rays. Neutrinos (ν) escape without heating the gas.

and collide with other particles. This energy of motion helps raise the temperature of the gas. The neutrinos, on the other hand, don't heat the gas. Neutrinos are particles that almost never interact with other particles. The average neutrino could pass unhindered through a lead wall more than a light-year thick. Consequently, the neutrinos do not warm the gas but race out of the Sun at nearly the speed of light, carrying away approximately 2 percent of the energy produced by the fusion reactions.

Creating one helium nucleus makes only a small amount of energy, not enough to raise a housefly one-thousandth of a millimeter. Because one reaction produces such a small amount of energy, it is obvious that reactions must occur at a tremendous rate to supply the energy output of a star. The Sun, for example, completes 10^{38} fusion reactions every second, transforming about 4 million tons of matter into energy. It might sound as if the Sun is losing mass at a furious rate, but in its entire estimated 12-billion-year lifetime, the Sun will convert only about 0.1 percent of its mass into energy.

It is a **Common Misconception** that nuclear fusion in the Sun is tremendously powerful. After all, the fusion of a milligram of hydrogen (roughly the mass of a match head) produces as much energy as burning 5 gallons of gasoline. However, at any one time, only a tiny fraction of the hydrogen atoms are fusing into helium, and the nuclear reactions in the Sun are spread through a large volume in its core. Any single gram of matter produces only a little energy. A person of normal mass eating a normal diet produces about 3000 times more heat per gram than the matter in the core of the Sun. Gram for gram, you are a much more efficient heat producer than the Sun. The Sun produces a lot of energy because it contains many grams of matter in its core.

Fusion reactions can occur only when the nuclei of two atoms get very close to each other. Because atomic nuclei carry positive charges, they repel each other with an electrostatic force called the *Coulomb force* (Chapter 7). Physicists commonly refer to this electrical resistance to nuclear collisions as the **Coulomb barrier**. To overcome this barrier and get close together, atomic nuclei must collide violently. Sufficiently violent collisions are rare unless the gas is very hot, so that the nuclei move at high enough speeds. (Recall that an object's temperature is related to the speed with which its particles move.) Even so, the fusion of two protons is a highly unlikely process. If you could follow a single proton in the Sun's core, you would see it encountering and bouncing off other protons millions of times a second, but you would have to follow it around for many billions of years before it would have a 50/50 chance of penetrating the Coulomb barrier and combining with another proton.

Because of the dependence of nuclear reactions on particle collisions, the reactions in the Sun take place only near its center, where the gas is hot and dense. A high temperature ensures that collisions between nuclei are violent, and a high density ensures that there are enough collisions, and thus enough reactions per second, to make energy at the Sun's rate. The proton–proton chain requires temperatures above about 4 million K.

Energy Transport in the Sun

Now you are ready to follow the energy from the core of the Sun to the surface. You will learn in a later chapter that astronomers have computed models indicating that the temperature at the center of the Sun must be about 16 million K for the Sun to be stable. Compared with that, the Sun's surface is very cool, only about 5800 K. Heat always moves from hot regions to cool regions, so energy must flow from the Sun's high temperature core outward to the cooler surface where it is radiated into space.

Because the core is so hot, the photons there are gamma-rays. Each time a gamma-ray encounters one of the matter particles—electrons or nuclei—it is deflected or scattered in a random direction, and, as it bounces around, it slowly drifts outward toward the surface while being converted into several photons of lower energy. The net outward motion of energy in the inner parts of the Sun takes the form of radiation, so astronomers refer to that region as the **radiative zone**.

Energy originally produced in the core of the Sun and traveling outward as radiation eventually reaches the outer layers of the Sun where the gas is cool enough that it is not completely ionized. Partially ionized gas is much less transparent to radiation than is completely ionized gas. So, at that point, the energy flowing toward the Sun's surface backs up like water behind a dam, and the gas begins to churn in convection currents. Hot blobs of gas rise, and cool blobs sink. In this region, known as the **convective zone**, the energy is carried outward not as photons but as circulating gas (**Figure 8-16**). Rising hot gas carries energy outward, but sinking cool gas is a necessary part of the cycle. The result is net transport of energy continuing outward. Previously in this chapter you learned about granulation and supergranulation features observed on the Sun's photosphere; those are the visible effects of energy arriving at the Sun's surface from its interior by convection.

It can take millions of years for the energy that began in the form of a single gamma-ray produced in the center of the Sun to work its way outward first as radiation and then by convection. When that energy finally reaches the photosphere, it is radiated into space as about 2000 photons of visible light.

Counting Solar Neutrinos

It is time to ask the critical question that lies at the heart of science. What is the evidence to support this theoretical explanation of how the Sun shines?

Nuclear reactions in the Sun's core produce floods of neutrinos that rush out of the Sun and off into space. More than 10^{14} (100 trillion) solar neutrinos flow through your body every second, but you never feel them because you are almost perfectly transparent to neutrinos. If you could detect these neutrinos, you could probe the Sun's interior. You can't focus neutrinos with a lens or mirror, and they zip right through detectors used to count other atomic particles, but neutrinos of certain energies can trigger the radioactive decay of some atoms. That gives astronomers a way to detect neutrinos.

In the 1960s, chemist Raymond Davis Jr. created a device that could count neutrinos with energies produced by hydrogen fusion in the Sun. He buried a 100,000-gallon tank of cleaning fluid (perchloroethylene [C_2Cl_4]) in the bottom of a South Dakota gold mine where other types of cosmic rays could not reach it (**Figure 8-17a**) and invented a way to count individual argon atoms that were produced by neutrinos colliding with chlorine atoms in the tank.

The Davis neutrino experiment created a huge controversy. It was expected to detect one neutrino a day, but it actually counted one-third as many: Only one solar neutrino was captured in that 100,000-gallon tank every three days. Were scientists wrong about nuclear fusion in the Sun? Did they misunderstand how neutrinos behave? Was the detector not working properly? Because astronomers had reason for confidence in their understanding of the solar interior, they didn't immediately abandon their hypotheses (**How Do We Know? 8-2**). It took more than 30 years, but eventually physicists were able to build better and different neutrino detectors (Figure 8-17b). They discovered that neutrinos change back and forth among three different types, which physicists call "flavors." Nuclear reactions in the Sun produce just one flavor, and the Davis experiment was designed to detect (taste!) only that flavor. But during the 8-minute journey from the Sun's core to Earth, the neutrinos changed flavor so many times that they were distributed evenly among the three different flavors by the time they arrived at Earth. That's why the Davis experiment detected only one-third of the number originally predicted. Models of nuclear fusion in the Sun are now confirmed once the actual properties of neutrinos are taken into account.

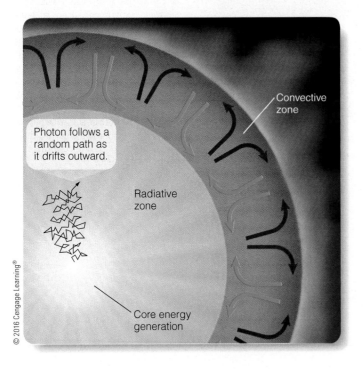

◀ **Figure 8-16** A cross-section of the Sun's interior. Near the center, nuclear fusion reactions sustain high temperatures. Energy flows outward through the radiative zone as photons that gradually make their way to the surface as they are randomly deflected over and over by collisions with electrons. In cooler, more opaque outer layers the energy is carried by rising convection currents of hot gas (*red arrows*) and sinking currents of cooler gas (*blue arrows*).

Photon follows a random path as it drifts outward.

Radiative zone

Convective zone

Core energy generation

© 2016 Cengage Learning®

How Do We Know? 8-2

Scientific Confidence

How can scientists be certain of something?
Sometimes scientists stick so firmly to their ideas in the face of contradictory claims that it almost sounds as if they are merely stubbornly refusing to consider alternatives. To understand what's actually going on, you might consider the perpetual motion machine, which is a device that supposedly runs continuously with no source of energy. If you could invent a real perpetual motion machine, you could make cars that would run without any fuel. That's good mileage.

For centuries many people have claimed to have invented perpetual motion machines, and for just as long scientists have been dismissing these claims as impossible. The problem with a perpetual motion machine is that it violates the law of conservation of energy, and scientists are not willing to accept that the law could be wrong. In fact, the Royal Academy of Sciences in Paris was so sure that a perpetual motion machine is impossible, and so tired of debunking hoaxes, that in 1775 they issued a formal statement refusing to deal with them. The U.S. Patent Office policy is that it won't even consider starting the patent process for one without seeing a working model first.

Why do scientists seem so stubborn and close-minded on this issue? Why isn't one person's belief in perpetual motion just as valid as another person's belief in the law of conservation of energy? In fact, the two positions are not equally valid. The confidence physicists have in that law is not a belief or even an opinion; it is an understanding founded on the fact that the law has been tested uncountable times and has never failed. In contrast, no one has ever successfully demonstrated a perpetual motion machine. The law of conservation of energy is a fundamental truth about nature and can be used to understand what is possible and what is impossible.

When the first observations of solar neutrinos detected fewer than were predicted, some scientists speculated that astronomers misunderstood how the Sun makes its energy or that they misunderstood the internal structure of the Sun. But astronomers stubbornly refused to reject their model because the nuclear physics of the proton–proton chain is well understood, and models of the Sun's structure have been tested successfully by other measurements many times. The confidence astronomers felt in their understanding of the Sun was an example of scientific certainty, and that confidence in basic natural laws prevented them from abandoning decades of work in the face of a single contradictory observation.

What seems to be stubbornness among scientists is really their confidence in basic principles that have been tested over and over. Those principles are the keel that keeps the ship of science from rocking before every little breeze. Without even looking at that perpetual motion machine, your physicist friends can warn you not to invest in it.

For centuries, people have tried to design a perpetual motion machine, but not a single one has ever worked. Scientists understand why.

◄ **Figure 8-17** (a) The Davis solar neutrino experiment used a large tank of cleaning fluid as a detector and could detect only one of the three flavors of neutrinos. (b) The Sudbury Neutrino Observatory is a 12-meter-diameter globe containing water rich in deuterium (heavy hydrogen) in place of ordinary hydrogen. Buried 2100 m (6800 ft) down in an Ontario mine, it can detect all three flavors of neutrinos and confirms that neutrinos oscillate.

The center of the Sun seems forever beyond human experience, but counting solar neutrinos provides evidence to confirm the theories. The Sun makes its energy through nuclear fusion.

DOING SCIENCE

Why does nuclear fusion require that the gas be very hot? Is there any alternative? Answering these questions requires scientists to rehearse what is known about the basic physics of atoms and thermal energy. Occasional reconsidering and questioning of basic principles are important parts of doing science.

Inside a star, the gas is so hot it is ionized, which means the electrons have been stripped off the atoms, leaving bare, positively charged nuclei. In the case of hydrogen, the nuclei are single protons. These atomic nuclei repel each other because of their positive charges, so they must collide with each other at high speed if they are to overcome that repulsion and get close enough together to fuse. If the atoms in a gas are moving rapidly, then the gas must have a high temperature, so nuclear fusion requires that the gas be very hot. If the gas is cooler than about 4 million K, hydrogen can't fuse because the protons don't collide violently enough to overcome the repulsion of their positive charges. In spite of rumors to the contrary, there does not seem to be any "short-cut" allowing fusion to happen at much lower temperatures.

It is easy to see why nuclear fusion in the Sun requires a high temperature, but now consider a related question about basic physical laws: **Why is fusion helped by high density?**

What Are We? Sunlight

We live very close to a star and depend on it for survival. All of our food comes from sunlight that was captured by plants on land or in the oceans. We either eat those plants directly or eat the animals that feed on those plants. Whether you had salad, seafood, or a cheeseburger for supper last night, you dined on sunlight, thanks to photosynthesis.

Almost all of the energy that powers human civilization came from the Sun through photosynthesis by ancient plants that were buried and converted to coal, oil, and natural gas. New technology is making energy from plant products like corn, soybeans, and sugar. It is all stored sunlight. Windmills generate electrical power, but the wind blows because of heat from the Sun. Photocells make electricity directly from sunlight. Even our bodies have adapted to use sunlight to help manufacture vitamin D.

Our planet is warmed by the Sun; without that warmth the oceans would be ice, and the atmosphere would be a coating of frost. Books often refer to the Sun as "our Sun" or "our star." It is ours in the sense that we are creatures of its light.

Study and Review

Summary

▶ The Sun's visible surface, the **photosphere (p. 148)**, is the layer in the Sun from which visible photons most easily escape. The solar atmosphere can be considered to consist of the photosphere plus two layers of hotter, lower-density gas above the photosphere: the **chromosphere (p. 148)** and **corona (p. 148)**.

▶ **Granulation (p. 149)** of the photosphere is produced by **convection (p. 149)** currents of hot gas rising from below, which cool and sink below the visible surface. **Supergranules (p. 149)** appear to be caused by larger convection currents rising from deeper depths than those currents at shallower depths producing the granules.

▶ The edge or **limb (p. 150)** of the solar disk is dimmer than the center. This **limb darkening (p. 150)** effect is evidence that the temperature in the solar photosphere increases with depth.

▶ The chromosphere is most easily visible during total solar eclipses, when it flashes into view for a few seconds. It is a thin, hot layer of gas just above the photosphere. Its pink color is from the Balmer lines in the Sun's emission spectrum.

▶ **Filtergrams (p. 150)** of the chromosphere reveal **spicules (p. 150)**, which are flamelike structures that extend upward into the lower corona.

▶ The corona is the Sun's outermost atmospheric layer and can be imaged using a **coronagraph (p. 151)**. It is composed of a very low-density, very hot gas extending many solar radii from the visible Sun. The current hypothesis is that the corona's high temperature—more than 2 million K—is maintained by energy transported via motions of the magnetic field extending up through the photosphere—the **magnetic carpet (p. 152)**—and by magnetic waves coming from below the photosphere.

▶ Parts of the corona give rise to the **solar wind (p. 153)**, a breeze of low-density ionized gas streaming away from the Sun. The solar wind extends to the **heliopause (p. 153)**, marking an outer boundary of the Solar System.

▶ The strength of spectral lines in the Sun's spectrum depends partly on the temperature of its atmosphere. Some elements that are abundant have weak spectral lines, and some that are not very abundant have strong spectral lines. The mathematical techniques for deriving true abundances of elements from the solar spectrum were worked out by Cecilia Payne in the 1920s.

- Solar astronomers can study the motion, density, and temperature of gases inside the Sun by analyzing the way the solar photosphere oscillates. Known as **helioseismology (p. 154)**, this field of study requires large amounts of data and extensive computer analysis.

- The Sun's light and infrared radiation can burn your eyes, so you must take great care in observing it. **Sunspots (p. 156)** come and go on the Sun, but only rarely are they large enough to be visible to the unaided eye.

- Sunspots seem dark to our eyes because they are slightly cooler than the rest of the photosphere. The average sunspot is about twice the size of Earth. Sunspots appear for a month or so and then fade away, and the number and location of spots on the Sun vary with an 11-year sunspot cycle.

- Early in a sunspot cycle, spots appear farther from the Sun's equator, and later in the cycle they appear closer to the equator. This repetitive pattern of sunspot locations over time is shown in the **Maunder butterfly diagram (p. 158)**.

- Astronomers can use the **Zeeman effect (p. 159)** to measure magnetic fields on the Sun. The average sunspot contains magnetic fields a few thousand times stronger than Earth's field. This increase in strength is part of the evidence that magnetic fields are involved in the sunspot cycle. Two sunspot cycles occur for every one solar magnetic cycle, which is 22 years long.

- The sunspot cycle does not always have the same pattern each cycle. For example, during the **Maunder minimum (p. 159)** from 1645 to 1715, the number of sunspots was significantly fewer, solar activity was very low, and Earth's climate was slightly colder.

- Sunspots are the visible consequences of **active regions (p. 159)** where the Sun's magnetic field is strong. Arches of magnetic field can produce sunspots where the field passes through the photosphere.

- The Sun's magnetic field is produced by the **dynamo effect (p. 157)** operating at the base of the zone of convection currents.

- Alternate sunspot cycles have reversed magnetic **polarity (p. 160)**, which has been explained by the **Babcock model (p. 157)**, in which the Sun's **differential rotation (p. 157)** and convection currents tangle the magnetic field. The magnetic field's tangles arch through the photosphere and cause active regions visible to your eyes as sunspot pairs. As the Sun's magnetic field becomes strongly tangled, **meridional flow (p. 161)** currents carry portions of the field from the equator toward the poles at the surface and more slowly from the poles to the equator below the surface until the magnetic field finally reorders itself into a simpler but reversed field, and the sunspot cycle starts over.

- Arches of magnetic field are visible as **prominences (p. 162)** in the chromosphere and corona. Seen from above in filtergrams, prominences are visible as dark **filaments (p. 162)** silhouetted against the bright chromosphere.

- Magnetic field **reconnection events (p. 163)** can produce powerful **flares (p. 163)**, which are sudden eruptions of X-ray, ultraviolet, and visible radiation plus high-energy atomic particles. Flares are important because they can have dramatic effects on Earth, such as communications blackouts.

- The solar wind originates in regions on the solar surface called **coronal holes (p. 163)**, where the Sun's magnetic field leads out into space and does not loop back to the Sun. **Coronal mass ejections, or CMEs (p. 163)**, occur when magnetic fields on the surface of the Sun eject bursts of ionized gas that flow outward in the solar wind. If they strike Earth, such bursts can produce **auroras (p. 163)** and other phenomena.

- Other stars are too far away to observe visible star spots, which are the equivalents of sunspots. However, some stars vary in brightness in ways that reveal they do indeed have spots on their surfaces. Spectroscopic observations reveal that many other stars have spots and magnetic fields that follow long-term cycles like the Sun's.

- The **solar constant (p. 165)** is a measure of all the electromagnetic radiation reaching Earth from the Sun; its value is about 1370 joules/m²/s. Sensitive measurements show that the solar constant is not truly constant but varies during a sunspot cycle and perhaps on longer time scales.

- There are only four fundamental forces in nature: the electromagnetic force, the gravitational force, the **weak nuclear force (p. 166)**, and the **strong nuclear force (p. 166)**. The strong force binds atomic nuclei together. The weak force is involved in radioactive decay and other interactions of certain kinds of nuclear particles.

- Nuclear reactors on Earth generate energy through **nuclear fission (p. 166)**, in which large nuclei such as uranium break into smaller fragments and generate energy in the process. The Sun generates its energy through **nuclear fusion (p. 166)**, in which small nuclei such as hydrogen fuse to form larger nuclei, such as helium, generating energy in the process.

- The fusion of hydrogen into helium in the Sun proceeds in three steps known as the **proton–proton chain (p. 167)**. The first step in the chain combines two hydrogen nuclei to produce a heavy hydrogen nucleus called **deuterium (p. 167)**. The second step forms light helium, and the third step combines the light helium nuclei to form normal helium. During the process, **positrons (p. 167)**, **neutrinos (p. 167)**, and gamma-rays are formed and energy is released as the particles fly away.

- Fusion can occur only in the core of the Sun where temperatures are pressures are high enough. Because particles of like charge repel one other, high temperatures are needed to give particles high enough velocities to overcome this **Coulomb barrier (p. 168)** and fuse together. High densities are needed to provide large numbers of reactions.

- Energy is transported from the core of the Sun's by photons traveling through the **radiative zone (p. 169)** to the base of the **convective zone (p. 169)**. Then energy is transported to the photosphere by rising currents of hot gas and sinking currents of cooler gas through this convective zone.

- Neutrinos escape from the Sun's core at nearly the speed of light, carrying away about 2 percent of the energy produced by fusion. Early experiments of detection rates reveal fewer neutrinos than expected coming from the Sun's core. This result is now known to be because neutrinos oscillate among three different types (called "flavors") while traveling to Earth, and experiments were designed to detect only one of these three flavors. Later experiments detected all solar neutrino flavors, confirming the hypothesis that much of the Sun's energy comes from the proton–proton chain.

Review Questions

1. Why can't you see deeper into the Sun than the photosphere?
2. What color is the photosphere as viewed from the ground on a clear, cloudless day when the Sun is highest overhead? When the Sun has sunk to just above the ocean's horizon? When the Sun has sunk to half below the ocean's horizon? Does the photosphere really change colors during this sunset? Why or why not?
3. You stayed a little too long outside in the sunshine, and now you have a sunburn. From which atmospheric layer of the Sun did the photons originate that resulted in your sunburn? How do you know?
4. The average temperature of the photosphere is 5800 K. What color is the maximum intensity of a 5800 K blackbody? Is this

the color we normally associate with the photosphere? Why or why not? (*Hint:* Refer to Figure 7-6 and Section 8-1.)

5. What information can we obtain from the Sun's absorption line spectrum shown in Figure 7-7a?

6. Which atmospheric layer is associated with the Sun's continuous spectrum? With its absorption spectrum? With its emission spectrum?

7. What evidence can you give that granulation is caused by convection?

8. How are granules and supergranules alike? How are they different?

9. How can astronomers detect structure in the chromosphere?

10. What evidence can you give that the corona has a very high temperature?

11. What heats the chromosphere and corona to maintain such high temperatures?

12. Why does hydrogen, which is abundant in the Sun's atmosphere, have relatively weak spectral lines, whereas calcium, which is not abundant, has very strong spectral lines?

13. What is the shape of the heliopause? Is it a point, line, circle, sphere, box, or something else? (*Hint:* Think about what defines the heliopause and try to draw it.)

14. How are astronomers able to explore the layers of the Sun below the photosphere?

15. Energy can be transported by convection, conduction, and radiation. Which of these is (or are) associated with the interior of the Sun?

16. What evidence can you give that sunspots are magnetic?

17. When in the cycle does the maximum number of sunspots occur? Is it at the beginning, in the middle, or at the end of a sunspot cycle? Is this time in the cycle reflected in the Maunder butterfly diagram? Why or why not?

18. How does the Babcock model explain the sunspot cycle?

19. How is meridional flow related to the Sun's magnetic dynamo, the sunspot cycle, and the Maunder butterfly diagram?

20. *Meridional* is derived from *meridian*. Your local meridian is from the north point on your horizon through your zenith to the south point on your horizon. Based on the definition of *meridian,* what direction is meridional flow?

21. What does the spectrum of a prominence reveal? What does its shape reveal?

22. Do prominences affect Earth? If so, how? If not, why not?

23. How can solar flares affect Earth?

24. Which has a more tightly bound nucleus, uranium, or helium? How do you know?

25. Why does nuclear fusion require high temperatures and high densities?

26. Why does nuclear fusion in the Sun occur only near the center?

27. How many protons are ultimately involved in the fusion to helium by the proton–proton fusion chain?

28. Give an example of a charged subatomic particle and a neutral subatomic particle discussed in this chapter.

29. If the Sun began its life with 75 percent H and 25 percent He by mass, what has happened since then that resulted in the percentages by mass listed in Table 8-1?

30. How can astronomers detect neutrinos from the Sun?

31. How did neutrino oscillation affect the detection of solar neutrinos by the Davis experiment?

32. **How Do We Know?** How do confirmation and consolidation extend scientific understanding?

33. **How Do We Know?** What does it mean when scientists say they are certain? What does scientific certainty really mean?

Discussion Questions

1. Some clothing now comes with ultraviolet protection factor (UPF) ratings like sun protection factor (SPF) ratings for sunscreens. A UPF rating of 50 is considered excellent because it only allows 1/50 of available UV radiation to pass through it. Construction, dye, chemical treatment, fiber type, stretching ability, efficiency when wet, and condition of garment affect the UPF rating. Can you assume a black shirt has a UPF of 50+? Why or why not? Should all clothing be required to have UPF ratings listed?

2. Have *Voyager 1* and *Voyager 2* passed through the heliopause yet? How would we know that they have? What is on the other side of the heliopause as viewed from Earth?

3. Explain why the presence of spectral lines of a given element in the solar spectrum tells you that element is present in the Sun, but the absence of the lines would not necessarily mean the element is absent from the Sun.

4. What information can we gain from a comparison of the Sun's spectrum with that of a spectrum from a nebula as shown in Figures 7-7a and 7-7b?

5. What energy sources on Earth *cannot* be thought of as stored sunlight?

6. You step outside into the sunshine and feel the warmth of the Sun. Which of the three ways to transport energy—conduction, convection, or radiation—applies to this scenario?

7. What would the spectrum of an auroral display look like? Why?

8. You live in Phoenix, Arizona. A news announcement breaks into your regularly scheduled TV program to let you know that a large coronal mass ejection is going to collide with Earth tomorrow. What do you do?

9. What observations would you make if you were ordered to set up a system that could warn orbiting astronauts of dangerous solar flares? (Such a warning system actually exists.)

10. Which energy generation process—chemical burning, fusion, gravitational contraction, or fission—do you suppose generated the elements in the Sun other than H and He that are listed in Table 8-1?

Problems

1. The radius of the Sun is 0.7 million km. What percentage of the radius is taken up by the chromosphere?

2. What fraction of the Sun's interior is the core? The radiative zone? The convective zone? Which layer occupies most of the volume of the Sun's interior? (*Hint:* Refer to Figure 8-16.)

3. The smallest detail visible with ground-based solar telescopes is about 1 arc second. How large a region does this represent on the Sun? (*Hint:* Use the small-angle formula, Chapter 3.)

4. What is the angular diameter of a star the same size as the Sun located 5 light-years (ly) from Earth? Is the *Hubble Space Telescope* able to detect detail on the surface of such a star? (*Hint:* Use the small-angle formula, Chapter 3.)

5. If a sunspot has a temperature of 4200 K and the sunspot can be considered a blackbody, what is the wavelength of maximum intensity in nm units and what color is associated with this wavelength? Is this the color we see the sunspot as from Earth? Why or why not? (*Hint:* Refer to Wien's law, Chapter 7.)

6. How many watts of radiation does a 1-meter-square region of the Sun's photosphere emit, at a temperature of 5800 K? How much would the wattage increase if the temperature were twice as much, 11,600 K? (*Hint:* Use the Stefan-Boltzmann law, Chapter 7.)

7. If a sunspot has a temperature of 4200 K and the average solar photosphere has a temperature of 5800 K, how much more energy is emitted in 1 second from a square meter of the photosphere compared to a square meter of the sunspot? (*Hint:* Use the Stefan-Boltzmann law, Chapter 7.)

8. The radial velocity of a granule's center is found to be −0.4 km/s. If the observed spectral line is Balmer H-alpha at a laboratory wavelength of 656.300 nm, at what wavelength is the line observed? Is that a blueshift or a redshift? Does that mean the gas is rising, sinking, or moving laterally across the line of sight? (*Hint:* Use the Doppler formula, Chapter 7.)

9. Gusts of the solar wind travel as fast as 1000 km/s. How many days would the solar wind take to reach Earth at this speed? (*Hint:* Refer to the Appendix A tables for the distance between the Sun and Earth.)

10. If the Sun rotates once every 24.5 days at the equator, once every 27.8 days at latitude 45 degrees, and once every 34.3 days at the poles, what is the average number of days the Sun takes to rotate once based on these three numbers?

11. How much energy is produced when the Sun converts 1 kg of mass into energy?

12. How much energy is produced when the Sun converts 1 kg of hydrogen into helium? (*Hint:* How does this problem differ from Problem 11?)

13. A 1-megaton nuclear weapon produces about 4×10^{15} J of energy. How much mass must vanish when a 1-megaton weapon explodes?

14. A solar flare can release 10^{25} J. How many megatons of TNT would be equivalent? (See Problem 13 for conversion between megatons and joules.)

15. The United States consumes about 2.5×10^{19} J of energy in all forms in a year. How many years could you run the United States on the energy released by the solar flare in Problem 14?

16. Use the luminosity of the Sun (the total amount of energy the Sun emits each second) to calculate how much mass the Sun converts into energy each second.

17. If the Sun began its life with 75 percent H and 25 percent He by mass, by how much did the respective percentages by mass change to result in the percentages listed in Table 8-1? Why are these changes in percentages not equal?

Learning to Look

1. Whenever there is a total solar eclipse, you can see something like the image shown at right. Explain why the shape and extent of the glowing gases are observed to be different for each eclipse.

Visual

NOAO/AURA/NSF

2. Look at Figure 8-3. To what height and which atmospheric layer does the red color correspond? How about the yellow color? Which color corresponds to the transition zone? Where in the image is the transition zone?

3. Refer to Figure 7-7 and Figure 8-3. What kinds of spectra are shown in Figure 7-7? What atmospheric layers of the Sun are associated with these spectra? At what height in km is the gas that generated these spectra?

4. Refer to labels C, F, and G in the solar spectrum shown in Figure 7-7a. What are the wavelengths of those lines? Which hydrogen spectral line series are these lines in? (*Hint:* Refer to Atomic Spectra, page 141.)

5. Look at the stars in image shown in How Do We Know? 4-1 (page 64). Can you see limb darkening? Why or why not? What about the full moon images in Figure 3-1? Does limb darkening apply?

6. The two images here show two solar phenomena. What are they, and how are they related? How do they differ?

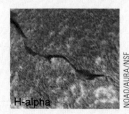

H-alpha

NOAO/AURA/NSF

Visual

NOAO/AURA/NSF

7. This image of the Sun was recorded in the extreme ultraviolet by the *SOHO* spacecraft. Explain the features you see.

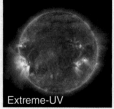

Extreme-UV

ESA/NASA/SOHO

Perspective: Origins

Guidepost As you begin your study of Earth and its sibling planets, you should first get some perspective on where and what our Solar System is. That means you need to take a quick tour of stars, the Milky Way Galaxy, other galaxies, and the Universe as a whole to help you view the Solar System and Earth in context. After that, you will understand better how Earth—and you—fit into the history and evolution of the Universe. Your tour of the cosmos will address four important questions:

► **How are stars born, and how do they die?**

► **What are galaxies, and how did they form and evolve?**

► **How did the Universe begin?**

► **How were the atoms in our bodies formed?**

Most of all, your tour of the Universe will show how galaxies, stars, the Sun, and other planets are made of the same elements of which Earth and our bodies are composed, and that those atoms were made long ago and far away.

The Universe, as has been observed before, is an unsettlingly big place, a fact which for the sake of a quiet life most people tend to ignore.

DOUGLAS ADAMS, *THE RESTAURANT AT THE END OF THE UNIVERSE*

Igor Chekalin/Moment/Getty Images

M78

Messier 78 (M78) is a cloud of interstellar gas and dust 1600 light-years from Earth in the direction of Orion. Some parts of the cloud are dense and opaque at visual wavelengths, appearing as dark regions devoid of stars. Other regions are illuminated by neighboring stars; the reflected light appears blue. The image was created by a private citizen, Igor Chekalin of Russia, for a worldwide contest sponsored by the European Southern Observatory (ESO).

Visual

EVERYTHING HAS TO COME FROM SOMEWHERE. Look at your thumb. The atoms in your thumb are billions of years old. Astronomers have compelling evidence that five billion years ago, those atoms were part of a cloud of gas floating in space, and before that those atoms were inside stars. The atoms inside your body are old, but the matter the atoms are made of is even older. That matter had its origin within minutes of the beginning of the Universe 13.8 billion years ago. If your atoms could tell their stories, you would be amazed to know where they came from and where they have been.

As you study the Solar System, remember the history of the matter it is made of. The iron inside Earth and in your blood, the oxygen, nitrogen and carbon in Earth's atmosphere that you breathe in and out, the calcium in your bones—all exist because stars have lived and died. You are no more isolated from the rest of the Universe than a raindrop is isolated from the sea.

9-1 The Birth of Stars

The stars above seem permanent fixtures of the sky, but astronomers know that stars are born and stars die. Their lives are long compared with a human life; however, if you know where to look, you can see all of the stages of stellar birth, aging, and death represented in the sky. Astronomers have put those stages into the proper order and can tell the life story of the stars, a story that begins in the darkness of interstellar space.

Although space seems empty, it is actually filled with thinly spread gas and dust, the **interstellar medium**. The gas atoms are mostly hydrogen, a few centimeters apart on average, and the dust is made of microscopic grains of heavier atoms such as carbon and iron. The dust makes up only about one percent of the mass of matter between the stars.

The interstellar medium is tenuous in the extreme, yet you can see clear evidence that it exists. In some places, the interstellar medium is collected into great dark clouds of dusty gas that obscure the stars beyond (see, for example, the image on page 175). In other cases, a nearby hot star can ionize the gas and create a glowing cloud (Figure 9-1). Astronomers refer to both dark and glowing interstellar clouds as **nebulae** (singular, nebula), from the Latin word for cloud or mist.

These nebulae not only adorn the sky, they also mark the birthplace of stars. As the cold gases of a nebula grow denser, gravity can pull parts of it together to form warmer, denser bodies that eventually become **protostars**—objects destined to become stars.

Astronomers can't see these protostars easily because they are hidden deep inside the dusty gas clouds from which they form, but they are easily detected at infrared wavelengths. As the more massive protostars become hot, luminous stars, the light and gas flowing away from the newborn stars blow the nebula away to reveal the new stars (Figure 9-2). In this way, a single gas cloud can give birth to a cluster of stars. Astronomers have evidence that our Sun was born in such a nebula almost 5 billion years ago.

Panel a: NOAO/AURA/NSF/N. Sharp; Panel b: ESO

Horsehead Nebula

Dusty foreground gas silhouetted against glowing gas illuminated by young, high-temperature stars.

Young star buried in the nebula

5 ly

Nebula N44

More than 40 massive young stars are inflating a bubble of hot gas inside the nebula from which they formed.

100 ly

Visual

▲ Figure 9-1 (a) The Horsehead Nebula is within the Milky Way Galaxy, about 1500 ly distant in the direction of Orion. (b) Young stars are found in and near clouds of gas and dust from which they have been born. Nebula N44 is 170,000 ly from Earth in a nearby galaxy. Gas in both nebulae is excited to glow by ultraviolet radiation from the most massive and luminous of the young stars. Interstellar dust is visible as dark clouds seen against the bright background gas.

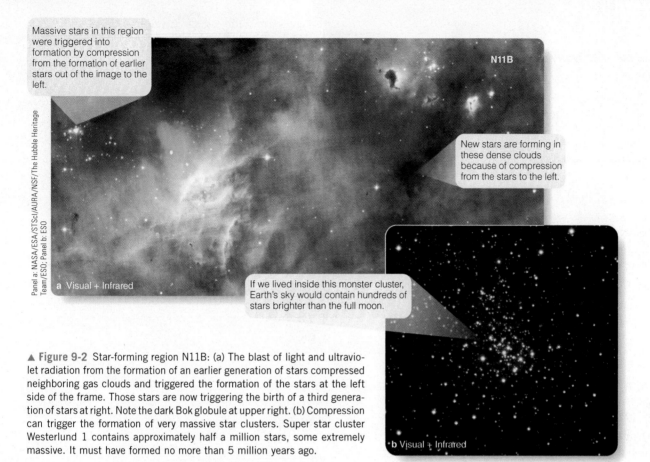

Massive stars in this region were triggered into formation by compression from the formation of earlier stars out of the image to the left.

N11B

New stars are forming in these dense clouds because of compression from the stars to the left.

Panel a: NASA/ESA/ESA/STScI/AURA/NSF/The Hubble Heritage Team/ESO; Panel b: ESO

a Visual + Infrared

If we lived inside this monster cluster, Earth's sky would contain hundreds of stars brighter than the full moon.

b Visual + Infrared

▲ **Figure 9-2** Star-forming region N11B: (a) The blast of light and ultraviolet radiation from the formation of an earlier generation of stars compressed neighboring gas clouds and triggered the formation of the stars at the left side of the frame. Those stars are now triggering the birth of a third generation of stars at right. Note the dark Bok globule at upper right. (b) Compression can trigger the formation of very massive star clusters. Super star cluster Westerlund 1 contains approximately half a million stars, some extremely massive. It must have formed no more than 5 million years ago.

If you look at the constellation Orion, you can see one of these regions with your unaided eye. The patch of haze in Orion's sword is the Great Nebula in Orion. Read **Star Formation in the Orion Nebula** on pages 178–179 and notice four points and one new term:

1 The visible nebula is only a small part of a vast, dusty cloud. You see the nebula because the stars born within it have ionized the gas and driven it outward, breaking out of the cloud. In some cases, the turbulence has produced small, dense clouds of gas and dust called *Bok globules* that may be in the process of forming stars.

2 Also notice that a single very hot star is almost entirely responsible for producing the ultraviolet photons that ionize the gas and make the nebula glow.

3 Infrared observations reveal clear evidence of active star formation deeper in the cloud behind the visible nebula.

4 Many of the young stars in the Orion Nebula and other star-forming regions are surrounded by disks of gas and dust. Astronomers have abundant evidence that planets can form in such disks.

As protostars contract, something happens that is quite important to planet-walkers like you—planets form. You will see in the next chapter how our Solar System formed from the disk of gas and dust that orbited the protostar that became our Sun.

You could identify the birth of a star as the time when most of its energy output comes from fusing hydrogen into helium instead of from contraction. Nuclear fusion reactions release energy, support the star, and stop its contraction. Stars that generate energy in their cores by hydrogen fusion are called **main-sequence stars**.

9-2 The Deaths of Stars

Stars spend most of their lives fusing hydrogen into helium, and, when the hydrogen is exhausted, they fuse helium into carbon. The more massive stars can fuse carbon into even heavier atoms, but iron fusion is the limit beyond which no star can go and remain stable.

How long a star can live depends on its mass. The smallest stars are very common, but they are not very luminous. These cool **red dwarf** stars are hardly massive enough to fuse

Star Formation in the Orion Nebula

1 The visible Orion Nebula shown below is a pocket of ionized gas on the near side of a vast, dusty molecular cloud that fills much of the southern part of the constellation Orion. The molecular cloud can be mapped by radio telescopes. To scale, the cloud would be many times larger than this page. As the stars of the Trapezium were born in the cloud, their radiation has ionized the gas and pushed it away. Where the expanding nebula pushes into the larger molecular cloud, it is compressing the gas (see diagram at right) and may be triggering the formation of the protostars that can be detected at infrared wavelengths within the molecular cloud.

Hundreds of stars lie within the nebula, but only the four brightest, those in the Trapezium, are easy to see with a small telescope. A fifth star, at the narrow end of the Trapezium, can be visible on nights of good seeing.

The cluster of stars in the nebula is less than 2 million years old. This means the nebula is similarly young.

Side view of Orion Nebula

Hot Trapezium stars   Protostars

To Earth

Expanding ionized hydrogen

Molecular cloud

Artist's conception

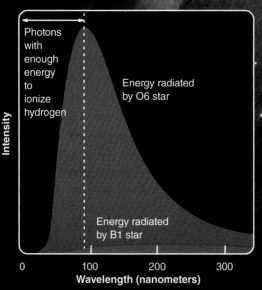

Trapezium

Visual

Credit: NASA, ESA, M. Robberto, STScI and the
Hubble Space Telescope Orion Treasury Project Team

Infrared

The near-infrared image above reveals more than 50 low-mass, cool protostars.

NASA/JPL-Caltech/N. Flagey (IAS), A. Noriega-Crespo (SSC)

Visual

NASA/ESA/STScI/AURA/NSF/The Hubble Heritage Team

Small dark clouds called **Bok globules**, named after astronomer Bart Bok, are found in and near star forming regions. The one pictured here is part of nebula NGC 1999 near the Orion Nebula. Typically about 1 light-year in diameter, they contain from 10 to 1000 solar masses.

Intensity (vertical axis)

Photons with enough energy to ionize hydrogen

Energy radiated by O6 star

Energy radiated by B1 star

0   100   200   300
Wavelength (nanometers)

2 Of all the stars in the Orion Nebula, just one is hot enough to ionize the gas. Only photons with wavelengths shorter than 91.2 nm can ionize hydrogen. The second-hottest stars in the nebula are B1 stars, and they emit little of this ionizing radiation. The hottest star, however, is an O6 star that has 30 times the mass of the Sun. At a temperature of 40,000 K, it emits plenty of photons with wavelengths short enough to ionize hydrogen. Remove that one star, and the nebula's emission would turn off.

3 The infrared image from the *Spitzer Space Telescope* reveals extensive nebulosity surrounding the visible Orion Nebula. Red and orange show the locations of warm dust that has been heated by starlight, green shows hot dust and ionized gas, and blue shows light coming directly from stars.

AAO/David Malin Images

In this near-infrared image, image, fingers of gas (nicknamed the "Hand of God") rush away from of a region full of infrared protostars.

Infrared

NASA/IRTF/D. Gezari (GSFC), D. Backman (Franklin & Marshall College), M. Werner (JPL-Caltech)

Infrared image

BN

KL

The Becklin-Neugebauer (BN) object is a B star just reaching the main sequence. The Kleinmann-Low (KL) Nebula is a cluster of cool, young protostars. BN and KL are detectable only at infrared wavelengths.

NASA

Trapezium cluster

The gas behind and around the Trapezium stars appears green because of the filters used to record this image.

Visual

500 AU

4 As many as 85 percent of the stars in the Orion Nebula are surrounded by disks of gas and dust. One such disk is seen at the upper right of this *Hubble Space Telescope* image, magnified in the inset. Radiation from the nearby hot, luminous Trapezium stars is evaporating gas from the disk and driving it away to form an elongated nebula.

Visual

NASA

Infrared

NASA/JPL-Caltech/T. Megeath (University of Toledo, Ohio); © 2016 Cengage Learning®

How Do We Know? 9-1

Mathematical Models

How can scientists study aspects of nature that cannot be observed directly? One of the most powerful methods in science is the mathematical model, a group of equations carefully designed to describe the behavior of objects and processes that scientists want to study. Astronomers build mathematical models of stars to study the structure hidden deep inside them. Models can allow you to imagine speeding up the slow evolution of stars or slowing down the rapid processes that generate energy. Stellar models are based on only four equations, but other models are much more complicated and may require many more equations.

For example, scientists and engineers designing a new airplane don't just build it, cross their fingers, and ask a test pilot to try it out. Long before any metal parts are made, mathematical models are created to test whether the wing design will generate enough lift, whether the fuselage can support the strain, and whether the rudder and ailerons can safely control the plane during takeoff, flight, and landing. Those mathematical models are put through all kinds of tests: Can a pilot fly with one engine shut down? Can the pilot recover from sudden turbulence? Can the pilot land in a crosswind? By the time the test pilot rolls the plane down the runway for the first time, the mathematical models have flown many thousands of miles.

Scientific models are only as good as the assumptions that go into them and must be compared with the real world at every opportunity. If you are an engineer designing a new airplane, you can test your mathematical models by making measurements in a wind tunnel. Models of stars are much harder to test against reality, but they do predict some observable things. Stellar models predict the existence of a main sequence, the mass–luminosity relation, the observed numbers of giant and supergiant stars, and the evolution of star clusters. Without mathematical models, astronomers would know little about the lives of the stars, and flying new airplanes would be a very dangerous business.

Before any new airplane flies, engineers build mathematical models to test its stability.

The Boeing Company

---

hydrogen, and they can survive for over 100 billion years. The Sun, a medium-mass star, will live a total of about 11 billion years as a main-sequence star. In contrast, the most massive stars, up to almost 100 times the mass of the Sun, are tremendously luminous and fuse their fuels so rapidly that they can live only a few million years.

This chapter tells the story of the birth and death of stars in a few paragraphs, but modern astronomers know a great deal about the lives of the stars. How could mere humans understand the formation, development, and deaths of objects that can survive for millions or even billions of years? The answers lie in mathematical models (**How Do We Know? 9-1**). Astronomers can express the physical laws that govern the gas and energy inside a star as equations that can be solved in computer programs. The resulting mathematical model describes the internal structure of a star and allows astronomers to follow its evolution as it ages (**Figure 9-3**). Solving the mystery of the evolution of stars is one of the greatest accomplishments of modern astronomy.

How a star dies depends on its mass. A medium-mass star like the Sun has enough hydrogen fuel to survive for billions of years, but it must eventually exhaust its fuel, swell to become a **giant** star, and then expel its outer layers in an expanding nebula. These **planetary nebulae** were named for their planet-like appearance in small telescopes, but they are, in fact, the remains of dying stars. Once the outer layers of a star are ejected into space, the hot core contracts to form a small, dense, cooling star—a **white dwarf**.

Study **Formation of Planetary Nebulae and White Dwarfs** on pages 182–183 and notice four important points:

1 You can understand what planetary nebulae are with simple analyses using Kirchhoff's laws and the Doppler effect that you learned about in Chapter 7.

2 Astronomers have developed a model to explain planetary nebulae. Real nebulae are more complex than is implied by the simple model of a slow wind and a fast wind, but that model provides a way to understand observations of these phenomena.

3 Oppositely directed jets and multiple shells produce many of the complicated shapes of planetary nebulae.

4 After the star has lost much of its outer layers, the core of the star contracts to become a remnant called a white dwarf.

When medium-mass stars like the Sun die, they expel their outer layers back into space, and some of the atoms that have been cooked up inside the stars get mixed back into the interstellar medium.

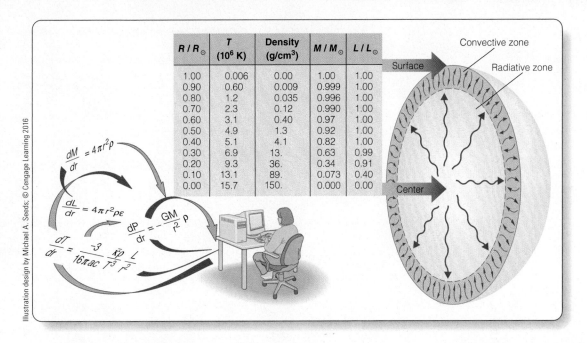

$R/R_\odot$	T (10^6 K)	Density (g/cm³)	$M/M_\odot$	$L/L_\odot$
1.00	0.006	0.00	1.00	1.00
0.90	0.60	0.009	0.999	1.00
0.80	1.2	0.035	0.996	1.00
0.70	2.3	0.12	0.990	1.00
0.60	3.1	0.40	0.97	1.00
0.50	4.9	1.3	0.92	1.00
0.40	5.1	4.1	0.82	1.00
0.30	6.9	13.	0.63	0.99
0.20	9.3	36.	0.34	0.91
0.10	13.1	89.	0.073	0.40
0.00	15.7	150.	0.000	0.00

The equations shown:

$$\frac{dM}{dr} = 4\pi r^2 \rho$$

$$\frac{dL}{dr} = 4\pi r^2 \rho \varepsilon$$

$$\frac{dP}{dr} = -\frac{GM}{r^2}\rho$$

$$\frac{dT}{dr} = \frac{-3}{16\pi ac}\frac{\bar{\kappa}\rho}{T^3}\frac{L}{r^2}$$

▲ **Figure 9-3** A stellar model is a table of numbers that represent conditions inside a star. Such tables can be computed using basic laws of physics, shown here in mathematical form. The table in this figure describes the Sun.

In contrast with stars like the Sun, massive stars become very large giants or even larger **supergiants**. These stars are so luminous that they exhale gas back into the interstellar medium (**Figure 9-4**). They live very short lives, perhaps only millions of years, before they develop iron cores and explode as **supernovae** (singular, supernova) (**Figure 9-5**). The core of such a dying massive star may form a **neutron star** or a **black hole**, but the outer parts of the star, newly enriched with the atoms cooked up inside the star, are returned to the interstellar medium.

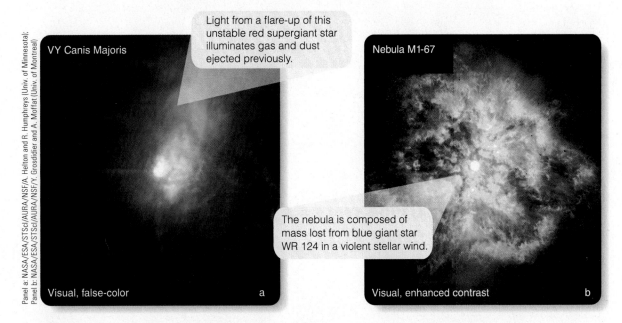

▲ **Figure 9-4** Stars can lose mass if they are very hot or very luminous. (a) The red supergiant VY Canis Majoris is ejecting loops, arcs, and knots of gas as it ages. (b) A massive hot star such as WR 124 constantly loses mass into space. This star is surrounded by nebula M1-67 composed of material it has expelled in the past.

Formation of Planetary Nebulae and White Dwarfs

1 Simple observations tell astronomers about the nature of planetary nebulae. Their angular size and distances indicate that their radii range from 0.2 to 3 ly. The presence of emission lines in their spectra implies that they are excited, low-density gas. Doppler shifts show they are expanding at 10 to 20 km/s. If you divide radius by velocity, you find that planetary nebulae are no more than about 10,000 years old. Older nebulae evidently become mixed into the interstellar medium and disappear.

Astronomers find about 1500 planetary nebulae in the sky. Because planetary nebulae are short-lived formations, you can conclude that they must be a common part of stellar evolution. Medium-mass stars up to a mass of about 8 to 10 $M_\odot$ are destined to die by forming planetary nebulae.

The Helix Nebula is 2.5 ly in diameter, and the radial texture shows how light and winds from the central star are pushing outward.

NASA/ESA/STScI/AURA/NSF/JPL-Caltech

Visual + Infrared

2 The process that produces planetary nebulae involves two stellar winds. First, as an aging giant, the star gradually blows away its outer layers in a slow breeze of low-excitation gas that is not easily visible. Once the hot interior of the star is exposed, it ejects a high-speed wind that overtakes and compresses the gas of the slow wind like a snowplow, while ultraviolet radiation from the hot remains of the central star excites the gases to glow like a giant neon sign.

2a The Cat's Eye, below, lies at the center of an extended nebula that must have been exhaled from the star long before the fast wind began forming the visible planetary nebula. See other images of this nebula on the opposite page

Slow stellar wind from a red giant

Fast wind from exposed interior

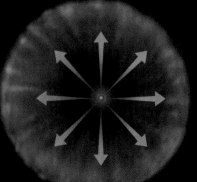

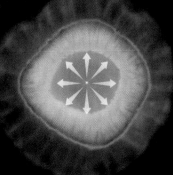

The gases of the slow wind are not easily detectable.

You see a planetary nebula where the fast wind compresses the slow wind.

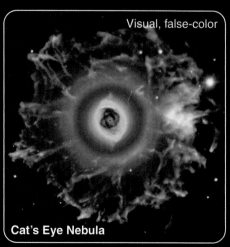

Visual, false-color

Cat's Eye Nebula

Nordic Optical Telescope/R. Corradi

3 Images from the *Hubble Space Telescope* reveal that asymmetry is the rule in planetary nebulae rather than the exception. A number of causes have been suggested. A disk of gas around a star's equator might form during the slow-wind stage and then deflect the fast wind into oppositely directed flows. Another star or planets orbiting the dying star, rapid rotation, or magnetic fields might cause these peculiar shapes. The Hour Glass Nebula seems to have formed when a fast wind overtook an equatorial disk (white in the image). The nebula Menzel 3 shows evidence of multiple ejections, as do many other planetary nebulae.

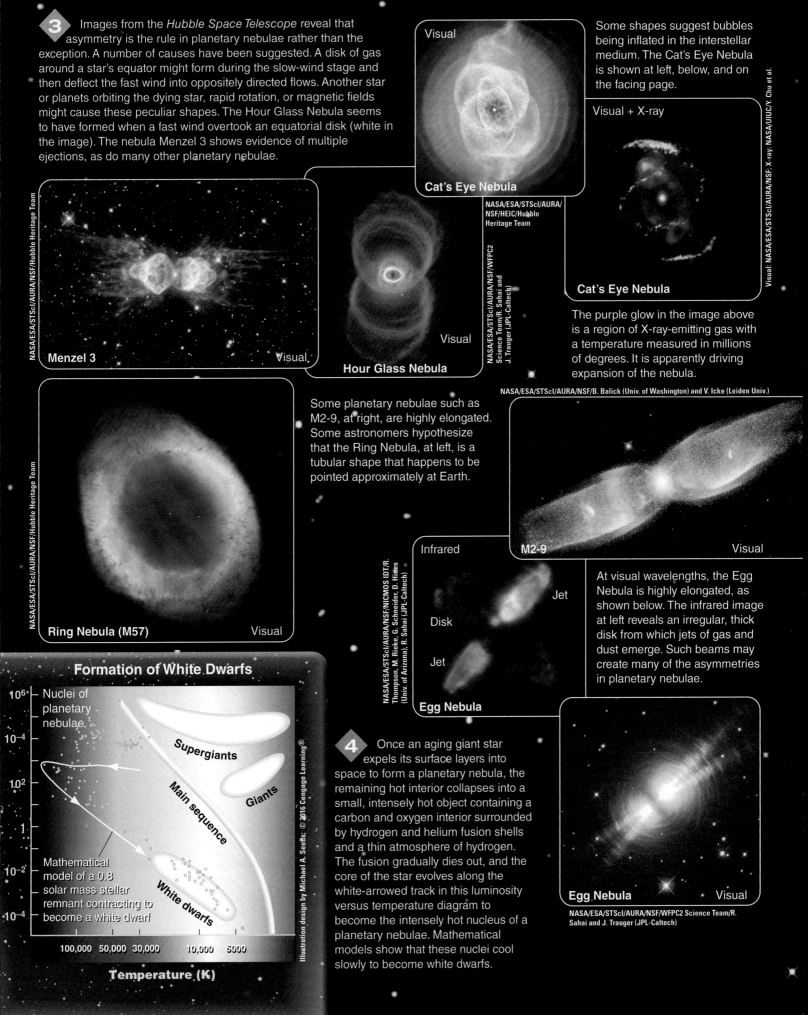

Some shapes suggest bubbles being inflated in the interstellar medium. The Cat's Eye Nebula is shown at left, below, and on the facing page.

Visual

Cat's Eye Nebula

NASA/ESA/STScI/AURA/NSF/HEIC/Hubble Heritage Team

Visual + X-ray

Cat's Eye Nebula

Visual: NASA/ESA/STScI/AURA/NSF X-ray: NASA/UIUC/Y. Chu et al.

The purple glow in the image above is a region of X-ray-emitting gas with a temperature measured in millions of degrees. It is apparently driving expansion of the nebula.

NASA/ESA/STScI/AURA/NSF/B. Balick (Univ. of Washington) and V. Icke (Leiden Univ.)

Menzel 3 Visual

NASA/ESA/STScI/AURA/NSF/Hubble Heritage Team

Hour Glass Nebula Visual

NASA/ESA/STScI/AURA/NSF/WFPC2 Science Team/R. Sahai and J. Trauger (JPL-Caltech)

Some planetary nebulae such as M2-9, at right, are highly elongated. Some astronomers hypothesize that the Ring Nebula, at left, is a tubular shape that happens to be pointed approximately at Earth.

Ring Nebula (M57) Visual

NASA/ESA/STScI/AURA/NSF/Hubble Heritage Team

M2-9 Visual

Infrared

Jet

Disk

Jet

Egg Nebula

NASA/ESA/STScI/AURA/NSF/NICMOS IDT/R. Thompson, M. Rieke, G. Schneider, D. Hines (Univ. of Arizona), R. Sahai (JPL-Caltech)

At visual wavelengths, the Egg Nebula is highly elongated, as shown below. The infrared image at left reveals an irregular, thick disk from which jets of gas and dust emerge. Such beams may create many of the asymmetries in planetary nebulae.

Egg Nebula Visual

NASA/ESA/STScI/AURA/NSF/WFPC2 Science Team/R. Sahai and J. Trauger (JPL-Caltech)

Formation of White Dwarfs

10^6	Nuclei of planetary nebulae
10^{-4}	Supergiants
10^2	Main sequence / Giants
1	
10^{-2}	Mathematical model of a 0.8 solar mass stellar remnant contracting to become a white dwarf / White dwarfs
10^{-4}	

100,000 50,000 30,000   10,000   5000

Temperature (K)

Illustration design by Michael A. Seeds; © 2016 Cengage Learning®

4 Once an aging giant star expels its surface layers into space to form a planetary nebula, the remaining hot interior collapses into a small, intensely hot object containing a carbon and oxygen interior surrounded by hydrogen and helium fusion shells and a thin atmosphere of hydrogen. The fusion gradually dies out, and the core of the star evolves along the white-arrowed track in this luminosity versus temperature diagram to become the intensely hot nucleus of a planetary nebulae. Mathematical models show that these nuclei cool slowly to become white dwarfs.

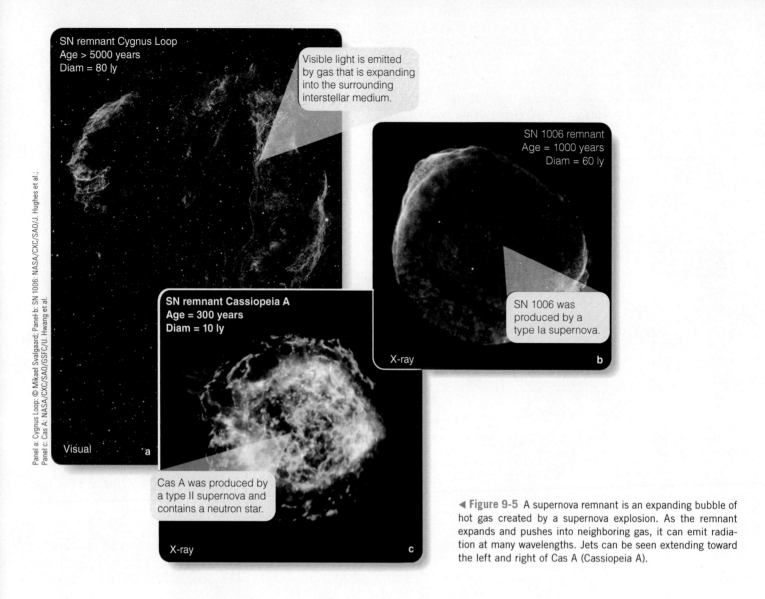

Panel a: Cygnus Loop: © Mikael Svalgaard; Panel b: SN 1006: NASA/CXC/SAO/J. Hughes et al.; Panel c: Cas A: NASA/CXC/SAO/GSFC/U. Hwang et al.

SN remnant Cygnus Loop
Age > 5000 years
Diam = 80 ly

Visible light is emitted by gas that is expanding into the surrounding interstellar medium.

Visual   a

SN 1006 remnant
Age = 1000 years
Diam = 60 ly

SN 1006 was produced by a type Ia supernova.

X-ray   b

SN remnant Cassiopeia A
Age = 300 years
Diam = 10 ly

Cas A was produced by a type II supernova and contains a neutron star.

X-ray   c

◀ **Figure 9-5** A supernova remnant is an expanding bubble of hot gas created by a supernova explosion. As the remnant expands and pushes into neighboring gas, it can emit radiation at many wavelengths. Jets can be seen extending toward the left and right of Cas A (Cassiopeia A).

The violence of a supernova explosion can fuse atoms together to build elements heavier than iron. Gold, platinum, uranium, and other elements heavier than iron are rare and valuable because they are made only in the moments of a supernova explosion. The iodine atoms in your thyroid gland and the gold atoms in your class ring were made in supernova explosions.

The atoms of which you are made had their birth inside stars. That process is common in the Universe because stars are common. Our galaxy contains billions of them.

9-3 Our Home Galaxy

From a dark location away from city lights, you can see the **Milky Way** stretching across the sky. The winter Milky Way is especially dramatic (**Figure 9-6a**). It is actually the disk of the galaxy that we live in—the **Milky Way Galaxy**—seen from the inside. Of course, no one has ever journeyed out into space to look back and take a picture of our galaxy, but astronomers have evidence that the nearby Andromeda Galaxy, shown in **Figure 9-6b**, looks much like our own. Our galaxy contains roughly 100 billion stars, and we live about two-thirds of the way from the center to the edge.

Astronomers estimate that our galaxy is about 80,000 light-years in diameter. That is, light takes 80,000 years to travel from one edge to the other. If you had a model of the Milky Way Galaxy as big as North America, the entire Solar System would be about the size of a small cookie, and the Sun and planets would all be too small to see without a powerful microscope.

Large astronomical telescopes reveal other galaxies scattered across the sky. You have already met the Andromeda Galaxy in Figure 9-6. It is so close you can see its nucleus with the unaided eye as a hazy patch in the constellation Andromeda. Other galaxies are all around us, and some are dramatically beautiful (**Figure 9-7**).

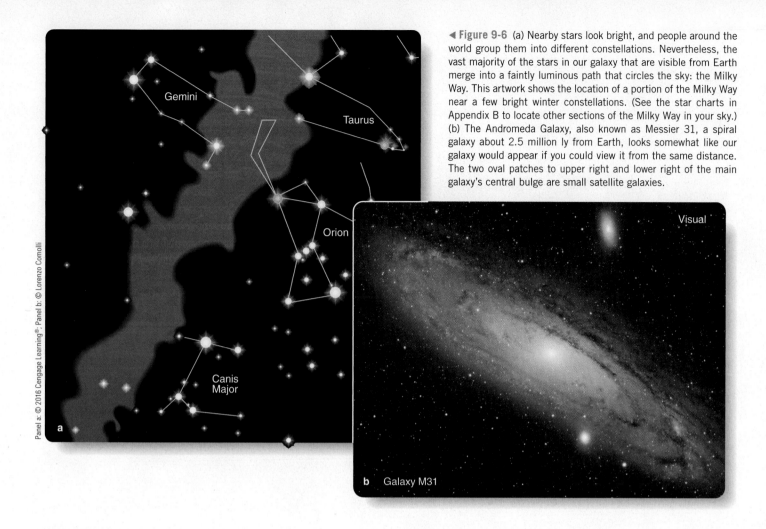

◄ Figure 9-6 (a) Nearby stars look bright, and people around the world group them into different constellations. Nevertheless, the vast majority of the stars in our galaxy that are visible from Earth merge into a faintly luminous path that circles the sky: the Milky Way. This artwork shows the location of a portion of the Milky Way near a few bright winter constellations. (See the star charts in Appendix B to locate other sections of the Milky Way in your sky.) (b) The Andromeda Galaxy, also known as Messier 31, a spiral galaxy about 2.5 million ly from Earth, looks somewhat like our galaxy would appear if you could view it from the same distance. The two oval patches to upper right and lower right of the main galaxy's central bulge are small satellite galaxies.

Gemini

Taurus

Orion

Canis Major

a

Visual

b   Galaxy M31

9-4 The Universe of Galaxies

Our galaxy is our home, but ours is only one of billions of galaxies. Some, like our Milky Way Galaxy, are disk shaped with graceful spiral arms marked by clouds of gas and bright newborn stars. But many galaxies are great swarms of stars with relatively little gas and dust.

Study **Galaxy Classification** on pages 186–187 and notice three important points and four new terms:

1 Many galaxies have no disk, no spiral arms, and almost no gas and dust. These *elliptical galaxies* (class E) range from huge giants to small dwarfs.

2 Disk-shaped galaxies usually have spiral arms and contain gas and dust that is the raw material for star formation. Many of these *spiral galaxies* (class S) have a central region shaped like an elongated bar and are called *barred spiral galaxies* (class SB). The Milky Way Galaxy is a barred spiral. A few disk galaxies contain relatively little gas and dust; these are called *lenticular galaxies* (class S0).

3 *Irregular galaxies* (class Irr) are generally shapeless and tend to be rich in gas and dust.

You might expect such titanic objects as galaxies to be rare, but large telescopes reveal that the sky is filled with galaxies. Like leaves on the forest floor, galaxies carpet the sky. They fill the Universe in every direction as far as telescopes can see (**Figure 9-8**). Grouped in clusters and superclusters, galaxies are the homes of the billions of stars that illuminate the Universe and create the chemical elements.

When two galaxies collide, the stars swirl past each other without bumping, but the great clouds of gas and the magnetic fields inside the galaxies do collide and compress each other. The compression of the gas clouds can stimulate two colliding galaxies to form vast numbers of new stars and massive star clusters. Furthermore, tidal forces twist and distort the colliding galaxies. Images of such colliding galaxies show their twisted shapes, far-flung streamers of stars and gas, and extensive regions of star formation (**Figure 9-9**). Such bursts of star formation salt the galaxies with newly formed heavy atoms. Our own

1 **Elliptical galaxies** are round or elliptical, contain no visible gas and dust, and lack hot, bright stars. They are classified with a numerical index ranging from 1 to 7; E0s are round, and E7s are highly elliptical. The index is calculated from the largest and smallest diameter of the galaxy using the following formula, rounding to the nearest integer:

$$\frac{10(a-b)}{a}$$

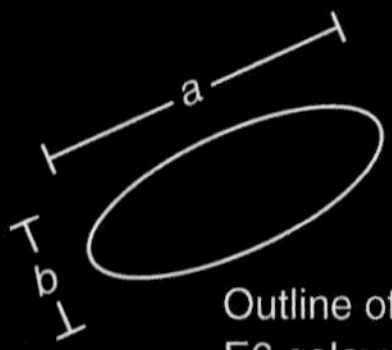

Outline of an E6 galaxy

NOAO/AURA/NSF

AATB/David Malin Images

Visual

The Leo 1 dwarf elliptical galaxy is not much bigger than a globular cluster.

Visual

M87 is a giant elliptical galaxy classified E1. It is several times larger in diameter than our own galaxy and is surrounded by a swarm of over 500 globular clusters.

2 **Spiral galaxies** contain a disk and spiral arms. Their halo stars are not visible, but presumably all spiral galaxies have halos. Spirals contain gas and dust and hot, bright O and B stars, as shown at right and below. The presence of short-lived O and B stars alerts us that star formation is occurring in these galaxies. Sa galaxies have larger nuclei, less gas and dust, and fewer hot, bright stars. Sc galaxies have small nuclei, lots of gas and dust, and many hot, bright stars. Sb galaxies are intermediate. The Milky Way Galaxy is classified as Sbc, between Sb and Sc.

AATB/David Malin Images

Sa

Visual   NGC 3623

AATB/David Malin Images

BAR

Sb

AATB/David Malin Images

NGC 3627   Visual

Sc

AATB/David Malin Images

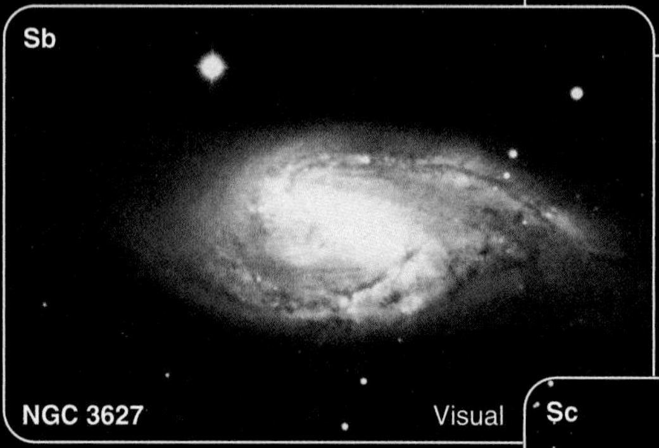

2a Roughly two-thirds of all spiral galaxies are **barred spiral galaxies** classified SBa, SBb, and SBc. They have an elongated nucleus with spiral arms springing from the ends of the bar, as shown at left. The Milky Way Galaxy is a barred spiral, so its complete classification is SBbc.

NGC 1365   Visual

NGC 2997   Visual

2b Some disk galaxies are rich in dust, which is concentrated along their spiral arms. NGC 4013, shown below, is a galaxy much like ours, but seen edge-on its dust is readily apparent.

Dust visible in spiral arm crossing in front of more distant galaxy

Visual

NGC 2207 and IC 2163

Dust in spiral galaxies is most common in the spiral arms. Here the spiral arms of one galaxy are silhouetted in front of a more distant galaxy.

Visual

The galaxy IC 4182 is a dwarf irregular galaxy about 4 million pc from our galaxy.

2c Galaxies with an obvious disk and nuclear bulge but little or no visible gas and dust and no hot, bright stars are called **lenticular galaxies**, classified as S0 (pronounced "Ess Zero"). Compare this galaxy with the edge-on spiral above.

Visual

Visual

3 **Irregular galaxies** (classified Irr) are a chaotic mix of gas, dust, and stars with no obvious nuclear bulge or spiral arms. The Large and Small Magellanic Clouds are visible to the unaided eye as hazy patches in the Southern Hemisphere sky. Telescopic images show that they are irregular galaxies that are interacting gravitationally with our own much larger galaxy. Star formation is rapid in the Magellanic Clouds. The bright pink regions are emission nebulae excited by newborn massive stars. The brightest nebula in the Large Magellanic Cloud is called the Tarantula Nebula.

Tarantula Nebula

Small Magellanic Cloud

Large Magellanic Cloud

Visual

Visual

NGC 1300

Visual

NASA/ESA/STScl/AURA/NSF/The Hubble Heritage Team

▲ Figure 9-7 The beautiful symmetry of the spiral pattern is clear in this image of the barred spiral galaxy NGC 1300. Star-forming regions containing young star clusters with hot, luminous stars and clouds of ionized hydrogen gas are located along the spiral arms, while dark lanes of dust mark the inner edges of the arms.

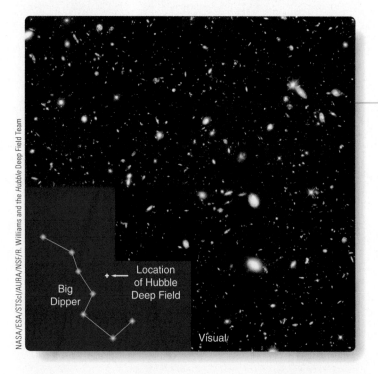

Big Dipper

+ ← Location of Hubble Deep Field

Visual

NASA/ESA/STScl/AURA/NSF/R. Williams and the Hubble Deep Field Team

▲ Figure 9-8 An apparently empty spot on the sky only 1/30 the diameter of the full moon contains over 1500 galaxies in this extremely long time exposure known as the Northern Hubble Deep Field. Presumably the entire sky is similarly filled with galaxies.

Milky Way Galaxy has probably collided with more than one smaller galaxy, and some of the atoms of which you are made may have been formed because of those collisions.

9-5 The Origin of the Universe

A little less than a century ago, astronomers made an astonishing discovery. The galaxies in the Universe are moving away from each other. The spectrum of a galaxy is the combined spectrum of billions of stars, and the spectral lines visible in galaxy spectra are shifted slightly toward the red end of the spectrum. This redshift is proportional to distance. That is, the farther away a galaxy is, the larger is the redshift that is visible in its spectrum (Figure 9-10). This result, first discovered by the American astronomer Edwin Hubble in 1929, is clear evidence that the Universe is expanding.

If the galaxies are all rushing away from each other, then you can imagine viewing a video running backward. You would see the galaxies moving toward each other, and eventually you would see the galaxies pushing into each other, compressing and heating the gas until your video screen was filled with the glare of an intensely hot, fantastically dense gas filling the Universe. This, the beginning of the Universe, is a state called the **big bang**.

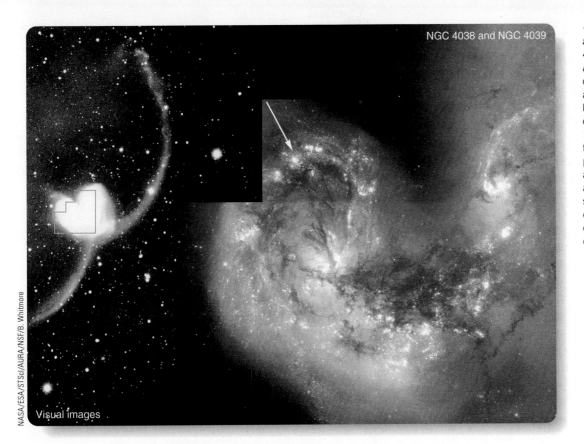

NGC 4038 and NGC 4039

Visual images

NASA/ESA/STScI/AURA/NSF/B. Whitmore

◄ **Figure 9-9** The colliding galaxies NGC 4038 and NGC 4039 are known as the Antennae because their long curving tails resemble the antennae of an insect. Earth-based photos (left) show little detail, but a *Hubble Space Telescope* image (right) reveals the collision of the two galaxies producing thick clouds of dust and raging star formation, creating roughly a thousand massive star clusters such as the one at the top (arrow). Such collisions between galaxies are common.

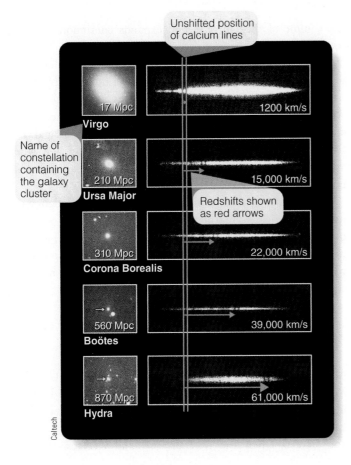

Unshifted position of calcium lines

Name of constellation containing the galaxy cluster

Redshifts shown as red arrows

17 Mpc	1200 km/s	
Virgo		
210 Mpc	15,000 km/s	
Ursa Major		
310 Mpc	22,000 km/s	
Corona Borealis		
560 Mpc	39,000 km/s	
Boötes		
870 Mpc	61,000 km/s	
Hydra		

Caltech

How can anyone know the big bang really happened? The evidence is conclusive—astronomers can see it (**How Do We Know? 9-2**). To understand how you might see an event that occurred billions of years ago, you need to look once again at the galaxies. Nearby galaxies, such as the Andromeda Galaxy, are only a few million light-years away. The hazy light you see in the sky coming from that direction in the constellation Andromeda has been traveling through space for only 2.5 million years, and you see the Andromeda Galaxy as it was 2.5 million years ago when the light began its journey. If you used a telescope and looked at more distant galaxies, you would see them as they were a billion years ago or more; light from those galaxies has been traveling that long. If you used a big enough telescope and looked at the most distant galaxies, you would be looking back

◄ **Figure 9-10** These galaxy spectra extend from near-ultraviolet wavelengths at left to the blue part of the visible spectrum at right. The two dark absorption lines of once-ionized calcium are prominent in the ultraviolet. The redshifts in galaxy spectra are expressed here as apparent velocities of recession. Note that the apparent velocity of recession is proportional to distance, which is known as the Hubble law.

How Do We Know? 9-2

Science: A System of Knowledge

What is the difference between believing in the big bang and understanding it? If you ask a scientist, "Do you believe in the big bang?" she or he may hesitate before responding. The question implies something incorrect about the way science works.

The goal of science is to understand nature. Science is a logical process based on observations and experiments used to test and confirm hypotheses and theories. A scientist does not really believe in even a well-confirmed theory in the way people normally use the word *believe*. Rather, the scientist understands the theory and recognizes how different pieces of evidence support or contradict the theory.

There are other ways to know things, and there are many systems of belief. Religions, for example, are systems of belief that are not entirely based on observation. In some cases,

a political system is also a system of belief; many people believe that democracy is the best form of government and do not ask for, or expect, evidence supporting that belief. A system of belief can be powerful and lead to deep insights, but it is different from science.

Scientists try to be careful with words, so thoughtful scientists would not say they *believe* in the big bang. They would say that the evidence is overwhelming that the big bang really did occur and that they are compelled by a logical analysis of both the observations and the theory to conclude that the theory is very likely correct. In this way scientists try to be objective and reason without distortion by personal feelings and prejudices.

A scientist once referred to "the terrible rule of evidence." Sometimes the evidence forces a scientist to a conclusion she or he does not like, but the personal preferences of

each scientist must take second place to the rule of evidence.

Do you believe in the big bang? Or, instead, do you have confidence that the theory is right because of the evidence? There is a big difference.

Scientific knowledge is based objectively on evidence such as that gathered by spacecraft.

NASA/WMAP Science Team

---

in time and seeing them as they were over 10 billion years ago when the Universe was young.

What would you see if you looked at the empty places on the sky between the most distant galaxies? You could detect a glow that was emitted by the hot, dense clouds of gas of the big bang, but you couldn't see that glow with your unaided eyes because the redshift is so great that the photons of light are shifted into the long-wavelength infrared and radio parts of the spectrum. Nevertheless, astronomers can "see" the radiation from the big bang with infrared and radio detectors. This is called the **cosmic microwave background radiation**, and it fills the Universe, pouring in on Earth from all directions, and telling you that you are part of a Universe that was very hot and very dense about 13.8 billion years ago (**Figure 9-11**). Your atoms and Earth's atoms were part of the big bang.

Notice that the background radiation is visible in any direction in the sky. The big bang did not occur in a specific place, but it filled the entire volume of the Universe. As the Universe expanded, the total volume of space increased, and the hot gases of the big bang cooled and formed galaxies. Except for relatively small random motions, galaxies do not move as the Universe expands, but are carried away from each other as the volume of space continues to increase. Galaxies are not fragments ejected from an explosion but rather are parts of a whole that fills all of

the volume of the Universe and expands continuously as the total volume increases.

The big bang theory seems fantastic, but it has been tested and confirmed over and over, and modern astronomers have great confidence that there really was a big bang (**How Do We Know? 9-3**). Of course, there are more details to understand. How did the first galaxies form? Why did the galaxies form in giant clusters? You live in an exciting age when astronomers are able to ask these questions and expect to discover answers.

These ideas strain human imagination and challenge the most sophisticated mathematics, but the lesson is a simple one. The Universe had a beginning not so long ago, and the matter you are made of had its birthday in that moment of cosmic beginnings. You are small, but you are part of something vast.

9-6 The Story of Matter

Astronomers can tell the story of the matter in the Universe. Mathematical models of the big bang show that the first few minutes in the history of the Universe were unimaginably hot. Energy was so intense that it could form particles of matter, and that is when the first protons and electrons formed.

How Do We Know? 9-3

Theories and Proof

How do astronomers know the Sun isn't made of burning coal? People say dismissively of a scientific explanation they dislike, "That's only a theory," as if a theory were just a random guess. A hypothesis is like a guess in some ways, although of course not a random one. What scientists mean by the word *theory* is a hypothesis that has "graduated" to being confidently considered a well-tested truth. You can think of a hypothesis as equivalent to having a suspect in a criminal case, whereas a theory is equivalent to finishing a trial and convicting someone of the crime.

Of course, no matter how many tests and experiments you conduct, you can never prove that any scientific theory is absolutely true. It is always possible that the next observation you make will disprove the theory. And it is unfortunately sometimes true that innocent people go to jail and guilty people are free, but, occasionally, with further evidence, those legal mistakes can be fixed.

There have always been hypotheses about why the Sun is hot. It was once thought that the Sun is a ball of burning material. Only a century ago, most astronomers accepted the hypothesis that the Sun is hot because gravity was making it contract. In the late 19th century, geologists showed that Earth was

much older than the Sun could be if it was powered by gravity, so the "gravity hypothesis" had to be wrong. It wasn't until 1920 that a new hypothesis was proposed by Sir Arthur Eddington, who suggested that the Sun is powered somehow by the energy contained in atomic nuclei. In 1938 the German-American astrophysicist Hans Bethe showed how nuclear fusion could power the Sun. He won the Nobel Prize for that work in 1967.

The fusion hypothesis is now so completely confirmed that it is fair to call it a theory. No one will ever go to the center of the Sun, so you can't *prove* the fusion theory is right. Many observations and model calculations support this theory, and in Chapter 8 you saw further evidence in the neutrinos that have been detected coming from the Sun's core. Nevertheless there remains some tiny possibility that all the observations and models are misunderstood and that the theory will be overturned by some future discovery. Astronomers have tremendous confidence that the Sun is powered by fusion and not gravity or coal, but a scientific theory can never be proved conclusively correct.

There is a great difference between a theory in the colloquial sense of a far-fetched guess and a scientific theory that has

undergone decades of testing and confirmation with observations, experiments, and models. However, no theory can ever be proved absolutely true. It is up to you as a consumer of knowledge and a responsible citizen to distinguish between a flimsy guess and a well-tested theory that deserves to be treated like truth—at least pending substantial further information.

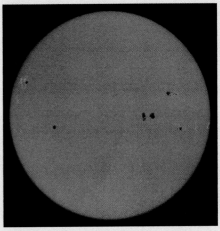

ESA/NASA/SOHO/MDI

Technically it is still a theory, but astronomers have tremendous confidence that the Sun gets its power from nuclear fusion and not from burning coal.

NASA/WMAP Science Team

▲ **Figure 9-11** Data from the *Wilkinson Microwave Anisotropy Probe (WMAP)* space telescope were used to make this microwave-wavelength map of the sky. This radiation was emitted when the Universe was very young and filled with the hot gas of the big bang. The statistical distribution of the tiny irregularities in brightness allows astronomers to determine the age of the Universe and its rate of expansion.

Protons and electrons, even when they are not attached to each other, are hydrogen, so the first gas of the big bang was hydrogen. The temperature and density were so high that nuclear fusion reactions could occur, fusing some of the hydrogen into helium.

By the time the Universe was a few minutes old, expansion had cooled its gases, and nuclear fusion stopped. In those first few minutes, about 25 percent of the mass had been converted into helium. But because there are no stable atomic nuclei with masses of five or eight times that of hydrogen, the fusion process could not go past helium. Few heavier atoms could be made.

As the Universe expanded and cooled, it became dark. Gravity drew matter together to form great clouds that contracted to form the first galaxies. Eventually, stars began to form; and, as those first stars began to shine, they lit up the Universe and ionized the hydrogen gas in great shells around the galaxies (Figure 9-12).

Since the formation of the first stars, galaxies have collided and merged; generation after generation of stars have lived and died. Mass loss from aging stars, planetary nebulae, and supernova explosions has spread heavy elements out into the thin gas in space where they have become incorporated into new stars in a continuous gas-star-gas cycle (Figure 9-13). Except for hydrogen, all of the atoms of which you are made were cooked up inside stars before the Sun and Earth were born (Figure 9-14). When the Sun began to form from a cloud of gas, the atoms now inside your body were there. They were part of Earth as it formed, and now they are part of you. Every atom has a history that stretches back through the birth and death of stars to the first moment in time. You exist because stars have died.

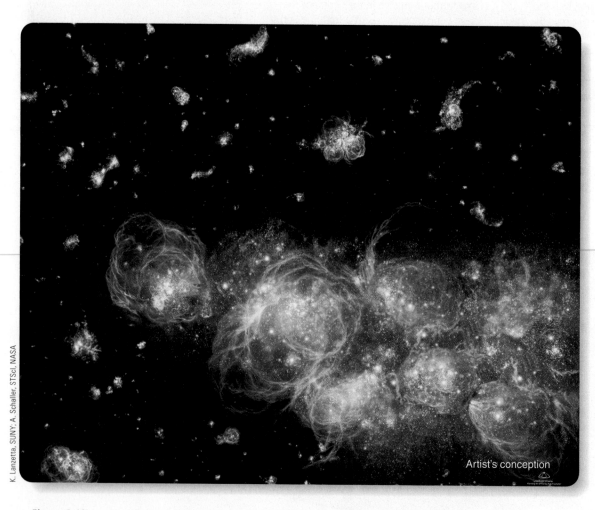

K. Lanzetta, SUNY; A. Schaller, STScI, NASA

Artist's conception

▲ **Figure 9-12** In this artist's conception, the first stars that formed after the big bang produced floods of ultraviolet photons that ionized the gas in expanding bubbles around young galaxies. Such a storm of star formation ended an era lasting hundreds of millions of years during which the Universe expanded in darkness. As star formation began, so did the production of chemical elements heavier than helium.

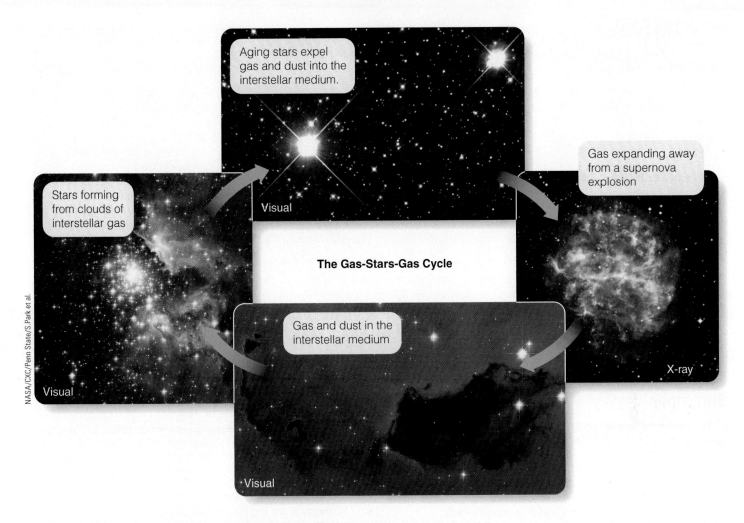

▲ **Figure 9-13** Matter cycles from the interstellar medium to form stars and back into the interstellar medium when stars die. Our galaxy is slowly but steadily using up its star-making supply of hydrogen and increasing the amount of heavier elements.

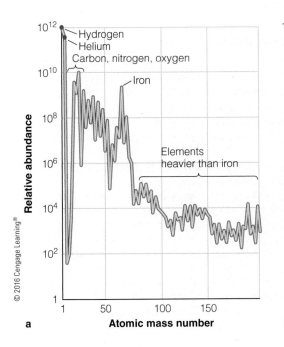

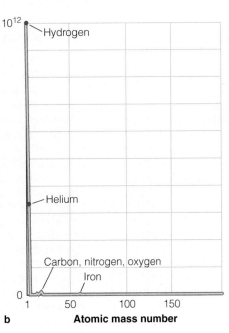

◀ **Figure 9-14** The abundances of chemical elements in the Universe. (a) When the elemental abundances are plotted on an exponential scale, you see that elements heavier than iron are about a million times less common than iron and that all elements heavier than helium (the "metals") are quite rare. (b) The same data plotted on a linear scale provide a more realistic impression of how rare the metals are. Carbon, nitrogen, and oxygen make small peaks near atomic mass 15, and iron is just visible in the graph.

Hang on tight. The Sun, with Earth in its clutch, is ripping along at about 225 km/s (that's 500,000 mph) as it orbits the center of the Milky Way Galaxy. We live on a wildly moving ball of rock in a large galaxy that humanity calls home, but the Milky Way is more than just our home. Perhaps "parent galaxy" would be a better name.

Except for hydrogen atoms, which have survived unchanged since the Universe began, you and Earth are made of metals—atoms heavier than helium. There is no helium in your body, but there is plenty of carbon, nitrogen, and oxygen. There is calcium in your bones and iron in your blood. All of those atoms and

more were cooked up inside stars or during their supernova death throes.

Stars are born when clouds of gas orbiting the center of our galaxy collide with the gas in spiral arms and are compressed. That process has given birth to generations of stars, and each generation has produced elements heavier than helium and spread them back into the interstellar medium. The abundance of metals has grown slowly in our galaxy. About 4.6 billion years ago, a cloud of gas enriched in those heavy atoms slammed into a spiral arm and produced the Sun, Earth, and, eventually, you. You have been created by the Milky Way—your parent galaxy.

Study and Review

Summary

► The **interstellar medium (p. 176)**, the gas and dust between the stars, can be seen as **nebulae (p. 176)**. Small, dense, dark clouds are called **Bok globules (p. 178)**. Stars are born when clouds of gas and dust contract to form **protostars (p. 176)**.

► Protostars are cool and hidden inside clouds of gas and dust, so they are most easily observed at infrared wavelengths.

► Planets form in the disks of gas and dust around protostars.

► Stars that generate energy by hydrogen fusion, including the Sun, are called **main-sequence stars (p. 177)**.

► The lowest-mass stars, the **red dwarfs (p. 177)**, can survive for many billions of years.

► The Sun is a medium-mass star and can live for about 11 billion years before it expands to become a **giant (p. 180)**, expels its outer layers to form a **planetary nebula (p. 180)**, and collapses to form a **white dwarf (p. 180)**.

► The most massive stars live only a few million years and expand to become **supergiants (p. 181)**. Giant and supergiant stars exhale gas back into the interstellar medium. The most massive stars die in explosions called **supernovae (p. 181)** that can form elements heavier than iron and blow them plus atoms of lighter elements made earlier in the star's life back into the interstellar medium. Supernova explosions can leave behind **neutron stars (p. 181)** or **black holes (p. 181)** as stellar remnants.

► Stars cook hydrogen atoms into atoms of heavier elements and return those atoms to the interstellar medium when the star dies.

► The **Milky Way (p. 184)**, the hazy band of light across the night sky, is our galaxy, the **Milky Way Galaxy (p. 184)**, seen from the inside. It contains about 100 billion stars and is about 80,000 ly in diameter. We live about two-thirds of the way from the center to the edge.

► Billions of galaxies fill the sky. Some are **elliptical (p. 186)**, some are **lenticular (p. 187)**, and some are **irregular (p. 187)**. Some spiral galaxies, including the Milky Way Galaxy, contain a bar-shaped nucleus and are called **barred-spiral galaxies (p. 186)**.

► Collisions between galaxies are common and can trigger star formation.

► The redshifts of the galaxies show that the Universe is expanding, and imagining running the expansion backward in time leads to

inference of the high-temperature, high-density beginning of the Universe called the **big bang (p. 188)**.

► The **cosmic microwave background radiation (p. 190)** is light released from matter soon after the big bang that has been redshifted by the expansion of the Universe into the far-infrared, microwave, and radio parts of the electromagnetic spectrum. It is direct evidence that the big bang really happened. Because Earth is part of the big bang, the radiation is detectable in every direction in the sky.

► The energy of the big bang produced protons and electrons (hydrogen gas), and some of that hydrogen fused into helium during the first few minutes, but no heavier elements could form.

► Gravity drew the gas into great clouds that formed galaxies, and stars fused hydrogen into heavier elements. Matter passes through a gas–star–gas cycle and is enriched in elements heavier than helium. A number of generations of stars have produced the chemical elements in Earth and in you.

Review Questions

1. What evidence could you cite that there is an interstellar medium?
2. Why are protostars difficult to observe at visible wavelengths?
3. How are protostars related to planet formation?
4. How do stars like the Sun die?
5. How do massive stars die?
6. What is the difference between the Milky Way and the Milky Way Galaxy?
7. Describe how star formation is related to gas and dust in galaxies such as ellipticals and spirals.
8. What evidence can you site to show that the Universe is expanding? That there really was a big bang?
9. Why would it be accurate to say that the big bang formed the matter in your body, but the stars formed the atoms?
10. **How Do We Know?** How can scientists use mathematical models to understand processes that they cannot observe directly?
11. **How Do We Know?** Why do scientists hesitate to say they "believe" in a certain theory?
12. **How Do We Know?** How would you respond to someone who said, "Oh, that's only a theory"?

Origin of the Solar System and Extrasolar Planets

10

Guidepost You have studied the appearance, origin, structure, and evolution of stars, galaxies, and the entire Universe. So far, though, your studies have left out one important type of object—planets. Now it is time for you to fill in that blank. In this chapter, once you learn about the general characteristics of the Solar System and the evidence for how it formed, you can understand how the processes you have been studying produced Earth, your home planet.

As you explore our Solar System in space and time, you will find answers to four important questions:

▶ **What are the observed properties of the Solar System?**

▶ **What is the theory for the origin of the Solar System that explains the observed properties?**

▶ **How did Earth and the other planets form?**

▶ **What do astronomers know about extrasolar planets orbiting other stars?**

In the following six chapters, you will explore in more detail each of the planets, plus asteroids, comets, and meteoroids. By studying the origin of the Solar System before studying the individual objects in it, you give yourself a better framework for understanding these fascinating worlds.

What place is this?
Where are we now?

CARL SANDBURG, *GRASS*

NASA/JPL-Caltech

Artist's conception of a "heavy bombardment" episode in a young planetary system around the star Eta Corvi. Evidence for this barrage comes from spectroscopic observations by the *Spitzer* infrared space telescope that detected fragments of both rocky and icy bodies in that system.

$\mathbf{M}$ICROSCOPIC CREATURES LIVE in the roots of your eyelashes. Don't worry. Everyone has them, and they are harmless. (*Demodex folliculorum* has been found in 97 percent of individuals and is a characteristic of healthy skin.) They hatch, fight for survival, mate, lay eggs, and die in the tiny spaces around the roots of your eyelashes without doing any harm. Some live in renowned places—the eyelashes of a glamorous movie star, for example—but the tiny beasts are not self-aware; they never stop to say, "Where are we?"

There are many good reasons to study the Solar System. You should study Earth and its sibling planets because, as you are about to discover, there are almost certainly more planets in the Universe than stars. Above all, you should study the Solar System because it is your home in the Universe. Humans are an intelligent species, so we have the ability and the responsibility to wonder where we are and what we are. Our kind have inhabited this Solar System for at least a hundred thousand years, but only within the past few hundred years have we begun to understand what the Solar System is.

10-1 A Survey of the Solar System

Over the course of decades astronomers have been searching the present Solar System for evidence of its past. In this section, you will survey the Solar System and compile a list of its most significant characteristics that are potential clues to how it formed.

You can begin with the most general view of the Solar System (Figure 10-1). It is, in fact, almost entirely empty space (look back to Figure 1-7, page 4). Imagine making a model of the Solar System in which 1 AU, the average distance between Sun and Earth, is represented by 4 m (13 ft). Then the Sun would be the size of a plum, Earth a grain of table salt, and the Moon a speck of pepper about 1 cm (0.4 in.) from Earth. Jupiter would be represented by an apple seed 21 m (69 ft) from the Sun, and Neptune, at the edge of the planetary zone, would be a large grain of sand 120 m (400 ft) from the central plum. Your model Solar System would be larger than a football field, but you would need a magnifying glass to detect even the largest asteroids orbiting between Mars and Jupiter. The planets and other Solar System objects are relatively tiny specks of matter scattered around the Sun.

Revolution and Rotation

The planets revolve around the Sun in orbits that lie close to a common plane. (Recall from Chapter 2 that the words *revolve* and *rotate* refer to different types of motion. A planet revolves around the Sun but rotates on its axis. Cowboys in the Old West didn't carry revolvers; those guns should have been called rotators.) The orbit of Mercury, the closest planet to the Sun, is tipped 7.0 degrees to Earth's orbit. The rest of the planets' orbital planes are inclined by no more than 3.4 degrees. As you can see, the Solar System is basically flat and disk shaped.

© 2016 Cengage Learning®

▲ **Figure 10-1** A fanciful conception of the Solar System as if seen from a nearby vantage point. All the planets orbit in the same direction, in one plane, in approximately circular orbits. Comets, in contrast, normally have very eccentric orbits that are often inclined to the plane of the planets' orbits. These are all clues to how the Solar System formed. The planets are shown here more than 1000 times larger than their true diameters relative to the sizes of their orbits.

The rotation of the Sun and planets on their axes also seems related to the rotation of the disk. The Sun rotates with its equator inclined only 7.2 degrees to Earth's orbit, and most of the other planets' equators are tipped less than 30 degrees to their respective orbits. The rotations of Venus and Uranus are peculiar, however. Venus rotates backward compared with the other planets, whereas Uranus rotates on its side with its equator almost perpendicular to its orbit. Later in this chapter you will be able to understand how they might have acquired their peculiar rotations; you will explore those planets in detail in subsequent chapters.

There is a preferred direction of motion in the Solar System—counterclockwise as seen from the north. All the planets revolve around the Sun in that direction. With the exception of Venus and Uranus, all the planets also rotate on their axes in that direction. Furthermore, nearly all of the moons in the Solar System, including Earth's Moon, orbit around their respective planets in that same direction. With only a few exceptions, revolution and rotation in the Solar System follow a single theme. Apparently, these motions today are related to the original rotation of a disk of Solar System construction material.

Two Kinds of Planets

Perhaps the most striking clue to the origin of the Solar System comes from the obvious division of the planets into two groups, the small Earth-like worlds and the giant Jupiter-like worlds. The difference is so dramatic that you are led to say, "Aha, this must mean something!" Study **Terrestrial and Jovian Planets** on pages 198–199, notice three important points, and learn two new terms:

1 The two kinds of planets are distinguished by their location. The four inner *Terrestrial planets* are quite different from the four outer *Jovian planets*.

2 Craters are common. Almost every solid surface in the Solar System is covered with craters.

3 The two groups of planets are also distinguished by properties such as number of moons and presence or absence of rings. A theory of the origin of the planets needs to explain those properties.

The division of the planets into two groups is a clue to how our Solar System formed. The present properties of individual planets, however, don't tell everything you need to know about their origins. The planets have all evolved since they formed. For further clues about the origin of the planets, you can look at smaller objects that have remained largely unchanged since soon after the birth of the Solar System.

Cosmic Debris

The Sun and planets are not the only objects in the Solar System; it is littered with several kinds of space debris that are a rich source of information about the origin of the planets. They will be described briefly in the following pages, and you will learn more about them in later chapters.

The **asteroids** are small, rocky worlds, most of which orbit the Sun in a belt between the orbits of Mars and Jupiter. The term *asteroid* means "starlike," but of course they are not anything like stars. Planetoid, meaning "planetlike," would be a more accurate term; sometimes they are called *minor planets*. As of July 2014, almost 400,000 asteroids have orbits that are charted. It is a **Common Misconception** that the asteroids are the remains of a planet that broke apart. In fact, planets are held together very tightly by their gravity and do not "break apart." Astronomers recognize the asteroids as debris left over from the failure of a planet to form at a distance of about 3 AU from the Sun.

Only about 200 asteroids are more than 100 km (60 mi) in diameter, and tens of thousands are estimated to be more than 10 km (6 mi) in diameter. There are probably more than a million that are larger than 1 km (0.6 mi), and billions that are smaller. Because the largest are only a few hundred kilometers in size, Earth-based telescopes can detect no details on their surfaces, and even the *Hubble Space Telescope* can image only the largest features. Photographs returned by robotic spacecraft show that asteroids are generally irregular in shape and covered with craters (**Figure 10-2**). Spectroscopic observations indicate that asteroid surfaces are made up of a variety of rocky and metallic materials. (Note that metal is used here in the familiar sense, referring to substances like iron, rather than the stellar astronomer's sense meaning any element other than hydrogen and helium.) Observations of asteroids will be discussed in detail in a later chapter, but in this quick survey you have enough information to conclude that when the Solar System formed it included elements that compose rock and metal and also that collisions have played an important role in the Solar System's history.

Since 1992, astronomers have discovered more than a thousand small, icy bodies orbiting in the outer fringes of the Solar System beyond Neptune. This collection of objects is called the **Kuiper Belt** after astronomer Gerard Kuiper (*KYE-per*), who predicted their existence in the 1950s. There are probably 100 million objects larger than 1 km in the Kuiper Belt—many more than in the asteroid belt. A successful theory for the formation of the Solar System should include an explanation for how the **Kuiper Belt Objects (KBOs)** came to be where they are. You will find out more about the Kuiper Belt and its origin in a later chapter regarding the outer Solar System.

Terrestrial and Jovian Planets

1 The distinction between the Terrestrial planets and the Jovian planets is dramatic. The inner four planets, Mercury, Venus, Earth, and Mars, are **Terrestrial** (Earth-like) **planets**, meaning they are small, dense, rocky worlds with little or no atmosphere. The outer four planets, Jupiter, Saturn, Uranus, and Neptune, are **Jovian** (Jupiter-like) **planets**, meaning they are large, low-density worlds with thick atmospheres and liquid or ice interiors.

The planets and the Sun to scale. Saturn's rings would reach more than halfway from Earth to the Moon.

Planetary orbits to scale. The Terrestrial planets lie close to the Sun, whereas the Jovian planets are spread far from the Sun.

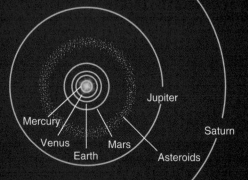

1a Of the Terrestrial planets, Earth is the most massive, but the Jovian planets are much more massive. Jupiter contains over 300 Earth masses, Saturn nearly 100 Earth masses. Uranus and Neptune each contain about 15 Earth masses.

Mercury is only 40 percent larger than Earth's Moon, and its weak gravity cannot retain a permanent atmosphere. Like the Moon, it is covered with craters from meteorite impacts.

Mercury

Earth's Moon

2 Craters are common on all of the surfaces in the Solar System that are strong enough to retain them. Earth has about 150 impact craters, but many more have been erased by erosion. Terrestrial planets, asteroids, comet nuclei, and nearly all of the moons in the Solar System are scarred by craters. Ranging from microscopic to hundreds of kilometers in diameter, these craters have been produced over the ages by meteorite impacts. When astronomers see a rocky or icy surface that contains few craters, they know that the surface is young.

Mercury is so close to the Sun it is difficult to study from Earth. The MESSENGER spacecraft went into orbit around Mercury in 2011 and was able to take detailed photos of the planet's entire surface.

Moon

Mercury

These five worlds are shown in proper relative size.

Earth

3 The Terrestrial planets have densities like that of rock or metal. The Jovian planets all have low densities, and Saturn's average density is only about 70 percent that of water.

The atmospheres of the Jovian planets are turbulent, and some are marked by great storms such as the Great Red Spot on Jupiter, but the atmospheres are not deep. If Jupiter were shrunk to the size of a tennis ball, its atmosphere would be no deeper than the fuzz.

Mars

Mars has a thin atmosphere and little water. Craters and volcanoes are common on its desert surface.

Venus (radar image)

These Jovian worlds are shown in proper relative size.

3a The interiors of the Jovian planets contain small cores of heavy elements such as metals, surrounded by a liquid. Jupiter and Saturn contain hydrogen forced into a liquid state by the high pressure. Less-massive Uranus and Neptune contain heavy-element cores surrounded by partially solid water mixed with some rocks and minerals.

Venus at visual wavelengths

Jupiter

Great Red Spot

The Terrestrial planets are drawn here to the same scale as the Jovian planets.

The Jovian planets have extensive systems of satellites. For example, Jupiter is orbited by four large moons which were discovered by Galileo in 1610, and dozens of smaller moons discovered up to the present day.

Saturn's rings seen through a small telescope

Neptune

Uranus

Saturn

3b All four Jovian planets have ring systems. Saturn's rings are made of ice particles. The rings of Jupiter, Uranus, and Neptune are made of dark rocky particles. Terrestrial planets have no rings.

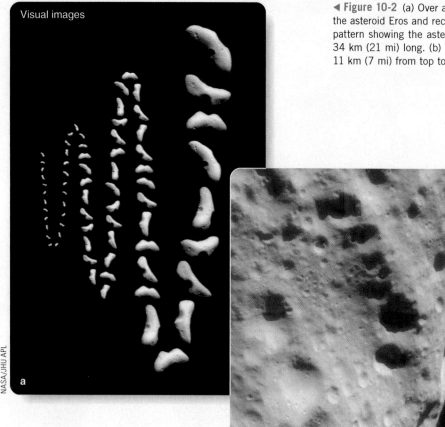

Visual images

a

b

NASA/JHU APL

◄ **Figure 10-2** (a) Over a period of three weeks, the *NEAR* spacecraft approached the asteroid Eros and recorded a series of images arranged here in an entertaining pattern showing the asteroid's irregular shape and 5-hour rotation period. Eros is 34 km (21 mi) long. (b) This close-up of the surface of Eros shows an area about 11 km (7 mi) from top to bottom.

In contrast to the small asteroids and distant Kuiper Belt objects, the brightest **comets** can be seen with the naked eye and are impressively beautiful (Figure 10-3). A comet may be visible for months as it sweeps through the inner Solar System. Most comets are faint, however, and difficult to locate even at their brightest.

The nuclei of comets are ice-rich bodies a few kilometers or tens of kilometers in diameter, similar in size to asteroids. Comets provide evidence that at least some parts of the Solar System had abundant icy material when it formed. You will discover more about the composition and history of comets in a later chapter.

A comet nucleus remains frozen and inactive while it is far from the Sun. If the comet's orbit carries it into the inner Solar System, the Sun's heat begins to vaporize the ices, releasing gas and dust. The flow of solar wind (look back to Chapter 8, page 153) plus **radiation pressure** exerted by sunlight pushes the gas and dust away, forming a long tail. As a result, the tail of a comet always points approximately away from the Sun (Figure 10-3b), no matter what direction the comet itself is moving in. The beautiful tail of a comet can be longer than an AU, although is produced by a relatively tiny nucleus only a few kilometers in diameter.

Unlike the stately comets, **meteors** flash across the sky in momentary streaks of light (Figure 10-4). They are commonly called *shooting stars*. Of course, they are not stars but small bits of rock and metal colliding with Earth's atmosphere and bursting into incandescent vapor because of friction with the air about 80 km (50 mi) above the ground. This vapor condenses to form dust that settles slowly to Earth, adding about 40,000 tons per year to our planet's mass.

Technically, the word *meteor* refers to the streak of light in the sky. In space, before its fiery plunge, the object is called a **meteoroid**, and any part of it that survives its fiery passage to Earth's surface is called a **meteorite**. Most meteoroids are specks of dust, grains of sand, or tiny pebbles. Almost all the meteors you see in the sky are produced by meteoroids that weigh less than 1 gram. Only rarely is a meteoroid massive and strong enough to survive its plunge, reach Earth's surface, and become a meteorite.

Thousands of meteorites have been found, and you will learn more about their various types in a later chapter. Meteorites are mentioned here for one specific reason: They can reveal the age of the Solar System.

Age of the Solar System

The most accurate way to find the age of a rocky body is to bring a sample into the laboratory and analyze the radioactive elements it contains. When a rock solidifies, it incorporates known percentages of the chemical elements. A few of these elements have forms called isotopes (Chapter 7, page 132) that are radioactive, meaning they gradually decay into other isotopes. For example, potassium-40, called a *parent isotope*, decays into calcium-40 and argon-40, called *daughter isotopes*. The **half-life** of a radioactive substance is the time it takes for half of the parent isotope atoms to decay into daughter isotope atoms. The abundance of a radioactive substance gradually decreases as it decays, and the abundances of the daughter substances gradually

Panel a: Kent Wood/Science Source, Inc.; Panel b: © 2016 Cengage Learning®

Comet's orbit

a   Visual

b

▲ **Figure 10-3** (a) A comet may remain visible in the sky for weeks as it passes through the inner Solar System. Although comets are actually moving rapidly along their orbits, they are so distant that, on any particular evening, a comet seems to hang motionless in the sky relative to the background constellations. Comet Hyakutake is shown here near Polaris in 1996. (b) A comet in a long, elliptical orbit becomes visible when the Sun's heat vaporizes its ices and pushes the gas and dust away in a tail.

increase (**Figure 10-5**). The half-life of potassium-40 is 1.3 billion years. If you also have information about the abundances of the elements in the original rock, you can compare those with the present abundances and find the age of the rock. For example, if you study a rock and find that only 50 percent of the potassium-40 remains and the rest has become a mixture of daughter isotopes, you could conclude that one half-life must have passed and that the rock is 1.3 billion years old.

Potassium isn't the only radioactive element used in radioactive dating. Uranium-238 decays with a half-life of 4.5 billion years to form lead-206 and other isotopes. Rubidium-87 decays into strontium-87 with a half-life of 47 billion years. Any of these substances can be used as a radioactive clock to find the age of mineral samples.

Daniel Good

Visual

▲ **Figure 10-4** A meteor is a sudden streak of glowing gases produced by a meteoroid, a bit of solid material, colliding with Earth's atmosphere. Friction with the air vaporizes the material about 80 km (50 mi) above Earth's surface. This meteor is seen against the background of part of the Milky Way.

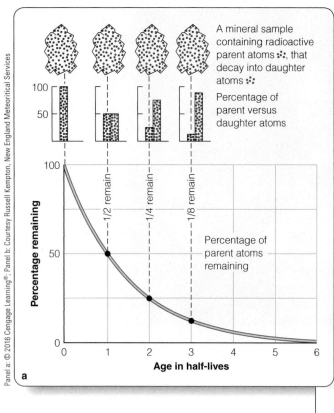

A mineral sample containing radioactive parent atoms ∴, that decay into daughter atoms ∵

Percentage of parent versus daughter atoms

Percentage of parent atoms remaining

a

b

◄ **Figure 10-5** (a) The radioactive parent atoms (*red*) in a mineral sample decay into daughter atoms (*blue*). Half the radioactive atoms are left after one half-life, a fourth after two half-lives, an eighth after three half-lives, and so on. (b) Radioactive dating shows that this fragment of the Allende meteorite is 4.56 billion years old. It contains a few even older interstellar grains, which formed long before our Solar System did.

Of course, to find a radioactive age, you need to get a sample into the laboratory, and the only celestial bodies of which scientists have samples for which ages have been determined are Earth, the Moon, Mars, and meteorites. The oldest Earth rocks so far discovered and dated are tiny zircon crystals from Australia that are 4.4 billion years old. That does not mean that Earth formed 4.4 billion years ago. As you will see in the next chapter, the surface of Earth is active, and the crust is continuously destroyed and replaced with material welling up from beneath the crust. Those types of processes tend to dilute the daughter atoms and spread them away from the parent atoms, effectively causing the radioactive clocks to reset to zero. Thus, the radioactive age of a rock is actually the length of time since the material in that rock was last melted. Consequently, the dates of these oldest rocks tell you only a lower limit to the age of Earth, in other words, that Earth is at least 4.4 billion years old.

One of the most important scientific goals of the Apollo lunar landings was to bring lunar rocks back to Earth's laboratories where their ages could be measured. Because the Moon's surface is not geologically active like Earth's surface, some Moon rocks might have survived unaltered since early in the history of the Solar System. In fact, the oldest Moon rocks are 4.5 billion years old. That means the Moon must be at least 4.5 billion years old.

Although no one has yet been to Mars, more than a dozen meteorites found on Earth have been identified by their chemical composition as having come from Mars. Most of these have ages of only a billion years or so, but one has an age of approximately 4.5 billion years. Mars must be at least that old.

Meteorites are actually the primary source for determining the age of the Solar System. Radioactive dating of meteorites yields a range of ages, but there is a fairly precise upper limit—many meteorite samples have ages of 4.56 billion years old, and none is older. That figure is widely accepted as the age of the Solar System and is often rounded to 4.6 billion years. The true ages of Earth, the Moon, and Mars are also assumed to be 4.6 billion years, although no rocks from those bodies have yet been found that have remained unaltered for that entire stretch of time.

One last celestial body deserves mention: the Sun. Astronomers estimate the age of the Sun to be about 5 billion years, but that is not a radioactive date because we have no samples of radioactive material from the Sun. Instead, an independent estimate for the age of the Sun can be made using helioseismological observations and mathematical models of the Sun's interior (Chapter 8). This yields a value of about 5 billion years, plus or minus 1.5 billion years, a number that is in agreement with the age of the Solar System independently derived from the age of meteorites. The evidence is consistent with all the bodies of the Solar System forming at about the same time, some 4.6 billion years ago.

10-2 The Great Chain of Origins

You are linked through a great chain of origins that leads backward through time to the instant when the Universe began, 13.8 billion years ago. The gradual discovery of the links in that chain has been one of the most exciting adventures of the human intellect. In previous chapters, you studied some of that story: the formation of stars, the growth of chemical elements in stellar furnaces, the formation of galaxies, and the origin of the Universe in the big bang. Now you have enough information to understand the origin of planets.

History of the Atoms in Your Body

Astronomers have compelling evidence that the Universe began in the big bang. By the time the Universe was a few minutes old, the protons, neutrons, and electrons in your body had come into existence. You are made of very old matter.

Although those particles formed quickly, they were not linked together to form many of the atoms that are common today. Most of the atoms in the early Universe were hydrogen, and about 10 percent were helium. Although your body does not contain helium, it does contain many of those ancient hydrogen atoms unchanged since the Universe began. Evidence indicates that almost no atoms heavier than helium were made in the big bang.

Within a few hundred million years after the big bang, matter began to collect to form galaxies containing billions of stars. You have learned that nuclear reactions inside stars are where low-mass atoms such as hydrogen are combined to make heavier atoms. Generation after generation of stars cooked the original particles, fusing them into atoms such as carbon, nitrogen, and oxygen that are common in your body. Even the calcium atoms in your bones were assembled inside stars.

Massive stars produce iron in their cores, but much of that iron is destroyed when the core collapses and the star explodes as a supernova. Model calculations indicate that most of the iron on Earth and in your body was produced instead by carbon fusion in type Ia supernova explosions and by the decay of radioactive atoms in the expanding matter ejected by type II supernovae. Atoms heavier than iron, such as gold, silver, and iodine, are also created by rapid nuclear reactions that occur during supernova explosions. Iodine is critical to the function of your thyroid gland, and you probably have gold and silver jewelry or dental fillings. Realize that atoms of these types, which are part of your life on Earth, were made during violent stellar explosions billions of years ago.

Our galaxy contains at least 100 billion stars, of which the Sun is one. Astronomers have a variety of evidence that the Sun formed from a cloud of gas and dust about 5 billion years ago, and the atoms in your body were part of that cloud. This chapter explains how the cloud gave birth to the planets and how the atoms in your body found their way onto Earth and into you. As you explore the origin of the Solar System, keep in mind the great chain of origins that created the atoms. As the geologist Preston Cloud remarked, "Stars have died that we might live."

Early Hypotheses for the Origin of Earth and the Solar System

The earliest descriptions of Earth's origin are myths and folktales that go back beyond the beginning of recorded history. From the time of Galileo, telescopes yielded observational evidence on which to base rational explanations for celestial phenomena. Although people like Copernicus, Kepler, and Galileo worked to find logical explanations for the motions of Earth and the other planets, other scholars began thinking about the origin of Earth and the Solar System.

The first physical theory of the Solar System's origin was proposed by the French philosopher and mathematician René Descartes in 1644. Because he lived and wrote before the time of Isaac Newton, Descartes did not recognize that gravity is the dominant force in the Universe. Rather, he believed that forces are transmitted by contact between bodies and that the Universe is filled with vortices of whirling invisible particles. Descartes proposed that the Sun and planets formed when a large vortex contracted and condensed. His hypothesis explained the general properties of the Solar System known at the time.

A century later, in 1745, the French naturalist Georges-Louis Leclerc proposed an alternate hypothesis that the planets were formed when a passing comet collided with or passed close to the Sun and pulled matter out of the Sun gravitationally. Working after the publication of Newton's *Principia*, Buffon was aware of the power of gravity. However, he did not know that the solid parts of comets are small, insubstantial bodies. Later astronomers modified Buffon's hypothesis to propose that

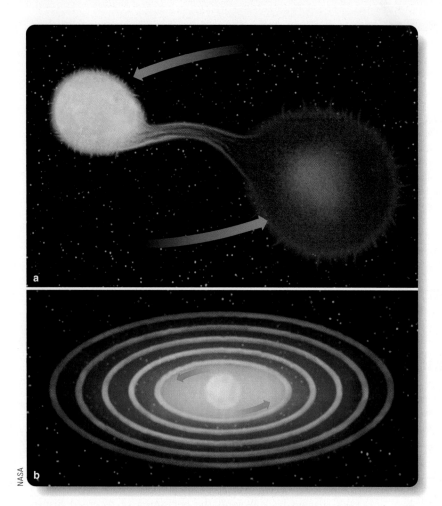

NASA

▲ Figure 10-6 (a) The passing star hypothesis proposed that the Sun was hit by, or had a very close encounter with, another star and that matter torn from the Sun and the other star formed planets orbiting the Sun, and perhaps the other star. This is an example of a catastrophic hypothesis. (b) Originally suggested in the 18th century, the nebular hypothesis proposed that a contracting disk of matter around the Sun spun faster as it conserved angular momentum and shed rings of matter that then formed planets. This is an example of an evolutionary hypothesis.

another star, rather than a comet, interacted with the Sun. According to the modified hypothesis, matter ripped from the Sun and the other star condensed to form the planets, which were driven into orbit around the Sun by the motion of the two stars' collision (Figure 10-6a).

This **passing star hypothesis** was popular off and on for two centuries, but it was problematic for several reasons. First, stars are very far apart in relation to their sizes and their relative velocities, so they collide extremely infrequently. Only a small number of stars in our galaxy have ever suffered a collision or close encounter with another star. Also, the gas pulled from the Sun and the other star would have been much too hot to condense to make planets, and would have dispersed instead. Furthermore, even if planets did form from that gas, they would not have gone into stable orbits around the Sun.

The hypotheses of Descartes and Buffon fall into two different categories. Descartes proposed an **evolutionary hypothesis** involving common, gradual processes to produce the Sun and planets. If it were correct, stars with planets would be common. Buffon's idea, on the other hand, is a **catastrophic hypothesis**. It involves an unlikely, sudden event to produce the Solar System, and thus implies that planetary systems are very rare. Although your imagination may enjoy picturing the spectacle of colliding stars, modern scientists have observed that changes in nature are usually gradual, occurring in small steps rather than sudden, dramatic events. The modern theory for the origin of the planets, based on many independent types of evidence, is evolutionary rather than catastrophic (**How Do We Know? 10-1**).

Early versions of the present-day theory of the Solar System's origin were suggested by philosophers Emanuel Swedenborg and Immanuel Kant, respectively, in 1734 and 1755. They reasoned, qualitatively, that a spinning cloud could contract under the influence of its own gravity and produce a disk of material that might condense into planets orbiting a central mass, the Sun. In 1796, Pierre-Simon de Laplace, a brilliant French astronomer and mathematician, put that theory on a mathematical basis to produce the **nebular hypothesis**. Laplace knew that as the disk grew smaller, it had to conserve angular momentum and spin faster and faster. (Recall from Chapter 5 that angular momentum is the tendency of a rotating object to continue rotating.) Laplace reasoned that, when the disk was spinning as fast as it could, it would shed its outer edge to leave behind a ring of matter. Then the disk could contract further, speed up again, and leave another ring. In this way, he imagined, the contracting disk would leave behind a series of rings, each of which could become a planet circling the newborn Sun at the center of the disk (Figure 10-6b).

According to the nebular hypothesis, the Sun should be spinning very rapidly, or, to put it another way, the Sun should have most of the Solar System's total angular momentum. As astronomers studied the planets and the Sun, however, they found that the Sun rotates relatively slowly and that the planets moving in their orbits actually have most of the angular momentum in the Solar System. In fact, although the Sun contains 99.9 percent of the Solar System's mass, the Sun's rotation represents less than 0.5 percent of the Solar System's angular momentum. Because the nebular hypothesis could not explain this **angular momentum problem**, it was never fully successful, so 19th- and early 20th-century astronomers instead considered various versions of the passing star hypothesis. Since around the middle of the 20th century, evidence has become overwhelming

How Do We Know? (10-1)

Two Kinds of Hypotheses: Catastrophic and Evolutionary

How big a role have sudden, catastrophic events played in the history of the Solar System? Many hypotheses in science can be classified as either evolutionary, in that they involve gradual processes, or catastrophic, in that they depend on specific, unlikely events. Scientists generally prefer evolutionary hypotheses. Nevertheless, catastrophic events do occur.

Some people prefer catastrophic hypotheses, perhaps because they like to see spectacular violence from a safe distance, which may explain the success of movies that include lots of car crashes and explosions. Also, catastrophic hypotheses resonate with scriptural accounts of cataclysmic events and special acts of creation. Thus, many people have an interest in catastrophic hypotheses.

Nevertheless, evidence shows that nearly all natural processes are gradual, and thus evolutionary. Scientific hypotheses almost never depend on unlikely events or special acts. For example, geologists study hypotheses about

mountain building processes that are evolutionary and describe mountains being pushed up slowly as millions of years pass. The evidence of erosion and the folded rock layers show that the process is gradual. Because most such natural processes are evolutionary, scientists are generally reluctant to accept hypotheses that depend on catastrophic events.

You will see in this and later chapters that catastrophes do occur. The planets, for example, are bombarded by debris from space, and some of those impacts are very large. Also, there is not a strict dividing line between evolutionary and catastrophic processes. Plate tectonics is usually a gradual, evolutionary process, but occasionally a Richter 9.0 earthquake happens that seems like a catastrophe and can make large and instantaneous changes to the landscape. As you study astronomy or any other natural science, notice that most hypotheses are evolutionary but that you need to allow for the possibility of unpredictable catastrophic events.

Mountains evolve to great heights by rising slowly, not suddenly and catastrophically.

---

for the nebular hypothesis. In fact, the nebular hypothesis is so comprehensive and explains so many of the observations that it can be considered to have "graduated" from being just a hypothesis to being properly called a theory. Today, astronomers are continuing to refine the details of that theory.

The Solar Nebula Theory

By about 1940, astronomers were beginning to understand how stars form and how they generate their energy, and it became clear that the origin of the Solar System was linked to that story. The **solar nebula theory** supposes that planets form in rotating disks of gas and dust around young stars (**Figure 10-7**). You have seen clear evidence that such disks of gas and dust are common around young stars. Bipolar flows from protostars were the first evidence of such disks; now, space telescopes and ground-based interferometers (look back to Chapter 6) can image those disks directly (**Figure 10-8**).

The evidence strongly supports the solar nebula theory: Earth and the other planets of the Solar System formed in a disk of material around the Sun as the Sun condensed from a cloud

Solar Nebula Theory

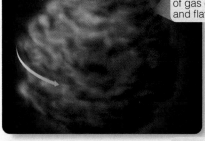

A rotating cloud of gas contracts and flattens...

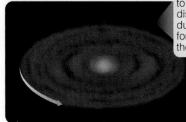

to become a thin disk of gas and dust around the forming Sun at the center.

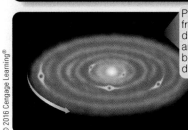

Planets grow from gas and dust in the disk and are left behind when the disk clears.

▶ **Figure 10-7** The solar nebula theory implies that the planets formed along with the Sun.

© 2016 Cengage Learning®

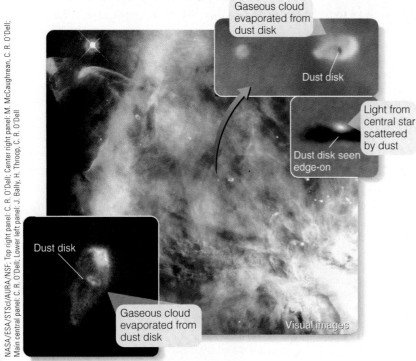

Gaseous cloud
evaporated from
dust disk

Dust disk

Light from
central star
scattered
by dust

Dust disk seen
edge-on

Dust disk

Gaseous cloud
evaporated from
dust disk

Visual images

NASA/ESA/STScI/AURA/NSF: Top right panel: C. R. O'Dell; Center right panel: M. McCaughrean, C. R. O'Dell;
Main central panel: C. R. O'Dell; Lower left panel: J. Bally, H. Throop, C. R. O'Dell

◀ **Figure 10-8** Many of the young stars in the Orion Nebula are surrounded by disks of gas and dust, but intense light from the brightest star in the neighborhood is evaporating the disks to form expanding clouds of gas. These particular disks may evaporate before they can form planets, but the large number of such disks shows that planet construction material around young stars is common.

of interstellar gas and dust. Therefore, if planet formation is a natural part of star formation, most stars should have planets.

10-3 Building Planets

The challenge for modern planetary scientists is to compare the characteristics of the Solar System with predictions of the solar nebula theory so they can work out details of how the planets formed.

Chemical Composition of the Solar Nebula

Everything astronomers know about the Solar System and star formation suggests that the solar nebula was a fragment of an interstellar gas cloud. Such a cloud would have been mostly hydrogen with some helium and small amounts of the heavier elements.

That is precisely what you see in the composition of the Sun (look back at Table 8-1, page 154). Analysis of the solar spectrum shows that the Sun is mostly hydrogen, with a quarter of its mass being helium and only about 2 percent being heavier elements. Of course, nuclear reactions have fused some hydrogen into helium, but this happens in the Sun's core and has not affected the composition of its surface and atmosphere, which are the parts you can observe directly. That means the composition revealed in the Sun's spectrum is essentially the composition of the gases from which the Sun formed.

This must have been the composition of the solar nebula, and you can also see that composition reflected in the chemical compositions of the planets. The inner planets are composed of rock and metal, and the outer planets are rich in low-density gases such as hydrogen and helium. The chemical composition of Jupiter resembles the composition of the Sun, but if you allowed hydrogen, helium, and hydrogen-bearing compounds to escape from a blob of stuff with the same overall composition as the Sun or Jupiter, the remainder would be more like the chemical composition of Earth and the other Terrestrial planets.

Condensation of Solids

An important clue to understanding the process that converted the nebular gas into solid matter is the variation in density among objects in the Solar System. You have already noted that the four inner planets are small and have high density, resembling Earth, whereas the outermost planets are large and have low density, resembling Jupiter.

Even among the four Terrestrial planets, you will find a pattern of slight differences in density. Merely listing the observed densities of the Terrestrial planets does not reveal the pattern clearly because Earth and Venus, being more massive, have stronger gravity and have squeezed their interiors to higher densities. The **uncompressed densities**—the densities the planets would have if their gravity did not compress them, or, to put it another way, the average densities of their original construction materials—can be calculated from the actual densities and masses of each planet (Table 10-1). In general, the closer a planet is to the Sun, the higher its uncompressed density.

According to the solar nebula theory, the observed pattern of planet densities originated when solid grains first formed

TABLE 10-1 Observed and Uncompressed Densities

Planet	Observed Density (g/cm³)	Uncompressed Density (g/cm³)
Mercury	5.43	5.0
Venus	5.24	3.9
Earth	5.51	3.96
Mars	3.93	3.70

© 2016 Cengage Learning®

from the gas of the nebula as it cooled, a process called **condensation**. The kind of matter that could condense in a particular region depended on the temperature of the gas there. In the inner regions of the nebula, close to the Sun, the temperature was evidently 1500 K or so. The only materials that can form grains at that temperature are compounds with high melting points, such as metal oxides and pure metals, which are very dense. Farther out in the nebula it was cooler, and silicates (rocky material) could also condense in addition to metal. Silicates are less dense than metal oxides and metals. Mercury, Venus, Earth, and Mars are evidently composed of a mixture of metals, metal oxides, and silicates, with proportionately more metals close to the Sun and more silicates farther from the Sun.

Even farther from the Sun there was a boundary called the **frost line** beyond which water vapor could freeze to form icy particles. Yet a little farther from the Sun, compounds such as methane and ammonia could condense to form other types of ice. Water vapor, methane, and ammonia were abundant in the solar nebula, so beyond the frost line the nebula would have been filled with a blizzard of ice particles, mixed with small amounts of silicate and metal particles that could also condense there. Those ices are low-density materials. The densities of Jupiter and the other outer planets correspond to a mix of ices plus relatively small amounts of silicates and metal.

The sequence in which the different materials would condense from the gas as a function of nebular temperature is called the **condensation sequence** (Table 10-2). It suggests that planets forming at different distances from the Sun should have accumulated from different kinds of materials in a predictable way.

People who have read a little bit about the origin of the Solar System may hold the **Common Misconception** that the matter in the solar nebula was sorted by density, with the heavy rock and metal sinking toward the Sun and low-density gases being blown outward. That is not the case. The chemical composition of the solar nebula should originally have been roughly the same throughout the disk when it was hot enough to be entirely gas. Later, as the disk cooled down, the inner parts close to the Sun would have had higher temperatures so that only metals and rock could condense there, whereas lots of ices along with metals and rock could condense in the cooler outer parts of the disk, far from the Sun. The frost line, beyond which ice could condense into solid particles, seems to have been between Mars and Jupiter, in the outer part of what is now the asteroid belt. That line separates the region for formation of the high-density Terrestrial planets from that of the low-density Jovian planets.

Formation of Planetesimals

In the development of the planets from the material of the solar nebula disk, three processes operated to collect solid bits of matter—metal, rock, ice—into larger bodies called **planetesimals**, which eventually made the planets. The study of planet building is the study of these processes: condensation, accretion, and gravitational collapse.

In the previous section, you learned about the condensation sequence. Planetary development in the solar nebula began with the formation of dust grains by condensation. A particle grows by condensation when it adds matter one atom or molecule at a time from a surrounding gas. Snowflakes, for example, grow by condensation in Earth's atmosphere. In the solar nebula, dust grains would have been bombarded continuously by atoms of gas, and some of those stuck to the grains. A microscopic grain capturing a layer of gas molecules on its surface increases its mass by a much larger fraction than a gigantic boulder capturing a single layer of molecules. For that reason, condensation can increase the mass of a small grain rapidly, but as the grain grows larger, condensation becomes less effective and other processes became more important.

The second process of planetesimal formation is **accretion**, which is the sticking together of solid particles. You may have seen accretion in action if you have walked through a snowstorm with big, fluffy flakes. If you caught one of those "flakes" on your glove and looked closely, you saw that it was actually made up of many tiny, individual flakes that had collided as they fell and accreted to form larger particles. Model calculations indicate that dust grains in the solar nebula were, on average, less than a meter apart, so they collided frequently and could accrete into larger particles.

When the particles grew to sizes larger than a centimeter, they would have been subject to new processes that tended to concentrate them. One important effect was that the growing solid objects would have collected into the plane of the solar nebula. Small dust grains could not fall into the plane because the turbulent motions of the gas kept them stirred up, but more massive objects would have been able to move through the gas and settle in the disk midplane.

TABLE 10-2 The Condensation Sequence

Temperature (K)	Condensate	Object; Estimated Temperature of Formation (K)
1500	Metal oxides	Mercury; 1400
1300	Metallic iron and nickel	
1200	Silicates	
1000	Feldspars	Venus; 900
680	Troilite (FeS)	Earth; 600
		Mars; 450
175	H_2O ice	Jovian; 175
150	Ammonia-water ice	
120	Methane-water ice	
65	Argon-neon ice	Pluto; 65

© 2016 Cengage Learning®

NASA/JPL-Caltech

Visual

▲ **Figure 10-9** What did the planetesimals look like? You can get a clue from this photo of the 5-km-wide nucleus of Comet Wild 2 (pronounced *Vildt-two*). Whether rocky or icy, the planetesimals must have been small, irregular bodies, scarred by craters from collisions with other planetesimals.

Astronomers calculate this process would have concentrated the larger solid particles into a relatively thin layer about 0.01 AU thick that would have allowed further rapid growth, resulting in the formation of planetesimals. There is no clear distinction between a very large grain and a very small planetesimal, but you might consider an object to be a planetesimal when its diameter approaches a kilometer (0.6 mi) or so (**Figure 10-9**).

The process of concentrating large particles and planetesimals into the plane of the solar nebula is analogous to the flattening of a forming galaxy, and a process found in galaxies might also have become important in the young Solar System once the plane of planetesimals formed. Calculations indicate that the rotating disk of particles should have been gravitationally unstable and would have been disturbed by spiral density waves resembling the much larger ones found in spiral galaxies. Those waves could have further concentrated the planetesimals and helped them coalesce into objects up to 100 km (60 mi) in diameter.

Through these processes, according to the solar nebula theory, the disk of gas and dust around the forming Sun became filled with trillions of solid particles ranging in size from pebbles to mini-planets. As the largest began to exceed 100 km in diameter, a third process began to affect them, and a new stage in planet building began, the formation of protoplanets.

Growth of Protoplanets

Collisions and coalescing of planetesimals eventually produced **protoplanets**, the name for massive objects destined to become planets. As these larger bodies grew, a new process helped them grow faster and altered their physical structure.

If planetesimals had collided at orbital velocities, they would have been unable to stick together. Typical orbital velocities in the Solar System are many kilometers per second. Head-on collisions at these speeds would vaporize any solid material. However, the planetesimals were all moving in the same direction in the nebular plane and didn't collide head-on. Instead, they merely "rubbed shoulders," so to speak, at low relative velocities. Such gentle collisions would have been more likely to combine planetesimals than to shatter them.

The largest planetesimals would grow the fastest because they had the strongest gravitational field. Their stronger gravity would attract additional material and could also hold on to a cushioning layer of dust capable of trapping incoming fragments. Astronomers calculate that the largest planetesimals would have grown quickly to protoplanetary dimensions, sweeping up more and more material.

Protoplanets initially grew only by attracting and accumulating solid bits of rock, metal, and ice because they did not have enough gravity to capture and hold large amounts of gas. In the warm solar nebula, the atoms and molecules of gas were traveling at velocities much larger than the escape velocities of modest-size protoplanets. However, once a protoplanet approached a size of 15 Earth masses or so, it could begin to grow by **gravitational collapse**, which is the rapid accumulation of large amounts of infalling gas from the nebula.

The theory of protoplanet growth into planets supposes that all the planetesimals had about the same chemical composition. The planetesimals accumulated to form planet-size balls of material with homogeneous composition throughout. Once a planet formed, heat would begin to accumulate in its interior from the decay of short-lived radioactive elements. The violent impacts of infalling particles would also have released energy called **heat of formation**. These two heating sources would eventually have melted the planet and allowed it to differentiate.

Differentiation is the separation of material according to density. After a planet melted, the heavy metals such as iron and nickel, plus elements chemically attracted to them, would settle to the core, while the lighter silicates and related materials floated to the surface to form a low-density crust. The scenario of planetesimals combining into planets that subsequently differentiated is shown in **Figure 10-10**.

Astronomers know that radioactive elements capable of releasing enough heat to melt the interiors of planets were present because the oldest meteorites contain daughter isotopes such as magnesium-26. That isotope is produced by the decay of aluminum-26 with a half-life of only 0.73 million years. The aluminum-26 and similar short-lived radioactive isotopes are gone now, but they must have been created in a supernova explosion that occurred shortly before the formation of the solar nebula. In fact, supernova explosions can trigger the formation of stars by compressing interstellar clouds. Some astronomers

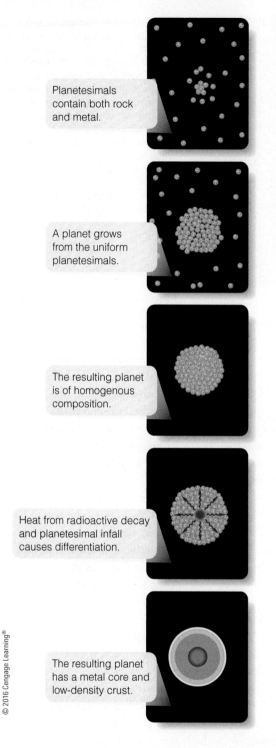

Planetesimals contain both rock and metal.

A planet grows from the uniform planetesimals.

The resulting planet is of homogenous composition.

Heat from radioactive decay and planetesimal infall causes differentiation.

The resulting planet has a metal core and low-density crust.

© 2016 Cengage Learning®

▲ **Figure 10-10** This simple model of planet building assumes planets formed from accretion and collision of planetesimals that were of uniform composition, containing both metals and rocky material, and that the planets later differentiated, meaning they melted and separated into layers by density and composition.

hypothesize that our Solar System may exist because of a supernova explosion that occurred about 4.6 billion years ago.

If planets formed by accretion of planetesimals and were later melted by radioactive decay and heat of formation, then

Earth's early atmosphere probably consisted of a combination of gases delivered by planetesimal impacts and released from the planet's interior during differentiation. The accumulation of gases from a planet's interior to create an atmosphere is called **outgassing**. Given the location of Earth in the solar nebula, planetary scientists calculate that outgassing during differentiation would not have included as much water as Earth now has. Therefore, astronomers hypothesize that some of Earth's water and atmosphere might have accumulated late in the formation of the planet as Earth swept up volatile-rich planetesimals, supplementing outgassing. Those icy planetesimals would have formed in the cool outer parts of the solar nebula, at or beyond the frost line, and could have been scattered toward the Terrestrial planets by the gravitational influence of the Jovian planets.

According to the solar nebula theory, the Jovian planets could begin growing by the same processes that built the Terrestrial planets. However, in the inner solar nebula, only metals and silicates could form solids, so the Terrestrial planets grew slowly. In contrast, the outer solar nebula contained not just solid bits of metals and silicates but also ices that included plentiful hydrogen. Model calculations show that the Jovian planets would have grown faster than the Terrestrial planets and quickly become massive enough to begin even faster growth by gravitational collapse, drawing in large amounts of gas from the solar nebula. The Terrestrial planet zone did not include ice particles, so those planets developed relatively slowly and never became massive enough to grow further by gravitational collapse.

The Jovian planets must have reached their present size in no more than about 10 million years, before the Sun became hot and luminous enough to blow away the remaining gas in the solar nebula, removing the raw material and preventing further Jovian growth. As you will learn in the next section, disturbances from outside the forming Solar System might have reduced the time available for Jovian planet formation even more severely. The Terrestrial planets, in comparison, grew from solids and not from the gas, so they could have continued to grow by accretion from solid debris left behind after the gas was removed. Computer models indicate that the Terrestrial planets were at least half finished within 10 million years but probably continued to grow for another 20 million years or so.

The solar nebula theory has been very successful overall in explaining the formation of the Solar System. But there are some problems with the theory, and the Jovian planets are the main troublemakers.

The Jovian Problem

Recent observations of star formation make it difficult to understand how the Jovian planets could have assembled during the lifetime of the solar nebula, and this has caused astronomers to expand and revise the theory of planet formation (**Figure 10-11**).

The new information is that gas and dust disks around newborn stars don't last long. You have seen images of dusty gas

Visual

NASA/JPL-Caltech/Univ. of Arizona

◄ **Figure 10-11** Image of Jupiter made by the *Cassini* spacecraft as it flew by on the way to Saturn; the shadow of one moon is visible near the left limb. Jovian worlds pose a problem for astronomers investigating the origin of planets. Planet-forming nebulae are blown away in only a few million years by nearby luminous stars, so Jovian planets must form more quickly than initial calculations predicted. Newer research suggests that accretion followed by gravitational collapse can build Jovian planets in about a million years. Under certain conditions, direct gravitational collapse may form some large planets in just thousands of years.

disks around the young stars in the Orion star-forming region (Figure 10-8). Those disks are being evaporated by intense ultraviolet radiation from hot O and B stars forming nearby. Astronomers have calculated that most stars form in clusters containing sibling O and B stars, so this evaporation is hypothesized to happen to most disks. Even if a disk did not evaporate quickly, the gravitational influence of the crowded stars in a cluster could strip away the outer parts of the disk. Those are troublesome observations because they seem to indicate that disks usually don't last as long as 10 million years; some may even evaporate within the astronomically short span of 100,000 years or so. That's not long enough to grow a Jovian planet by the combination of condensation, accretion, and gravitational collapse proposed in the standard solar nebular theory.

Yet Jovian planets are common. In the final section of this chapter, you will see evidence that astronomers have found planets orbiting other stars, and almost all of the planets discovered so far have the mass of Jovian planets. There may also be many Terrestrial planets orbiting those stars that are too small to be detected at present, but the important point is that there are lots of Jovian planets around. How they can form quickly enough, before the disks of raw material evaporate, is referred to as the **Jovian problem**.

Detailed mathematical models of the solar nebula have been produced using computers running specially constructed programs that take days to finish a calculation. The results show that the rotating gas and dust of the solar nebula could have become unstable and rapidly collapse gravitationally to make Jovian planets. That is, massive planets may have been able to form by **direct collapse**, skipping the slower step of forming a dense core by condensation and accretion of solid material. Jupiters and Saturns can form in these direct collapse models in only a few hundred years. If the Jovian planets formed in this way, they could have formed long before the solar nebula disappeared, even if the nebula was eroded quickly by neighboring massive, hot stars.

Another possible answer to the Jovian problem is a revised estimate of the opacity of the solar nebula, which was probably much less opaque than the interstellar medium. That would have allowed the gravitational collapse of nebular material onto the solid proto-Jovian solid cores to proceed much faster than in previous models, taking only one million instead of 10 million years to make the Jovian planets. Both proposed modifications to the solar nebula theory suggest that the outer planets could have formed within the likely lifetime of the gas and dust disk around the forming Sun.

These new insights into the formation of the outer planets may also help explain a puzzle about the formation of Uranus and Neptune. Those planets are so far from the Sun that accretion could not have built them rapidly. The gas and dust of the solar nebula must have been sparse out there, and Uranus and Neptune orbit so slowly they would not have swept up material very rapidly. The conventional view has been that they grew by accretion so slowly that they never became quite massive enough to begin accelerated growth by gravitational collapse. In fact, it is hard to understand how they could have reached even their present sizes if they started growing by accretion so far from the Sun. Theoretical calculations show that Uranus and Neptune might instead have formed closer to the Sun, in the region of Jupiter and Saturn, and then could have been shifted outward by gravitational interactions with the bigger planets. In any case, explaining the formation of Uranus and Neptune is part of the Jovian problem.

Explaining the Characteristics of the Solar System

Now you have learned enough to put all the pieces of the puzzle together and explain the distinguishing characteristics of the Solar System in Table 10-3.

The disk shape of the Solar System is inherited from the motion of material in the solar nebula. The Sun and planets and moons mostly revolve and rotate in the same direction (Figure 10-1) because they formed from the same rotating gas cloud. The orbits of the planets lie in the same plane because the rotating solar nebula collapsed into a disk, and the planets formed in that disk.

The solar nebula theory is evolutionary in that it involves continuing processes to gradually build the planets. To explain the odd rotations of Venus and Uranus, however, you might need to consider catastrophic events. Uranus rotates on its side. This might have been caused by an off-center collision with a massive planetesimal when the planet was nearly formed. Two hypotheses have been proposed to explain the backward rotation of Venus. Theoretical models suggest that the Sun produced tides in the thick atmosphere of Venus that could have eventually reversed the planet's rotation—an evolutionary hypothesis. It is also possible that the rotation of Venus was altered by an impact late in the planet's formation, and that is a catastrophic hypothesis. Both hypotheses may be true.

The second item in Table 10-3—the division of the planets into Terrestrial and Jovian worlds—can be understood through the condensation sequence. The Terrestrial planets formed in the inner part of the solar nebula, where the temperature was high and only substances such as silicates and metals could condense to form solid particles. That produced the small, dense Terrestrial planets. In contrast, the Jovian planets formed in the outer solar nebula, where the lower temperature allowed the gas to condense into large amounts of ices that included plentiful hydrogen in addition to silicates and metal. That allowed the Jovian planets to grow rapidly and become massive, low-density worlds. Also, Jupiter and Saturn are so massive they were able to grow by drawing in cool gas by gravitational collapse from the solar nebula. The Terrestrial planets could not do this because they never became massive enough.

The heat of formation released by infalling matter was tremendous for these massive planets. Jupiter must have grown hot enough to glow with a luminosity of about 1 percent that of the present Sun. As a result, Jupiter is still hot inside. In fact, both Jupiter and Saturn radiate more heat than they absorb from the Sun, so they are evidently still cooling. (Note that model calculations indicate neither Jupiter nor Saturn ever became hot enough in their cores to generate nuclear energy as a star would.)

A glance at the Solar System suggests that you should expect to find a planet between Mars and Jupiter at the present location of the asteroid belt. Mathematical models indicate that the reason asteroids are there rather than a planet is that Jupiter grew into such a massive body so quickly that it was able to gravitationally disturb the motion of nearby planetesimals. The bodies that could have formed a planet between Mars and Jupiter instead collided at high speeds and shattered rather than combining, were thrown into the Sun, or were ejected from the Solar System. The asteroids seen today are the last remains of those rocky planetesimals.

The comets, in contrast, are evidently the last of the icy planetesimals. Some may have formed in the outer solar nebula beyond Neptune, but many probably formed in the denser part of the nebula among the Jovian planets where ices could condense easily. Mathematical models show that the massive Jovian planets could have ejected some of these icy planetesimals into the far outer Solar System. In a later chapter, you will see evidence that some comets are icy bodies coming from those distant locations, falling back into the inner Solar System.

The icy KBOs also appear to be ancient planetesimals that formed in the outer Solar System but were never incorporated into a planet. They orbit slowly far from the light and warmth of the Sun and, except for occasional collisions, have not changed much since the Solar System was young. The gravitational influence of the planets can deflect some KBOs into the inner Solar System where they also are seen as comets.

The large satellite systems of the Jovian worlds may contain two kinds of moons. Some moons may have formed in orbit around forming planets in a miniature version of the solar nebula. In contrast, some of the smaller moons, especially those in eccentric, inclined, or retrograde orbits, may be captured planetesimals, asteroids, and comets. The large masses of the Jovian planets would have made it easier for them to capture satellites.

You see in Table 10-3 that all four Jovian worlds have ring systems, and that makes sense if you consider the large masses of those worlds and their locations in the outer Solar System. A massive planet can more easily hold onto small orbiting ring particles that are strongly affected by radiation pressure and the solar wind. It is hardly surprising, then, that the Terrestrial planets, low-mass worlds located near the Sun, have no planetary rings.

The last entry in Table 10-3 is the common ages of Solar System bodies, and the solar nebula theory has no difficulty explaining that characteristic. If the theory is correct, then the planets formed at the same time as the Sun and should have the same age.

Clearing the Nebula

Evidently the Sun formed, along with many other stars, in a cloud of interstellar material. You have already learned that observations of young stars suggest that radiation and gravitational effects from the Sun's siblings, especially the larger and nearer ones, would have tended to disturb and erode the disk of planet construction material around the Sun. Even without those external effects, four internal processes would have gradually destroyed the solar nebula.

The two most important of these internal processes were radiation pressure and the solar wind. Earlier in this chapter you learned that those two forces create and determine the shapes of comet tails. Once the Sun became a luminous object, light streaming from its photosphere exerted radiation pressure on the particles of the solar nebula. Large bits of matter like planetesimals and planets were not affected, but low-mass specks of dust and individual atoms and molecules were pushed outward and driven from the Solar System. Due to a subtle side effect of radiation pressure, medium-mass specks of dust would actually spiral toward the Sun, but that would also remove them from the nebula.

The second process that helped clear the nebula was pressure from the solar wind, the flow of ionized hydrogen and other atoms away from the Sun's upper atmosphere (Chapter 8). This flow is a steady breeze that rushes past Earth at about 400 km/s (250 mi/s). Young stars have even stronger winds than stars of the Sun's age and also irregular fluctuations in luminosity, like those observed in young stars such as T Tauri stars, which can accelerate the wind.

The strong surging wind from the young Sun would have helped radiation pressure remove dust and gas from the nebula.

The third effect that helped clear the nebula was the sweeping up of space debris by the planets. All of the old, solid surfaces in the Solar System are heavily cratered by meteorite impacts (Figure 10-12). Earth's Moon, Mercury, Venus, Mars, and most of the moons in the Solar System are covered with craters. A few of these craters have been formed recently by the steady rain of meteorites that falls on all the planets in the Solar System, but most of the craters appear to have been formed roughly 4 billion years ago in what is called the **heavy bombardment**, as the last of the debris in the solar nebula was swept up by the planets. The image that opens this chapter (page 195) is an artist's conception of the heavy bombardment phase in another planetary system.

The fourth effect that would have cleared the nebula was the ejection of material from the Solar System by close encounters with planets. If a small object such as a planetesimal passes close to a planet, the small object's path will be affected by the planet's gravitational field. In some cases, the small object can gain energy from the planet's motion and be thrown out of the Solar System. Ejection is most probable in encounters with massive planets, so the Jovian planets were probably very efficient at ejecting the icy planetesimals that formed in their region of the nebula.

Attacked by the radiation and gravity of nearby stars and racked by internal processes, the solar nebula could not survive very long in astronomical terms. Once the gas and dust were gone and most of the planetesimals were swept up, the planets could no longer gain significant mass, and the era of planet building ended.

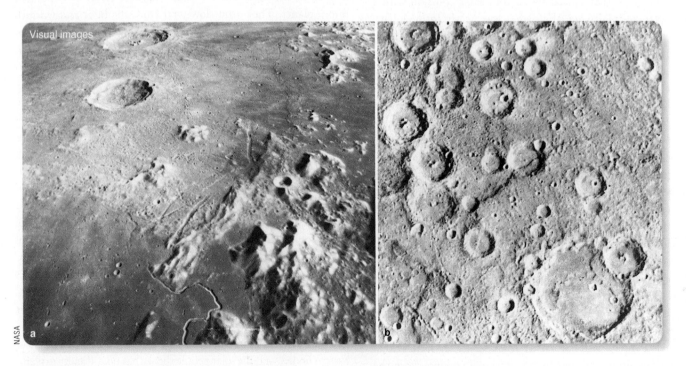

▲ Figure 10-12 Every old, solid surface in the Solar System is scarred by craters. (a) Earth's Moon has craters ranging from basins hundreds of kilometers in diameter down to microscopic pits. (b) The surface of Mercury, as photographed by a passing spacecraft, shows vast numbers of overlapping craters.

larger in diameter than our Solar System. The Orion star-forming region is only a few million years old, so planets may not have finished forming in these disks yet. Furthermore, the intense radiation from nearby hot stars is evaporating the disks so quickly that planets may never have a chance to grow large. The important point is that disks of gas and dust that could become planetary systems are a common feature around stars that are forming.

The *Hubble Space Telescope* can detect dense disks around young stars in another way. Some disks show up in silhouette against the nebulae that surround the newborn stars (**Figure 10-13**). These disks are related to the formation of bipolar flows in that they focus the gas flowing away from a young star into two jets shooting in opposite directions.

In addition to these dense, hot disks forming planets around young stars, astronomers have found cold, low-density dust disks around stars much older than the newborn stars in Orion, old enough to have reached the main sequence. These tenuous dust disks are sometimes called **debris disks** because they are evidently made of dusty debris produced in collisions among small bodies such as comets, asteroids, and KBOs rather than dust left over from an original protostellar disk. That conclusion is based on calculations showing that the observed dust would be removed by

10-4 Planets Orbiting Other Stars

Do other planetary systems exist? The evidence says yes. Do they contain planets like Earth? The first such objects have now been discovered.

Planet-Forming Disks

Both visible and radio-wavelength observations detect dense disks of gas and dust orbiting young stars. For example, at least 50 percent of protostars in the Orion Nebula are surrounded by such disks. A young star is surrounded by such disks. Astronomers can determine that the disks contain many Earth masses of material in a region a few times

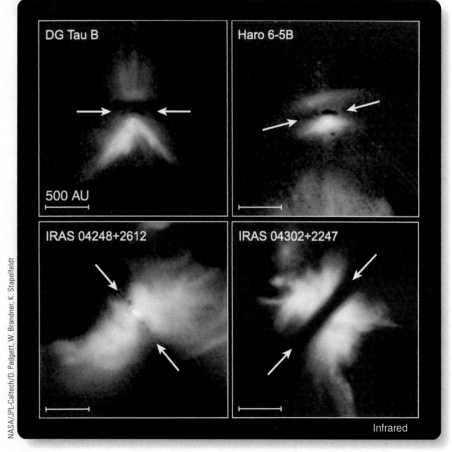

NASA/JPL-Caltech/D. Padgett, W. Brandner, K. Stapelfeldt

DG Tau B

Haro 6-5B

500 AU

IRAS 04248+2612

IRAS 04302+2247

Infrared

▶ Figure 10-13_ Dark bands (indicated by arrows) are edge-on disks of gas and dust around young stars seen in *Hubble Space Telescope* near-infrared images. Planets may eventually form in these disks. These systems are so young that material is still falling inward and being illuminated by light from the stars.

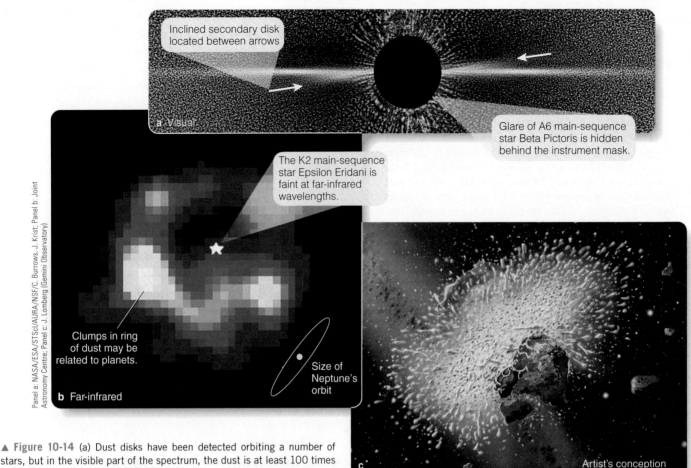

Inclined secondary disk located between arrows

a Visual

Glare of A6 main-sequence star Beta Pictoris is hidden behind the instrument mask.

The K2 main-sequence star Epsilon Eridani is faint at far-infrared wavelengths.

Clumps in ring of dust may be related to planets.

b Far-infrared

Size of Neptune's orbit

Panel a: NASA/ESA/STScI/AURA/NSF/C. Burrows, J. Krist; Panel b: Joint Astronomy Centre; Panel c: J. Lomberg (Gemini Observatory)

c   Artist's conception

▲ **Figure 10-14** (a) Dust disks have been detected orbiting a number of stars, but in the visible part of the spectrum, the dust is at least 100 times fainter than the stars, which must be hidden behind masks to make the dust detectable. The second faint inclined disk in the Beta Pictoris system may show the orbital plane of a massive planet. (b) At far-infrared wavelengths, the dust in debris disks can be much brighter than the central star. Warps and clumps in these disks suggest the gravitational influence of planets. (c) Collisions between asteroids are rare events, but they generate lots of dust and huge numbers of fragments, as in this artist's conception. Further collisions between fragments can continue to produce dust. Because such dust is blown away quickly, astronomers treat the presence of dust as evidence that objects of at least planetesimal size are also present.

radiation pressure in a much shorter time than the ages of those stars, meaning the dust there now must have been created relatively recently. The presence of dust with short lifetimes around mature main-sequence stars indicates that larger bodies such as asteroids and comets must be present as reservoirs for the dust. Our own Solar System contains such "second-generation" dust produced by asteroids, comets, and KBOs. Astronomers consider the Solar System's Kuiper Belt extending beyond the orbit of Neptune as an example of an old debris disk.

Some examples of debris disks are around the stars Beta Pictoris and Epsilon Eridani (**Figure 10-14**). The dust disk around Beta Pictoris, an A-type star more massive and luminous than the Sun, is about 20 times the diameter of our Solar System. The dust disk around Epsilon Eridani, which is a K-type star

somewhat smaller than the Sun, is similar in size to the Solar System's Kuiper Belt. Like most of the other known low-density disks, both of these examples have central zones with even lower density. Those inner regions are understood to be places in which planets have finished forming and have swept up most of the construction material.

Infrared observations reveal that Favorite Star Vega, easily visible in the Northern Hemisphere summer sky, also has a debris disk, and detailed studies show that most of the dust particles in that disk are tiny. Radiation pressure from Vega should blow away small dust particles quickly, so astronomers conclude that the dust being observed now must have been produced by a big event like the collision of two large planetesimals within the past million years (Figure 10-14c). Fragments from that collision are still smashing into each other and producing more dust, continuing to enhance the debris disk. This effect has also been found in the disks around other relatively old stars. Such smashups probably happen rarely in a dust disk, but when they happen they make the disk easy to detect.

Notice the difference between the two kinds of planet-related disks that astronomers have found. The low-density dust disks such as the ones around Beta Pictoris, Epsilon Eridani,

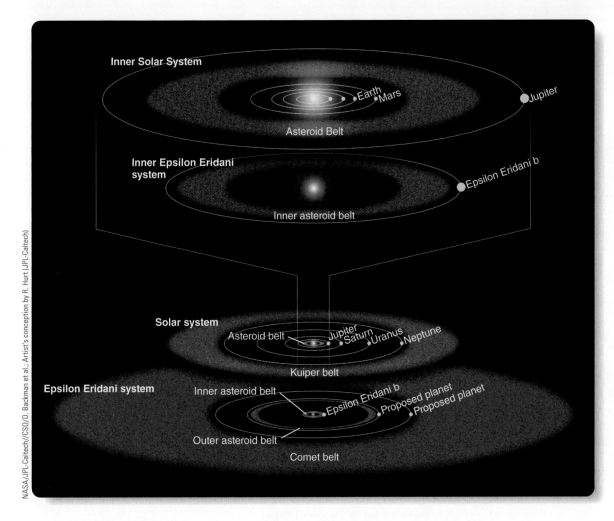

NASA/JPL-Caltech//CSO/D. Backman et al.: Artist's conception by R. Hurt (JPL-Caltech)

▲ **Figure 10-15** Dust in debris belts around older main-sequence stars indicates ongoing collisions of asteroids and comets. Such activity in our Solar System is ultimately driven by the gravitational influence of planets. Debris belt edges may be defined by adjacent orbits of planets. The inferred architecture of the inner (*top*) and outer (*bottom*) parts of the Epsilon Eridani planetary system are shown in comparison with the corresponding parts of our Solar System.

and Vega are produced by dust from collisions among comets, asteroids, and KBOs. Such disks are evidence that planetary systems have already formed around those stars (**Figure 10-15**). In comparison, the dense disks of gas and dust such as those seen round the stars in Orion are sites where planets could be forming right now.

Observing Extrasolar Planets

A planet orbiting another star is called an **extrasolar planet** or exoplanet. Such planets are quite faint and difficult to see close to the glare of their parent stars, but there are ways to find them. To understand one important way, all you have to do is imagine walking a dog.

You will remember that Earth and the Moon orbit around their common center of mass, and two stars in a binary system orbit around their center of mass. When a planet orbits a star, the star moves very slightly as it orbits the center of mass of the

planet-star system. Think of someone walking a poorly trained dog on a leash; the dog runs around pulling on the leash, and even if it were an invisible dog, you could plot its path by watching how its owner was jerked back and forth (**Figure 10-16a**). Astronomers can detect a planet orbiting another star by watching how the star moves as the planet tugs on it. As the planet circles the star, the star wobbles slightly, and that very small motion of the star is detectable by Doppler shifts in the star's spectrum (Chapter 7).

The first planet detected this way was discovered in 1995 orbiting the sunlike star 51 Pegasi (Figure 10-16b). From the observed motion of the star and an estimate of its mass from its spectral type, astronomers deduced that the planet has at least half the mass of Jupiter and orbits only 0.05 AU from the star. Half the mass of Jupiter amounts to 160 Earth masses, so this is a large planet—bigger than Saturn. Note that it orbits very close to its star, much closer than Mercury orbits around our Sun.

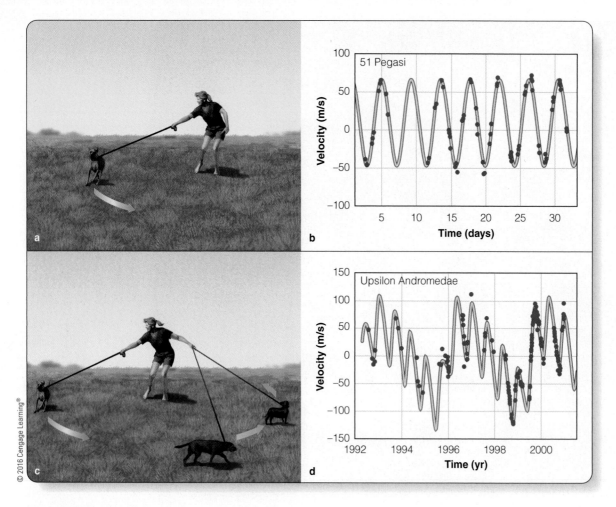

▲ **Figure 10-16** (a) A person walking a lively dog is tugged off course by the dog. (b) The star 51 Pegasi is pulled back and forth by the gravity of the planet that orbits it every 4.2 days. The wobble is detectable in precision observations of the star's Doppler shifts. (c) A person walking multiple dogs has a complicated motion. (d) Doppler shifts of the star Upsilon Andromedae show the combined effects of at least four planets orbiting it. The influence of its shortest-period planet has been removed in this graph to reveal more clearly the orbital influences of the other three planets.

Astronomers were not surprised by the announcement that a planet had been found orbiting a sunlike star. For years, astronomers had assumed that many stars had planets. Nevertheless, they acted as professional skeptics, carefully testing the data and making further observations that confirmed the discovery (**How Do We Know? 10-2**). In fact, as of July 2014, almost 600 planets had been discovered by the Doppler radial velocity method, including at least four planets orbiting the star Upsilon Andromedae (Figure 10-16d), four orbiting Gliese 581, and five orbiting 55 Cancri—true planetary systems.

Another way to search for planets is to look for changes in the brightness of a star when an orbiting planet crosses in front of it, called a **transit**. The decrease in light during a planetary transit is very small, but it is detectable, and astronomers have used this technique to find more than a thousand extrasolar planets. From the amount of light lost, astronomers can tell that the transiting planets that have Jovian masses also have Jovian diameters and thus Jovian densities and compositions. In other words, hot Jupiters really are Jovian planets instead of monster Terrestrial planets. The transit method will be described in further detail in the next section regarding results from the *Kepler* mission.

The *Spitzer Space Telescope* detected infrared radiation from more than a dozen planets already known from Doppler shifts or transits. As these planets orbit their parent stars, the amount of infrared radiation from each system varies. When the planets pass behind their parent stars, the total infrared brightness of the systems noticeably decreases. These measurements confirmed the existence of those planets, but, more importantly, allow determination of their temperatures and sizes.

Notice how the techniques used to detect extrasolar planets resemble techniques used to study binary stars.

How Do We Know? (10-2)

Scientists: Courteous Skeptics

What does it mean to be skeptical, yet also open to new ideas? "Scientists are just a bunch of skeptics who don't believe in anything." That is a **Common Misconception** among people who don't understand the methods and goals of science. Yes, scientists are skeptical about new ideas and discoveries, but they do hold strong beliefs about how nature works. Scientists are skeptical not because they want to disprove everything but because they are searching for the truth and want to be sure that a new description of nature is reliable before it is accepted.

Another **Common Misconception** is that scientists automatically accept the work of other scientists. On the contrary, scientists skeptically question every aspect of a new discovery. They may wonder if another scientist's instruments were properly calibrated or whether the scientist's mathematical models are correct. Other scientists will want to repeat the work themselves using their own instruments to see if they can obtain the same results. Every observation is tested, every discovery is confirmed, and only an idea that survives many of these tests begins to be accepted as a scientific truth.

Scientists are prepared for this kind of treatment at the hands of other scientists. In fact, they expect it. Among scientists it is not bad manners to say, "Really, how do you know that?" or "Why do you think that?" or "Show me the evidence!" And it is not just new or surprising claims that are subject to such scrutiny. Even though astronomers had long expected to discover planets orbiting other stars, when a planet was finally discovered circling 51 Pegasi, astronomers were skeptical. That was not because they thought the observations were necessarily flawed but because this is how science works.

The goal of science is to tell stories about nature. Some people use the phrase "telling a story" to describe someone who is telling a fib. But the stories that scientists tell are exactly the opposite; perhaps you could call them "antifibs" because they are as true as scientists can make them. Skepticism eliminates stories with logical errors, flawed observations, or misunderstood evidence and tends eventually to leave standing only the stories that best describe nature.

Skepticism is not a refusal to hold beliefs. Rather, it is a way for scientists to find and keep those natural principles that are worthy of trust.

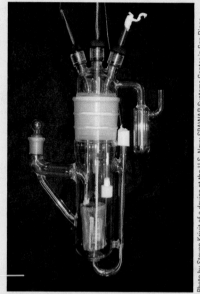

<!-- vertical credit -->Photo by Steven Krivit of a device at the U.S. Navy SPAWAR Systems Center in San Diego

A laboratory cell for the investigation of cold fusion claims that were thereby found to be false.

Almost all of the extrasolar planets discovered so far were found with the same observational methods used to study eclipsing binary and spectroscopic binary star systems, pushed to the limits of instrumental capabilities.

About 30 extrasolar planets have been found by a technique called **microlensing**. In those cases, an extrasolar planet passed precisely between Earth and a background star, briefly magnifying the distant star's brightness by gravitational lensing.

The extrasolar planets discovered so far tend to be massive and have short orbital periods because lower-mass planets or longer-period planets are harder to detect. Low-mass planets don't tug on their stars very much, and long-period planets produce only slow motions of their stars. Only specially designed spectrographs can detect the very small stellar velocity changes that these gentle and/or slow tugs produce. Planets with longer periods are also harder to detect because astronomers have not been making high-precision observations for a long enough time. Jupiter takes 11 years to circle the Sun once, so it will take decades for astronomers to find and confirm the longer-period

wobbles produced by planets lying that far, or even farther, from their parent stars. You should therefore not be surprised that the first extrasolar planets discovered are massive and have short orbital periods.

The new planets seem puzzling for several reasons. As you have learned regarding our Solar System, the large planets formed farther from the Sun where the solar nebula was colder and ices could condense. So, how could big planets near their stars (called "**hot Jupiters**") have formed? Theoretical calculations indicate that planets forming in an especially dense disk of matter could spiral inward as they sweep up gas, planetesimals, and even smaller planets. That means it is possible for a few planets to become the massive, short-period planets that are detected most easily.

Another puzzle is that many of the newly discovered extrasolar planets have eccentric orbits, and a few others appear to have orbits inclined at large angles to the equators of their parent stars. Simple interpretation of the solar nebular theory would predict that planets generally should have nearly circular orbits that lie approximately in their star's equatorial planes, as do the

planets in our Solar System. Theorists point out, however, that planets in some young planetary systems can interact with each other and can be thrown into eccentric or highly inclined orbits. This effect is probably rare in planetary systems, but astronomers have found some of these extreme systems more easily because they have easily detected wobbles.

You might be wondering if astronomers have been able to make images of any extrasolar planets. Getting an image of a planet orbiting another star is about as easy as photographing a bug crawling on the bulb of a searchlight miles away; planets are small and dim and get lost in the glare of the stars they orbit. Nevertheless, astronomers managed to image a planet orbiting at the inner edge of the debris disk around the star Fomalhaut (Alpha Piscis Austrinis) using the *Hubble Space Telescope*'s near-infrared camera, and around the star Beta Pictoris using a specially designed near-infrared camera plus adaptive optics on the Gemini telescope atop Mauna Kea. Both of those discoveries further confirmed the predicted connection between debris disks and finished planetary systems (Figure 10-17).

The *Kepler* Planet-Finding Mission

The *Kepler* space telescope, launched into solar orbit in 2009, searched for planets by the transit method mentioned in the previous section. *Kepler*'s 42 visual-wavelength charge-coupled device (CCD) detectors (Chapter 6, page 121) monitored the brightness of 150,000 solar-type stars during its 4-year primary mission, detecting slight decreases in brightness caused by planets passing in front of some of those stars. *Kepler*'s original search field, between the constellations of Cygnus and Lyra, was chosen to be close to, but not in, the plane of the Milky Way so that the number of solar-type stars was maximized relative to other types of stars.

As you know, the Sun's diameter is about 100 times larger than Earth's. Therefore, when an Earth-size planet transits its parent star (meaning, moves between you and the star), the planet covers up 1/10,000 of the surface area of the star, causing a decrease of 1 part in 10,000 (1/100 of a percent) in the star's brightness. Remarkably, *Kepler*'s detectors are sensitive and stable enough to detect transits that small (Figure 10-18). However, only a few percent of planetary systems can be expected to have orbits inclined parallel enough to our line of sight so that transits can be observed. That is the main reason why the target star sample was so large.

The *Kepler* search scheme involved very precisely measuring the brightness of each of the 150,000 target stars at least once every 30 minutes, producing a tremendous amount of data that had to be carefully analyzed. The *Kepler* team was especially interested in finding Earth-like planets, meaning planets that are not only about the size of Earth but also have Earth-like temperatures. Earth-like temperatures require an orbit about 1 AU in radius if the parent star has about the same mass and luminosity as our Sun.

You can see that an Earth-like planet around a Sun-like star will produce a transit only once per year. Each transit will last only 13 hours if the planet's orbit is exactly parallel to the line of sight and even less time if the orbit is inclined. Thus, the *Kepler* investigators had the amazingly tough task of searching

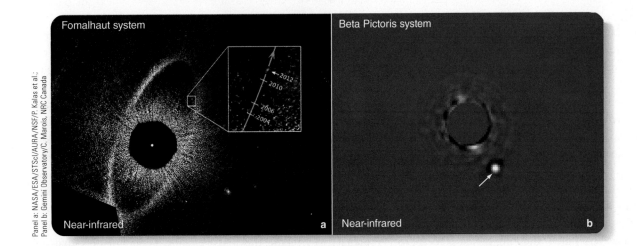

▲ **Figure 10-17** (a) Images of a Jovian-sized planet orbiting about 120 AU from the star Fomalhaut (Alpha Piscis Austrinus), at the inside edge of that star's previously known debris disk/ring. The inset shows motion of the object between 2004 and 2012, at the correct rate for a planet at that position orbiting a star of that mass. (b) Image of a Jovian-size planet orbiting 9 AU from the star Beta Pictoris. In both images, the central circular blank region represents a combination of hardware and software masks implemented to block light from the central stars that are much brighter than the planets.

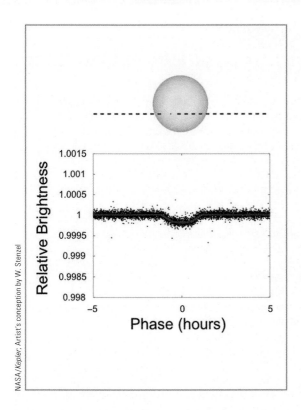

▲ **Figure 10-18** Transit light curve of the first confirmed extrasolar Terrestrial planet, Kepler-10b. The planet has an orbital period of 0.84 day. The depth of the transit "dip" indicates that the planet has 1.4 times the diameter of Earth. Spectroscopic measurements with ground-based telescopes of the parent star's radial velocity variation give a mass for the planet of 4.6 times Earth's mass. The planet's diameter and mass yield a density of 8.8 g/cm³, higher than Earth's, indicating a mostly metallic composition.

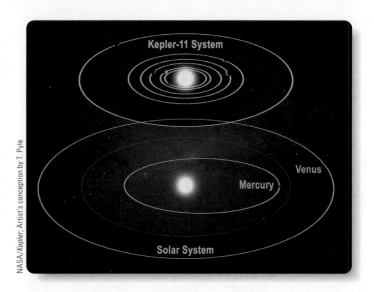

▲ **Figure 10-19** Artist's conception comparing the Kepler-11 planetary system with our Solar System from a tilted perspective, showing that in both systems the orbits of the planets lie in single planes.

in their data for occasions when some of their 150,000 target stars (nobody knew ahead of time which ones) become 0.01 percent dimmer than usual, for a few hours, once per year. And, to be sure that the transits actually represented orbiting planets, at least three transits spaced equally in time had to be observed. Unless transits repeated on schedule, any observed variations in a star's brightness could have been caused by other phenomena. Thus, the *Kepler* investigators required at least three years of observations to discover and confirm each Earth-like planet.

Of course, finding and confirming planets orbiting closer than 1 AU to their parent stars, with orbital periods shorter than one year, is quicker. Also, you can see that finding planets bigger than Earth will be easier because they will cover a larger fraction of their parent star during transit, causing a more noticeable "dip" in the light curve. Thus, *Kepler* easily found many hot Jupiters within a few weeks of the start of the mission.

After four years of observations and another year of data processing, the total number of confirmed extrasolar planets discovered by the *Kepler* mission reached almost 1000, with 4000 more candidates waiting for confirmation. The majority of those planets are hot Jupiters or "hot Neptunes," larger than Earth but smaller than Jupiter; about 150 of them are "hot Earths." One star, designated Kepler-11, has 6 planets, all somewhat larger than Earth, all orbiting closer to their parent star than Venus does to our Sun (**Figure 10-19**). Interestingly, because the Kepler-11 planets perturb each other gravitationally, making successive transits come a tiny bit earlier or later than would happen if they were orbiting alone, their masses can be estimated. The amount of light lost during the transits gives the diameters of the planets, and you know that combining the masses and the diameters lets you calculate their densities. All 6 of Kepler-11's known planets turn out to have densities like Earth or higher. Thus, they are definitely Terrestrial planets.

A few of the extrasolar planets and planet candidates identified by *Kepler* have transit light curves indicating that they are smaller than Earth in diameter. A handful of these systems have masses determined directly by radial velocity Doppler shift measurements; so far, three extrasolar planets are known to be less massive than Earth. These are all "hot Earths," in orbits so close to their parent stars that their surfaces are probably molten rock and metal. (Note that several other objects known to have smaller masses than Earth have been found orbiting pulsars.)

In 2014, the *Kepler* team announced confirmation of the discovery of the first known Earth-size, Earth-temperature extrasolar planet, designated Kepler-186f. That planet orbits an M dwarf star with 4 other planets previously discovered by *Kepler*.

You can expect that space observatories will someday be able to image even Terrestrial planets directly around Sun-like stars. The discovery of extrasolar planets gives astronomers added confidence in the solar nebula theory. The theory predicts that planets are common, and astronomers are finding them orbiting many stars.

DOING SCIENCE

Why are debris disks evidence that planets have already formed? A scientist often reaches conclusions using a combination of indirect evidence, theory, and past experiences, a kind of scientific common sense.

Certainly the cold debris disks seen around stars like Vega are not places where planets are forming. They are not hot enough or dense enough to be young disks. Rather, the debris disks must be older, and the dust is being produced by collisions among comets, asteroids, and KBOs. Small dust particles would be blown away or destroyed relatively quickly, so these collisions must be a continuing process. The successful solar nebula theory gives astronomers reason to believe that where you find comets, asteroids, and KBOs, you should also find planets, so the debris disks are probably evidence that planets have already formed in such systems.

Now try reaching a conclusion based on direct rather than indirect evidence. **What is the evidence that planets orbit other stars?**

What Are We? Planet-Walkers

The matter you are made of came from the big bang, and it has been cooked into a wide variety of atoms inside stars. Now you can see how those atoms came to be part of Earth. Your atoms were in the cloud of gas that formed the Solar System 4.6 billion years ago, and nearly all of that matter contracted to form the Sun, but a small amount left behind in a disk formed planets. In the process, your atoms became part of Earth.

You are a planet-walker, and you have evolved to live on the surface of Earth. Are there other beings like you in the Universe? Now that you know planets are common, you can reasonably suppose that there are more planets in the Universe than there are stars. However complicated the formation of the Solar System was, it is a common process, so there may indeed be planet-walkers living on other worlds.

But what are those distant planets like? Before you can go very far in your search for life beyond Earth, you need to explore the range of planetary types. It is time to pack your spacesuit and voyage out among the planets of our Solar System, visit them one by one, and search for the natural principles that relate planets to each other. That journey begins in the next chapter.

Study and Review

Summary

▶ Descartes proposed that the Solar System formed from a contracting vortex of matter—an **evolutionary hypothesis (p. 204)**. Buffon later suggested that a passing comet pulled matter out of the Sun to form the planets—a **catastrophic hypothesis (p. 204)**. Later astronomers replaced the comet with a star to produce the **passing star hypothesis (p. 204)**.

▶ Laplace's **nebular hypothesis (p. 204)** required a contracting nebula to leave behind rings that formed each planet. This hypothesis could not explain the Sun's low angular momentum relative to the planets, a puzzle known as the **angular momentum problem (p. 204)**.

▶ The evidence now strongly favors an evolutionary scenario for the origin of the Solar System. The **solar nebula theory (p. 205)** is a more extensive and mathematically sophisticated version of the nebular hypothesis.

▶ Modern astronomy reveals that nearly all the matter in the Universe, including our Solar System, was originally formed from hydrogen and helium that was present at the time of the big bang. Atoms heavier than helium (which astronomers refer to as "metals") were generated from nuclear reactions that occurred in the cores of subsequent generations of stars. The Sun and planets evidently formed from an interstellar cloud of gas and dust.

▶ The solar nebula theory proposes that the planets formed in a disk of gas and dust orbiting around the protostar that evolved to became the Sun. Hot, dense disks of gas and dust have been detected around many protostars and are understood to be the kind of disks in which planets could form.

▶ The Solar System is disk shaped, with all the planets orbiting nearly in the same plane. The orbital revolution of all the planets, the rotation of most of the planets on their rotation axes, and the orbital revolution of most of their moons are in the same direction, counterclockwise as seen from the north.

▶ The planets are divided into two major groups. The inner four planets are **Terrestrial planets (p. 198)**—small, rocky, and dense worlds, some of which have little or no atmosphere. The next four outward are **Jovian planets (p. 198)**—large, liquid or icy, and low-density worlds which all have thick atmospheres.

▶ All four of the Jovian worlds have ring systems and large families of moons. The Terrestrial planets have no ring systems and few moons.

▶ Most of the **asteroids (p. 197)**, also known as planetoids or minor planets, are small, irregular, rocky bodies, and most are located between the orbits of Mars and Jupiter.

▶ The **Kuiper Belt (p. 197)** is composed of small, icy bodies called **Kuiper Belt Objects (KBOs; p. 197)**. KBOs orbit the Sun beyond Neptune.

- **Comets (p. 200)** are icy bodies that pass through the inner Solar System on long elliptical orbits. As the ices vaporize and release dust, the solar wind and **radiation pressure (p. 200)** push the gas and dust away from the comet, creating a straight ionized gas tail and a curved dust tail that both point approximately away from the Sun.

- **Meteoroids (p. 200)** that fall into Earth's atmosphere are vaporized by friction and are visible as streaks of light called **meteors (p. 200)**. Larger and stronger meteoroids may survive passage through the atmosphere and reach the ground, where they are called **meteorites (p. 200)**.

- The age of a rock can be found by radioactive dating, which is based on the decay **half-life (p. 200)** of radioactive atoms in the rock. Radioactive dating of the oldest rocks from Earth, Moon, and Mars indicate ages of more than 4 billion years. The oldest meteorites that formed with our Solar System have ages of 4.56 billion years, or 4.6 billion years to a precision of two digits, which is taken to be the age of the Solar System.

- **Condensation (p. 207)** in the solar nebula converted some of the gas into solid grains as the nebula cooled. According to the **condensation sequence (p. 207)**, the inner part of the solar nebula's disk was so hot that only metals and rocky particles could form solid grains. The dense Terrestrial planets grew from those solid metal and rocky grains but did not include ices or any significant quantity of gas.

- A comparison between the **uncompressed densities (p. 206)** of the Terrestrial planets shows a trend with distance from the Sun running from high density to low density. This trend is understood to result from the condensation sequence. Further evidence that the condensation sequence was important in the formation of the Solar System is found in a comparison between the high densities of the Terrestrial planets relative to the low densities of the Jovian planets.

- The outer solar nebula's disk, beyond the **frost line (p. 207)**, was cold enough that significant quantities of ices (frozen gases) as well as metals and rocky minerals could form solid particles. The Jovian planets therefore grew rapidly, incorporating large amounts of low-density ices and gases.

- The solar nebula theory predicts that as solid particles became larger they would no longer grow efficiently by condensation. Instead, they could continue to grow by **accretion (p. 207)**, resulting in the formation of billions of **planetesimals (p. 207)** a few km in diameter.

- The collision and coalescing of planetesimals eventually formed **protoplanets (p. 208)**. If a protoplanet grows to about 15 Earth masses, models indicate it can begin further rapid growth by **gravitational collapse (p. 208)**, pulling in gas from the solar nebula.

- The Terrestrial planets may have formed slowly from the combining of planetesimals of similar composition. Radioactive decay plus **heat of formation (p. 208)** evidently melted each planet's interior, causing them to **differentiate (p. 208)** into layers of differing density.

- Earth's early atmosphere was probably produced by a combination of **outgassing (p. 209)** from the planet's interior during differentiation plus planetesimal impacts.

- Observations of disks around protostars suggest that the solar nebula might not have survived long enough for the Jovian planets to form the way the solar nebula theory predicts, by condensation and accretion of solids followed by gravitational collapse of gas. This discrepancy is referred to as the **Jovian problem (p. 210)**. Some models indicate that Jovian planets could have formed very rapidly by **direct collapse (p. 210)**, skipping the condensation and accretion steps.

- Several processes were involved in the clearing of the solar nebula, including solar radiation and solar wind pressure, evaporation by the light from hot nearby stars, the gravitational influence of passing stars, the sweeping up of space debris by the planets, and the ejection of debris from the Solar System by the gravitational influence of the planets.

- All of the solid surfaces in the Solar System were heavily cratered during the **heavy bombardment (p. 218)** period by impacts of planetesimals left over from the formation of the planets.

- **Debris disks (p. 213)** are cold dust disks around main-sequence stars. They represent dust produced by collisions among comets, asteroids, and KBOs. Such disks are probably signs that planets have already formed in those systems.

- Planets orbiting other stars, called **extrasolar planets (p. 215)** or exoplanets, have been detected by many methods, including measuring cyclical Doppler shifts in a star's spectrum produced by an orbiting planet gravitationally tugging on the star. Other discovery methods include discovering repeated **transits (p. 216)** during which an extrasolar planet crosses in front of its parent star and partially blocks the light, or observing eclipses during which the infrared light from a system is temporarily reduced as an extrasolar planet passes behind its host star. A few extrasolar planets have been detected by gravitational **microlensing (p. 217)**.

- Many massive, extrasolar Jovian worlds have been found orbiting close to their parent stars and are therefore called **hot Jupiters (p. 217)**. A hot Jupiter may have formed beyond the frost line and then spiraled inward to its current location as it was swept up protoplanetary disk material.

- The *Kepler* space telescope used the transit method to examine 150,000 stars and, as of July 2014, had found almost a thousand confirmed extrasolar planets ranging in size from larger than Jupiter to smaller than Earth.

- Extrasolar planets more massive than Neptune but less massive than Jupiter are the most common type found so far, combining results from all detection methods.

Review Questions

1. Why is the solar nebula theory considered a theory rather than a hypothesis?

2. Why was the nebular hypothesis never fully accepted by astronomers of the day?

3. What produced the helium now present in the Sun's atmosphere? In Jupiter's atmosphere? In the Sun's core?

4. What produced the iron and heavier elements such as gold and silver in Earth's core and crust?

5. Where did the atoms in your body originate? How does the answer depend on which kind of atom?

6. What evidence can you cite that disks of gas and dust are common around young stars?

7. According to the solar nebula theory, why is Earth's orbit nearly in the plane of the Sun's equator?

8. Look at the orbit inclination values in Table A-10 for all the planets in the Solar System. Explain those values using the solar nebula theory. (*Hint:* See the previous question.)

9. What is the common direction of rotation and orbital motion of celestial objects in the Solar System as viewed from the north? What is the cause of this motion?

10. Why does the solar nebula theory predict that planetary systems are common?

11. What evidence can you cite that the Solar System formed about 4.6 billion years ago?

12. If you could visit an extrasolar planetary system, would you be surprised to find planets older than Earth? Why or why not?

13. Which planet in the Solar System orbits with its equator nearly perpendicular to its orbit? (*Note:* Necessary data can be found in Appendix Table A-10.)

14. I have mostly an iron core, I am heavily cratered, and my orbit has a high inclination angle to Earth's orbit. Which planet am I?

15. I have a ring system, I have many moons, I am outside the frost line, and I am the least dense planet. Which planet am I?

16. I am sometimes called a minor planet. What am I?

17. Between which two planets does the asteroid belt lie?

18. I am a frozen, inactive comet. What am I called, and where am I located?

19. Asteroids are more numerous than KBOs. True or false?

20. I am pea-sized and I am passing through Earth's atmosphere, creating a streak in the sky. Am I a meteoroid, a meteor, or a meteorite?

21. I am massive but not dense, and I have a dense atmosphere. Am I a Terrestrial or a Jovian planet?

22. I am larger than the Moon but smaller than Earth. I have no permanent atmosphere. Which planet am I?

23. Why is almost every solid surface in the Solar System scarred by craters?

24. I am the most massive Terrestrial planet. Which planet am I?

25. I am a Jovian planet, and my color is more blue than green. Which planet am I?

26. I go through phases like the Moon. I am usually easy to observe with an unaided eye. Sometimes I am seen from Earth as a morning star. Which planet am I?

27. What is the difference between condensation and accretion?

28. Why don't Terrestrial planets have ring systems like the Jovian planets?

29. How does the solar nebula theory help you understand the location of the asteroid belt?

30. The oldest rocks on Earth contain zircon crystals. These rocks are 4.4 billion years old. Does this mean that Earth is 4.4 billion years old? Why or why not?

31. If rocks obtained from the Moon indicate an age of 4.5 billion years and the oldest rocks from Earth indicate an age if 4.4 billion years, is the Moon necessarily older than Earth? Why or why not?

32. Which is older, the Moon or the Sun? How do you know?

33. How does the solar nebula theory explain the significant density difference between the Terrestrial and Jovian planets?

34. Did hydrogen gas condense from the nebula as the nebula cooled? What about helium gas? How do you know?

35. Differences in density among various materials in the condensation sequence was the most important factor in determining which materials each planet is made of. True or false?

36. What happens if a planet has differentiated? Would you expect differentiation to be common among the planets? Why or why not?

37. Order the following steps in the formation of a Terrestrial planet chronologically: gravitational collapse, accretion, outgassing, condensation, and differentiation.

38. Which of the step(s) listed in the previous question can be eliminated in models that form Jovian planets in thousands of years, a time frame that solves the Jovian problem?

39. Describe two methods to warm the interior of a planet.

40. Jupiter and Saturn are planets that evidently are still cooling, because they radiate more energy than they receive from the Sun. True or false?

41. Describe two processes that cleared the solar nebula and ended planet formation.

42. A protoplanet is a planetesimal. True or false?

43. Uranus has enough mass to have formed by gravitational collapse. True or false?

44. What properties of the gas and dust disks observed around many protostars indicate they could evolve into planetary systems?

45. Why would the astronomically short lifetime of gas and dust disks around protostars pose a problem in understanding how the Jovian planets formed? What modification of the solar nebula theory might solve this problem?

46. What is the difference between the dense, hot disks seen around some stars and the low-density, cold disks seen around other stars?

47. What evidence can you cite that there are extrasolar planets orbit host stars?

48. Describe three methods to find extrasolar planets.

49. Why is the existence of hot Jupiters puzzling? What is the current hypothesis for how they formed?

50. What is the difference between a hot Jupiter and Jupiter?

51. **How Do We Know?** The evidence is overwhelming that the Grand Canyon was dug over a span of millions of years by the erosive power of the Colorado River and that river's tributary streams. Does this evidence support a catastrophic theory or an evolutionary theory?

52. **How Do We Know?** Why must scientists be skeptical of new discoveries even when they think that the discoveries are correct?

Discussion Questions

1. Why is studying discarded models for the formation of the Solar System important?

2. If you visited some extrasolar planetary systems while the planets were forming, would you expect to see the condensation sequence at work, or is the condensation sequence probably unique to our Solar System? How do the properties of the extrasolar planets discovered so far affect your answer?

3. Should most extrasolar planetary systems have asteroid belts? Should all extrasolar planetary systems show evidence of an age of heavy bombardment? Why do you think so?

4. If the solar nebula hypothesis is correct, then extrasolar planets may be more common in the Universe than stars. Do you agree? Why or why not?

5. Do you think our Solar System is unique? Why or why not?

6. In your opinion, could Earth have a twin—a planet orbiting a star like the Sun, with liquid water on its surface, a biosphere, and an intelligent species?

Problems

1. If you observed the Solar System from the vantage point of the nearest star, at a distance of 1.3 pc, what would the maximum angular separation be between Earth and the Sun? (*Note:* 1 pc = 2.1×10^5 AU. *Hint:* Use the small-angle formula, Chapter 3.)

2. Venus can be as bright as apparent magnitude −4.7 when at a distance of about 1 AU. How many times fainter would Venus look from a distance of 1 pc? What would its apparent magnitude be? Assume Venus has the same illumination phase from your new vantage point. (*Note:* 1 pc = 2.1×10^5 AU. *Hint:* Review the definition of apparent visual magnitudes, Chapter 2.)

3. What is the smallest-diameter crater you can identify in the photo of Mercury on page 198? (*Hint:* See Appendix Table A-10, Properties of the Planets, to find the diameter of Mercury in kilometers for scale.)

4. Refer to the masses of the planets in Table A-10. If the mass of Jupiter is defined as 1 M_{Jup}, what is the mass of Saturn in units of M_{Jup}? Of Uranus? Of Neptune? Of Earth? Of Mercury? Compare your results and decide the number of distinct categories of planet masses. Does your determination agree with the text?

5. A sample from a meteorite that landed on Earth has been analyzed, and the result shows that out of every 1000 nuclei of potassium-40 originally in the meteorite, only 200 are still present, have not yet decayed. How old is the meteorite? (*Hint:* See Figure 10-5.)

6. You analyze a sample of a meteorite that landed on Earth and find 87.5 percent of the uranium-238 radioactive atoms have decayed into lead-206. What percentage of the sample are daughter isotopes and what percentage are uranium-238 radioactive atoms? What is the age of the sample? What do you conclude about the sample?

7. You analyze a sample of a meteorite that landed on Earth and find 93.75 percent of a certain type of radioactive atoms have decayed into the corresponding daughter atom. Calculate the number of half-lives that have occurred.

8. Examine the values in Table 10-1. Which object's observed density differs least from its uncompressed density? Why?

9. Examine Table 10-2. What might a planet's composition be if the planet formed in a region of the solar nebula where the temperature was about 100 K?

10. Examine Table 10-2. What might a planet's composition be if the planet formed in a region of the solar nebula where the temperature was about 1200 K?

11. Suppose that Earth grew to its present size in 10 million years from particles averaging 100 grams each. On average, how many particles did Earth capture per second? (*Notes:* 1 yr = 3.2×10^7 s. Earth's mass can be found in Appendix Table A-10.)

12. How many impacts would you expect to strike a 100 m² region in one hour during Earth's formation as described in Problem 11? (*Notes:* The surface area of a sphere is $4\pi r^2$; 1 yr = 3.2×10^7 s. *Hint:* Assume that Earth had its current radius; see Appendix Table A-10.)

13. The speed of the solar wind is approximately 400 km/s. How many days does the solar wind take to travel from the Sun to Earth? (*Notes:* 1 AU = 1.5×10^8 km and 1 day = 86,400 s.)

Learning to Look

1. What do you see in the image at right that indicates this planet formed far from the Sun?

Visual

NASA/JPL-Caltech/SSI

2. Why do astronomers conclude that the surface of Mercury, shown at right, is old? When did the majority of those craters form?

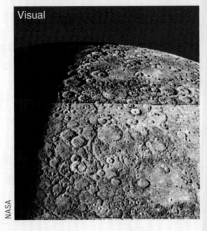

Visual

NASA

3. In the mineral specimen represented to the right, radioactive parent atoms (*red*) have decayed to form daughter atoms (*blue*). How old is this specimen in half-lives? (See Figure 10-5).

4. Redraw the mineral specimen of the previous problem, with the correct ratio of red to blue dots, after three more half-lives have passed.

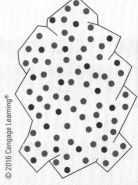

© 2016 Cengage Learning®

11 Earth: The Active Planet

Guidepost In the preceding chapter, you learned how our Solar System formed as a by-product of the formation of the Sun. You also saw how distance from the Sun determined the general composition of each planet. In this chapter, you begin your study of individual planets with Earth. You will come to see Earth not only as your home but also as a planet among other planets. On the way, you will answer four important questions:

▶ **How does Earth compare with the other Terrestrial planets?**

▶ **How has Earth changed since it formed?**

▶ **What is the interior of Earth like?**

▶ **How has Earth's atmosphere formed and evolved?**

Like a mountain climber establishing a base camp before attempting the summit, you can establish your basis of comparison on Earth in this chapter. In the following chapters, you will visit worlds that are un-Earthly but, in some ways, still familiar.

Nature evolves. The world was different yesterday.

PRESTON CLOUD, *COSMOS, EARTH AND MAN*

Mount Etna erupting in 2001 above the Sicilian city of Catania. Humanity, a relative newcomer to Earth, has so far witnessed only a narrow range of the types of powerful activity that a rocky planet with a hot interior can display.

P LANETS, LIKE PEOPLE, are more alike than they are different. They are described by the same basic principles, and their differences arise mostly because of small differences in background. To understand the planets, you can compare and contrast them to identify those common basic principles, an approach called **comparative planetology**.

Earth is the ideal starting point for your study because it is the planet you know best. It is also a complex planet. Earth's core is partly liquid and generates a magnetic field. Its crust is active, with moving continents, earthquakes, volcanoes, and mountain building. Earth's oxygen-rich atmosphere is unique in the Solar System. The properties of Earth will give you perspective on the other planets in our Solar System.

11-1 A Travel Guide to the Terrestrial Planets

If you visit the city of Granada in Spain, you will probably consult a travel guide. If it is a good guide, it will do more than tell you where to find museums and restrooms. It will give you a preview of what to expect. You are beginning a journey to the Earth-like worlds, so you should consult a travel guide and see what is in store for you when you get there.

Five Worlds

In this chapter and the next two chapters, you are about to visit Earth, Earth's Moon, Mercury, Venus, and Mars. It may surprise you that the Moon is included in your itinerary. It is, after all, just a natural satellite orbiting Earth and isn't one of the planets. But the Moon is a fascinating world of its own, one of the largest moons in the Solar System. It makes a striking comparison with the other worlds on your list, and its properties gives you important information about the history of Earth and the other planets.

Figure 11-1 compares the five worlds. The first feature you might notice is diameter. The Moon is small, and Mercury is not much bigger. Earth and Venus are large and quite similar in size, but Mars is a medium-sized world. You will discover that size is a critical factor in determining a world's personality. Small worlds tend to be geologically inactive, whereas larger worlds tend to be active.

Core, Mantle, and Crust

The Terrestrial worlds are made up of rock and metal. They are all differentiated, which means they are each separated into layers of different density, with dense metallic cores surrounded by less dense rocky **mantles**, and low-density crusts on their exteriors.

You learned in the preceding chapter that when the planets formed, their surfaces were subjected to heavy bombardment by leftover planetesimals and debris in the young Solar System. You will see lots of craters on these worlds, especially on Mercury and the Moon, many of them dating back to the heavy bombardment era. Notice that cratered surfaces are old. For example, if a lava flow covered up some cratered landscape after the end of the heavy bombardment, few craters could be formed later on that surface because most of the debris in the Solar System was gone. When you see a smooth plain on a planet, you can guess that surface is younger than the heavily cratered areas.

Another important way you can study a planet is by following the energy flow. In the preceding chapter you learned that the heat in the interior of a planet may be partly from radioactive decay and partly left over from the planet's formation, but in any case it must flow outward toward the cooler surface where it is radiated into space. In the process of flowing outward, the heat can cause convection currents, magnetic fields, plate motions, quakes, faults, volcanism, mountain building, and more. Heat flowing outward through the cooler crust makes a large world like Earth geologically active (**How Do We Know? 11-1**). In contrast, the Moon and Mercury—both small worlds—cooled quickly, so they have little heat flowing outward now and are relatively inactive.

Atmospheres

When you look at Mercury and the Moon in Figure 11-1, you can see their craters, plains, and mountains clearly; they each have little or no atmosphere to obscure your view. In comparison, the surface of Venus is completely hidden by a cloudy atmosphere even thicker than Earth's. Mars, the medium-size planet, has a relatively thin atmosphere.

You might ponder two questions. First, why do some worlds have atmospheres whereas others do not? You will discover that both size and temperature are important. The second question is more complex. Where did those atmospheres come from? To answer that question in later chapters, you will have to study the geological histories of these worlds.

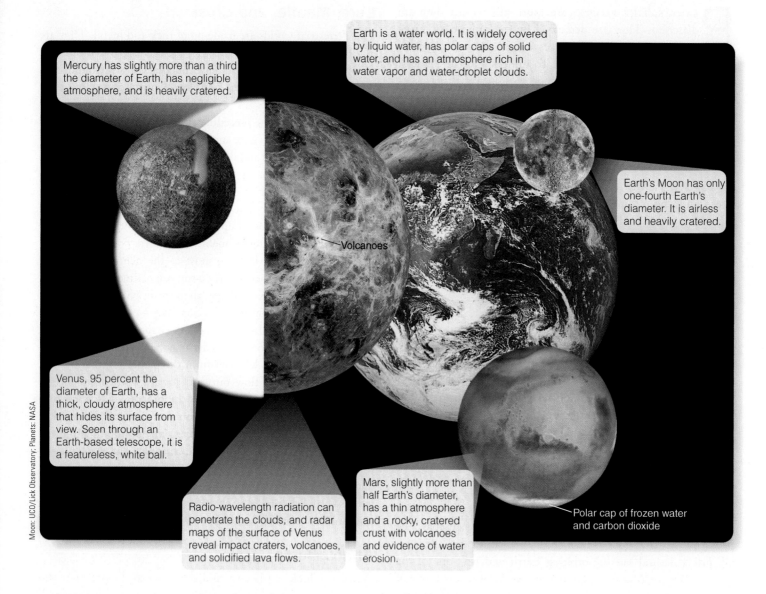

Moon: UCO/Lick Observatory; Planets: NASA

Mercury has slightly more than a third the diameter of Earth, has negligible atmosphere, and is heavily cratered.

Earth is a water world. It is widely covered by liquid water, has polar caps of solid water, and has an atmosphere rich in water vapor and water-droplet clouds.

Volcanoes

Earth's Moon has only one-fourth Earth's diameter. It is airless and heavily cratered.

Venus, 95 percent the diameter of Earth, has a thick, cloudy atmosphere that hides its surface from view. Seen through an Earth-based telescope, it is a featureless, white ball.

Radio-wavelength radiation can penetrate the clouds, and radar maps of the surface of Venus reveal impact craters, volcanoes, and solidified lava flows.

Mars, slightly more than half Earth's diameter, has a thin atmosphere and a rocky, cratered crust with volcanoes and evidence of water erosion.

Polar cap of frozen water and carbon dioxide

▲ **Figure 11-1** Planets in comparison: Earth and Venus are similar in size, but their atmospheres and surfaces are different. The Moon and Mercury are much smaller, and Mars is intermediate in size.

11-2 Earth as a Planet

Like all the Terrestrial planets, Earth formed from the inner solar nebula about 4.6 billion years ago. Even as it took form, it began to change.

Four Stages of Planetary Development

There is evidence that Earth and the other Terrestrial planets, plus Earth's Moon, passed through four developmental stages (Figure 11-2).

The first stage of planetary evolution is *differentiation*, the separation of material according to density. As you have already learned, Earth is differentiated, meaning separated into layers of different density. That differentiation is understood to have occurred as a result of melting of Earth's interior caused by heat from a combination of radioactive decay plus energy released by infalling matter during the planet's formation. Once the interior of Earth melted, the densest materials were able to sink to the core.

The second stage, *cratering*, could not begin until a solid surface formed. The heavy bombardment of the early Solar System made craters on Earth just as it did on the Moon and other planets. As the debris in the young Solar System cleared away, the rate of cratering impacts fell gradually to its present low rate. You will learn in the next chapter that evidence provided by lunar crater counts and rock samples indicates there

How Do We Know? 11-1

Understanding Planets: Follow the Energy

What causes change? One of the best ways to think about a scientific problem is to follow the energy. According to the principle of cause and effect, every effect must have a cause, and every cause must involve energy. Energy moves from regions of high concentration to regions of low concentration and, in doing so, produces changes. For example, coal burns to make steam in a power plant, and the steam passes through a turbine and then escapes into the air. In flowing from the burning coal to the atmosphere, the heat spins the turbine that spins a generator to make electricity.

Scientists commonly use energy as a key to understanding nature. A biologist might ask where certain birds get the energy to fly thousands of miles, and a geologist might ask where the energy comes from to power a volcano. Energy is everywhere, and when it moves, whether it is in birds or molten magma, it causes change. Energy is the "cause" in "cause and effect."

In previous chapters, keeping track of the flow of energy from the inside of a star to its surface helped you understand how the Sun and other stars work. You saw that the outward flow of energy supports the star against its own weight, drives convection currents that produce magnetic fields, and causes surface activity such as spots, prominences, and flares. You were able to understand stars because you could follow the flow of energy outward from their interiors.

You can also think of a planet by following the energy. The heat in the interior of a planet may be left over from the formation of the planet, or it may be heat generated by radioactive decay, but it must flow outward toward the cooler surface, where it is radiated into space. In flowing outward, the heat can cause convection currents in the mantle, magnetic fields, plate motions, quakes, faults, volcanism, mountain building, and more.

When you think about any world, be it a small asteroid or a giant planet, think of it as a source of heat that flows outward through that object's surface into space. If you can follow that energy flow, you can understand a great deal about the world. A planetary astronomer once said, "The most interesting thing about any planet is how its heat gets out."

Heat flowing out of Earth's interior generates geological activity such as that at Yellowstone National Park.

was a temporary large increase in the impact cratering rate near the end of the heavy bombardment, and this violent event likely would have affected all the planets.

The third stage, *basin flooding*, began as radioactive decay continued to heat Earth's interior and caused rock to melt in the upper mantle, where the pressure was lower than in the deep interior. Some of that molten rock welled up through cracks in the crust and flooded the deeper impact basins. Later, as the environment cooled, water fell as rain and flooded the basins to form the first oceans. Note that on Earth, basin flooding was first by lava and later by water.

The fourth stage, *slow surface evolution*, has continued for at least the past 3.5 billion years. Earth's surface is constantly changing as sections of crust slide over and against each other, push up mountains, and shift continents. In addition, moving air and water erode the surface and wear away geological features. Almost all traces of the first billion years of Earth's history have been destroyed by the active crust and erosion.

Terrestrial planets pass through these four stages, but differences in mass, temperature, and composition among the planets can emphasize some of those stages over others and produce surprisingly different worlds.

Exceptional Earth

Earth is a good standard for comparative planetology (■ Celestial Profile 2). Every major process on any rocky world in our Solar System is represented in some form on Earth. Nevertheless, Earth is unusual, if not unique, among the planets in our Solar System in two ways: the presence of abundant liquid water and the presence of life.

First, 75 percent of Earth's surface is covered by liquid water. No other planet in our Solar System has liquid water on its surface, although, as you will learn in later chapters, Mars had surface water long ago, Venus probably did, and some moons in the outer Solar System show evidence of having liquid water under their surfaces. Water fills Earth's oceans, evaporates into the atmosphere, forms clouds, and then falls as rain. Water falling on the continents flows downhill to form rivers that flow back to the sea and, in doing so, produce vigorous erosion. Entire mountain ranges can literally dissolve and

Four Stages of Planetary Development

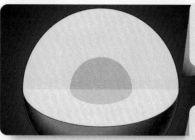

Differentiation produces a dense core, thick mantle, and low-density crust.

The young Earth was heavily bombarded in the debris-filled early solar system.

Flooding by molten rock and later by water can fill lowlands.

Slow surface evolution continues due to geological processes, including erosion.

© 2016 Cengage Learning®

▲ **Figure 11-2** The four stages of Terrestrial planet development are illustrated for Earth.

wash away in only a few tens of millions of years, less than 1 percent of Earth's total age. You will not see such rapid erosion on most worlds.

Your home planet is special in a second way. Some of the matter on the surface of this world is alive, and a small part of that living matter, including you, is aware. No one is sure how the presence of living matter has affected the evolution of Earth, but this process seems to be totally missing from other worlds in our Solar System. Furthermore, as you will learn later in this chapter, the thinking part of the life on Earth—humankind—is actively altering our planet.

NGDC

Earth's surface is marked by high continents and low seafloors. The crust is only 10 to 60 km thick, and interior to that are a deep mantle and an inner core.

Celestial Profile 2 Earth

Motion:

Average distance from the Sun	1.00 AU (1.50×10^8 km)
Eccentricity of orbit	0.017
Maximum distance from the Sun	1.017 AU (1.52×10^8 km)
Minimum distance from the Sun	0.983 AU (1.47×10^8 km)
Inclination of orbit to ecliptic	0° (by definition)
Orbital period	1.0000 y (365.26 days)
Period of rotation	24.00 h (with respect to the Sun)
Period of rotation	23.93 h (with respect to the stars)
Inclination of equator to orbit	23.4°

Characteristics:

Equatorial diameter	1.28×10^4 km
Mass	5.97×10^{24} kg
Average density	5.51 g/cm³ (3.96 g/cm³ uncompressed)
Surface gravity	1.00 Earth gravity
Escape velocity	11.2 km/s
Surface temperature	−90° to +60°C (−130° to +140°F)
Average albedo	0.31
Oblateness	0.0034

Personality Point:

The name *Earth* comes, through Old English *eorthe* and Greek *Eraze*, from the Hebrew *erez*, which means *ground*. *Terra* comes from the Roman goddess of fertility and growth; thus, Terra Mater, Mother Earth.

11-3 The Solid Earth

Although you might think of Earth as solid rock, it is in fact neither entirely solid nor entirely rock. The thin crust seems solid, but it floats and shifts on a semiliquid layer of molten rock just below the crust. Below that lies a deep, rocky mantle surrounding a core of liquid metal. Much of what you see on Earth's surface is determined by conditions and processes in its interior.

Earth's Interior

The theory of the origin of planets from the solar nebula predicts that Earth should have melted and differentiated into a dense metallic core and a dense mantle with a low-density silicate crust. But did it? Where's the evidence? Earth's average density can be calculated easily from its known mass and density. Clearly, the silicate rocks on Earth's surface have lower density than material inside the planet. But what more can be determined about Earth's interior?

High temperature and tremendous pressure in Earth's interior make any direct exploration impossible. Even the deepest oil wells extend only a few kilometers down and don't reach through the crust. It is impossible to drill far enough to sample Earth's core. Yet Earth scientists have studied the interior and found clear evidence that Earth did differentiate (**How Do We Know? 11-2**).

This exploration of Earth's interior is possible because earthquakes produce vibrations called **seismic waves** that travel through the crust and interior and eventually register on sensitive detectors called **seismographs** all over the world (**Figure 11-3**). Two kinds of seismic waves are important to this discussion. The **pressure (P) waves** are much like sound waves in that they travel as a sequence of compressions and decompressions. As a *P* wave passes, particles of matter vibrate back and forth parallel to the direction of wave travel (**Figure 11-4a**). In contrast, the **shear (S) waves** move as displacement of particles perpendicular to the waves' direction of travel (Figure 11-4b). That means that *S* waves distort the material but do not compress it. Normal sound waves are pressure waves, but the vibrations you see in a bowl of jelly are shear waves. Because *P* waves are compression waves, they can move through a liquid. *S* waves can move along the surface of a liquid but not through it. A glass of water can't shimmy like jelly because a liquid does not have the rigidity required to transmit *S* waves.

The *P* and *S* waves caused by an earthquake do not travel in straight lines or at constant speed within Earth. The waves may reflect from boundaries between layers of different density, or they may be refracted as they pass through a boundary. In addition, the gradual increase in temperature and density toward Earth's center causes the speed of sound to increase as well. These changes cause seismic waves to be refracted as they travel through Earth's interior, meaning that, instead of following straight lines, the waves curve away from the denser, hotter central regions. Geoscientists can use the arrival times of reflected and refracted seismic waves from distant earthquakes to construct a model of Earth's interior.

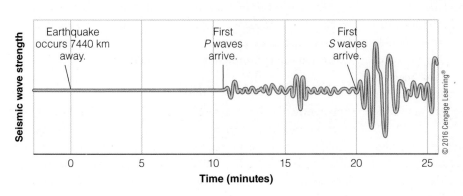

◀ **Figure 11-3** A seismograph in northern Canada made this record of seismic waves from an earthquake in Mexico. The first vibrations, *P* waves, arrived 11 minutes after the quake, but the slower *S* waves took 20 minutes to make the journey.

© 2016 Cengage Learning®

How Do We Know? 11-2

Studying an Unseen World

How can studying what can't be seen save your life? Science tells us how nature works, and the basis for scientific knowledge is evidence gathered through observation. But much of the natural world can't be observed directly because it is too small, or far away, or deep underground. Yet geologists describe molten rock deep inside Earth, and biologists discuss the structure of genetic molecules. So how can these scientists know about things they can't observe directly?

A virus, which can be as common as a cold or as deadly as Ebola, contains a tiny bit of genetic information in the form of a DNA molecule. You have surely had a virus, but you've never seen one. Even under the best electron microscopes, a virus can be seen only as a hazy pattern of shadows. Nevertheless scientists know enough about them to devise ingenious ways to protect us from viral disease.

A virus is DNA hidden inside a protective coat of protein molecules, which is a rigid molecular lattice almost like a mineral. In fact, a culture of viruses can be crystallized, and the shapes of the crystals reveal the shape and structure of the virus. Unlike a crystal of calcite, however, a crystal of viruses also contains genetic information.

Scientists can make a vaccine to protect against a certain virus if they can identify a unique molecular pattern on the protein coat. The vaccine is harmless but contains that same pattern as the active virus and thereby trains your body's immune system to recognize the pattern and attack it. Vaccines significantly reduce the danger of common illness such as chicken pox and influenza, and they have virtually wiped out devastating diseases like polio and smallpox in the developed world. Researchers are currently working on a vaccine for HIV/AIDS that would potentially save millions of lives.

Even though viruses are too small to see, scientists can use chains of inference and the interaction of theory and evidence to deduce the structure of a virus. Whether it is a virus or the roots of a volcano, science takes us into realms beyond human experience and allows us to see the unseen.

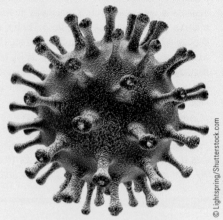

Electron microscopes allow biologists to deduce the elegant structures of viruses.

© Lightspring/Shutterstock.com

Such studies confirm that the interior consists of three parts: a central core, a thick mantle, and a thin crust. *S* waves provide an important clue to the nature of the core. When an earthquake occurs, no direct *S* waves pass through the core to register on seismographs on the opposite side of Earth, as if the core were casting a shadow (**Figure 11-5**). The absence of *S* waves shows that the core is mostly liquid, and the size of the *S*-wave shadow fixes the radius of the core at about 55 percent of Earth's radius. Mathematical models based on the seismographic measurements and other data predict that the core is also hot (about 5000 K), dense (about 14 g/cm³), and composed of iron and nickel.

Earth's core is as hot as the surface of the Sun, but it is under such tremendous pressure that the material cannot vaporize. Because of its high temperature, most of the core is a liquid. Nearer the center, the material is under even higher pressure, which in turn raises the melting point so high that the material cannot melt (**Figure 11-6**). That is why there is an inner core of solid iron and nickel. Estimates suggest the inner core's radius is about 22 percent that of Earth.

The paths of seismic waves in the mantle, the layer of dense rock that lies between the molten core and the crust, show that it is not molten, but it is not precisely solid either. Mantle material behaves like a **plastic**, meaning a material with the properties of a solid but capable of flowing under pressure. The asphalt used in paving roads is a common example of a plastic. It shatters if struck with a sledgehammer, but it bends under the steady weight of a heavy truck. Just below Earth's crust, where the pressure is less than at greater depths, the mantle is most plastic.

Earth's rocky crust is made up of low-density rocks and floats on the denser mantle. The crust is thickest under the continents, up to 60 km (35 mi) thick, and thinnest under the oceans, where it is only about 10 km (6 mi) thick. Unlike the mantle, the crust is brittle and can break when it is stressed.

Although Earth's core is only 2000 miles from you, it is completely inaccessible. Earth's seismic activity reveals some of Earth's innermost secrets. But there is another source of evidence about Earth's interior—its magnetic field.

Earth's Magnetic Field

Apparently, Earth's magnetic field is a direct result of its rapid rotation and its molten metallic core. Internal heat forces the liquid core to circulate with convection, while Earth's rotation turns it about an axis. The core is a highly conductive iron–nickel alloy, an even better electrical conductor than copper, the

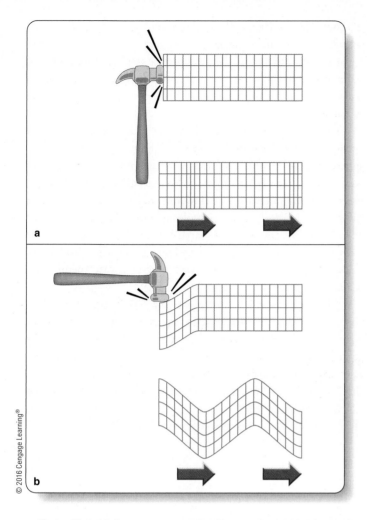

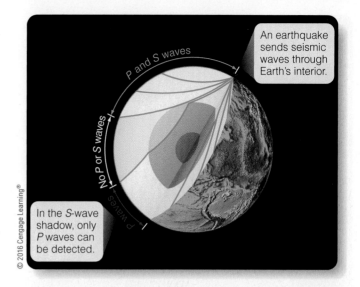

▲ Figure 11-5 *P* and *S* waves give you clues to the structure of Earth's interior. No direct S waves from an earthquake reach the side of Earth opposite their source, indicating that Earth's core is liquid. The size of the *S*-wave "shadow" tells you the size of the liquid outer part of the core.

▲ Figure 11-4 (a) *P*, or pressure, waves, like sound waves in air, travel as a region of compression. (b) *S*, or shear, waves, like vibrations in a bowl of jelly, travel as displacements perpendicular to the direction of travel. *S* waves tend to travel more slowly than *P* waves and cannot travel through liquids.

material commonly used for electrical wiring. The rotation of this convecting, conducting liquid generates Earth's magnetic field in a process called the *dynamo effect* (Figure 11-7). That is the same process that generates the solar magnetic field in the Sun's convective layers (Chapter 8, pages 154–155), and you will see it again when you explore other planets.

Earth's magnetic field protects it from the solar wind. Blowing outward from the Sun at about 400 km/s, the solar wind consists of ionized gases carrying a small part of the Sun's magnetic field (Chapter 8, pages 153 and 163). When the solar wind encounters Earth's magnetic field, it is deflected like water flowing around a boulder in a stream. The surface where the solar wind is first deflected is called the **bow shock**, and the cavity dominated by Earth's magnetic field is called the

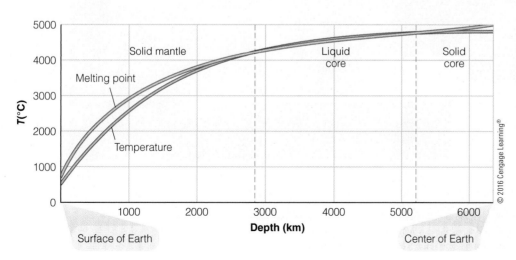

◀ Figure 11-6 Theoretical models combined with observations of the velocity of seismic waves reveal the temperature inside Earth (*blue line*). The melting point of the material (*red line*) is determined by its composition and by the pressure. In the mantle and in the inner core, the melting point is higher than the existing temperature, and the material is not molten.

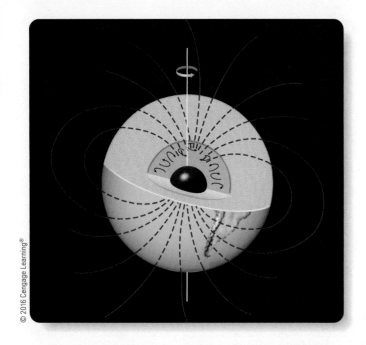

Figure 11-7 The dynamo effect couples convection in the liquid core with Earth's rotation to produce electric currents that are understood to be responsible for Earth's magnetic field.

magnetosphere (Figure 11-8a). High-energy particles from the solar wind leak into the magnetosphere and become trapped within Earth's magnetic field to produce the **Van Allen belts** of radiation. You will see in later chapters that other planets that have magnetic fields have bow shocks, magnetospheres, and radiation belts.

Earth's magnetic field produces the dramatic and beautiful auroras, glowing rays and curtains of light in the upper atmosphere (Figure 11-8b). The solar wind carries charged particles past Earth's extended magnetic field, and this generates tremendous electrical currents that flow into Earth's atmosphere near the north and south magnetic poles (Chapter 8, page 163). The currents ionize gas atoms in Earth's atmosphere, and when the ionized atoms capture electrons and recombine, they produce an emission spectrum as if they were part of a vast "neon" sign (Chapter 7, pages 133 and 140).

Although you can be confident that Earth's magnetic field is generated within its molten core, many mysteries remain. For example, rocks retain traces of the magnetic field in which they solidify, and some contain fields that point backward. That is, they imply that Earth's magnetic field was reversed at the time they solidified. Careful analysis of such rocks indicates that Earth's field has reversed itself in a very irregular pattern, on average every 700,000 years or so during the past 330 million years,

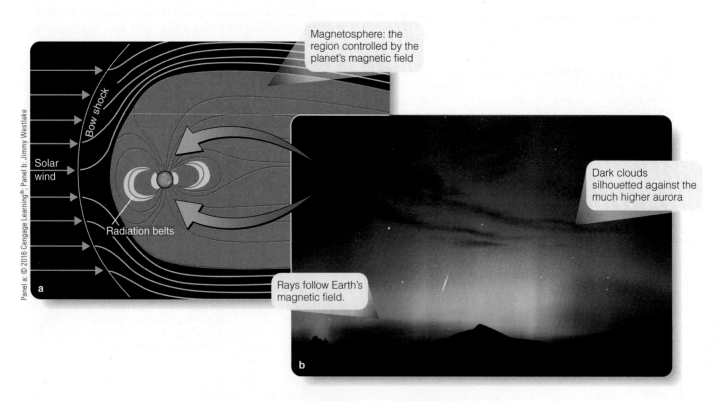

Figure 11-8 (a) Earth's magnetic field dominates space around Earth by deflecting the solar wind and trapping high-energy particles in radiation belts. The magnetic field lines enter Earth's atmosphere around the north and south magnetic poles. (b) Powerful currents flow down along the magnetic field lines near the poles and excite gas atoms to emit photons, creating auroras. Colors are produced as different atoms are excited to produce characteristic spectral emission lines. Note the meteor (shooting star).

with the north magnetic pole becoming the south magnetic pole and vice versa. The reversals are poorly understood, but they are probably related to changes in the core's convection.

Convection in Earth's core is important because it generates the magnetic field. As you will see in the next section, convection in the mantle constantly remakes Earth's surface.

Earth's Active Crust

Earth's crust is composed of low-density rock that floats on the mantle. The image of a rock floating may seem odd, but recall that the rock of the mantle is very dense. Also, just below the crust, the mantle rock tends to be highly plastic, so great sections of low-density crust do indeed float on the semiliquid mantle like great lily pads floating on a pond.

The motion of the crust and the erosive action of water make Earth's crust highly active. Read **The Active Earth** on pages 234–235 and notice three important points and six new terms:

1. *Plate tectonics*, which is the motion of crustal plates, produces much of the geological activity on Earth. Plates spreading apart can form *rift valleys* or, on the ocean floor, *midocean rises* where molten rock solidifies to form *basalt*. A plate sliding into a *subduction zone* can trigger volcanism, and the collision of plates can produce *folded mountain ranges*. Chains of volcanoes such as the Hawaiian Islands can result when a plate moves horizontally across a hot spot.

2. Notice how the continents on Earth's surface have moved and changed over periods of hundreds of millions of years. A hundred million years is only 0.1 billion years, so sections of Earth's crust are in rapid motion relative to the 4.6-billion-year age of the planet.

3. Most of the geological features you know—mountain ranges, the Grand Canyon, and even the familiar outlines of the continents—are recent products of Earth's active surface. Earth's surface is constantly renewed. The oldest rocks on Earth, which are small crystals of the mineral zircon from western Australia, are 4.4 billion years old. Most of the crust is much younger than that. Most of the mountains and valleys you see around you are no more than a few tens of millions of years old.

The average speed of plate movement is slow, but sudden movements do occur. Plate margins can stick, accumulate stress, and then release it suddenly. That's what happened in 2011 along a major subduction zone in the Pacific Ocean off the coast of Japan. The total motion was as much as 15 meters (50 ft), and the resulting earthquake caused devastating tsunamis (tidal waves). Every day, minor earthquakes occur on moving faults, and the stress that builds in those faults that are sticking will eventually be released in major earthquakes.

Earth's active crust explains why Earth contains so few impact craters. The Moon is richly cratered, but Earth's surface

has only about 150 impact craters. Plate tectonics and erosion have destroyed all but the most recent craters on Earth.

You can see that Earth's geology is dominated by two processes. Heat rising from the interior drives plate tectonics. Just below the thin crust of solid rock lies a churning molten layer that rips the crust to fragments and pushes the pieces about like rafts of wood on a pond. The second process modifying the crust is erosion by water. Water falls as rain and snow and tears down mountains, erodes river valleys, and washes any raised ground into the sea. Tectonics builds up mountains and continents, and then erosion rips them down.

DOING SCIENCE

What evidence indicates that Earth has a liquid metal core? Scientists often do not have the luxury of direct observations to settle a question.

In the case of Earth's core, the evidence is indirect because you can never visit Earth's core. Seismic waves from distant earthquakes pass through Earth, but a certain kind of wave, the *S* type, does not pass through the core. Because the *S* waves cannot move through a liquid, scientists conclude that Earth's core is partly liquid. Earth's magnetic field is further evidence of a liquid metallic core. The theory for the generation of magnetic fields, the dynamo effect, requires a moving, conducting liquid (for a planet) or gas (for a star) in the interior. If Earth's core were not partly a liquid metal, it would not be able to generate a magnetic field.

Two different kinds of evidence tell you that our planet has a liquid core. Now investigate another important characteristics of Earth: **What evidence can you cite to support the theory of plate tectonics?**

11-4 Earth's Atmosphere

You can't tell the story of Earth without mentioning its atmosphere. It is not only necessary for life but also intimately related to the crust. It affects the surface through erosion by wind and water, and in turn the chemistry of Earth's surface affects the composition of the atmosphere.

Origin of the Atmosphere

Until a few decades ago, planetary scientists thought the early Earth might have attracted small amounts of gases such as hydrogen, helium, and hydrogen compounds from the solar nebula to form a **primeval atmosphere**. According to that old hypothesis, slow decay of radioactive elements eventually heated Earth's interior; melted it; caused it to differentiate into core, mantle, and crust; and triggered widespread volcanism.

When a volcano erupts, 50 to 80 percent of the gas released is water vapor. The rest is mostly carbon dioxide, nitrogen, and

The Active Earth

1   Our world is an astonishingly active planet. Not only is it rich in water and therefore subject to rapid erosion, but its crust is divided into moving sections called *plates*. Where plates spread apart, lava wells up to form new crust; where plates push against each other, they crumple the crust to form mountains. Where one plate slides over another, you see volcanism. This process is called **plate tectonics**, referring to the Greek word for "builder." (An architect is literally an arch builder.)

A typical view of planet Earth

Mountains are common on Earth, but they erode away rapidly because of the abundant water.

Janet Seeds

William K. Hartmann

A **rift valley** forms where continental plates begin to pull apart. The Red Sea is forming as Africa starts to separate from the Arabian peninsula.

Midocean rise

Red Sea

Midocean rise

NGDC

1a   Evidence of plate tectonics was first found in ocean floors, where plates spread apart and magma rises to form **midocean rises** made of rock called **basalt** that is typical of solidified lava. Radioactive dating shows that the basalt is younger near the midocean rise. Also, the ocean floor carries less sediment near the midocean rise. As Earth's magnetic field reverses back and forth, it is recorded in the magnetic fields frozen into the basalt. This produces a magnetic pattern in the basalt that shows that the seafloor is spreading away from the midocean rise.

1b   A **subduction zone** is a deep trench where one plate slides under another. Melting releases low-density magma that rises to form volcanoes such as those along the northwest coast of North America, including Mount St. Helens.

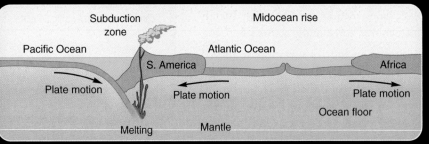

Subduction zone

Midocean rise

Pacific Ocean

Atlantic Ocean

S. America

Africa

Plate motion

Plate motion

Plate motion

Ocean floor

Melting

Mantle

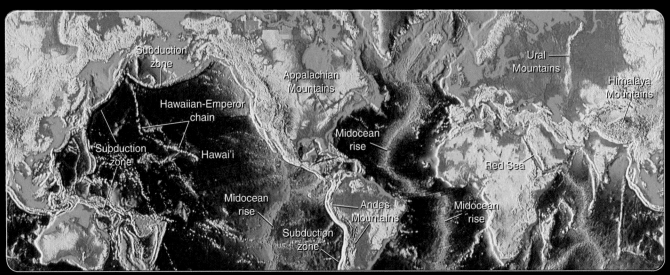

Subduction zone

Appalachian Mountains

Ural Mountains

Himalaya Mountains

Hawaiian-Emperor chain

Midocean rise

Subduction zone

Hawai'i

Red Sea

Midocean rise

Andes Mountains

Midocean rise

Subduction zone

1c   Hot spots caused by rising magma in the mantle can poke through a plate and cause volcanism such as that in Hawai'i. As the Pacific plate has moved northwestward, the hot spot has punched through to form a chain of volcanic islands, now mostly worn below sea level. **Folded mountain ranges** can form where plates push against each other. For example, the Ural Mountains lie between Europe and Asia, and the Himalaya Mountains are formed by India pushing north into Asia. The Appalachian Mountains are the remains of a mountain range thrust up when North America was pushed against Europe and Africa.

1d The floor of the Pacific Ocean is sliding into subduction zones in many places around its perimeter. That process pushes up mountains such as the Andes and triggers earthquakes and active volcanism all around the Pacific in what is called the Ring of Fire. In places such as Southern California, the plates slide past each other, causing frequent earthquakes.

NGDC

Hawai'i

Yellow lines on this globe mark plate boundaries. Red dots mark earthquakes since 1980. Earthquakes within the plate, such as those at Hawai'i, are related to volcanism over hot spots in the mantle.

Continental Drift

Not long ago, Earth's continents came together to form one continent dubbed Pangaea by geoscientists.

Pangaea

200 million years ago

Pangaea broke into a northern and a southern continent.

Laurasia

Gondwanaland

135 million years ago

Notice India moving north toward Asia.

65 million years ago

The continents are still drifting on the highly plastic upper mantle.

Today

2 The floor of the Atlantic Ocean is not being subducted. It is locked to the continents and is pushing North and South America away from Europe and Africa at about 3 cm per year in a motion called *continental drift*. Scientists can measure the motion by, for example, comparing the time for laser flashes sent from European and American observatories to bounce off mirrors left on the moon by the Apollo astronauts. Roughly 200 million years ago, North and South America were joined to Europe and Africa. Evidence of that lies in similar fossils and similar rocks and minerals found in the matching parts of the continents. Notice how North and South America fit against Europe and Africa like a puzzle.

Michael A. Seeds

Formation of Grand Canyon

Formation of Earth

Age of dinosaurs

Heavy bombardment

Breakup of Pangaea

? Oldest fossil life

First animals emerge on land

4.6 4 3 2 1 Now

Billions of years ago

3 Plate tectonics pushes up mountain ranges and causes bulges in the crust, and water erosion wears the rock away. The Colorado River began cutting the Grand Canyon only about 10 million years ago when the Colorado plateau warped upward under the pressure of moving plates. That sounds like a long time ago, but it is only 0.01 billion years. A mile down, at the bottom of the canyon, lie rocks 0.57 billion years old, the roots of an earlier mountain range that stood as high as the Himalayas. It was pushed up, worn away to nothing, and covered with sediment long ago. Many of the geological features we know on Earth have been produced by relatively recent events.

smaller amounts of sulfur gases such as hydrogen sulfide—the rotten-egg gas that you smell if you visit geothermal pools and geysers such as those at Yellowstone National Park. These gases could have diluted the primeval atmosphere and eventually produced a **secondary atmosphere** rich in carbon dioxide, nitrogen, and water vapor.

In contrast, a modern understanding of planet building implies that Earth formed so rapidly that it was substantially heated by the impacts of infalling material, as well as by radioactive decay. If Earth's surface was molten as it formed, then outgassing would have been continuous, and the early atmosphere would have been rich in carbon dioxide, nitrogen, and water vapor from the beginning. In other words, planetary scientists now think Earth went straight to the volcanic "secondary atmosphere" and never had a hydrogen-rich primeval atmosphere. You will find in the final chapter of the book that the lack of hydrogen in Earth's original atmosphere has important implications for how life began on our planet.

Astronomers also have suspected that some of the abundant water on Earth arrived late in the formation process as a bombardment of volatile-rich planetesimals. These icy bodies, the theory goes, were scattered by the growing mass of the outer planets and by the outward migrations of Saturn, Uranus, and Neptune. The inner Solar System, including Earth, would then have been bombarded by a storm of comets, some of which could have supplied some or all of Earth's water. That hypothesis once faced a serious objection. Spectroscopic studies of a few comets revealed that the ratio of deuterium to hydrogen in comets does not match the ratio in the water on Earth. Some astronomers thought this meant that the water now on Earth could not have arrived in cometlike planetesimals.

However, studies of Comet LINEAR, which broke up in 1999 as it passed near the Sun, show that the water in that comet had a ratio of isotopes similar to water on Earth. Those data suggest that there may be major differences in composition among comets. Icy planetesimals that formed far from the Sun may be richer in deuterium, whereas those that formed closer to the orbit of Jupiter may contain water with isotope ratios more like those in Earth's water. This is a subject of continuing research, and it shows that the origins of Earth's atmosphere and oceans are yet to be fully understood.

In whatever way Earth's atmosphere originated, the mix of gases must have changed over time. The young atmosphere would have been rich in water vapor, carbon dioxide, and other gases. As it cooled, the water condensed to form the first oceans. Carbon dioxide is easily soluble in water—which is why carbonated beverages are so easy to manufacture—and the first oceans began to absorb atmospheric carbon dioxide. Once in solution, the carbon dioxide reacted with dissolved substances in the seawater to form silicon dioxide, calcium carbonate (limestone), and other mineral sediments in the ocean floor, freeing the seawater to absorb more carbon dioxide. Thanks to those chemical reactions in the oceans, the carbon dioxide was transferred from the atmosphere to seafloor sediments.

When Earth was young, its atmosphere had no free oxygen, that is, oxygen not combined with other elements. Oxygen is very reactive and quickly forms oxides in the soil or combines with iron and other substances dissolved in water. Only the action of plant life keeps a steady supply of oxygen in Earth's atmosphere via photosynthesis, which makes energy for plants by absorbing carbon dioxide and releasing oxygen. Beginning about 2 billion to 2.5 billion years ago, photosynthetic plants in the oceans had multiplied to the point where they made oxygen at a rate faster than chemical reactions could remove it from the atmosphere. After that time, atmospheric oxygen increased rapidly. (This topic will be discussed again in the final chapter regarding life in the universe.) It is a **Common Misconception** that there is life on Earth because of oxygen. The truth is exactly the opposite: There is oxygen in Earth's atmosphere because of life. Except for the minority of creatures, including us, that are animals, most life forms on Earth do not need oxygen, and some are even poisoned by it.

An ozone molecule consists of three oxygen atoms linked together (O_3). Ozone molecules are very good at absorbing ultraviolet photons. Earth's lower atmosphere is now protected from ultraviolet radiation by an **ozone layer** about 15 to 30 km above the surface that exists because the atmosphere contains abundant ordinary oxygen (O_2) from which ozone is made. Because the atmosphere of the young Earth did not contain oxygen, an ozone layer could not form, and the Sun's ultraviolet radiation was able to penetrate deep into the atmosphere. There the energetic ultraviolet photons would have broken up weaker molecules such as water (H_2O). The hydrogen from the water then escaped to space, and the oxygen formed oxides in the crust. Earth's atmosphere could not reach its present composition (Table 11-1) until it was protected by an ozone layer, and that required oxygen.

TABLE 11-1 Earth's Atmosphere

Gas	Percent by Weight*
N_2	75.5
O_2	23.1
Ar	1.3
CO_2	0.060
Ne	0.0013
He	0.00007
CH_4	0.0001
Kr	0.0003
H_2O (vapor)	0.006–1.7

*The values in the first eight rows of the table are for a dry atmosphere excluding water vapor.

© 2016 Cengage Learning®

Climate and Human Effects on the Atmosphere

If you climb to the top of a high mountain, you will find the temperature to be much lower than at sea level (**Figure 11-9**). Most clouds form at such altitudes. Higher still, you would find the air much colder and so thin it could not protect you from the intense ultraviolet radiation in sunlight.

You can live on Earth's surface in safety because of Earth's atmosphere, but modern civilization is altering Earth's atmosphere in at least two different serious ways, by adding carbon dioxide (CO_2) and by destroying ozone.

The concentration of CO_2 in Earth's atmosphere is important because CO_2 can trap heat in a process called the **greenhouse effect** (**Figure 11-10a**). When sunlight shines through the glass roof of a greenhouse, it heats the benches and plants inside. The warmed interior radiates infrared radiation, but the glass is opaque to infrared. Warm air in the greenhouse cannot mix with cooler air outside, so heat is trapped within the greenhouse, and the temperature climbs until the glass itself grows warm enough to radiate heat away as fast as the sunlight enters. This is the same process that heats a car when it is parked in the sunlight with the windows rolled up.

Earth's atmosphere is transparent to sunlight, and when the ground absorbs the sunlight, it grows warmer and radiates at infrared wavelengths. However, CO_2 makes the atmosphere less transparent to infrared radiation, so infrared radiation from the warm surface is absorbed by the atmosphere and cannot escape back into space. That traps heat and makes Earth warmer (Figure 11-10b).

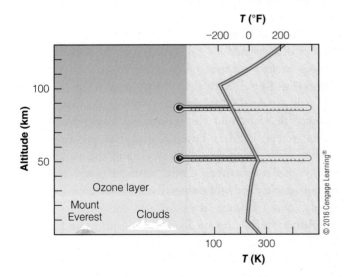

▲ **Figure 11-9** Thermometers placed in Earth's atmosphere at different levels would register the temperatures shown in the graph at the right. The lower few kilometers where you live are comfortable, but higher in the atmosphere the temperature is quite low. The ozone layer lies about 15 to 30 km above Earth's surface.

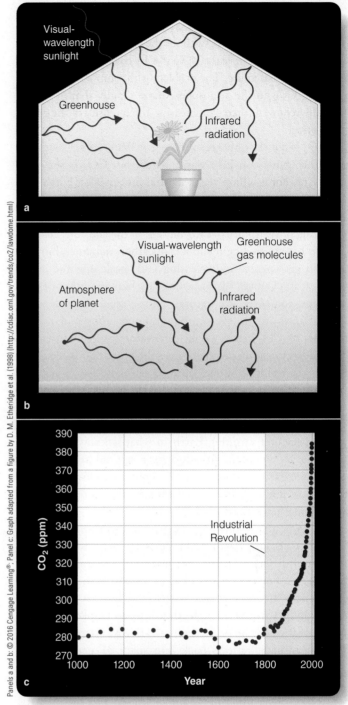

Panels a and b: © 2016 Cengage Learning®. Panel c: Graph adapted from a figure by D. M. Etheridge et al. (1998) (http://cdiac.oml.gov/trends/co2/lawdome.html)

▲ **Figure 11-10** The greenhouse effect: (a) Visual-wavelength sunlight can enter a greenhouse and heat its contents, but the longer-wavelength infrared radiation cannot get out. (b) The same process can heat a planet's surface if its atmosphere contains greenhouse gases such as CO_2. (c) The concentration of CO_2 in Earth's atmosphere as measured in Antarctic ice cores remained roughly constant for thousands of years until the beginning of the Industrial Revolution around the year 1800. Since then it has increased by more than 40 percent. Evidence from proportions of carbon isotopes and oxygen in the atmosphere proves that most of the added CO_2 is the result of burning fossil fuels.

It is a **Common Misconception** that the greenhouse effect is only bad. Evidence indicates that Earth has had a greenhouse effect for its entire history. Without the greenhouse effect, Earth presently would be at least 30°C (54°F) colder and uninhabitable for water-based life. The problem is that human civilization is adding CO_2 to the atmosphere more rapidly than it can be naturally removed, thereby increasing the intensity of the greenhouse effect.

CO_2 is not the only greenhouse gas. Water vapor, methane, and other gases also help warm Earth, but CO_2 is the most important. For 4 billion years, natural processes on Earth have removed CO_2 from the atmosphere and buried the carbon in the form of limestone, coal, oil, and natural gas. Since the beginning of the Industrial Revolution in the late 18th century, humans have been digging up lots of carbon-rich fuels, burning them to get energy, and releasing CO_2 back into the atmosphere. It is a **Common Misconception** that human output of CO_2 is minor compared to natural sources such as volcanoes. Careful measurements of carbon isotope ratios and relative amounts of CO_2 versus O_2 in the atmosphere show that the CO_2 added to the atmosphere since the year 1800 is mostly or entirely the result of human burning of fossil fuels. Some estimates are that the amount of CO_2 in the atmosphere could be twice preindustrial levels by the year 2100.

The increased concentration of CO_2 is increasing the greenhouse effect and warming Earth in what is known as **global warming**. Studies of the growth rings in very old trees show that the average Earth climate had been cooling for most of the past 1000 years, but the 20th century reversed that trend with a rise of 0.6 to 0.9°C (1.0 to 1.7°F).

The amount of warming to expect in the future is difficult to predict because Earth's climate is critically sensitive to a number of different factors, not just the abundance of greenhouse gases. For example, a planet's **albedo** is the fraction of the sunlight hitting it that gets reflected away. A planet with an albedo of 1 would be perfectly white, and a planet with an albedo of 0 would be perfectly black. Earth's overall albedo is 0.31, meaning it reflects back into space 31 percent of the sunlight that hits it. Much of that reflection is caused by clouds, and the formation of clouds depends critically on the presence of water vapor in the upper atmosphere, the temperature of the upper atmosphere, and the patterns of atmospheric circulation.

Even a small change in any of those factors could change Earth's albedo and thus its climate. For example, a slight warming should increase water vapor in the atmosphere, and water vapor is another greenhouse gas that would enhance the warming. But increased water vapor could increase cloud cover, increase Earth's albedo, and partially reduce the warming. On the other hand, high icy clouds tend to enhance the greenhouse effect. The situation is complex, and therefore precise calculations of future warming are not easy to make. Also, even small changes in temperature can alter circulation patterns in the atmosphere and in the oceans, and the consequences of such changes are very difficult to model.

Even though the future is uncertain, general trends now point to substantial continued warming. Mountain glaciers have melted back dramatically since the 19th century. Measurements show that polar ice in the form of permafrost, ice shelves, and ice on the open Arctic Ocean is melting. It is a **Common Misconception** that the observed warming of Earth is the result of natural causes rather than the greenhouse effect. Regular and predictable changes in Earth's axis inclination and orientation and in the shape of its orbit, called Milankovitch cycles (Chapter 2, pages 27–29), currently would be driving Earth's climate slowly toward lower, not higher, temperatures. Also, careful observations during recent decades by space probes indicate the Sun's luminosity averaged over its activity cycles has been constant or decreasing very slightly. The observed warming must be strong to be occurring in the face of opposing astronomical effects.

Although changes are small now, it is a serious issue for the future. Even a small rise in temperatures will dramatically affect agriculture, not only through rising temperatures but also through changes in rainfall. It is a **Common Misconception** that all of Earth will warm at the same rate if there is global warming. Models predict that although most of North America will grow warmer and dryer, Europe initially will become cooler and wetter. Also, the melting of ice on polar landmasses such as Greenland can cause a rise in sea levels that will flood coastal regions and alter shore environments. A modest rise will cover huge low-lying areas such as nearly all of Florida.

There is no doubt that civilization is warming Earth through an enhanced greenhouse effect, but a remedy is difficult to imagine. Reducing the amount of CO_2 and other greenhouse gases released to the atmosphere is difficult because modern society depends on burning fossil fuels for energy. Conserving forests is difficult because growing populations, especially in developing countries with large forest reserves, demand the wood and the agricultural land produced when forests are cut. Political, business, and economic leaders argue that the issue is uncertain, but all around the world scientists have reached a consensus: Global warming is real, is driven by human activity, and will change Earth. What humanity can or will do about it is uncertain.

Human influences on Earth's atmosphere go beyond the greenhouse effect. Our modern industrial civilization is also reducing ozone in Earth's atmosphere. Many people have a **Common Misconception** that ozone is bad because they hear it mentioned as a pollutant of city air, produced by auto emissions. Breathing ozone is bad for you, but, as you learned earlier in this chapter, the ozone layer in the upper atmosphere protects the lower atmosphere and Earth's surface

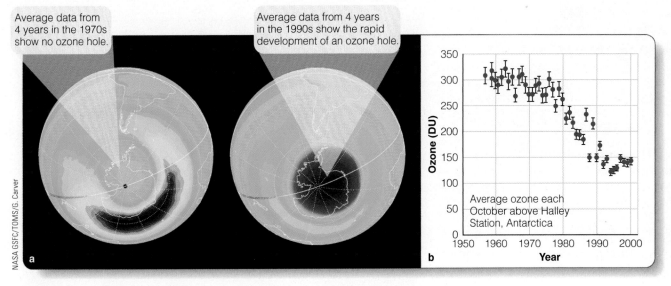

Average data from 4 years in the 1970s show no ozone hole.

Average data from 4 years in the 1990s show the rapid development of an ozone hole.

NASA GSFC/TOMS/G. Carver

a

b

▲ **Figure 11-11** (a) Satellite observations of ozone concentrations over Antarctica are shown here as red for highest concentration and violet for lowest. Since the 1970s, a hole in the ozone layer has developed over the South Pole. (b) Although ozone depletion is most dramatic above the South Pole, ozone concentrations have declined at all latitudes.

from harmful solar ultraviolet photons. Ozone is an unstable molecule and is chemically active. Certain chemicals called chlorofluorocarbons (CFCs), used for refrigeration, air conditioning, and some industrial processes, can destroy ozone. As these CFCs escape into the atmosphere, they become mixed into the ozone layer and convert the ozone (O_3) back into normal oxygen (O_2) molecules. Ordinary oxygen does not block ultraviolet radiation, so depleting the ozone layer causes an increase in ultraviolet radiation at Earth's surface. In small doses, ultraviolet radiation can produce a suntan, but in larger doses it can cause skin cancers.

The ozone layer over the Antarctic is especially sensitive to CFCs. Starting in the late 1970s, the ozone concentration fell significantly over the Antarctic, and a hole in the ozone layer developed over the continent each October at the time of the Antarctic spring (**Figure 11-11**). Satellite and ground-based measurements showed the same thing beginning to happen at higher northern latitudes, with the amount of ultraviolet radiation reaching the ground increasing. This was an early warning that human activity is modifying Earth's atmosphere in a potentially dangerous way. Fortunately, as a result of that warning, international agreements banned most uses of CFCs, and the trend of ozone hole expansion seems to have slowed and may be reversing.

There is yet another **Common Misconception** that global warming and ozone depletion are two names for the same thing. Take careful note that the ozone hole is a second Earth environmental issue that is basically separate from global warming. The CO_2 and ozone problems in Earth's atmosphere are paralleled on Venus and Mars. When you study Venus in a later chapter

you will discover a runaway greenhouse effect that has made the surface of the planet hot enough to melt lead. On Mars you will discover an atmosphere without an ozone layer. A few minutes of sunbathing on Mars would kill you. Once again, you can learn more about your own planet by studying extreme conditions on other planets.

DOING SCIENCE

Why does Earth's atmosphere contain little carbon dioxide and lots of oxygen? As a scientist, you must learn to expect the unexpected.

Because volcanic outgassing releases mostly CO_2, N_2, and water vapor, you might expect Earth's atmosphere to be very rich in CO_2. However, CO_2 is highly soluble in water, and Earth's surface temperature allows most of the surface to be covered with liquid water. The CO_2 dissolves in the oceans and combines with minerals in seawater to form deposits of silicon dioxide, limestone, and other mineral deposits. In this way, the CO_2 is removed from the atmosphere and buried in Earth's crust. Oxygen, in contrast, is highly reactive and forms oxides so easily you might expect it to be rare in the atmosphere. Happily for us animals, it is continuously replenished as green plants release oxygen into Earth's atmosphere faster than chemical reactions can remove it. Were it not for liquid water oceans and plant life, Earth would have a thick CO_2 atmosphere with no free oxygen.

Now follow up on your discovery. **Why would an excess of CO_2 and a deficiency of free oxygen be harmful to all life on Earth in ways that go beyond mere respiration?**

One of the most fascinating aspects of science is its power to reveal the unseen. That is, it reveals regions you can never visit. You saw this in previous chapters when you studied the inside of the Sun and stars, the surface of neutron stars, the event horizon around black holes, the cores of active galaxies, and more. In this chapter, you have "seen" Earth's core.

An engineer is a person who builds things, so you might call a person who imagines things an *imagineer*. Most creatures on Earth cannot imagine situations that do not exist, but humans have evolved the ability to say, "What if?" Our ancient ancestors could imagine what would happen if a tiger was hiding in the grass, and we can imagine the inside of Earth.

A poet can imagine the heart of Earth, and a great writer can imagine a journey to the center of Earth. Scientists learn to use their imaginations in carefully controlled ways. Guided by evidence and theory, they can imagine the molten core of our planet in detail. As you read this chapter, if you could see the yellow-orange glow and feel the heat of the liquid iron, you were a scientific imagineer.

Human imagination makes science possible and provides one of the great thrills of science—exploring beyond the limits of normal human experience.

Study and Review

Summary

- Earth is the standard used in **comparative planetology (p. 225)** for the Terrestrial planets primarily because Earth contains nearly all of the phenomena found on the other Terrestrial planets.

- The Terrestrial worlds are Earth, the Moon, Mercury, Venus, and Mars. Earth's Moon is included because it is a complex world, one of the largest moons in the Solar System, and its characteristics give insight about the other worlds.

- The Terrestrial worlds differ mainly in size, but they all have dense metallic cores, less dense rocky **mantles (p. 225)**, and low-density crusts.

- Comparative planetology leads you to expect that cratered surfaces are old, that heat flowing out of a planet drives geological activity, and that the nature of a planet's atmosphere depends on both the size of the planet and the planet's temperature.

- At some point early in its history, Earth was hot enough to be completely molten, which resulted in the planet differentiating into layers of different density.

- Earth has passed through four stages as it developed: (1) differentiation, (2) cratering, (3) basin flooding, and (4) slow surface evolution. The other Terrestrial worlds also passed through the same stages. However, the effects and durations of the respective stages differed because of the specific properties of each world.

- Earth is unique in that large amounts of liquid water have existed on its surface for most of the history of the Solar System. Among other effects, water drives strong erosion that alters the planet's surface features on relatively short time scales.

- Earth is also unique in that so far it is the only known home for life.

- **Seismic waves (p. 229)** generated by earthquakes can be detected by **seismographs (p. 229)** all over the planet and can be used to construct a model of Earth's internal structure.

- **Pressure (P) waves (p. 229)** can travel through a liquid, but **shear (S) waves (p. 229)** cannot. Observations show that *S* waves cannot pass through Earth's core. This is evidence that a significant portion of Earth's core must be liquid. Measurements of heat flowing outward from the interior, combined with mathematical models, reveal that the core is very hot and composed of iron and nickel.

- Although Earth's crust is brittle and breaks under stress, the mantle behaves like a **plastic (p. 230)**, able to deform and flow under pressure.

- Earth's magnetic field is generated by the dynamo effect in the mostly liquid core, which is convecting, rotating, and conducting. This magnetic field defines a **magnetosphere (p. 232)** around the planet, which mostly shields Earth from the solar wind encountering the planet at the **bow shock (p. 231)**. Radiation belts called the **Van Allen belts (p. 232)**, as well as aurora displays, are products of the interaction of the solar wind with Earth's magnetic field.

- Earth is dominated by **plate tectonics (p. 234)**, with the crust divided into moving sections. Heat flowing upward from the interior drives plate tectonics.

- Earth's crustal plates are made of low-density, brittle rock that floats on the hotter plastic upper layers of the mantle. **Rift valleys (p. 234)** can be produced where plates begin pulling away from each other. New crust is formed in rift valleys but also, especially, along **midocean rises (p. 234)**: Plates spread apart and magma rises and solidifies to form **basalt (p. 234)** rocks.

- Crust is destroyed when one plate slides under another, sinking into the mantle along **subduction zones (p. 234)**. Volcanism is common in subduction zones, and earthquakes are common at plate boundaries where plates move relative to each other.

- Volcanism at a hot spot can produce a series of volcanic islands such as the Hawaiian Island chain. Hot-spot volcanism is not related to volcanism in plate-boundary subduction zones.

- The continents drift slowly on the plastic mantle, and their arrangement changes with time. Where plates collide, they can buckle and form **folded mountain ranges (p. 234)**.

- Most prominent geological features such as mountain ranges and the Grand Canyon have been formed recently in Earth's history. The first billion years of Earth's history have been almost entirely erased by plate tectonics and erosion.

- Because Earth has a relatively low mass and formed at a high temperature, it likely never had a **primeval atmosphere (p. 233)** consisting of hydrogen, helium, and hydrogen compounds captured from the solar nebula. Instead, the forming Earth evidently went straight to a **secondary atmosphere (p. 236)**, one that was a combination of gases baked out of the interior and carried in by volatile-rich planetesimals.

- Because Earth formed in a molten state, its original atmosphere was probably mostly carbon dioxide, nitrogen, and water vapor. Most of the atmospheric carbon dioxide eventually became absorbed in seawater, forming mineral sediments on the sea floor. Subduction carries such sediments to the mantle. Volcanism returns the carbon dioxide back to the atmosphere, continuing the carbon dioxide cycle. Plant life added oxygen to Earth's atmosphere via photosynthesis.

- When enough oxygen built up in Earth's atmosphere, an **ozone layer (p. 236)** (O_3 molecules) could form at high altitudes. Ultraviolet photons can break up water molecules in a planet's atmosphere but ozone absorbs ultraviolet photons, so Earth's ozone layer preserves Earth's water.

- The **albedo (p. 238)** of a planet is the fraction of sunlight hitting the planet that is reflected into space. Small changes in the albedo of Earth caused by changes in clouds, snow and ice cover, and vegetation can have a dramatic effect on Earth's climate.

- The **greenhouse effect (p. 237)** warms the surface of a planet when atmospheric greenhouse gases such as carbon dioxide are transparent to incoming sunlight but opaque to outgoing infrared light. The greenhouse effect warms Earth's surface, keeping the average temperature above freezing. Greenhouse gases added by industrial civilization are responsible for an enhanced greenhouse effect and **global warming (p. 238)**.

- Measurements of carbon isotope ratios and carbon dioxide versus oxygen abundances in the atmosphere prove that the carbon dioxide added to Earth's atmosphere since the early 1800s comes from burning of fossil fuels.

- Observations and model calculations have eliminated possibilities such as natural climate cycles or variations in the Sun's output as causes for the current warming. CO_2 produced by human burning of fossil fuels is indicated as the primary driver of the warming.

- The ozone layer high in Earth's atmosphere protects the surface from ultraviolet radiation. However, chlorofluorocarbons (CFCs) released by industrial processes have attacked the ozone layer and thinned it, especially near the poles, allowing more of this harmful ultraviolet radiation to reach Earth's surface. International intervention to eliminate the use of CFCs resulted in partial reversal of the ozone layer's destruction, suggesting that humans are capable of changing behavior that is potentially dangerous to us all.

Review Questions

1. Why would you include the Moon in a comparison of the Terrestrial planets?
2. Which of the five Terrestrial worlds has an oxygen-rich atmosphere?
3. Which is the most geologically active Terrestrial world? Why?
4. In what ways is Earth unique among the Terrestrial worlds?
5. Which Terrestrial worlds have thin or no atmospheres?
6. Describe the four stages of Terrestrial planet development.
7. The Moon did not pass through all of the four stages of planetary development. True or false?
8. Earth shows few craters on its surface. What is the explanation for this? Did Earth somehow avoid being hit during the heavy bombardment period?
9. How do you know that Earth differentiated?
10. What keeps Earth's interior warm today?
11. Lava flows today are examples of basin flooding. True or false?
12. Describe three forms of erosion that cause slow evolution of Earth's surface.
13. Earth's interior is separated into core, mantle, and crust. Order the interior layers by increasing density.
14. Which type of seismic wave cannot pass through Earth's core? What does that indicate about the composition of the core?
15. What property makes Earth's mantle behave like a plastic? How do you know?
16. All five of the Terrestrial worlds have bow shocks, magnetospheres, and radiation belts. True or false? How do you know?
17. How is the root cause of earthquakes in Hawai'i different from earthquakes in Southern California?
18. What characteristics must Earth's core have to generate a magnetic field?
19. All five of the Terrestrial worlds have plate tectonics. True or false?
20. What characteristic does a Terrestrial planet's interior need for it to be a geologically active world?
21. How do island chains located in the centers of tectonic plates, such as the Hawaiian-Emperor chain, indicate ongoing plate tectonic activity?
22. What evidence can you cite that the Atlantic Ocean is growing wider?
23. How are the inferred properties of Earth's original atmosphere related to the location and timescale of Earth's formation from the solar nebula?
24. What produced the oxygen in Earth's atmosphere?
25. Explain the natural carbon dioxide cycle on Earth. Start by explaining how carbon dioxide is removed from Earth's atmosphere and end with how carbon dioxide is returned to Earth's atmosphere.
26. Life on Earth exists because of oxygen in Earth's atmosphere. True or false?
27. Where is the ozone layer? Where are the ozone holes?
28. How does the increasing abundance of CO_2 in Earth's atmosphere cause a rise in Earth's temperature?
29. The greenhouse effect is bad for Earth's climate. True or false?
30. Name three greenhouse gases in Earth's atmosphere.
31. Where is most of Earth's carbon dioxide located?
32. In three sentences or fewer, explain global warming using the following vocabulary words: temperature, ground, sunlight, atmosphere, infrared light, greenhouse gases, global warming.

33. Why would a decrease in the density of the ozone layer in Earth's atmosphere cause public health problems?

34. **How Do We Know?** How can the flow of energy out of a planet's interior affect its surface and atmosphere?

35. **How Do We Know?** How is deducing the structure of a virus like finding the composition of Earth's core?

Discussion Questions

1. Why might it be more accurate to divide the Solar System into three sets of planets instead of two: (i) the Terrestrial planets, (ii) Jupiter & Saturn, and (iii) Neptune & Uranus?

2. Is the life on Earth definitely unique in our Solar System?

3. If you orbited a planet in another planetary system and discovered oxygen in its atmosphere, what might you expect to find on its surface?

4. If you wanted to find evidence of *intelligent* life on an extrasolar planet, what signatures might you look for in the planet's atmosphere?

5. If sustained presence of abundant liquid water is rare on the surface of planets, then most Terrestrial planets in the Universe must have CO_2-rich atmospheres. Correct or incorrect? Why or why not?

6. Should we replace the words *global warming* with *climate change* when discussing effects of human activity on Earth's climate? Are these two different, or equivalent, terms?

Problems

1. Examine Figure 11-1. Using a ruler, measure the diameters of the Terrestrial planets on the figure in units of millimeters. Plot planet diameter on the horizontal x-axis and the observed densities from Table 19-1 on the vertical y-axis, with Mercury's data located nearest the origin, followed by the data points for Mars, Venus, and Earth. Which planet does not fit the trend line set using the other three planets' data? If the diameter of the nonconforming planet is assumed to be "correct," what should the nonconforming planet's density be, and which other planet's density should the nonconforming planet more closely resemble?

2. Look at Figure 11-3. The earthquake occurred 7440 km from the seismograph. How fast did the P waves travel in km/s? How fast did the S waves travel? How long did the P waves and the S waves take to travel 100 km from the epicenter? Assume the wave speeds are constant.

3. Look at Figure 11-3. The lag time is the difference between when the P waves arrived and when the S waves arrived. Using the earthquake data shown in the figure, what is the lag time? Form a general conclusion about the relationship between lag times and locations of earthquakes.

4. What percentage of Earth's volume is the metallic core? (*Note:* The volume of a sphere is $\frac{4}{3}\pi r^3$.)

5. How many magnetic pole reversals has Earth endured in the last 330 million years if the average time between reversals is 700,000 years?

6. If the Atlantic seafloor is spreading at 3.0 cm/year and is now 6400 km wide, how long ago were the continents in contact? How does that time span compare to the age of Earth?

7. The Hawaiian-Emperor chain of undersea volcanoes is about 7500 km long, and the Pacific plate is moving 9.2 cm a year. How old is the oldest detectable volcano in the chain? What has happened to older volcanoes in the chain?

8. From Hawai'i to the bend in the Hawaiian-Emperor chain is about 4000 km. Use the speed of Pacific plate motion given in Problem 7 to estimate how long ago the direction of plate motion changed. (*Note:* It may not be a coincidence that the San Andreas fault became active in what is now Southern California at about the same time.)

9. Calculate the age of the Grand Canyon as a fraction of Earth's age.

Learning to Look

1. Look at the hemisphere of Earth shown on the right-hand side of **The Active Earth** and find volcanoes scattered over the Pacific Ocean. What is producing those volcanoes?

2. Look at the hemispheres of Earth shown on the two pages of **The Active Earth.** Name a folded mountain range. Describe the locations of one subduction zone and one midocean rise.

3. Look at the series of figures depicting continental drift on the right-hand side of **The Active Earth.** Based on the amount of motion shown, develop a hypothesis about where the continents will be 200 million years from now. Draw the next picture in the series.

4. Look at Figure 11-9. Rising from Earth's surface to the cloud layer shown, does the temperature increase, decrease, or stay the same? How about from the clouds to the ozone layer? At about what altitude does the temperature change most abruptly, almost 400°F in 30 km?

5. In what ways is the photo at the right a typical view of the surface of planet Earth? How is this view unusual among the other Terrestrial worlds?

William K. Hartmann

6. What do you see in the photo at right that suggests heat is flowing out of Earth's interior?

USGS

The Moon and Mercury: Comparing Airless Worlds

Guidepost Want to fly to the Moon? You will need to pack more than a lunch. There is no air and no water, and the sunlight is strong enough to kill you. If you take shelter in the shade, you might freeze to death in moments. Mercury is the same kind of world. Earth seems normal to you, and other worlds that are, well, un-Earthly, are nevertheless related to Earth in surprising ways. Exploring these two airless worlds will answer three important questions:

► **How did the Moon form and evolve?**

► **In what ways is Mercury similar to, and different from, the Moon?**

► **How are the histories of the Moon and Mercury connected to Earth's history?**

You are beginning your detailed study of planets by exploring airless worlds; in the next chapter, you will move on to bigger planets with atmospheres. They are not necessarily more interesting places, but they are less un-Earthly.

---

*That's one small step for [a] man . . .
one giant leap for mankind.*

NEIL ARMSTRONG, FIRST HUMAN TO WALK ON THE MOON

Beautiful, beautiful. Magnificent desolation.

EDWIN ("BUZZ") ALDRIN, SECOND HUMAN TO WALK ON THE MOON

NASA/JHU APL/CIW

Artist's conception of the *MESSENGER* spacecraft against a real image of a portion of the planet Mercury's surface with enhanced color contrast. The enhancement emphasizes variation in composition among different regions on the surface.

Enhanced-contrast visual image
plus artist's conception

F YOU HAD BEEN ONE of the first people to walk on the Moon, what would you have said? Neil Armstrong responded to the historic significance of being the first human to step onto the surface of another world. Buzz Aldrin was second, and he responded to the Moon itself. It *is* desolate, and it *is* magnificent. But it is not unusual. Many planets in the Universe probably look like Earth's Moon, and astronauts may someday walk on such worlds and compare them with our Moon.

In this chapter, you will use comparative planetology to study the Moon and Mercury, and continue following three important themes of planetary astronomy: internal heat flow, cratering, and giant impacts. These three themes will help you organize the flood of details astronomers have learned about the Moon and Mercury.

12-1 The Moon

Only 12 people have stood on the Moon, but planetary scientists know it well. The photographs, measurements, and samples brought back to Earth paint a picture of the airless, ancient, battered crust of a world created by a planetary catastrophe.

The View from Earth

A few billion years ago, the Moon probably rotated faster than it does today, but Earth is more than 80 times more massive than the Moon (■ Celestial Profile 3, p. 256), and its tidal forces on the Moon are strong. Earth's gravity raised tidal bulges on the Moon, and friction in the bulges slowed the Moon until it now rotates once each orbit, keeping the same side facing Earth. A moon with rotation locked to its planet is said to be **tidally coupled**. That is why we always see the same side of the Moon; the back of the Moon is never visible from Earth. The Moon's familiar face has shone down on Earth since long before there were humans (**Figure 12-1**).

Based on what you already know, you can predict that the Moon should have no atmosphere. It is a small world with an escape velocity too low to keep gas atoms and molecules from departing into space. You can confirm your hypothesis with even a small telescope. The Moon has no clouds or other obvious traces of an atmosphere. With a small telescope you could watch stars disappear behind the Moon's limb (edge of its disk) without first being dimmed by an atmosphere. Also, shadows near the **terminator**, the dividing line between daylight and darkness, are sharp and black, indicating there is no air on the Moon to scatter light and soften shadows. Clearly, the Moon is an airless, and therefore soundless, world.

The surface of the Moon is divided into two dramatically different kinds of terrain. The lunar highlands are filled with jumbled mountains, but there are no folded mountain ranges like the ones on Earth. This shows that the Moon has no plate

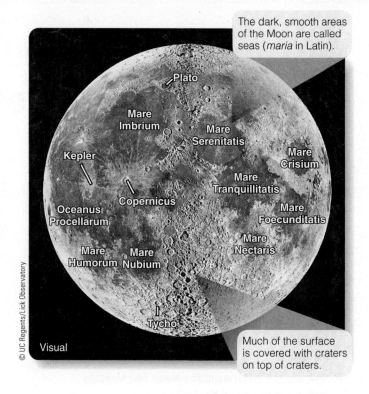

The dark, smooth areas of the Moon are called seas (*maria* in Latin).

Plato
Mare Imbrium
Mare Serenitatis
Kepler
Mare Crisium
Mare Tranquillitatis
Copernicus
Oceanus Procellarum
Mare Foecunditatis
Mare Nectaris
Mare Humorum
Mare Nubium
Tycho

Visual

© UC Regents/Lick Observatory

Much of the surface is covered with craters on top of craters.

▲ Figure 12-1 The side of the Moon that faces Earth is a familiar sight. Craters have been named for famous scientists and philosophers, and the so-called seas have been given romantic names. Mare Imbrium is the Sea of Rains, and Mare Tranquillitatis is the Sea of Tranquillity. There is, in fact, no water on the Moon.

tectonics. Instead, the Moon's mountains are pushed up by millions of overlapping impact craters. In fact, the highlands are saturated with craters, meaning that it would be impossible to form a new crater without destroying the equivalent of one old crater. In contrast, the lowlands, about 3 km (2 mi) lower than the highlands, are smooth, dark plains called **maria**, the Latin word for "seas." (The singular of *maria* is **mare**, pronounced *MAH-ray*.) The first observers using telescopes thought these were bodies of water, but further examination showed that the maria are marked by ridges, faults, and scattered craters, so they can't be water. Rather, the maria are ancient lava flows that apparently have covered older, cratered lowlands.

Those lava flows suggest volcanism, but no major volcanic peaks are visible on the Moon, and no active volcanism has ever been detected. The lava flows that created the maria happened long ago and were much too fluid to build peaks. However, with a good telescope and some diligent searching, you can find a few small domes pushed up by lava below the surface, as well as some long, winding channels called **sinuous rilles** (**Figure 12-2**). These channels are often found near the edges of the maria and were evidently cut by flowing lava. In some cases, such a channel may once have had a roof of solid rock,

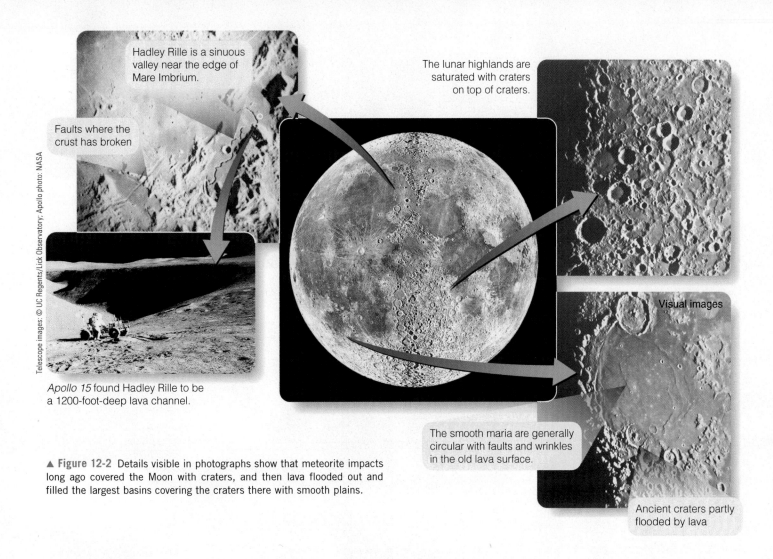

Hadley Rille is a sinuous valley near the edge of Mare Imbrium.

Faults where the crust has broken

Apollo 15 found Hadley Rille to be a 1200-foot-deep lava channel.

The lunar highlands are saturated with craters on top of craters.

Visual images

The smooth maria are generally circular with faults and wrinkles in the old lava surface.

Ancient craters partly flooded by lava

Telescope images: © UC Regents/Lick Observatory; Apollo photo: NASA

▲ **Figure 12-2** Details visible in photographs show that meteorite impacts long ago covered the Moon with craters, and then lava flooded out and filled the largest basins covering the craters there with smooth plains.

forming a lava tube. After the lava drained away, meteorite impacts collapsed the roof to form a sinuous rille. The view from Earth provides just a few hints of ancient volcanic activity associated with the maria.

Lava flows and impact cratering have dominated the history of the Moon. Study **Impact Cratering** on pages 246–247 and notice three important points and five new terms:

1 Impact craters have certain distinguishing characteristics, such as their shape and the *ejecta*, *rays*, and *secondary craters* around them.

2 Lunar impact craters range from tiny pits formed by *micrometeorites* to giant *multiringed basins*.

3 Most of the craters on the Moon are old; they were formed long ago when the Solar System was young.

Meteorites strike the Moon all the time, but large impacts are rare today. In 2014 a flash was detected by Earth-based observers, later calculated to be the result of an impact by an object a meter in diameter that would have dug a crater at least 10 meters (33 ft) across. Astronomers estimate that meteorites

with diameters of tens of meters strike the Moon every few decades, but no one has ever seen such an impact with certainty. No significant change has been seen on the Moon since the invention of the telescope. As you will learn in a later chapter, large impacts do happen on the Moon and Earth, but nearly all of the lunar craters seen through telescopes date from the Solar System's youth.

The lunar features visible from Earth allowed astronomers to construct a hypothetical history of the Moon that could not be confirmed until astronauts went to the Moon, made on-site measurements and observations, and brought rocks back for analysis. That history goes like this: As the Moon formed, its crust would have been heavily cratered by debris left over from the formation of the planets. Sometime after the cratering subsided, lava welled up from below the crust and flooded the lowlands, covering the craters there and forming the smooth maria. The maria are only lightly scarred by impacts and must be younger than the cratered highlands. You can locate a few large craters on the maria such as Kepler and Copernicus in Figure 12-1, which therefore must be younger than the maria. This hypothetical history provides a

Impact Cratering

1 The craters that cover the Moon and many other bodies in the Solar System were produced by the high-speed impact of meteorites of all sizes. Meteorites striking the Moon travel 10 to 70 km/s.

A meteorite striking the Moon at those speeds can produce an impact crater 10 or more times larger in diameter than the meteorite. The vertical scale is exaggerated at right for clarity.

Impact Cratering

An object approaches the lunar surface at high velocity.

On impact, the meteorite is deformed, heated, and vaporized.

The resulting explosion blasts out a round crater.

Slumping produces terraces in crater walls, and rebound can raise a central peak.

1a Lunar craters such as Euler, 27 km (17 mi) in diameter, look deep when you see them near the terminator where shadows are long, but a typical crater is only a fifth to a tenth as deep as its diameter, and the largest craters are even shallower.

Because craters are formed by shock waves rushing outward, by the rebound of the rock, and by the expansion of hot vapors, craters are almost always round, even when the meteorite strikes at a steep angle.

Debris blasted out of a crater is called **ejecta**, and it falls back to blanket the surface around the crater. Ejecta shot out along specific directions can form bright rays.

Euler

NASA

Visual

Rays

Tycho

Visual

Visual

NASA

NASA

1b Rock ejected from distant impacts can fall back to the surface and form smaller craters called **secondary craters.** The chain of craters here is a 45-km (28-mi)-long chain of secondary craters produced by ejecta from the large crater Copernicus 200 km (125 mi) out of the frame to the lower right.

Bright ejecta blankets and rays gradually darken as sunlight alters minerals and small meteorites stir the dusty surface. Bright rays are signs of youth. Rays from the crater Tycho, perhaps only 100 million years old, extend halfway around the Moon.

2 Shown at right, Plum Crater, 40 m (130 ft) in diameter, was visited by *Apollo 16* astronauts. Note the many smaller craters visible. Lunar craters range from giant impact basins to tiny pits in rocks struck by **micrometeorites**, meteorites of microscopic size.

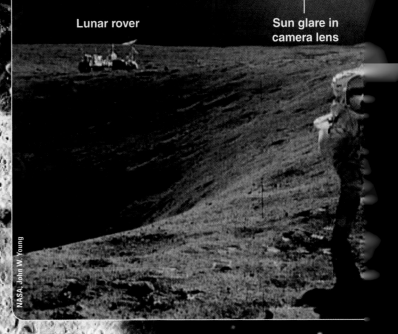

Lunar rover

Sun glare in camera lens

NASA, John W. Young

Mare Orientale

Solidified lava

Visual

NASA

2a In larger craters, the deformation of the rock can form one or more inner rings concentric with the outer rim. The largest of these craters are called **multiringed basins**. In Mare Orientale (shown at left) on the west edge of the visible Moon, the outermost ring is almost 900 km (550 mi) in diameter.

2b The energy of an impact can melt rock, some of which falls back into the crater and solidifies. When the Moon was young, craters could also be flooded by lava welling up from below the crust.

A few meteorites found on Earth have been identified chemically as fragments of the Moon's surface blasted into space by cratering impacts. The fragmented nature of these meteorites indicates that the Moon's surface has been battered by countless impacts.

3 Most of the craters on the Moon were produced long ago when the Solar System was filled with debris from planet building. As that debris was swept up, the cratering rate fell rapidly, shown schematically below.

Meteorite from Moon

Rate of Crater Formation

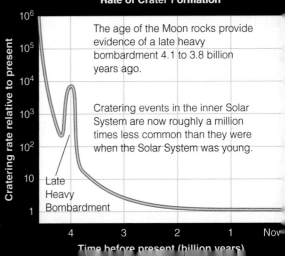

The age of the Moon rocks provide evidence of a late heavy bombardment 4.1 to 3.8 billion years ago.

Cratering events in the inner Solar System are now roughly a million times less common than they were when the Solar System was young.

Cratering rate relative to present

10^6
10^5
10^4
10^3
10^2
10
1

Late Heavy Bombardment

4   3   2   1   Nov

Time before present (billion years)

How Do We Know? 12-1

How Hypotheses and Theories Unify the Details

Why is playing catch more than just looking at the ball? Like any technical subject, science includes a mass of details, facts, figures, measurements, and observations. The flood of details can be overwhelming, but one of the most important characteristics of science comes to your rescue. The goal of science is not to discover more details but to explain the details with a unifying hypothesis or theory. A good theory is like a basket that makes it easier for you to carry a large assortment of details.

For example, when a psychologist begins studying the way the human eye and brain respond to moving objects, the data are a sea of detailed measurements and observations. Infants look at a moving ball for only moments, but older children look longer. Adults can concentrate longer on the moving ball, but their eyes move differently if they are given a stick to point with. Scans of brain activity show that different areas of the brain are active in subjects of different ages and under different circumstances.

From the data, the psychologist might form the hypothesis that the human brain processes visual information differently depending on its intended use. If you look at a baseball being rubbed in the hands of a pitcher, your brain processes the visual information one way. If you see a baseball flying at you and you have to catch it, your brain processes the information in a different way. The psychologist's hypothesis brings all of the details into place as parts of a logical argument about the ability and necessity of action. Babies don't catch balls. Sometimes a ball is an object that might be rough or smooth, but sometimes it is an object to be caught. The brain responds appropriately.

The goal of science is to understand nature, not to memorize details. Whether scientists are psychologists studying brain functions or astronomers studying the formation of other worlds, they are trying to unify their data and explain it with a single hypothesis or theory.

When scientists create a hypothesis, it draws together a great many observations and measurements.

Franklin and Marshall College/Phyllis Leber

---

framework that organizes the available details and observations (**How Do We Know? 12-1**).

It is difficult to estimate the true age of any specific crater. In some cases, you can find **relative ages** by noting that a crater or its rays partially cover other craters. Clearly the crater on top must be younger than the craters on the bottom. Once lunar samples were available, those relative ages could be calibrated using radioactive dating. The result would be an indication of how the cratering rate changed over time. Combining all this information, astronomers can study the size and number of craters on a section of the Moon's surface not visited by astronauts and still be able to estimate that section's **absolute age** in years. The maria are 2 to 4 billion years old, and the highlands are older. You can see that to really understand the history of the lunar surface humans had to go there and bring back samples.

The Apollo Missions

In 1961, President Kennedy committed the United States to landing a human being on the Moon by 1970. Although the reasons for that decision related more to economics, international politics, and the stimulation of technology than to science, the Apollo program became a fantastic scientific adventure, including six expeditions to the surface of the Moon that changed how humans think about Earth.

Flying to the Moon is not particularly difficult. With powerful enough rockets and enough food, water, and air, it is a straightforward trip. Landing on the Moon is more difficult but not impossible. The Moon's gravity is only one-sixth that of Earth, and there is no atmosphere to disturb the trajectory of the spaceship. The difficulty is in getting to the Moon, landing, taking off, and returning to Earth all in one trip. The craft must carry food, water, and air for a number of days in space plus fuel and rockets for midcourse corrections, landing, and launch back to Earth. This would require a vehicle that is too massive to make a safe landing on the lunar surface. The solution was to take two spaceships to the Moon, one for the round trip and one to land in (**Figure 12-3**).

The command module was the long-term living space and command center for the trip. Three astronauts had to live in it for a week, and it had to carry all the life-support equipment, supplies, navigation instruments, computers, and so on for a week's journey in space. The small lunar landing module (LM) was tacked to the front of the command module like a bicycle strapped to the front of the family camper. It carried only enough fuel and supplies for the short trip from lunar orbit to the lunar surface, and it was built to minimize weight and maximize maneuverability.

The weaker gravity of the Moon made the design of the LM relatively simple. Landing on Earth requires reclining couches for the astronauts, but the trip to the lunar surface involved smaller accelerations. In an early version of the LM, the astronauts sat on what looked like bicycle seats, but these were later

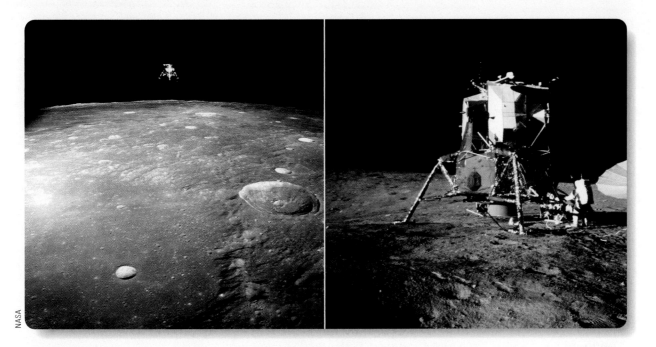

▲ **Figure 12-3** Inside the lunar module, the two astronauts stood in a space hardly bigger than two telephone booths. The metal skin was so thin it was easily flexible, like metal foil, and the legs of the module, designed specifically for the Moon's weak gravity, could not support the lander's weight on Earth. Only the upper half of the lander blasted off from the surface to return the astronauts to the command module in orbit around the Moon.

scrapped to save weight. The astronauts had no seats at all in the LM, and once they began their descent and acquired weight, they stood at the controls supported by straps, riding the LM like daredevils on a rocket surfboard. One astronaut stayed in lunar orbit in the command module while the other two landed and returned to orbit in the LM.

When the LM lifted off from the lunar surface, the larger descent rocket and support stage were left behind to save weight. Only the compartment containing the two astronauts, their instruments, and their cargo of rocks returned to the command module orbiting above. The astronauts in the LM blasting up from the lunar surface were again standing at the controls. The rocket engine that lifted them back into orbit around the Moon was not much bigger than a dishwasher.

The most complicated part of the trip was the rendezvous and docking between the tiny ascent stage of the LM and the command module. Aided by radar systems and computers, the two astronauts docked with the command module, transferred their Moon rocks, and jettisoned the remains of the LM. Only the command module returned to Earth.

The first human-piloted lunar landing was made on July 20, 1969. While Michael Collins waited in orbit around the Moon, Neil Armstrong and Edwin Aldrin took the LM down to the surface. Although computers controlled much of the descent, the astronauts had to override a number of computer alarms and take control of the LM to avoid a boulder-strewn crater bigger than a football field.

Between July 1969 and December 1972, 12 people reached the lunar surface and collected 380 kg (840 lb) of rocks and soil (Table 12-1). The flights were carefully planned to visit different regions and develop a comprehensive understanding of the lunar surface.

The first flights went to relatively safe landing sites (Figure 12-4)—Mare Tranquillitatis for *Apollo 11* and Oceanus Procellarum for *Apollo 12*. *Apollo 13* was aimed at a more complicated site, but an explosion in an oxygen tank on the way to the Moon ended all chances of a landing and nearly cost the astronauts their lives. They succeeded in using the life support in the LM to survive, looping around the back of the Moon and returning to Earth safely a few days later in the crippled command module.

The last four Apollo missions, 14 through 17, sampled geologically important places on the Moon. *Apollo 14* visited the Fra Mauro region, which is covered by ejecta from the impact that dug the multiringed basin now filled by Mare Imbrium. *Apollo 15* visited the edge of Mare Imbrium at the foot of the Apennine Mountains and examined Hadley Rille (see Figure 12-2). *Apollo 16* and *Apollo 17* visited highland regions to sample older parts of the lunar crust (Figure 12-4). Almost all of the lunar samples from these six landings are now held at the Planetary Materials Laboratory at the Johnson Space Center in Houston, although one has been embedded in a stained glass window in the National Cathedral in Washington, DC. These lunar samples are a national treasure containing clues to the beginnings of our Solar System.

TABLE 12-1 Apollo Lunar Landings

Apollo Mission*	Astronauts: Commander LM Pilot CM Pilot	Date	Mission Goals	Sample Mass (kg)	Typical Samples	Ages (10^9 y)
11	Armstrong Aldrin Collins	July 1969	First human landing; Mare Tranquillitatis	22	Mare basalts	3.48–3.72
12	Conrad Bean Gordon	Nov. 1969	Visit Surveyor 3; sample Oceanus Procellarum (mare)	34	Mare basalts	3.15–3.37
14	Shepard Mitchell Roosa	Feb. 1971	Sample Imbrium ejecta sheet; Fra Mauro hills	43	Breccia	3.85–3.96
15	Scott Irwin Worden	July 1971	Sample edge of Mare Imbrium; Appenine Mountains; Hadley Rille	77	Mare basalts Highland anorthosite	3.28–3.44 4.09
16	Young Duke Mattingly	April 1972	Sample highland crust; Cayley formation (ejecta); Descartes region	95	Highland basalts Breccia	3.84 3.92
17	Cernan Schmitt Evans	Dec. 1972	Sample highland crust; Dark halo craters; Taurus–Littrow region	110	Mare basalt Highland breccia Fractured dunite	3.77 3.86 4.48

*The *Apollo 13* mission suffered an explosion on the way to the Moon and did not land.

Apollo 17, the last Apollo mission to the moon, landed in the highlands in December 1972.

Apollo 11 landed in the lunar lowlands in July 1969.

◄ **Figure 12-4** *Apollo 11*, the first mission to the Moon, landed on the smooth surface of Mare Tranquillitatis in the lunar lowlands, and the horizon was straight and level. When *Apollo 17* landed at Taurus–Littrow in the lunar highlands, the astronauts found the horizon mountainous and the terrain rugged. The large boulder is a piece of ejecta that, at some time in the past, was thrown here from an impact far beyond the horizon.

Left panel: © UC Regents/Lick Observatory; Right and bottom panel: NASA

The Apollo astronauts found that all Moon rocks are igneous, meaning they solidified from molten rock.

Rocks exposed on the lunar surface become pitted by micrometeorites.

Vesicular basalt contains bubbles frozen into the rock when it was molten.

A breccia is formed by rock fragments bonded together by heat and pressure.

NASA

◄ **Figure 12-5** Rocks returned from the Moon show that the Moon formed in a molten state, that it was heavily fractured by cratering when it was young, and that it is now affected mainly by micrometeorites grinding away at surface rock.

Lunar Geology

Scientists eagerly awaited return of Moon rocks to Earth. Analysis could reveal clues to the chemical and physical history of the Moon, the origin and evolution of Earth, and the conditions in the solar nebula from which the planets formed. Those studies, combined with the measurements made on site by the Apollo astronauts and with data radioed to Earth by the seismographs they left behind, allowed a giant leap for humanity's understanding of the Moon. Since the last astronauts walked on the Moon in 1972, a number of robot probes have continued our exploration of Earth's companion.

Of the many rock samples that the Apollo astronauts carried back to Earth, every one is igneous. That is, they formed by the cooling and solidification of molten rock. No sedimentary rocks were found, consistent with the understanding that the Moon has never had liquid water on its surface. In addition, the rocks were extremely dry. Almost all Earth rocks contain 1 to 2 percent water, either as free water trapped in the rock or as water molecules chemically bonded with certain minerals. In contrast, the Moon rocks brought back by the Apollo astronauts contain very little water.

Rocks from the lunar maria are dark-colored, dense basalts much like the solidified lava produced by the Hawaiian

volcanoes (**Figure 12-5**). These rocks are rich in heavy elements such as iron, manganese, and titanium, which give them their dark color. Some of the basalts are **vesicular**, meaning that they contain holes caused by bubbles of gas in the molten rock. Like bubbles in a carbonated beverage, these bubbles do not form while the magma is under pressure. Only when the molten rock flows out onto the surface, where the pressure is low, do bubbles appear. The vesicular nature of some of the basalts shows that these rocks formed in lava flows that reached the surface and did not solidify underground.

Absolute ages of the mare basalts, measured by radioactive dating, range from about 2 to 4 billion years. These ages confirm that the lava flows happened after the end of the heavy bombardment (look back to Chapter 10).

The highlands are composed of low-density rock containing calcium-, aluminum-, and oxygen-rich minerals that would have been among the first to solidify and float to the top of molten rock. Some of this rock is **anorthosite**, a light-colored rock that contributes to the highlands' bright contrast with the dark, iron-rich basalts of the lowlands. The rocks of the highlands, although badly shattered by impacts, represent the Moon's original low-density crust, whereas the mare basalts rose as molten rock from the deep crust and upper mantle. The crustal rocks range in age from 4.0 to 4.5 billion years old, significantly older than the mare basalts.

Moon rocks are igneous, but many are classified as **breccias**, rocks that are made up of fragments of earlier rocks cemented together by heat and pressure. Evidently, after the molten rock solidified, meteorite impacts broke up the rocks and fused them together time after time.

Both the highlands and the lowlands of the Moon are covered by a layer of powdered rock and crushed fragments called the **regolith**. It is about 10 m deep on the maria but more than 100 m (330 ft) deep in certain places in the highlands. Impacts dominate the lunar surface and are responsible for the lunar regolith. About 1 percent of the regolith is meteoric fragments; the rest is the smashed remains of Moon rocks that have been pulverized by the constant rain of meteorites. The smallest meteorites—micrometeorites—do the most damage by constantly sandblasting the lunar surface, grinding the rock down to fine dust. The Apollo astronauts found that the dust coated their spacesuits and equipment, and then the interior of the lunar module after they climbed back inside.

In 2009, part of the *Lunar Crater Observation and Sensing Satellite (LCROSS)* probe was purposely crashed into the permanently dark floor of crater Cabeus near the Moon's south pole to eject a plume of crater floor material into the sky. This was to test the hypothesis that water ice lies buried in the soil there as permafrost. Instruments on another portion of the *LCROSS* probe flying behind the impactor detected water ice and vapor in the plume. The water in the Moon's polar region is more likely to have been delivered by comets and other water-bearing planetesimals over the history of the Solar System, rather than being "native" to the Moon. Nevertheless, its presence increases the possibility that a human lunar base might be established some day.

The Moon rocks are old, dry, igneous, and badly shattered by impacts, but in some protected regions water lurks as permafrost. You can use these facts, combined with what you know about lunar features, to fill in and revise the story of the Moon.

A History of the Moon

Evidence preserved in the Apollo Moon rocks shows that the Moon must have formed in a molten state. Planetary geologists now refer to the newborn Moon as a magma ocean. Evidently, denser materials sank to form a core, and, as the magma ocean cooled, low-density minerals crystallized and floated to the top to form a low-density crust. As a result the Moon differentiated into core, mantle, and crust. This corresponds to the first of the four stages of Terrestrial planet development displayed in Figure 11-2. The radioactive ages of the Moon rocks show that the surface solidified between 4.6 billion and 4.1 billion years ago. The Moon has a low average density and no magnetic field, so its dense core must be small. The core may still retain enough heat to be partially molten, but it can't contain much molten iron, or the dynamo effect would produce a magnetic field.

The second stage of development—cratering—began as soon as the crust solidified, and the older highlands show that the cratering was intense during the first 0.5 billion years or so—during the heavy bombardment at the end of planet building. The cratering rate should have fallen rapidly as the Solar System was cleared of debris. However, there is evidence from lunar crater counts and rock sample ages that, near the end of the heavy bombardment era about 4 billion years ago, there was a temporary surge in the impact rate, called the **late heavy bombardment** (see part 3 of **Impact Cratering**, page 247).

Models of the Solar System's evolution described in a later chapter indicate that Jupiter and Saturn may have migrated and temporarily moved into a mutual resonance during which their orbital periods had exactly a 2:1 ratio. The result would have been that, for a few million years, the eccentricities of Jupiter's and Saturn's orbits would have increased, and their gravity would have scattered remnant planetesimals into collisions with all the Solar System's planets and moons. Later, after the Solar System settled down again, continued bombardment by comets and icy asteroids could have brought in the water that has been detected in the form of permafrost at the Moon's south pole.

The Moon's crust was shattered to a depth of 10 km (6 mi) or so, and the largest impacts during the heavy bombardment and late heavy bombardment formed giant multiringed crater basins hundreds of kilometers in diameter, such as Mare Imbrium and Mare Orientale. This led to the third stage of Terrestrial planet development—basin flooding. Although most of the Moon cooled rapidly after its formation, radioactive decay heated material deep in the crust, and part of it melted. Molten rock followed the cracks up to the surface and flooded the giant basins with successive lava flows of dark basalts from about 4 billion to about 2 billion years ago. This formed the maria (**Figure 12-6**).

Evidence confirming the lunar basin flooding scenario came from the two *GRAIL* (*Gravity Recovery And Interior Laboratory*) spacecraft dubbed *Ebb* and *Flow* that worked in tandem to make detailed measurements of the Moon's gravity field for a year after their launch in 2011. The *GRAIL* data indicate that mass concentrations dubbed "mascons" under most of the mare basins are composed of a combination of surface rocks melted by the impact plus denser mantle rock that rose up to fill in the initial impact crater.

It is a **Common Misconception** that the lava flooding out on the surfaces of Earth and other planets comes from their molten cores. The lava actually comes from the lower crust and upper mantle. The pressure is low enough there that the melting temperature of the rock is lowered and heat flowing out of the interior is sufficient to melt portions of the rock. If there are faults and cracks, the magma can reach the surface and form volcanoes and lava flows. Whenever you see lava flows on a planet, you can be sure heat is flowing out of the interior, but the lava did not come all the way from the core.

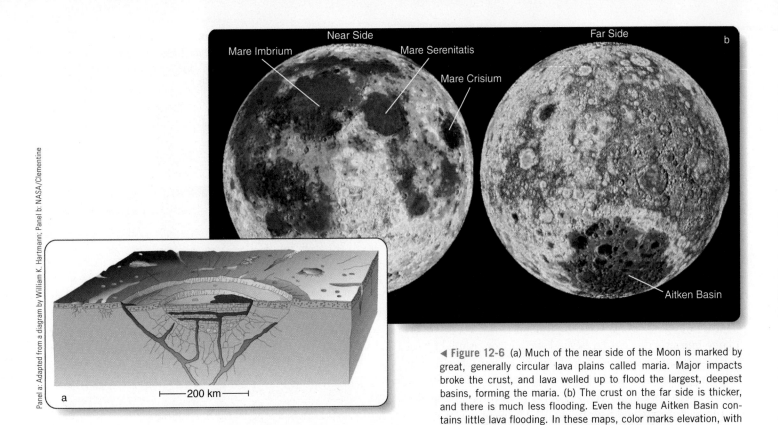

Panel a: Adapted from a diagram by William K. Hartmann; Panel b: NASA/Clementine

Near Side

Far Side   b

Mare Imbrium   Mare Serenitatis

Mare Crisium

Aitken Basin

├─── 200 km ───┤

a

◄ **Figure 12-6** (a) Much of the near side of the Moon is marked by great, generally circular lava plains called maria. Major impacts broke the crust, and lava welled up to flood the largest, deepest basins, forming the maria. (b) The crust on the far side is thicker, and there is much less flooding. Even the huge Aitken Basin contains little lava flooding. In these maps, color marks elevation, with red the highest regions and purple the lowest.

Some maria on the Moon, such as Mare Imbrium, Mare Serenitatis, Mare Humorum, and Mare Crisium, retain their round impact-crater shapes, but others are irregular because lava overflowed the edges of the basin or because the shape of the basin was modified by further cratering. The floods of lava left other characteristic features frozen into the maria. As you learned previously, in some places the lava formed channels that are seen from Earth as sinuous rilles. Also, the weight of the maria pressed the crater basins downward, and the solidified lava was compressed and formed wrinkle ridges visible even in small telescopes. The tension at the edges of the maria broke the hard lava to produce straight fractures and faults. (All of these features are visible in the different panels of Figure 12-2.) As time passed, further cratering and overlapping lava floods modified the maria. Consequently, you should think of the maria as accumulations of features reflecting multiple events during the Moon's complex history.

Mare Imbrium is a dramatic example of how the great basins became the maria. Its story can be told in detail in part because of evidence gathered by the *Apollo 14* astronauts, who landed on ejecta from the Imbrium impact (**Figure 12-7**), and by the *Apollo 15* astronauts, who landed at the edge of the mare itself.

NASA

▲ **Figure 12-7** *Apollo 14* landed on rolling terrain covered with ejecta from the Imbrium impact.

Near the end of the heavy bombardment, roughly 4 billion years ago, a planetesimal estimated to have been as much as 275 km (170 mi) across (about the size of Massachusetts, Rhode Island, and Connecticut combined) struck the Moon and blasted out a giant multiringed basin. The impact was so violent the ejecta blanketed 16 percent of the Moon's surface. After the cratering rate fell at the end of the heavy bombardment, lava flows welled up time after time and flooded the Imbrium Basin, burying all but the highest parts of the giant multiringed basin. The Imbrium Basin is now a large, generally round mare marked by only a few craters that have formed since the last of the lava flows (Figure 12-8).

This story of the Moon might suggest that it was a violent place during the cratering phase, but large impacts were in fact rare; the Moon was, for the most part, a peaceful place even during the heavy bombardment. Had you stood on the Moon at that time, you would have experienced a continuous rain of micrometeorites and much less common pebble-size impacts. Centuries might pass between major impacts. Of course, when a large impact did occur far beyond the horizon, it might have buried you under ejecta or jolted you by seismic shocks. You could have felt the Imbrium impact anywhere on the Moon, but had you been standing on the side of the Moon directly opposite that impact, you would have been at the focus of seismic waves traveling around the Moon from different directions. When the waves met under your feet, the surface would have jerked up and down by as much as 10 m (30 ft). The place on the Moon opposite the Imbrium Basin that was subjected to that effect is a strangely disturbed landscape called **jumbled terrain**. You will see similar effects of large impacts on other worlds.

Studies of our Moon show that its crust is thinner on the side facing Earth, perhaps because of tidal effects. Consequently, although lava flooded the basins on the Earth-facing side, lava was unable to rise through the thicker crust to flood the lowlands on the far side. One of the largest impact basins in the Solar System is the Aitken Basin near the Moon's south pole (Figure 12-6). It is about 2600 km (1600 mi) in diameter and as deep as 13 km (8 mi) in places, but flooding has never filled it with smooth lava flows.

The Moon is small, and small worlds cool rapidly because they have a large ratio of surface area to volume. The rate of heat loss is proportional to the surface area, and the amount of heat in a world is proportional to the volume. The smaller a world is, the easier it is for the heat to escape. That is why a small cupcake fresh from the oven cools more rapidly than a large cake. The Moon lost much of its internal heat when it was young, but the outward flow of heat is what drives geological activity, so the Moon is mostly inactive today. The crust of the Moon rapidly grew thick and never divided into moving plates. There are no rift valleys or folded mountain chains on the Moon. The last lava flows on the Moon ended about 2 billion years ago when the Moon's internal temperature fell too low to maintain subsurface lava.

The overall terrain on the Moon is almost unchanging. On Earth a billion years from now, plate tectonics will have totally

Origin of Mare Imbrium

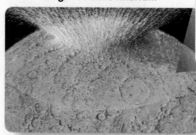

Four billion years ago, an impact forms a multiringed basin over 1000 km (600 mi) in diameter.

Continuing impacts crater the surface but do not erase the high walls of the multiringed basin.

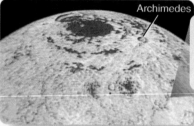

Archimedes

Impacts form a few large craters, and, starting about 3.8 billion years ago, lava floods low regions.

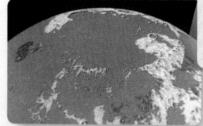

Repeated lava flows cover most of the inner rings and overflow the basin to merge with other flows.

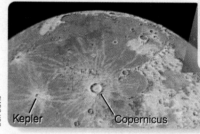

Impacts continue, including those that formed the relatively young craters Copernicus and Kepler.

Kepler   Copernicus

Don Davis

▲ Figure 12-8 Lava flooding after the end of the heavy bombardment filled a giant, multiringed basin and formed Mare Imbrium.

altered the shapes of the continents, and erosion will have long ago worn away the mountain ranges you see today. On the Moon, with no atmosphere and no water, there is no Earth-like erosion. Over the next billion years, impacts will have formed only a few more large craters, and nearly all of the lunar scenery will be unchanged. Micrometeorites are the biggest influence; they will have blasted the soil, erasing the footprints left by the Apollo astronauts and reducing the equipment they left behind to peculiar chemical contamination in the soil at the six Apollo landing sites.

You have studied the story of the Moon's evolution in detail for later comparison with other planets and moons in our Solar System, but the story has skipped one important question: Where did Earth get such a large satellite?

Origin of Earth's Moon

Over the past two centuries, astronomers developed three different hypotheses for the origin of Earth's Moon. The *fission hypothesis* proposed that the Moon broke from a rapidly spinning young Earth. The *condensation hypothesis* suggested that Earth and the Moon condensed together from the same cloud of matter in the solar nebula. The *capture hypothesis* suggested that the Moon formed elsewhere in the solar nebula and was later captured by Earth. Each of these older ideas had problems and failed to survive comparison with all the evidence.

In the 1970s, after Moon rocks were returned to Earth and studied in detail, a new hypothesis originated that combined some aspects of the three older hypotheses. The **large-impact hypothesis** proposes that the Moon formed when a large planetesimal, estimated to have been at least as massive as Mars (1/10 the mass of Earth), smashed into the proto-Earth. Model calculations indicate that this collision would have ejected a disk of debris into orbit around Earth that would have quickly formed the Moon (**Figure 12-9**).

This hypothesis does a good job of explaining many puzzling characteristics of both the Moon and Earth. If the two colliding planetesimals had already differentiated, the ejected material would be mostly iron-poor mantle and crust. Calculations indicate that the iron core of the impacting body would have combined with the larger body that became Earth. This would explain the Moon's overall low density relative to Earth, why the Moon is so poor in iron, and why the abundances of other elements are so similar to those in Earth's mantle. The collision must have occurred at a steep angle to eject enough matter to make the Moon, so the objects could not have collided head-on. A glancing collision would have spun the material rapidly enough to explain the observed angular momentum in the Earth–Moon system. Furthermore, the collision-heated material that eventually became the Moon would have lost its volatile components during the time it was in a disk of heated debris in space, so the Moon would have formed lacking volatiles. Such an impact would have melted the proto-Earth, and the material falling together to form the Moon would also have been heated hot enough to melt completely. This fits the evidence that the highland anorthosite in the Moon's oldest rocks formed by differentiation of large quantities of molten material. The large-impact hypothesis survives comparison with the known evidence and is now considered likely to be correct.

The Moon is evidently the result of a giant impact. Until recently, astronomers have been reluctant to consider such

The Large-Impact Hypothesis

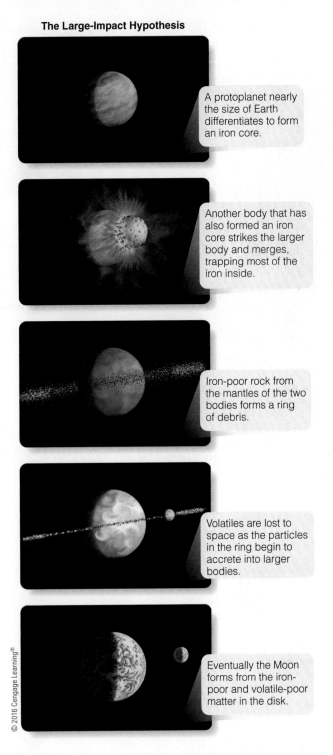

A protoplanet nearly the size of Earth differentiates to form an iron core.

Another body that has also formed an iron core strikes the larger body and merges, trapping most of the iron inside.

Iron-poor rock from the mantles of the two bodies forms a ring of debris.

Volatiles are lost to space as the particles in the ring begin to accrete into larger bodies.

Eventually the Moon forms from the iron-poor and volatile-poor matter in the disk.

© 2016 Cengage Learning®

▲ **Figure 12-9** Sometime before the Solar System was 50 million years old, a collision produced Earth and the Moon in its orbit inclined to Earth's equator.

catastrophic events, but a number of lines of evidence suggest that other planets also may have been affected by giant impacts. Consequently, the third theme identified in the introduction to this chapter, giant impacts, has the potential to help you understand other worlds. Catastrophic events are rare, but they can occur.

Visual

The Moon

Earth

Earth's Moon has about one-fourth the diameter of Earth. Its low density indicates that it contains little iron, but the size of its iron core and the amount of remaining heat are unknown.

Celestial Profile 3 The Moon

Motion:

Average distance from Earth	3.84×10^5 km
Eccentricity of orbit	0.055
Inclination of orbit to ecliptic	5.1°
Orbital period (sidereal)	27.3 d
Orbital period (synodic)	29.5 d (phase cycle)
Inclination of equator to orbit	6.7°

Characteristics:

Equatorial diameter	3.48×10^3 km (0.273 $D_\oplus$)
Mass	7.35×10^{22} kg (0.0123 $M_\oplus$)
Average density	3.35 g/cm³ (3.27 g/cm³ uncompressed)
Surface gravity	0.17 Earth gravity
Escape velocity	2.4 km/s (0.21 $V_\oplus$)
Surface temperature	$-170°$ to $+130°C$ ($-275°$ to $+265°F$)
Average albedo	0.12
Oblateness	0

Personality Point:

Lunar superstitions are common. The words *lunatic* and *lunacy* come from *Luna,* the Moon. Someone who is moonstruck is supposed to be a bit nutty. Because the Moon affects the ocean tides, many superstitions link the Moon to water, to weather, and to women's cycle of fertility. According to legend, moonlight is supposed to be harmful to unborn children, but, on the plus side, moonlight rituals are said to remove warts.

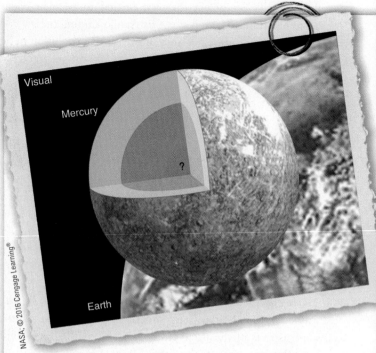

Visual

Mercury

?

Earth

Mercury is a bit more than one-third the diameter of Earth. Its high density must mean it has a large iron core. The amount of heat it retains is unknown.

Celestial Profile 4 Mercury

Motion:

Average distance from the Sun	0.387 AU (5.79×10^7 km)
Eccentricity of orbit	0.206
Inclination of orbit to ecliptic	7.0°
Orbital period	0.241 y (88.0d)
Period of rotation	58.6 d (direct)
Inclination of equator to orbit	0.0°

Characteristics:

Equatorial diameter	4.88×10^3 km (0.383 $D_\oplus$)
Mass	3.30×10^{23} kg (0.0553 $M_\oplus$)
Average density	5.43 g/cm³ (5.0 g/cm³ uncompressed)
Surface gravity	0.38 Earth gravity
Escape velocity	4.3 km/s (0.38 $V_\oplus$)
Surface temperature	$-170°$ to $+430°C$ ($-275°$ to $+805°F$)
Average albedo	0.12
Oblateness	0

Personality Point:

Mercury lies very close to the Sun and completes an orbit in only 88 days. For this reason, the ancients named the planet after Mercury, the fleet-footed messenger of the gods. The name is also applied to the element mercury, which is also known as *quicksilver* because it is a heavy, quickly flowing, silvery liquid at room temperatures.

DOING SCIENCE

If the Moon was intensely cratered by the heavy bombardment and then formed great lava plains, why didn't the same thing happen on Earth? Even though the answer to this question seems obvious, scientists still review the logic to test their understanding of basic concepts.

In fact, the same thing (heavy bombardment and cratering) did happen on Earth. Although the Moon has more craters than Earth, the Moon and Earth are the same age, and both were battered by meteorites during the heavy bombardment. Some of those impacts on Earth must have been large and dug giant, multiringed basins. Lava flows must have welled up through Earth's crust and flooded the lowlands to form great lava plains much like the lunar maria.

Earth, however, is a larger world and has more internal heat, which escapes more slowly than the Moon's heat did. The Moon is now geologically dead, but Earth is very active, with heat flowing outward from the interior to drive plate tectonics. The moving plates long ago erased all evidence of the cratering and lava flows dating from Earth's youth.

Comparative planetology is a powerful conceptual tool in that it allows you to see similar processes occurring under different circumstances. Now, employ comparative planetology to explain a different phenomenon: ***Why doesn't the Moon have a magnetic field?***

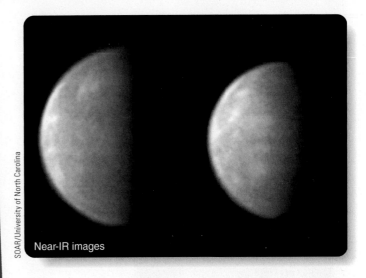

Near-IR images

▲ **Figure 12-10** Using the 4.1-m SOAR telescope in Chile by remote control from the University of North Carolina, astronomer Gerald Cecil and undergraduate student Dmitry Rashkeev resolved details on Mercury as small as 15 km (10 mi), the highest-quality images of that planet ever made from Earth. A series of very rapid exposures was taken using a digital camera, with most images being discarded, saving only the highest-resolution data.

12-2 Mercury

Earth's Moon and Mercury are good subjects for comparative planetology. They are similar in a number of ways. Most important, they are both small worlds (■ Celestial Profile 4); the Moon is only a fourth of Earth's diameter, and Mercury is slightly more than a third of Earth's diameter. They have negligible atmospheres, their rotation has been altered by tides, their surfaces are heavily cratered, their lowlands are flooded in places by ancient lava flows, and both now have ancient, inactive surfaces. The impressive differences between them also will help you understand the nature of these airless worlds.

Mercury is the innermost planet in the Solar System, and thus its orbit keeps it near the Sun in the sky as viewed from Earth. It is sometimes visible near the horizon in the evening sky after sunset or in the dawn sky just before sunrise. Astronomers using Earth-based telescopes have a difficult time discerning any surface features on the planet (**Figure 12-10**). The *Mariner 10* spacecraft looped through the inner Solar System in 1974 and 1975, taking photographs and other measurements during three flybys of Mercury. A new spacecraft called *MErcury Surface, Space ENvironment, GEochemistry, and Ranging (MESSENGER)* passed Mercury three times and then entered orbit around the planet in 2011 to begin a multiyear close-up study that is helping planetary scientists build a comprehensive understanding of Mercury's surface, interior, and history (see the image that opens this chapter, page 243).

Rotation and Revolution

During the 1880s, Italian astronomer Giovanni Schiaparelli sketched the faint features he thought he saw on the disk of Mercury and concluded that the planet was tidally locked to the Sun and kept the same side facing the Sun throughout its orbit. This was actually a very good guess because, as you will notice in the next few chapters, tidal coupling between rotation and revolution is common in the Solar System. You have already learned that the Moon is tidally locked to Earth. The rotation of Mercury is more complicated than Schiaparelli thought, however.

In 1962, radio astronomers detected blackbody emission from the planet and concluded that the dark side was not as cold as it should have been if the planet kept one side in perpetual darkness. In 1965, radio astronomers used the 305-m Arecibo dish (look back to Figure 6-17) to transmit a pulse of radio energy at Mercury and then waited for the reflected signal to return. Doppler shifts in the reflected radio pulse showed that the planet was rotating with a period of about 59 days, noticeably shorter than the orbital period of 88 days.

Mercury is tidally coupled to the Sun but in a different way than the Moon is coupled to Earth. Mercury rotates not once per orbit but 1.5 times per orbit. That is, its period of rotation is exactly 2/3 of its orbital period. This means that a mountain on Mercury directly below the Sun at one place in its orbit will point away from the Sun one orbit later and toward the Sun after the next orbit (**Figure 12-11**).

If you flew to Mercury and landed your spaceship in the middle of the day side, the Sun would be high overhead, and it

The Rotation of Mercury

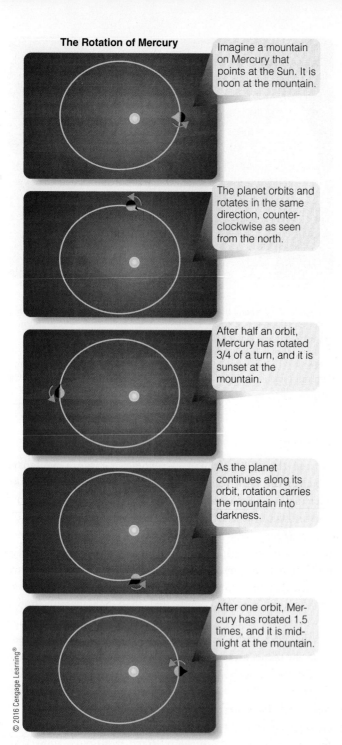

Imagine a mountain on Mercury that points at the Sun. It is noon at the mountain.

The planet orbits and rotates in the same direction, counter-clockwise as seen from the north.

After half an orbit, Mercury has rotated 3/4 of a turn, and it is sunset at the mountain.

As the planet continues along its orbit, rotation carries the mountain into darkness.

After one orbit, Mercury has rotated 1.5 times, and it is midnight at the mountain.

© 2016 Cengage Learning®

▲ **Figure 12-11** Mercury's rotation is in resonance with its orbital motion. It orbits the Sun in 88 days and rotates on its axis in two-thirds of that time. One full day on Mercury from noon to noon takes two orbits.

would be noon. Your clock would show almost 44 Earth days passing before the Sun set in the west, and a total of 88 Earth days would pass before the Sun reached the midnight position. In those 88 Earth days, Mercury would have completed one orbit around the Sun (Figure 12-11). It would require another entire orbit of Mercury for the Sun to return to the noon position overhead. So a full day on Mercury is two Mercury years long!

The complex tidal coupling between the rotation and revolution of Mercury is an important illustration of the power of tides. Just as the tides in the Earth–Moon system have slowed the Moon's rotation and locked it to Earth, so have the Sun–Mercury tides slowed the rotation of Mercury and coupled its rotation to its revolution. Astronomers refer to such a relationship as a **resonance**. You will see other such resonances as you continue to explore the Solar System. Model calculations indicate that, because of Mercury's moderately eccentric orbit, the 3:2 resonance is more stable than the 1:1 resonance that is exhibited by Earth's Moon.

Like its rotation, Mercury's orbital motion is complex. Recall from Chapter 5 that Mercury's elliptical orbit precesses (twists) faster than can be explained by Isaac Newton's laws, but at precisely the rate predicted by Einstein's theory of general relativity. The orbital motion of Mercury is taken as strong confirmation of the curvature of space-time as predicted by general relativity.

Mercury's Surface

Because Mercury is close to the Sun, the temperatures on Mercury are extreme. If you stood in direct sunlight on Mercury, you would hear your spacesuit's cooling system cranking up to high power as it tried to keep you cool. Daytime temperatures can exceed 700 K (800°F), although about 500 K (440°F) is a more usual high temperature. If you stepped into shadow on Mercury or took a walk at night, your spacesuit heaters would struggle to keep you warm. The surface can cool to −170°C (−275°F), so nights on Mercury are bitter cold.

Nights are cold on Mercury because it has almost no atmosphere. It borrows hydrogen and helium atoms from the solar wind, and atoms such as oxygen, sodium, potassium, and calcium have been detected in a cloud above the planet's surface. Some of these atoms are probably baked out of the crust, or possibly produced by very low-level remnant volcanic venting. Mercury's "atmosphere" has such a low density that the atoms do not collide with each other but just bounce from place to place on the surface and, because of the low escape velocity, eventually disappear into space.

In photographs, Mercury looks much like Earth's Moon (page 259). It is heavily battered, with craters of all sizes, including some large basins. Some craters are obviously old and degraded; others seem quite young and have bright rays of ejecta.

When planetary scientists began looking at Mercury close-up photographs in detail, they discovered something not seen on the Moon. Mercury is marked by great curved cliffs called **lobate scarps** (Figure 12-12). These are now understood to have formed when the planet cooled and shrank in diameter by a few kilometers, wrinkling its crust as a drying apple wrinkles its skin. Some of these scarps are as high as 3 km (2 mi) and reach hundreds of kilometers across the surface. Other faults in Mercury's crust are straight rather than curved, and model calculations indicate they were produced by tidal stresses generated when the Sun slowed Mercury's rotation.

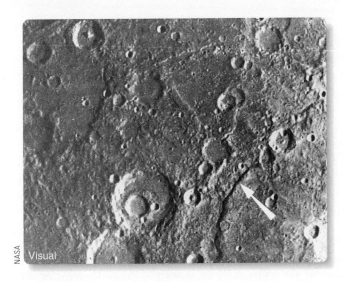

NASA
Visual

▲ **Figure 12-12** A lobate scarp (*arrow*) crosses craters, indicating that Mercury cooled and shrank, wrinkling its crust, after many of its craters had formed.

The largest basin on Mercury (**Figure 12-13**) is called Caloris Basin after the Latin word for "heat," recognition of its location at one of the two "hot poles" that face the Sun at alternate perihelions. At the times of the *Mariner* encounters, the Caloris Basin was half in shadow. Although half cannot be seen, the low angle of illumination was ideal for the study of the lighted half because it produced dramatic shadows. Caloris is a gigantic multiringed impact basin about 1550 km (950 mi)

in diameter with concentric mountain rings up to 3 km high. The impact threw ejecta more than 1000 km (600 mi) across the planet, and the focusing of seismic waves on the far side produced peculiar terrain that looks much like the jumbled area on the Moon's surface that lies opposite the Imbrium Basin (**Figure 12-14**).

The Caloris Basin is partially filled with lava flows. Some of this lava may be material melted by the energy of the impact, but some may be lava from below the crust that leaked up through cracks. The weight of this lava and the sagging of the crust have produced deep cracks in the central lava plains. Caloris Basin seems to be the same kind of structure as the Imbrium Basin on the Moon, although it has not been as deeply flooded with lava. The geophysics of such large, multiringed impact basins is still not well understood.

Mercury's Plains

The most striking difference between Mercury and the Moon is that Mercury lacks the great dark lava plains so obvious on the Moon. Under careful examination, the *Mariner 10* photographs show that Mercury has plains, two different kinds in fact, but they are different from the Moon's. Understanding these differences is the key to understanding the history of Mercury.

Much of Mercury's surface is old, cratered terrain, but other areas called **intercrater plains** are less heavily cratered. Those plains are marked by meteorite craters less than 15 km in diameter, plus secondary craters produced by chunks of ejecta

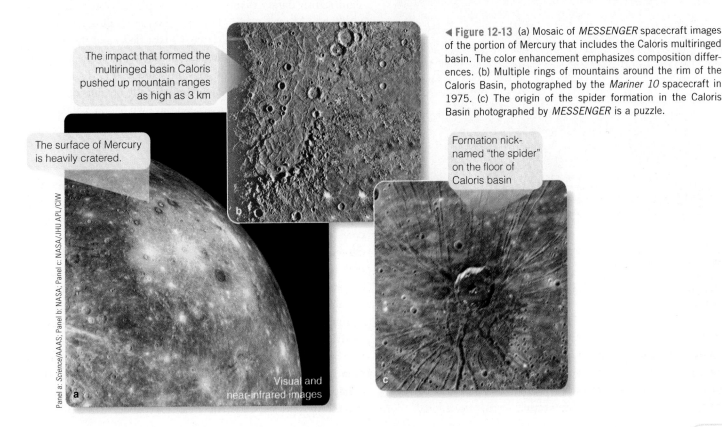

The impact that formed the multiringed basin Caloris pushed up mountain ranges as high as 3 km

The surface of Mercury is heavily cratered.

◄ **Figure 12-13** (a) Mosaic of *MESSENGER* spacecraft images of the portion of Mercury that includes the Caloris multiringed basin. The color enhancement emphasizes composition differences. (b) Multiple rings of mountains around the rim of the Caloris Basin, photographed by the *Mariner 10* spacecraft in 1975. (c) The origin of the spider formation in the Caloris Basin photographed by *MESSENGER* is a puzzle.

Formation nick-named "the spider" on the floor of Caloris basin

Panel a: *Science*/AAAS; Panel b: NASA; Panel c: NASA/JHU APL/CIW

Visual and near-infrared images

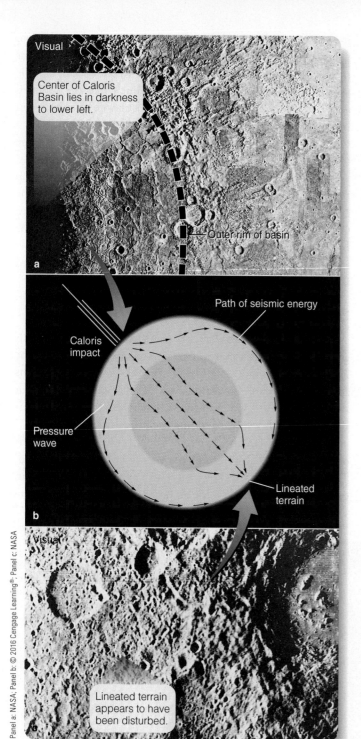

Center of Caloris Basin lies in darkness to lower left.

Outer rim of basin

a

Path of seismic energy

Caloris impact

Pressure wave

Lineated terrain

b

Visual

Lineated terrain appears to have been disturbed.

c

▲ **Figure 12-14** (a) The huge impact that formed the Caloris Basin on Mercury sent seismic waves through the planet. (b) and (c) Where the waves came together on the far side, they produced this lineated terrain, which resembles the jumbled terrain on Earth's Moon opposite the Imbrium impact.

from larger impacts. Unlike the heavily cratered regions, the intercrater plains are not totally saturated with craters. With your knowledge of comparative planetology, you can recognize that this means the intercrater plains were produced by later lava flows, which buried older terrain.

Enhanced-color visual

▲ **Figure 12-15** Some of Mercury's smooth plains are shown in the right half of this image made by the *MESSENGER* spacecraft. The surface is not saturated with craters as are the lunar highlands, but the crater density is higher than in the lunar maria. This shows that the smooth plains formed after most of the heavy bombardment in the early Solar System.

Smaller regions called **smooth plains** are evidently even younger than the intercrater plains. They have even fewer craters and appear to be lava flows that occurred after most cratering had ended. Much of the region around the Caloris Basin is composed of these smooth plains (Figure 12-15), and they appear to have formed soon after the Caloris impact.

Given the available evidence, planetary astronomers conclude that the plains of Mercury are solidified lava flows much like the maria on the Moon, but Mercury's lava plains are not significantly darker than the rest of the planet's crust, strikingly unlike the Moon's maria. This may be the result of a compositional difference between Mercury's lava flows and the Moon's. Except for a few bright crater rays, Mercury's surface is a uniform gray with an albedo of only about 0.1. That means Mercury's lava plains are not as dramatically obvious on photographs as the much darker maria on our own Moon, which show up in contrast to the lighter highlands.

Mercury's Interior

One of the most striking differences between Mercury and the Moon is the composition of their interiors. You have seen that the Moon is a low-density world that contains at most a small core of metals. In contrast, Mercury is more than 60 percent denser than the Moon, yet Mercury's surface appears to be normal, relatively low-density crustal rock. From these facts you can conclude that Mercury's interior contains a large core of dense metals, mostly iron. In proportion to its size, Mercury must have a larger metallic core than Earth (see the diagram in **Celestial Profile 4**, page 256).

If Mercury had a large metallic core that remained molten, then the dynamo effect would generate a magnetic field (Chapter 11, page 231). The *Mariner 10* and *MESSENGER* spacecraft found a magnetic field only about 1.1 percent as strong as Earth's, and this weak field made it difficult to understand the planet's interior. Because Mercury is a small world, it should have lost most of its internal heat long ago and should not have a molten core. Nevertheless, radar observations of Mercury's rotation show that the surface of the planet is shifting back and forth slightly in response to the Sun's tidal influence as the planet moves in its elliptical orbit. That must mean that at least the outer core is molten. If Mercury's iron core contains a higher-than-Earthly concentration of sulfur, the melting point would be lowered, and the outer core, where the pressure is lower, could be molten. It is not clear how a planet that formed so close to the Sun could contain so much sulfur, which is a volatile material. Continued measurements of Mercury's gravitational and magnetic fields by the *MESSENGER* spacecraft may help planetary astronomers finally understand its core.

It is also difficult to explain the disproportionately large size of the metallic core inside Mercury. You learned in Chapter 10 that the condensation sequence predicts planets forming near the Sun should incorporate more metals, but Mercury's metallic core is even larger than expected from theory. One hypothesis involves a giant impact when Mercury was young, an impact much like the planet-shattering impact proposed to explain the origin of Earth's Moon. If the forming planet had differentiated and was then struck by a large planetesimal, the impact could have shattered the crust and mantle and blasted much of the lower-density material into space. The denser core could have survived, re-formed, and then swept up some of the lower-density debris to form a thin mantle and crust. This scenario would leave Mercury with a deficiency of low-density crustal rock. It is possible that Mercury, like the Moon, is the product of a giant impact. However, *MESSENGER* imaged surface hollows in many locations (Figure 12-16) that may indicate a substantial amount of volatiles were incorporated in the crust and subsequently vaporized. Abundant volatiles would not have survived a giant impact, so an impact may not, after all, be the explanation for the size of Mercury's metal core. Because of that, some astronomers have proposed an alternate hypothesis to explain Mercury's overly massive core, which is that heat from

gigantic solar flares during the Sun's temperamental youth vaporized and drove away some of the rock-forming elements in the inner solar nebula.

The *MESSENGER* spacecraft data should allow testing of those hypotheses. Studies of Mercury's crust and interior can not only reveal much about the history of that planet, they can yield clues about the formation and histories of the other planets, including Earth.

A History of Mercury

Can you combine evidence and theory to tell the story of Mercury? It formed in the innermost part of the solar nebula, and, as you have seen, a giant impact may have robbed it of some of its lower-density rock and left it a small, dense world with a surprisingly large metallic core.

Like the Moon, Mercury suffered heavy cratering by debris in the young Solar System. Planetary scientists don't know accurate absolute ages for features on Mercury because they do not have rock samples to subject to radioactive dating, but you can safely assume that cratering, the second stage of planetary development, occurred over the about same period as the cratering on the Moon. This intense cratering declined rapidly as the planets swept up the last of the debris left over from planet building.

The cratered surface of Mercury is not exactly like that of the Moon. Because of Mercury's stronger gravity, the ejecta from an impact on Mercury are thrown only about 65 percent as far as on the Moon, and that means the ejecta from an impact on Mercury do not blanket as much of the surface. Also, the intercrater plains appear to have formed when lava flows occurred during the heavy bombardment, burying the older surface, and then accumulated more craters. Sometime near the end of cratering, a planetesimal more than 100 km in diameter smashed into the planet and blasted out the great multiringed Caloris Basin. Only parts of that basin have been flooded by lava flows.

The smooth plains contain fewer craters and may date from the time of the Caloris impact. The impact may have been so big it fractured the crust and allowed lava flows to resurface wide areas. Because this happened near the end of cratering, the smooth plains have few craters.

Enhanced-color Visual

◄ **Figure 12-16** Mosaic of images from *MESSENGER* of a section of the peak-ring mountains and floor of the Raditladi impact basin on Mercury. The individual frames in the mosaic are about 20 km (12 mi) wide. The rounded depressions called "hollows," seen in many locations on Mercury, may have been formed by sublimation of a volatile component in the surface material.

Finally, the cooling interior contracted, and the crust broke to form the lobate scarps. Lava flooding ended quickly, perhaps because the shrinking planet squeezed off the lava channels to the surface. Mercury lacks a true atmosphere, so you would not expect flooding by water, but radar images show a bright spot at the planet's north pole that may be caused by ice trapped in perpetually shaded crater floors where the temperature never exceeds 60 K (–350°F). This may be water from comets that occasionally collide with Mercury and deliver bursts of water vapor. As you read earlier in this chapter, water deposits of this type have been identified at the Moon's south pole.

The fourth stage in the story of Mercury, slow surface evolution, is now limited to micrometeorites, which grind the surface to dust; rare larger meteorites, which leave bright-rayed craters; and the slow but intense cycle of heat and cold, which weakens the rock at the surface. The planet's crust is now thick, and although its core may be partially molten, the heat flowing outward is unable to drive plate tectonics that would actively erase craters and build folded mountain ranges.

DOING SCIENCE

Why don't Earth and the Moon have lobate scarps? Answering this question calls for another use of the principle of comparative planetology.

You might expect that any world with a large metallic interior should have lobate scarps. When the metallic core cools and contracts, the world should shrink, and the contraction should wrinkle and fracture the brittle crust to form lobate scarps. But there are other factors to consider. Earth has a fairly large metallic core, but it has not cooled very much, and the crust is thin, flexible, and active. If any lobate scarps ever did form on Earth, they would have been quickly destroyed by plate tectonics.

The Moon does not have a large metal core. The rocky interior might have contract slightly as the Moon lost its internal heat, but that slight contraction of a much smaller core may not have produced major lobate scarps.

You can, in a general way, understand lobate scarps, but now consider them using a geologist's basic method of interpreting time sequences: *How do you know the lobate scarps on Mercury formed after most of the heavy bombardment was over?*

What Are We? Comfortable

Many planets in the Universe probably look like the Moon and Mercury—small, airless, and cratered. Some are made of stone; and some, because they formed farther from their star, are made mostly of ice. If you randomly visited a planet anywhere in the Universe, you might find yourself standing on a cratered moonscape.

Earth-like worlds are unusual but perhaps not rare. The Milky Way Galaxy contains more than 100 billion stars, and more than 100 billion galaxies are visible with existing telescopes. Most of those 10^{22} stars probably have planets, and although many planets look like Earth's Moon and Mercury, there are also probably plenty of Earth-like worlds.

As you look around your planet, you should feel comfortable living on such a beautiful planet, but it was not always such a nice place. The craters on the Moon and the Moon rocks returned by astronauts show that the Moon formed as a sea of magma. Mercury seems to have had a similar history, so the Earth likely formed the same way. It was once a seething ocean of liquid rock swathed in a hot, thick atmosphere, torn by explosions of gas from the interior, and occasional impacts from space. The Moon and Mercury show that that is the way Terrestrial planets begin. Earth has evolved to become your home world, but Mother Earth has had a violent past.

Study and Review

Summary

▶ The Moon is **tidally coupled (p. 244)** to Earth and rotates on its axis once each orbit, keeping the same side facing Earth.

▶ The Moon has only one-sixth the gravity of Earth. It has such a low escape velocity that it is unable to retain an atmosphere. For that reason observers on Earth see sharp shadows on the Moon's surface, especially near the **terminator (p. 244)**, the dividing line between daylight and darkness, and stars disappear behind the limb of the Moon without dimming. Astronauts visiting the Moon verified that the Moon has no measurable atmosphere.

▶ Large, smooth, dark plains on the Moon called **maria (singular, mare) (p. 244)** are old lava flows that filled lowlands. Evidence of lava flows includes **sinuous rilles (p. 244)** that once carried flowing lava, faults where lava plains cracked, and wrinkle ridges.

- When a meteorite strikes the Moon, it digs an impact crater and throws out debris that falls back as **ejecta (p. 246)**, which can form **rays (p. 246)** and **secondary craters (p. 246)**. The largest impacts on the Moon dug huge **multiringed basins (p. 247)**.

- Astronomers can find **relative ages (p. 248)** of lunar features by looking to see which features lie on top of or underneath other features, and by counting craters. Those relative ages can then be calibrated using **absolute ages (p. 248)** of lunar rock samples determined by radioactive dating.

- The highlands on the Moon are saturated with craters. Most of the craters on the Moon were formed during the heavy bombardment era, which occurred at the end of planet building, between about 4 and 4.5 billion years ago. Occasional meteorite impacts continue to form new craters, although no large crater is known with certainty to have been formed on the Moon in historic times. **Micrometeorites (p. 247)** are currently the main source of lunar erosion, constantly grinding the surface down to dust.

- Between 1969 and 1972, 12 Apollo astronauts set foot on the Moon and returned specimens to Earth.

- Moon rocks returned to Earth are all igneous, showing that they solidified from molten rock. Some of the basalt rocks are **vesicular (p. 251)**, indicating that they formed in surface lava flows. Light-colored **anorthosite (p. 251)** rocks are part of the old crust and help make the highlands brighter than the lowland maria. Many of the rocks are **breccias (p. 252)**, indicating that much of the lunar crust was fractured by meteorite impacts.

- The surface of the Moon is covered by **regolith (p. 252)**, a form of soil that was created from meteorite impacts that crushed and powdered surface Moon rocks.

- Evidence from lunar crater counts and rock sample ages indicate an episode of increased impact rate. This **late heavy bombardment (p. 252)** occurred about 4 billion years ago, near the end of the heavy bombardment era. Many of the giant impact basins containing mare were formed during this time. The late heavy bombardment may have been caused by impacts of remnant planetesimals that were scattered when Jupiter and Saturn were migrating.

- The Imbrium Basin formed about 4 billion years ago by the impact of a planetesimal estimated to have been 275 km (170 mi) in diameter. Seismic waves traveling through the Moon focused on the far side, producing **jumbled terrain (p. 254)**. Later, flooding nearly buried the original basin.

- The fission, condensation, and capture hypotheses for the origin of the Moon have all been abandoned. The commonly accepted hypothesis is the **large-impact hypothesis (p. 255)**, which proposes that the Moon forming from a ring of debris that was ejected into space when a large planetesimal struck the proto-Earth after the planet had differentiated. This formation process would explain the Moon's orbit inclination, low density, and lack of volatiles.

- The composition of lunar rocks show that the Moon formed in a molten state referred to as a magma ocean. Although the Moon differentiated, it contains little iron and its core is not massive. Low-density rock rose to the surface to form a crust that later was heavily cratered and shattered to great depth.

- Lava, welling up through the cracked crust, filled the lowlands to form the smooth maria plains. The maria formed after the end of the heavy bombardment and contain few craters.

- The near side of the Moon has a thin crust, possibly because of tidal forces. The far side of the Moon has a thicker crust and not much lava flooding.

- Because the Moon is small, it has lost its internal heat and is geologically dead today. The only slow surface evolution occurring now is the blasting by micrometeorites.

- Mercury rotates 1.5 times per orbit in a **resonance (p. 258)** relationship between its rotation and its revolution.

- Mercury is a small world that has been unable to retain an atmosphere and has lost most of its internal heat. It is geologically inactive today and covered with craters.

- As Mercury's large metallic core cooled and contracted, the brittle crust broke. **Lobate scarps (p. 258)** formed, analogous to wrinkles in the skin of a drying apple.

- The Caloris Basin is a large, multiringed basin on Mercury that has been partially flooded by lava flows.

- Mercury was heavily cratered during the heavy bombardment. Lava flows covered some of those craters, and new craters formed the intercrater plains. Fractures produced by the Caloris impact may have triggered additional lava flows that formed the smooth plains.

- The **intercrater plains (p. 259)** on Mercury may have been formed by lava flows that occurred later in Mercury's evolution. These lava flows covered older craters and then accumulated newer craters. **The smooth plains (p. 260)** contain few craters and evidently were formed by more recent lava flows. Unlike the lunar maria, all of the lava flows on Mercury's plains are a similar shade of gray and thus are not easily visible in photographs of Mercury.

- Mercury formed at high temperature in the inner solar system and therefore contains a large proportion of dense metals. In fact Mercury has a larger metallic core than predicted by the condensation sequence. One hypothesis is that a large impact shattered and drove off some of the planet's low-density crust, increasing the proportion of metallic core.

- The amount of heat remaining in Mercury's interior is not well known. The planet has a weak but detectable magnetic field. Radar observations of Mercury's rotation suggest that the outer layers of Mercury's core remain molten.

- The *MESSENGER* spacecraft went into orbit around Mercury in 2011. It has provided more extensive photography of Mercury's surface as well as detailed measurements of Mercury's physical properties.

Review Questions

1. Which of the four fundamental forces results in tidally coupled celestial objects?

2. As viewed from Earth, how many times does the Moon rotate during one orbit? As viewed from outside the Earth–Moon system, how many times does the Moon rotate in one orbit? How do you know?

3. If the Moon is tidally coupled to Earth, is Earth tidally coupled to the Moon? How do you know?

4. How can you determine the relative ages of the Moon's maria and highlands?

5. From looking at images of the Moon's near side, how can you tell that Copernicus is a young crater?

6. Why did the first Apollo missions land on the maria? Why were the other areas of more scientific interest?

7. Why do planetary scientists hypothesize that the Moon formed with a molten surface?

8. A Moon rock classified as a breccia does not necessarily have to be also igneous. True or false?

9. Why are so many lunar samples classified as breccias?

10. What do the vesicular basalts tell you about the evolution of the lunar surface?

11. What is the most significant kind of erosion that occurs on the Moon today?

12. List any observational evidence of tectonics on the Moon's surface.

13. What evidence can you cite that the Moon had volcanism? Does the Moon have volcanism today? How do you know?

14. What evidence would you expect to find on the Moon if the Moon had been subjected to plate tectonics? Is there such evidence?

15. How does the large-impact hypothesis explain the Moon's lack of iron?

16. Look at **Celestial Profiles 2**, **3**, and **4**. Compare the average albedo values of the Moon, Mercury, and Earth. Can you explain their different values?

17. Look at **Celestial Profile 3** and **4**. Which has the more elliptical orbit, Mercury or the Moon? How do the shapes of their respective orbits explain the rotation–orbit resonances of the Moon and Mercury?

18. Look at **Celestial Profile 3** and **4**. Which has the higher density, Mercury or the Moon? Why?

19. Look at **Celestial Profiles 2**, **3**, and **4**. Which is most oblate: Mercury, the Moon, or Earth? Why?

20. Look at **Celestial Profile 3** and **4**. Explain the differences in surface temperatures between Mercury and the Moon.

21. Maria are observed on Mercury. True or false?

22. What are the relative ages of the intercrater plains versus the smooth plains on Mercury?

23. What evidence can you cite that Mercury has a partially molten, metallic core?

24. Describe any observational evidence of tectonics on Mercury's surface.

25. What evidence can you cite that Mercury had volcanism? Does Mercury have volcanism today? How do you know?

26. How are the histories of the Moon and Mercury similar? How are they different?

27. What property of the Moon and Mercury has resulted in almost complete cessation of surface evolution on both those worlds, whereas Earth's surface evolution continues?

28. **How Do We Know?** How do scientists use a hypothesis or theory to remember an assortment of details?

Discussion Questions

1. Why does the flag in Figure 12-4 appear to be waving in a breeze?

2. Old science fiction paintings and drawings of colonies on the Moon often show very steep and jagged peaks on the Moon's highlands. Why did artists assume that lunar mountains would be more steep and jagged than Earth's mountains? Why are the lunar highlands actually less jagged than mountains on Earth?

3. From your knowledge of comparative planetology, propose a description of the view that astronauts would have if they landed on the surface of Mercury.

4. From your knowledge of Mercury, where would you choose to land your spacecraft and walk around outside?

Problems

1. Look at the right top and bottom images in Figure 12-2. Count the number of craters in each image. Based on your result, which portion of the Moon's surface do you judge to be relatively older, and why?

2. Calculate the escape velocity of the Moon from its mass and diameter. (*Note:* Relevant information can be found in **Celestial Profile 3**. *Hint:* Use the formula for escape velocity, Chapter 5.)

3. If you transmitted radio signals to the Moon and waited to receive the echo, how long would you wait? (*Notes:* The speed of light is 3.0×10^5 km/s. Other relevant information can be found in **Celestial Profile 3**.)

4. Why do small planets cool faster than large planets? Choose any two of the five Terrestrial worlds and calculate for each one the ratio of its surface area to its volume. Why is this ratio important? (*Note:* The surface area of a sphere is $4\pi r^2$, and the volume of a sphere is $\frac{4}{3}\pi r^3$. *Hint:* Does this ratio have anything to do with the ability of a planet to lose internal heat?)

5. The smallest detail visible through Earth-based telescopes is about 1 arc second in diameter. What linear size is this on the Moon? (*Hint:* Use the small-angle formula, Chapter 3.)

6. Review part 1a of **Impact Cratering**. Estimate the approximate size, in km, of the meteorite that struck the Moon to form Euler crater. Estimate the approximate depth of Euler crater in km.

7. The trenches where Earth's seafloor slips downward are 1 km or less wide. Could Earth-based telescopes resolve such features on the Moon? Why can you be sure that such features are not present on the Moon? (*Note:* Relevant information can be found in **Celestial Profile 3**. *Hint:* Use the small-angle formula, Chapter 3.)

8. An *Apollo* command module orbited the Moon about 100 km above the surface. What was its orbital velocity? What was its orbital period? (*Note:* Relevant information can be found in **Celestial Profile 3**. *Hint:* Use the formula for circular orbit velocity, Chapter 5.)

9. From a distance of 100 km above the surface of the Moon, what is the angular diameter of an astronaut in a spacesuit who has a linear diameter of 0.7 m as viewed from above? The unaided human eye has a resolution of about 100 arc seconds in bright lighting conditions. Could someone looking out the command module window have seen the astronauts on the Moon? (*Hint:* Use the small-angle formula, Chapter 3.)

10. What is the angular diameter of Mercury when it is closest to Earth? How does that compare with the angular diameter of the Moon? (*Note:* Relevant information can be found in **Celestial Profile 3** and **4** and Appendix Table A-10. *Hint:* Use the small-angle formula, Chapter 3.)

11. If you transmit radio signals to Mercury when Mercury is closest to Earth and wait to hear the radar echo, how long will you wait? (*Note:* The speed of light is 3.0×10^5 km/s. Other relevant information can be found in **Celestial Profile 4** and Appendix Table A-10.)

12. What is the wavelength of the most intense radiation emitted from the surface of Mercury at high noon? In which band of the electromagnetic spectrum is that wavelength? (*Hints:* Use Wien's law, Chapter 7, and see Figure 6-3.)

13. Suppose you send a probe to land on Mercury, and the probe transmits radio signals to Earth at a wavelength of 10.0000 cm. You listen for the probe when Mercury is moving away from Earth at its full orbital velocity of 48 km/s around the Sun. What wavelength would have to tune your radio telescope to detect that signal? (*Note:* The speed of light is 3.0×10^5 km/s. *Hint:* Use the Doppler shift formula in Chapter 7.)

14. The smallest detail visible through Earth-based telescopes is about 1 arc second in diameter. What linear size does that correspond to on Mercury when Mercury is at a distance of 1 AU? Can Caloris Basin be resolved? (*Note:* 1 AU is 1.5×10^8 km. *Hint:* Use the small-angle formula, Chapter 3.)

Learning to Look

1. Look at the image of the astronaut on the Moon at the upper right of the right-hand page of **Impact Cratering.** Can you tell whether the Sun's location in toward the upper right, upper left, lower right, or lower left of the image? Is the relative time of day more like sunrise/sunset or high noon?

2. Examine the shape of the horizon at the *Apollo 17* landing site (Figure 12-4, upper right panel). What processes shape mountains on Earth that have not affected mountains on the Moon?

3. In the photo at right, astronaut Alan Bean works at the *Apollo 12* lander. Describe the horizon and the surface you see. What kind of terrain did they land on for this, the second human Moon landing, and why?

NASA

13 Venus and Mars

Guidepost After visiting the Moon and Mercury, you will find Venus and Mars dramatically different from those small, geologically inactive, airless worlds. Venus and Mars have internal heat and atmospheres. The internal heat means they are geologically active, and the atmospheres mean they have weather. As you explore, you will discover answers to four important questions:

▶ **What is the evidence that Venus's surface conditions were originally more Earth-like than they are now?**

▶ **How did Venus form and evolve?**

▶ **What is the evidence that Mars's surface conditions were originally more Earth-like than they are now?**

▶ **How did Mars form and evolve?**

The comparative planetology questions that you need to keep in mind when you explore another world are these: How and why is this world similar to Earth? How and why is this world different from Earth? You will see that small initial differences can have big effects.

You are a planet-walker and are becoming an expert on the kind of planets you can imagine walking on. But there are other worlds beyond Mars in our Solar System so peculiar they have no surfaces to walk on, even in your imagination. You will explore them in the next two chapters.

The only truly alien planet is Earth.

J. G. BALLARD

NASA's *Curiosity* rover took a set of 55 high-resolution images that were combined to create this full-color self-portrait during the rover's 84th Martian day on the surface. Four scoop-sample scars can be seen in front of the rover. The base of Mount Sharp, Gale Crater's 5-km (3-mi) high central mountain, rises on the right side of the frame.

NASA/JPL-Caltech/MSSS

Visual

THE TEMPERATURE ON MARS on a hot summer day at noon might feel pretty pleasant at around 15°C (60°F). But without a spacesuit, you could survive there for only a few moments because the air is mostly carbon dioxide with almost no oxygen. Even more important, the air pressure is less than 1 percent that at the surface of Earth, so your exposed body fluids such as tears and saliva would boil if you stepped outside your spaceship unprotected.

Not even a spacesuit would save you on Venus. The surface is hot enough to melt lead, the air pressure is like being a half-mile underwater, and the air is almost entirely carbon dioxide with traces of various acids.

Venus and Mars resemble Earth in some ways, so why are they such unfriendly places to visit? Comparative planetology will give you some clues.

13-1 Venus

Venus is almost a twin of Earth, and you might expect surface conditions on the two planets to be quite similar. Venus and Earth are almost exactly the same size and mass, with Venus having 95 percent the diameter of Earth (■Celestial Profile 5, p. 277). Venus and Earth have similar average densities, and they formed in the same part of the solar nebula; Venus's orbit is the closest to Earth's of all the planets, so Venus's overall composition is similar to Earth's (look back to Chapter 10). Also, planets the size of Earth and Venus cool slowly, so you might predict that Venus, like Earth, would have a molten metallic core, a convective mantle, and an active crust with plate tectonics.

Until fairly recently, those predictions could not be checked because the surface of Venus is perpetually hidden by thick clouds that completely envelop the planet, preventing us from easily observing conditions there. From the time of Galileo until the early 1960s, astronomers could only speculate about Earth's twin. Science fiction writers imagined that Venus was a steamy swamp inhabited by strange creatures, or a windblown sandy desert, or completely covered by an ocean.

In the 1960s, astronomers used measurements of microwave blackbody emission from Venus to determine its surface temperature, and radar to penetrate the clouds, making images of the surface as well as measuring the planet's rotation. More than 25 spacecraft have flown past or orbited Venus, and more than a dozen have landed on its surface. The resulting picture of Venus is different from any of the fiction writers' visions. In fact, the surface of Venus is drier than any desert on Earth and twice as hot as a kitchen oven set to its highest temperature, with an atmosphere 100 times denser than Earth's. Other surprising contrasts to Earth are that Venus rotates slowly, backward, and the planet has no measurable magnetic field.

Venus has certainly gone down a different evolutionary path than our home planet. How did Earth's near-twin in terms of location, size, and density become so different? There are as yet only partial answers to this question.

Venus's Rotation

Nearly all of the planets in our Solar System rotate counterclockwise as seen from the north. Uranus is an exception, and so is Venus. Doppler shifts of radar pulses reflected off the planet reveal that Venus is rotating once every 243 Earth days. Furthermore, because the western edge of Venus produces a blueshifted signal, we know that limb is moving toward us, meaning the planet is rotating retrograde relative to most other motions in the solar system (Chapter 10, page 197).

Why does Venus rotate so slowly and backward? For decades, textbooks have suggested that proto-Venus was set spinning backward when it was struck off-center by a large planetesimal. That is a reasonable possibility; you have learned that a similar collision probably gave birth to Earth's Moon and may have caused Mercury's high density. But there is an alternative. Mathematical models suggest that the rotation of a Terrestrial planet orbiting close to the Sun with a molten core and a dense atmosphere can be gradually reversed by solar tides in the atmosphere. Notice the contrast between the catastrophic theory of a giant impact and the evolutionary theory of atmospheric tides. It is possible that both mechanisms played a role in causing Venus's peculiar rotation.

Venus's Interior

You might expect from the condensation sequence and Venus's position in the Solar System that it should have a similar overall composition with perhaps slightly higher metal content than Earth. Instead, Venus's uncompressed density is slightly *less* than Earth's (look back to Table 10-1, page 206). The density and size of Venus indicate that it has a dense metallic interior much like Earth's. However, no spacecraft has detected a magnetic field around Venus; if the planet has a magnetic field it must be at least 25,000 times weaker than Earth's.

Although the planet's rotation is very slow, if Venus's metal core is liquid, you would expect the dynamo effect to generate some magnetic field. (Note that Mercury has a significant magnetic field even though it rotates almost as slowly as Venus.) Some theorists wonder if the core of the planet is solid, but if it is solid, planetary scientists do not have a good explanation for how Venus got rid of its internal heat so much faster than Earth did.

Venus's Atmosphere

Venus's atmosphere is truly un-Earthly. The composition, temperature, and density of Venus's atmosphere make the planet's surface entirely inhospitable. About 96 percent of its atmosphere is carbon dioxide, and 3.5 percent is nitrogen. The remaining 0.5 percent is water vapor, sulfuric acid (H_2SO_4), hydrochloric acid (HCl), and hydrofluoric acid (HF). In fact, the thick clouds that hide the surface are composed of sulfuric acid droplets and microscopic sulfur crystals.

Soviet and U.S. spacecraft dropped probes into the atmosphere of Venus, and those probes radioed data back to Earth as they fell toward the surface. Those studies show that Venus's cloud layers are much higher and more stable than those on Earth. The highest layer of clouds—the layer visible from Earth—extends from about 60 to 70 km (about 40 to 45 mi) above the surface (**Figure 13-1**). For comparison, clouds in Earth's atmosphere normally do not extend higher than about 15 km (10 mi).

The Venusian cloud layers are highly stable because the atmospheric circulation on Venus is much more regular than that on Earth. The heated atmosphere at the **subsolar point**, the point on the planet where the Sun is directly overhead, rises and spreads out in the upper atmosphere. Convection circulates this gas toward the dark side of the planet and the poles, where it cools and sinks. This circulation produces 300-km/hour jet streams in the upper atmosphere, which move from east to west (the same direction the planet rotates) so rapidly that the entire atmosphere rotates with a period of only 4 days.

The details of this atmospheric circulation are not well understood, but it seems that the slow rotation of the planet is an important factor. On Earth, large-scale circulation patterns are broken up into cyclonic (spiral) disturbances by Earth's rapid rotation. Because Venus rotates more slowly, its atmospheric circulation is not broken up into small cyclonic storms but instead is organized as a planetwide wind pattern.

Although Venus's upper atmosphere is cool, the lower atmosphere is quite hot (Figure 13-1b). Instrumented probes that reached the surface reported that the temperature is 470°C (880°F), and the atmospheric pressure is 90 times that of Earth. Earth's atmosphere is 1000 times less dense than water, but on Venus the air is only 10 times less dense than water. If you could survive the unpleasant atmospheric composition, intense heat, and high pressure, you could strap wings to your arms and fly.

The present atmosphere of Venus is extremely dry, but there is evidence that it once had significant amounts of water. The ratio of abundances of deuterium, the heavy isotope of hydrogen, to ordinary hydrogen, as measured by several different probes descending through the atmosphere or orbiting Venus, is about 150 times higher than in Earth's atmosphere. Planetary scientists hypothesize that this high abundance of deuterium is the remnant of destroyed water. Venus has no ozone layer to absorb the ultraviolet (UV) radiation in sunlight. As a result, solar UV photons can easily break atmospheric water vapor molecules into hydrogen and oxygen. The oxygen forms oxides in the soil, and the hydrogen leaks away into space. The heavier deuterium atoms would leak away more slowly than normal hydrogen atoms, which would increase the ratio of deuterium to normal hydrogen.

Venus has essentially no water now, but the amount of deuterium in the atmosphere suggests that it may have once had enough water to make a planetwide ocean at least 25 m (80 ft) deep. (For comparison, the water on Earth would make a uniform ocean about 3000 m deep.) Now Venus only has enough

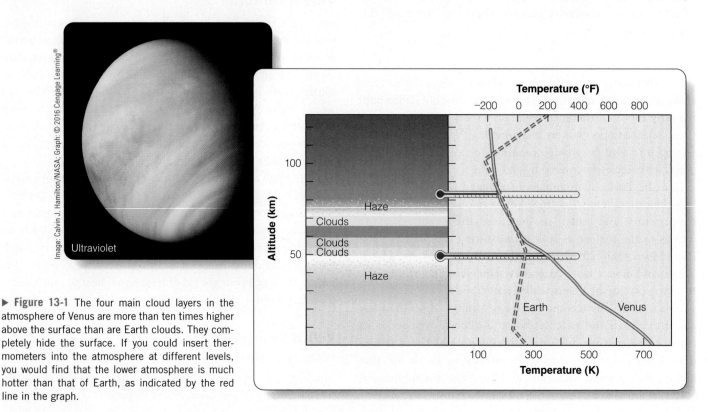

▶ **Figure 13-1** The four main cloud layers in the atmosphere of Venus are more than ten times higher above the surface than are Earth clouds. They completely hide the surface. If you could insert thermometers into the atmosphere at different levels, you would find that the lower atmosphere is much hotter than that of Earth, as indicated by the red line in the graph.

water vapor in its atmosphere to make a planetwide water layer 0.3 m (1 ft) deep. Venus's current lack of water is one of the biggest differences between that planet and Earth.

The Venusian Greenhouse

You learned in Chapter 11 how the greenhouse effect warms Earth. Carbon dioxide (CO_2) is transparent to light but opaque to infrared (heat) radiation. That means energy can enter the atmosphere as light and warm the surface, but the surface cannot radiate the energy back to space easily because the atmospheric CO_2 is opaque to infrared radiation. Venus also has a greenhouse effect, but on Venus the effect is fearsomely strong; that is the explanation for why Venus, although it is farther from the Sun than Mercury, is actually hotter. Whereas Earth's atmosphere contains only about 0.06 percent CO_2, the atmosphere of Venus contains 96 percent CO_2, and as a result temperatures on the surface of Venus are more than hot enough to melt lead. The thick atmosphere and its high winds carry heat around the planet efficiently enough to make surface temperatures on Venus nearly the same everywhere. This evidently offsets the effect of the planet's slow rotation that would otherwise cause a large temperature difference between the day side and the night side.

Planetary scientists think they know how Venus got into such a jam. When Venus was young, it may have been cooler than it is now, but because it formed 30 percent closer to the Sun than did Earth, it was always warmer than Earth, and that unleashed what scientists refer to as a **runaway greenhouse effect** that made it much hotter. Model calculations indicate that Venus and Earth should have outgassed about the same amount of CO_2, but Earth's oceans have dissolved most of Earth's CO_2 and converted it to sediments such as limestone. If all of Earth's crustal carbon were dug up and converted back to CO_2, our atmosphere would be about as dense as Venus's atmosphere and also composed mostly of CO_2.

The evidence provided by the atmospheric deuterium excess indicates that Venus once had substantial amounts of water on its surface, but that water would have begun to evaporate at the temperatures of early Venus. CO_2 is highly soluble in water, but as surface water disappeared on Venus, the ability to dissolve CO_2 and remove it from the atmosphere also would have disappeared. The surface of Venus is now so hot that even sulfur, chlorine, and fluorine have baked out of the rock and formed sulfuric, hydrochloric, and hydrofluoric acid vapors.

Venus's Surface

Given that the surface of Venus is perpetually hidden by clouds, is hot enough to melt lead, and suffers under crushing atmospheric pressure, it is surprising how much planetary scientists know about the geology of Venus. Early radar maps made from Earth penetrated the clouds and showed that it had mountains, plains, and some craters. The Soviet Union launched a number of spacecraft that landed on Venus; although the surface conditions caused the spacecraft to fail within an hour or so of landing, they did analyze some rocks and transmit a few images back to Earth. The rocks seem to be basalt, a typical product of volcanism. The images revealed dark-gray rocky plains bathed in a deep-orange glow caused by sunlight filtering down through the thick atmosphere (Figure 13-2).

Beginning in 1978, a series of U.S. and Soviet probes orbited Venus and made close-up radar maps of the surface. Maps made by the *Magellan* spacecraft between 1992 and 1994 showed details as small as 100 m (330 ft) in diameter. These radar maps provide a comprehensive look below the clouds.

The color of Venus radar maps is mostly arbitrary. *Magellan* scientists chose to use yellows and oranges for their radar maps in an effort to mimic the orange color of daylight caused by the thick

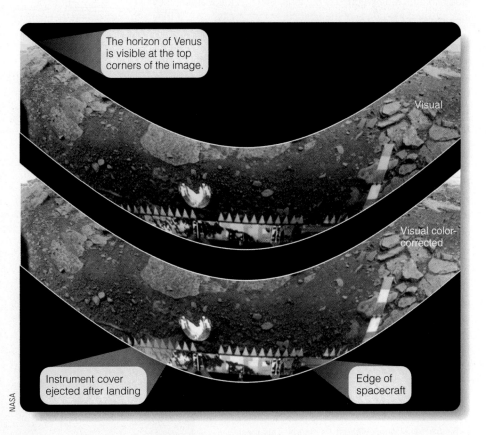

NASA

The horizon of Venus is visible at the top corners of the image.

Visual

Visual color-corrected

Instrument cover ejected after landing

Edge of spacecraft

▲ Figure 13-2 The *Venera 13* lander touched down on Venus in 1982 and carried a camera that swiveled from side to side to photograph the surface. The orange glow is produced by the thick atmosphere; when that is corrected to produce the view as it would be under white light, you can see that the rocks are dark gray. Isotopic analysis suggests they are basalts.

How Do We Know? 13-1

Data Manipulation

Why do scientists think it is OK to visually enhance their data? Planetary astronomers studying Venus change the colors of radar maps and stretch the height of mountains. If they were making political TV commercials and were caught digitally enhancing a politician's voice, they would be called dishonest, but scientists often manipulate and enhance their data. It's not dishonest because the scientists are their own audience, and the data are not altered, just presented differently. The goal is to reveal truth rather than conceal it.

Research physiologists studying knee injuries, for instance, can use magnetic resonance imaging (MRI) data to study both healthy and damaged knees. By placing a patient in a powerful magnetic field and irradiating his or her knee with precisely tuned radio frequency pulses, the MRI machine can force one in a million hydrogen atoms to emit radio frequency photons. The intensity and frequency of the emitted photons depend on how the hydrogen atoms are bonded to other atoms, so bone, muscle, and cartilage emit different signals. An antenna in the machine picks up the emitted signals and stores huge masses of data in computer memory as tables of numbers.

The tables of numbers are meaningless in that form to the physiologists, but by manipulating the data they can produce images that reveal the anatomy of a knee. By enhancing the data, they can distinguish between bone and cartilage and see how tendons are attached. They can filter the data to see fine detail or smooth the data to eliminate distracting textures. Because the physiologists are their own audience, they know how they have manipulated the data and can use it to devise better ways to treat knee injuries.

When scientists say they are "massaging the data," they mean they are filtering, enhancing, and manipulating it to bring out the features they need to study—but not altering it. If they were presenting that data to a television audience to promote a cause or sell a product, it would be dishonest, but scientists' manipulation of the data allows them to better understand how nature works.

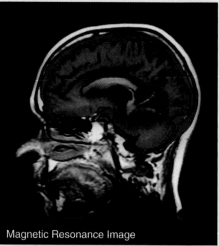

Magnetic Resonance Image

You are accustomed to seeing data manipulated and presented in convenient ways.

Jim Wehtje/Photodisc/Getty Images

---

atmosphere (**Figure 13-3**). When you look at these orange images, you need to remind yourself that the true color of the rock would be dark gray if illuminated by white light (**How Do We Know? 13-1**).

Radar maps do not show how the surface would look to human eyes but rather provide information about altitude, roughness, and, in some cases, chemical composition. For example, if you transmit a radio signal down through the clouds and measure the time until you hear the echo coming back up, you can measure the altitude of the surface. As a result, part of the *Magellan* data is a detailed altitude map. You can also measure the amount of the signal that is reflected from each spot on the surface. Much of the surface of Venus is made up of old, smooth lava flows that do not look bright in radar maps, but faults and uneven terrain look brighter. Young

Radar map

NASA/JPL-Caltech

▶ **Figure 13-3** Venus without its clouds: This mosaic of *Magellan* radar maps has been given an orange color to mimic the coloration of daylight at the surface of the planet. The image shows scattered impact craters and volcanic regions such as Beta Regio and Atla Regio.

▲ **Figure 13-4** Although it is nearly 1000 years old, this lava flow near Flagstaff, Arizona, is still such a rough jumble of sharp rock that it is dangerous to venture onto its surface. Rough surfaces are very good reflectors of radio waves and look bright in radar maps. Solidified lava flows on Venus show up as bright regions in the radar maps because they are rough.

lava flows are generally very rough (**Figure 13-4**), containing billions of tiny crevices that bounce the radar signal around and shoot it back the way it came. Those rough lava flows look very bright in radar maps. Deposits of certain minerals also cause bright radar echoes. All of these radar maps paint a picture of a hot, violent, desert world with recent volcanic activity.

The map in **Figure 13-5** covers all of Venus except the polar regions. By international agreement, the names of celestial bodies and features on celestial bodies are assigned by the IAU, which has decided that all names on Venus should be feminine. Examples are the highland regions Ishtar Terra and Aphrodite Terra, named for the Babylonian and Greek goddesses of love.

There are only a few exceptions, features that were discovered during early Earth-based radar mapping before the naming convention was adopted. These include the mountain Maxwell Montes (named for James Clerk Maxwell, the 19th-century physicist who first described electromagnetic radiation), which is 50 percent higher than Mount Everest, and the volcanic peaks Alpha Regio and Beta Regio.

Radar maps show that most of the surface of Venus consists of low, rolling plains and highland regions. Those rolling plains appear to be large-scale smooth lava flows, whereas the highlands are regions of deformed crust.

Just as in the case of the lunar landscape, craters are the key to figuring out the age of the surface. With nearly 1000 impact craters on its surface, Venus has many more craters than Earth but not nearly as many as the Moon. The craters are uniformly scattered over the surface and look sharp and fresh (**Figure 13-6**). With no water and a thick, slow-moving lower atmosphere, there is little erosion on Venus, and the thick atmosphere protects the surface from small meteorites. Consequently, there are no small craters. Planetary scientists conclude that the surface of Venus is older than Earth's surface but not as old as the Moon's. Unlike the Moon, there are no very old, cratered highlands. Lava flows seem to have completely resurfaced Venus within approximately the past half-billion years.

Volcanism on Venus

Signs of volcanism dominate the surface of Venus. As you just learned, much of Venus is covered by lava flows such as those photographed close-up by the *Venera* landers (Figure 13-2). Also, volcanic peaks and other volcanic features are evident in radar maps.

Comparing volcanism on Venus, Earth, and Mars tells you something about all three worlds. Look through **Volcanoes** on pages 272–273 and notice three important ideas and two new terms:

1 There are two main types of volcanoes found on Earth. *Composite volcanoes* tend to be associated with plate motion and located near the edges of plates, whereas *shield volcanoes* are associated with hot spots caused by columns of magma rising from deep in the mantle and are generally not near the edges of plates.

Radar maps

Lakshmi Planum   Maxwell Montes

ISHTAR TERRA

Atalanta Planitia

Beta Regio

Lowlands are colored blue, and highlands yellow and red.

Atla Regio

APHRODITE TERRA

Phoebe Regio

Alpha Regio

Themis Regio

Artemis Chasma

Maxwell Montes is very rough.

Lakshmi Planum

Maxwell Montes

Colette   Sacajawea

Cleopatra

The top of this volcano is coated with a mineral that is not stable on the warmer lower slopes.

The old lava plains of Lakshmi Planum are smooth and marked by two large volcanic calderas.

◄ **Figure 13-5** Notice how these three radar maps show different things. The main map here shows elevation over most of the surface of Venus. Only the polar areas are not shown. The inset map at left shows an electrical property of surface minerals related to chemical composition. The detailed map of Maxwell Montes and Lakshmi Planum on the right is color coded to show degree of roughness, with purple smooth and orange rough.

Volcanoes

1 Molten rock (magma) is less dense than the surrounding rock and tends to rise. Where it bursts through Earth's crust, you see volcanism. The two main types of volcanoes on Earth provide good examples for comparison with those on Venus and Mars.

On Earth, **composite volcanoes** form above subduction zones where the descending crust melts and the magma rises to the surface. This forms chains of volcanoes along the subduction zone, such as the Andes along the west coast of South America.

Magma rising above subduction zones is not very fluid, and it produces explosive volcanoes with sides as steep as 30°.

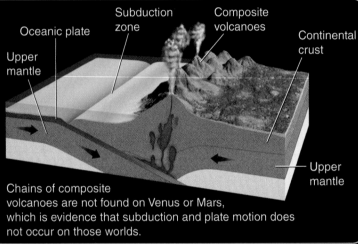

Oceanic plate · Subduction zone · Composite volcanoes · Continental crust · Upper mantle · Upper mantle

Chains of composite volcanoes are not found on Venus or Mars, which is evidence that subduction and plate motion does not occur on those worlds.

Based on *Physical Geology*, 4th edition, James S. Monroe and Reed Wicander, Wadsworth Publishing Company. Used with permission.

Mount St. Helens exploded northward in 1980, killing 63 people and destroying 600 km² (230 mi²) of forest with a blast of winds and suspended rock fragments that moved as fast as 480 km/hr (300 mph) and had temperatures as hot as 350°C (660°F). Note the steep slope of this composite volcano.

Shield volcano · Lava flow

Oceanic crust · Magma chamber

1a A shield volcano is formed by highly fluid lava (basalt) that flows easily and creates low-profile volcanic peaks with slopes of 3 to 10 degrees. The volcanoes of Hawai'i are shield volcanoes that originated over a hot spot in the middle of the Pacific plate.

Magma collects in a chamber in the crust and finds its way to the surface through cracks.

Magma forces its way upward through cracks in the upper mantle and causes small, deep earthquakes.

A hot spot is formed by a rising convection current of magma moving upward through the hot, deformable (plastic) rock of the mantle.

The Cascade Range composite volcanoes are produced by an oceanic plate being subducted below North America and partially melting.

Seattle · Washington · St Helens · Rainier · Portland · Hood · Oregon · Pacific Ocean · Shasta · Nevada · California · Lassen

USGS

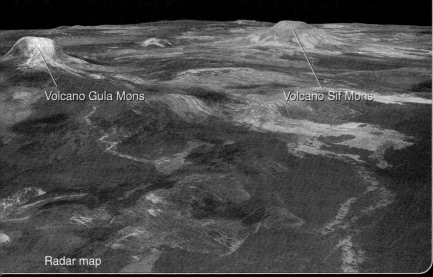

Volcano Gula Mons

Volcano Sif Mons

Radar map

2 Volcanoes on Venus are shield volcanoes. They appear to be steep sided in some images created from Magellan radar maps, but that is because the vertical scale has been exaggerated to enhance detail. Venusian volcanoes are actually shallow-sloped shield volcanoes.

3 Volcanism over a hot spot results in repeated eruptions that build up a shield volcano of many layers. Such volcanoes can grow very large.

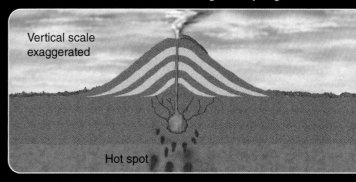

Vertical scale exaggerated

Hot spot

2a This computer model of a mountain with the vertical scale magnified 10 times appears to have steep slopes such as those of a composite volcano.

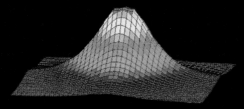

A true profile of the computer model shows the mountain has very shallow slopes typical of shield volcanoes.

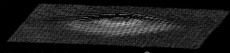

Old volcanic island eroded below sea level

Plate motion

Hot spot

If the crustal plate is moving, magma generated by the hot spot can repeatedly penetrate the crust to build a chain of volcanoes. Only the volcanoes over the hot spot are active. Older volcanoes slowly erode away. Such volcanoes cannot grow large because the moving plate carries them away from the hot spot.

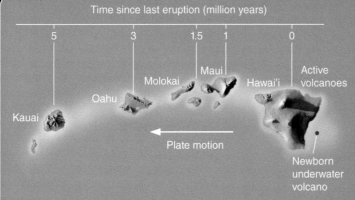

Time since last eruption (million years)

5   3   1.5   1   0

Kauai   Oahu   Molokai   Maui   Hawai'i   Active volcanoes

Plate motion

Newborn underwater volcano

3a The volcanoes that make up the Hawaiian Islands have been produced by a hot spot poking upward through the middle of the moving Pacific plate, as shown at left.

3b The plate moves about 9 cm/yr and carries older volcanic islands northwest, away from the hot spot. The volcanoes are carried away from the hot spot before they can become extremely large. New islands form to the southeast over the stationary hot spot.

Olympus Mons, shown at right, is the largest volcano in the Solar System. It is a shield volcano 25 km (16 mi) high and more than 700 km (430 mi) in diameter at its base. Its vast size is evidence that the crust under the mountain must have remained stationary over the hot spot. This is evidence that Mars has not had plate tectonics.

Olympus Mons on Mars contains 95 times more volume than the largest volcano on Earth, Mauna Loa in Hawai'i.

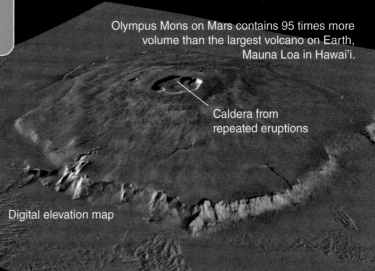

Caldera from repeated eruptions

Digital elevation map

▲ **Figure 13-6** Impact crater Howe in the foreground of this *Magellan* radar image is 37 km (23 mi) in diameter. Craters in the background are 47 km (29 mi) and 63 km (39 mi) in diameter. This radar map has been digitally modified to represent the view as if from a spacecraft flying over the craters.

② Volcanoes on Venus and Mars are all shield volcanoes produced by hot-spot volcanism rather than by plate tectonics.

③ Volcanoes on Venus and Mars have grown very large because of repeated eruptions at the same place in the crust. This is clear evidence that, unlike Earth, neither Venus nor Mars has been dominated by plate tectonics and horizontal crust motions.

The radar image of Sapas Mons in **Figure 13-7** shows a dramatic overhead view of that volcano, which is 400 km (250 mi) in diameter at its base and 1.5 km (about 1 mi) high. Many radar-bright and therefore presumably young lava flows extend outward from the center, covering older, darker flows. Remember that the colors in this image are artificial; if you could walk across these lava flows and shine your spacesuit's white headlight on them, you would find them solid, dark gray stone.

In addition to the volcanoes, radar images reveal other volcanic features on the planet's surface. Lava channels are common, and they appear similar to the sinuous rills visible on Earth's Moon. The longest channel on Venus is also the longest known channel in the Solar System. It stretches 6800 km (4200 mi), roughly twice the distance from Chicago to Los

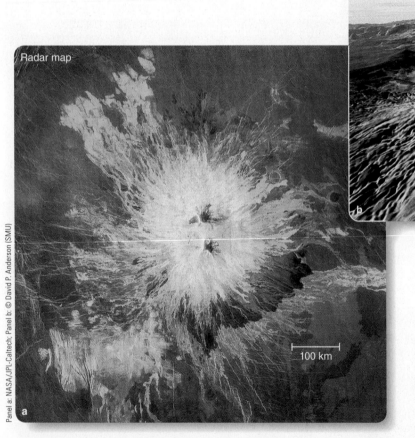

100 km

◀ **Figure 13-7** (a) Volcano Sapas Mons, lying along a major fracture zone, is topped by two lava-filled calderas and flanked by rough lava flows. The orange color of this radar map mimics the orange light that filters through the thick atmosphere. (b) Seen by light typical of Earth's surface, Sapas Mons might look more like this computer-generated landscape. Volcano Maat Mons rises in the background. The vertical scale has been exaggerated by a factor of 15 to reveal the shape of the volcanoes and lava flows.

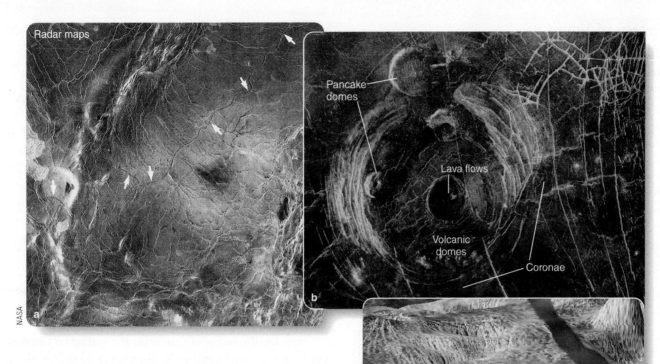

Radar maps

Pancake domes

Lava flows

Volcanic domes

Coronae

NASA

a

b

c

50 km (30 mi)

d

▶ **Figure 13-8** Volcanic features on Venus: (a) Arrows point to a 600-km (370-mi) segment of Baltis Vallis, the longest lava-flow channel in the Solar System. It is at least 6800 km (4200 mi) long. (b) Aine Corona, about 200 km (125 mi) in diameter, is marked by faults, lava flows, small volcanic domes, and pancake domes of solidified lava. (c) This perspective view shows a hill typical of those associated with a corona. Molten lava below the surface appears to have pushed up the hill, causing the network of radial faults. (d) Other regions of the crust appear to have collapsed as subsurface magma drained away.

Angeles. These channels are 1 to 2 km wide and can sometimes be traced back to collapsed areas where lava appears to have drained from beneath the crust.

For further evidence of volcanism on Venus, you can consider features called **coronae**, circular bulges up to 2000 km (1250 mi) in diameter containing volcanic peaks and lava flows. The coronae appear to be caused by rising currents of molten magma below the surface that uplifted a dome of crust and then withdrew, allowing the surface to subside and fracture. Coronae are sometimes accompanied by features called *pancake domes* that are understood to be solidified outpourings of viscous lava. These volcanic features and others are shown in **Figure 13-8**.

There is no reason to suppose that all of the volcanoes found on Venus are extinct, so volcanoes might be erupting on Venus right now. However, there are no confirmed observations of actual eruptions in progress.

The history of Venus apparently has been a fiery tale, dominated by volcanism, but widespread signs of plate tectonics do not appear on the surface of Venus as they do on Earth. For example, the only obvious example of wrinkled mountains

on Venus are in the areas north and west of the volcanic plateau Lakshmi Planum (inset at the right of Figure 13-5) where some type of horizontal crust motion has pushed up against the big land mass named Ishtar Terra (upper left of Figure 13-5). In contrast, plate collisions have produced many wrinkled and folded mountain ranges on Earth. Faults and deep chasms are found in places on the surface of Venus, suggesting that the crust has stretched in those areas. Thus, there are signs of some limited crustal motion on that planet. Nevertheless, Earth's dominant geological process appears mostly missing from Venus. You will find a hypothesis about why the histories of the planets have been so different in the next section.

A History of Venus

Earth passed through four major stages in its history (Chapter 11, especially Figure 11-2), and you have seen how the Moon and Mercury were affected by their own versions of the same stages. Venus, however, seems to have followed a peculiar path through planetary development, and its history is difficult to understand. Planetary scientists are not sure of all the details about how the planet formed and differentiated, how it was cratered and flooded, or how its surface has continued to evolve.

Presumably, Venus and Earth formed in the same way and outgassed atmospheres rich in CO_2 as they differentiated into dense cores and less dense mantles and crusts. Venus and Earth should have outgassed approximately the same amount of CO_2, but Earth's oceans have dissolved that CO_2 and converted it to sediments such as limestone. The main cause of the difference between surface conditions on Earth and Venus is the lack of water on Venus. There is evidence that Venus might have had oceans when it was young, but because Venus is closer to the Sun than Earth, it was initially warmer, so substantial amounts of water would have evaporated, increasing the greenhouse effect and making the planet even warmer. That process would have been a runaway greenhouse, a vicious cycle that dried up any oceans that did exist and thereby severely reduced the ability of the planet to clear its atmosphere of CO_2. As more CO_2 was outgassed, the greenhouse effect grew even more severe. Eventually, solar UV radiation destroyed the atmospheric water vapor, leaving an overabundance of deuterium as a fossil clue to the fate of Venus's oceans.

In comparison, Earth avoided a runaway greenhouse effect because it was farther from the Sun and always cooler than Venus. Consequently, it could form and preserve liquid-water oceans, which absorbed the CO_2 and left an atmosphere of mostly nitrogen that was relatively transparent at infrared wavelengths. As you learned previously, if all of the carbon in Earth's sediments was put back into the atmosphere as CO_2, our air would be as dense as that of Venus, and Earth would suffer from a tremendous greenhouse effect, much worse than anything humanity would be able to cause (Chapter 11, pages 237–238).

True plate tectonics are apparently not important on Venus. Although measurements by lander probes show that the surface rock on Venus is the same kind of dark-gray basalt found in Earth ocean crust, they also reveal that the crust is very dry and therefore about 12 percent less dense than Earth's crust. Venus's low-density crust is more buoyant than Earth's crust and would resist being pushed into the interior. Also, model calculations indicate that water embedded in rocks helps lubricate plate motion, so Venus's dry crustal rocks would not slide past each other easily. Finally, Venus's crust is so hot that it is halfway to its melting point. Such hot rock is not very stiff, so it cannot form the rigid plates typical of plate tectonics on Earth. Planetary scientists hypothesize that the low density, dryness,

and pliability of Venus's crustal rocks are the reasons that planet lacks plate tectonics.

Fully 70 percent of the heat from Earth's interior flows outward through volcanism along mid-ocean ridges, the places where crustal plates are spreading apart, but Venus lacks tectonic crustal rifts. The planet's numerous volcanoes are insufficient to carry most of the heat out of the interior. Rather, Venus seems to get rid of its interior heat through large convection currents of hot magma that rise beneath the crust. Those currents evidently deform the surface to create coronae and related features. Also, detailed maps of gravity strength made by orbiting spacecraft show that Maxwell Montes and other mountains must be supported by rising currents of magma rather than having deep roots like Earth's mountains. Thus, although features such as the folded mountains around Ishtar Terra provide evidence of limited horizontal crustal motion, most of Venus's tectonic processes seem to be vertical rather than horizontal.

The small number of craters on the surface of Venus indicates that the entire crust has been replaced within the past half-billion years or so, which is only 10 percent of the age of Venus. This may have occurred in a planetwide overturning as the old crust broke up and sank and lava flows created a new crust. Such global catastrophes could happen periodically on Venus, or the planet may have had geological processes more like Earth's until a single resurfacing geologically recently. In any case, studying un-Earthly Venus may eventually reveal more about how our own world works.

DOING SCIENCE

What evidence can you point to that Venus does not have plate tectonics? A planetary scientist would focus immediately on this as one of the most significant differences between Venus and Earth.

On Earth, plate tectonics is identifiable by the worldwide network of faults, subduction zones, volcanism, and folded mountain chains that outline the plates. Although some of these features are visible on Venus, they do not occur in a planetwide network of plate boundaries. Volcanism is widespread, but folded mountain ranges occur in only a few places, such as near Lakshmi Planum and Maxwell Montes; unlike on Earth, they do not make up long mountain chains. Also, the large size of the shield volcanoes on Venus shows that the crust is not moving over the hot spots in the way the Pacific seafloor is moving over the Hawaiian hot spot.

At first glance, you might think that Earth and Venus should be as similar as siblings, but comparative planetology reveals that they are more like cousins. You can blame the thick atmosphere of Venus for altering its geology, but that calls for focusing on another big difference between the planets: ***Why isn't Earth's atmosphere as thick as Venus's?***

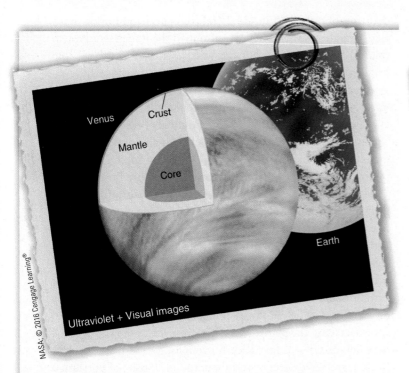

NASA: © 2016 Cengage Learning®

Venus is only 5 percent smaller than Earth, but its atmosphere is perpetually cloudy, and its surface is hot enough to melt lead. It may have a metal core about the size of Earth's.

Celestial Profile 5 Venus

Motion:

Average distance from the Sun	0.723 AU (1.08×10^8 km)
Eccentricity of orbit	0.007
Inclination of orbit to ecliptic	3.4°
Orbital period	0.6152 y (224.7 d)
Period of rotation (sidereal)	243.0 d
Period of rotation (solar)	116.8 d
Inclination of equator to orbit	177.3° (retrograde rotation)

Characteristics:

Equatorial diameter	1.21×10^4 km (0.945 $D_\oplus$)
Mass	4.87×10^{24} kg (0.815 $M_\oplus$)
Average density	5.20 g/cm³ (4.2 g/cm³ uncompressed)
Surface gravity	0.90 Earth gravity
Escape velocity	10.4 km/s (0.93 $V_\oplus$)
Surface temperature	+470C° (+880°F)
Albedo (cloud tops)	0.90
Oblateness	0

Personality Point:

Venus is named for the Roman goddess of love, perhaps because the planet often shines so beautifully in the evening or dawn sky. In contrast, the ancient Maya identified Venus as their war god Kukulkan and sacrificed human victims to the planet when it rose in the dawn sky.

NASA: © 2016 Cengage Learning®

Mars is only half the diameter of Earth and probably retains some internal heat, but the size and composition of its core are not well known.

Celestial Profile 6 Mars

Motion:

Average distance from the Sun	1.52 AU (2.28×10^8 km)
Eccentricity of orbit	0.093
Inclination of orbit to ecliptic	1.8°
Orbital period	1.881 y (687.0 d)
Period of sidereal rotation	24.62 h
Inclination of equator to orbit	25.2°

Characteristics:

Equatorial diameter	6.79×10^3 km (0.533 $D_\oplus$)
Mass	6.42×10^{23} kg (0.107 $M_\oplus$)
Average density	3.93 g/cm³ (3.70 g/cm³ uncompressed)
Surface gravity	0.38 Earth gravity
Escape velocity	5.0 km/s (0.45 $V_\oplus$)
Surface temperature	−140°C to + 15°C (−220° to + 60°F)
Average albedo	0.25
Oblateness	0.009

Personality Point:

Mars is named for the god of war. Minerva was the goddess of defensive war, but Bullfinch's *Mythology* refers to Mars's "savage love of violence and bloodshed." You can see how the planet glows blood red in the evening sky because of iron oxides in its soil.

13-2 Mars

Mercury and the Moon are small. Venus and Earth are the largest of the Terrestrial planets. Mars has an intermediate size. It has twice the diameter of the Moon but only a little more than half of Earth's diameter (■ Celestial Profile 6, page 277). Mars's small size has allowed it to cool faster than Earth, and much of its atmosphere has leaked away. Its present CO_2 atmosphere is a bit less than 1 percent as dense as Earth's.

No Canals on Mars

Long before the space age, the planet Mars was a mysterious landscape in the public mind. In the century following Galileo's first astronomical use of the telescope, astronomers discovered dark markings on Mars as well as bright polar caps. By timing the motions of the markings, they concluded that a Martian day was about 24 hours 40 minutes long, only slightly longer than Earth's day. Mars's axis is tipped 25.2 degrees to its orbit, almost exactly the same as Earth's 23.4-degree tilt, so Mars has seasons with about the same winter–summer contrast as on Earth. Mars's year is about 1.88 Earth years long. These similarities with Earth encouraged the belief that Mars might, like Earth, be inhabited.

In 1858, astronomer Angelo Secchi referred to a region he had viewed on Mars as a *canale*, the Italian word for "channel," a narrow body of water that is a natural geological feature. Two decades later, Giovanni Schiaparelli, using a telescope with a diameter of only 8.75 in (22.2 cm), thought he glimpsed many fine, straight lines on Mars. He also used the Italian word *canali* (plural) for these lines, but the word was then translated into English not as "channel," but as "canal," an artificially constructed channel. Thus, the "canals of Mars" were born from a mistranslation. Many astronomers could not see the canals at all, but others drew maps showing hundreds (Figure 13-9a).

In the decades that followed Schiaparelli's announcement, excitement about the possibility of intelligent life on Mars was promoted especially by Percival Lowell, a wealthy Bostonian. In 1894 Lowell founded Lowell Observatory in Flagstaff, Arizona, principally for the study of Mars. He not only mapped hundreds of canals but also popularized his results in books and lectures. Although some astronomers continued to say that the canals were merely illusions, by 1907 the general public was so sure life existed on Mars that *The Wall Street Journal* suggested the most extraordinary event of the previous year had been "the proof by astronomical observations … that conscious, intelligent human life exists upon the planet Mars." Adding fuel to the fire over the next few decades, Edgar Rice Burroughs, author of the Tarzan stories, wrote a series of novels about the adventures of Earthman John Carter lost on Mars. (Note that Burroughs decided to portray Martians as small, with green skin.)

People became so familiar with the idea of intelligent life on Mars that they were ready to believe that Earth could be invaded. When a radio announcer repeatedly interrupted a dance music program on Halloween night in 1938 to report the

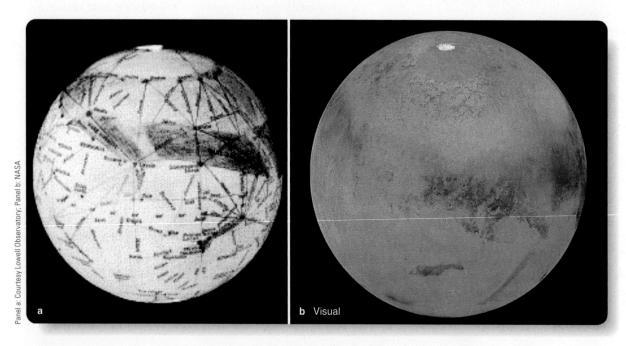

Panel a: Courtesy Lowell Observatory; Panel b: NASA

a

b Visual

▲ **Figure 13-9** (a) Early in the 20th century, Percival Lowell mapped canals over the face of Mars and concluded that intelligent life must reside there. (b) Modern images recorded by spacecraft reveal a globe of Mars with no canals. Instead, the planet is marked by craters and, in some places, volcanoes. Both of these images are reproduced with south at the top, as they appear in telescopes. Lowell's globe is inclined more nearly vertically and is rotated slightly to the right compared with the modern globe.

landing of a spaceship in New Jersey, the emergence of monstrous creatures, and the subsequent destruction of entire cities, thousands of otherwise sensible people fled in panic, not knowing that Orson Welles and other actors were dramatizing H. G. Wells's book *The War of the Worlds*.

Public fascination with Mars, its canals, and its little green men lasted right up until 1965, when *Mariner 4*, the first spacecraft to fly past Mars, radioed back photos of a dry, cratered surface and showed that there are no canals and no Martian civilization. The canals are optical illusions produced by the human brain's powerful ability to assemble a field of disconnected marks into a coherent image. If your brain could not do this, the photo on this page would be nothing but swarms of dots, and the images on a TV screen would never make sense. The downside of this is that the brain of an astronomer, looking for something at the edge of visibility, is capable of connecting faint, random markings on Mars into the straight lines of canals, and the brain of another astronomer might also see canals in the same places.

Even today, Mars holds some fascination for the general public. Grocery store tabloids regularly run stories about a giant face carved on Mars by an ancient race. Although planetary scientists recognize that as nothing more than chance shadows in a photograph and dismiss the issue as a silly hoax, the stories persist. A hundred years of speculation have raised high expectations for Mars.

Mars's Interior

Observations made from orbit show that Mars has no overall magnetic field, but it does have traces of magnetism frozen into some sections of old crust. This probably means that soon after Mars formed it had a liquid metallic core in which the dynamo effect generated a magnetic field strong enough to magnetize surface rocks. Evidence that, like Earth, Mars is differentiated comes from exquisitely sensitive Doppler-shift measurements of radio signals from orbiting spacecraft. Detecting those shifts allowed planetary scientists to map the gravitational field and study the shape of Mars in such detail that tides in the body of Mars caused by the Sun's gravity can be detected. Those tides are less than a centimeter high, but by comparing them with models of the interior of Mars, it can be shown that Mars has a very dense core, a less dense mantle, and a low-density crust.

Because Mars is small, it lost its heat rapidly, and most of its core gradually froze solid. That may be why the dynamo finally shut down. Today Mars probably has a large solid core surrounded by a shell of liquid metal that is too thin for the dynamo effect to be able to generate a magnetic field.

Mars's Atmosphere

For a planetary scientist comparing Mars to other planets, the atmosphere of Mars is of major interest. The gases that cloak a planet are critical to understanding that planet's history.

The air on Mars is 95 percent CO_2, with a few percent each of nitrogen and argon, quite similar to the composition of Venus's atmosphere. The reddish color of the Martian soil is caused by oxides (rusts), meaning that the oxygen humans would prefer to find in the atmosphere is locked in chemical compounds in the soil. The Martian atmospheric density at the surface of the planet is only about 1 percent that of Earth's atmosphere. This does not provide enough pressure to prevent liquid water from boiling into vapor. Water can exist at the Martian surface only as ice or vapor.

Although the air is thin, it is dense enough to be visible in photographs (Figure 13-10). Haze and clouds come and go, and occasional weather patterns are visible. Winds on Mars can be strong enough to produce dust storms that envelop the entire planet. The polar caps visible in photos are also related to the Martian atmosphere. The ices in the polar caps are frozen CO_2 ("dry ice") with frozen water underneath.

To understand Mars more fully, you can ask why its atmosphere is so thin and dry and why the surface is rich in oxides. To find the answers, you need to consider the atmosphere's origin and evolution.

Presumably, the gases in Mars's atmosphere were mostly outgassed from its interior. Volcanism on Terrestrial planets typically releases CO_2 and water vapor plus smaller amounts of other gases. Because Mars formed farther from the Sun, you might expect that it would have incorporated more volatiles than Earth when it formed. On the other hand, Mars is smaller than Earth, so it has had less internal heat to drive geological activity, and that would lead you to suspect that it has not outgassed as much as Earth. In either case, whatever outgassing

NASA/ESA/STScI/AURA/NSF/P. James (Univ. of Toledo, Ohio) and S. Lee (Univ. of Colorado–Boulder)

Visual

▲ Figure 13-10 The atmosphere of Mars is evident in this image made by the *Hubble Space Telescope*. The haze is made up of high, water-ice crystals in the thin CO_2 atmosphere. The spot at extreme left is the volcano Ascraeus Mons, 15 km (10 mi) high, poking up through the morning clouds. Note Mars's north polar cap at the top of the image.

took place occurred early in the planet's history. Mars is small, so it cooled rapidly, became nearly inactive geologically, and now releases little gas.

The ability of a planet to hold onto an atmosphere depends on the planet's mass and temperature. The more massive the planet, the higher its escape velocity (look back to Chapter 5), and the more difficult it is for gas atoms to leak into space. The temperature of a planet's atmosphere is also important. If a gas is hot, its molecules have a higher average velocity and are more likely to exceed escape velocity. That means a planet near the Sun is less likely to retain an atmosphere than a more distant, cooler planet of the same size. The velocity of a gas molecule, however, also depends on the mass of the molecule. On average, a low-mass molecule travels faster than a massive molecule. For that reason, a planet loses its lowest-mass gases more easily because those molecules travel fastest.

You can see this principle of comparative planetology if you plot a diagram such as that in **Figure 13-11**. The data points show the escape velocity versus temperature for the larger objects in our Solar System. The temperature used in the diagram is the temperature of the gas in a position to escape. For the Moon, which has essentially no atmosphere, that is the temperature of the sunlit surface. For Mars, the important temperature is that at the top of the atmosphere.

The curves in Figure 13-11 show the velocities of the fastest-traveling subsets of various molecules as a function of temperature. At any given temperature, some CO_2 molecules, for example, travel faster than others, and it is the highest-velocity molecules that escape from a planet. The diagram shows that Earth and Venus can't hold hydrogen; Mars, even though it is colder, is so small it can hold only the more massive molecules. Earth's Moon is too small to keep any gases from leaking away. You can refer back to this diagram when you study the atmospheres of other worlds in later chapters.

Over the 4.6 billion years since Mars formed, it has lost some of its lower-mass gases. Water molecules are massive enough that Mars should have been able to keep them, but solar ultraviolet radiation can break them up. Recall that on Earth the

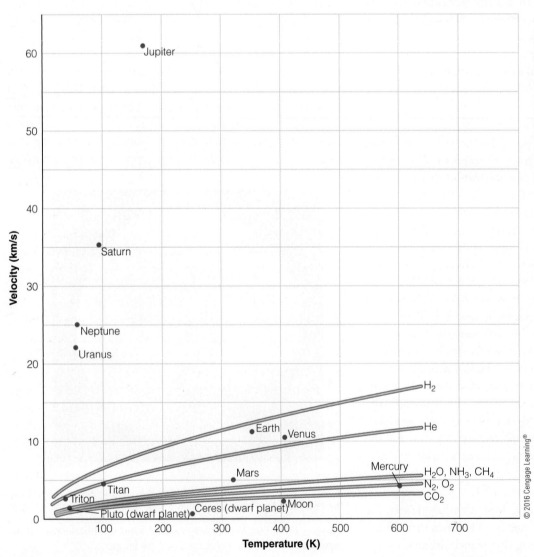

◄ **Figure 13-11** Loss of planetary atmosphere gases. Dots represent the escape velocity and temperature of various Solar System bodies. The lines represent the typical highest velocities of molecules of various masses. The Jovian planets have high escape velocities and can hold onto even the lowest-mass molecules. Mars can hold only the more massive molecules, and the Moon has such a low escape velocity that even massive molecules can escape.

© 2016 Cengage Learning®

ozone layer protects water vapor from UV radiation, but Mars never had an oxygen-rich atmosphere, so it never had an ozone layer. UV photons from the Sun can penetrate deep into the atmosphere and break up water molecules. Then, the hydrogen escapes, and the oxygen, a very reactive element, forms oxides in the soil. This is the explanation for the oxides that make Mars the "Red Planet."

The argon in the Martian atmosphere is evidence that there once was a denser blanket of air. Argon atoms are massive, almost as massive as a CO_2 molecule, and would not be lost easily. Also, argon is inert and cannot form compounds in the soil. The 1.6 percent argon in Mars's atmosphere, almost twice the fraction in Earth's, is understood to be remnant of an ancient Martian atmosphere that may have been 10 to 100 times denser than currently, perhaps nearly as dense as Earth's atmosphere.

Planetary scientists debate how important the solar wind has been in the evolution of Mars's atmosphere. Mars now has no planetwide magnetic field, so the solar wind interacts directly with the upper atmosphere. Detailed calculations show that significant amounts of CO_2 could have been carried away by the solar wind over the history of the planet. However, Mars might have had a magnetic field when it was younger with significant internal heat and a liquid metal core. A magnetic field would have protected its atmosphere from the solar wind. More observations are needed to determine how long that protection might have lasted.

There is abundant evidence that Mars's polar caps and atmosphere are intimately connected. The polar caps contain large amounts of CO_2 ice; as spring comes to a hemisphere, that ice begins to vaporize and returns to the atmosphere. Meanwhile, at the other pole, CO_2 is freezing out and adding to the polar cap there. Dramatic evidence of this cycle appeared when the camera aboard the *Mars Odyssey* probe sent back images of dark markings on the south polar cap. Evidently, as spring comes to the polar cap and the Sun begins to peek above the horizon, sunlight penetrates the meter-thick ice and vaporizes CO_2, which bursts out in geysers tens of meters high carrying dust and sand. Local winds push the debris downwind to form the fan-shaped markings (**Figure 13-12**).

Although planetary scientists remain uncertain as to how much of an atmosphere Mars had in its past and how much it has lost, it is a good example for your study of comparative planetology. When you look at Mars, you see what can happen to the atmosphere of a medium-size world. And, like its atmosphere, the geology of Mars is probably typical of those worlds.

Mars's Surface

If you ever decide to visit another world, Mars may be your best choice. As you've already learned, it is a cold, reddish desert with very thin air (**Figure 13-13**), but even so Mars's surface is much more Earth-like than the Moon, Mercury, or Venus. Mars has

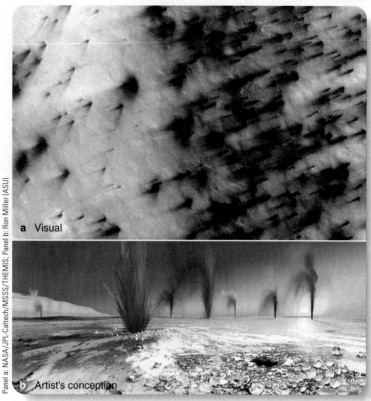

Panel a: NASA/JPL-Caltech/MSSS/THEMIS; Panel b: Ron Miller (ASU)

a Visual

b Artist's conception

▲ Figure 13-12 (a) Each spring, spots and fans appear on the ice of the south polar cap on Mars. (b) Studies show the ice is frozen CO_2 in a nearly clear layer about a meter thick. High-pressure CO_2 gas vaporized by spring sunlight bursts out of the ice in geysers. The gas carries sand and dust hundreds of meters into the air.

weather, complex geology, and signs that water once flowed there. You might even hope to find traces of ancient life hidden in the rocks.

Spacecraft have been visiting Mars for almost 50 years. A small armada of spacecraft has gone into orbit around Mars to photograph and analyze its surface. The first successful landings were made in 1976 by *Viking 1* and *2*. The *Phoenix* probe landed in the north polar region of Mars in 2008. Rovers *Spirit* and *Opportunity* landed in 2004 and *Curiosity* in 2012, carrying sophisticated instruments to explore the rocky surface. Rovers have an advantage because they are wheeled robots that can be controlled from Earth and directed to travel from place to place, making detailed measurements.

Photographs made by rovers and landers on the surface of Mars such as Figure 13-13 and the image on page 266 that opens this chapter, show stretches of broken and oxidized rock. These appear to be rocky plains fractured by meteorite impacts, but they don't look much like the meteorite-blasted surface of Earth's Moon. The atmosphere of Mars, thin though it is, protects the surface from the constant rain of micrometeorites that grinds Moon rocks to dust. Also, Martian dust storms may sweep fine dust away from some areas, leaving larger rocks exposed.

◀ Figure 13-13 This image taken by the Mast Camera on the *Curiosity* rover highlights the interesting geology of Mount Sharp, located inside Gale Crater where the rover landed. Prior to the rover's landing, observations from orbiting satellites indicated that the lower reaches of the mountain are rock layers containing water-bearing minerals.

Spacecraft orbiting Mars have imaged the surface and measured elevations to reveal that all of Mars is divided into two parts. The southern highlands are heavily cratered, and the number of craters shows that they must be old. In contrast, the northern lowlands are smooth (Figure 13-14) and so remarkably free of craters that they must have been resurfaced no more than a billion years ago. Some astronomers suggested that volcanic floods filled the northern lowlands and buried the craters there, or that the lowlands are actually a giant impact basin. However, consensus is growing that the northern lowlands were once filled with an ocean of liquid water that would have been about the size of the Mediterranean

Sea. This is an exciting hypothesis and will be mentioned again later in this chapter when you consider the history of water on Mars.

The cratering and volcanism on Mars fit with what you already know of comparative planetology. Mars is larger than Earth's Moon, so it cooled more slowly, and its volcanism has continued longer. But Mars is smaller than Earth and less geologically active, so some of its ancient cratered terrain has survived undamaged by volcanism or plate tectonics.

Martian volcanoes are shield volcanoes with shallow slopes, showing that the lava flowed easily. As you learned from the examples of Earth and Venus, shield volcanoes occur over hot

▲ Figure 13-14 These globes of Mars are color coded to show elevation. The northern lowlands lie about 4 km (2.5 mi) below the southern highlands. Volcanoes are very high (white), and the giant impact basins, Hellas and Argyre, are low. Note the depth of the canyon Valles Marineris. The two *Viking* spacecraft landed on Mars in 1976, *Pathfinder* in 1997, rovers *Spirit* and *Opportunity* in 2004, and rover *Curiosity* in 2012.

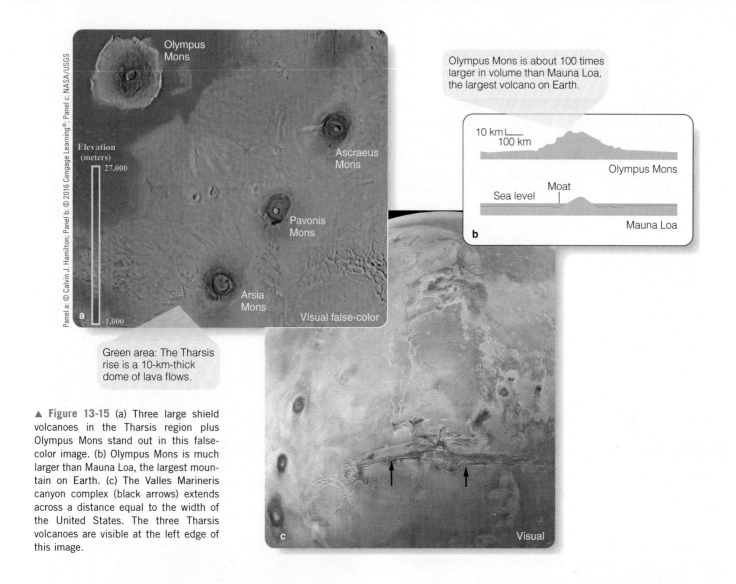

Olympus Mons

Ascraeus Mons

Pavonis Mons

Arsia Mons

Elevation (meters)

27,000

-1,000

a

Visual false-color

Panel a: © Calvin J. Hamilton; Panel b: © 2016 Cengage Learning®; Panel c: NASA/USGS

Green area: The Tharsis rise is a 10-km-thick dome of lava flows.

Olympus Mons is about 100 times larger in volume than Mauna Loa, the largest volcano on Earth.

10 km

100 km

Olympus Mons

Sea level

Moat

Mauna Loa

b

c

Visual

▲ **Figure 13-15** (a) Three large shield volcanoes in the Tharsis region plus Olympus Mons stand out in this false-color image. (b) Olympus Mons is much larger than Mauna Loa, the largest mountain on Earth. (c) The Valles Marineris canyon complex (black arrows) extends across a distance equal to the width of the United States. The three Tharsis volcanoes are visible at the left edge of this image.

spots of rising magma below the crust and are not related to plate tectonics. The largest volcano in the Solar System is Mars's Olympus Mons (**Figure 13-15**). Figure 13-15b compares Olympus Mons to the most massive mountain on Earth, the shield volcano Mauna Loa in Hawai'i. Mauna Loa is so heavy it has sunk into Earth's crust to form an undersea depression like a moat around a castle. Olympus Mons is 100 times larger than Mauna Loa but has not sunk into the crust of Mars. This shows that the crust of Mars is much thicker than the crust of Earth. Olympus Mons and other Martian volcanoes are relatively free of impact craters, indicating that volcanic activity continued on the planet long past the end of the heavy bombardment period, the age estimated for the southern highlands.

Other evidence indicates that the Martian surface has been much more active than the surfaces of the Moon and Mercury. Valles Marineris is a network of canyons 4000 km (2500 mi) long, enough to stretch from New York to Los Angeles, and up to 600 km (370 mi) wide (Figure 13-15c). At its deepest, it is four times deeper than the Grand Canyon on Earth. Analysis of

images from orbiting spacecraft indicate that the canyon was produced originally by faults that allowed great blocks of crust to sink, and it was enlarged and modified by later landslides and erosion. Crater counts show that Valles Marineris, like Olympus Mons, is the product of Martian geological activity more recent than the heavy bombardment era.

The faults that created Valles Marineris seem to be linked at its western end to a great volcanic bulge in the crust of Mars called the Tharsis rise. Nearly as large as the continental United States, the Tharsis rise extends 10 km (6 mi) above the mean radius of Mars. Tharsis is home to many smaller volcanoes, but on its summit lie three giants, and just off of its northwest edge lies huge Olympus Mons (Figure 13-15a). The origin of the Tharsis rise is not well understood, but it appears that magma rising from below has pushed up the crust and broken through repeatedly to build a giant bulge of volcanic deposits. This bulge is large enough to have modified the climate and seasons on Mars and may be critical in understanding the history of the planet.

A similar uplifted volcanic bulge—the Elysium region, visible in Figure 13-14—lies halfway around the planet. It appears to be similar to the Tharsis rise, but it is more heavily cratered and so must be older.

The vast sizes of features like the Tharsis rise and Olympus Mons show that the crust of Mars is not broken into mobile plates. If a plate were moving over a hot spot, the rising magma would produce a long chain of shield volcanoes and not a single large peak. On Earth, the hot spot that created the Hawaiian Islands has punched through the moving Pacific plate repeatedly to produce the Hawaiian-Emperor island chain extending 7500 km (4700 mi) northwest across the Pacific seafloor (Chapter 11, page 234). No such chains of volcanoes are evident on Mars, so you can conclude that the crust is not moving horizontally.

No spacecraft has ever photographed an erupting volcano on Mars, but it is possible that some of the volcanoes are still active. Lack of impact craters in the youngest lava flows in the Tharsis and Elysium regions shows that some of the volcanoes may have been active as recently as a few million years ago, which, geologically speaking, is only yesterday. Mars may still retain enough heat to trigger an eruption, but the interval between eruptions could be very long.

Finding the Water on Mars

The quest for water on Mars is interesting because water has been deeply involved in the histories of both Earth and Mars. The quest for water on Mars is exciting because biologists conclude that life on Earth depends on water. As you will learn in this section, Mars evidently had liquid water on its surface for at least as long as it took life to arise on Earth. If conditions under which life started on Earth existed on Mars at the same time, then Mars provides a real opportunity for scientists to check their hypotheses about how life began on Earth.

The two *Viking* spacecraft reached orbit around Mars in 1976 and photographed compelling hints that water once flowed over the surface. As you have learned, liquid water cannot exist on the Martian surface now because it would boil away in the extremely low atmospheric pressure, so the *Viking* photos indicating water once flowed on Mars meant that surface conditions there must have been quite different long ago. More recent missions to Mars such as *Mars Odyssey* (which reached Mars in 2001), *Mars Express* (2003), and *Mars Reconnaissance Orbiter* (2006) have identified numerous additional features related to water.

Several types of formations hint at water flowing over the Martian surface. **Outflow channels** appear to have been cut by massive floods carrying as much as 10,000 times the volume of water flowing down the Mississippi River (Figure 13-16a). Perhaps in just a matter of hours or days, such floods swept away landscape features and eroded deep channels. The number of craters formed on top of the outflow channels show that they are

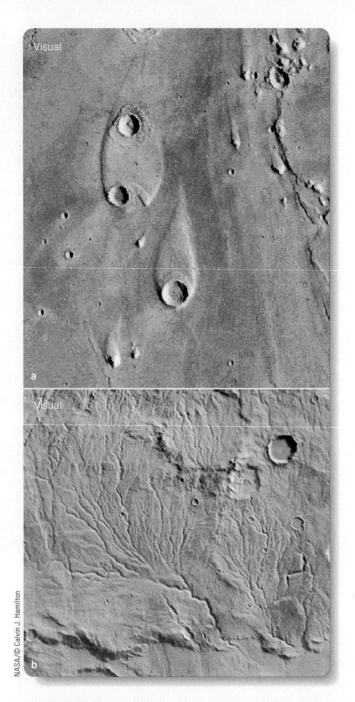

▲ Figure 13-16 These images made by the *Viking* orbiters show some of the features that suggest liquid water on Mars. (a) Outflow channels are broad and shallow and deflect around obstructions such as craters. They appear to have been produced by sudden floods. (b) Valley networks resemble drainage patterns and suggest water flowing over long periods. Crater counts show that both formations are old, but valley networks are older than outflow channels.

billions of years old. The **valley networks** look like meandering riverbeds that probably formed over long periods (Figure 13-16b). The valley networks are also located in the old, cratered southern hemisphere, so they must be very old as well. There are other signs that there has been water on the Martian surface, some of which may be geologically recent (Figure 13-17).

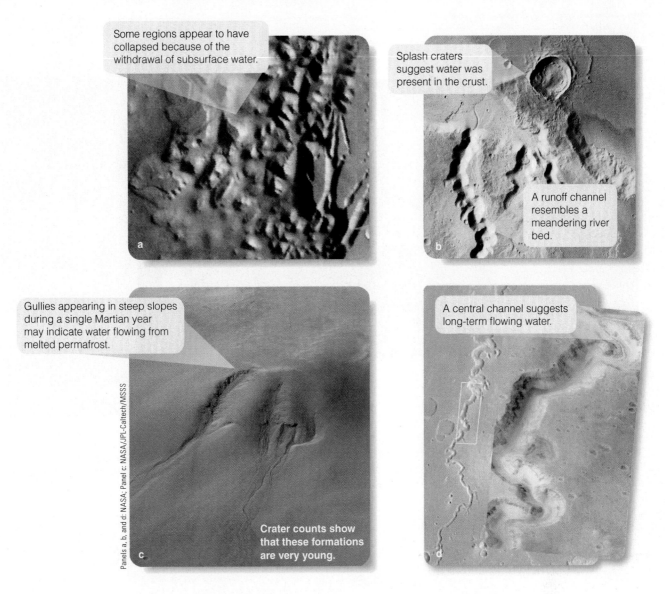

Some regions appear to have collapsed because of the withdrawal of subsurface water.

Splash craters suggest water was present in the crust.

A runoff channel resembles a meandering river bed.

Gullies appearing in steep slopes during a single Martian year may indicate water flowing from melted permafrost.

A central channel suggests long-term flowing water.

Crater counts show that these formations are very young.

Panels a, b, and d: NASA; Panel c: NASA/JPL-Caltech/MSSS

▲ **Figure 13-17** These visual-wavelength images made by *Viking* and *Mars Global Surveyor* orbiters show more features that suggest liquid water once flowed on the surface of Mars. Runoff channels are billions of years old, but some of the gullies have appeared since the first Earth-built spacecraft arrived.

Many flow features lead into the northern lowlands, and the smooth terrain there has been interpreted as ancient ocean floor. Features along the edges of the lowlands have been compared to shorelines, and many planetary scientists conclude that the northern lowlands contained an ocean when Mars was younger. Large, generally circular depressions such as Hellas and Argyre that appear to be impact basins might also have been flooded by water.

Mars Reconnaissance Orbiter photographed the eroded remains of a river delta in an unnamed crater in the ancient highlands. Detailed examination of features in the images indicate that the river flowed for long periods of time, shifting its channel to form meanders and braided channels as rivers on Earth do. The shape of the delta suggests it formed when the river flowed into deeper water and dropped its sediment, much as the Mississippi drops its sediment and builds its delta in the Gulf of Mexico.

Did Mars once have that enough water to fill a sea? Deuterium is 5.5 times more abundant than normal (light) hydrogen in the Martian atmosphere, suggesting that Mars once had about 20 times more water than it has now. Presumably, much of the water was broken up by solar UV and the normal hydrogen mostly lost to space, like the process hypothesized to have destroyed Venus's water.

The remaining water on Mars could survive if it is frozen in the crust. High-resolution images and measurements made from orbit reveal features that suggest subsurface ice. Some regions of collapsed terrain appear to be places where subsurface water has

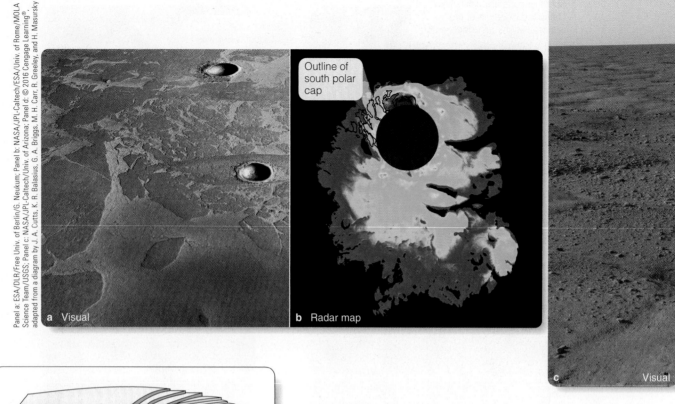

Panel a: ESA/DLR/Free Univ. of Berlin/G. Neukum; Panel b: NASA/JPL-Caltech/ESA/Univ. of Rome/MOLA Science Team/USGS; Panel c: NASA/JPL-Caltech/Univ. of Arizona; Panel d: © 2016 Cengage Learning®, adapted from a diagram by J. A. Cutts, K. R. Balasius, G. A. Briggs, M. H. Carr, R. Greeley, and H. Masursky

a Visual

Outline of south polar cap

b Radar map

c Visual

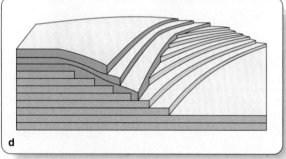

d

▲ **Figure 13-18** (a) Like broken pack ice, these formations near the equator of Mars suggest floating ice that broke up and drifted apart. Scientists propose that the ice was covered by a protective layer of dust and volcanic ash and may still be present. (b) Radar aboard the *Mars Express* satellite probed beneath the surface to image water ice below the south polar cap. The black circle is the area that could not be studied from the satellite's orbit. (c) This view from the *Phoenix* lander shows the landscape of Mars's north polar plain, including polygonal cracks understood to result from seasonal expansion and contraction of ice under the surface. (d) In some regions of the north polar cap, layers of ice plus dust with different orientations are superimposed, suggesting periodic changes in the Martian climate.

drained away (Figure 13-17a). Gullies leading downhill appear to have been eroded recently, judging from their lack of craters, and a few gullies have been seen to appear between one Martian year and the next (Figure 13-17c).

Instruments aboard the *Mars Odyssey* spacecraft detected water frozen in the soil over large areas of the planet. At latitudes farther than 60 degrees from the equator, water ice may make up more than 50 percent of the surface soil. If you added a polar bear, changed the colors, and hid the craters in **Figure 13-18a**, it would look like the broken pack ice on Earth's Arctic Ocean. *Mars Express* photographed these dust-covered formations near the Martian equator, and the shallow depth of the craters suggests that the ice is still there, just below the surface.

Much of the ice on Mars may be hidden below the polar caps. Radar aboard the *Mars Express* orbiter was able to penetrate 3.7 km (2.3 mi) below the surface and map ice deposits hidden below the planet's south polar region (Figure 13-18b). There is enough water there, at least 90 percent pure, to cover

the entire planet to a depth of 11 m. The *Phoenix* probe, which landed in 2008 near Mars's north pole, found water ice mixed with the soil not only as permafrost but also as small chunks of pure ice, indicating that there once was standing water that froze in place.

Rovers *Spirit* and *Opportunity* were targeted to land in areas suspected of having had water on their surfaces. (As of August 2014, *Opportunity* was still roving and communicating with Earth, more than 10 years after it landed.) Both rovers found evidence of past water including small spherical concretions of the mineral hematite (dubbed "blueberries") that require abundant water to form (**Figure 13-19a**). In other rocks, *Opportunity* found layers of sediments with ripple marks and crossed layers showing they were deposited in moving water. Chemical analysis of the rocks at both the *Opportunity* and *Spirit* sites detected sulfates much like Epsom salts plus bromides and chlorides. On Earth, these compounds are left behind when bodies of water such as desert lakes dry up. In its

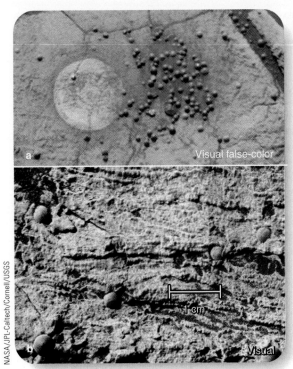

Rover *Spirit's* tire tracks exposed bright sulfate deposits.

NASA/JPL-Caltech/Cornell/USGS

◀ **Figure 13-19** (a) Rover *Opportunity* photographed these hematite concretions ("blueberries") in a rock near its landing site. The spheres appear to have grown as minerals collected around small crystals in the presence of water. Similar concretions are found on Earth. (b) The layers in this rock were deposited as sand and silt in rapidly flowing water. From the way the layers curve and cross each other, geologists can estimate that the water was at least 10 cm deep. A few blueberries and one small pebble are also visible in this image. (c) Rover *Spirit*'s discovery of bright sulfate deposits in soil excavated by its tire tracks confirmed other evidence that a body of salty water once stood in that region and then evaporated, leaving the sulfates behind.

first year at work, the *Curiosity* rover discovered rounded rocks of the sort produced by running streams.

Mars has water, but it is hidden. When humans reach Mars, they will not need to dig far to find water in the form of ice. They can use solar power to break the water into hydrogen and oxygen. Hydrogen is fuel, and oxygen is the breath of life, so the water on Mars may prove to be buried treasure. Even more exciting is the knowledge that Mars definitely once had bodies of liquid water on its surface. It is a desert world now, but someday an astronaut may scramble down an ancient Martian streambed, turn over a rock, and find a fossil. It is even possible that life started on Mars and has managed to persist as the planet gradually became less hospitable. For example, life may have hung on by retreating to limited warm and wet oases underground. You will learn in the final chapter about possible evidence of ancient life in one of the Martian meteorites.

A History of Mars

The history of Mars is like a play where all the exciting stuff happens in the first act. After the first 2 billion years on Mars, most of the activity was over, and it has gone downhill ever since.

As you have already learned, there is evidence that Mars differentiated and once had a liquid metal core capable of producing a planetary magnetic field that has left traces in some surface rocks. The planet no longer has a detectable magnetic field, which probably means the core has solidified and cannot support the dynamo effect.

Planetary scientists divide the history of Mars into three periods. The first—the **Noachian period**—extended from the formation of the crust about 4.3 billion years ago until roughly 3.7 billion years ago. During this time, the crust was battered by the heavy bombardment as the last of the debris in the young Solar System was swept up. The old southern hemisphere survives from this period. The largest impacts blasted out the great basins like Hellas and Argyre very late in the cratering, and there is no trace of magnetic field in those basins. Evidently the dynamo had shut down by then.

The Noachian period included flooding by great lava flows that smoothed some regions. Volcanism in the Tharsis and Elysium regions was very active, and the Tharsis rise grew into a huge bulge on the side of Mars (Figure 13-15). Some of the oldest lava flows on Mars are in the Tharsis rise, but it also contains some of the most recent lava flows. It has evidently been a major volcanic area for most of the planet's history.

The valley networks found in the southern highlands were formed during the Noachian period when water fell as rain or snow and drained down slopes. For that water to remain liquid required a higher temperature and higher atmospheric pressure than is present on Mars today. Violent volcanism could have vented gases, including more water vapor that kept the air pressure high. This may have produced episodes in which water flowed over the surface and collected in the northern lowlands and in the deep basins to form oceans and lakes, but it's not known how long those bodies of water survived. The original planetwide magnetic field may have protected the atmosphere from the solar wind.

Because Mars is small, it lost its internal heat quickly, and atmospheric gases escaped into space. The **Hesperian period** extended from roughly 3.7 billion years ago to about 3 billion years ago. During this time, massive lava flows covered some sections of the surface. Most of the outflow channels date from this period, which suggests that the loss of atmosphere drove Mars to become a deadly cold desert world with its water frozen in the crust. When volcanic heat or large impacts melted subsurface ice, the water could have produced violent floods and shaped the outflow channels.

The history of Mars may hinge partly on climate variations. Models calculated by planetary scientists suggest that Mars may have once had a much steeper rotation axis inclination, as much as 45 degrees. This could have resulted in a generally warmer climate and kept more of the CO_2 from freezing out at the poles. Model calculations indicate that the rise of the Tharsis bulge could have tipped the axis to its present 25 degrees and permanently cooled the climate. Other models suggest that Mars should go through cycles similar to Earth's Milankovitch cycles (Chapter 2, pages 27–29), in which solar heating at different latitudes varies widely as the planet's rotation inclination, axis orientation, and orbit parameters fluctuate. This could cause variations in climate on time scales of thousands to millions of years, much like the ice ages on Earth. Perhaps that is the explanation for the changing orientation of chronologic layers of polar ice and dust deposits (Figure 13-18d).

The third period in the history of Mars—the **Amazonian period**—extended from about 3 billion years ago to the present and has been a period of slow evolution. The planet has lost much of its internal heat, and the core no longer generates a global magnetic field. The crust of Mars is too thick to be active with plate tectonics, and consequently there are no folded mountain ranges on Mars resembling the ones on Earth. The huge size of the Martian volcanoes clearly shows that crustal plates have not moved horizontally on Mars. Volcanism may still occur occasionally on Mars, but the crust has grown too thick for much geological activity beyond slow erosion by wind-borne dust, rare flows of water onto the surface making small gullies, and the occasional meteorite impact.

Planetary scientists cannot tell the story of Mars in complete detail, but it is clear that the size of Mars has influenced both its atmosphere and its geology. Medium-sized Mars has characteristics intermediate between smaller Mercury and Earth's Moon on one hand and larger Venus and Earth on the other.

Comparative Planetology, Once Again

Venus and Mars share at least one characteristic: They have evolved since they formed and are now quite different than they were when the Solar System was young. Moreover, as you have learned, planetary scientists have evidence that surface conditions on Venus, Earth, and Mars were much more similar long ago than they are now. Study **When Good Planets Go Bad** on pages 290–291 and notice four important points:

1 The difference between Venus and Earth is not in the amount of CO_2 they have outgassed but in the amount they have removed from their atmospheres. Being warmer and consequently losing liquid water from its surface sealed Venus's fate as a runaway greenhouse.

2 Venus is highly volcanic with a crust made up of lava flows that have covered over any older crust. The entire planet has been resurfaced within recent geological history.

3 Mars is significantly smaller than Earth, with gravity too weak to prevent much of its atmosphere from leaking away. Evidence shows that Mars once had liquid water on its surface, but much of its water has been lost to space. The low atmospheric pressure means that the remaining water is frozen in the soil or the polar caps.

4 Because Mars is small, its interior cooled relatively quickly and volcanism has died down. The lack of volcanism means the escaping atmosphere is not replenished. The planet's crust was thinner when the planet was young but has grown thick and never broke into moving plates as did Earth's crust.

DOING SCIENCE

Why doesn't Mars have coronae like those on Venus? This question provides another opportunity for a scientist to employ comparative planetology to understand the underlying reasons for similarities and differences among the planets.

The coronae on Venus are caused by rising currents of molten magma in the mantle pushing upward under the crust and then withdrawing to leave circular scars. Earth, Venus, and Mars have had significant amounts of internal heat, and plenty of evidence exists that they have had rising convection currents of magma under their crusts. But you wouldn't expect to see coronae on Earth; its surface is rapidly modified by erosion and plate tectonics. Furthermore, the mantle convection on Earth seems to produce plate tectonics rather than coronae.

In contrast, Mars is a smaller world and must have cooled faster. There is no evidence of plate tectonics on Mars, and giant volcanoes suggest rising plumes of magma erupting up through the crust at the same point over and over. Perhaps there are no coronae on Mars because the crust of Mars rapidly grew too thick to deform easily over a rising plume. On the other hand, perhaps you could think of the entire Tharsis bulge as a single, giant corona.

Of the Terrestrial planets, Mars is the farthest from the Sun. Like a scientist would, imagine the effect of altering one important variable. ***Would Mars's atmosphere have evolved differently if the planet had been much closer to the Sun, for example, in Venus's orbit?***

13-3 Mars's Moons

If you could camp overnight on Mars, you might notice its two small moons, Phobos and Deimos. Phobos, shaped like a flattened loaf of bread measuring 18 by 22 by 27 km in size, would appear less than half as large in angular diameter as Earth's full moon. Deimos, only 13 km in diameter and three times farther from Mars, would look only one-fourteenth the diameter of Earth's Moon.

Both moons are tidally locked to Mars, keeping the same side facing the planet as they orbit. Also, both moons revolve around Mars in the same direction that Mars rotates, but Phobos follows such a close orbit that it revolves faster than Mars rotates. From the surface you would see Phobos rise in the west, drift eastward across the sky, and set in the east 6 hours later.

Origin and Evolution of Phobos and Deimos

Phobos and Deimos are typical of the smaller rocky moons in our Solar System (Figure 13-20). Their albedos are only about 0.07, making them look as dark as coal. They have low densities, less than 2 g/cm³.

Many of the properties of these moons suggest that they are captured asteroids. In the outer parts of the asteroid belt, almost all asteroids are dark, low-density objects like Phobos and Deimos. Massive Jupiter, orbiting just outside the asteroid belt, can scatter such bodies throughout the Solar System, so a number of them might have encountered Mars, the closest Terrestrial planet to the asteroid belt.

However, capturing a passing asteroid into a permanent orbit is not so easy that it can be expected to happen often. An asteroid approaches a planet along a hyperbolic (open) orbit and, if it is unimpeded, swings around the planet and heads back into space. To change the hyperbolic orbit into a closed orbit, the asteroid must somehow be slowed down as it passes so it can be captured. Tidal forces, interactions with other moons, or a grazing collision with a thick planetary atmosphere might do the trick. It is interesting in this context to note that Phobos is so close to Mars that tides currently are making its orbit shrink, and it will fall into Mars or be ripped apart by tidal forces within about 100 million years.

Both Phobos and Deimos have been photographed by passing spacecraft, and those photos show that the satellites are heavily cratered. The heavy battering has broken the satellites

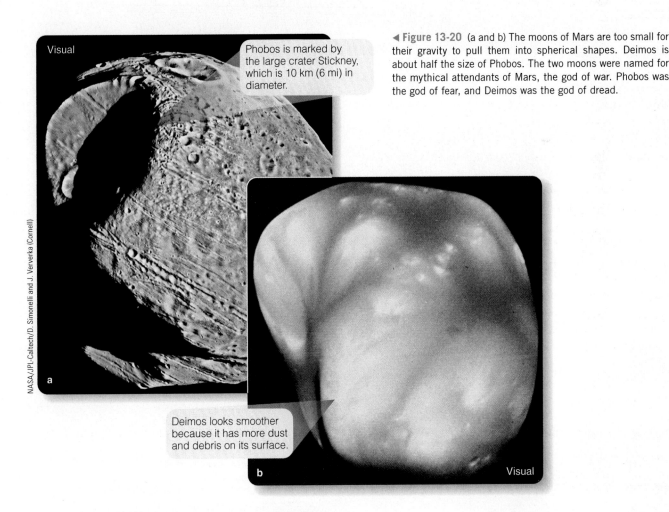

Visual

Phobos is marked by the large crater Stickney, which is 10 km (6 mi) in diameter.

Deimos looks smoother because it has more dust and debris on its surface.

a

b

Visual

NASA/JPL-Caltech/D. Simonelli and J. Ververka (Cornell)

◀ Figure 13-20 (a and b) The moons of Mars are too small for their gravity to pull them into spherical shapes. Deimos is about half the size of Phobos. The two moons were named for the mythical attendants of Mars, the god of war. Phobos was the god of fear, and Deimos was the god of dread.

Venus and Mars haven't really gone wrong, but they have changed since they formed, and those changes can help you understand your own world.

1 Venus and Earth have outgassed about the same amount of CO_2, but Earth's oceans have dissolved most of Earth's CO_2 and converted it to sediments such as limestone. If all of Earth's sedimentary carbon were dug up and converted back to CO_2, our atmosphere would be much like Venus's.

Because Venus was warmer when it formed, it had little if any liquid water to dissolve CO_2, and that produced a greenhouse effect that made the planet even warmer. The planet could not purge its atmosphere of CO_2, and as more was outgassed, Venus was trapped in a runaway greenhouse effect.

NASA

Only 0.7 AU from the Sun, Venus receives almost twice the solar energy per square meter that Earth does. Moved to Venus's orbit, Earth's surface would be 50°C (90°F) hotter.

VENUS

Ultraviolet

2 Lava covers Venus in layers of basalt, and volcanoes are common. Some of those volcanoes may be active right now whereas others are dormant. Traces of past lava flows show up in radar images including long narrow valleys cut by moving lava.

Even its thick atmosphere cannot protect Venus from larger meteorites, yet it has few craters. That must mean the surface is not old.

Baltis Vallis (*arrows*), at least 6800 km (4200 mi) long, is the longest lava flow channel in the solar system.

Radar map

NASA/JPL-Caltech

Radar map

NASA/JPL-Caltech

2a The crust of Venus must be no older than 0.5 billion years. One hypothesis is that an earlier crust broke up and sank in a sea of magma as fresh lava flows formed a new crust. Such resurfacing events might occur periodically on a hot, volcanic planet like Venus.

What could cause a resurfacing event? Models of the climate on Venus show that an outburst of volcanism could increase the greenhouse effect and drive the surface temperature up by as much as 100°C. This could soften the crust, increase the volcanism, and push the planet into a resurfacing episode. This type of catastrophe may happen periodically on Venus, or the planet may have had just one resurfacing event about half a billion years ago.

Even if bodies of water existed when Venus was young, they could not have survived long.

Michael A. Seeds; © 2016 Cengage Learning®

Mars: Runaway Refrigerator

3 When Mars was young, water was abundant enough to flow over the surface in streams and floods, and may have filled oceans. That age of liquid surface water ended more than 3 billion years ago. The climate on Mars has changed as atmospheric gases and water were lost to space and as water was frozen into the soil as permafrost.

3a This distributary fan formed where an ancient stream flowed into a lake within a crater. Sediment deposited in the still lake water formed a lobed delta that later became sedimentary rock. Detailed analysis reveals that the stream changed course repeatedly, showing that the stream was not just a short-term flood.

Visual

NASA/JPL-Caltech/Malin Space Science Systems

MARS

The Northern Lowlands of Mars and the Hellas Basin may once have been filled with water.

Hellas Basin

NASA

Radar map

4 Early in the history of Mars when its crust was thin, convection in the mantle could have pushed up volcanic regions such as Tharsis and Elysium, and limited plate motion could have produced Valles Marineris, but Mars cooled too fast, and its crust never broke into moving tectonic plates as did Earth's crust.

As its crust thickened, volcanism abated, and Mars lost the ability to replenish its waning atmosphere.

Olympus Mons volcano on Mars

4a Mars has very large volcanoes, and their size shows that the crust has grown thick as the little planet has lost its heat to space. A thinner crust could not support such large volcanoes.

If the crust of Mars were made up of moving plates, its volcanoes could not have grown so big.

Computer-enhanced visual image

NASA GSFC Scientific Visualization Studio

into irregular chunks of rock, and they cannot pull themselves into smooth spheres because their gravity is too weak to overcome the structural strength of the rock. You will discover in the next chapters that low-mass moons are typically irregular in shape, whereas massive moons are normally spherical.

Images of Phobos reveal a unique set of narrow, parallel grooves (Figure 13-20a). Averaging 150 m (500 ft) wide and 25 m (80 ft) deep, the grooves run from Stickney, the largest crater, to an oddly featureless region on the opposite side of the satellite. One hypothesis is that the grooves are deep fractures produced by the impact that formed Stickney. The featureless region opposite Stickney may be similar to the jumbled terrains found on Earth's Moon and on Mercury that are thought to have been produced by the focusing of seismic waves from major impacts on the far sides of each body.

Observations made with the *Mars Global Surveyor*'s infrared spectrometer show that Phobos's surface cools quickly from −4°C to −112°C (25°F to −170°F) as it passes from sunlight into the shadow of Mars. Solid rock would retain heat and cool more slowly. To cool as quickly as it does, most of Phobos must be covered with very fine dust at least a meter deep. Deimos looks even smoother than Phobos, probably because of an even thicker layer of surface dust (Figure 13-20b).

The debris on the surfaces of the moons raises an interesting question: How can the weak gravity of small bodies hold on to fragments from meteorite impacts? Escape velocity on Phobos is only 11 m/s (25 mph). An athletic astronaut could almost jump into space. Certainly, most fragments from impacts should escape, but evidently some do fall back and accumulate on the surface.

Deimos and Phobos illustrate three principles of comparative planetology that you will find helpful as you explore farther from the Sun. First, some satellites are probably captured asteroids. Second, small satellites tend to be irregular in shape and heavily cratered. And third, tidal forces can affect small moons and gradually change their orbits. You will find even stronger tidal effects in Jupiter's satellite system in the next chapter.

DOING SCIENCE

Why would you be surprised if you found volcanism on Phobos or Deimos? The answer may seem obvious, but a good scientist rethinks supposedly obvious answers to check whether there might be alternatives.

In studying the Moon, Mercury, Venus, Earth, and Mars, you have seen illustrations of the principle that the larger a world is, the more slowly it loses its internal heat. It is the flow of that heat from the interior through the surface into space that drives geological activity such as volcanism and plate motion. A small world, like Earth's Moon, cools quickly and remains geologically active for a shorter time than a larger world like Earth. Phobos and Deimos are not just small; they are tiny. However they formed, any interior heat should have leaked away very quickly; with no energy flowing outward, there can be no volcanism.

Some futurists suggest that the first human missions to Mars will not land on the surface of the planet but will build a colony on Phobos or Deimos. These plans are based on speculation that there may be water deep inside the moons that colonists could use.

Now practice understanding the unexpected. *What might be the explanation if objects as small as Phobos and Deimos are found to be geologically active?*

What Are We? Earth-Folk

Space travel isn't easy. We humans made it to the Moon, but it took everything we had in the late 1960s. Going back to the Moon will be easier next time because the technology will be better, but it will still be expensive and will require people with heroic talent to design, build, and fly the spaceships. Going beyond the Moon will be even more difficult.

Going to Mercury or Venus doesn't seem worth the effort. Mercury is barren and dangerous, and the heat and air pressure on Venus may prevent any astronaut from ever visiting its surface. In the next two chapters, you will discover that the Jovian planets and their moons also are not places humans are likely to visit soon. The stars are so far away they may be forever beyond the reach of human spaceships. But Earth has a neighbor.

Astronomically, Mars is just up the street, and it isn't such a bad place. You would need a good spacesuit and a pressurized colony to live there, but it isn't impossible. Solar energy and water are abundant. It seems likely that humans will not only walk on Mars but someday live there. We Earth-folk have an exciting future. Eventually we will be the Martians.

Study and Review

Summary

▶ Venus is nearly Earth's twin in size and density (overall composition), but Venus's surface conditions have evolved very differently from those of Earth.

▶ Venus rotates in the retrograde (backward) direction. This retrograde rotation may have been caused by an off-center impact of a very large planetesimal as Venus was forming, by solar tides in the planet's thick atmosphere, or both.

▶ Venus has no detectable magnetic field. That means the planet's core does not support the dynamo effect and may be completely solid. The reason for this is not understood.

▶ The atmosphere of Venus is 95 times thicker than Earth's and composed almost entirely of CO_2. Venus rotates so slowly that solar heat at the **subsolar point (p. 268)** produces strong atmospheric currents that circle the planet in about 4 days.

▶ The temperature at the surface of Venus is about 470°C (880°F), hotter than Mercury even though Venus is farther from the Sun.

▶ Because Venus formed closer to the Sun than Earth did, it was initially warmer. Any oceans on early Venus originally began to evaporate and so were less able to remove CO_2 from the atmosphere. Accumulating CO_2 increased the greenhouse effect. The increasing surface temperature of the planet evaporated the remaining surface water in a **runaway greenhouse effect (p. 269)**.

▶ The surface of Venus is so hot that compounds have cooked out of the crust to form traces of sulfuric, hydrochloric, and hydrofluoric acids in the atmosphere. The very high clouds on Venus are composed of small droplets of sulfuric acid and sulfur crystals.

▶ Although perpetually hidden below thick clouds, the surface of Venus can be studied using radar. Radar mapping reveals rugged highlands and low rolling plains. Impact craters, volcanoes, lava flows, and channels similar to the Moon's sinuous rilles are also detectable. Radar maps can measure roughness and, in some cases, mineral composition of the surface.

▶ Signs of volcanism are common on Venus, and much of the surface appears to be solidified lava flows. Landers have analyzed the surface rocks at several locations on Venus and found these rocks to be similar to Earth's volcanic basalt rocks.

▶ **Composite volcanoes (p. 272)** on Earth have steep slopes and are associated with subduction zones at the edges of tectonic plates. **Shield volcanoes (p. 272)** on Earth have shallow slopes and are associated with hot spots where magma rises from the mantle. The volcanoes on Venus are of the shield type. Volcanoes are probably still active on Venus.

▶ On Earth, plate motion across a hot spot produces chains of volcanic peaks. The large size of the shield volcanoes on Venus and the fact that there are no volcanoes of the composite type indicate that Venus does not have plate tectonics.

▶ Circular **coronae (p. 275)** hundreds of kilometers across form where rising currents of magma push the crust up and then withdraw. Coronae are associated with signs of surface volcanism.

▶ The surface of Venus appears to be about a half billion years old. Planetary scientists suspect that the entire planet was resurfaced by numerous lava flows.

▶ Mars is smaller than Earth but larger than the Moon. Mars has lost lower-mass atoms from its atmosphere because of its low escape velocity. As a result the atmosphere on Mars has a very low pressure and consists mostly of CO_2. The pressure at the surface is too low to allow liquid water to exist there.

▶ Although 19th-century astronomers thought they saw networks of canals of Mars, images from spacecraft show that Mars is a dry desert world with no surface water. The canals are an optical illusion.

▶ The southern hemisphere of Mars is old and heavily cratered. In contrast, the northern lowlands are smooth and mostly free of craters.

▶ Images from spacecraft orbiting Mars reveal **valley networks (p. 284)** and **outflow channels (p. 284)**. The valley networks resemble dry riverbeds. The outflow channels appear to have been formed by massive floods. Crater counts show that the valley networks and outflow channels are in old terrain.

▶ The smooth lowlands of Mars's northern hemisphere may have once contained a liquid water ocean. Some outflow channels lead into the lowlands, and features resembling shorelines have been found.

▶ Evidence suggests that Mars has been able to retain water frozen in the crust as permafrost and as large deposits in the polar caps. When liquid water is able to melt and seep out, it can cut gullies in the terrain and form other flow features. However, rivers, lakes, and oceans containing sustained liquid water cannot currently exist because of Mars's low atmospheric pressure.

▶ There is evidence that Mars once had a partly liquid metal core that was able to generate a magnetic field. Today, Mars has no detectable planetwide magnetic field. Planetary scientists hypothesize that most of the core has solidified and that any remaining molten material is not enough to produce a magnetic field.

▶ The **Noachian period (p. 287)** extended from the formation of Mars's crust to the end of heavy cratering, about 3.7 billion years ago. The valley networks formed during this period, suggesting that Mars's atmosphere was denser then and water that fell as rain was able to flow across the surface.

▶ The **Hesperian period (p. 288)** began as cratering declined and massive lava flows resurfaced some regions. The climate was colder and the atmosphere thinner than during the Noachian period. During this period, water began to freeze in the crust. Massive floods and outflow channels seem to have been produced by sudden episodes of subsurface ice melting.

▶ The **Amazonian period (p. 288)** extended from about 3 billion years ago to the present day. This period is marked by continued impact cratering, but at a slower rate, plus erosion by wind and by small amounts of water seeping from subsurface ice.

▶ Volcanism has been important throughout the history of Mars. Volcanoes were active during the Noachian period. The Tharsis rise is a huge volcanic uplift and the region contains some of the oldest and youngest lava flows.

▶ Near the Tharsis rise is Olympus Mons, the largest volcano in the Solar System. Because that enormous mountain has not sunk into the crust, the crust of Mars must be very thick to be able to support it.

▶ The development of the Tharsis rise, which grew into a huge bulge on one side of the planet, may have resulted in solar tidal forces reducing the inclination of Mars's rotation axis, making its climate generally colder.

▶ Mars has two moons named Phobos and Deimos that are hypothesized to be captured asteroids. They are small, irregularly shaped, and cratered. Both moons have tidally locked rotation with respect to Mars.

Review Questions

1. Describe four ways Venus is similar to Earth today. Describe four ways Venus is different from Earth today.
2. Why might you expect that Venus's surface conditions should resemble Earth's more than they do?
3. Describe and explain changes in Venus's surface temperature during the planet's history.
4. Describe sources and "sinks" of CO_2, if any, on Venus today.
5. Does Venus's surface experience meteorite impacts today? How do you know?
6. Describe two different tectonic features (horizontal or vertical) observed on Venus.
7. Why isn't the crust of Venus broken into mobile plates as Earth's crust is? How do you know?
8. There are some composite volcanoes and some shield volcanoes on both Venus and Mars. True or false?
9. What evidence can you cite that Venus once had significant amounts of water? Where did that water come from? Where did it go?
10. What evidence shows that Venus has been resurfaced within the past half-billion years?
11. Describe four ways Mars is similar to Earth today. Describe four ways Mars is different from Earth today.
12. How are today's atmospheres of Venus and Mars similar? How are they different?
13. Where is the oxygen on Mars today? How do you know?
14. Why doesn't Mars have folded mountain ranges like the ones on Earth? Why doesn't Earth have large volcanoes like those on Mars?
15. Why isn't the crust of Mars broken into mobile plates as Earth's crust is? How do you know?
16. What were the canals on Mars eventually found to be? How do they differ from the outflow channels and valley networks on Mars?
17. How can planetary scientists estimate the ages of the outflow channels and valley networks on Mars?
18. Propose an explanation for the nearly pure CO_2 atmospheres of Venus and Mars. Why is Earth's atmosphere different?
19. Describe and explain changes Mars's surface temperature during the planet's history. What evidence can you cite that the climate on Mars has changed?
20. Describe sources and "sinks" of CO_2, if any, on Mars today.
21. Does Mars's surface experience any meteorite impacts today? How do you know?
22. Describe two different tectonic features (horizontal or vertical) observed on Mars.
23. What surface features on Mars today indicate that there was significant water erosion in the past?
24. Why are Phobos and Deimos nonspherical? Why is Earth's Moon much more spherical?
25. **How Do We Know?** How are a weather radar map and an image of a highland on Venus related?

Discussion Questions

1. Earthlings automatically think of a "day" as an Earth day, or the average amount of time Earth takes to rotate once on its axis with respect to the Sun. However, if you lived on Venus, the length of a Venusian solar day is 117 Earth days. The length of a solar day on Mercury is 176 Earth days. Should Earthlings be so biased as to assign the time unit of "day" to that of an Earth day, or should we establish a less biased time unit?

2. Explain the challenges that humans would need to overcome to colonize Venus.
3. If you were able to stand on the surface of Venus in the daytime, what would it look like? For example, would it be raining? Would there be high winds? Would there be any light? What color would the sky be?
4. From what you know about Earth, Venus, and Mars, do you expect the volcanoes on Venus and Mars to be active or extinct? Why or why not?
5. If you had a time machine, plus superpowers sufficient to modify or move entire planets, what would you change about Venus during its formation, to make the surface environment remain more Earth-like up to the present day? How about Mars?
6. Propose a hypothesis about a single event that could explain both Venus's slow rotation and Venus's geologically recent resurfacing.
7. The largest challenges humans must overcome to colonize Mars probably would be finding water and oxygen. With plenty of solar energy available, how might colonizers extract water and oxygen from the Martian environment?
8. If you were able to stand on the surface of Mars in the daytime, what would it look like? For example, would it be raining? Would there be high winds? Would there be any light? What color would the sky be?

Problems

1. Atmospheric jet streams on Venus travel at about 300 km/hr. How long does it take a jet stream to circle the planet once? How many times does the jet stream circle the planet during one solar rotation of the planet? (*Notes:* The circumference of a sphere is $c = \pi d$, where d is diameter. The diameter and solar rotation period of Venus are given in **Celestial Profile 5**.)
2. How long would radio signals take to travel from Earth to Venus and back if Venus were at its nearest point to Earth? At its farthest point from Earth? (*Notes:* The speed of light is 3.00×10^8 m/s. Necessary data to derive the distances between the objects in those two situations are given in **Celestial Profile 2** and **Celestial Profile 5**.)
3. What is the maximum angular diameter of Venus as seen from Earth? (*Hint:* Use the small-angle formula, Chapter 3.) (*Note:* Necessary data to derive the distance between the objects in that situation are given in **Celestial Profile 2** and **Celestial Profile 5**.)
4. The *Pioneer Venus* orbiter circled Venus with a period of 24 hours. What was its average distance above the *surface* of Venus? (*Hints:* Use the formula for circular orbital velocity, Chapter 5. Remember to convert quantities to kg, m, and s.) (*Note:* Necessary data are given in **Celestial Profile 5**.)
5. Calculate the velocity of Venus as it orbits the Sun. (*Hint:* Use the formula for circular orbital velocity, Chapter 5.) (*Note:* Necessary data are given in **Celestial Profile 1** and **Celestial Profile 5**.)
6. The distance between Haleakala, the inactive volcano on Maui, and Kilauea, the active volcano on the Big Island of Hawai'i, is 171 km. If the plate under Hawai'i moves at 9 cm/yr, how long did the plate take to travel this distance? Speculate why your calculated value does not agree with the age of Maui given in the figure on the right-hand page of **Volcanoes**.
7. If the *Magellan* spacecraft transmitted radio signals down through the clouds on Venus and heard an echo from a certain spot 0.000133 second before the main echo, how high is the spot above the average surface of Venus, in m and km units? (*Note:* The speed of light is 3.00×10^8 m/s.)
8. Examine Figure 13-11. What is the ratio of the velocity of molecular hydrogen at Venus's temperature to the escape velocity from? What does that ratio tell you?

9. The smallest feature visible through an Earth-based telescope has an angular diameter of about 1 arc second. Can Mars's moon Phobos be resolved in Earth-based observations? (*Hint:* Use the small-angle formula, Chapter 3.) (*Notes:* Necessary data to derive the distance between the objects in that situation are given in **Celestial Profile 2** and **Celestial Profile 6**. For the size of Phobos, use the maximum dimension given in the text.)

10. Examine Figure 13-11. What is the ratio of the velocity of CO_2 at Mars's temperature to the escape velocity from? What does that ratio tell you?

11. How fast does Phobos travel in its orbit around Mars? (*Hint:* Use the formula for circular orbital velocity, Chapter 5. Remember to convert quantities to kg, m, and s.) (*Note:* Necessary data are given in **Celestial Profile 6** and Appendix Table A-11.)

12. Deimos is about 13 km in diameter and has a density of 2 g/cm^3. Assume Deimos is a sphere. What is its mass? (*Note:* The volume of a sphere is $\frac{4}{3}\pi r^3$.)

Learning to Look

1. Look at Figure 13-1. Compare temperature profiles of Venus's and Earth's atmospheres. Describe the differences between the two profiles.

2. Look at the map of the Hawaiian chain of islands on the right-hand page of **Volcanoes.** Which island formed most recently? How do you know? Is the newly formed volcano of a type found on Venus, on Mars, on both planets, or on neither?

3. Look at Figure 13-11. Which molecule(s) can escape from Earth's gravity? From Mars? From Venus?

4. Volcano Sif Mons on Venus is shown in this radar image. What kind of volcano is it, and why is it orange in this image? What color would the rock be if you could see it with your own eyes, and why the difference?

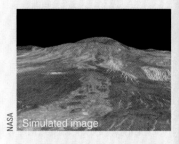
Simulated image

5. Olympus Mons on Mars is an enormous volcano. In this image, you can see multiple calderas (craters) at the top. What do the numbers of calderas and the immense size of the volcano indicate about the geology of Mars?

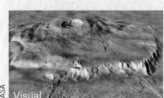
Visual

14 Jupiter and Saturn

Guidepost As you begin this chapter, you leave behind the psychological security of planetary surfaces. You can imagine standing on the Moon, on Mars, or even on Venus, but Jupiter and Saturn have no surfaces. Here you face a new challenge: to use comparative planetology to study worlds so un-Earthly you cannot imagine really being there. On the other hand, Jupiter and Saturn also have extensive systems of moons and rings. Someday humans may walk on some of the moons and watch erupting volcanoes or stroll through methane rain storms, then journey to the rings and float among the ring particles. As you study these worlds, you will find answers to four important questions:

▶ **How do the Jovian planets compare with the Terrestrial planets?**

▶ **How did Jupiter and Saturn form and evolve?**

▶ **How did Jupiter's and Saturn's systems of moons and rings form and evolve?**

▶ **What is the evidence that some of Jupiter's and Saturn's moons have been geologically active?**

After learning about the two largest Jovian planets, in the next chapter you will continue your trip away from the Sun and visit their two smaller, and in some ways even stranger, siblings, Uranus and Neptune, plus the dwarf planets in the Kuiper Belt at the outer fringe of the Solar System. It will be interesting, but there is no place like home.

There is something fascinating about science. One gets such wholesale returns of conjecture out of such a trifling investment of fact.

MARK TWAIN, *LIFE ON THE MISSISSIPPI*

NASA/JPL-Caltech/SSI

Saturn's moon Enceladus backlit by the Sun. This greatly enhanced image shows the enormous extent of the fountain-like plume of material that towers hundreds of kilometers above the moon's south polar region. The plume is salty liquid water bearing organic compounds, erupting from pressurized subsurface reservoirs.

Visual, false-color, enhanced contrast

When Mark Twain wrote the sentences that open this chapter, he was poking gentle fun at science, but he was right. The exciting thing about science isn't the so-called facts, that is, the observations in which scientists have greatest confidence. Rather, the excitement lies in the understanding that scientists get by rubbing a few facts together. Science can take you to strange new worlds such as Jupiter and Saturn, and you can get to know them by combining the available observations with known principles of comparative planetology.

14-1 A Travel Guide to the Outer Solar System

If you travel much, you know that some cities make you feel at home, and some do not. In this chapter and the next one, you will visit worlds that are truly un-Earthly. This travel guide will warn you about what to expect.

The Outer Planets Plus Pluto

The worlds of the outer Solar System can be studied from Earth, but much of what astronomers know has been radioed back to Earth from robot spacecraft. The *Pioneer* and *Voyager* missions flew past the outer planets in the 1970s and 1980s, *Galileo* orbited Jupiter and dropped a probe into the planet's atmosphere in the late 1990s, and the *Cassini* orbiter plus *Huygens* probe (pronounced, approximately, *HOWK-ginz*) arrived at Saturn and its moon Titan in 2004 and 2005, respectively. The *New Horizons* craft will pass Pluto in 2015 and then sail deeper into the Kuiper Belt. Throughout this discussion, you will find images and data returned by these robotic explorers.

The outermost planets in our Solar System are Jupiter, Saturn, Uranus, and Neptune—all classified as Jovian planets, meaning they resemble Jupiter. In fact, they are each individuals with separate personalities. **Figure 14-1** compares the four outer worlds to each other. One striking feature, of course, is their sizes. Figure 14-1 also shows Earth in scale to the Jovian planets, and it seems tiny in comparison; Jupiter, the largest of the Jovian worlds, is more than 11 times the diameter of Earth. You can also see that the four Jovian planets can be divided into two pairs, with Jupiter and Saturn being large and nearly the same size and Uranus and Neptune being smaller but still four times the size of Earth.

Pluto, not pictured in the figure, is smaller than Earth's Moon but was considered a planet at the time of its discovery in 1930. In 2006 it was reclassified as a dwarf planet. You will learn about Pluto's characteristics, and the reasons for that decision, in the next chapter.

The other feature you will notice immediately when you look at Figure 14-1 is Saturn's rings. They are bright and beautiful and composed of billions of ice particles, each particle following its own orbit around the planet. Astronomers have discovered that Jupiter, Uranus, and Neptune also have rings, but they are not easily detected from Earth and are not visible in this figure. As you visit those worlds in this chapter and the next, you will be able to compare and contrast four giant planets, four moon systems, and four different sets of planetary rings.

Atmospheres and Interiors

All the Jovian worlds have hydrogen-rich atmospheres filled with clouds. On Jupiter and Saturn, you can see that the clouds form dark belts and light zones that circle the planets like the stripes on a child's ball. This form of atmospheric structure is called **belt–zone circulation**. You will find traces of belts and zones on Uranus and Neptune, but they are much less distinct. All the Jovian worlds also have giant circulating storms, the primary example of which is Jupiter's Great Red Spot, which is more than twice the size of Earth. These Jovian storms are comparable to Terrestrial hurricanes, but they can last for centuries. The Great Red Spot storm has been going strong at least since it was first noticed by early telescope observers 350 years ago.

The gaseous atmospheres of the Jovian planets are not very deep. Jupiter's atmosphere makes up only about 1 percent of its radius. Below that, Jupiter and Saturn are composed of liquid hydrogen, so the conventional term for these planets, the *gas giants*, should probably be changed to *liquid giants*. Uranus and Neptune are sometimes called the *ice giants* because they contain abundant water in solid forms. Only near their centers do the Jovian planets have cores of dense material with the composition of rock and metal. None of the Jovian worlds has a definite solid surface on which you could walk.

Satellite Systems

All of the Jovian worlds have extensive satellite systems. Those moons can be classified into two groups: (1) the **regular satellites**, which tend to be large and orbit relatively close to their parent planet, with low inclinations to the planet's equator, moving in the **prograde** direction along with most of the objects in the Solar System, versus (2) the **irregular satellites**, which tend to be smaller than the regular satellites, sometimes have retrograde and/or highly inclined orbits, and are generally far from their parent planet. Astronomers have evidence that the regular satellites formed approximately where they are now as the planets formed but that the irregular satellites are mostly, if not all, captured objects.

As you focus on the moons of the Jovian worlds, look for evidence of two processes. The orbits of some moons may have been modified by interactions with other moons so that they now revolve around their planet in mutual resonances. The same process may allow moons to affect the orbital motions of particles in planetary rings.

The second process allows tides to heat the interiors of some moons and produce geological activity on their surfaces, including

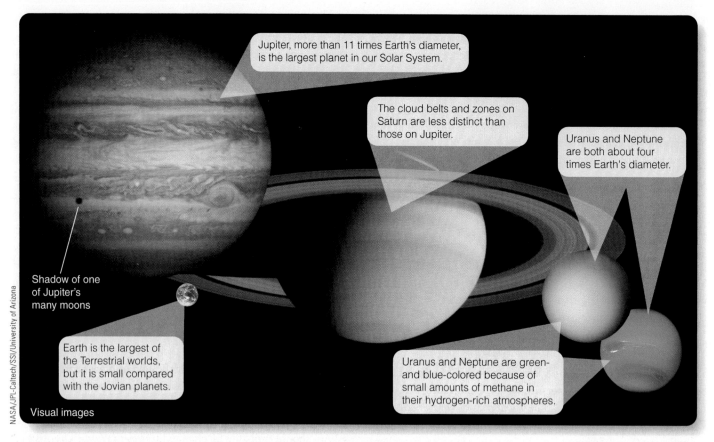

Jupiter, more than 11 times Earth's diameter, is the largest planet in our Solar System.

The cloud belts and zones on Saturn are less distinct than those on Jupiter.

Uranus and Neptune are both about four times Earth's diameter.

Shadow of one of Jupiter's many moons

Earth is the largest of the Terrestrial worlds, but it is small compared with the Jovian planets.

Uranus and Neptune are green- and blue-colored because of small amounts of methane in their hydrogen-rich atmospheres.

Visual images

NASA/JPL-Caltech/SSI/University of Arizona

▲ **Figure 14-1** The principal worlds of the outer Solar System are the four massive but low-density Jovian planets, each much larger than Earth.

volcanoes and lava flows. You have learned that the entire Solar System received a heavy bombardment after the planets formed, and heavily cratered surfaces are old, so when you see a section of a moon's surface, or an entire moon, that has few craters, you know that moon must have been geologically active since the end of the heavy bombardment.

14-2 Jupiter

Jupiter is the largest and most massive of the Jovian planets, containing 71 percent of all the planetary matter in the entire Solar System. Just as you used Earth, the largest of the Terrestrial planets, as the basis for comparison with the others, you can examine Jupiter in detail as a standard in your comparative study of the other Jovian planets.

Surveying Jupiter

Jupiter is extreme because it is big, massive, mostly liquid hydrogen, and very hot inside. The preceding facts are common knowledge among astronomers, but you should demand an explanation of how they know these facts. Often the most interesting thing about a fact isn't the fact itself but how it is known.

At its closest point to Earth, Jupiter is about eight times farther away than Mars, but even a small telescope will reveal that the disk of Jupiter appears more than twice as big as the disk of Mars. From its apparent size and distance, and using the small-angle formula (look back to Chapter 3), you can compute the diameter of Jupiter—1.4×10^5 km, which is about 11 times Earth's diameter (■ Celestial Profile 7, page 313).

You can see that Jupiter is massive by watching its moons race around it at high speeds. Io is the innermost of the four **Galilean moons** (named after their discoverer, the astronomer Galileo Galilei), and its orbit is just a bit larger than the orbit of our Moon around Earth. Io streaks around its orbit in less than two days, whereas Earth's Moon takes a month. Jupiter has to be a massive world to hold on to such a rapidly moving moon (Figure 14-2). In fact, you can use the radius of Io's orbit and its orbital period in Isaac Newton's version of Johannes Kepler's third law (look back to Chapter 5) to calculate the mass of Jupiter, which is 1.9×10^{27} kg, 318 times Earth's mass.

Learning the size and mass of Jupiter is relatively easy, but you might wonder how astronomers know that it is made mostly of hydrogen. The first step is to divide mass by volume to find Jupiter's average density, 1.3 g/cm³. Of course, it is denser at the center and less dense near the surface, but this average density reveals that it can't contain much rock. Rock has a density of 2.5

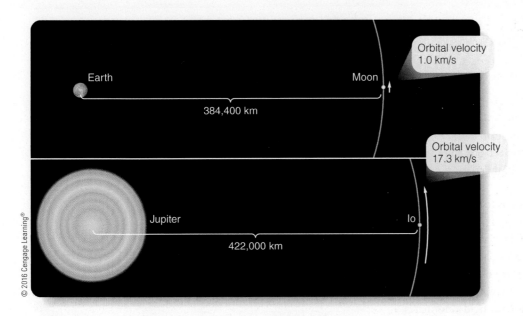

Orbital velocity
1.0 km/s

Orbital velocity
17.3 km/s

Earth

Moon

384,400 km

Jupiter

Io

422,000 km

© 2016 Cengage Learning®

▲ **Figure 14-2** It is obvious that Jupiter is a very massive planet when you compare the motion of Jupiter's moon Io with Earth's Moon. Although Io is 10 percent farther from Jupiter, it travels 17 times faster in its orbit than does Earth's Moon around Earth. Clearly, Jupiter's gravitational field is much stronger than Earth's, and that means Jupiter must be very massive.

to 5 g/cm³, so Jupiter must contain material mostly of lower density, such as hydrogen.

Spectra recorded from Earth and from spacecraft visiting Jupiter show that the composition of Jupiter is much like that of the Sun. This was confirmed in 1995 when a probe from the *Galileo* spacecraft parachuted into the atmosphere and radioed its results back to Earth. Jupiter is mostly hydrogen and helium, with traces of heavier atoms that form molecules such as methane (CH_4), ammonia (NH_3), and water (Table 14-1).

Jupiter's Interior

Just as astronomers can build mathematical models of the interiors of stars, they can use the equations that describe gravity, energy, and the compressibility of matter to build mathematical models of the interior of Jupiter. These models reveal that the

TABLE 14-1 **Composition of Jupiter and Saturn's Upper Atmospheres (by number of molecules)**

Molecule	Jupiter (%)	Saturn (%)
H_2	90	96
He	10	3
H_2O	0.0004	0.0004
CH_4	0.3	0.4
NH_3	0.03	0.01

© 2016 Cengage Learning®

interior of the planet is mostly liquid hydrogen containing small amounts of heavier elements. The pressure and temperature are higher than the **critical point** for hydrogen, and that means there is no difference between gaseous hydrogen and liquid hydrogen. If you parachuted into Jupiter, you would fall through the gaseous atmosphere and notice the density of the surrounding fluid gradually increasing until you were in a liquid, but you would never splash into a liquid surface.

Roughly a quarter of the way to the center, the pressure is high enough to force the hydrogen into being **liquid metallic hydrogen**, which is a very good electrical conductor. Because liquid metallic hydrogen has been very difficult to create and study in the laboratory so far, its properties are poorly understood. That is the reason why the models are uncertain about the depth of the transition from normal to metallic liquid hydrogen.

The models are also uncertain about the presence of a heavy-element core in Jupiter. The planet contains about 30 Earth masses of elements heavier than helium, but much of that may be suspended in the convectively stirred liquid hydrogen. Measurements by orbiting spacecraft indicate that no more than 10 Earth masses are included in a heavy-element core. The *Juno* probe is expected to begin orbiting Jupiter in 2016 to further investigate the mass of the planet's core. Some astronomy books refer to this as a rocky core, but if it exists it cannot be anything like the rock you know on Earth. The center of Jupiter is five or six times hotter than the surface of the Sun and is prevented from exploding into vapor only by the tremendous pressure. If there is a core, it is "rocky" only in the sense that it contains heavy elements.

How do astronomers know Jupiter is hot inside? Infrared observations show that Jupiter is glowing strongly in the infrared, radiating 1.7 times more energy than it receives from the Sun (look at the *SOFIA* infrared image of Jupiter, Figure 6-18b). That observation, combined with models of its interior, provides an estimate of its internal temperature. You will learn later in this chapter that Saturn resembles Jupiter in this respect.

You can tell that Jupiter is mostly a liquid just by looking at it. If you measure a photograph of Jupiter, you will discover that it is noticeably flattened; it is a bit more than 6 percent larger in diameter through its equator than through its poles. This is referred to as Jupiter's **oblateness**. The amount of flattening depends on the speed of rotation and on the rigidity of the planet. Jupiter's flattened shape shows that the planet cannot be as rigid as a Terrestrial planet and must have a liquid interior.

Basic observations and the known laws of physics can tell you a great deal about Jupiter's interior. Its vast magnetic field can tell you even more.

Jupiter's Magnetic Field

As early as the 1950s, astronomers detected radio noise coming from Jupiter and recognized it as synchrotron radiation. That type of radio energy is produced by fast electrons spiraling in a magnetic field, so it was obvious that Jupiter has a magnetic field.

In 1973 and 1974, two *Pioneer* spacecraft flew past Jupiter, followed in 1979 by two *Voyager* spacecraft. Those probes found that Jupiter has a magnetic field about 14 times stronger than Earth's field. Apparently, the field is produced by the dynamo effect operating in the highly conductive liquid metallic hydrogen as it is circulated by convection and spun by the rapid rotation of the planet. This powerful magnetic field dominates a huge magnetosphere around the planet. Compare the size of Jupiter's field with that of Earth in Figure 14-3.

Jupiter's magnetic field deflects the solar wind and traps high-energy particles in radiation belts much more intense than Earth's. The radiation is more intense because Jupiter's magnetic field is stronger and can trap and hold more particles, and higher-energy particles, than Earth's field can. The spacecraft

passing through the radiation belts received radiation doses equivalent to a billion chest X-rays—at least 4000 times the lethal dose for a human. Some of the electronics on the spacecraft were damaged by the radiation.

You will recall that Earth's magnetosphere interacts with the solar wind to produce auroras, and the same process occurs on Jupiter. Charged particles in the magnetosphere leak downward along the magnetic field, and where they enter the atmosphere they produce auroras 1000 times more luminous than those on Earth (look back to Chapter 8, page 163). The auroras on Jupiter, like those on Earth, occur in rings around the magnetic poles (Figure 14-4).

The four Galilean moons of Jupiter orbit inside the magnetosphere, and some of the heavier ions in the radiation belts come from the innermost moon, Io. As you will learn later in this chapter, Io has active volcanoes that spew gas and ash. Because Io orbits with a period of 1.8 days, and Jupiter's magnetic field rotates in only 10 hours, the wobbling magnetic field rushes past Io at high speed, sweeping up stray particles, accelerating them to high energy, and spreading them around Io's orbit in a doughnut of ionized gas called the **Io plasma torus**.

Jupiter's magnetic field interacts with Io to produce a powerful electric current (about a million amperes) that flows through a curving path called the **Io flux tube** from Jupiter out to Io and back to Jupiter. Small spots of bright auroras lie at the two points where the Io flux tube enters Jupiter's atmosphere (Figure 14-4).

Fluctuations in the auroras reveal that the solar wind buffets Jupiter's magnetosphere, but some of the fluctuations seem to be caused by changes in the magnetic dynamo deep inside the planet. In this way, studies of the auroras on Jupiter can help astronomers learn more about its liquid depths.

Jupiter's powerful magnetic field is invisible to your eyes, but the swirling cloud belts are beautifully visible in their complexity.

Jupiter's Atmosphere

You now know that Jupiter is a liquid world that has no surface. The gaseous atmosphere blends gradually with the liquid hydrogen interior. Below the clouds of Jupiter lies the largest ocean in the Solar System—an ocean that has no surface and no waves.

When you look at Jupiter, all you see are clouds. When you look near the limb of Jupiter (the edge of its disk), the clouds are much dimmer (Figure 14-1) because it is nearly sunset or sunrise along the limb. If you were on Jupiter at that location, you would see the Sun just above the horizon, and sunlight arriving there would be dimmed by passing through the planet's atmosphere. In addition, sunlight reflected from the clouds must travel back out through the atmosphere at a steep angle to reach Earth, dimming the light further. Jupiter is brighter near the center of the disk because the sunlight shines nearly straight down on the clouds.

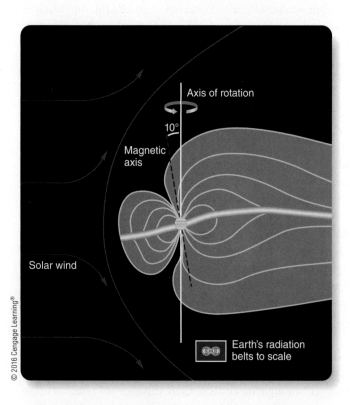

▲ Figure 14-3 Jupiter's magnetic field is large and powerful. It traps particles from the solar wind to form powerful radiation belts. The rapid rotation of the planet forces the slightly inclined magnetic field to wobble up and down as the planet rotates. Earth's magnetosphere and radiation belts are shown to scale.

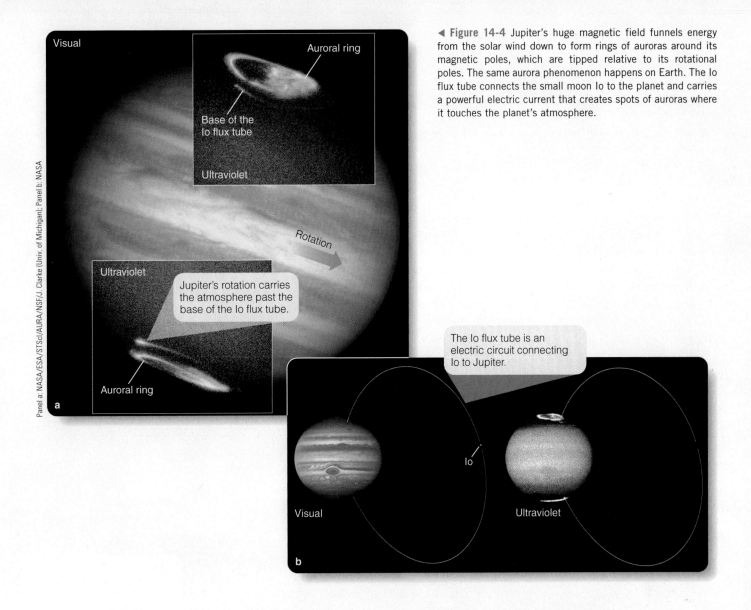

Panel a: NASA/ESA/STScI/AURA/NSF/J. Clarke (Univ. of Michigan); Panel b: NASA

◄ Figure 14-4 Jupiter's huge magnetic field funnels energy from the solar wind down to form rings of auroras around its magnetic poles, which are tipped relative to its rotational poles. The same aurora phenomenon happens on Earth. The Io flux tube connects the small moon Io to the planet and carries a powerful electric current that creates spots of auroras where it touches the planet's atmosphere.

Visual

Auroral ring

Base of the Io flux tube

Ultraviolet

Rotation

Ultraviolet

Jupiter's rotation carries the atmosphere past the base of the Io flux tube.

Auroral ring

a

The Io flux tube is an electric circuit connecting Io to Jupiter.

Visual

Io

Ultraviolet

b

Study **Jupiter's Atmosphere** on pages 302–303 and notice four important ideas:

1 The atmosphere is hydrogen rich, and the clouds are confined to a shallow layer.

2 The cloud layers are located at certain levels within the atmosphere where the temperatures are such that ammonia (NH_3), ammonium hydrosulfide (NH_4SH), and water (H_2O) can condense to form ice particles.

3 The belt–zone circulation is driven by high- and low-pressure areas related to those on Earth.

4 The large circular or oval spots seen in Jupiter's clouds are circulating storms that can remain stable for decades or even centuries.

Circulation in Jupiter's atmosphere is not totally understood. Observations made by the *Cassini* spacecraft as it raced past Jupiter on its way to Saturn revealed that the dark belts, which were thought to be entirely regions of sinking gas, contain small rising storm systems too small to have been seen in images by previous probes. Evidently, the general circulation usually attributed to the belts and zones is much more complex when it is observed in more detail. Further understanding of the small-scale motions in Jupiter's atmosphere may have to await future planetary probes.

A History of Jupiter

Your goal in studying any planet is to be able to tell its story—to describe how it got to be the way it is. Although you can understand part of the story of Jupiter, there is still much to learn.

If the solar nebula theory for the origin of the Solar System is correct, then Jupiter formed from the colder gases of the outer solar nebula, where ices of water and other molecules were able to condense. Thus, Jupiter grew rapidly and became massive enough to capture hydrogen and helium gas from the

Jupiter's Atmosphere

1 Humans will probably never visit Jupiter's atmosphere. Its cloud layers are deathly cold, and the deeper layers that are warmer have a crushingly high pressure. There is no free oxygen to breathe; the composition is roughly three-quarters hydrogen and a quarter helium by mass, plus small amounts of water vapor, methane, ammonia, and similar molecules. Traces of sulfur and molecules containing sulfur probably make it smell bad. Of course, Jupiter has no surface, so there isn't even a place to stand.

Belts are dark bands of clouds.

Zones are bright bands of clouds.

Shadow of Europa

Jupiter's moon Europa

NASA/JPL/University of Arizona

1a The first spacecraft to enter Jupiter's atmosphere was the *Galileo* probe. Released from the main *Galileo* spacecraft, the probe entered Jupiter's atmosphere in 1995. It parachuted through the upper atmosphere of clear hydrogen, released its heat shield, and then sent measurements back to Earth as it descended through layers of increasing pressure in Jupiter's stormy atmosphere until it was finally crushed.

Lightning bolts are common in Jupiter's turbulent clouds.

Artist's conception

Hughes Aircraft Co.

Jupiter's atmosphere is a thin layer of turbulent gas above the liquid interior. It makes up only about 1 percent of the radius of the planet.

The Great Red Spot at right is a giant circulating storm in one of the southern zones. It has lasted at least 350 years since astronomers first noticed it after the invention of the telescope. Smaller spots are also circulating storms.

NASA/JPL-Caltech/Univ. of Arizona

2 The visible clouds on Jupiter are composed of ammonia crystals, but models predict that deeper layers of clouds contain ammonium hydrosulfide crystals, and deeper still lies a cloud layer of water droplets. These compounds are normally white, so planetary scientists think the colors arise from small amounts of other molecules formed in reactions powered by lightning or sunlight.

If you could put thermometers in Jupiter's atmosphere at different levels, you would discover that the temperature rises below the uppermost clouds.

Far below the clouds, the temperature and pressure climb so high that the gaseous atmosphere merges gradually with the liquid hydrogen interior and there is no surface.

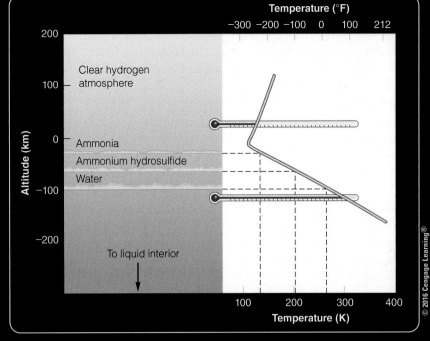

Clear hydrogen atmosphere

Ammonia
Ammonium hydrosulfide
Water

To liquid interior

Temperature (°F)
−300 −200 −100 0 100 212

Altitude (km)
200 100 0 −100 −200

Temperature (K)
100 200 300 400

© 2016 Cengage Learning®

3 On Earth, the temperature difference between the poles and equator drives a wave-shaped, high-speed wind that organizes the high- and low-pressure areas into cyclonic circulations familiar from weather maps.

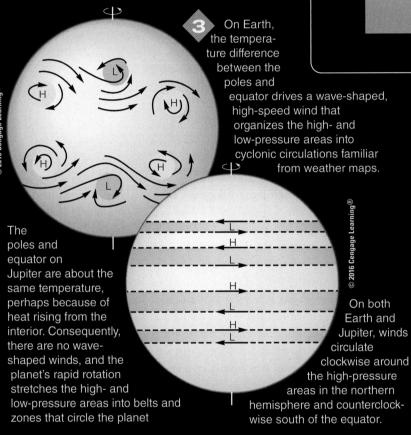

© 2016 Cengage Learning®

The poles and equator on Jupiter are about the same temperature, perhaps because of heat rising from the interior. Consequently, there are no wave-shaped winds, and the planet's rapid rotation stretches the high- and low-pressure areas into belts and zones that circle the planet

On both Earth and Jupiter, winds circulate clockwise around the high-pressure areas in the northern hemisphere and counterclockwise south of the equator.

Zones are brighter than belts because rising gas forms clouds high in the atmosphere, where sunlight is strong.

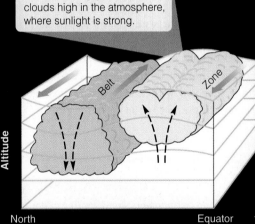

Belt
Zone
Altitude
North
Equator

4 Three circulating storms visible as white ovals since the 1930s merged in 1998 to form a single white oval. In 2006, the storm intensified and turned red like the Great Red Spot. The reason for the red color is unknown, but it may show that the storm is bringing material up from lower in the atmosphere.

Storms in Jupiter's atmosphere may be stable for decades or centuries, but astronomers had never before witnessed the appearance of a new red spot. The development of such storms is unpredictable, and most eventually disappear. Even the Great Red Spot may someday vanish.

Great Red Spot

Red Jr.

Visual + Infrared, enhanced contrast

solar nebula and form a deep liquid hydrogen envelope. Model calculations yield conflicting results as to whether a heavy-element core survives; it may have been mixed in with the convecting liquid hydrogen envelope. Astronomers estimate that the mass of Jupiter's heavy-element core is no more than 10 Earth masses, but it could be zero.

In the interior of Jupiter, hydrogen exists as liquid metallic hydrogen, a very good electrical conductor. The planet's rapid rotation, coupled with the outward flow of heat from its hot interior, drives a dynamo effect that produces a powerful magnetic field. That vast magnetic field traps high-energy particles from the solar wind to form intense radiation belts and auroras.

The rapid rotation and large size of Jupiter cause belt–zone circulation in its atmosphere. Heat flowing upward from the interior causes rising currents in the bright zones, and cooler gas sinks in the dark belts. As on Earth, winds blow at the margins of these regions, and large spots appear to be cyclonic disturbances. Internal heat has been escaping since Jupiter formed, so you can guess that Jupiter's atmospheric circulation and storms were stronger in the distant past and will diminish in the future.

Your study of Jupiter has been challenging because Jupiter is so unlike Earth. Most of the features and processes you found on the Terrestrial planets are missing on Jupiter, but as the prototype of the Jovian worlds, it earns its place as the ruler of the Solar System.

The highly complex spacecraft that have visited Jupiter are examples of how technology can give scientists the raw data they need to form their understanding of nature. Science is about understanding nature, and Jupiter is an entirely new kind of planet in your study. In fact, Jupiter has another feature you did not find anywhere among the Terrestrial planets. Jupiter has rings that you will explore in the next section along with the planet's impressive moons.

DOING SCIENCE

How do astronomers know Jupiter is hot inside? Scientists routinely review and check even the most basic information.

You can tell that something is hot if you can feel heat when you hold your hand near it. That is, you can detect infrared radiation with your skin. In the case of Jupiter, you would need greater sensitivity than the back of your hand, but infrared telescopes reveal that Jupiter is a source of infrared radiation; it is glowing in the infrared. Sunlight would warm Jupiter a little bit, but it is emitting 1.7 times as much energy as it receives from the Sun. That means it must be hot inside. From models of the interior, astronomers conclude that the center must be five or six times hotter than the surface of the Sun to cause the surface of the planet to glow as much as it does in the infrared.

Now review another simple but profound bit of information. **How do astronomers know that Jupiter has a low density?**

14-3 Jupiter's Moons and Rings

How many moons does Jupiter have? Astronomers are finding many small moons, and the count is now more than 60. (You will have to check the Internet to get the latest figure because more moons are discovered every year.) Most of these moons are small and rocky, and many are probably captured asteroids. The four Galilean moons are large and have interesting geologies (Figure 14-5).

Your study of the moons of Jupiter will illustrate three important principles in comparative planetology. First, a body's composition depends on the temperature of the material from which it formed. This is illustrated by the prevalence of ice as a building material in the outer Solar System, where sunlight is weak. You are already familiar with the second principle: that cratering can reveal the age of a surface. Also, as you have seen in your study of the Terrestrial planets, internal heat has a powerful influence over the geology of these larger moons.

Callisto: An Ancient Surface

The outermost of Jupiter's four large moons, Callisto, is half again as large in diameter as Earth's Moon. Like all of Jupiter's larger satellites, Callisto is tidally locked to its planet, keeping the same side forever facing Jupiter. From its gravitational influence on other moons and passing spacecraft, astronomers can calculate Callisto's mass, and dividing that mass by its volume shows that its density is 1.8 g/cm^3. Ice has a density of about 1 and rock 2.5 to 5 g/cm^3, so Callisto must be a mixture of rock and ice.

Images from the *Voyager* and *Galileo* spacecraft show that the surface of Callisto is dark, dirty ice heavily pocked with craters (Figure 14-6). Old, icy surfaces in the Solar System become dark because solar UV radiation and solar wind particles cause chemical changes in the ice and also because meteorite impacts deposit dust and vaporize water, leaving any dust and rock in the ice behind to form a dirty crust. If you live in a city in a cold climate, you may have seen the latter process happen to an urban snowbank. As the snow evaporates over a few days, the crud in the snow is left behind to form a dirty rind. If you break through that dirty surface, you find much cleaner snow underneath.

Spectra of Callisto's surface show that in most places it is a 50:50 mix of ice and rock, but some areas are ice free. Nevertheless, the slumped shapes of craters suggest that the outer 10 km (6 mi) of this moon is mostly frozen water; ice isn't very strong, so big piles of it tend to slump under their own weight. The disagreement between the spectra and the shapes of craters can be understood when you recall that the spectra contain information about only the outer 1 mm of the surface, which can contain lots of dirt, whereas the shapes of craters tell you about the outermost 10 km, which appear to be rich in ice.

Visual

▲ **Figure 14-5** The Galilean moons of Jupiter from left to right are Io, Europa, Ganymede, and Callisto. The white circle around Europa shows the size of Earth's Moon.

Careful measurements of the shape of Callisto's gravitational field were made by the *Galileo* spacecraft as it flew by. Those measurements show that Callisto has never fully differentiated to form a dense core and a lower-density mantle. Its interior is a mixture of rock and ice rather than having distinct layers of different composition. This is consistent with the observation that Callisto has only a weak magnetic field of its own. A strong magnetic field could be generated by the dynamo effect in a liquid convecting core, and Callisto has no core. It does, however, interact with Jupiter's magnetic field in a way that suggests it has a layer of salty liquid water roughly 10 km thick about 100 km (60 mi) below its icy surface. Slow radioactive decay in Callisto's interior may produce enough heat to keep this layer of water from freezing.

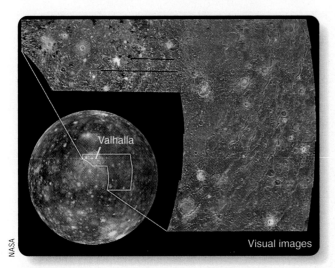

Valhalla

Visual images

▲ **Figure 14-6** The dark surface of Callisto is dirty ice marked by craters. The youngest craters look bright because they have dug down to cleaner ice. Valhalla, with a diameter of 3800 km (2400 mi), is the scar of a giant impact, the largest multiringed basin in the Solar System. Valhalla is so large and old that the icy crust has flowed back to partially heal itself, and the outer rings of Valhalla are shallow troughs marking fractures in the crust.

Ganymede: A Puzzling Past

The next Galilean moon inward is Ganymede, larger than Mercury and more than three-quarters the diameter of Mars. In fact, Ganymede is the largest moon in the Solar System. Its density is 1.9 g/cm^3, and its influence on the *Galileo* spacecraft reveals that it is differentiated into a rock and metal core, an ice-rich mantle, and a crust of ice 500 km (300 mi) thick. It may even have a small inner iron core. Evidently Ganymede is large enough for radioactive decay to have melted its interior after it formed, allowing rock and metal to sink to its center.

Ganymede's surface hints at an active past. Although a third of the surface is old, dark, and cratered like Callisto's, the rest is marked by bright parallel grooves. Because this bright **grooved terrain** (Figure 14-7a) contains fewer craters, it must be younger.

Observations show that the bright terrain was produced when the icy crust broke and water flooded up from below and froze. As the surface broke over and over, sets of parallel grooves were formed. Some low-lying regions are smooth and appear to have been flooded by water. Spectra reveal concentrations of salts such as those that would be left behind by the evaporation of mineral-rich water. Also, some features in or near the bright terrain appear to be calderas formed when subsurface water drained away and the surface collapsed (Figure 14-7b).

The *Galileo* spacecraft found that Ganymede has a magnetic field about 10 percent as strong as Earth's. It even has its own magnetosphere inside the larger magnetosphere of Jupiter. Mathematical models calculated by planetary scientists do not predict that a magnetic field this strong should arise from the dynamo effect in a liquid water mantle layer with the size and location of the one in Ganymede, and there does not appear to be enough heat in Ganymede for it to have a molten metallic core. Thus, the cause of Ganymede's unique magnetic field remains a puzzle. One hypothesis is that the magnetic field is left over and frozen into the rock from a time when Ganymede was hotter and more active.

Ganymede's magnetic field fluctuates with the 10-hour period of Jupiter's rotation. The rotation of the planet sweeps its

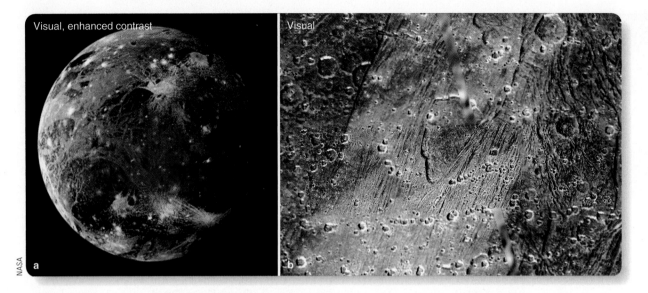

▲ **Figure 14-7** (a) This color-enhanced image of Ganymede shows the frosty poles at top and bottom, the old dark terrain, and the brighter grooved terrain. (b) A band of bright terrain runs from lower left to upper right, and a collapsed area, possibly a caldera, lies at the center in this image. Calderas form where subsurface liquid has drained away, and the bright areas do contain other features probably a result of flooding by water.

tilted magnetic field past the moon, and the two fields interact. That interaction reveals that the moon has a layer of liquid water about 170 km (110 mi) below its surface. The data indicate that the water layer is about 5 km (3 mi) thick. It is possible that the water layer was thicker and closer to the surface long ago when the interior of the moon was warmer. That might explain the flooding that appears to have formed the bright grooved terrain.

Ganymede orbits close enough to massive Jupiter that this moon is exposed to two unusual processes that many worlds never experience. **Tidal heating**, the frictional heating of a body by changing tides (**Figure 14-8a**), could have heated Ganymede's interior and added to the heat generated by radioactive decay. In its current nearly circular orbit, this moon experiences little or no tidal heating. But at some point in the past, interactions with the other moons could have pushed Ganymede into a more eccentric orbit. Tidal forces resulting from Jupiter's gravity would have deformed the moon; as Ganymede followed its orbit, varying in distance from Jupiter, tides would have flexed it, and friction would have heated it. Such an episode of tidal heating might have been enough to drive a dynamo to produce a magnetic field and break the crust to make the bright terrain.

The second process that affects Ganymede is the inward focusing of meteorites. Because massive planets like Jupiter draw debris inward, the closer a moon orbits to the planet, the more often it will be struck by meteorites (Figure 14-8b). You should expect such a moon to have lots of craters, but the bright terrain on Ganymede has few craters. That part of Ganymede's surface must be only about 1 billion years old, and this should alert you that the Galilean moons are not just dead lumps of rock and ice. The closer you get to Jupiter, the more active the moons are.

Europa: A Hidden Ocean

The next Galilean moon inward is Europa, which is a bit smaller than Earth's Moon (Figure 14-5). Europa has a density of 3.0 g/cm^3, so it must be mostly rock and metal, yet its surface is ice.

Europa lies closer to Jupiter than Ganymede does, so it should be exposed to more meteorite impacts than Callisto or Ganymede, yet the icy crust of Europa is almost free of craters. A few craters such as Pwyll are prominent, but most are hardly more than blemishes in the ice (**Figures 14-9**). Evidently, the

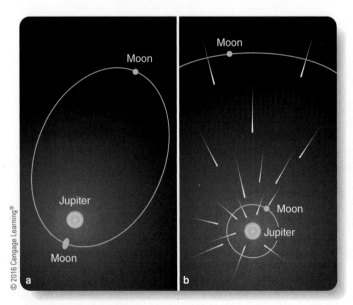

▲ **Figure 14-8** Two effects on planetary satellites. (a) Tidal heating occurs when changing tides cause friction within a moon. (b) The focusing of meteoroids exposes satellites in inner orbits to more impacts than satellites in outer orbits receive.

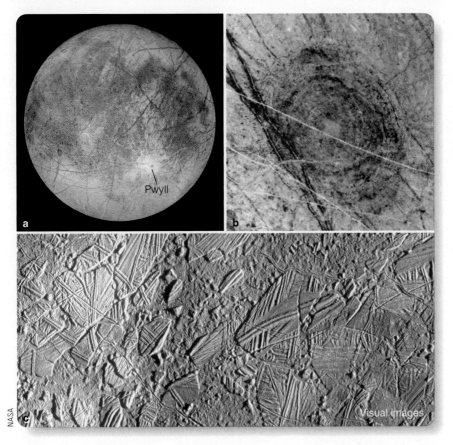

surface of Europa is active and erases craters almost as fast as they form. The number of impact scars on Europa suggests that the average age of its surface is only 10 million years. Other signs of activity include long cracks in the icy crust and regions where the crust has broken into sections that have moved apart as if they were icebergs floating on water (Figure 14-9c).

Europa's clean, bright face is another clue that its surface is young. The surface reflects an average of 67 percent of the sunlight that hits it. This reflectivity is produced by clean ice. You have learned that old icy surfaces tend to be very dark, so Europa's high reflectivity means the surface is active, covering older surfaces with fresh ice.

Europa is too small to have retained much heat from its formation or from radioactive decay, and the *Galileo* spacecraft found that Europa has no magnetic field of its own. It cannot have a molten conducting core. Tidal heating, however, is important for Europa and apparently provides enough heat to keep the little moon active. In fact, the curving cracks in its crust reveal the shape of the tidal forces that flex it as Europa orbits Jupiter.

If you hiked on Europa with a compass in your hand, you would detect a magnetic field, but not from Europa itself. Jupiter rotates rapidly and drags its strong magnetic field past the little moon. That induces a fluctuating magnetic field at Europa that would make your compass wander uselessly. Europa's interaction with Jupiter's magnetic field reveals the presence of a liquid water ocean lying only 15 km (about 10 mi) below the icy surface. The ocean might be as deep as 150 km (100 mi; **Figure 14-10**), containing twice as much water as all the oceans on Earth. It is likely to

▲ **Figure 14-9** (a) The icy surface of Europa is shown here in natural color. Many faults are visible on its surface, but very few craters. The bright crater is Pwyll, a young impact feature. (b) This circular bull's-eye is 140 km (90 mi) in diameter. It is the remains of an impact by an object estimated to have been about 10 km (6 mi) in diameter. Notice the younger cracks and faults that cross the older impact feature. (c) Like icebergs on the Arctic Ocean, blocks of crust on Europa appear to have floated apart. Spectra show that the blue ice is stained by salts such as those that would be left behind by mineral-rich water welling up from below and evaporating. White areas are ejecta from the impact that formed crater Pwyll.

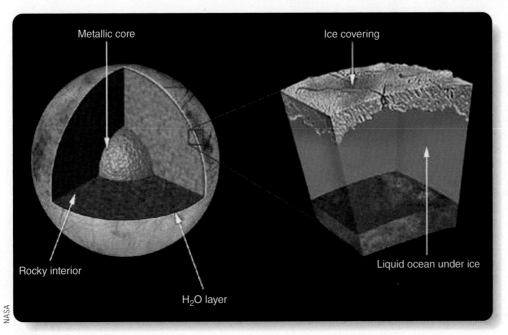

◄ **Figure 14-10** The gravitational influence of Europa on the passing *Galileo* spacecraft shows that this moon has differentiated into a dense core and rocky mantle. Magnetic interactions with Jupiter show that it has a liquid water ocean below its icy crust. Heat produced by tidal heating could flow outward as convection in such an ocean and drive geological activity in the icy crust.

be rich in dissolved minerals, which make the water a good electrical conductor and allow it to interact with Jupiter's magnetic field. No one knows what, if anything, might be swimming through such an ocean, and many scientists hope for a future mission to Europa to drill through the ice crust and explore the ocean below for signs of life.

Tidal heating makes Europa geologically active. Apparently, rising currents of water can break through the icy crust or melt surface patches. Many of the cracks show evidence that they have spread apart and that fresh water has welled up and frozen between the walls of the cracks. In other regions, compression of Europa's crust is revealed by networks of faults and low ridges. Compression on Earth pushes up mountain ranges, but no such ranges appear on Europa. The icy crust isn't strong enough to support ridges higher than a kilometer or so.

Orbiting deep inside Jupiter's radiation belts, Europa is bombarded by high-energy particles that alter the icy surface. Water molecules are freed and broken up, then dispersed into a doughnut-shaped cloud spread round Jupiter and enclosing Europa's orbit. Flying past Jupiter in 2002 on its way to Saturn, the *Cassini* spacecraft was able to image this cloud of glowing gas. Europa's gas cloud is evidence that moons orbiting deep inside a massive planet's radiation belts are exposed to a form of erosion that is entirely lacking on Earth's Moon.

Io: Roaring Volcanoes

Geological activity is driven by heat flowing out of a planet's interior, and nothing could illustrate this principle better than Io, the innermost of Jupiter's Galilean moons. Photographs from the *Voyager* and *Galileo* spacecraft show no impact craters at all—surprising considering Jupiter's power to focus meteoroids inward. But there is no difficulty explaining the missing craters. More than 150 active volcanoes are visible on Io's surface, blasting enough ash out over the surface to quickly bury any newly formed craters (Figure 14-11). Io is more geologically active than any other object in the Solar System, even more than Earth.

Spectra reveal that Io has a tenuous atmosphere of gaseous sulfur and oxygen, but those gases can't be permanent. Even though the erupting volcanoes pour out about 1 ton of gases per second, the gases leak into space easily because of Io's low escape velocity. Also, any gas atoms that become ionized are swept away by Jupiter's rapidly rotating magnetic field. The ions produce a cloud of sulfur and sodium ions in a torus (doughnut shape) enclosing Io's orbit (Figure 14-12).

The temperature at the surface averages 130 K (−225°F) and the atmospheric pressure is very low. Because of the continuous volcanism and the sulfurous gases, Io's thin atmosphere is smelly with sulfur. In fact, the reddish color of Jupiter's small

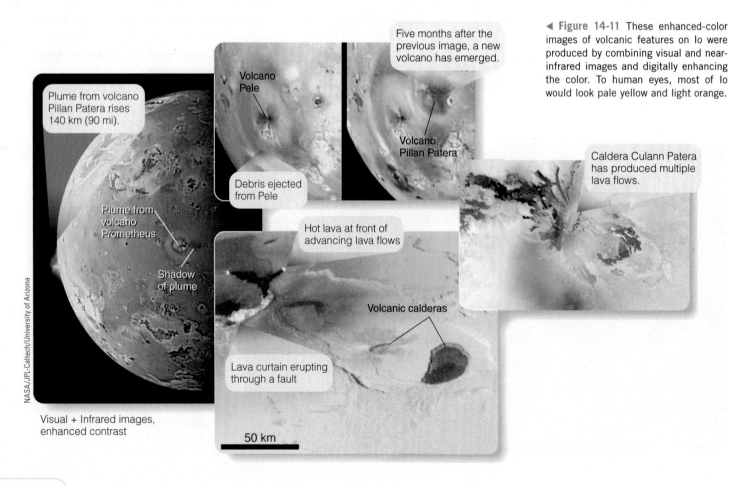

◀ Figure 14-11 These enhanced-color images of volcanic features on Io were produced by combining visual and near-infrared images and digitally enhancing the color. To human eyes, most of Io would look pale yellow and light orange.

Plume from volcano Pillan Patera rises 140 km (90 mi).

Plume from volcano Prometheus

Shadow of plume

Volcano Pele

Debris ejected from Pele

Five months after the previous image, a new volcano has emerged.

Volcano Pillan Patera

Caldera Culann Patera has produced multiple lava flows.

Hot lava at front of advancing lava flows

Volcanic calderas

Lava curtain erupting through a fault

Visual + Infrared images, enhanced contrast

50 km

NASA/JPL-Caltech/University of Arizona

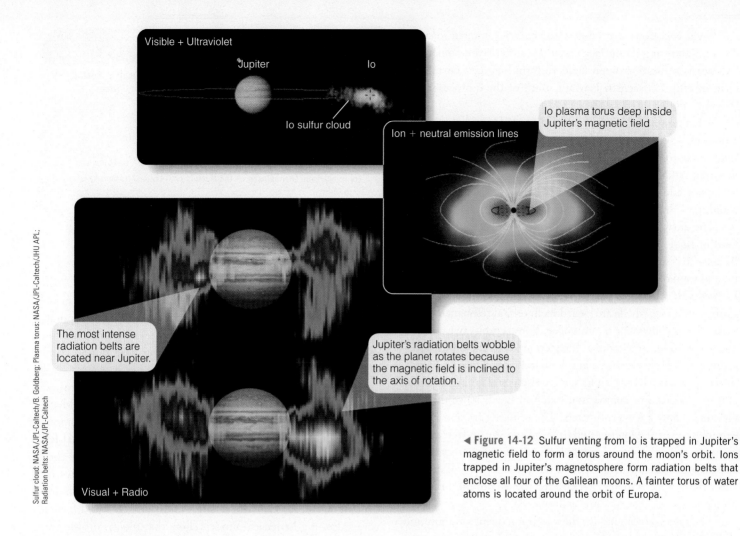

Visible + Ultraviolet

Jupiter

Io

Io sulfur cloud

Ion + neutral emission lines

Io plasma torus deep inside
Jupiter's magnetic field

The most intense
radiation belts are
located near Jupiter.

Jupiter's radiation belts wobble
as the planet rotates because
the magnetic field is inclined to
the axis of rotation.

Visual + Radio

Sulfur cloud: NASA/JPL-Caltech/B. Goldberg; Plasma torus: NASA/JPL-Caltech/JHU APL;
Radiation belts: NASA/JPL-Caltech

◀ **Figure 14-12** Sulfur venting from Io is trapped in Jupiter's magnetic field to form a torus around the moon's orbit. Ions trapped in Jupiter's magnetosphere form radiation belts that enclose all four of the Galilean moons. A fainter torus of water atoms is located around the orbit of Europa.

inner moon Amalthea may be caused by sulfur escaping from Io. The main problem for you to consider before walking across the surface of Io would be radiation. Io is deep inside Jupiter's magnetosphere and radiation belts. Unless your spacesuit had impressive shielding, the radiation would be lethal. Io, like Venus, may be a place that humans will never visit in person.

You can use basic observations to deduce the nature of Io's interior. From its density, 3.5 g/cm³, you can conclude that it is rocky. Spectra reveal no trace of water at all, so there is no ice on Io. In fact, it is the driest world in our Solar System. The oblateness of Io caused by its rotation and by the slight distortion produced by Jupiter's gravity gives astronomers more clues to the properties of its interior. Model calculations suggest it contains a modest core of iron or iron mixed with sulfur, a deep rocky mantle that is partially molten, and a thin, rocky crust.

The colors of Io have been compared to those of a badly made pizza. The reds, oranges, and browns of Io are caused by sulfur and sulfur compounds, and an early hypothesis proposed that the crust is mostly sulfur. New evidence says otherwise. Infrared measurements show that volcanoes on Io erupt lava with a temperature of more than 1500°C (2700°F), about 300°C hotter than lavas on Earth. Sulfur on Io would boil at only

550°C, so the volcanoes must be erupting molten rock and not just liquid sulfur. Also, a few isolated mountains exist that are as high as 18 km, twice the height of Mount Everest. Sulfur is not strong enough to support such high mountains. These are all indications that the crust of Io is probably silicate rock.

Volcanism is continuous on Io. Plumes come and go over periods of months, but some volcanic vents, such as Pele, have been active since the *Voyager* spacecraft first visited Io in 1979 (Figure 14-11). Earth's explosive volcanoes eject lava and ash because of water dissolved in the lava. As rising lava reaches Earth's surface, the sudden decrease in pressure allows the water to come out of solution in the lava, like popping the cork on a bottle of champagne: The water flashes into vapor and blasts material out of the volcano, the process that was responsible for the Mount St. Helens explosion in 1980. But Io is dry. Instead, its volcanoes appear to be powered by sulfur dioxide dissolved in the magma. When the pressure on the magma is released, the sulfur dioxide boils out of solution and blasts gas and ash high above the surface in plumes up to 500 km (300 mi) high. Ash falling back to the surface produces debris layers around the volcanoes, such as those around Pele in Figure 14-11. Whitish areas on the surface are frosts of sulfur dioxide.

Great lava flows can be detected carrying molten material downhill, burying the surface under layer after layer. Sometimes lava bursts upward through faults to form long lava curtains, a form of eruption seen in Hawai'i. Both of these processes are shown in Figure 14-11.

What powers Io? It has abundant internal heat, but it is only 5 percent bigger than Earth's Moon, which is cold and dead. Io is too small to have retained heat from its formation or to remain hot from radioactive decay. In fact, the energy blasting out of its volcanoes adds up to about three times more energy than it could make by radioactive decay in its interior.

The answer is that Io is heated by a stronger version of the kind of tidal heating that has affected Ganymede and Europa. Because Io is so close to Jupiter, the tides it experiences are powerful and should have forced Io's orbit to become circular long ago. Io, however, is strongly influenced by its neighboring moons. Io, Europa, and Ganymede are locked in an orbital resonance; in the time it takes Ganymede to orbit once, Europa orbits twice and Io four times. This gravitational interaction keeps the orbits, especially Io's, slightly eccentric; and Io, also being closest to Jupiter, suffers dramatic tides, with its surface rising and falling by about 100 m (330 ft). For comparison, tides on Earth move the solid ground by only a few centimeters. The resulting friction in Io is enough to melt the interior and drive volcanism. In fact, there is enough energy flowing outward to continuously recycle Io's crust: Deep layers melt, are spewed out through the volcanoes to cover the surface, and are later covered themselves until they are buried so deeply that they are again melted.

The four Galilean moons show a clear sequence of more and more tidal heating the nearer they are to Jupiter. The more distant moons have geologies dominated by impacts, whereas the closer moons are dominated by heat flow from inside and have few craters. What a difference a few hundred thousand miles makes!

The History of the Galilean Moons

Each time you have finished studying a world, you have tried to summarize its history. Now you have studied a system of four small worlds. Can you tell their story? To do that you need to draw on what you have learned about the moons and also on what you have learned about Jupiter and the origin of the Solar System (look back to Chapter 10).

The minor, irregular moons of Jupiter are probably captured asteroids, but the regular Galilean moons seem to be primordial. That is, they formed with Jupiter. Also, they seem to be interrelated in that their densities decrease with their distance from Jupiter (Table 14-2).

From all the evidence, astronomers propose that the four moons formed in a disk-shaped nebula around Jupiter—a mini solar nebula—in much the same way the planets formed from the solar nebula around the Sun. As Jupiter grew massive, it would have formed a hot, dense disk of matter around its

TABLE 14-2 The Galilean Moons*

Name	Radius (km)	Density (g/cm³)	Orbital Period (days)
Io	1820	3.5	1.77
Europa	1560	3.0	3.55
Ganymede	2630	1.9	7.15
Callisto	2410	1.8	16.69

*For comparison, the radius of Earth's Moon is 1740 km, and its density is 3.3 g/cm³.
© 2016 Cengage Learning®

equator. The moons could have condensed inside that disk with the innermost moons, Io and Europa, forming from rocky material and the outer moons, Ganymede and Callisto, incorporating more ice. This hypothesis follows the same condensation sequence that led to rocky planets forming near the Sun and ice-rich worlds forming farther away.

There are objections to this hypothesis. The disk around Jupiter would have been dense and hot, and moons would have formed rapidly, perhaps in only 1000 years. If the moons formed quickly, the heat of formation released as material fell into the moons would not have leaked away quickly, and they would have grown so hot they would have lost their water. Ganymede and Callisto are rich in water. Furthermore, Callisto has never been hot enough to differentiate and form a core. Also, mathematical models show that moons orbiting in the dense disk would have swept up debris and lost orbital momentum; they would have spiraled into Jupiter within a century.

A newer hypothesis proposes that Jupiter's early disk was indeed dense and hot and may have created moons, but those moons spiraled into the planet and were lost. Only later, as the disk grew thinner and cooler, did the present Galilean moons begin to form. Additional material may have dribbled slowly into the disk, and the moons could have formed slowly enough to retain their water and avoid spiraling into Jupiter. In this scenario, many large moons may have accreted around Jupiter. The Galilean moons you observe now would thus be only the last batch of moons, formed when the disk of construction material around Jupiter had become thin enough that they did not fall into Jupiter and disappear. In Chapter 10 you learned that the same migration and destruction processes may apply, on a larger scale, to planets forming in extrasolar planetary systems.

You can combine this hypothesis with what you know about tidal heating to understand the interiors of the moons. The moons formed relatively slowly, over perhaps 100,000 years, and were not heated severely by infalling material. The inner moons, however, were cooked by tidal heating—possibly enhanced when an orbital resonance developed between Ganymede, Europa, and Io. The innermost moon, Io, was heated so much it lost all of its water, and Europa retained only a small amount. Ganymede was heated enough to differentiate but retained much of its water. Callisto, orbiting far from Jupiter and avoiding orbital resonances, was never heated enough to differentiate. The Galilean moons as they appear today seem to be the result of a combination of slow formation in a hot nebula and tidal heating.

Jupiter's satellite system is full of clues to the history of the Solar System. And, as it turns out, there is an intricate and close relationship between the moons and rings of the Jovian planets.

Jupiter's Rings

Astronomers have known for centuries that Saturn has rings, but Jupiter's rings were not discovered until 1979 when the *Voyager 1* spacecraft sent back photos. The discovery was confirmed soon after by difficult ground-based measurements. Less than 1 percent as bright as Saturn's rings, the ghostly rings around Jupiter are a puzzle. What are they made of? Why are they there? A few simple observations will help you solve some of these puzzles.

Saturn's rings are made of bright ice chunks, but the particles in Jupiter's rings are very dark and reddish. This is evidence that those rings are rocky rather than icy. You can also conclude that the ring particles are mostly microscopic. Photos show that Jupiter's rings are very bright when illuminated from behind (Figure 14-13). In other words, they are scattering light forward. Efficient **forward scattering** occurs when particles have diameters roughly the same as the wavelength of light, a few millionths of a meter. Large particles do not scatter light forward, so a ring filled with basketball-size particles would look dark when illuminated from behind. The forward scattering tells you that Jupiter's rings are made mostly of particles about the size of those in smoke.

Larger particles are not entirely ruled out. A sparse component of rocky objects ranging from pieces of gravel to boulders is possible, but objects larger than 1 km would have been detected in spacecraft photos. The vast majority of the ring particles are microscopic dust.

The size of the ring particles is a clue to their origin, and so is their location. They orbit inside the **Roche limit**, the distance from a planet within which a moon cannot hold itself together by its own gravity. If a moon orbits relatively far from its planet, then the moon's gravity will be much greater than the tidal forces caused by the planet, and the moon will be able to hold itself together. If, however, a planet's moon comes inside the Roche limit, the tidal forces can overcome its gravity and pull the moon apart. The International Space Station can orbit inside Earth's Roche limit because it is welded and

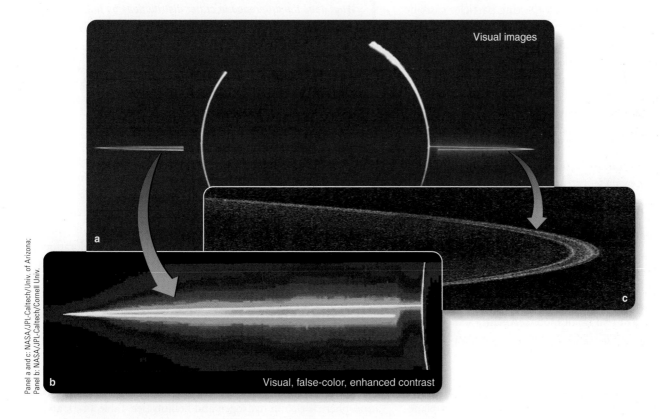

Visual images

Visual, false-color, enhanced contrast

▲ **Figure 14-13** (a) The main ring of Jupiter, illuminated from behind, glows brightly in this visual image made by the *Galileo* spacecraft while it was within Jupiter's shadow. (b) Digital enhancement and false color reveal the halo of ring particles that extends above and below the main ring. The halo is just visible in panel (a). (c) Structure in the ring is probably caused by the gravitational influence of Jupiter's inner moons.

bolted together, and a single large rock can survive inside the Roche limit if it is strong enough not to break. However, a moon composed of separate rocks and particles held together by their mutual gravity could not survive inside a planet's Roche limit. Tidal forces would destroy such a moon. If a planet and its moon have the same average densities, the Roche limit is at 2.44 times the planet's radius. Jupiter's main ring has an outer radius of 130,000 km (1.8 Jupiter radii) and therefore lies inside Jupiter's Roche limit. The rings of Saturn, Uranus, and Neptune also lie within those planets' respective Roche limits.

Now you can understand the dust in Jupiter's rings. If a dust speck gets knocked loose from a larger rock orbiting inside the Roche limit, the rock's gravity cannot hold the dust speck. And the billions of dust specks in the rings can't pull themselves together to make a larger body—a moon—because of the tidal forces inside the Roche limit.

You can also be sure that the ring particles are not old. The pressure of sunlight and Jupiter's powerful magnetic field alter the orbits of the particles, and they gradually spiral into the planet. Images show faint ring material extending down toward Jupiter's cloud tops, and this is evidently dust grains spiraling inward. Dust is also lost from the rings as electromagnetic effects force the particles out of the ring plane to form a low-density halo above and below the rings (Figure 14-13b). Yet another reason the ring particles can't be old is that the intense radiation around Jupiter can grind dust specks down to nothing in a century or so. For all these reasons, the rings seen today can't be made up of material that has been in the form of small particles for the entire time since the formation of Jupiter.

Obviously, the rings of Jupiter must be continuously resupplied with new material. Dust particles can be chipped off rocks ranging in size from gravel to boulders within the rings, and small moons that orbit near the outer edge of the rings lose particles as they are hit by meteorite impacts. Observations made by the *Galileo* spacecraft show that the main ring is densest at its outer edge, where the small moon Adrastea orbits, and that another small moon, Metis, orbits inside the ring. Clearly these moons must be structurally strong to withstand Jupiter's tidal forces. Images from the *Voyager* and *Galileo* probes also reveal much fainter rings, called the **gossamer rings**, extending twice as far from the planet as the main ring. These gossamer rings are densest at the orbits of two small moons, Amalthea and Thebe, more evidence that ring particles are being blasted into space by impacts on the moons.

Besides supplying the rings with particles, the moons help confine the ring particles and keep them from spreading outward. You will find that this is an important process in planetary rings when you study the rings of Saturn later in this chapter.

Your exploration of Jupiter reveals that it is much more than just a big planet. It is the gravitational and magnetic center of an entire community of objects. In the next section you will study Saturn, the ruler of another large celestial community.

DOING SCIENCE

What produces Io's internal heat? Io ought not to have substantial internal heat, but evidently it does. Finding explanations for unexpected phenomena is an especially fun part of doing science.

In this case, you understand that small worlds lose their internal heat quickly and become geologically inactive. Io is only slightly larger than Earth's Moon, which is cold and dead, but Io is full of energy flowing outward. Clearly, Io must have a powerful source of heat inside, and that heat source is tides. Io's orbit is slightly elliptical, so it is sometimes closer to Jupiter and sometimes farther away. This means that Jupiter's powerful gravity sometimes squeezes Io more than at other times, and the flexing of the moon's interior produces heat through friction. Such tides would rapidly force Io's orbit to become circular, and then tidal heating would end and the planet would become inactive—except that the gravitational tugs of the other moons keep Io's orbit eccentric. Thus, it is the influence of its companions that keeps Io in such an active state.

Continue your exploration of Jupiter's moons by using comparative planetology. Io has almost no impact craters, but Callisto has many. *What does the difference in crater distributions on the four Galilean moons tell you about their histories?*

14-4 Saturn

Saturn has played second fiddle to its own rings since Galileo first saw them in 1610. He didn't recognize the rings for what they are, but today they are instantly recognizable as one of the wonders of the Solar System. Nevertheless, Saturn itself, not quite ten times Earth's diameter (■ Celestial Profile 8), is a fascinating planet with a few mysteries of its own. Your exploration of Saturn and its rings can make use of the principles you have learned from Jupiter.

Surveying Saturn

The basic characteristics of Saturn reveal its composition and interior. Only about a third of the mass of Jupiter and 15 percent smaller in diameter, Saturn has an average density of 0.7 g/cm³. It is less dense than water; it would float! Spectra show that its atmosphere is rich in hydrogen and helium (see Table 14-1), and models predict that it is mostly liquid hydrogen with a core of heavy elements.

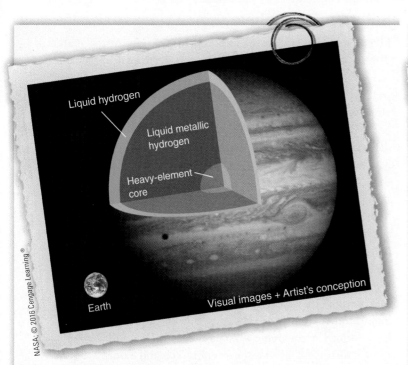

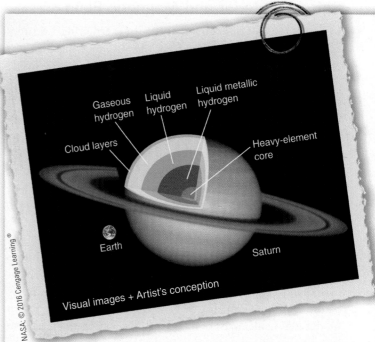

Jupiter is mostly a liquid planet. It may have a small core of heavy elements not much bigger than Earth.

Saturn's atmosphere blends gradually into its liquid interior. The size of its core is uncertain.

Celestial Profile 7 Jupiter

Motion:

Average distance from the Sun	5.20 AU (7.79 × 10⁸ km)
Eccentricity of orbit	0.048
Inclination of orbit to ecliptic	1.3°
Orbital period	11.9 y
Period of rotation	9.92 h
Inclination of equator to orbit	3.1°

Characteristics:

Equatorial diameter	1.43 × 10⁵ km (11.2 $D_\oplus$)
Mass	1.90 × 10²⁷ kg (318 $M_\oplus$)
Average density	1.33 g/cm³
Gravity at cloud tops	2.5 Earth gravities
Escape velocity	59.5 km/s (5.3 $V_\oplus$)
Temperature at cloud tops	145 K (−200°F)
Albedo	0.34
Oblateness	0.065

Personality Point:

Jupiter is named for the Roman king of the gods and is the largest planet in our Solar System. It can be very bright in the night sky, and its cloud belts and four largest moons can be seen through even a small telescope or a good pair of binoculars mounted on a tripod. Its moons are visible even with a good pair of binoculars mounted on a tripod or braced against a wall.

Celestial Profile 8 Saturn

Motion:

Average distance from the Sun	9.58 AU (1.43 × 10⁹ km)
Eccentricity of orbit	0.056
Inclination of orbit to ecliptic	2.5°
Orbital period	29.5 y
Period of rotation (sidereal)	10.57 h
Inclination of equator to orbit	26.7°

Characteristics:

Equatorial diameter	1.21 × 10⁵ km (9.45 $D_\oplus$)
Mass	5.68 × 10²⁶ kg (95.2 $M_\oplus$)
Average density	0.69 g/cm³
Gravity at cloud tops	1.1 Earth gravities
Escape velocity	35.5 km/s (3.2 $V_\oplus$)
Temperature at cloud tops	95 K (−290°F)
Albedo	0.34
Oblateness	0.098

Personality Point:

The Greek god Cronus was forced to flee when his son Zeus took power. Cronus fled to Italy, where the Romans called him Saturn, protector of the sowing of seed. He was celebrated in a weeklong wild party called the Saturnalia at the time of the winter solstice. Early Christians took over the holiday to celebrate Christmas.

Saturn's Interior and Magnetic Field

Infrared observations show that Saturn is radiating 1.8 times as much energy as it receives from the Sun, showing that heat is flowing out of its interior. It must be very hot inside Saturn, as is also true for Jupiter. In fact, Saturn's interior is too hot. It should have lost more heat since it formed. Astronomers have calculated models indicating that helium in the liquid hydrogen interior is condensing into droplets and falling inward. The falling droplets, releasing energy as they pick up speed, heat the planet. This heating is similar to the heating produced when a star contracts and may also occur to some extent in the atmospheres of Jupiter, Uranus, and Neptune.

Just as for Jupiter, you can learn more about Saturn's interior from its magnetic field. Spacecraft have found that Saturn's magnetic field is about 20 times weaker than Jupiter's. It also has correspondingly weaker radiation belts. Models comparing Saturn with Jupiter predict that the lower pressure inside Saturn produces a smaller mass of liquid metallic hydrogen. Heat flowing outward causes convection in this conducting layer, and the rapid rotation drives a dynamo effect that produces the magnetic field. Unlike most magnetic fields, Saturn's is not inclined to its axis of rotation, something you can see in UV images that show rings of auroras around Saturn's poles (Figure 14-14). This perfect alignment between Saturn's magnetic axis and the axis of rotation is peculiar, is not observed for any other planet, and is not understood.

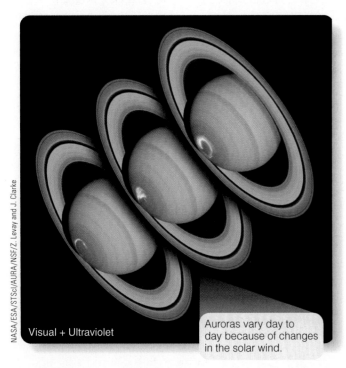

Visual + Ultraviolet

Auroras vary day to day because of changes in the solar wind.

▲ **Figure 14-14** Auroras on Saturn occur in rings around the planet's magnetic poles. Because the magnetic field is not inclined to the axis of rotation, the auroral rings occur nearly at the planet's geometrical poles. (Compare with Figure 14-4.)

Saturn's Atmosphere

Like that of Jupiter, Saturn's atmosphere is rich in hydrogen and displays belt–zone circulation, which appears to arise in the same way as the circulation patterns on Jupiter. The light-colored zones are higher clouds formed by rising gas, and the darker belts are lower clouds formed by sinking gas.

Notice, however, that the zones and belt clouds are not very distinct on Saturn compared with Jupiter (Figure 14-15a). Measurements from the *Voyager* and *Cassini* spacecraft indicate that Saturn's atmosphere is much colder than Jupiter's, something you would expect because Saturn is twice as far from the Sun and receives only one-fourth as much solar energy per square meter. The clouds on Saturn form at about the same temperature as the clouds on Jupiter, but those temperature levels are deeper in Saturn's cold atmosphere. Compare the cloud layers in Figure 14-15b with those shown in the diagram on page 303. Because they are deeper in the atmosphere, the cloud layers look dimmer from your viewpoint outside, and a high layer of haze formed by methane crystals makes the cloud layers even more indistinct. The atmospheres of Jupiter and Saturn are quite similar once you account for the fact that Saturn is colder.

One dramatic difference between Jupiter and Saturn concerns the winds. On Jupiter, winds form the boundaries for each of the belts and zones, but on Saturn the pattern is not the same. Saturn has fewer such winds, but they are much stronger. The eastward wind at the equator of Saturn, for example, blows at 500 m/s (1100 mph), roughly five times faster than the eastward wind at Jupiter's equator. The reason for this difference is not clear.

> ### DOING SCIENCE
>
> ***Why do the belts and zones on Saturn look so much fainter than the ones on Jupiter?*** One of the most powerful tools of critical thought available to scientists, and to people in general, is simple comparing and contrasting.
>
> In the atmosphere of Jupiter, the dark belts form in regions where gas sinks, and zones form where gas rises. The rising gas cools and condenses to form icy crystals of ammonia, which are visible as bright clouds. Clouds of ammonium hydrosulfide and water form deeper, below the ammonia clouds, and are not as visible.
>
> Saturn is twice as far from the Sun as Jupiter, so sunlight is four times dimmer. The atmosphere is colder, and gas currents do not have to rise as far to reach cold levels and form clouds. This means that the clouds are deeper in Saturn's atmosphere than they are in Jupiter's atmosphere. Because the clouds are deeper, they are not as brightly illuminated by sunlight and look dimmer. Also, a layer of methane-ice-crystal haze high above the ammonia clouds makes the clouds even less distinct.
>
> Now compare and contrast the interiors of the two planets. ***How is Saturn's magnetic field similar to, and different from, Jupiter's magnetic field?***

NASA/ESA/STScI/AURA/NSF/Z. Levay and J. Clarke

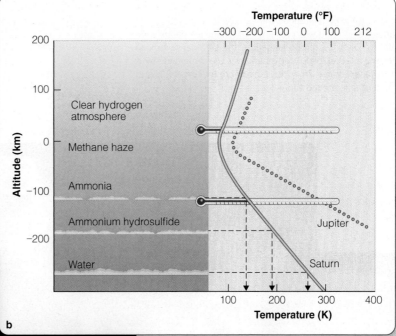

▼ **Figure 14-15** (a) Saturn's belt–zone circulation is not very distinct at visible wavelengths. These images were recorded when Saturn's southern hemisphere was tipped toward Earth. (b) Because Saturn is colder than Jupiter, the clouds form deeper in the hazy atmosphere. Notice that the three cloud layers on Saturn form at about the same respective temperatures as do the three cloud layers on Jupiter.

Ultraviolet

Visual

a Infrared

14-5 Saturn's Moons and Rings

Saturn has more than 60 moons with charted orbits—far too many to examine individually—but these moons share characteristics common to icy worlds. Most of them are small and dead, but one is big enough to have an atmosphere and perhaps even oceans or lakes—but not of water.

Titan

Saturn's largest satellite is a giant ice moon with a thick atmosphere and a mysterious surface. From Earth it is only a dot of light, with no visible detail. Nevertheless, a few basic observations can tell you a great deal about this strange world.

Titan's mass can be estimated from its influence on both passing spacecraft and on other moons. Its mass divided by its volume reveals that its density is 1.9 g/cm³, which means it is about a 60/40 mixture of rock and ice. Although its core

must be rocky, its mantle and crust contain a large amount of ice.

Titan is a bit larger than the planet Mercury and almost as large as Jupiter's moon Ganymede. Unlike those worlds, Titan has a thick atmosphere. Its escape velocity is low, but it is so far from the Sun that it is very cold, and most gas atoms don't move fast enough to escape (review Figure 13-11). Most of Titan's atmosphere is nitrogen with about 1.6 percent methane. A variety of organic compounds more complex than methane such as acetylene, propane, and hydrogen cyanide have been detected in observations from Earth as well as from the *Cassini* orbiter and *Huygens* probe, (Note that, although organic molecules are common in living things on Earth, they are not necessarily derived from living things. One chemist defined an organic molecule as "any molecule with a carbon backbone.")

When the *Voyager 1* and *Voyager 2* spacecraft flew past Saturn in the early 1980s, their cameras could not penetrate Titan's hazy atmosphere (**Figure 14-16**). Measurements showed that the average surface temperature is 94 K (−290°F), and the surface atmospheric pressure is 50 percent greater than on Earth. Model calculations show that in the conditions on Titan methane could condense from the atmosphere and fall as rain, so planetary scientists hypothesized that Titan should have rivers, lakes, and possibly oceans of methane.

Sunlight converts methane (CH_4) into the gas ethane (C_2H_6) plus a collection of other organic molecules.* Some of these molecules produce the smoglike haze, and as the smog particles gradually settle, they were predicted to deposit smelly,

▼ **Figure 14-16** (a) As the *Huygens* probe descended through Titan's smoggy atmosphere, (b) it photographed the surface from an altitude of 8 km (5 mi). Although no liquid was present, dark drainage channels led into the lowlands. (c) Once the probe landed on the surface, it radioed back photos showing a level plain and chunks of ice smoothed by a moving liquid. (d) Radar images from the *Cassini* orbiter reveal lakes of liquid methane and ethane around the poles.

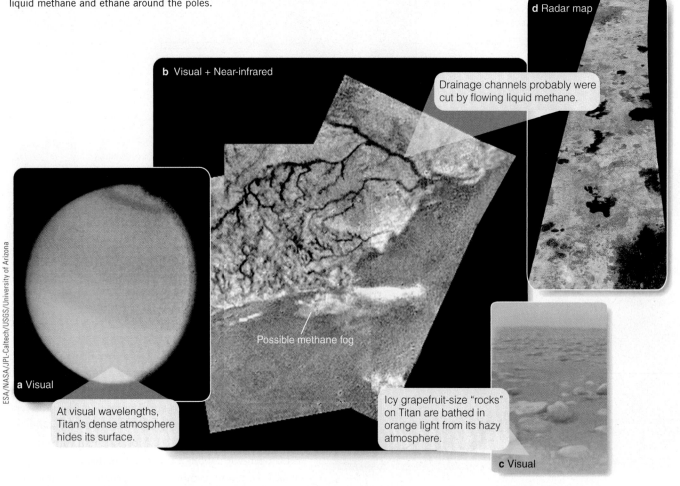

Cassini radar image of methane lakes that look dark because they do not reflect radio waves

d Radar map

b Visual + Near-infrared

Drainage channels probably were cut by flowing liquid methane.

Possible methane fog

a Visual

At visual wavelengths, Titan's dense atmosphere hides its surface.

Icy grapefruit-size "rocks" on Titan are bathed in orange light from its hazy atmosphere.

c Visual

ESA/NASA/JPL-Caltech/USGS/University of Arizona

organic goo on the surface. This goo is important because similar organic molecules may have been the precursors of life on Earth. You will examine this idea further in a later chapter.

Infrared cameras and radar instruments on *Cassini* spacecraft, which began exploring Saturn and its moons in 2004, have been able to see through the hazy atmosphere. The surface consists of icy, irregular highlands and smoother dark lowland areas. There are only a few craters, suggesting that geological activity is erasing craters almost as quickly as they are formed.

The *Cassini* spacecraft released the *Huygens* probe that parachuted down through the atmosphere of Titan and eventually landed on the surface. *Huygens* radioed back images of the surface as it descended under its parachute, and those images show dark drainage networks that lead into dark, smooth areas (Figure 14-16). Those dark regions, which look superficially like bodies of liquid, are actually dry or mostly dry. Precipitation may have washed the black goo off the highlands into the stream channels and lowlands so that they look smooth and

dark even though the liquid had temporarily evaporated. If you visit Titan, you could get caught in a shower of methane rain, but it probably won't rain often at your landing site.

When the *Huygens* probe landed on Titan's frigid surface, it radioed back measurements and images. The surface is mostly frozen water ice with some methane mixed in. The sunlight is orange because it has filtered down through the orange haze. Rocks littering the ground are actually steel-hard chunks of supercold water ice smoothed by erosion. Some rest in small depressions, suggesting that a liquid has flowed around them. You can see these depressions around the rocks in Figure 14-16c.

Radar observations made as the *Cassini* probe passed by Titan several times revealed lakes of liquid methane in its polar regions, confirming the earlier hypotheses about surface conditions. Some of those lakes are as large as Lake Superior. Evaporation from the lakes can maintain the 1.4 percent methane gas in the atmosphere, but sunlight eventually destroys methane, so Titan must have a large supply of methane ice.

Planetary scientists hypothesize that ice volcanoes on Titan may occasionally vent methane into the atmosphere.

By the way, before you go to Titan, check your spacesuit for leaks. Nitrogen is not a reactive gas, but methane is used as cooking gas on Earth and is highly flammable. Of course, there is no free oxygen on Titan, so you are safe so long as your spacesuit does not leak oxygen.

Saturn's Smaller Moons

In addition to Titan, Saturn has a large family of smaller moons. They are mixtures of rock and ice and are heavily cratered. Some of the smallest are probably captured objects and are geologically dead, but some of the larger moons show traces of geological activity. You can compare the sizes of a few of these moons in **Figure 14-17**.

Phoebe is at the outer fringes of Saturn's satellite family, and it moves in the retrograde direction; that is, it orbits backward. It is quite small, only about 210 km (130 mi) in diameter, but is nevertheless the largest of Saturn's irregular satellites. Phoebe's surface is dark, with an albedo of only 6 percent, and heavily cratered (**Figure 14-18**). Traces of ice are detected where impacts have excavated deeper layers or where landslides have exposed

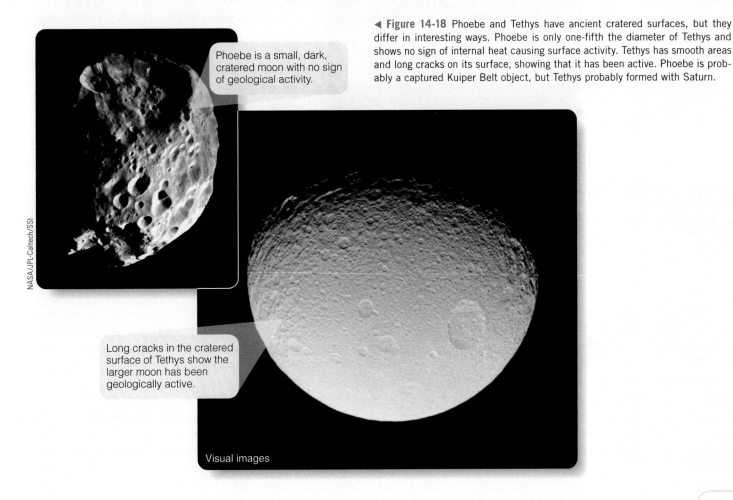

▲ **Figure 14-17** A few of Saturn's moons compared with Earth's Moon at the right. In general, larger moons are round and more likely to show signs of geological activity. Small moons such as Phoebe and Hyperion are cratered and do not have enough gravity to overcome the strength of their own material and squeeze themselves into a spherical shape.

fresh material. The density of Phoebe is 1.6 g/cm³, which is high enough to show that it contains a significant amount of rock. It seems unlikely that Phoebe came from the asteroid belt, where ice is relatively rare. It is more likely to be a captured Kuiper Belt object that was originally in an orbit beyond Neptune.

Other regular moons such as Tethys, which has a diameter of more than 1060 km (660 mi), are icy and cratered, but they show some signs of geological activity. Some smooth areas on

◄ **Figure 14-18** Phoebe and Tethys have ancient cratered surfaces, but they differ in interesting ways. Phoebe is only one-fifth the diameter of Tethys and shows no sign of internal heat causing surface activity. Tethys has smooth areas and long cracks on its surface, showing that it has been active. Phoebe is probably a captured Kuiper Belt object, but Tethys probably formed with Saturn.

Phoebe is a small, dark, cratered moon with no sign of geological activity.

Long cracks in the cratered surface of Tethys show the larger moon has been geologically active.

Visual images

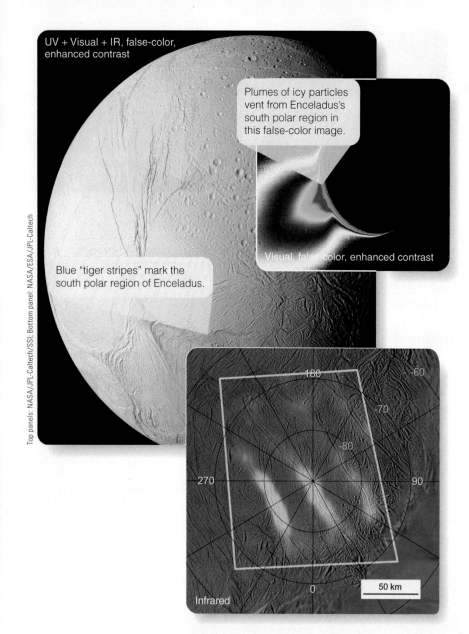

UV + Visual + IR, false-color, enhanced contrast

Plumes of icy particles vent from Enceladus's south polar region in this false-color image.

Blue "tiger stripes" mark the south polar region of Enceladus.

Visual, false-color, enhanced contrast

Infrared

50 km

Top panels: NASA/JPL-Caltech/SSI; Bottom panel: NASA/ESA/JPL-Caltech

▲ **Figure 14-19** The bright, clean, icy surface of Enceladus does not look old. Some areas have few craters, and the numerous cracks and lanes of grooved terrain resemble the surface of Jupiter's moon Ganymede. Enceladus is venting water, ice, and organic molecules from geysers near its south pole. A thermal infrared image reveals internal heat leaking to space from the "tiger stripe" cracks where the geysers are located.

Tethys appear to have been resurfaced by flowing water "lava," and long cracks and grooves may have formed when geological activity strained the icy crust (Figure 14-18).

With a diameter of 520 km (320 mi), the small moon Enceladus isn't much larger than Phoebe, but Enceladus shows dramatic signs of geological activity (**Figure 14-19**). For one thing, Enceladus has an albedo of 0.99. That is, it reflects 99 percent of the sunlight that hits it, and that makes it the most reflective object in the Solar System. You know that old icy surfaces become dark, so the surface of Enceladus must be quite young.

Look closely at the surface and you will see that some regions have few craters and also that grooves and cracks are common. Observations made by the *Cassini* spacecraft show that Enceladus has a tenuous atmosphere of water vapor and nitrogen. It is too small to keep such an atmosphere, so it must be releasing gas continuously. *Cassini* detected a large cloud of water vapor over the moon's south pole where water vents through cracks and produces ice-crystal jets extending hundreds of kilometers above the surface. Infrared images made by *Cassini* show significant amounts of heat escaping to space through the same cracks from which the water is venting (Figure 14-19).

The possibility of liquid water below the icy crust of Enceladus has excited those scientists searching for life on other worlds. You will read more about this possibility in the final chapter of this book. Nevertheless, it will be a long time before explorers can drill through the crust and analyze the water below for signs of living things.

Of course, you are wondering how a little moon like Enceladus can have heat flowing up from its interior. With a density of 1.6 g/cm^3, Enceladus must contain a significant rocky core, but radioactive decay is not enough to keep it active. A clue lies in the moon's orbit. Enceladus orbits Saturn in a resonance with the larger moon Dione. Each time Dione orbits Saturn once, Enceladus orbits twice. That means Dione's gravitational tugs on Enceladus always occur in the same places and make the orbit of the little moon slightly eccentric. As Enceladus follows that eccentric orbit around Saturn, tides flex it, and tidal heating warms the interior. You saw how resonances and tidal heating keep some of Jupiter's moons active; now you can add Enceladus to the list.

The *Voyager* spacecraft discovered small moonlets trapped at the L4 and L5 Lagrange points in the orbits of Dione and Tethys. These points of stability lie 60 degrees ahead of and 60 degrees behind the two moons, and small moonlets can become trapped in these regions. (You will see in a later chapter that some asteroids are trapped in the Lagrange points of Jupiter's and Neptune's orbits around the Sun.) This gravitational curiosity is therefore not unique to the Saturn system.

Saturn has too many moons to discuss in detail here, but you should meet at least one more, Iapetus (pronounced *ee-YAP-eh-tus*). It is literally an odd ball: Iapetus is an asymmetric moon. Its trailing side, the side that always faces backward as it orbits Saturn, is old, cratered, icy, and about as bright as dirty snow.

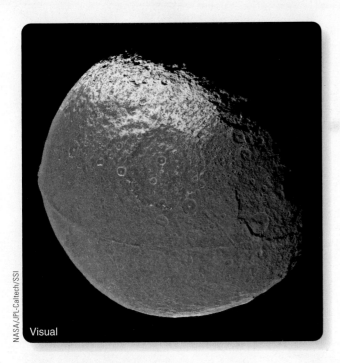

NASA/JPL-Caltech/SSI

Visual

▲ **Figure 14-20** Like the windshield of a speeding car, the leading side of Saturn's moon Iapetus seems to have accumulated a coating of dark material. The poles and trailing side of the moon have much cleaner ice. The equatorial ridge is 20 km (12 mi) wide and up to 13 km (8 mi) high. It stretches roughly 1300 km (800 mi) along the moon's equator.

Its leading side, the side that always faces forward in its orbit, is also old and cratered, but it is much darker than you would expect. It has an albedo of only 4 percent—about as dark as fresh asphalt on a highway (Figure 14-20). The origin of this dark material is unknown, but theorists suspect that the little moon has swept up dark, silicon- and carbon-rich material on its leading side. The source of the material covering the leading side of Iapetus could be meteorites striking and eroding the carbonaceous surface of the outermost moon, Phoebe, and tossing the resulting dust into space to be scooped up eventually by Iapetus.

Another odd feature on Iapetus shows up in *Cassini* images—an equatorial ridge that stands as high as 13 km (8 mi) in some places. You can see the ridge clearly in Figure 14-20. The origin of this ridge is unknown, but it is not a minor feature. It is more than 50 percent higher than Mount Everest, and it extends for a long distance across the surface. That is one big pile of rock and ice. The ridge sits atop an equatorial bulge, and both ridge and bulge may have formed when Iapetus was young, spun rapidly, and was still mostly molten.

Saturn's moons illustrate a number of principles of comparative planetology. Small moons are irregular in shape, and old surfaces are dark and cratered. Resonances can trigger tidal heating, and that can in turn resurface moons and outgas atmospheres. Small moons can't keep atmospheres, but big, cold moons can. You are an expert in all of this, so you are ready to wonder where the moons came from.

Origin of Saturn's Moons

Jupiter's four Galilean moons seem clearly related to one another, and you can safely conclude that they formed with Jupiter. No such simple relationships link Saturn's satellites. That seems to indicate that, unlike Jupiter, Saturn was not enough of a heat source during that system's formation to cause the densities of its regular moons to follow the condensation sequence. Planetary scientists also suspect that comet impacts have so badly fractured the regular moons that they no longer show much evidence of their common origin. Understanding the origin of Saturn's moons is also difficult because the moons interact gravitationally so that the orbits they now occupy may differ significantly from their earlier orbits.

The complex orbital relationships of Saturn's moons and their evidently intense cratering suggest that the moons have interacted and may have collided with each other, with comets, and with large planetesimals in the past. Nevertheless, as with Jupiter's moons, astronomers hypothesize that most or all of Saturn's regular moons formed with the planet and that the irregular moons are captured. As you continue your exploration of the outer Solar System, you can be alert for the presence of more such small, icy worlds.

Saturn's Rings

Looking at the beauty and complexity of Saturn's rings, an astronomer once said, "The rings are made of beautiful physics." You could add that the physics is actually rather simple, but the result is one of the most amazing sights in our Solar System.

In 1610, Galileo became the first human to see the rings of Saturn, but perhaps because of the poor optics in his telescopes, he did not recognize the rings as a disk. He drew Saturn as three objects—a central body and two smaller ones on either side. In 1659, Christiaan Huygens (after whom the 2005 Titan probe was named) realized that the rings form a disk surrounding but not touching the planet.

Understanding Saturn's rings has required human ingenuity continuing to the present day. In 1859, James Clerk Maxwell (for whom the large mountain on Venus is named) proved mathematically that solid rings would be unstable. Saturn's rings, he concluded, had to be made of separated particles. In 1867, Daniel Kirkwood demonstrated that gaps in the rings were caused by resonances with some of Saturn's moons. Spectra of the rings eventually showed that the particles were mostly water ice.

Study **The Ice Rings of Saturn** on pages 320–321 and notice three points and a new term:

❶ The rings are made up of billions of ice particles, each in its own orbit around the planet. But, just as for Jupiter's rings, the particles observed now in Saturn's rings can't have been there since the planet formed. The rings must be replenished now and then by impacts on Saturn's icy moons or by the disruption of a small moon that moves too close to the planet.

The brilliant rings of Saturn are made up of billions of ice particles ranging from microscopic specks to chunks bigger than a house. Each particle orbits Saturn in its own circular orbit. Much of what planetary scientists know about the rings has been learned from the *Voyager 1* and *2* flybys and the *Cassini* orbiter. From Earth, astronomers see three rings labeled A, B, and C. *Voyager* and *Cassini* images reveal over a thousand ringlets within the rings.

Saturn's rings can't be leftover material from the formation of Saturn. The rings are made of ice particles, and the planet would have been so hot when it formed that it would have vaporized and driven away any icy material. Rather, the rings must be debris from collisions between Saturn's icy moons and passing comets or asteroids. Impacts large enough to scatter ice throughout the Saturn system are estimated to occur every 100 million years or so. The ice would quickly settle into the equatorial plane, and some would become trapped in rings.

Although the ice will tend to waste away because of meteorite impacts and damage from radiation in Saturn's magnetosphere, new impacts could replenish the rings with fresh ice. The bright, beautiful rings you see today may be only a temporary enhancement caused by an impact that occurred since the extinction of the dinosaurs.

Cassini Division

Gap

A ring

B ring

C ring

As in the case of Jupiter's ring, Saturn's rings lie inside the planet's Roche limit where the ring particles cannot pull themselves together to form a moon.

Because it is so dark, the C ring was once called the crepe ring.

Earth to scale

Visual

The C ring contains boulder-size chunks of ice, whereas most particles in the A and B rings are more like golf balls, down to dust-size ice crystals. Further, C ring particles are less than half as bright as particles in the A and B rings. *Cassini* observations show that the C ring particles contain less ice and more minerals.

1a An astronaut could swim through the rings. Although the particles orbit Saturn at high velocity, all particles at the same distance from the planet orbit at about the same velocity, so they collide gently at low relative speeds. If you could visit the rings, you could push your way from one icy particle to the next. This artwork is based on a model of particle sizes in the A ring.

NASA; Line art: © 2016 Cengage Learning®

© 2016 Cengage Learning®

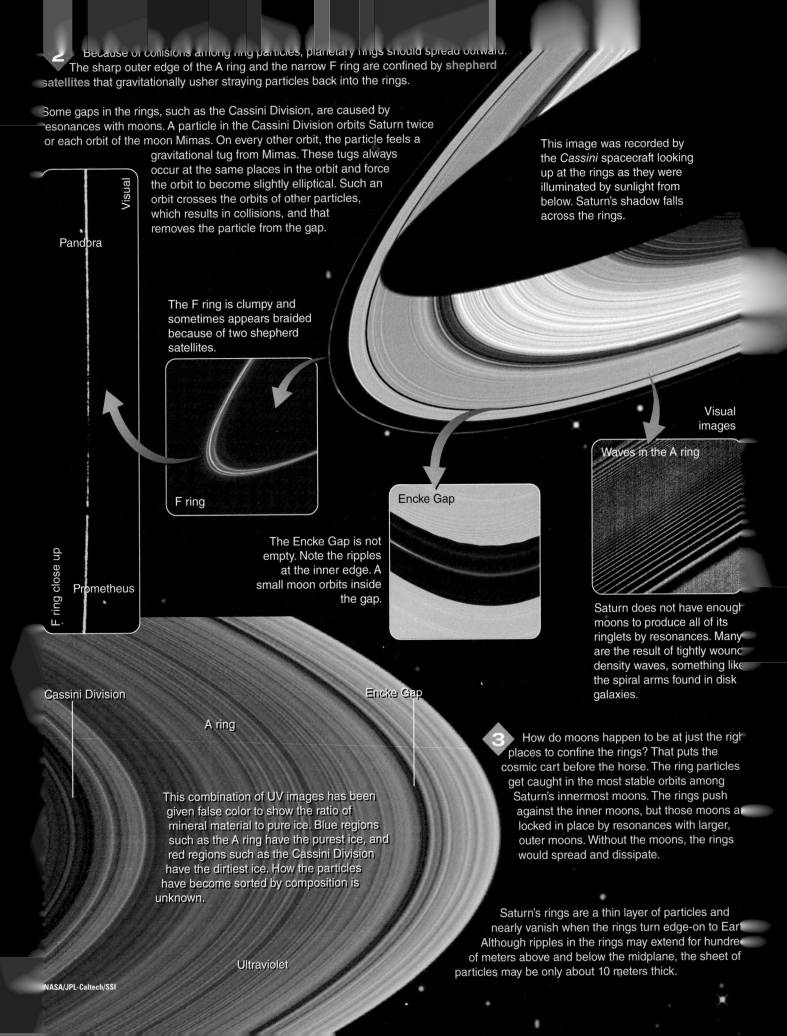

2 Because of collisions among ring particles, planetary rings should spread outward. The sharp outer edge of the A ring and the narrow F ring are confined by **shepherd satellites** that gravitationally usher straying particles back into the rings.

Some gaps in the rings, such as the Cassini Division, are caused by resonances with moons. A particle in the Cassini Division orbits Saturn twice for each orbit of the moon Mimas. On every other orbit, the particle feels a gravitational tug from Mimas. These tugs always occur at the same places in the orbit and force the orbit to become slightly elliptical. Such an orbit crosses the orbits of other particles, which results in collisions, and that removes the particle from the gap.

This image was recorded by the *Cassini* spacecraft looking up at the rings as they were illuminated by sunlight from below. Saturn's shadow falls across the rings.

Visual

Pandora

The F ring is clumpy and sometimes appears braided because of two shepherd satellites.

F ring

F ring close up

Prometheus

Visual images

Waves in the A ring

The Encke Gap is not empty. Note the ripples at the inner edge. A small moon orbits inside the gap.

Encke Gap

Saturn does not have enough moons to produce all of its ringlets by resonances. Many are the result of tightly wound density waves, something like the spiral arms found in disk galaxies.

Cassini Division

A ring

Encke Gap

This combination of UV images has been given false color to show the ratio of mineral material to pure ice. Blue regions such as the A ring have the purest ice, and red regions such as the Cassini Division have the dirtiest ice. How the particles have become sorted by composition is unknown.

3 How do moons happen to be at just the right places to confine the rings? That puts the cosmic cart before the horse. The ring particles get caught in the most stable orbits among Saturn's innermost moons. The rings push against the inner moons, but those moons are locked in place by resonances with larger, outer moons. Without the moons, the rings would spread and dissipate.

Saturn's rings are a thin layer of particles and nearly vanish when the rings turn edge-on to Earth. Although ripples in the rings may extend for hundreds of meters above and below the midplane, the sheet of particles may be only about 10 meters thick.

Ultraviolet

2 The gravitational effects of small moons called *shepherd satellites* can confine some rings in narrow strands or keep the edges of rings sharp. Moons can also produce waves in the rings that are visible as tightly wound ringlets.

3 The ring particles lie in a thin layer in Saturn's equatorial plane and are prevented from spreading outward by the gravity of small moons. The small moons in turn are controlled by gravitational interactions with larger, more distant moons. The rings of Saturn, and the rings of the other Jovian worlds, are created from and controlled by the planet's moons. Without the moons, there would be no rings.

Modern astronomers find simple gravitational interactions producing even more complex processes in the rings. Where particles orbit in resonance with a moon, the moon's gravity triggers spiral density waves in much the same way that spiral arms are produced in galaxies. The spiral density waves spread outward through the rings. If the moon follows an orbit that is inclined to the ring plane, the moon's gravity causes a different kind of wave—spiral bending waves—with ripples extending above and below the ring plane and spreading inward. Both of these kinds of processes are shown in the inset ring images on page 321.

Many other processes occur in the rings. Specks of dust become electrically charged by sunlight, and Saturn's magnetic field lifts them out of the ring plane. Small moonlets embedded in the rings produce gaps, waves, and scallops in the rings. The *Cassini* spacecraft has recorded dramatic images (**Figure 14-21**) of the Saturn ring system, including two faint outer rings (E and G) that are rarely detectable from Earth. The E ring appears to be replenished at least in part by ice crystals blasted into space by the geysers on Enceladus, and the source of the G ring seems to be a tiny embedded moonlet discovered in *Cassini* orbiter images.

The word *particle* in colloquial language connotes tiny specks, but in the context of Saturn's rings astronomers use that term to refer to any object from snowlike powder grains up to building-size icy minimoons (look again at page 320). The larger objects are understood to be aggregates of the smaller ones. The subtle colors of the rings arise from contamination in the ice, and some areas have unusual compositions. The Cassini Division, for instance, contains particles that are richer in rock than most of the ring. No one knows how these differences in composition arise, but they must be related to the way the rings are formed and replenished.

Like a beautiful flower, the rings of Saturn are controlled by many different natural processes. Observations from spacecraft such as *Voyager* and *Cassini* will continue to reveal even more about the rings. Such missions are expensive, of course, but they are helping us understand what we are (**How Do We Know? 14-1**).

A History of the Saturn System

The farther you journey from the Sun, the more difficult it is to understand the history of the planets. Any fully successful history of Saturn should explain its low density, its peculiar magnetic field, and its beautiful rings. Planetary scientists can't

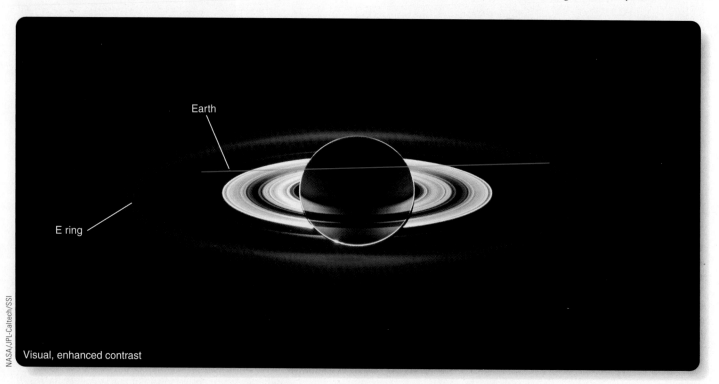

NASA/JPL-Caltech/SSI

Visual, enhanced contrast

Earth

E ring

▲ **Figure 14-21** The *Cassini* spacecraft recorded this image as it passed through Saturn's shadow. Earth is visible as a faint blue dot just inside the G ring, and jets of ice particles vented from the moon Enceladus are visible at the left extreme of the larger E ring. Two faint rings associated with small moons were discovered in this image.

How Do We Know? 14-1

Who Pays for Science?

Why shouldn't you plan for a career as an industrial paleontologist? Searching out scientific knowledge can be expensive, and that raises the question of funding. Some science has direct applications, and industry supports such research. For example, pharmaceutical companies have large budgets for scientific research leading to the creation of new drugs. But some basic science is of no immediate practical value. Who pays the bill?

A paleontologist is a scientist who studies ancient life forms by examining fossils of plant and animal remains, and such research does not have commercial applications. Except for the rare Hollywood producer about to release a dinosaur movie, corporations can't make a profit from the discovery of a new dinosaur. The practical-minded stockholders of a company will not approve major investments in such research. Consequently, digging up dinosaurs, like astronomy, is poorly funded by industry.

It falls to government institutions and private foundations to pay the bill for this kind of research. The Keck Foundation has built two giant telescopes with no expectation of financial return, and the National Science Foundation has funded thousands of astronomy research projects for the benefit of society.

The discovery of a new dinosaur or a new galaxy is of no great financial value, but such scientific knowledge is not worthless. Its value lies in what it tells us about the world we live in. Such scientific research enriches our lives by helping us understand what we are. Ultimately, funding basic scientific research is a public responsibility that society must balance against other needs. There isn't anyone else to pick up the tab.

Sending the Cassini *spacecraft to Saturn costs each U.S. citizen 56¢ per year over the life of the project.*

---

tell a complete story yet, but you can understand a few of the principles that affected the formation of Saturn and its rings.

Most of Saturn's story parallels that of Jupiter. Saturn formed in the outer solar nebula, where ice particles were stable. It grew rapidly, becoming massive enough to capture hydrogen and helium gas directly from the nebula. The heavier elements probably form a denser core, and the hydrogen forms a liquid mantle containing liquid metallic hydrogen. The outward flow of heat from the core drives convection currents in this mantle that, coupled with the rapid rotation of the planet, produce its magnetic field. Because Saturn is smaller than Jupiter, it has less liquid metallic hydrogen, and its magnetic field is weaker.

The rings of Saturn definitely are not primordial, meaning the material in them now has not been in its current form since the formation of the planet. Saturn, like Jupiter, would have been very hot when it formed, and that heat would have vaporized and driven off any nearby small, icy particles of leftover material. Also, such a hot Saturn would have had a very distended atmosphere, which would have slowed ring particles by friction and caused the particles to fall into the planet. Finally, the processes that tend to destroy Jupiter's ring particles also apply to Saturn's rings.

Planetary rings do not seem to be stable over 4.6 billion years, so the ring material must have been produced more recently. Saturn's beautiful rings may have been created within just the past 100 million years, an astronomically short time. One suggestion is that a large comet, asteroid, or Kuiper Belt object struck one of Saturn's moons. Such a collision would produce a mix of icy and rocky debris, some of which would have settled into the ring plane. Bright planetary rings such as Saturn's may be temporary phenomena, forming when violent events produce fresh ice debris and then wasting away as the ice is gradually lost.

DOING SCIENCE

What features on Enceladus suggest that it has been active? Answering this type of question requires a scientist to extend comparative planetology to moons.

The smaller moons of Saturn are icy worlds mostly battered by impact craters, and you might suspect that they are all internally cold, with old surfaces. Small worlds lose their heat quickly; with no internal heat, there is no geological activity to erase impact craters. Enceladus, however, is peculiar. Although it is small and icy, its surface is highly reflective, and some areas seem almost free of craters. Grooves and faults mark some regions of the little moon and suggest motion in the crust. These features should have been destroyed long ago by impact cratering, so you must suppose that the moon has been geologically active at some time since the end of the heavy bombardment when planet building finished. But tall geyser plumes plus infrared emission from "tiger stripes" discovered near the south pole of Enceladus show that the moon is still active.

Now apply a different principle of comparative planetology to another moon. ***How can a world as small as Titan keep a thick atmosphere?***

People often describe science that has no known practical value as basic science or basic research. The exploration of distant worlds would be called *basic science,* and it is easy to argue that basic science is not worth the effort and expense because it has no known practical use. Of course, the problem is that no one has any way of knowing what knowledge will be of use until that knowledge is acquired.

In the middle of the 19th century, Queen Victoria asked physicist Michael Faraday what good his experiments with electricity and magnetism were. He answered, "Madam, what good is a baby?" Of course, Faraday's experiments were the beginning of the electronic age. Many of the practical uses of scientific knowledge that fill your world—digital electronics, synthetic materials, and modern vaccines—began as basic research. Basic scientific research provides the raw materials that technology and engineering use to solve problems; so, to protect its future, the human race must continue its struggle to understand how nature works.

Basic scientific research has yet one more important use that is so valuable it seems an insult to refer to it as *merely practical.* Science is the study of nature, and as you learn more about how nature works, you learn more about what your existence in this Universe means. The seemingly impractical knowledge gained from space probes visiting other worlds tells you about your own planet and your own role in the scheme of nature. Science tells us where we are and what we are, and that knowledge is beyond value.

Study and Review

Summary

▶ The outer planets in the Solar System—Jupiter, Saturn, Uranus, and Neptune—are much larger than Earth and lower in density. All the Jovian planets are rich in hydrogen and have rings, multiple satellite systems, and shallow atmospheres above liquid hydrogen mantles.

▶ Strong **belt–zone circulation (p. 297)** is seen in the atmospheres of Jupiter and Saturn. Belt-zone circulation is present but difficult to discern in the atmospheres of Uranus and Neptune. Belts are low-pressure areas where gas is sinking. Zones are high-pressure areas where gas is rising.

▶ Moons in the Jovian satellite systems interact gravitationally. Some moons are, or have been, heated internally by tides to produce geological activity. Most moons are old and cratered. **Regular satellites (p. 297)** are generally larger, orbit closer to the parent planet, orbit in the **prograde (p. 297)** direction, and have low eccentricities orbital inclinations. **Irregular satellites (p. 297)** are generally smaller, orbit further from the parent planet, and have high eccentricities and orbital inclinations.

▶ Observations taken from Earth show that Jupiter is 11 times Earth's diameter and 318 times Earth's mass. As Jupiter's density is much lower than Earth's and Jupiter is rich in hydrogen and helium, Jupiter cannot contain more than a small core of heavy elements.

▶ Not far below Jupiter's clouds, the temperature and pressure increase beyond the **critical point (p. 299)** at which gas, liquid, and solid phases can coexist. Thus, the transition from gaseous hydrogen to liquid hydrogen is gradual, and the liquid hydrogen layer has no definite surface.

▶ Jupiter's atmospheric composition is much like that of the Sun—mostly hydrogen and helium with smaller amounts of heavier elements.

▶ Infrared observations show that Jupiter radiates more heat than Jupiter receives from the Sun. Model calculations indicate that Jupiter's interior must be five or six times hotter than the Sun's surface, and its interior is prevented from flashing into vapor by its high pressure.

▶ Jupiter's high internal pressure results in a **liquid metallic hydrogen (p. 299)** interior. The **oblateness (p. 299)** of Jupiter arises because its interior is liquid and the planet rotates rapidly.

▶ The liquid metallic hydrogen supports a dynamo effect that generates a powerful magnetic field contributing, in part, to the rings of auroras around the planet's magnetic poles. The magnetic field also traps high-energy solar wind particles to form the planet's intense radiation belts.

▶ Ionized atoms from Jupiter's inner moon Io are swept up by Jupiter's rapidly orbiting and wobbling magnetic field to form the **Io plasma torus (p. 300)** that encloses the orbit of Io. Powerful electrical currents flow through the **Io flux tube (p. 300)** and produce enhanced aurora spots where the flux tube enters Jupiter's atmosphere.

▶ Jupiter's shallow atmosphere is rich in hydrogen, with three layers of clouds—ammonia, ammonium hydrosulfide, and water—that condense at different temperatures.

▶ Spots on Jupiter, such as the Great Red Spot, are long-lasting, cyclonic storm systems analogous to hurricanes on Earth.

▶ The four **Galilean moons (p. 298)** appear to have formed with Jupiter. Their densities generally decrease with distance from Jupiter, similar to the general decrease of planet densities with distance from the Sun that is caused by the condensation sequence. This is evidence that Jupiter was a strong luminosity source when those moons were forming.

▶ Callisto, the outermost Galilean moon, is composed of ice and rock and has an old and cratered surface. Unlike the three inner

Galilean moons, Callisto is not caught in an orbital resonance and does not appear to be active.

▶ Galilean moons Ganymede, Europa, and Io are locked in mutual orbital resonances, which results in **tidal heating (p. 306)** by Jupiter's tidal force that decreases with distance from Jupiter. Ganymede's surface is old and cratered in some areas, but bright **grooved terrain (p. 305)** must have been produced by a past episode of geological activity.

▶ The gravitational pull focusing of incoming objects by massive Jovian planets should result in more cratering impacts on inner moons. Hence, it is surprising to find that Europa and Io have almost no craters. This lack of craters is explained by geological activity caused by tidal heating.

▶ Europa is mostly rock with a thin, icy crust that contains only a few scars caused by past impact craters. Cracks and lines show that the crust has broken repeatedly. A subsurface ocean evidently vents through the crust and deposits ice to cover the craters as fast as the impact craters form. Europa's subsurface ocean could conceivably harbor life.

▶ Io is strongly heated by tides and has no water at all. More than 150 volcanoes erupt molten rock and throw ash high above the surface. No impact craters are visible because they have been destroyed or buried as fast as the craters can form. Sulfur compounds color the surface yellow and orange and vent into space to be swept up by Jupiter's magnetic field.

▶ At least some of Jupiter's irregular moons are probably captured asteroids.

▶ **Forward scattering (p. 311)** shows that Jupiter's ring is composed of tiny dust specks orbiting inside Jupiter's **Roche limit (p. 311)**. The dust particles cannot have survived since the formation of the planet. Rather, they are thought to be produced by meteorite impacts on some of Jupiter's inner moons, spraying fragments from the surfaces of the moons into space to be captured by Jupiter.

▶ A small moon can orbit inside a planet's Roche limit and survive if it is a solid piece of rock strong enough to endure the tidal forces tending to pull it apart.

▶ The dimmer **gossamer rings (p. 312)** lie near the orbits of two moons, further evidence that the rings are sustained by particles from moons.

▶ Saturn must have formed much as Jupiter did, but it has less than one-third of Jupiter's mass. Its average density is less than that of water. Saturn has a hot interior but model calculations indicate it contains less liquid metallic hydrogen than Jupiter, which explains why its magnetic field is weaker than Jupiter's. For some unknown reason, the magnetic fields axis is almost exactly aligned with the rotation axis of Saturn, unlike the magnetic field of any other planet.

▶ Saturn is twice as far from the Sun as Jupiter and is thus significantly colder. Saturn has the same three cloud layers as Jupiter, but because of the colder temperatures these cloud layers form deeper in Saturn's hazy atmosphere and are not as clearly visible from Earth.

▶ Saturn's rings are composed of ice particles and cannot have lasted since the formation of the planet. The rings must receive occasional additions of ice particles, perhaps when asteroids or comets hit the planet's icy moons, scattering ice particles into space that then settle into orbit around Saturn.

▶ Icy particles can become trapped in stable bands among the orbits of the innermost small moons that are within Saturn's Roche limit. Resonances with outer moons can produce gaps in the rings and generate waves that move like ripples through the rings. Small **shepherd satellites (p. 321)** can confine sections of the ring to produce sharp edges, ripples, and/or narrow ringlets.

Without moons to confine them, the rings would have spread outward and dissipated long ago.

▶ All of Saturn's moons have densities indicating they are mixtures of rock and ice. Some of the smaller moons, such as Phoebe, are probably captured asteroids or Kuiper Belt Objects.

▶ Titan, the largest moon, is so massive and cold that it can retain a dense atmosphere of nitrogen mixed with a small amount of methane. Models plus observations from the *Huygens* probe indicate that methane condenses from the atmosphere, falls as rain, and drains downslope, washing dark, organic material into lowland basins.

▶ The methane in Titan's atmosphere is destroyed by the UV component of sunlight, so it must be continuously replenished. Methane probably evaporates into the atmosphere from the lakes of liquid methane observed on in the moon's polar regions, and might also vent from volcanoes.

▶ Some of the Saturn's other moons, such as Tethys, have old, dark, cratered surfaces with cracks and smoothed areas that suggest past geological activity.

▶ Enceladus, a rather small moon, is the most reflective object in the Solar System. It has large smooth areas, so it must have been very geologically active recently. Organic-laden water vapor has been observed venting from cracks near the south pole of Enceladus. The water vents into space and forms ice crystals, which apparently resupply Saturn's E ring. Enceladus orbits in a resonance with the moon Dione, which may cause tidal heating of Enceladus's interior.

▶ The moon Iapetus has a bright, icy surface on its trailing side relative to its orbit around Saturn but a dark surface on its leading side. The leading side may be a coat of debris from the next moon out, Phoebe. Iapetus has a long equatorial ridge higher than Mount Everest on Earth. The ridge and the moon's equatorial bulge may have formed when the moon was young, molten, and spinning rapidly.

▶ The origin and evolution of Saturn's moons are not as clear as for Jupiter's Galilean moons. Orbital interactions and impacts have been important to these moons' evolution.

Review Questions

1. Describe four differences between the Jovian planets and the Terrestrial planets.

2. Why is Jupiter more oblate than Earth? Just because a planet is a Jovian planet, would it necessarily be more oblate than Earth? Why or why not?

3. Which molecules and atoms are Jupiter and Saturn able to retain in their atmospheres that can't be retained in Earth's atmosphere? (*Hints:* See Table 14-1 and Figure 13-11)

4. The ammonia hydrosulfide layer in Jupiter's and Saturn's atmospheres is cooler and at a lower altitude than the ammonium layer. True or false?

5. Jupiter radiates more energy than received from the Sun and so has nuclear fusion occurring in its core. True or false?

6. How does belt–zone circulation transport energy—by radiation, conduction, or convection? Explain your answer.

7. Why are belts and zones wrapped entirely around the planet?

8. What ingredients are needed to power a dynamo effect inside a planet?

9. Why are magnetic phenomena such as extensive radiation belts and auroras so strong around Jupiter?

10. How do the interiors of Jupiter and Saturn differ? How does this difference affect the magnetic fields of Jupiter and Saturn?

11. Which planet formation step did the Jovian planets undergo that the Terrestrial planets did not? Why?

12. Io is an example of a regular satellite. True or false?

13. Phoebe is an example of an irregular satellite. True or false?

14. If Jupiter had a satellite the size of our own Moon orbiting outside the orbit of Callisto, what would you predict for the satellite's density and surface features?

15. The density of Earth's Moon is 3.35 g/cm³. Which of Jupiter's moons has a density closest to Earth's Moon? What does this tell you about that moon?

16. Ganymede was once completely molten on the inside. True or false? How do you know?

17. Describe evidence of tectonic features seen on Jovian moons.

18. Why are no craters seen on Io and few seen on Europa?

19. Why should you expect Io to suffer more impacts per square kilometer than Callisto?

20. How can you be certain that Jupiter's ring does not date from the formation of the planet? Where do the ring particles come from?

21. Why are the belts and zones in the atmosphere of Saturn less distinct than those in the atmosphere of Jupiter?

22. Describe the composition of Saturn from its center outward. What causes these different internal layers?

23. If Saturn had no moons, do you think it would have rings?

24. How can Titan keep an atmosphere when Titan is smaller than airless Ganymede?

25. What should the interior composition of Titan be if its density is 1.9 g/cm³?

26. If you were able to stand on the surface of Titan in the daytime, what would you see? For example, would it be raining? Would there be high winds? Would there be lots of sunlight? What color would the sky be?

27. Does Titan experience volcanism today? Impact cratering? How do you know?

28. Describe the types of geological activity observed on the moons of Saturn.

29. More Jovian moons are geologically active than Terrestrial planets. True or false? How would you explain this?

30. Saturn's moons formed in the same way as the Galilean moons formed around Jupiter. True or false? How do you know?

31. If you piloted a spacecraft to visit Saturn's moons and wanted to land on a geologically old surface, what moon would you choose? Why?

32. The ring systems around Jupiter and Saturn lie outside those planet's respective Roche limits. True or false? How do you know?

33. Saturn's rings are primordial, meaning that they originated when the planet formed. True or false? How do you know?

34. Ripples in ring systems are likely caused by moons orbiting within the rings. True or false?

35. Gaps in ring systems are likely caused by moons orbiting within the rings. True or false?

36. **How Do We Know?** Why would you expect research in archaeology to be not as well funded as research in chemistry?

Discussion Questions

1. In Chapter 11, Earth was presented as the standard for comparative planetology of the Terrestrial planets. However, Earth is not an average Terrestrial planet. In fact, it's the largest one, which makes it special in several ways. Jupiter is the largest Jovian planet. Do you think Jupiter makes a good standard for comparative planetology of the Jovian planets? Why or why not?

2. Earth's diameter is about 4 times the diameter of the Moon. Jupiter's diameter is about 40 times the diameter of its moon Io.

(Io is almost exactly the same size as Earth's Moon.) Given what you know about how Earth's Moon formed versus how Jupiter's moons formed, do you think a planet/moon size ratio of 4, or 40, is more typical in the Universe?

3. Look back to sections 3 and 4 of Chapter 10 regarding the condensation sequence and extrasolar planets. Do you think Jupiter's and Saturn's positions in our Solar System, in orbits distant from the Sun, are typical of Jovian planets in the Universe? Why or why not? Support your argument with evidence.

4. Why don't the Terrestrial planets have ring systems and large numbers of moons?

Problems

1. What is the angular diameter of Jupiter as seen from Earth when the two planets are closest together? When the two planets are farthest apart? (*Hint:* Use the small-angle formula, Chapter 3.) (*Note:* Necessary data to calculate the two distances are given in **Celestial Profiles 2** and **7**.)

2. How fast is Io moving in orbiting around Jupiter? (*Hint:* Use the formula for circular orbit velocity, Chapter 5. The formula requires input quantities in kg and m.) (*Note:* Necessary data are given in **Celestial Profile 7** and Appendix Table A-11.)

3. What is the angular diameter of Jupiter as seen from the surface of Callisto? *Hint:* Use the small-angle formula, Chapter 3.) (*Note:* Necessary data are given in **Celestial Profile 7** and Appendix Table A-11.)

4. What is the escape velocity from the surface of Ganymede? Ganymede's mass is 1.5×10^{23} kg and its radius is 2.6×10^3 km. (*Hint:* Use the formula for escape velocity, Chapter 5. The formula requires input quantities in kg and m.)

5. Calculate the mass of Callisto using a value for its density of 1.8 g/cm³. Convert your answer to units of kg, and compare to the mass of Ganymede given in Problem 5. (*Notes:* Density is mass divided by volume, and the volume of a sphere is $\frac{4}{3}\pi r^3$. Necessary data are given in Appendix Table A-11.)

6. Using the data in Table 14-2, determine which Galilean moons have orbital periods that are approximately integer multiples of Io's period. These moons are in mutual orbital resonances.

7. Calculate the radius of Jupiter's Roche limit and decide which moons are likely candidates to contribute to the rings around Jupiter. Are any of the Galilean moons candidates? Necessary data are given in **Celestial Profile 7** and Appendix Table A-11.

8. How long does the eastward wind at the equator of Saturn take to circle the planet once at a speed of 500 m/s? Compare this value with the rotation period of the planet. (*Note:* Necessary data are given in **Celestial Profile 8**.)

9. What is the orbital velocity and period of a ring particle at the outer edge of Saturn's A ring? (*Hint:* Use the formula for circular orbital velocity, Chapter 5. The formula requires input quantities in kg and m.) (*Note:* The radius of the outer edge of the A ring is 136,500 km.)

10. If you were to record the spectrum of Saturn as well as the A ring, you would find light from one edge of the rings redshifted and light from the other edge blueshifted. If you observed a spectral line at a wavelength of 500.000 nm, what difference in wavelength should you expect between the opposite edges of the rings? (*Hints:* See Problem 9, and use the formula for Doppler shift, Chapter 7.)

11. What is the difference in orbital velocity between Saturn's two co-orbital satellites if the semimajor axes of their orbits are 151,400 km and 151,500 km? (*Hint:* Use the formula for orbital velocity, Chapter 5. The formula requires input quantities in kg and m.)

Learning to Look

1. Look at Figure 14-4b. Compare the visual and UV images of Jupiter. What do you notice? What does it mean?

2. Examine the planetary atmosphere profile plots in Figures 11-9, 13-1, and 14-15. How do the temperature profiles of the atmospheres of Jupiter and Saturn compare with those of Earth and Venus?

3. This image to the right shows a segment of the surface of Jupiter's moon Callisto. Why are portions of the surface dark? Why are some craters dark and some white? What does this image tell you about the history of Callisto?

4. The *Cassini* spacecraft recorded the image below of Saturn's A ring and the Encke Gap. What do you see in this photo that tells you about processes that confine and shape planetary rings?

15 Uranus, Neptune, and the Kuiper Belt

Guidepost Two planets circle the Sun in the twilight beyond Saturn. You will find Uranus and Neptune substantially different from Jupiter and Saturn but still recognizable as Jovian planets. As you explore further, you will also discover a family of smaller bodies, including dwarf planets Pluto, Eris, Haumea, and Makemake, which evidently are leftover planet construction material. This chapter will help you answer three important questions:

▶ **How are Uranus and Neptune similar to, and different from, Jupiter and Saturn?**

▶ **How did Uranus and Neptune form and evolve?**

▶ **What do Pluto and the other Kuiper Belt Objects tell you about the origin and evolution of the Solar System?**

When you finish this chapter, you will have visited all of the major worlds in our Solar System and finished with a pass through a zone of leftover planetary construction material beyond the planets. But there is more to see. Vast numbers of small rocky and icy bodies orbit among the planets, and the next chapter will introduce you to these messengers from the age of planet building.

A good many things go around in the dark besides Santa Claus.

HERBERT HOOVER

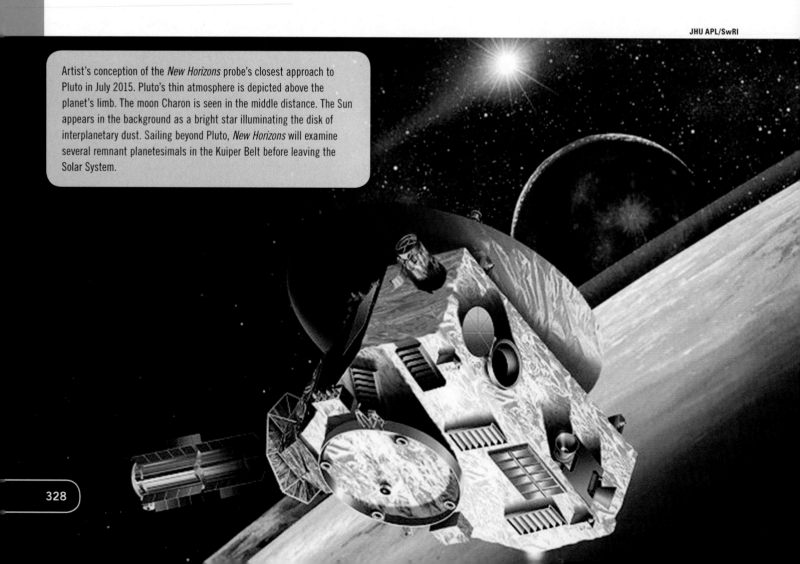

JHU APL/SwRI

Artist's conception of the *New Horizons* probe's closest approach to Pluto in July 2015. Pluto's thin atmosphere is depicted above the planet's limb. The moon Charon is seen in the middle distance. The Sun appears in the background as a bright star illuminating the disk of interplanetary dust. Sailing beyond Pluto, *New Horizons* will examine several remnant planetesimals in the Kuiper Belt before leaving the Solar System.

Out in the darkness beyond Saturn, out where sunlight is 100 to 1000 times fainter than on Earth, there are objects orbiting the Sun that Aristotle, Galileo, and Newton never imagined. They knew about Mercury, Venus, Mars, Jupiter, and Saturn, but our Solar System also includes worlds that were not discovered until after the invention of the telescope. The stories of those discoveries highlight the process of scientific discovery, and the characteristics of these dimly lit worlds will reveal more of how Earth and the rest of the Solar System formed.

15-1 Uranus

In March 1781, Benjamin Franklin was in France raising money, troops, and arms for the American Revolution. George Washington and his colonial army were only six to seven months away from the defeat of Lord Cornwallis at Yorktown and the end of the war. In England, King George III was beginning to show signs of madness. And a German-born music teacher in the English resort city of Bath was about to discover the planet Uranus.

Discovery of Uranus

William Herschel (Figure 15-1a) came from a musical family in Hanover, Germany, but emigrated to England as a young man and eventually obtained a prestigious job as the organist at the Octagon Chapel in Bath.

While working as a musician and music teacher, Herschel studied the mathematical principles of musical harmony from a book by Professor Robert Smith of Cambridge. The mathematics in the book was so interesting that Herschel searched out other works by Smith, including a book on optics. Soon, Herschel and his brother Alexander began building telescopes. Herschel developed ways of making exceptionally large mirrors for that time: One of his favorite telescopes was a bit more than 2 m (7 ft) long and had a mirror 16 cm (6.2 in.) in diameter. Using this telescope, he began the research project that led to the discovery of Uranus.

One night in late winter 1781, Herschel set up the 7-foot telescope in his back garden to continue a 2-year project detecting and cataloging binary stars. He later wrote, "In examining the small stars in the neighborhood of H Geminorum, I perceived one that appeared visibly larger than the rest." As seen from Earth, Uranus is never larger in angular diameter than 3.7 arc seconds, so Herschel's detection of the disk indicates the quality of his telescope and his eye. At first he suspected that the object was a comet, but other astronomers quickly realized that it was a planet orbiting the Sun beyond Saturn.

The discovery of Uranus made Herschel world famous. Since antiquity, astronomers had known of five planets—Mercury, Venus, Mars, Jupiter, and Saturn—but had never

▲ Figure 15-1 (a) When William Herschel discovered Uranus in 1781, he saw only a tiny green-blue dot. He never knew how interesting that planet is. (b) An image of Uranus recorded by a near-infrared camera with an adaptive optics (AO) system on the Keck telescope shows banding and cloud features. The rings look red in this view because of image processing.

Infrared, enhanced contrast

Keck Observatory/CARA/L. Sromovsky (UW-Madison)

© Georgios Kollidas/Shutterstock.com

imagined there could be more. Herschel's discovery extended the classical universe by adding a new planet. The English public accepted Herschel as their astronomer-hero, and, having named the new planet Georgium Sidus (George's Star) after King George III, Herschel received a royal pension. Years later, German astronomer Johann Bode suggested the name Uranus, after the father of Cronus, the Greek name for the god Saturn. That name for the planet is the one we use today because it proved much more popular with astronomers in other countries than did Herschel's choice.

Herschel's new financial position allowed him to build large telescopes on his estate, and with his sister Caroline, also a talented astronomer, he attempted to map the extent of the Universe. You met Herschel in Chapter 6 as the discoverer of infrared radiation.

Continental astronomers were less than thrilled that an Englishman had made such a great discovery, and even some professional English astronomers thought Herschel a mere amateur. They called his discovery a lucky accident. But, as a musician, Herschel knew the value of practice and applied it to the

How Do We Know? 15-1

Scientific Discoveries

Why didn't Galileo expect to discover Jupiter's moons? In 1928, Alexander Fleming noticed that bacteria in a culture dish were avoiding a spot of mold. He went on to discover penicillin. In 1895, Conrad Roentgen noticed a fluorescent screen glowing in his laboratory when he experimented with other equipment. He discovered X-rays. In 1896, Henri Becquerel stored a uranium mineral on a photographic plate safely wrapped in black paper. The plate was later found to have been fogged, and Becquerel discovered natural radioactivity. Like many discoveries in science, these seem to be accidental, but, as you have seen in this chapter, "accidental" doesn't quite describe what happened.

The most important discoveries in science are those that totally change the way people think about nature, and it is very unlikely that anyone would predict such discoveries. For the most part, scientists work within a paradigm (look back at **How Do We Know? 4-1**, page 64), a set of models, hypotheses, theories, and expectations about nature, and it is difficult to imagine natural events that lie beyond that paradigm. Ptolemy, for example,

could not have imagined galaxies because they were not part of his geocentric paradigm. That means that the most important discoveries in science are almost always unexpected.

An unexpected discovery, however, is not the same as an accidental discovery. Fleming discovered penicillin in his culture dish not because he was the first to see it, but because he had studied bacterial growth for many years, so when he saw what many others must have seen before, he recognized it as important. Roentgen realized that the glowing screen in his lab was important, and Becquerel didn't discard that fogged photographic plate. Long years of experience prepared them to recognize the significance of what they saw.

A historical study has shown that each time astronomers build a telescope that significantly surpasses the capabilities of existing telescopes, their most important discoveries are unexpected. Herschel didn't expect to discover Uranus with his 7-foot focal-length telescope, and modern astronomers didn't expect to discover evidence of dark energy with the *Hubble Space Telescope*.

Scientists pursuing basic research are rarely able to explain the potential value of their work, but that doesn't make their discoveries accidental. They earn their right to those lucky accidents.

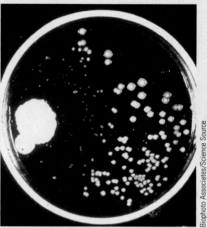

Alexander Fleming's photo of a laboratory disk with bacteria and Penicillin mold.

---

business of astronomical observing. In fact, records show that other astronomers had seen Uranus at least 17 times before Herschel, but each time they failed to notice that it was not a star. They plotted Uranus on their charts as if it were just another faint star.

This illustrates one of the ways in which scientific discoveries are made. Often, discoveries seem accidental, but on closer examination you find that the scientist has earned the right to the discovery through many years of study and preparation (**How Do We Know? 15-1**). To quote a common saying, "Luck is what happens to people who work hard."

Over the half-century following the discovery of Uranus, astronomers noted that Newton's laws did not exactly predict the observed position of the planet. Tiny variations in the orbital motion of Uranus eventually led to the discovery of Neptune, a controversial story you will read later in this chapter.

Uranus's Motion

Uranus orbits nearly 20 AU from the Sun and takes 84 years to go around once (■ Celestial Profile 9, page 341). The ancients thought of Saturn as the slowest of the planets, but Saturn

orbits in slightly more than 29 years. Uranus, being farther from the Sun, moves even slower than Saturn and has a longer orbital period.

The rotation of Uranus is peculiar. Earth rotates approximately upright in its orbit. That is, Earth's axis of rotation is inclined only 23 degrees from the perpendicular to its orbit. The other planets have similarly moderate axial inclinations. Uranus, in contrast, rotates on an axis that is inclined 98 degrees from the perpendicular to its orbit. It rotates on its side; in other words, the ecliptic on Uranus passes very near the planet's celestial poles (**Figure 15-2**).

Because of its odd axial tilt, seasons on Uranus are extreme. The first good photographs of Uranus were taken in 1986, when the *Voyager 2* spacecraft flew past. At that time, Uranus was in the segment of its orbit in which its south pole faces the Sun. Consequently, its southern hemisphere was bathed in continuous sunlight, and an observer there would have seen the Sun near the planet's south celestial pole. The Sun was at southern solstice on Uranus in 1986, as you can see at the lower left in Figure 15-2.

Over the next two decades, Uranus moved about a quarter of the way around its orbit, and, with the Sun shining down

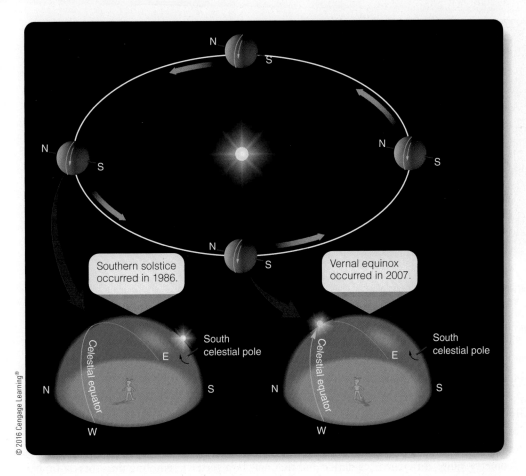

◄ **Figure 15-2** Uranus rotates on an axis that is tipped 98 degrees from the perpendicular to its orbit, so its seasons are extreme. When one of its poles is pointed nearly at the Sun (a solstice), a citizen of Uranus would see the Sun near a celestial pole, and it would never rise or set. As it orbits the Sun, the planet maintains the direction of its axis in space, and thus the Sun would apparently move from pole to pole. At the time of an equinox on Uranus, the Sun would be on the celestial equator and would rise and set with each rotation of the planet. Compare with similar diagrams for Earth on page 25.

from above the planet's equator, a citizen of Uranus would see the Sun rise and set with the rotation of the planet. The Sun reached equinox on Uranus in December 2007, and you can see that geometry at the lower right in Figure 15-2. As Uranus continues along its orbit, the Sun will approach the planet's north celestial pole, and the southern hemisphere of the planet will experience a lightless winter lasting 21 Earth years.

Uranus's Atmosphere

Like Jupiter and Saturn, Uranus has no surface. The gases of its atmosphere—mostly hydrogen, 26 percent helium (by mass), and a few percent methane, ammonia, and water vapor—blend gradually into a fluid interior.

Seen through Earth-based telescopes, Uranus is a small, featureless, greenish-blue disk. The green-blue color arises because the atmosphere contains methane, a good absorber of longer-wavelength photons. As sunlight penetrates into the atmosphere and is scattered back out, the longer-wavelength (red) photons are more likely to be absorbed. That means that the sunlight reflecting off Uranus and then entering your eye has proportionately more blue photons, giving the planet a blue color.

As *Voyager 2* drew close to the planet in late 1985, astronomers studied the images radioed back to Earth. Uranus was a pale green-blue ball with no obvious clouds, and only when the images were carefully computer enhanced was any banded structure detected (**Figure 15-3**). A few very high clouds of methane ice particles were detected, and their motions allowed astronomers to make the first good measurement of the planet's rotation period.

You can understand the nearly featureless appearance of the atmosphere by studying the temperature profile of Uranus shown in **Figure 15-4**. The atmosphere of Uranus is much colder than that of Saturn or Jupiter. Consequently, the three cloud layers of ammonia, ammonium hydrosulfide, and water that form the belts and zones in the atmospheres of Jupiter and Saturn lie very deep in the atmosphere of Uranus. These cloud layers, if they exist at all in Uranus, are not visible because of the thick atmosphere of hydrogen through which an observer has to look. The clouds that are visible on Uranus are clouds of methane ice crystals, which form at such a low temperature that they occur high in the atmosphere of Uranus. Figure 15-4 shows that there can be no methane clouds on Jupiter because that planet is too warm. The coldest part of Saturn's atmosphere is just cold enough to form a thin

© 2016 Cengage Learning®

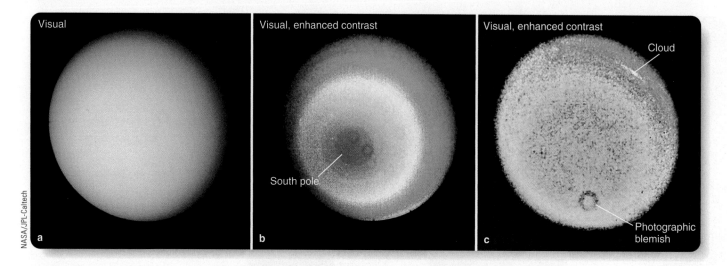

NASA/JPL-Caltech

▲ Figure 15-3 (a) This *Voyager 2* image of Uranus made in 1986 shows no clouds. Only when computer processing enhances the contrast, as in (b), is a banded structure visible. At the time of this image, the planet's axis of rotation was pointed nearly at the Sun. (c) With extreme computer-enhanced contrast, small methane clouds become visible. The geometry of the banding and the clouds suggests belt–zone circulation analogous to that on Saturn and Jupiter.

methane haze high above its more visible cloud layers (look back to Figure 14-15).

The clouds and atmospheric banding that are faintly visible on Uranus appear to be the result of belt–zone circulation, which is a bit surprising. Because Uranus rotates on its "side," solar energy strikes its surface with geometry quite different than that for Jupiter and Saturn. Evidently, belt–zone circulation is dominated by the rotation of the planet and not by the direction of sunlight.

The *Voyager 2* images from 1986 caused astronomers to expect that Uranus was always a nearly featureless planet, but later observations made with the *Hubble Space Telescope* and giant Earth-based telescopes as spring came to the planet's northern hemisphere detected changing clouds on Uranus, including a dark cloud that may be a vortex resembling the spots on Jupiter (Figure 15-5). The clouds appear to be part of a seasonal cycle on Uranus, but its year lasts 84 Earth years, so you will have to be patient to see the effects of northern hemisphere summer.

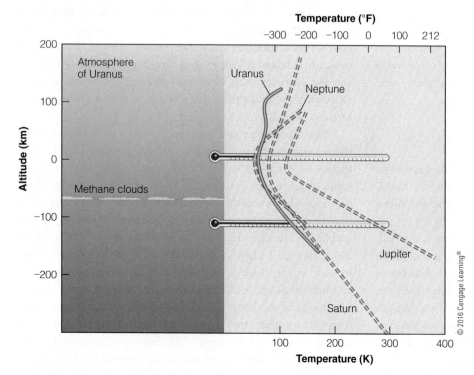

© 2016 Cengage Learning®

◀ Figure 15-4 The atmosphere of Uranus is much colder than that of Jupiter or Saturn, and the only visible cloud layer is one formed of methane ice crystals deep in the hydrogen atmosphere. Other cloud layers would be even deeper in the atmosphere and are not visible. The temperature profile of Neptune is similar to that of Uranus, and it has methane clouds at about the same place in its atmosphere.

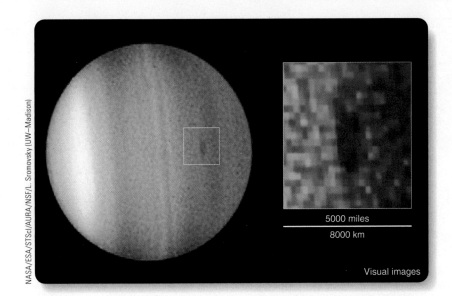

5000 miles

8000 km

Visual images

▲ **Figure 15-5** A dark cloud, possibly a circulating storm, is visible in this *Hubble Space Telescope* image of Uranus.

Uranus's Interior

Astronomers cannot describe the interiors of Uranus and Neptune as accurately as they can the interiors of Jupiter and Saturn. Observational data are sparse, and the materials inside these planets are not as easy to model as simple liquid hydrogen.

The average density of Uranus—1.3 g/cm³—tells you that the planet must contain a larger share of dense materials than Saturn. Nearly all models of the interior of Uranus contain three layers. The uppermost layer—the atmosphere—is rich in hydrogen and helium. Below the atmosphere, a deep mantle must contain large amounts of water, methane, and ammonia in a solid or slushy state, mixed with hydrogen and silicate matter. This mantle is sometimes described as "ice," but because of high pressure and a temperature of a few thousand degrees, it is quite unlike the Earthly material that word suggests. The third layer in the three-layer models is a small heavy-element core. Many books refer to the core as "rocky," but, again, because of the high pressure and high temperature, the material is not very rocklike. The term *rock* refers to its chemical composition and not to its other properties.

It is a **Common Misconception** to imagine that the four Jovian planets are gaseous. You have learned about the evidence and models indicating that Jupiter and Saturn are mostly liquid hydrogen. Uranus and Neptune are sometimes described as "ice giant planets," in recognition of the large proportion of solid water inferred to be in their interiors.

Because Uranus has a much lower mass than Jupiter, its internal pressure is not high enough to produce liquid metallic hydrogen. Consequently, you might expect it to lack a strong magnetic field, but the *Voyager 2* spacecraft found that Uranus has a magnetic field about 75 percent as strong as Earth's. Surprisingly, Uranus's field is tipped 59 degrees to the axis of rotation and is offset from the center of the planet by about 30 percent of the planet's radius (**Figure 15-6**). Theorists suggest that this oddly oriented magnetic field is produced by a dynamo effect operating not in the planet's core but nearer the surface in a layer of liquid water with dissolved ammonia and methane. Such a material would be a good conductor of electricity, and the rotation of the planet coupled with convection in the fluid could generate the magnetic field.

As it made its closest approach to Uranus, *Voyager 2* observed effects of the planet's magnetic field. This allowed a more precise measurement of Uranus's rotation period than was possible from motions of difficult-to-detect cloud features. The magnetic field deflects the solar wind and traps some charged particles to create weak radiation belts in the planet's magnetosphere. High-speed electrons spiraling along the magnetic field produce synchrotron radio emission just as around Jupiter, and the *Voyager 2* spacecraft recorded this radiation fluctuating with a period of 17.2 hours, the period of rotation of the magnetic field and, presumably, the planetary interior.

The magnetic field and the high inclination of the planet produce some peculiar effects. As is the case for all planets with magnetic fields, the solar wind deforms the magnetosphere and draws it out into a long tail extending away from the planet in the direction opposite the Sun. The rapid rotation of Uranus and its high inclination give the magnetosphere and its long extension a corkscrew shape. At the time *Voyager 2* flew past in 1986, the south pole of Uranus was pointed nearly at the Sun, and once during each rotation the solar wind poured down into the south magnetic pole. The resulting interaction produced strong auroras that *Voyager 2* detected in the ultraviolet at both magnetic poles (**Figure 15-7**).

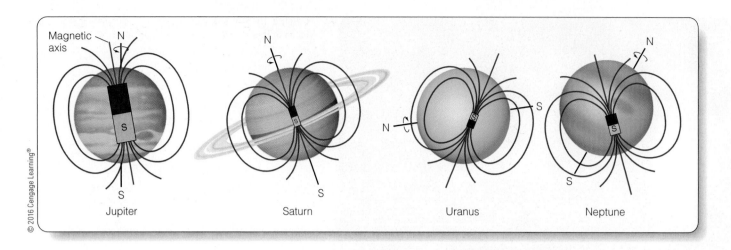

▲ **Figure 15-6** The magnetic fields of Uranus and Neptune are peculiar. Although the magnetic axis of Jupiter is tipped only 10 degrees from its axis of rotation, and the magnetic axis of Saturn is not tipped at all, the magnetic axes of Uranus and Neptune are tipped at large angles. Furthermore, the magnetic fields of Uranus and Neptune are offset from the centers of both planets. This suggests that the dynamo effect operates differently in Uranus and Neptune than in Jupiter, Saturn, and Earth.

In the years since, Uranus has moved around its orbit, and the geometry of its interaction with the solar wind has changed. Unfortunately, no spacecraft is currently anywhere near Uranus, so there is no way to observe the effects of these changes in detail.

Like the magnetic field, the temperature of Uranus can reveal something about its interior. Jupiter and Saturn are warmer than you would expect, given the amount of energy they receive from the Sun. This means that heat is leaking out from their hot interiors. Uranus, in contrast, is about the temperature you would expect for a world at its distance from the Sun; the planet radiates less than 10 percent more heat than it receives from the Sun. Apparently, Uranus has lost much of its interior heat. Nevertheless, there must be enough internal heat to cause convection in the fluid mantle, drive the dynamo effect, and create the magnetic field. The temperature in its core is estimated from models to be more than 5000 K. The decay of natural radioactive elements would generate heat, but some

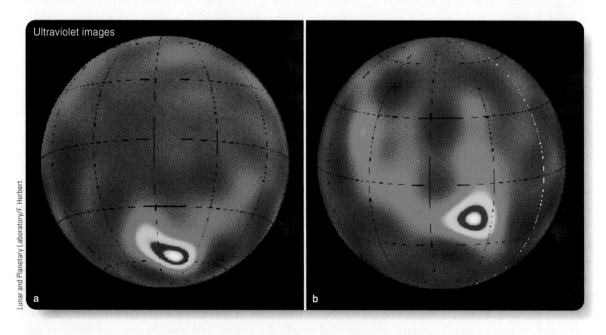

▲ **Figure 15-7** Auroras on Uranus were detected in the ultraviolet by the *Voyager 2* spacecraft when it flew by in 1986. These maps of opposite sides of Uranus show the location of auroras near the magnetic poles. Recall that the magnetic field is highly inclined and offset from the planet's center, so the magnetic poles do not lie near the poles defined by rotation. The white dashed line marks zero longitude.

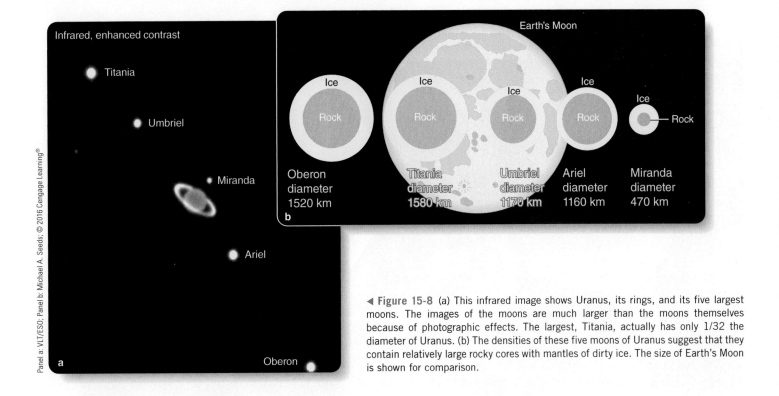

Infrared, enhanced contrast

Titania

Umbriel

Miranda

Ariel

Oberon

Earth's Moon

Oberon	Titania	Umbriel	Ariel	Miranda
Ice Rock	Ice Rock	Ice Rock	Ice Rock	Ice Rock
Oberon diameter 1520 km	Titania diameter 1580 km	Umbriel diameter 1170 km	Ariel diameter 1160 km	Miranda diameter 470 km

◄ **Figure 15-8** (a) This infrared image shows Uranus, its rings, and its five largest moons. The images of the moons are much larger than the moons themselves because of photographic effects. The largest, Titania, actually has only 1/32 the diameter of Uranus. (b) The densities of these five moons of Uranus suggest that they contain relatively large rocky cores with mantles of dirty ice. The size of Earth's Moon is shown for comparison.

Panel a: VLT/ESO; Panel b: Michael A. Seeds; © 2016 Cengage Learning®

astronomers have suggested that the slow settling of heavier elements through the fluid mantle could also release energy to warm the interior.

Laboratory studies of methane show that it can break down under the temperature and pressure inside Uranus and form various compounds plus pure carbon in the form of diamonds. If this happens in Uranus, the diamond crystals would fall inward, warming the interior through friction. Determining for sure whether a planetwide rain of diamonds actually exists inside Uranus is probably forever beyond human reach.

For a Jovian world, Uranus seems small and mostly featureless. But now you are ready to visit some of its best attractions—its moons.

Uranus's Moons

Uranus has five large regular moons that were discovered from Earth-based observations (**Figure 15-8**). Those five moons, from the outermost inward, are Oberon, Titania, Umbriel, Ariel, and Miranda. The names Umbriel and Ariel are names from Alexander Pope's *The Rape of the Lock,* and the rest are from Shakespeare's *A Midsummer Night's Dream* and *The Tempest* (an Ariel also appears in *The Tempest*). Spectra show that the moons contain frozen water, although their surfaces are dark. Planetary scientists assumed they were made of ices mixed with dirt, but little more was known of the moons before *Voyager 2* flew through the system.

In addition to imaging the known moons, the *Voyager 2* cameras discovered ten more moons too small to have been seen from Earth. Since then, the construction of new-generation telescopes and the development of new imaging techniques (Chapter 6, pages 115 and 124) have allowed astronomers to find even more small moons orbiting Uranus. Currently 27 moons are known: 13 small objects orbiting among the planet's rings, the five large regular moons, and nine small moons in large, irregular orbits. There are almost certainly more to be found.

The smaller inner moons are all as dark as coal. They are icy worlds with surfaces that have been darkened by impacts vaporizing ice and concentrating embedded dirt. Additionally, they orbit inside the planet's radiation belts, and the radiation can convert methane ice into dark carbon deposits to further darken their surfaces.

The five large moons all have rotations tidally locked to Uranus, which means their south poles were pointed toward the Sun in 1986 so *Voyager 2* could not photograph their northern hemispheres. Thus, current analysis of their geology must depend on images of only half their surfaces. The densities of the moons suggest that they contain relatively large rock cores surrounded by icy mantles, as shown in Figure 15-8.

Oberon, the outermost of the large moons, has a cratered surface, but visible evidence indicates that it was once geologically active (**Figure 15-9**). A large fault crosses the sunlit hemisphere, and dark material—"lava" perhaps composed of dirty water—appears to have flooded the floors of some craters.

Visual images

Oberon has an old, icy, heavily cratered surface.

Deep valleys up to 100 km across suggest Titania has had an active past.

The surface of Ariel is crossed by faults and broad valleys.

Umbriel is a dark, cratered world with no sign of faults or valleys.

Miranda, the innermost and smallest of the five main moons, has the most extreme signs of past activity.

Ariel and Miranda: USGS, Flagstaff, Arizona; other images: NASA/JPL-Caltech

▲ **Figure 15-9** The five largest moons of Uranus, shown here in correct relative size scale, range from the largest, Titania, 45 percent the diameter of Earth's Moon, down to Miranda, only 14 percent the diameter of Earth's Moon. For a better view of Miranda, see Figure 15-10.

Titania is the largest of the five moons and has a heavily cratered surface, but it has no large craters (Figure 15-9). This suggests that after the end of the heavy bombardment, the young Titania underwent an active phase in which its surface was flooded with water that covered early craters with fresh ice. Since then, the craters that have formed are not as large as the largest of those that were erased. The network of faults that crosses Titania's surface is another sign of past activity.

Umbriel, the next moon inward, is a dark, cratered world with no sign of faults or surface activity (Figure 15-9). It is the darkest of Uranus's major moons, with an albedo of only 0.10 compared with 0.14 to 0.23 for the other moons. Its crust is apparently a mixture of rock and ice. A bright crater floor in one region suggests that clean ice may lie at shallow depths in some regions.

Ariel has the brightest surface of the five major Uranian satellites and shows clear signs of geological activity. It is crossed by faults more than 10 km (6 mi) deep, and some regions appear to have been smoothed by resurfacing, as you can see in Figure 15-9. Crater counts show that the smoothed regions are in fact younger than the other regions. Ariel may have been subject to tidal heating caused by orbital resonances with Miranda and Umbriel.

Visual

USGS FSC

▲ **Figure 15-10** Miranda, the smallest of the five major moons of Uranus, is only 470 km (290 mi) in diameter, but its surface shows signs of activity. This photomosaic of *Voyager 2* images reveals that it is marked by great oval systems of grooves. The smallest features detected in this image are about 1.5 km (1 mi) in diameter.

Miranda is a mysterious moon. As you can see in Figure 15-9, it is the smallest of the five large moons, but it appears to have been the most active. In fact, its active past appears to have been quite unusual. Miranda is marked by oval patterns of grooves known as **ovoids** (**Figure 15-10**). Careful studies of the ovoids show that they are associated with faults, ice-lava flows, and rotated blocks of crust, suggesting that they may have been created by internal heat driving large, slow, convection currents in Miranda's icy mantle.

Near Miranda's equator, a huge cliff rises 20 km (12 mi). If you stood in your spacesuit at the top of the cliff and dropped a rock over the edge, it would fall for 12 minutes before hitting the bottom. Nevertheless, crater counts indicate that the cliff and the ovoids are old. Miranda is no longer geologically active, but you can read hints of its active past on its disturbed surface. Miranda is so small that its heat of formation must have been lost quickly, and there is no reason to expect it to have been strongly heated by radioactive decay. Your knowledge of tidal heating leads you to imagine that Miranda's activity probably resulted from its orbit being made temporarily slightly eccentric by a resonance with one or more of the other moons.

Uranus's Rings

Both Uranus and Neptune have rings that are more like Jupiter's than Saturn's. They are dark, faint, not easily visible from Earth, and confined by shepherd satellites.

Study **Uranus's and Neptune's Rings** on pages 338–339 and notice three important points about the rings of Uranus and one new term:

1 The rings of Uranus were discovered during an *occultation* when Uranus crossed in front of a star.

2 The rings are made up of a thin layer of very dark boulders. They are confined by small moons. Except for the outermost rings, the Uranian rings contain almost no small dust particles.

3 Like the rings of Jupiter and Saturn, the rings around Uranus and Neptune cannot survive for long periods and are not primordial. All the ring systems around the Jovian planets need to be resupplied continuously with new material, presumably from impact erosion of nearby moons.

When you read about Neptune's rings later in this chapter, you will return to this artwork and see how closely the two ring systems compare.

Images made with the *Hubble Space Telescope* in 2003 and 2005 revealed two larger, fainter, dustier rings lying outside the previously known ring system (**Figure 15-11**). The larger of these rings coincides with the orbit of the small moon Mab and is probably replenished by particles blasted off of that moon by meteorite impacts. The smaller of the rings is confined between the orbits of the moons Portia and Rosalind. (These moons are also named after Shakespearean characters.)

As you read about planetary rings, notice their close relationship with moons. Because of collisions among ring particles, planetary rings tend to spread outward, almost like an expanding gas. If a planet had no moons, its rings would spread out into a more and more tenuous sheet until they were gone. The spreading rings can be anchored by small shepherd moons, which interact gravitationally with wandering ring particles, absorb orbital energy, cause the particles to remain within the rings, and define ring outer boundaries. As these shepherd moons gain orbital energy they move slowly outward, but they can be anchored in turn by orbital resonances with larger, more distant moons that are so massive they do not get pushed outward significantly. In this way, a system of moons can confine and preserve a system of planetary rings.

A History of the Uranus System

The challenge of comparative astronomy is to tell the story of a world, and Uranus may present the biggest challenge of all the objects in our Solar System. Not only is it so distant that it is difficult to study, but it is also peculiar in many ways.

Uranus certainly formed from the solar nebula, as did the other Jovian planets, but calculations show that Uranus and

Uranus's and Neptune's Rings

1 The rings of Uranus were discovered in 1977, when Uranus crossed in front of a star being observed from NASA's *Kuiper Airborne Observatory*. During this **occultation**, astronomers saw the star dim a number of times before and again after the planet crossed over the star. The dips in brightness were caused by rings circling Uranus.

More rings were discovered by *Voyager* 2. The rings are identified in different ways depending on when and how they were discovered.

> Notice the eccentricity of the ε (epsilon) ring. It lies at different distances on opposite sides of the planet.

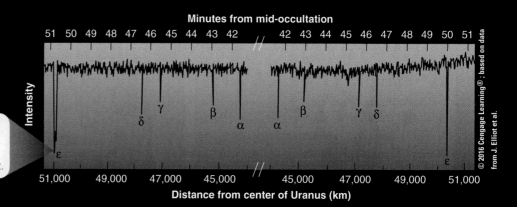

Minutes from mid-occultation

© 2016 Cengage Learning® ; based on data from J. Elliot et al.

Distance from center of Uranus (km)

2 The albedo of the ring particles is only about 0.015, darker than lumps of coal. If the ring particles are made of methane-rich ices, particle radiation from the planet's radiation belts could break the methane down to release carbon and darken the ices. The same process probably darkens the icy surfaces of Uranian moons.

The narrowness of the rings suggests they are shepherded by small moons. *Voyager* 2 found Ophelia and Cordelia shepherding the ε ring. Other small moons must be shepherding the other narrow rings. Such moons must be structurally strong to hold themselves together inside the planet's Roche limit.

The eccentricity of the ε ring is apparently caused by the eccentric orbits of Ophelia and Cordelia.

ε
λ
η
γ δ
β
α
6
5
4

1986
U2R

Ophelia

Cordelia

2a When the *Voyager* 2 spacecraft looked back at the rings illuminated from behind by the Sun, the rings were not bright. That is, the rings are not bright in forward-scattered light. That means they must not contain small dust particles. The nine main rings contain no particles smaller than meter-size boulders.

Uranus

3 Ring particles don't last forever because they collide with each other or are pushed away by radiation pressure. Uranus's rings are probably resupplied with fresh particles occasionally from impacts on moons that scatter debris.

Collisions among the large particles in the ring produce small dust grains. Friction with Uranus's tenuous upper atmosphere plus sunlight pressure act to slow the dust grains and make them fall into the planet. Uranus's rings actually contain very little dust.

Visual

The rings of Neptune are bright in forward-scattered light, as in the image above, and that indicates that the rings contain significant amounts of dust. The ring particles are as dark as those that circle Uranus, so they probably also contain methane-rich ice darkened by radiation.

4 Two narrow rings around Neptune are visible in this *Voyager* 2 image, as well as a wider, fainter ring lies closer to the planet. More ring material is visible between the two narrow rings. The black bar is an artifact of the imaging process, designed to eliminate most of the glare from the bright planet.

Arc

Arc

5 When Neptune occulted stars, astronomers observing from Earth sometimes detected rings and sometimes did not. From this, they concluded that Neptune might have incomplete rings, or ring arcs. Computer enhancement of this *Voyager* 2 visual-wavelength image shows arcs, regions of higher density, in the outer ring. The ring arcs visible in the outer ring appear to be generated by the gravitational influence of the moon Galatea, but other moons must also be present to confine the rings.

Visual, enhanced contrast

4a Neptune's rings lie in the plane of the planet's equator and inside the Roche limit. The narrowness of the rings suggests that shepherd moons must confine them, and a few such moons have been found among the rings. There must be more undiscovered small moons to confine the rings completely.

Naiad

Galetea

Thalassa

Adams

LeVerrier

Despina

Galle

4c Like the rings of the other Jovian planets, the ring particles that orbit Neptune cannot have survived since the formation of the planet. Presumably, occasional impacts on Neptune's moons scatter debris and resupply the rings with fresh particles.

Neptune

4b Neptune's rings have been given names associated with the planet's history. French astronomer Le Verrier and English astronomer Adams predicted the existence of Neptune from the motion of Uranus. The German astronomer Galle discovered the planet in 1846 based on Le Verrier's prediction.

Mab

Ring 2003 U1

Ring 2003 U2

Ring 2003 U1

Visual, enhanced contrast + Artist's conception

NASA/ESA/STScI/AURA/NSF/M. Showalter (SETI Institute)

▲ **Figure 15-11** Two newly discovered rings orbit Uranus far outside the previously known rings. The outermost ring follows the orbit of the small moon Mab, only 12 km (7 mi) in diameter. The short, bright arcs in this photo were caused by moons moving along their orbits during the long time exposure.

Uranus; the two planets later switched places as they moved out. As you will learn later in this chapter, the migration of giant planets also could explain the Solar System-wide late heavy bombardment episode. These are interesting hypotheses, and they illustrate how uncertain the histories of the outer planets really are.

The highly inclined axis of Uranus may have originated late in its formation when it was struck by a planetesimal perhaps as large as Earth. That impact could also have disturbed the interior of the world and caused it to lose much of its heat. There is now just enough heat flowing outward to drive circulation in its slushy mantle and generate its magnetic field. An alternate model suggests the possibility that tidal interactions with Saturn could have altered Uranus's axis of rotation as it was migrating outward. This is another example of a catastrophic hypothesis being challenged by an evolutionary hypothesis (look back to "How Do We Know?" 10-1, page 205).

Impacts have been important in the history of the moons and rings of Uranus. All of the moons are cratered, and some show signs of large impacts. Meteorites and the nuclei of comets striking the moons can create debris that becomes trapped among the orbits of the smaller moons to produce the narrow rings. The ring particles observed now could not have lasted since the formation of the planet, so they must be replenished with fresh material now and then. Like the rest of the Solar System, the full history of the Uranus system evidently includes the effects of multiple impact events.

Neptune could not have accumulated enough material to grow to their present size in the slow orbits they now occupy so far from the Sun. Computer models of planet formation suggest instead that Uranus and Neptune formed closer to the Sun, in the neighborhood of Jupiter and Saturn. Gravitational interactions among the Jovian planets could have eventually moved Uranus and Neptune outward to their present locations. One of those models even had Neptune form closer to the Sun than

DOING SCIENCE

How are conditions in Uranus's interior determined?
Obviously, we can't see inside a planet, so scientists are limited to a few basic observations that they must compare with model calculations in a chain of inference to describe the interior.

First, Uranus's distance is known from observations of the planet's motion plus Kepler's third law. Uranus's size can then be found from its angular diameter and distance. You can find its mass by observing the orbital radii and orbital periods of its moons, then using Newton's version of Kepler's third law. The planet's mass divided by its volume equals its density. Uranus's density of 1.3 g/cm^3 implies that it must contain a certain proportion of dense material such as ice and rock, more than there is inside Saturn. Add to that the chemical composition of its atmosphere obtained from spectra, and you have the data needed to build a mathematical model of Uranus's interior. Such models predict a core of heavy elements and a mantle of ices mixed with heavier material of rocky composition.

Going from observable properties to unobservable properties is the heart of what scientists do. Now try another chain of inference: *What observational properties of Uranus's rings show that small moons must orbit among the rings?*

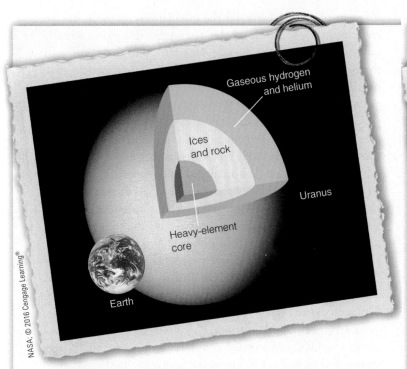

Uranus rotates on its side. When Voyager 2 flew past in 1986, the planet's south pole was pointed almost directly at the Sun.

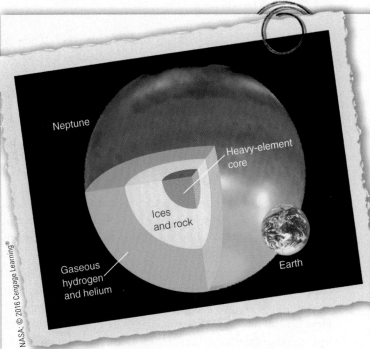

Neptune was tipped slightly away from the Sun when the Hubble Space Telescope recorded this image. The interior is much like Uranus's, but there is more outward heat flow.

Celestial Profile 9 Uranus

Motion:

Average distance from the Sun	19.2 AU (2.87×10^9 km)
Eccentricity of orbit	0.047
Inclination of orbit to ecliptic	0.8°
Orbital period	84.0 y
Period of rotation (sidereal)	17.23 h
Inclination of equator to orbit	97.8° (retrograde rotation)

Characteristics:

Equatorial diameter	5.11×10^4 km (4.01 $D_\oplus$)
Mass	8.68×10^{25} kg (14.5 $M_\oplus$)
Average density	1.27 g/cm^3
Gravity (at cloud tops)	0.9 Earth gravity
Escape velocity (at cloud tops)	21.3 km/s (1.9 $V_\oplus$)
Temperature (at cloud tops)	55 K (−360°F)
Albedo	0.30
Oblateness	0.023

Personality Point:

Most creation stories begin with a separation of opposites, and Greek mythology is no different. Uranus (the sky) separated from Gaia (Earth) who was born from the void, Chaos. They gave birth to the giant Cyclops, Cronos (Saturn, father of Zeus), and his fellow Titans. Uranus is sometimes called the starry sky, but the Sun (Helios), Moon (Selene), and the stars were born later, so Uranus, one of the most ancient gods, began as the empty, dark sky.

Celestial Profile 10 Neptune

Motion:

Average distance from the Sun	30.1 AU (4.50×10^9 km)
Eccentricity of orbit	0.009
Inclination of orbit to ecliptic	1.8°
Orbital period	164.8 y
Period of rotation (sidereal)	16.11 h
Inclination of equator to orbit	28.3°

Characteristics:

Equatorial diameter	4.95×10^4 km (3.88 $D_\oplus$)
Mass	1.02×10^{26} kg (17.1 $M_\oplus$)
Average density	1.64 g/cm^3
Gravity (at cloud tops)	1.1 Earth gravities
Escape velocity (at cloud tops)	23.5 km/s (2.1 $V_\oplus$)
Temperature (at cloud tops)	55 K (−360°F)
Albedo	0.29
Oblateness	0.017

Personality Point:

Because the planet Neptune looked so blue, astronomers named it after the Roman god of the sea, Neptune (Poseidon to the Greeks). His wife was Amphitrite, granddaughter of Ocean, who was one of the Titans. Neptune controlled the storms and waves and was a powerful god, not to be trifled with. His three-pronged trident became the symbol for the planet Neptune.

15-2 Neptune

Uranus and Neptune are often discussed together. They are about the same size and density. They do differ, however, in certain respects. Unlike Uranus, Neptune has a significant amount of heat flowing out from its interior. Also, Neptune has especially complicated ring and satellite systems. Even the discovery of Neptune was dramatically different from the discovery of Uranus.

Discovery of Neptune

The discovery of Neptune triggered one of the greatest controversies in the history of science. For more than 150 years, people told the story and took sides, but only in recent decades has the real story become known.

In 1843, the young English astronomer John Couch Adams completed his degree in astronomy and immediately began the analysis of one of the great problems of 19th-century astronomy. Herschel had discovered Uranus in 1781, but earlier astronomers had seen the planet as early as 1690 and had mistakenly plotted it on charts as a star. When 19th-century astronomers tried to combine all those data, they didn't quite fit together. No planet obeying Newton's laws of motion and controlled only by the gravity of the Sun and the other known planets could follow such an orbit.

Some astronomers suggested that the gravitational attraction of an undiscovered planet was causing the discrepancies. Adams began with the observed variations, never more than 2 arc minutes, from the predicted positions of Uranus, and by October 1845, he had, through a laborious and difficult calculation, computed the orbit of the undiscovered planet. He sent his prediction to the Astronomer Royal, Sir George Airy, who passed it on to an observer who began a painstaking search of the area star by star.

Meanwhile, the French astronomer Urbain Le Verrier made the same calculations and sent his predicted position of the planet to Johann Galle at the Berlin Observatory. Galle received Le Verrier's prediction on the afternoon of September 23, 1846, and after searching for 30 minutes that evening, found Neptune. It was only 1 degree from the position predicted by Le Verrier.

The discovery of a new planet caused a sensation, but English astronomers didn't like a Frenchman getting all the credit. After all, they said, the planet was found only 2 degrees from the position predicted by the orbit computed by their own astronomer, Adams. When the English announced Adams's work, the French suspected that he had plagiarized the calculations, and the controversy was bitter, as controversies between England and France often are. For more than a century, historians of science have repeated the story of the young English astronomer who missed his chance because the astronomers in charge of the search were careless and slow.

Original papers related to Adams's calculations were lost for decades, but when they were found in 1998, they painted a different picture. Adams did the calculations correctly, but he computed only the orbit for the new planet. He didn't actually calculate its position in the sky along the orbit; that was left to the English astronomers conducting the search. Once the new planet was discovered, the English astronomers, out of national pride, pressed Adams's case further than they should have.

It seems clear today that Le Verrier deserves credit for making an accurate, useful prediction of a position on the sky and then pressing it aggressively on an astronomer who had the right skills to make the search. You should still give both astronomers credit for solving a difficult problem that was one of the great challenges of the age. Modern analysis shows that both Le Verrier and Adams made unwarranted assumptions about the undiscovered planet's distance from the Sun. By accident their assumptions made no big difference, and the planet was close to the positions they predicted. Do you think they deserve less credit for that reason? They tried, and the other astronomers of the world didn't.

Le Verrier and Adams could have been beaten to the discovery had Galileo paid a little less attention to Jupiter and a little more attention to what he saw in the background. Modern studies of Galileo's notebooks show that he saw Neptune on December 24, 1612, and again on January 28, 1613, but he plotted it as a star in the background of drawings of Jupiter. It is interesting to speculate about the response of the Inquisition had Galileo proposed that a planet existed beyond Saturn. Unfortunately for history, but perhaps fortunately for Galileo, he did not recognize Neptune as a planet, and its discovery had to wait another 234 years.

Historians of science have recounted the discovery of Neptune as a triumph for Newtonian physics: The three laws of motion and the law of gravity had proved sufficient to predict the position and orbit of an unseen planet. Thus, the discovery of Neptune was fundamentally different from the discovery of Uranus. Uranus was discovered "accidentally" in the course of Herschel's attempt to systematically observe the entire sky, whereas the existence of Neptune was predicted by Le Verrier and Adams using basic laws of physics.

Neptune's Atmosphere and Interior

Little was known about Neptune before the *Voyager 2* spacecraft swept past it in 1989. Seen from Earth, Neptune is a tiny, blue-green dot never more than 2.3 arc seconds in diameter. Before *Voyager 2,* astronomers knew Neptune is almost four times the diameter of Earth, or about 4 percent smaller in diameter than Uranus (■ Celestial Profile 10, page 341), with a mass about 17 times that of Earth, which is 20 percent more than Uranus. Spectra revealed that, as in the case of Uranus, its blue-green color was caused by methane in its hydrogen-rich atmosphere absorbing red light. Neptune's density showed that it is a Jovian planet rich in hydrogen, but almost no detail was visible from Earth, so even its period of rotation was uncertain.

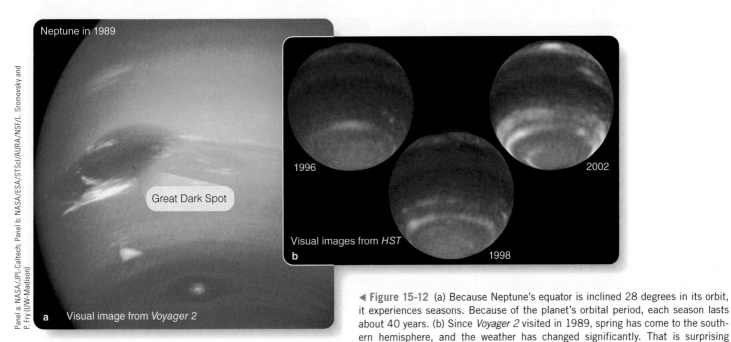

Panel a: NASA/JPL-Caltech; Panel b: NASA/ESA/STScI/AURA/NSF/L. Sromovsky and P. Fry (UW-Madison)

Neptune in 1989

Great Dark Spot

a Visual image from *Voyager 2*

1996

2002

Visual images from *HST*

b 1998

◄ **Figure 15-12** (a) Because Neptune's equator is inclined 28 degrees in its orbit, it experiences seasons. Because of the planet's orbital period, each season lasts about 40 years. (b) Since *Voyager 2* visited in 1989, spring has come to the southern hemisphere, and the weather has changed significantly. That is surprising because sunlight on Neptune is 900 times dimmer than on Earth.

Voyager 2 passed only 4900 km (3050 mi) above Neptune's cloud tops, closer than any spacecraft had ever come to one of the Jovian planets. The images it captured revealed that Neptune is marked by dramatic belt–zone circulation parallel to the planet's equator. *Voyager 2* also saw at least four cyclonic disturbances. The largest, dubbed the Great Dark Spot, looked similar to the Great Red Spot on Jupiter (**Figure 15-12**). Neptune's Great Dark Spot was located in the southern hemisphere and rotated around its center counterclockwise, with a period of about 16 days. Like the Great Red Spot, it appeared to be caused by gas rising from its planet's interior. Unexpectedly, when the *Hubble Space Telescope* began imaging Neptune in 1994, the Great Dark Spot was gone, and new cloud features were seen appearing and disappearing in Neptune's atmosphere (Figure 15-12). Evidently, the cyclonic disturbances on Neptune are not nearly as long lived as Jupiter's Great Red Spot, which has persisted for centuries.

The *Voyager 2* images reveal other cloud features standing out against the deep blue of the methane-rich atmosphere. The white clouds are made of methane ice particles and range up to 50 km above the deeper layers, just where the temperature in Neptune's atmosphere is low enough for rising methane to freeze into crystals (Figure 15-4). Presumably these features are related to rising convection currents that produce clouds high in Neptune's atmosphere where they catch sunlight and appear bright. Special filters can reveal these cloud belts in visual-wavelength images and at infrared wavelengths (**Figure 15-13**). Some observations made with the *Hubble Space Telescope* suggest that atmospheric activity on Neptune may be related to flares and other eruptions on the Sun, but more data are needed to explore this connection.

As on the other Jovian worlds, winds circle Neptune parallel to its equator, but Neptune's winds blow at very high speeds and tend to blow backward—against the rotation of the planet. Why Neptune should have such high-speed retrograde winds is not understood, and it is part of the larger problem of understanding belt–zone circulation.

Now that you have seen belts and zones on all four of the Jovian planets (assuming that the faint clouds observed on Uranus are in fact traces of belt–zone circulation), you can ask what drives this circulation. Because belts and zones remain parallel to a planet's equator even when the planet rotates at a high inclination, as in the case of Uranus, it seems reasonable to believe that the atmospheric circulation is dominated by the rotation of the planet, and perhaps also by heat flow and circulation currents in the liquid interior, but not by solar heating.

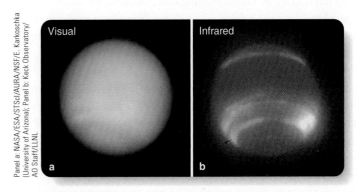

Panel a: NASA/ESA/STScI/AURA/NSF/E. Karkoschka (University of Arizona); Panel b: Keck Observatory/AO Staff/LLNL

Visual

Infrared

a **b**

▲ **Figure 15-13** (a) This visual-wavelength image of Neptune was taken through filters that cause belts of methane to stand out. (b) The infrared image at right was recorded using one of the Keck 10-m telescopes and adaptive optics. The locations of methane cloud belts around the planet are evident.

The same observations and calculations that allow planetary scientists to define the interior of Uranus can be applied to Neptune. Models suggest that Neptune has a small heavy-element core, surrounded by a deep mantle of slushy or solid water mixed with heavier material having a chemical composition resembling rock. Neptune's magnetic field is a bit less than half as strong as Earth's and is tipped 47 degrees from the axis of rotation. It is also offset 55 percent of the way to the surface (Figure 15-6). As in the case of Uranus, Neptune's field is probably generated by the dynamo effect operating in the conducting fluid mantle rather than in the planet's core.

Neptune has more internal heat than Uranus, and part of that heat may be generated by radioactive decay in the minerals in its interior. Some of the energy also may be released by denser material falling inward, including, as in the case of Uranus, diamond crystals formed by the disruption of methane. For whatever reason, Neptune has retained substantial internal heat whereas Uranus has not.

Neptune is a tantalizing world just big enough to be imaged by the *Hubble Space Telescope* but far enough away to make it difficult to study. The data from *Voyager 2* revealed that its moon system is surprisingly complex.

Neptune's Moons

Neptune has at least 14 moons. Before *Voyager 2* visited Neptune, only three moons were known. *Voyager 2* discovered six more small moons, and five additional small moons have been found since by observations from Earth. Two of the larger moons, Triton and Nereid, have been a puzzle for years because of their peculiar orbits.

Triton has a perfectly circular orbit, but it travels retrograde—orbiting Neptune clockwise as seen from the north. This makes Triton the only large satellite in the Solar System with a backward orbit; all other retrograde satellites are very small. Nereid moves in the prograde direction, but its orbit is highly elliptical and very large (Figure 15-14). Nereid takes 360 Earth days to orbit Neptune once. Some astronomers have speculated that the orbits of Triton and Nereid are evidence of a violent event long ago when Triton was captured into orbit during a close encounter with Neptune and scrambled the orbits of the other moons.

Eight moons orbit Neptune among the rings, in the prograde direction: the six moons found by *Voyager 2*, plus Larissa that was discovered by Earth-based stellar occultation observations, and a yet-unnamed moon identified in *Hubble Space Telescope* images. Triton's orbit lies outside the rings. Beyond the orbit of Triton there are six moons in irregular orbits, including Nereid that was discovered in 1949, and five others found since *Voyager 2* by observations using large Earth-based telescopes. Two of those moons have orbit semi-major axes of 0.3 AU, making them the most distant moons of any planet in the Solar System.

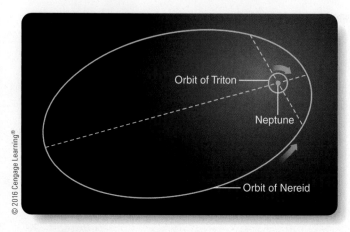

▲ **Figure 15-14** Triton follows a small, circular, retrograde orbit. At a distance of 30 AU from Earth, Triton is never seen farther than 16 arc seconds from the center of Neptune. Nereid has a large prograde elliptical orbit with a period of nearly 1 Earth year. There are also seven known small moons in prograde circular orbits close to Neptune than Triton, and five tiny moons in even larger and more eccentric orbits than Nereid, three retrograde and two prograde.

The *Voyager 2* images show that Triton is highly complex. Although it is only 2710 km (1680 mi) in diameter (78 percent the size of Earth's Moon), it is so cold (35 K, or –395°F) that it can hold a thin atmosphere—10^5 times less dense than Earth's atmosphere, composed of nitrogen and some methane. Although a few wisps of haze can be seen in the photographs, the atmosphere is transparent, and the surface is easily visible (Figure 15-15).

The surface of Triton is evidently composed of different types of ice. The surface ice is dominated by frozen nitrogen with some methane, carbon monoxide, and carbon dioxide. That is consistent with the nitrogen-rich atmosphere. Some regions of the surface that look dark may be slightly older terrain that darkened as sunlight converted methane into organic compounds. Triton's south pole had been turned toward the Sun for 30 years when *Voyager 2* flew past, and deposits of nitrogen frost in a large polar cap appeared to be vaporizing there and refreezing in the darkness of Triton's north pole (Figure 15-15a). The cycle of nitrogen on Triton resembles in some ways the cycle of carbon dioxide on Mars.

The surface of Triton contains evidence that the icy moon has been active recently and may still be active. Neptune and Triton are at the inner edge of the Kuiper Belt that you first read about in Chapter 10. You might predict that a moon located so close to the Kuiper Belt would have lots of craters, but Triton has few. The average age of the surface is no more than 100 million years, very young in astronomical terms. Some process has erased older craters. Evidence of geological activity includes long linear features that appear to be fractures in an icy crust and some roughly round basins that appear to have been flooded repeatedly by liquids from the interior (Figure 15-15b). Triton is much too cold for the liquid to be molten rock or even liquid water. Rather, the floods must have been composed of water that

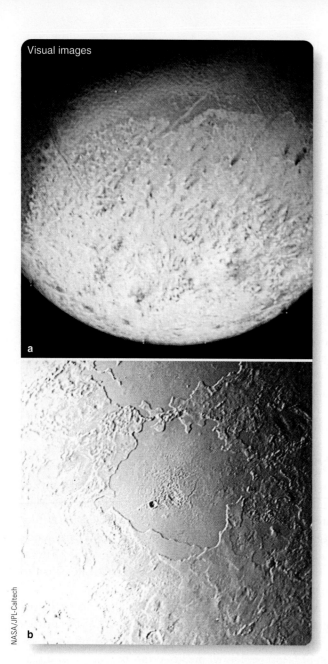

Visual images

a

b

NASA/JPL-Caltech

▲ **Figure 15-15** (a) Triton's south pole *(bottom)* had been in sunlight for 30 years when *Voyager 2* flew past in 1989. The frozen nitrogen in the polar cap appears to be vaporizing, perhaps to refreeze in the darkness at the planet's north pole. The dark smudges are produced when liquid nitrogen in the crust vaporizes and drives nitrogen geysers. (b) Roughly round basins on Triton may be old impact basins flooded repeatedly by liquids from the interior. Notice the small number of craters on Triton, a clue that it is a partially active world.

by the Sun, the nitrogen ice can change from one form of solid nitrogen to another and release heat that can vaporize some of the nitrogen. Heat rising from the interior can also vaporize nitrogen. This nitrogen vapor vents through the crust, forming nitrogen gas geysers up to 8 km (5 mi) high. The venting gas may carry dark material from below the ice that falls to the surface to form the dark smudges. Another possibility is that methane is carried along with the nitrogen, and sunlight converts some of the methane into dark organic material, which falls to the surface.

Active worlds must have a source of energy; Triton is big enough to retain some thermal energy from low-level radioactivity in its interior. That may be enough to melt some ices and cause flooding on the surface. The nitrogen geysers may be powered partly by heat from the interior and partly by sunlight. In fact, Triton is very efficient at absorbing sunlight. Its thin atmosphere does not dim sunlight, and its crust is composed of ices that are partially transparent to light. As the light penetrates into the ice and is absorbed, it warms the ices. However, the crust is not transparent to infrared radiation, so the heat is trapped in the ice. So the crust of Triton appears to be heated, in part, by a solid-ice version of the greenhouse effect.

This low-level heating would not be enough to erase nearly all of Triton's craters, so some planetary astronomers wonder if Triton might have been captured by Neptune relatively recently—meaning within the past billion years. Tidal forces from such a capture would have caused enough tidal heating to melt Triton and totally resurface it. Enough of that heat may remain to keep Triton active to this day.

Neptune's Rings

Astronomers on Earth saw hints of rings earlier when Neptune occulted stars, but the rings were not firmly identified until *Voyager 2* flew past the planet in 1989.

Look again at **Uranus's and Neptune's Rings** on pages 338–339 and compare the rings of Neptune with those of Uranus. Notice two additional points:

4 Neptune's rings, named after the astronomers involved in the discovery of the planet, are similar in some ways to those of Uranus but contain more small dust particles that forward-scatter light.

5 Also notice a new way that rings can interact with moons: One of Neptune's moons produces short arcs in the outermost ring. (A similar arc has been found in one of Saturn's rings.)

Like the rings of Uranus, Saturn, and Jupiter, the particles observed now in Neptune's rings cannot be primordial. Presumably, debris from impacts on the moons accumulates in places where the orbits of particles are most stable among the orbits of the moons.

contained agents such as ammonia, which would lower the freezing point of the liquid. It isn't possible to say now whether Triton is still active, but 100 million years isn't very long in the history of the Solar System. Triton may still suffer periodic eruptions and floods.

Another form of activity on Triton leaves dark smudges visible in the bright nitrogen ices near its south pole. These appear to be caused by nitrogen ice beneath the crust. Warmed slightly

A History of the Neptune System

Can you tell the story of Neptune? In some ways it seems to be a simpler, smaller version of Jupiter and Saturn, but, like that of Uranus, Neptune's magnetic field is peculiar, and its moons and rings also deserve careful attention.

You can assume that Neptune formed from the solar nebula in much the same way as did Uranus. Like Uranus, Neptune's hazy atmosphere, marked by changing cloud patterns, hides a mantle of partially frozen ices where astronomers suspect the dynamo effect generates its off-center magnetic field. Inside that mantle is a denser core. Neptune, like Jupiter and Saturn but unlike Uranus, has significant amounts of heat flowing out to space from its interior.

Neptune's satellite system suggests a peculiar history. Triton, the largest moon, revolves around Neptune in a retrograde orbit, whereas Nereid's long-period orbit is highly elliptical. These orbital oddities suggest that the satellite system may have been disturbed during the capture of Triton. You have seen evidence in other satellite systems of impacts with large objects, so hypothesizing such an event is not unreasonable.

A number of smaller moons orbit Neptune near its ring system. Because the rings are bright in forward scattering, you can conclude that they contain some dust, and the shepherding of small satellites must confine their width and produce the observed arcs. Evidently Neptune's rings must be occasionally supplied with fresh particles generated by impacts of meteorites and comets on Neptune's moons. Once again, the evidence of major impacts in the Solar System's history assures you that such impacts do occur.

15-3 Pluto and the Kuiper Belt

In 1930, Pluto was discovered orbiting beyond Neptune, and the public welcomed it as the ninth planet. To the surprise of astronomers, Pluto was found to be a solid object smaller than Earth's Moon rather than a low-density Jovian planet. At the end of the 20th century, with much-improved telescopes, astronomers found more small worlds in the same region, and it became clear that Pluto was just one of a large family of similar objects.

In 2006, the International Astronomical Union (IAU) voted to move Pluto out of the family of major planets and make it part of a larger family of small worlds. To understand this decision, which remains somewhat controversial, you can start with the details of Pluto's discovery.

Discovery of Pluto

Percival Lowell was fascinated with the idea that an intelligent race built the canals he thought he could see on Mars (Figure 13-9a). In 1894, Lowell founded Lowell Observatory in Flagstaff, Arizona, primarily for the study of Mars. Later, some say, motivated to improve the reputation of his observatory after the controversy about Martian canals, he began to search for a planet beyond Neptune.

Lowell used the same method that Adams and Le Verrier had used to predict the position of Neptune. Working from what were understood at the time to be irregularities in the motion of Neptune, Lowell predicted the location of an undiscovered planet beyond Neptune. He concluded it would contain about 7 Earth masses and would look like a 13th-magnitude object in eastern Taurus. Lowell and his staff searched for the planet photographically until his death in 1916.

In the late 1920s, 22-year-old amateur astronomer Clyde Tombaugh began using a homemade 9-in. telescope to sketch Jupiter and Mars from his family's wheat farm in western Kansas. He sent his drawings to Lowell Observatory, and the observatory director, Vesto Slipher, hired him without an interview. The young Tombaugh bought a one-way train ticket to Flagstaff without knowing what his new job would be like.

Slipher set Tombaugh to work photographing the sky along the ecliptic around the predicted position of the planet. The search technique was a classic method in astronomy. Tombaugh obtained pairs of 14-by-17-in. glass plates exposed 2 or 3 days

---

DOING SCIENCE

Why is Neptune blue but its clouds white? To answer this question, a scientist first needs to consider the route taken by light we receive from Neptune.

When you look at something, you turn your eyes toward it and receive light from the object. When you look at Neptune, the light you receive is sunlight that is reflected from various layers of Neptune and journeys to your eyes. Because sunlight contains a distribution of photons of all visible wavelengths, it looks white to human eyes, but sunlight entering Neptune's atmosphere must pass through hydrogen gas that contains a small amount of methane. Whereas hydrogen is almost completely transparent at visible wavelengths, methane is a good absorber of red wavelengths, so red photons are more likely to be absorbed than blue photons. Once the light is scattered back into space from deeper layers, it must run this methane gauntlet again to emerge from the atmosphere, and again red photons are more likely to be absorbed. The light that finally emerges from Neptune and eventually reaches your eyes is poor in longer wavelengths and thus looks blue.

The methane-ice-crystal clouds lie at high altitudes, so sunlight does not have to penetrate very far into Neptune's atmosphere to reflect off the clouds, and consequently it loses many fewer of its red photons. The clouds look white.

This discussion shows how a careful, step-by-step chain of inference can lead to a better understanding of how nature works. Now, build a another chain of inference to answer the following question: ***Where does the energy come from to power Triton's surface geysers?***

---

apart. To search a pair of plates, he mounted them in a blink comparator, a machine that allowed him to look through a microscope at a small spot on one plate and then at the flip of a lever see the same spot on the other plate. As he blinked back and forth, the star images did not move, but a planet would have moved along its orbit during the 2 or 3 days that elapsed before the second plate was exposed. So Tombaugh searched the giant plates, star image by star image, looking for an image that moved. A single pair of plates could contain 400,000 star images. He searched pair after pair and found nothing.

The observatory director turned to other projects, and Tombaugh, working alone, expanded his search to cover the entire ecliptic. For almost a year, Tombaugh exposed plates by night and blinked plates by day. Then, on February 18, 1930, nearly a year after he had left Kansas, a quarter of the way through a pair of plates, he found a 15th-magnitude image that moved (**Figure 15-16**). He later remembered the moment: "'Oh,' I thought, 'I had better look at my watch; this could be a historic moment. It was within about 2 minutes of 4 pm [MST].'" The discovery was announced on March 13, the 149th anniversary of the discovery of Uranus and the 75th anniversary of the birth of Percival Lowell. The object was named Pluto after the god of the underworld and also, in a way, after Lowell because the first two letters in Pluto are the initials of Percival Lowell.

The discovery of Pluto seemed a triumph of discovery by prediction, but Tombaugh sensed something was wrong from the first moment he saw the image. It was moving in the right direction by the right amount, but it was 2.5 magnitudes too faint. Clearly, Pluto was not the 7-Earth-mass planet that Lowell had predicted. The faint image implied that Pluto was a small world with a mass too low to seriously alter the motion of Neptune.

Later analysis has shown that the supposed variations in the motion of Neptune, which Lowell used to predict the location of Pluto, were random uncertainties of observation and could not have led to a trustworthy prediction. The discovery of the new planet only 6 degrees from Lowell's predicted position was apparently an accident.

▲ Figure 15-16 Pluto is small and far away, so its image is indistinguishable from that of a star on most photographs. Clyde Tombaugh discovered the planet in 1930 by looking for an object that moved relative to the stars on a pair of photographs taken a few days apart.

Pluto as a World

Pluto is difficult to observe from Earth. Only 68 percent the diameter of Earth's Moon, its angular size is only a bit larger than 0.1 arc second. Pluto shows little surface detail even when observed with the *Hubble Space Telescope,* although, as you will learn, astronomers have used a clever trick to make low-resolution maps of Pluto and its moon Charon. If all goes well, the *New Horizons* probe, due to arrive in 2015, will send the first close-up images of Pluto and its moons.

Most planetary orbits in our Solar System are nearly circular, but Pluto's is significantly eccentric. In fact, from 1979 to 1999, Pluto was closer to the Sun than Neptune. The two worlds will never collide, however, because Pluto's orbit is inclined 17 degrees to the ecliptic and also because Pluto and Neptune orbit in resonance with each other so they never come close together.

If you land on the surface of Pluto, your spacesuit will have to work hard to keep you warm. Orbiting so far from the Sun, Pluto is cold enough to freeze most compounds that you think of as gases; spectroscopic observations have found evidence of solid nitrogen ice on its surface with traces of frozen methane and carbon monoxide. The maximum daytime temperature of about 55 K (−360°F) is enough to vaporize some of the nitrogen and carbon monoxide and a little of the methane to form a thin atmosphere around Pluto. This atmosphere was detected in 1988 when Pluto occulted a distant star, and the starlight was observed to fade gradually as the atmosphere absorbed it rather than winking out suddenly at Pluto's solid edge.

Pluto's largest moon was discovered on photographs in 1978. It is very faint and about half the diameter of Pluto (**Figure 15-17**). The moon was named Charon after the mythological ferryman who transports souls across the river Styx into the underworld. Four smaller moons, named Stix, Nix, Kerberos, and Hydra, were found between 2005 and 2012 by several teams of astronomers using the *Hubble Space Telescope.*

The discovery of Pluto's moons is important for a number of reasons. Charon, the largest and easiest to track, orbits Pluto in a nearly circular orbit in the plane of Pluto's equator. Observations show that Charon and Pluto are tidally locked to each other and that Pluto's axis of rotation, like Uranus's, is highly inclined to its orbit around the Sun (**Figure 15-18**). Furthermore, tracking the orbital motion of Charon allowed the calculation of the mass of Pluto. Charon orbits 19,600 km from Pluto with an orbital period of 6.39 days. Kepler's third law reveals that the mass of the system is 7.3×10^{-9} solar mass or only about 0.0024 Earth mass. Most of that mass is Pluto, which is about 9 times more massive than Charon.

You know that finding the density of an object is important in astronomy because it gives an indication of the object's overall composition. From the mass and size of Pluto, its density is found to be about 2.0 g/cm³, and the density of Charon is about 1.6 g/cm³. Those densities indicate that Pluto and Charon must both contain mostly rock mixed with some ice.

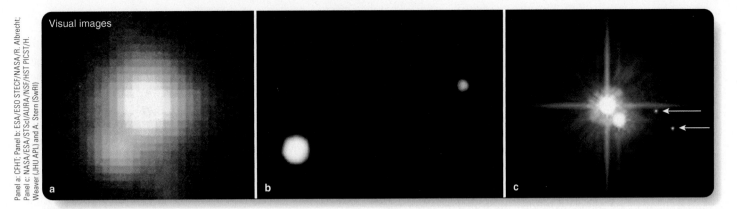

Panel a: CFHT; Panel b: ESA/ESO STECF/NASA/R. Albrecht; Panel c: NASA/ESA/STScI/AURA/NSF/HST PICST/H. Weaver (JHU APL) and A. Stern (SwRI)

Visual images

a b c

▲ **Figure 15-17** (a) A high-quality ground-based photo shows Pluto and its moon Charon badly blurred by seeing. (b) The *Hubble Space Telescope* image clearly separates the planet and its moon and allows more accurate measurements of the position of the moon. (c) A long-exposure photograph made in 2006 with the *Hubble Space Telescope* confirmed discovery of two more moons of Pluto that were named Nix and Hydra (indicated by arrows).

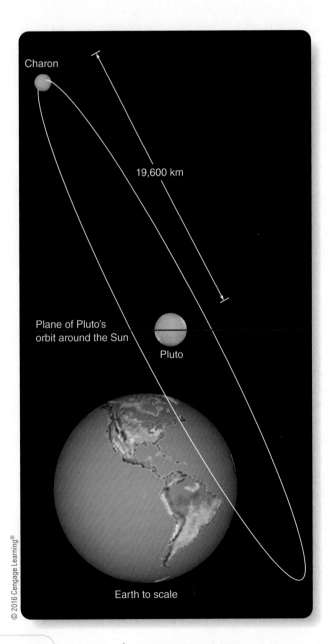

Charon

19,600 km

Plane of Pluto's orbit around the Sun

Pluto

Earth to scale

◀ **Figure 15-18** The nearly circular orbit of Charon is only a few times bigger than Earth's diameter. It is shown here nearly edge-on, and consequently it looks elliptical in this diagram. Charon's orbit and the equator of Pluto are inclined 120 degrees to the plane of Pluto's orbit around the Sun.

Spectra of Charon show that the small moon has a surface that is mostly water ice with not much evidence of other volatiles that are detected in spectra of Pluto. Perhaps Charon has lost its more volatile compounds because of its lower escape velocity. Water ice at Charon's surface temperature is no more volatile than is a piece of rock on Earth, so a water ice surface could last a long time.

Another reason Pluto's moon Charon has proved important is that because Charon and Pluto orbit the Sun their mutual orbit is occasionally seen edge-on from Earth. During those times, astronomers can watch Pluto and Charon eclipse each other, and by carefully measuring the combined light from the two objects, they can produce crude maps (**Figure 15-19**). Those observations reveal that Pluto has a surprising amount of albedo variation on its surface, but nobody knows what the dark and light features might be. Understanding that will have to wait until the *New Horizons* probe flies by in 2015.

Both Pluto and Charon go through dramatic seasons much like those on Uranus as they circle the Sun with their highly inclined rotation axes. This should cause large changes in Pluto's atmosphere as the planet grows warmer when it is closest to the Sun, as it was in the late 1980s, and then freezes as it draws away, as it is doing now.

The Family of Dwarf Planets

Perhaps the most interesting thing about Pluto is that it is not alone. More than a thousand objects have been discovered orbiting in the Kuiper Belt along with Pluto. At least one of them is larger than Pluto.

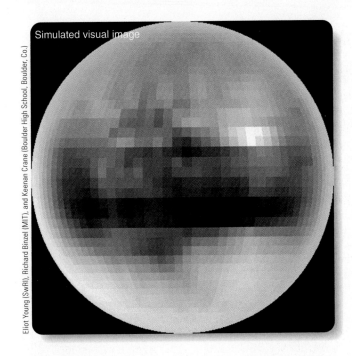

Simulated visual image

▲ **Figure 15-19** Observing Pluto's brightness variations as its moon Charon occasionally moves across the planet's disk allowed astronomers to make this low-resolution map of Pluto. These are approximately true colors.

The object known as Eris, discovered in 2003, has about the same diameter as Pluto but is 28 percent more massive. It orbits about 1.7 times farther away from the Sun than does Pluto. Eris's orbit is more eccentric and more highly inclined than Pluto's, but it seems to be a similar object. The discovery of Eris led the IAU to recognize a new class of Solar System objects called the **dwarf planets**—objects that orbit the Sun and are large enough to assume a spherical shape but not large enough to have sufficient gravitational influence to absorb or otherwise clear away remaining objects orbiting nearby. In contrast, the major planets, including Earth, were able to absorb or eject all of the nearby planetesimals as they were growing from protoplanets into planets.

So far, five Solar System objects have been designated by the IAU as dwarf planet: Pluto, Eris, Haumea (pronounced *how-MAY-ah*), and Makemake (pronounced *MAH-kay-MAH-kay*) in the Kuiper Belt, plus Ceres, the 950-km (590-mi)-diameter asteroid that orbits within the asteroid belt between Mars and Jupiter and is nearly twice as big as the next largest asteroids, Pallas and Vesta. (You will learn more about Ceres, Pallas, Vesta, and their other asteroid belt members in the next chapter.)

About ten other Kuiper Belt Objects are known to be almost as large as Pluto, Eris, Haumea, and Makemake, and thus are considered candidate dwarf planets, pending better determination of their properties. There are probably others yet to be discovered. Two large objects named Sedna and Orcus are each about two-thirds the diameter of Pluto. Another object called Quaoar (pronounced *KWAH-o-wahr*) is

half the diameter of Pluto. Haumea is a strange beast, with such a rapid spin—once every 4 hours—that it is shaped like a flattened loaf of bread.

Some astronomers argue that Charon, Pluto's big spherical moon, should be a member of the dwarf planets even though it orbits another world and not the Sun. Other astronomers are upset that Ceres, a rocky asteroid, is included. The classification may seem arbitrary until you begin thinking about how the dwarf planets formed. Some astronomers refer to them as oligarchs. The term *oligarch* is usually applied to business or political leaders who are the biggest, meanest dudes in town. They are not alone, but they are the bosses. The dwarf planets appear to have been bodies in the solar nebula that grew more rapidly than their neighbors and became dominant but never got big enough to take over completely and sweep up all objects orbiting nearby. Planets such as Earth and Jupiter cleared their orbital lanes around the Sun, but the dwarf planets never got quite big enough to do that. So they aren't planets; they are dwarf planets.

Pluto and the Plutinos

No, this section is not about a 1950s rock-and-roll band. It is about the history of the dwarf planets, and it will take you back billions of years to watch the outer planets form.

Hundreds of Kuiper Belt Objects are, like Pluto, caught in a 3:2 resonance with Neptune. That is, they orbit the Sun twice while Neptune orbits three times. You learned about orbital resonances when you studied Jupiter's Galilean moons. A 3:2 resonance with Neptune makes the orbiting bodies immune to any disturbing gravitational influence from Neptune, so their orbits are more stable. The orbits of Neptune and Pluto actually cross, although the resonance causes them to always be distant from each other so they will never collide. Because Pluto is one of the objects caught in the same 3:2 resonance, these Kuiper Belt Objects have been named **plutinos**.

How did the plutinos get caught in resonances with Neptune? You have already learned that computer models of the formation of the planets suggest that Uranus and Neptune may have formed closer to the Sun. Sometime later, gravitational interactions with Jupiter and Saturn gradually shifted the two ice giants outward, and, as Neptune migrated outward, its orbital resonances could have swept up small bodies like nets pushed in front of a fishing boat. Other Kuiper Belt Objects are caught in other stabilizing resonances, and they were apparently swept up in the same way. The plutinos in 3:2 resonance and other Kuiper Belt Objects in other resonances with Neptune are evidence that Neptune really did migrate outward. This planet migration could have scattered small bodies throughout the Solar System and caused the late heavy bombardment event during which Earth's Moon, presumably along with Earth and other Solar System bodies, suffered a brief but devastating increase in cratering about 4 billion years ago.

Some astronomers are still angry that the IAU "demoted" Pluto, but it is actually a more interesting world once you realize that it is the best studied of the dwarf planets. In the inner Solar System, only the asteroid Ceres was able to grow fast enough to become a dwarf planet, but in the outer Solar System huge numbers of icy bodies formed, ranging from pebbles to the oligarchs now recognized as dwarf planets. As they are understood better, the dwarf planets will reveal more secrets from the age of planet building.

DOING SCIENCE

Why is Earth a considered planet and not a dwarf planet?
One thing scientists normally do when beginning a new field of study is to make a classification scheme for the objects being examined.

To be useful, a classification scheme must be based on real characteristics, so you need consider the definition of the dwarf planets. According to the International Astronomical Union, a dwarf planet must orbit the Sun, not be a satellite of a planet, and be spherical. Earth meets these characteristics, but there is one more requirement.

By definition, a dwarf planet must be small enough to have been unable to clear out most of the smaller objects near its orbit. As Earth grew in the solar nebula, it accreted or ejected the small bodies that orbited the Sun in similar orbits. In other words, Earth was big enough that its gravity was able to clear its traffic lane around the Sun, so it is classified as a planet. Pluto never became massive enough to clear its lane (the Kuiper Belt), so Pluto is classified as a dwarf planet. (Note that a dwarf planet is a planet, just like a dwarf galaxy is a galaxy and a dwarf star is a star.)

What Are We? Trapped

No person has ever been farther from Earth than the Moon. We humans have sent robotic spacecraft to explore the worlds in our Solar System beyond Earth's Moon, but no human has ever set foot on them. We are trapped on Earth.

We lack the technology to leave Earth easily. Getting away from Earth's gravitational field calls for very large rockets. The United States built huge rockets in the 1960s and early 1970s to send astronauts to the Moon, but such rockets no longer exist. The best technology today can carry astronauts just a few hundred kilometers above Earth's surface to orbit above the atmosphere. We can probably reach Mars with a few decades of effort, but going beyond may take more resources than Earth can provide.

There is another reason we Earthlings are trapped on our planet. We have evolved to fit the environment on Earth. None of the planets or moons you explored beyond Earth would welcome you. Radiation belts, extreme heat or cold, and lack of air are obvious problems, but Earthlings have evolved to live with 1 Earth gravity. Astronauts in orbit for just a few weeks suffer biomedical problems because their muscles and bones no longer feel Earth's gravity. Could humans live for years in the weak gravity on Mars? We may be trapped on Earth not because we lack big rockets but because we need Earth's environment.

It seems likely that we need Earth more than it needs us. The human race is changing the world we live on at a terrific pace, and some of those changes are making Earth less hospitable. All of your exploring of un-Earthly worlds serves to remind you of the nurturing comfort and beauty of our home planet. It is probably the only one we will ever have.

Study and Review

Summary

▶ Discovered in 1781 by William Herschel, Uranus is the first planet to be found rather than known from prehistoric times.

▶ Although it is considered a Jovian planet, Uranus is significantly smaller and less massive than Jupiter, about four times the diameter of Earth.

▶ Uranus rotates "on its side," with its rotational axis nearly in the plane of its orbit. For this reason, it is the planet with the most extreme seasons.

▶ The atmosphere of Uranus is mostly hydrogen and helium with some methane. Methane absorbs longer-wavelength photons and thus gives Uranus a greenish-blue color.

▶ The atmosphere of Uranus is so cold that only methane ice particle clouds are visible. Signs of belt–zone circulation lie deep

in the atmosphere and are difficult to discern. In recent decades, spring has come to its northern hemisphere, and more cloud features have developed. This change suggests that Uranus has a seasonal cycle and is not always as bland as it was when *Voyager 2* flew past in 1986.

▶ Model calculations indicate that Uranus has a small core of dense matter and a deep slushy mantle of ice, water, and rock. Convection in the mantle may produce Uranus's magnetic field that is highly inclined and offset from its rotation axis.

▶ Uranus emits about the predicted amount of heat for a planet located at its distance from the Sun, which suggests that it has lost most of its internal heat.

▶ The rings of Uranus were discovered during an **occultation (p. 338)**, when the planet crossed in front of a star.

- The rings of Uranus are composed mostly of dark, boulder-size objects. The dark color may be from a surface coating of carbon caused by solar wind particles and UV breaking down methane ice. The same process may explain the dark surfaces on the Uranian moons. The rings are probably resupplied by debris from impacts on nearby icy moons.

- The five largest moons of Uranus appear to be icy with mostly old, cratered surfaces. However, Miranda appears to have had significant geological activity, as shown by the **ovoids (p. 337)** on its surface. Tidal heating is a likely source of the energy that drove this activity.

- Uranus appears to have formed slowly and was unable to capture significant quantities of hydrogen and helium from the solar nebula before the nebula dispersed. Uranus and Neptune may have formed closer to the Sun and migrated outward by gravitational interactions with Jupiter and Saturn.

- An impact by a large planetesimal while Uranus was forming, or tidal interactions with Jupiter and Saturn as Uranus migrated, may have caused its highly inclined rotation axis.

- Neptune was discovered in 1846 based on its predicted position in its orbit, computed from the discrepancies found in the orbital motion of Uranus.

- Neptune is slightly smaller in diameter but more massive than Uranus. Model calculations indicate that Neptune's core is denser than Uranus's core. Like Uranus, Neptune has a mantle of ice, water, and rock,

- Circulation in Neptune's conducting fluid mantle gives rise to the planet's magnetic field.

- Neptune has significantly more internal heat than Uranus. These two planets are otherwise quite similar, so the reason for this major difference is not understood.

- Neptune has an atmosphere of hydrogen and helium with traces of methane. The larger proportion of methane in Neptune's atmosphere results in the planet having a bluer color than Uranus. Methane clouds come and go in the cold atmosphere of Neptune and have a visible belt–zone circulation pattern.

- The rings of Neptune probably formed when impacts on moons scattered icy debris into stable bands among the moons' orbits. Observations of sunlight forward scattering show that Neptune's rings contain more small dust particles than the rings of Uranus. Short arcs in the outermost ring appear to be caused by the gravitational influence of a small moon or moons.

- Neptune has 14 known moons. The largest moon, Triton, follows a circular but retrograde orbit. Seven moons orbit closer than Triton in prograde orbits. Six moons orbit farther from the planet than Triton; Nereid and two other small moons have prograde orbits; three moons have high-inclination retrograde orbits, and are probably captured objects. Triton was most likely captured, and that event may have caused some of the unusual aspects of the Neptunian system.

- Triton has an icy surface and a thin atmosphere of nitrogen. The lack of many craters and the presence of flooded areas, cracks, and faults suggest that Triton may still be geologically active. Sunlight and heat from the interior appear to trigger nitrogen geysers in the crust. The interior heat may be leftover tidal heating from when Triton was captured into orbit.

- Neptune formed slowly, as did Uranus, and was unable to accumulate a deep atmosphere of hydrogen and helium before the solar nebula was blown away.

- Pluto was discovered in 1930 during a search for a large planet orbiting beyond Neptune that seemed to be disturbing Neptune's orbital motion. That discovery was in a sense an accident because Pluto is not massive enough or close enough to Neptune to affect Neptune's orbit. Later observations indicated that Neptune's orbital motion is actually undisturbed.

- Spectroscopic observations indicate that Pluto has a frigid crust of solid nitrogen ice with traces of frozen methane and carbon monoxide. Pluto's thin atmosphere is mostly nitrogen.

- Pluto is a small world with five known moons, one of which, Charon, is quite large in relation to Pluto. Observations of Charon and Pluto's mutual orbit indicate that both Pluto and Charon are composed of a mixture of rock and ice.

- The orbital plane of Pluto's moons is parallel to Pluto's equator and highly inclined to Pluto's orbit around the sun. Charon and Pluto are tidally locked to each other. Together, those facts mean that Pluto's rotation axis must be nearly parallel to its orbit, much like Uranus.

- Careful measurements of the brightness of Charon and Pluto on occasions when they move in front of each other have allowed astronomers to construct low-resolution maps of both objects.

- The densities of Pluto and Charon show that they must contain mostly rock mixed with substantial amounts of ice.

- The *New Horizons* probe will reach Pluto in 2015 and then continue on to fly by one or more Kuiper Belt Objects (KBOs). Currently, more than a thousand KBOs are known.

- The **dwarf planets (p. 349)** are small bodies that orbit the Sun, but they are not planets. Like the major planets, they are spherical. Unlike the major planets, they are not large enough to have cleared their orbital paths of other nearby objects.

- Pluto is now classified as a dwarf planet by the IAU. Other Kuiper Belt Objects (KBOs) Eris, Haumea, and Makemake, as well as the asteroid Ceres, are also classified as dwarf planets. At least ten other KBOs are candidate dwarf planets.

- The dwarf planets grew to be the largest objects in their respective regions of the solar nebula but never became large enough to capture or eject other objects orbiting nearby.

- Pluto is caught in a 3:2 stabilizing resonance with Neptune along with many other KBOs. This subset of Kuiper Belt Objects is called **plutinos (p. 349)**. Plutinos and any other Kuiper Belt Objects orbiting in other resonances with Neptune are evidence that Neptune migrated outward in the Solar System, sweeping up these remnant planetesimals.

Review Questions

1. Why didn't ancient astronomers know of Uranus's existence?
2. Describe the location of the equinoxes and solstices in the Uranian sky. What are the seasons like on Uranus?
3. What region(s) on Uranus, if any, experience(s) sunshine at least some portion of every day?
4. Methane emits blue light, hence Uranus and Neptune are bluish. True or false?
5. Why is belt–zone circulation difficult to detect on Uranus?
6. Belt–zone circulation seems to be more dependent on wind speeds in the atmosphere of a planet than the direction of sunlight. True or false?
7. With a density of 1.3 g/cm^3, what should the composition of Uranus be, in terms of proportions of icy materials versus rocky materials? (*Hint:* Refer to the discussions about density and the Galilean moons in Section 14-3.)
8. Describe four characteristics in common among all four Jovian planets. (*Hint:* Review **Celestial Profiles 7** through **10.**)
9. Describe four differences between the two ice giants, Uranus and Neptune, and the two gas giants, Jupiter and Saturn. (*Hint:* Review **Celestial Profiles 7** through **10.**)

10. Describe evidence of past geological activity on the major moons of Uranus. Which of the five major moons of Uranus has no indication of past geological activity?

11. What are hypotheses for the origin of the rings of Uranus and Neptune? Cite evidence to support these hypotheses.

12. Why do the characteristics of Uranus's and Neptune's magnetic fields suggest that the mantles of those planets are fluid?

13. If Uranus and Neptune had no satellites at all, would you expect them to have rings? Why or why not?

14. Why might the surface brightness of ring particles and small moons orbiting Uranus and Neptune depend on whether those planets have extensive, strong magnetic fields?

15. Both Uranus and Neptune have a blue-green tint when observed through a telescope. What does this color tell you about their atmospheric composition?

16. How are the atmospheres of Earth and Triton similar?

17. Describe evidence of geological activity on Triton.

18. Neptune's discovery was predicted. True or false?

19. How can small worlds like Triton and Pluto have atmospheres whereas a larger world such as Ganymede has none?

20. Why do you suspect that Triton had a geologically active past? What sources of energy could have powered such activity?

21. If you visited the surface of Pluto and found Charon as a full moon at your zenith, where would you be located on the surface of Pluto?

22. What evidence can you cite that Pluto and Charon are mixtures of rock and ice?

23. Why was Pluto reclassified to a dwarf planet?

24. **How Do We Know?** How was the discovery of Neptune not accidental?

Discussion Questions

1. William Herschel's discovery of Uranus was unexpected, but is it fairly described as an accident? What about the discovery of Uranus's rings?

2. Suggest a single phenomenon that could explain the inclination of the rotation axis of Uranus, the peculiar orbits of Neptune's major satellites, and the existence of Pluto's moons.

3. Pluto is now known to have five moons. Do these discoveries of moons suggest that Pluto should also have a ring system? Why or why not?

4. Pluto is a trans-Neptunian object, a plutino, a Kuiper Belt Object, and a dwarf planet. How can Pluto be classified as all of these? In your opinion, which should be the primary classification?

5. If we could only choose one Jovian moon to visit, which should we select? Why?

Problems

1. What is the maximum angular diameter of Uranus as seen from Earth? Of Neptune? (*Hint:* Use the small-angle formula, Chapter 3.) (*Note:* Necessary data are given in **Celestial Profiles 2, 9,** and **10.**)

2. One way to recognize a distant planet is by studying the planet's motion along its orbit. If Uranus circles the Sun in 84.0 years, how many arc seconds will it move in 1 Earth day? Assume a circular orbit for Uranus, and pretend that Earth is not moving.

3. What is the orbital velocity of Miranda around Uranus? (*Hint:* Use the formula for circular orbit velocity, Chapter 5. The formula requires input quantities in kg and m.) (*Note:* Necessary data are given in **Celestial Profile 9** and Appendix Table A-11.)

4. Calculate Uranus's Roche radius. Are all of Uranus's rings inside the Roche limit? Are any of the moons within the Roche limit? (*Notes:* The Roche limit is defined in Chapter 14. Necessary data are given in **Celestial Profile 9** and Appendix Table A-11. The structure of the Uranian system is displayed in **Uranus's and Neptune's Rings.**)

5. Use the data in Appendix Table A-11 to find which major moon of Uranus could have an orbital period resonance with Miranda or might have had one in the past.

6. What is the escape velocity from the surface of an icy moon that has a diameter of 20 km? (*Hint:* Use the formula for escape velocity, Chapter 5. The formula requires input quantities in kg and m.) (*Notes:* The density of ice is 1000 kg/m³. The volume of a sphere is $\frac{4}{3}\pi r^3$.)

7. What is the difference in the orbital velocities of the two shepherd satellites Cordelia and Ophelia, which have orbital radii of 49,800 km and 53,800 km, respectively (*Hint:* Use the formula for orbital velocity, Chapter 5. The formula requires input quantities in kg and m.)

8. Repeat Problem 2 for Pluto. In other words, ignoring the motion of Earth, how far across the sky would Pluto move in 1 Earth day? Assume Pluto is in a circular orbit around the Sun at its average distance of 39.3 AU.

9. Given the size of Triton's orbit ($r = 355,000$ km) and its orbital period ($P = 5.88$ days), calculate the mass of Neptune. (*Hint:* Use the formula for circular orbital velocity, Chapter 5. The formula requires input quantities in kg and m.)

Learning to Look

1. Look at Figure 15-8a. How are the orbits of Uranus's major moons arranged relative to the plane of the planet's equator and rings? Does that tell you anything about how the moons formed?

2. Compare Figure 15-8b with Figure 15-8a. Does the proportion of ice versus rock in the moons vary systematically with distance from Uranus? Does that tell you anything about how the moons formed?

3. Compare the interior cutaway sketches of the four Jovian planets in **Celestial Profiles 7** through **10.** What interior layer(s) is (are) shown in Jupiter and Saturn but not in Uranus and Neptune, and vice versa?

4. Look at Figure 15-19. Based your knowledge of comparative planetology, what do you think Pluto's albedo features might represent?

5. Review Figure 13-11. Which molecules can Triton retain in its atmosphere?

6. The image to the left shows how Uranus would look to the unaided human eye, whereas the right image shows how Uranus would look through a red filter, which enhances the methane clouds. What do the visible atmospheric features tell you about circulation on Uranus?

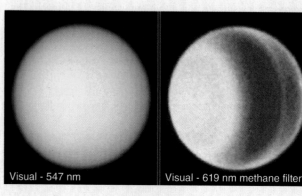

Visual - 547 nm   Visual - 619 nm methane filter

Meteorites, Asteroids, and Comets

16

Guidepost In Chapter 10, you began your study of planetary astronomy by considering evidence about how our Solar System formed. In the five chapters that followed, you surveyed the planets and found more clues about the origin of the Solar System but also learned that most traces of the early histories of the planets have been erased by geological activity or other processes. Now you can study smaller, less-altered objects that tell more about the era of planet building.

Asteroids and comets are unevolved objects, leftover planet construction "bricks." You will find them much as they were when they formed almost 4.6 billion years ago. Meteors and meteorites are fragments of comets and asteroids that arrive at Earth and can give you a close look at those ancient planetesimals. As you explore, you will find answers to four important questions:

▶ **Where do meteors and meteorites come from?**

▶ **What are asteroids?**

▶ **What are comets?**

▶ **What happens when asteroids and comets hit Earth and other planets?**

As you finish this chapter, you will have acquired real insight into your place in nature. You live on the surface of a planet. Are any other planets inhabited? That is the subject of the next, and final, chapter.

When they shall cry "PEACE, PEACE"
then cometh sudden destruction!
COMET'S CHAOS?—
What Terrible events will the Comet bring?

FROM A PAMPHLET PREDICTING THE END OF THE WORLD
BECAUSE OF THE APPEARANCE OF COMET KOHOUTEK IN 1973

Courtesy of Aleksandr Ivanov

A frame extracted from a car dashboard camera video of a large meteor (small asteroid) crossing the sky over the city of Chelyabinsk, Russia, in February 2013. The object was later estimated to be about 17 m (55 ft) in diameter, moving at 19 km/s (43,000 mph); its path indicates that it came from the asteroid belt. The object exploded at an altitude of about 30 km (20 mi) with the force of 25 Hiroshima-sized atomic bombs. The resulting shock wave broke many of the city's windows and sent more than 1500 people to seek medical treatment. Miraculously, nobody was killed.

One afternoon in 1954, while Mrs. E. Hulitt Hodges of Sylacauga, Alabama, lay napping on her living room couch, an explosion and a sharp pain jolted her awake. Analysis of the brick-size rock that smashed through the ceiling and bruised her left leg showed that it was a meteorite. In 2013, more than 1000 residents of the Russian city of Chelyabinsk had to be treated for glass cuts after a large meteorite exploded high overhead, shattering most of the windows in the city (page 353).

Meteorites arrive from space all over Earth every day, although normally not as spectacularly as in these two events. You will learn in this chapter that meteorites are fragments of asteroids and that asteroids, as well as their icy cousins the comets, carry precious clues about conditions in the solar nebula from which the Sun and planets formed. Because you cannot easily visit comets and asteroids, you can begin by learning about the pieces of those bodies that come to you.

16-1 Meteorites, Meteors, and Meteoroids

You first learned about meteorites in Chapter 10 when you studied evidence for the age of the Solar System. There you saw that the Solar System includes small particles called *meteoroids*. Some of them collide with Earth's atmosphere at speeds of 10 to 70 km/s. Friction with the air heats the meteoroids enough so that they glow, and you see them vaporize as streaks across the night sky. Those streaks are called *meteors* ("shooting stars"). If a meteoroid is big enough and holds together well enough, it can survive its plunge through the atmosphere and reach Earth's surface. Once the object strikes Earth's surface, it is called a *meteorite* ("–ite" being the Greek root for *rock*). As you will learn later in this chapter, the largest of those objects can blast out craters on Earth's surface, but such big impacts are extremely rare. The great majority of meteorites are too small to form craters.

What can meteorites and meteors tell you about the origin of the Solar System? To answer that question, you can consider their compositions and orbits.

Composition of Meteorites and Meteors

One of the best places to look for meteorites turns out to be certain parts of Antarctica—not because more meteorites fall there but because they are easy to recognize. No Earth rocks are on top of the Antarctic ice cap; the nearest native rocks are buried under the ice. Any rock you find there must have fallen from space. (For similar reasons, another good place to find meteorites is the Sahara desert, where deep layers of sand keep Earth rocks completely out of sight.) The slow flow of the Antarctic ice cap from the center of the continent toward the ocean concentrates meteorites in areas where the moving ice runs into mountain barriers, slows down, and evaporates. Teams of scientists travel to Antarctica and ride snowmobiles in systematic sweeps across the ice each Southern Hemisphere summer to recover meteorites (**Figure 16-1**). A four- to eight-person team can find a

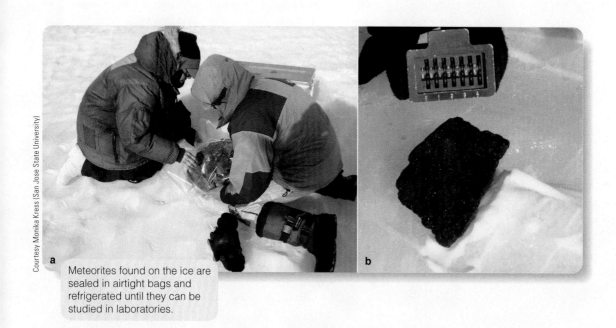

Meteorites found on the ice are sealed in airtight bags and refrigerated until they can be studied in laboratories.

Courtesy Monika Kress (San Jose State University)

a   b

▲ **Figure 16-1** (a) Braving bitter cold and high winds, teams of scientists riding snowmobiles search for meteorites that fell long ago in Antarctica and are exposed as the ice evaporates. When a meteorite is found, it is photographed where it lies, assigned a number, and placed in an airtight bag. (b) Thousands of meteorites have been collected in this way, including a few fragments from the Moon and Mars.

Iron meteorites are very dense and have a dark, irregular surface.

A stony-iron meteorite cut and polished reveals a mixture of iron and rock.

Stony meteorites tend to have a fusion crust caused by melting in Earth's atmosphere.

Chondrules are small, glassy spheres found in chondrites.

Lab photos courtesy of Russell Kempton, New England Meteoritical Services

◄ **Figure 16-2** The three main types of meteorites—(a) irons, (b) stony-irons, and (c) stones—are easily distinguished. (d) Carbonaceous chondrites are rare stony meteorites that are rich in carbon, making them very dark.

thousand meteorites during a single two-month field season. After 25 years of work, more than half of the 40,000 meteorites in human hands are from Antarctica.

Meteorites that are seen to fall are called **falls**; a fall is known to have occurred at a given time and place, and thus the meteorite is well documented. A meteorite that is discovered on or in the ground, but was not seen to fall, is called a **find**. Such a meteorite could have fallen thousands of years ago. The distinction between *falls* and *finds* will be important as you further analyze the different kinds of meteorites.

Meteorites can be divided into three broad composition categories. **Iron meteorites** (Figure 16-2a) are solid chunks of iron and nickel. **Stony-iron meteorites** (Figure 16-2b) are mixtures of iron and stone. **Stony meteorites** (Figure 16-2c) are silicate masses that resemble Earth rocks. **Carbonaceous chondrites** (pronounced *KON-drites*; Figure 16-2d) are a special type of stony meteorite.

Iron meteorites are easy to recognize because they are heavy, dense lumps of metal—a magnet will stick to them. That explains an important statistic. Iron meteorites make up 50 percent of finds (Table 16-1) but only 6 percent of falls. Why? Because an iron meteorite doesn't look like an ordinary rock.

If you trip over one on a hike, you are more likely to recognize it as something odd, carry it home, and show it to the local museum. Also, some stony meteorites deteriorate rapidly when exposed to weather; irons are made of stronger material and generally survive longer. The durability and recognizability of iron meteorites means there is a **selection effect** that makes it more likely they will be found than other types of meteorites (**How Do We Know? 16-1**). The fact that only 6 percent of falls are irons shows that iron meteoroids, although easier to find on Earth, are relatively rare in space.

TABLE 16-1 Proportions of Meteorites

Type	Falls (%)	Finds (%)
Iron	6	50
Stony-iron	1	5
Stony	93	45

© 2016 Cengage Learning®

How Do We Know? 16-1

Selection Effects

How is a red insect like a red car? Scientists must plan ahead and design their research projects with great care. Biologists studying insects in the rain forest, for example, must choose which ones to catch. They can't catch every insect they see, so they might decide to catch and study any insect that is red. If they are not careful, a selection effect could bias their data and lead them to incorrect conclusions without their ever knowing it.

For example, suppose you needed to measure the speed of cars on a highway. There are too many cars to measure every one, so you might reduce the workload and measure only red cars. It is quite possible that this selection criterion will mislead you because people who buy red cars may be more likely to be younger and drive faster. Should you instead measure only brown cars? No, because older, more sedate people might tend to buy brown cars. Only by very carefully designing your experiment can you be certain that the cars you measure are traveling at typical speeds.

Astronomers understand that what you see through a telescope depends on what you notice, and that is powerfully influenced by what are called "selection effects." The biologists in the rain forest, for example, should not catch and study only red insects. Often, the most brightly colored insects are poisonous or at least taste bad to predators. Catching only red insects could produce a result highly biased by a selection effect.

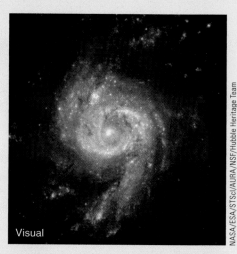
Visual

Things that are bright and beautiful, such as spiral galaxies, may attract a disproportionate amount of attention. Scientists must be aware of such selection effects.

NASA/ESA/STScI/AURA/NSF/Hubble Heritage Team

When iron meteorites are sliced open, polished, and etched with acid, they reveal regular bands called **Widmanstätten patterns** (pronounced *VEED-mahn-state-en*; Figure 16-3). Those patterns are caused by certain alloys of iron and nickel that formed crystals as the molten metal cooled and solidified long ago. The size and shape of the bands indicate that the molten metal cooled very slowly, no faster than 20 degrees Kelvin per million years.

The Widmanstätten pattern tells you that the metal in iron meteorites was once molten and must have been well insulated to cool so slowly. Such slow cooling indicates a location inside bodies at least 30 km (20 mi) in diameter. (In comparison, a small lump of molten metal exposed in space would cool in just a few hours.) On the other hand, the iron meteorites do not show effects of the very high pressures that would exist deep inside a planet. Evidently, iron meteorite material formed in the cooling interiors of planetesimal-size objects, smaller than planets. You will find this is one important clue to the origin of meteorites.

A small fraction of falls are meteorites made of mixed iron and stone (Figure 16-2b). These stony-iron meteorites appear to have solidified from both molten iron and rock—the kind of environment you might expect to find deep inside a planetesimal at the boundary between a liquid metal core and a rocky mantle.

In contrast to irons and stony-irons, stony meteorites (Figure 16-2c) are the most common type among falls (Table 16-1), meaning they are common in space near Earth. Although there are many different types of stony meteorites, you can classify them into two main categories, **chondrites** and **achondrites**, depending on their physical properties and chemical content.

Photo courtesy of Russell Kempton, New England Meteoritical Services

▲ **Figure 16-3** Sliced, polished, and etched with acid, iron meteorites show what is called a Widmanstätten pattern of large crystals, indicating that this material cooled very slowly from a molten state and must have been in the interior of a fairly large object.

Chondrites look like dark gray, granular rocks (see Figure 16-2d). The classification of meteorites has become quite complicated, and there are many types of chondrites. But in general they contain some volatiles, including water and organic (carbon) compounds. A few chondrites appear to have formed actually in the presence of liquid water.

Most types of chondrites also contain **chondrules**, small round bits of glassy rock only a few millimeters across. To be glassy rather than crystalline, the chondrules must have cooled from a molten state quickly, within a few hours. One hypothesis is that chondrules are bits of matter from the inner part of the solar nebula, near the Sun, that were blown outward by solar wind gusts or protostellar jets to cooler parts of the nebula where they condensed and were later incorporated into larger rocks. Another hypothesis is that the chondrules were once solid bits of matter that were melted by shock waves spreading through the solar nebula and then resolidified. The presence of chondrule particles inside chondrite meteorites indicates that those rocks have not melted since they formed because melting would have destroyed the chondrules.

Among the chondrites, the carbonaceous chondrites are rare but quite important. These dark gray, rocky meteorites are especially rich in water, other volatiles, and organic compounds. Those substances all would have been lost if the meteoroid had been heated even to room temperature.

One of the most important meteorites ever studied was a carbonaceous chondrite seen falling in 1969 near the little Mexican village of Allende (pronounced *ah-YEN-day*). About 2 tons of fragments were recovered. Studies of the Allende meteorite disclosed that it contained chondrules, water, complex organic compounds including amino acids, and a number of small, irregular inclusions rich in calcium, aluminum, and titanium (Figure 16-4). Now called **CAIs**, for calcium–aluminum-rich inclusions, these bits of matter are highly refractory; that is, they vaporize or condense only at high temperatures.

If you could scoop out a portion of the Sun's photosphere and cool it, the first particles to solidify would have the chemical composition of CAIs. As the temperature fell, other materials would become solid in accordance with the condensation sequence described in Chapter 10. When the material finally reached room temperature, you would find that almost all of the hydrogen, helium, and some other gases such as argon and neon had escaped and that the remaining lump would have almost exactly the same overall chemical composition as the Allende meteorite. This is evidence that the Allende meteorite is a very old sample of the solar nebula, confirmed by the fact that the CAIs have radioactive ages equal to the oldest of any other Solar System material.

Another large load of carbonaceous chondrite material arrived on Earth in the year 2000 at Tagish Lake in the

NASA

▲ **Figure 16-4** A sliced portion of the Allende carbonaceous chondrite meteorite, showing irregularly shaped white inclusions called CAIs that were probably the first solid material to condense as the Solar System formed.

Canadian Arctic. Analysis of that meteorite produced a surprise: It has noticeably less complex organics than Allende. Scientists are not sure whether this means that the Tagish organics formed so early in the Solar System's history that chemical reactions had not yet advanced to the stage of making Allende's complex compounds or whether the Tagish material was once heated just enough to break down big molecules into smaller ones.

The condensing solar nebula should have incorporated volatiles and organics into solid particles as they formed. If that material had later been heated, it would have lost the volatiles, and many of the organic compounds would have been destroyed. The chondrites show properties ranging from carbonaceous chondrites, most of which have avoided being heated or modified, to other chondrites in which the material was slightly heated and somewhat altered from the form in which it first solidified. Chondrites in general offer us the best direct information about conditions and processes occurring in the earliest days of the solar nebula when planetesimals and planets were forming.

Stony meteorites called achondrites contain no chondrules and also lack volatiles. These rocks appear to have been subjected to intense heat that melted the chondrules and completely drove off the volatiles, leaving behind rock with a composition similar to Earth's basalts.

The different types of meteorites evidently had a wide variety of histories. Some achondrites seem like pieces of lava flows, whereas stony-iron and iron meteorites apparently were once deep inside the molten interiors of differentiated objects, and carbonaceous chondrites seem to be unaltered lumps of condensed solar nebula. The differences in details between the compositions of various chondrites are thought to result from some locations in the solar nebula receiving material transported and mixed in from other locations. Meteorites provide evidence that the young Solar System was a complicated place.

Orbits of Meteors and Meteorites

Meteoroids are much too small to be visible through even the largest telescope. They are visible only when they fall into Earth's atmosphere and are heated by friction with the air. A typical meteoroid has roughly the mass of a paper clip and vaporizes at an altitude of about 80 km (50 mi) above Earth's surface. The meteor trail points back along the path of the meteoroid, so if you study the direction and speed of meteors, you can get clues to their orbits in the Solar System before they encountered Earth.

One way to backtrack meteor trails is to observe **meteor showers**. On any clear night, you can see 3 to 15 meteors per hour, but on some nights you can see a shower of hundreds of meteors per hour that are obviously related to each other. To confirm this, try observing a meteor shower. Pick a shower from Appendix Table A-12, and on an appropriate night stretch out in a lawn chair and watch a large area of the sky. When you see a meteor, sketch its path on the appropriate sky chart from the back of this book. In just an hour or so, you will discover that all or almost all of the meteors you see seem to come from a single area of the sky, called the **radiant** of the shower (Figure 16-5a). Meteor showers are generally named after the constellation or star from which they seem to radiate; for example, the Perseid shower seen in mid-August radiates from the constellation Perseus.

Observing a meteor shower is a natural fireworks show, but it is even more exciting when you understand what a meteor shower tells you. The fact that the meteors in a shower appear to come from a single point in the sky means that the meteoroids were traveling through space along parallel paths. When they encounter Earth and are vaporized in the upper atmosphere, you see their fiery tracks in perspective, so they appear to come from a single radiant point, just as railroad tracks seem to come from a single point on the horizon (Figure 16-5b).

Studies of meteor shower radiants reveal that those meteoroids are orbiting the Sun along the paths of comets. As you learned in Chapter 10, the vaporizing head of a comet releases bits of rock that eventually spread along its entire orbit (Figure 16-6). When Earth passes through this stream of material, you see a meteor shower. In some cases the comet has wasted away and is no longer visible, but in other cases the comet is still prominent, although located somewhere else along its orbit. For example, each May, Earth comes near the orbit of Comet Halley, and you can see the Eta Aquarid shower. Each October, Earth passes near the other side of the orbit of Comet Halley, and the Orionid shower appears.

Even when there is no shower, you can still occasionally see meteors that are called **sporadic meteors** because they are not part of specific showers. To determine their origin, scientists have photographed sporadic meteor trails from two or more locations on Earth a few miles apart. Then, they use triangulation to find the altitude, speed, and direction of the meteor as it moved through the atmosphere, and can work backward to calculate its orbit before it entered Earth's atmosphere. These studies confirm that some sporadic meteors, like shower meteors, have orbits that are similar to the orbits of comets. In contrast, a few sporadic meteors, including *all* observed meteorite falls, have orbits that lead back to the asteroid belt between Mars and Jupiter. From this you can conclude that meteors have a dual source: Many come from comets, but a few come from the asteroid belt. Meteors that are big and durable enough to become meteorites on the ground appear always to come from the asteroid belt.

It is a **Common Misconception** that a bright meteor disappearing behind a distant hill or line of trees probably landed just a mile or two away. This has triggered hilarious "wild goose chases" as police, fire companies, and TV crews try to find the impact site. Almost every meteor you see vaporizes high above Earth's surface. Only rarely does a meteor become a meteorite by reaching the ground, and it can land as far as 100 miles from where you are standing when you see it.

Origins of Meteoroids and Meteorites

Evidence you have already encountered suggests that many meteorites are fragments of parent bodies that were large enough to grow hot from radioactive decay or other processes. They then melted and differentiated to form iron-nickel cores plus rocky mantles and crusts. The molten iron cores would have been well insulated by the thick rocky mantles so that

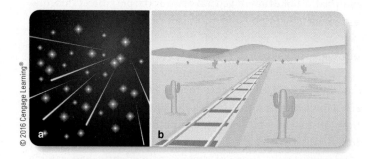

▲ **Figure 16-5** (a) Meteors in a meteor shower enter Earth's atmosphere along parallel paths, but perspective makes them appear to diverge from a single point in the sky. (b) Similarly, parallel railroad tracks appear to diverge from a point on the horizon.

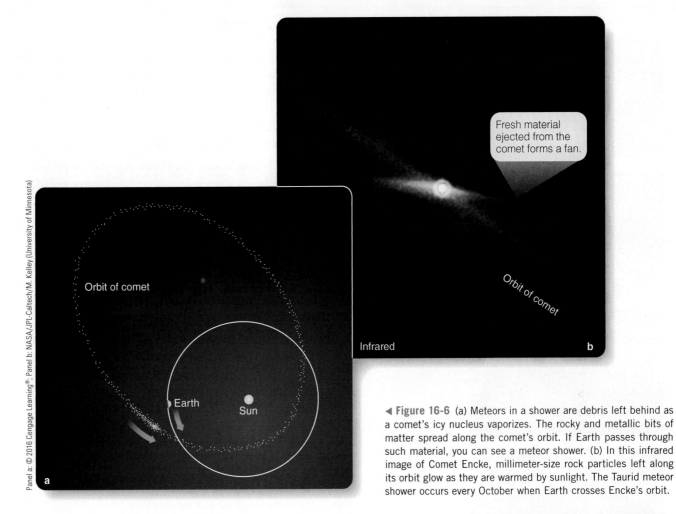

Panel a: © 2016 Cengage Learning®. Panel b: NASA/JPL-Caltech/M. Kelley (University of Minnesota)

Fresh material ejected from the comet forms a fan.

Orbit of comet

Orbit of comet

Orbit of comet

Earth

Sun

Infrared

◀ **Figure 16-6** (a) Meteors in a shower are debris left behind as a comet's icy nucleus vaporizes. The rocky and metallic bits of matter spread along the comet's orbit. If Earth passes through such material, you can see a meteor shower. (b) In this infrared image of Comet Encke, millimeter-size rock particles left along its orbit glow as they are warmed by sunlight. The Taurid meteor shower occurs every October when Earth crosses Encke's orbit.

the iron would have cooled slowly enough to produce big crystals that result in Widmanstätten patterns. Stony-iron meteorites apparently come from boundaries between stony mantles and iron cores. Stony meteorites that have been strongly heated evidently come from the mantles or surfaces of such bodies.

Collisions could break up such differentiated bodies and produce different kinds of meteorites (**Figure 16-7**). In contrast, many chondrites are probably fragments of smaller bodies that never melted, and carbonaceous chondrites may be from unaltered bodies that formed especially far from the Sun.

These hypotheses trace the origin of meteorites to plane-tesimal-like parent bodies, but they leave you with a puzzle. The small meteoroids now flying through the Solar System cannot have existed in their present form since the formation of the Solar System because they would have been swept up by the planets in a billion years or less. They could not have survived traveling in their current orbits for the full 4.6 billion year history of the Solar System. Nevertheless, when the orbits of meteorite falls are determined, those orbits lead back into

DOING SCIENCE

What is the evidence that meteors come from comets, but meteorites come from asteroids? The answer to this question is related to the distinction scientists have made between meteors and meteorites.

A meteor is the streak of light seen in the sky when a particle from space is heated by friction with Earth's atmosphere. A meteorite is a piece of space material that actually reaches the ground.

The difference between comet and asteroid sources must take into account two very strong effects that prevent you from finding meteorites that originated in comets. First, available evidence indicates that cometary material is physically weak, so comet particles vaporize in Earth's atmosphere easily. Very few ever reach the ground. Second, even if a comet particle reached the ground, it would be so fragile that it would weather away rapidly, and you would be unlikely to find it before it disappeared. Asteroidal particles, however, are made from rock or metal and are stronger. They are more likely to survive their plunge through the atmosphere, and afterward, more likely to survive erosion on the ground. Every known meteorite is from the asteroids; not a single meteorite is known to be cometary. This is true even though meteor tracks show that most meteors visible in the sky come from comets, and very few are from the asteroid belt.

Now consider a related question: ***What evidence suggests that meteorites were once part of larger bodies broken up by impacts?***

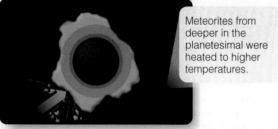

A large planetesimal can keep its internal heat long enough to differentiate.

Silicates

Cratering collisions

Iron

Collisions break up the layers of different composition. Meteorites from near the surface are rocky.

Meteorites from deeper in the planetesimal were heated to higher temperatures.

Fragments from near the core might have been melted entirely.

Fragments of the iron core can fall to Earth as iron meteorites.

Adapted from a diagram by Clark Chapman (PSI SAIC; SwRI); © 2016 Cengage Learning®

▲ **Figure 16-7** Some of the planetesimals that formed early in the Solar System's history differentiated, that is, melted and separated into layers of different density and composition, as did the Terrestrial planets. The fragmentation of such a planetesimal could produce different types of meteorites.

the asteroid belt. Thus, all the evidence together indicates that the meteorites now in museums around the world are fragments that were produced by asteroid collisions within the past billion years.

16-2 Asteroids

Asteroids are distant objects too small to study in detail with Earth-based telescopes. Astronomers nevertheless have learned a surprising amount about those little worlds using spacecraft and space telescopes.

Properties of Asteroids

Evidence from meteorites shows that the asteroids are the last remains of the population of rocky planetesimals from which the Terrestrial planets were built 4.6 billion years ago. Study **Observations of Asteroids** on pages 362–363 and notice four important points:

1 Most asteroids are irregular in shape and battered by impact cratering. Many asteroids seem to be rubble piles of broken fragments.

2 Some asteroids are double objects or have small moons in orbit around them. This is further evidence that asteroids have suffered collisions.

3 A few asteroids show signs of geological activity that probably happened on their surfaces when those asteroids were young.

4 Asteroids can be classified by their albedos, colors, and spectra to reveal clues to their compositions. This also allows them to be compared to meteorites in labs on Earth. *C-type, S-type,* and *M-type* asteroids are the main classes discovered by this method.

The Asteroid Belt

The first asteroid was discovered by the Sicilian monk Giuseppe Piazzi on January 1, 1801 (the first night of the 19th century). It was later named Ceres after the Roman goddess of the harvest (and source of our word *cereal*).

Astronomers were excited by Piazzi's discovery because there seemed to be a pattern to the location of planet orbits, except for a wide gap between Mars and Jupiter where the pattern implied a planet "ought" to exist at an average distance from the Sun of 2.8 AU. Ceres fit the pattern: Its average distance from the Sun is 2.77 AU. But Ceres is much smaller than the planets, and three even smaller objects—Pallas, Juno, and Vesta—were discovered within a few years, all orbiting between Mars and Jupiter, so astronomers decided that Ceres and the other asteroids should not be considered true planets. As you learned in Chapter 15, Ceres has now been reclassified as a dwarf planet because it has enough gravitational strength to squeeze itself into a spherical shape but not enough to have swept up or cleared away the rest of the asteroids.

Today, almost 400,000 asteroids have well-charted orbits. Other than the dwarf planet Ceres, only three are larger than 400 km (250 mi) in diameter (**Figure 16-8**), and most are much

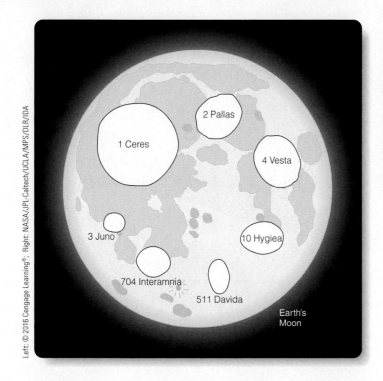

Vesta

Visual

◀ **Figure 16-8** (a) The relative size and approximate shape of Ceres, Vesta, and other large asteroids are shown here compared with the size of Earth's Moon. Smaller asteroids can be highly irregular in shape. (b) This image of the asteroid Vesta, made from a distance of 2700 km (1700 mi) by the *Dawn* spacecraft, shows Vesta's south polar region, dominated by the Rheasilvia impact basin.

smaller. There are probably a million or more asteroids larger in diameter than 1 km (0.6 mi). Astronomers are sure that all the large asteroids in the asteroid belt have been discovered but are also sure that many small asteroids remain undiscovered.

Movies and TV have created a **Common Misconception** that flying through an asteroid belt is a hair-raising plunge requiring constant dodging left and right. The asteroid belt between Mars and Jupiter is actually mostly empty space. In fact, if you were standing on an asteroid, it would be many months or years between sightings of other asteroids.

If you discover an asteroid, you are allowed to choose a name for it, and asteroids have been named for spouses, lovers, dogs, politicians, and others. Once an orbit has been calculated, the asteroid is assigned a number listing its order in the catalog known as the *Ephemerides of Minor Planets*. Thus, Ceres is officially known as 1 Ceres, Pallas as 2 Pallas, and so on. (Some sample asteroid names: Chicago, Vaticana, Noel, Tea, Hagar,

Tito, Zulu, Zappafrank, and Garcia; the latter two names honor the late musicians Frank Zappa and Jerry Garcia.)

The distribution of asteroids in the belt is strongly affected by Jupiter's gravitation. Certain orbits in the belt that are almost free of asteroids are called **Kirkwood gaps** after their discoverer, Daniel Kirkwood (**Figure 16-9**). These empty orbits have semi-major axes such that an object in them would have a resonance with Jupiter. For example, an asteroid with an average distance from the Sun of 3.28 AU will go exactly twice around the Sun in the time it takes Jupiter to go once. Such an asteroid would pass Jupiter at the same place in space every second orbit and be tugged outward. The cumulative perturbations would rapidly change the asteroid's orbit until it was no longer in resonance with Jupiter. Thus, Jupiter effectively eliminates objects from orbit resonances. The example given represents a 2:1 resonance, but gaps occur in the asteroid belt at many other resonances, including 3:1, 5:2, and 7:3. You will recognize that Kirkwood

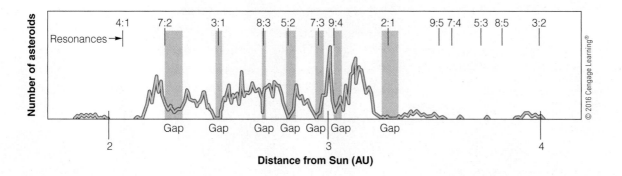

▲ **Figure 16-9** The red curve in this plot shows the number of asteroids versus orbit semi-major axis. Purple bars mark Kirkwood gaps, where there are few asteroids. Note that these gaps match resonances with the orbital motion of Jupiter.

Observations of Asteroids

1 Seen from Earth, asteroids look like faint points of light moving across the background of distant stars. Spacecraft have now visited asteroids, and the images radioed back to Earth show that the asteroids are mostly small, dark, irregular worlds heavily cratered by impacts.

The *Near Earth Asteroid Rendezvous (NEAR)* spacecraft visited asteroid Eros in 2000 and found it to be heavily cratered by collisions and covered by a layer of crushed rock ranging from dust to large boulders. The *NEAR* spacecraft eventually landed on Eros.

Data from *NEAR* indicate Eros is solid rock.

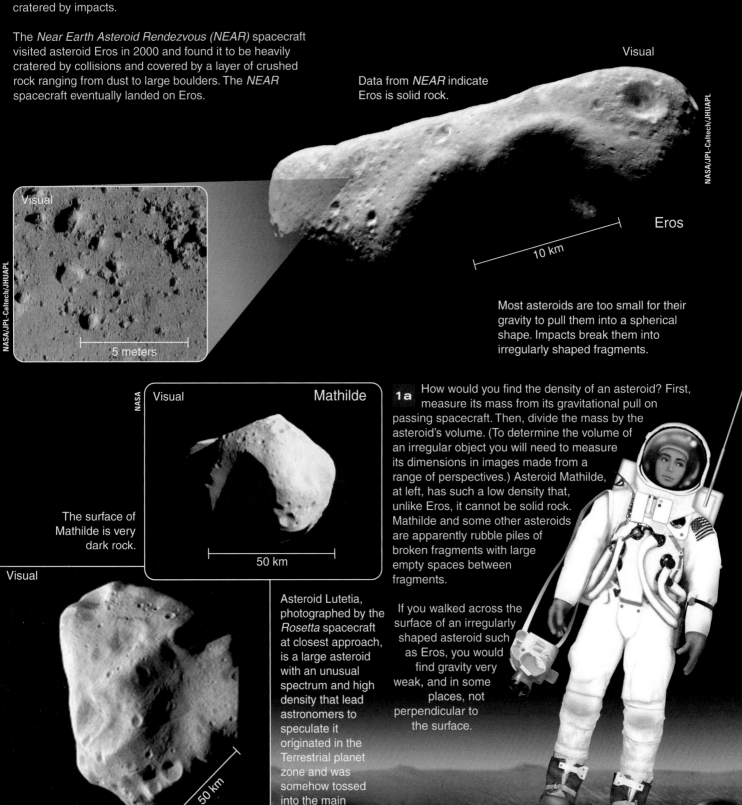

Visual

Eros

10 km

NASA/JPL-Caltech/JHUAPL

Visual

5 meters

NASA/JPL-Caltech/JHUAPL

Most asteroids are too small for their gravity to pull them into a spherical shape. Impacts break them into irregularly shaped fragments.

NASA

Visual — Mathilde

The surface of Mathilde is very dark rock.

50 km

Visual

Lutetia

50 km

ESA/MPS/UPD/LAM/IAA/RSSD/INTA/UPM/DASP/IDA

1a How would you find the density of an asteroid? First, measure its mass from its gravitational pull on passing spacecraft. Then, divide the mass by the asteroid's volume. (To determine the volume of an irregular object you will need to measure its dimensions in images made from a range of perspectives.) Asteroid Mathilde, at left, has such a low density that, unlike Eros, it cannot be solid rock. Mathilde and some other asteroids are apparently rubble piles of broken fragments with large empty spaces between fragments.

Asteroid Lutetia, photographed by the *Rosetta* spacecraft at closest approach, is a large asteroid with an unusual spectrum and high density that lead astronomers to speculate it originated in the Terrestrial planet zone and was somehow tossed into the main asteroid belt.

If you walked across the surface of an irregularly shaped asteroid such as Eros, you would find gravity very weak, and in some places, not perpendicular to the surface.

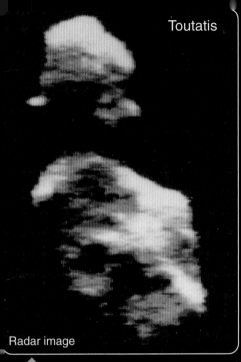

Toutatis

Radar image

NASA/JPL-Caltech

2 Asteroids that pass near Earth can be imaged by radar. Asteroid Toutatis is revealed to be a double object—two objects orbiting close to each other or actually in contact.

Ida

NASA

Dactyl

30 km

Visual + Infrared, enhanced contrast

Double asteroids are more common than was once thought, reflecting a history of collisions and fragmentation. Asteroid Ida is orbited by a moon, Dactyl, that is only about 1.5 km (1 mi) in diameter.

Occasional collisions among the asteroids rele fragments, and Jupiter's gr scatters them into the inner Sol System as a continuous supply of meteors.

3 The maps of asteroid Vesta's southern hemisphere in the panel at right are based on data acquired by the orbiting *Dawn* spacecraft. At upper left is a colored topographic (altitude) map; at upper right, a map of Vesta's gravity variations; at bottom center, a mineralogical map. The central peak of the large Rheasilvia basin, appearing as the yellow area just above and to the left of center in the gravity map, has a small positive gravity anomaly, indicating that the material there is denser, perhaps originating from deep within the asteroid. Geologists interpret the patterns in the mineralogical image to suggest that Vesta probably melted all the way through early in its history. (Compare with the visual-wavelength image of the same portion of Vesta in Figure 16-8b.)

NASA/JPL-Caltech/UCLA/INAF/MPS/DLR/IDA

Topography   Vesta   Gravity variations

Mineralogy (visual and near-infrared data)

Diogenite   Eucrite

-22.28   19.10
Elevation [km]

© 2016 Cengage Learning®

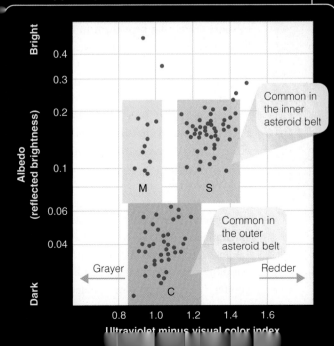

3a A class of meteorites that are spectroscopically identical to Vesta are thought to be fragments of that asteroid, perhaps blasted into space by the collision that created the big southern basin. Those meteorites appear to be solidified basalt lava, providing evidence that this asteroid once had geological activity.

Meteorite from Vesta

5 cm
2 in.

Lab photo courtesy of Russell Kempton,
New England Meteoritical Services

4 Although asteroids would look gray to your eyes, they can be classified according to their albedos (reflected brightness) and spectroscopic colors. For example, as shown in the plot at left, **S-types** have high albedos and tend to be reddish. They are the most common kind of asteroid and appear to be the source of the most common chondrite meteorites.

M-type asteroids are not too dark but are also not very red. They may be mostly iron-nickel alloys.

C-type asteroids are as dark as lumps of sooty coal and are

Common in the inner asteroid belt

Common in the outer asteroid belt

Grayer ← → Redder

Bright
Dark

Albedo (reflected brightness)

0.4
0.3
0.2
0.1
0.06
0.04

M   S
C

0.8   1.0   1.2   1.4   1.6

Ultraviolet minus visual color index

gaps in the asteroid belt are produced in the same way as some of the gaps in Saturn's rings (Chapter 14, page 321) that were also discovered by Kirkwood.

Computer models show that the motion of asteroids in Kirkwood gaps is described by a theory in mathematics that deals with chaotic behavior. As an example, consider how the smooth motion of water sliding over the edge of a waterfall decays rapidly into a chaotic jumble. The same theory of chaos that describes the motion of the water shows how the slowly changing orbit of an asteroid within one of the Kirkwood gaps can in a short time (astronomically speaking) become a long, eccentric orbit that carries the asteroid into the inner or outer Solar System.

Asteroids Outside the Main Belt

You don't have to go all the way to the asteroid belt if you want to visit an asteroid; some follow orbits that cross the orbits of the Terrestrial planets and come near Earth. Others wander far away, among the Jovian worlds. Some asteroids share orbits with the planets (Figure 16-10).

Main-belt asteroids lie between the orbits of Mars and Jupiter.

Asteroids that could approach Earth are shown in red.

Mars

Earth

Jupiter

Trojan asteroids orbit in two clouds 60° ahead of Jupiter and 60° behind.

▲ **Figure 16-10** This diagram plots the position of known asteroids between the Sun and the orbit of Jupiter on a specific day. Most asteroids are in the main belt. Squares, filled or empty, show the location of known comets. Although asteroids and comets are small bodies and lie far apart, there are a great many of them in the inner Solar System.

Apollo-Amor objects are asteroids with orbits that carry them into the inner Solar System. Amor objects follow orbits that cross the orbit of Mars but don't reach the orbit of Earth, whereas Apollo objects have Earth-crossing orbits. About 3000 Apollo and Amor objects have been found so far. The influences of Jupiter and other planets act to continuously change their orbits. Astronomers calculate that about one-third of Apollo-Amors will be thrown into the Sun, a few will be ejected from the Solar System, and, as you will discover later in this chapter, some are doomed to collide with one of the planets—perhaps Earth.

Several research teams are now intent on identifying **Near-Earth Objects (NEOs)**, including Apollo-Amor objects. For example, the Lowell Observatory Near-Earth Object Search (LONEOS) is searching the entire sky visible from northern Arizona once a month. The Lincoln Near-Earth Asteroid Research (LINEAR) project telescopes in New Mexico (Figure 16-11) and the Near-Earth Asteroids Tracking (NEAT) facilities in California and Hawai'i also have been successful in finding NEOs, as well as new main-belt asteroids and Kuiper Belt Objects. The combined searches are estimated to have found at least 90 percent of the Apollos and other NEOs larger than 1 km in diameter, and are now focused on finding a similar percentage of objects down to a size of about 150 meters (500 feet).

It is easy to hypothesize that the Apollo-Amor objects are rocky asteroids that have been sent into their unusual orbits by collisions in the main asteroid belt or by planetary perturbations, for example Jupiter's Kirkwood gap-clearing effect that you learned about in the previous section. There is evidence that a few of these objects instead may be comets that became trapped in short orbits that kept them in the inner Solar System so they have exhausted their volatiles. You can see from this that the distinction between comets and asteroids is not sharply defined.

Jupiter ushers two groups of nonbelt asteroids around its own orbit. These objects have become trapped in the L_4 and L_5 Lagrange points that lie in Jupiter's orbit 60 degrees ahead of and 60 degrees behind the planet. Lagrange points are regions like cosmic sinkholes where gravitational effects of the two larger bodies, in this case the Sun and Jupiter, combine to trap small bodies. The Jupiter Lagrange-point objects are

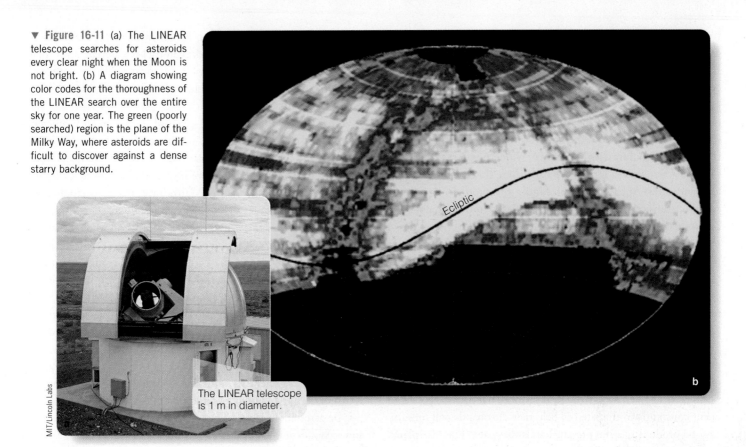

▼ **Figure 16-11** (a) The LINEAR telescope searches for asteroids every clear night when the Moon is not bright. (b) A diagram showing color codes for the thoroughness of the LINEAR search over the entire sky for one year. The green (poorly searched) region is the plane of the Milky Way, where asteroids are difficult to discover against a dense starry background.

Ecliptic

The LINEAR telescope is 1 m in diameter.

MIT/Lincoln Labs

called **Trojan asteroids** because individual asteroids have been named after the heroes of the Trojan War (e.g., 588 Achilles, 624 Hektor, 659 Nestor, and 1143 Odysseus). Almost 2000 Trojan asteroids are known, but only the brightest have been given names. Some astronomers speculate that there may be as many Trojan asteroids as asteroids in the main belt. A few objects have been found in the Lagrange points of the orbits of Mars, Uranus, and Neptune, and in 2011 astronomers announced detection by the *WISE* infrared space telescope of Earth's first known Lagrange-point asteroid, an object estimated to be about 300 m (1000 ft) in diameter.

There are other nonbelt asteroids beyond the main belt. The object Chiron, found in 1977, is about 180 km (110 mi) in diameter. Its orbit carries it from the orbit of Uranus to just inside the orbit of Saturn. Objects such as Chiron with orbits between, or crossing, orbits of the Jovian planets are called **centaurs**. Although it was first classified as an asteroid, Chiron surprised astronomers ten years after its discovery by suddenly brightening as it released jets of vapor and dust. Old photographs were found showing that Chiron had done this before. Astronomers now suspect Chiron has a rocky crust covering deposits of ices such as solid nitrogen, methane, and carbon monoxide. There is some similarity between Chiron and dwarf planet Ceres, both in overall composition and in the emission of water vapor. You will learn in the next section that these characteristics are more like comets than asteroids. The properties of centaurs are another reminder that the distinction between asteroids and comets is not clear-cut.

As technology allows astronomers to detect smaller and more distant objects, they are learning that our Solar System contains large numbers of these small bodies. The challenge is to explain their origin.

Origin and History of the Asteroids

An old hypothesis proposed that asteroids are the remains of a planet that exploded. Planet-shattering death rays may make for exciting science fiction movies, but in reality planets do not explode. The gravitational field of a planet holds the mass together so tightly that completely disrupting the planet would take tremendous energy. In addition, the current total mass of the asteroids is only about one-twentieth the mass of Earth's Moon, hardly enough to be the remains of a planet.

Astronomers have evidence that the asteroids are the remains of material lying 2 to 4 AU from the Sun that was prevented from assembling into a planet because of the gravitational influence of Jupiter, the next planet outward. Over the 4.6-billion-year history of the Solar System, most of the objects originally in the asteroid belt have collided, fragmented, and been covered with craters. Some asteroids were perturbed by the gravity of Jupiter and other planets into orbits that intersected planets or the Sun. Some were captured as planetary satellites or ejected from the Solar System. The present-day asteroids are understood to be a minor remnant of the original mass in that zone.

Collisions among asteroids must have been occurring since the formation of the Solar System (pages 362–363 as well as

Figures 16-7 and 16-8b). Astronomers have found evidence of catastrophic impacts powerful enough to shatter an asteroid. Early in the 20th century, astronomer Kiyotsugu Hirayama discovered that some groups of asteroids share similar orbits. Each group is distinct from other groups, but asteroids within a given group follow orbits with the same average distance from the Sun, the same eccentricity, and the same inclination. As many as 20 of these **Hirayama families** are known. Modern observations show that the asteroids in a family typically share similar spectroscopic characteristics. Apparently, each family was produced by a catastrophic collision that broke a single asteroid into fragments that continue traveling along similar orbits around the Sun. Studies of the fragment orbits in one family provide evidence that it was produced only 5.8 million years ago in a collision between asteroids estimated to have been 3 and 16 km (2 and 10 mi) in diameter, traveling at relative speeds of about 5 km/s (11,000 mph), typical for asteroid collisions. It seems that the fragmentation of asteroids is a continuing process.

In 1983, the *Infrared Astronomy Satellite* detected the infrared glow of Sun-warmed dust scattered in bands throughout the asteroid belt. These dust bands appear to be the products of past collisions. The dust will eventually be destroyed, but because collisions occur constantly in the asteroid belt, new dust bands will presumably be produced as the present bands dissipate. The interplanetary dust in our Solar System is analogous to dust in extrasolar planetary debris disks, produced astronomically recently by collisions of remnant planetesimals (Chapter 10, pages 214–215).

Even though most of the planetesimals originally in the main belt have been lost or destroyed, the objects left behind carry clues to their origin in their albedos and spectroscopic colors (p. 363). C-type asteroids have albedos less than 0.06 and would look very dark to your eyes. They are probably made of carbon-rich material similar to that in carbonaceous chondrite meteorites. C-type asteroids are more common in the outer asteroid belt. It is cooler there, and the condensation sequence (Chapter 10, pages 206–207) predicts that carbonaceous material would form more easily in the outer belt than in the inner belt.

S-type asteroids have albedos of 0.1 to 0.2, so they would look brighter and also redder than C-types; S-types may be composed of rocky material. M-type asteroids are also bright but not as red as the S-types; they seem to be metal rich and may be fragments from iron cores of differentiated asteroids. A few other types of asteroids are known, and a number of individual asteroids have been found that are unique, but these three classes include most of the known asteroids.

Although S-type asteroids are common in the inner asteroid belt, their colors and albedos are different from those of chondrites, the most common kind of meteorite. This represented a puzzle to astronomers; shouldn't the common type of asteroid nearest Earth be the source of the most common type of meteorite that hits Earth? New evidence from analysis of Moon rocks and from close-up observations of Eros, an S-type

asteroid, shows bombardment by micrometeorites and solar wind particles can redden and darken rocky materials until they have the colors and albedos of S-type asteroids. It therefore seems likely that chondrite meteorites are in fact fragments of S-type asteroids.

In October 2008, a small asteroid 2 to 3 m in diameter, about the size of a small truck, was spotted by the NEO detection network on a collision course with Earth. Astronomers were able to observe it in space before impact and discovered that its colors and albedo matched the fairly rare F-type asteroids that are mostly in the outer belt. The asteroid entered Earth's atmosphere over the desert of northern Sudan and was witnessed exploding. Scientists Peter Jenniskens from the SETI Institute and Muawia Shaddad of the University of Khartoum organized a team of Sudanese faculty and students to search for pieces of the object (**Figure 16-12**). They ultimately found about 4 kg (9 lb) of fragments corresponding to the rare ureilite meteorite type. For the first time, planetary scientists were able to make a definite connection between an asteroid observed in space and meteorites with properties measured in an Earth laboratory. In 2010, Japanese scientists announced that the *Hayabusa* probe returned to Earth with a few microscopic grains of soil from the S-type asteroid Itokawa. That material has a composition like some chondrite meteorites.

As you saw in the case of Vesta, a few asteroids may once have been geologically active, with lava flowing on their surfaces when they were young. Perhaps they incorporated especially large amounts of short-lived radioactive elements such as aluminum-26. Those radioactive elements were probably produced by a supernova explosion that could also have been the

Courtesy P. Jenniskens (Carl Sagan Center, SETI Institute)

▲ Figure 16-12 Search team members from the University of Khartoum and the SETI Institute pose with one of the first fragments found of the small asteroid that exploded over the Sudanese desert in 2009.

trigger for the formation of the Sun and planets while seeding the young Solar System with its nucleosynthesis products.

Ceres, 950 km (590 mi) in diameter, is almost twice as big as Vesta. The density of Ceres and reflection spectra of its surface indicate it is composed of ice-rich carbonaceous material but there are no signs of past silicate volcanic activity like Vesta's. However, observations of Ceres with the *Herschel* infrared space telescope revealed water vapor venting from two regions, creating a very thin atmosphere. The puzzling differences between those two large asteroids will be investigated by the *Dawn* spacecraft that spent 14 months orbiting Vesta in 2011–2012 and then headed for Ceres where it will arrive in 2015.

Although there are still mysteries to solve, you can understand the story of the asteroids. They are fragments of planetesimals, some of which differentiated, developed molten metal cores, in a few cases even had lava flows on their surfaces, and then cooled slowly. The largest asteroids astronomers see today may be nearly unbroken examples of original planetesimals, but the smaller asteroids are fragments produced by 4.6 billion years of collisions.

DOING SCIENCE

What is the evidence that asteroids have been fragmented?
This is the type of question that requires a scientist to keep in mind the difference between theory and evidence.

The solar nebula theory predicts that planetesimals collided and either stuck together or fragmented. It is called a "theory" because it is a comprehensive hypothesis for which there is abundant evidence. But even if you have strong confidence in a theory, it is not evidence. A theory can never be used as evidence to support some other theory or hypothesis. Evidence means observations or the results of experiments, so to answer this question requires citing observations and measurements.

Spacecraft photographs of asteroids show irregularly shaped little worlds heavily scarred by impact craters. Further evidence indicates some asteroids may be pairs of bodies split apart but still in contact, and images of asteroid Ida reveal a small satellite, Dactyl. Other asteroids with moons have been found. These double asteroids and asteroids with moons probably reveal the results of fragmenting collisions between asteroids. Furthermore, meteorites appear to have come from the asteroid belt astronomically recently, so fragmentation must be a continuing process there.

Now pursue another investigation into the history of asteroids: ***What evidence can you cite about the nature of the first planetesimals?***

16-3 Comets

Few sights in astronomy are more beautiful than a bright comet hanging in the night sky (**Figure 16-13**). It is a **Common Misconception** that comets whiz rapidly across the sky like meteors. Actually, comets move with the stately grace of great ships at sea—their motion hardly apparent. Night by night, they shift position slightly against the background stars, and they may remain visible for weeks.

Comets are now named after the person or persons who discover them. Comet Halley, in contrast, has been noted on every return for the past 2250 years and has no known "discoverer." It is named after Edmund Halley, an English astronomer who was a friend of Isaac Newton and who realized that certain comets appearing at 76-year intervals were actually the same comet on a repeating orbit. Halley correctly predicted that the comet would be seen in a certain year, and when it appeared, it was named after him posthumously.

Throughout history comets have been considered omens of doom. Comets may be beautiful, but they are also so strange in appearance that they can create some instinctive alarm. Even recent appearances of bright comets have caused predictions of the end of the world. In 1910, Comet Halley was spectacular, and it was frightening to some people. Comet Kohoutek in 1973, Comet Halley returning in 1986, and Comet Hale–Bopp in 1997 also caused concern among the superstitious.

Faint comets are common; several dozen are discovered each year. Truly bright comets appear about once per decade. Comet McNaught in 2007 was bright enough to be classed with other great comet appearances such as Comet Halley in 1910. An average person might see five or ten bright comets in a lifetime. Although everyone can enjoy the beauty of comets, astronomers study them because they are messengers from the past carrying cargos of information about the origin of our Solar System.

Properties of Comets

As always, you should begin your study of a new kind of object by summarizing its observational properties. What do comets look like, and how do they behave?

Study **Observations of Comets** on pages 368–369 and notice three important properties of comets plus three new terms:

1 Gas and dust released by a comet's icy nucleus produce a head or *coma* and are then blown outward, away from the Sun, responding differently to solar wind and solar radiation pressures to form a separate *gas tail* and *dust tail*.

2 Dust released from a comet's nucleus into the dust tail eventually spreads throughout the Solar System. Some of those comet dust particles later encounter the Earth and are seen as shower meteors and sporadic meteors.

3 Evidence shows that comet nuclei are fragile and can break into pieces easily.

Astronomers have combined these and other observations to understand the nature and structure of comet nuclei.

Observations of Comets

1 A comet's **gas tail** is produced by ionized gas carried away from the nucleus by the solar wind. The spectrum of a gas tail is an emission spectrum. The atoms are ionized by the ultraviolet component of sunlight. The wisps and kinks in gas tails are produced by the magnetic field embedded in the solar wind.

Spectra of gas tails reveal atoms and ions such as H_2O, CO_2, CO, H, OH, HCN, O, S, C, and so on. These are released by the vaporizing ices or produced by the breakdown of those molecules. Some gases such as hydrogen isocyanide (HNC) are apparently formed by chemical reactions in the coma.

Visual

Gas tail

Dust tail

1a A dust tail consists of dust that was contained in the vaporized ices of the nucleus. The dust is pushed gently outward by the pressure of sunlight but is not affected by the magnetic field of the solar wind, so dust tails are more uniform than gas tails. Dust tails are often curved because the dust particles follow individual orbits around the Sun once they leave the nucleus. Both gas and dust tails extend away from the Sun because of the forces acting on them.

Nucleus ———————————
(invisibly small at this scale)

1b The nucleus of a comet (too small to be visible here) is a lump of fragile, porous material containing ices of water, carbon dioxide, ammonia, and so on. Comet nuclei can be 1 to 100 km in diameter.

The **coma** of a comet is the cloud of gas and dust that surrounds the nucleus. It can be more than 1,000,000 km in diameter, as large as the Sun.

Coma

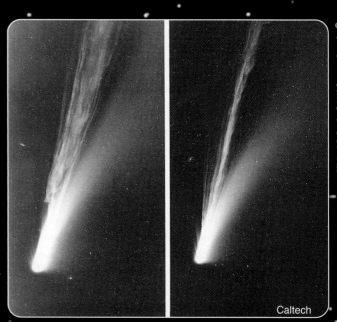

1c The gas tail of Comet Mrkos (1957; pronounced *MIHR-kosh*) showed significant changes from night to night caused by changes in the solar wind's magnetic fields.

Visual images

Caltech

The *Deep Impact* spacecraft released an instrumented probe into the path of Comet Tempel 1. When the comet slammed into the probe at 10 km/s as shown at right, huge amounts of gas and dust were released. From the results, scientists conclude that the nucleus of the comet is rich in dust finer than particles of talcum powder. The nucleus is marked by craters, but it is not solid rock. It is about the density of fresh-fallen snow.

Visual

NASA/JPL-Caltech/Univ. of Maryland

Craters on the dark surface of the comet were visible from the probe a few seconds before impact.

NASA/JPL-Caltech/Univ. of Maryland

Visual

Images of Comet Tempel 1 from the flyby probe 13 seconds after the impact probe hit. Gas and dust were thrown out of the impact crater.

Dust particles (*arrows*) were embedded in the collector when they struck at high velocity.

Direction of travel

NASA/JPL-Caltech

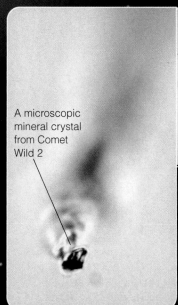

A microscopic mineral crystal from Comet Wild 2

NASA/JPL-Caltech

2a The *Stardust* spacecraft flew past the nucleus of Comet Wild 2 and collected dust particles (*as shown above*) in an exposed target that was later parachuted back to Earth. The dust particles hit the collector at high velocity and became embedded but can be extracted for study.

Some of the collected dust is made of high-temperature minerals that could only have formed near the Sun. This suggests that material from the inner solar nebula was mixed outward and became part of the forming comets in the outer Solar System.

Other minerals found include olivine, a common mineral on Earth but not one that scientists expected to find in a comet.

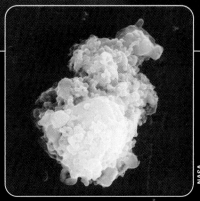

NASA

This dust particle was gathered by an aircraft flying at a very high altitude in the stratosphere. It is almost certainly from a comet.

Fragment B of Comet Schwassmann-Wachmann 3

3 Nuclei of comets are not strong. In 2006, Comet Schwassmann-Wachmann 3 broke into a number of fragments that themselves fragmented. Fragment B is shown at right breaking into smaller pieces. The gas and dust released by the breakup made the comet fragments brighten in the night sky; some were visible with binoculars. As its ices vaporize and its dust spreads, a comet may totally disintegrate and leave nothing but a stream of debris along its previous orbit.

Comets most often break up as they pass close to the Sun or close to a massive planet like Jupiter. Comet ISON came apart in 2012 as it passed near the Sun. Comet Shoemaker-Levy 9 that hit Jupiter in 1994 was first ripped to pieces by tidal stresses from Jupiter's gravity. Comets can also fragment far from planets, perhaps because of the collapse of cavities within the icy nucleus.

Fragments

NASA/ESA/STScI/AURA/NSF/H. Weaver (JHU APL), M. Mutchler and Z. Levay (STScI)

Visual

▲ Figure 16-13 Comet McNaught swept through the inner Solar System in 2007 and was a dramatic sight in the southern sky. Seen here from Australia, the comet was on its way back into deep space after making its closest approach to the Sun ten days earlier. Comet McNaught began this passage with a period of about 300,000 years, but gravitational perturbations by the planets changed its orbit shape from elliptical to hyperbolic, so it will never return but instead is leaving the Solar System to journey forever in interstellar space.

The Geology of Comet Nuclei

The nuclei of comets are quite small and cannot be studied in detail using Earth-based telescopes. Nevertheless, when a comet nucleus approaches the Sun, it emits material that forms into a coma and tails that can be millions of kilometers in size and are easily observed.

Spectra of comet comae (plural of *coma*) and tails indicate the nuclei must contain ices of water and other volatile compounds such as carbon dioxide, carbon monoxide, methane, ammonia, and so on. These are the kinds of compounds that would have condensed in cold regions of the solar nebula. This convinces astronomers that comets are ancient samples of the gases and dust from which the outer planets formed. As the ices absorb energy from sunlight, they sublime—change from a solid directly into a gas. The gases break down and also combine chemically, producing other substances found in comet spectra. For example, vast clouds of hydrogen gas observed around the heads of comets are understood to derive from the breakup of ice molecules.

Five spacecraft flew past the nucleus of Comet Halley when it visited the inner Solar System in 1985-1986. Other spacecraft flew past the nuclei of Comet Borrelly in 2001, Comet Wild 2 (pronounced *Vildt two*) in 2004, and Comet Tempel 1 in 2005. Images show that all these comet nuclei are 1 to 10 km across, similar in size to many asteroids, and also irregular in shape (Figure 16-14). In general, these nuclei are darker than a lump of coal, which suggests composition similar to carbon-rich carbonaceous chondrite meteorites. (At the time of this writing, the European probe *Rosetta* has begun orbiting the nucleus of Comet Churyumov-Gerasimenko, eventually to release a small landing probe.)

The mass and density of comet nuclei can be calculated from their gravitational influence on passing spacecraft. Comet nuclei appear to have densities between 0.1 and 0.25 g/cm^3, much less than the density of ice. Also, as you will learn later in this chapter, comets subjected to tidal stresses from Jupiter or the Sun come apart very easily. Comet nuclei have been described as

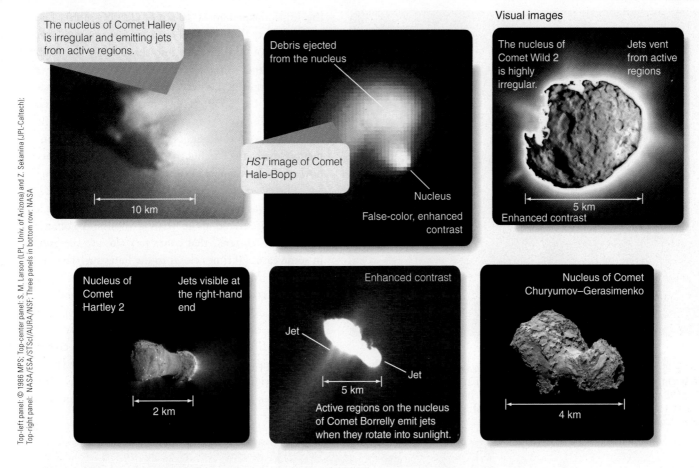

The nucleus of Comet Halley is irregular and emitting jets from active regions.

HST image of Comet Hale-Bopp

Debris ejected from the nucleus

Nucleus

False-color, enhanced contrast

The nucleus of Comet Wild 2 is highly irregular.

Jets vent from active regions

5 km

Enhanced contrast

10 km

Nucleus of Comet Hartley 2

Jets visible at the right-hand end

2 km

Enhanced contrast

Jet

Jet

5 km

Active regions on the nucleus of Comet Borrelly emit jets when they rotate into sunlight.

Nucleus of Comet Churyumov–Gerasimenko

4 km

Top-left panel: © 1986 MPS; Top-center panel: S. M. Larson (LPL, Univ. of Arizona) and Z. Sekanina (JPL-Caltech); Top-right panel: NASA/ESA/STScI/AURA/NSF; Three panels in bottom row: NASA

▲ **Figure 16-14** Visual-wavelength images made by the *Hubble Space Telescope* and from passing spacecraft show how the nuclei of comets produce jets of gases from regions where sunlight vaporizes ices.

dirty snowballs or icy mudballs, but that seems to be incorrect; their shapes, low densities, and lack of material strength suggest that comets are not solid objects. The evidence leads astronomers to conclude that most comet nuclei must be fluffy mixtures of ices and dust with significant amounts of empty space. On the other hand, images of the nucleus of Comet Wild 2 revealed cliffs, pinnacles, and other features that show the material has enough strength to stand against the weak gravity of the comet.

Photographs of comet comae (Figure 16-14) often show jets springing from the nucleus into the coma and being swept back by the pressures of sunlight and the solar wind to form the tail. Studies of the motions of these jets as the nucleus rotates reveal that they originate from small active regions that may be similar to volcanic faults or vents (Figure 16-15). As the rotation of a comet nucleus carries an active region into sunlight, it begins venting gas and dust, and as the active region rotates into darkness it shuts down.

The nuclei of comets appear to have a crust of rocky dust left behind as the ices vaporize. Breaks in that crust

NASA/NSSDC: Tom Herbst (MPIA), Doug Hamilton (MPIK), Hermann Böehnhardt, Universitäts-Sternwarte, and Jose Luis Ortiz Moreno, Instituto de Astrofisica

▲ **Figure 16-15** The crusts of comets are evidently delicate mixtures of rock, ice, and dust. The dust is ejected along with gases as the ices in a comet vaporize in sunlight, as shown in this artist's conception.

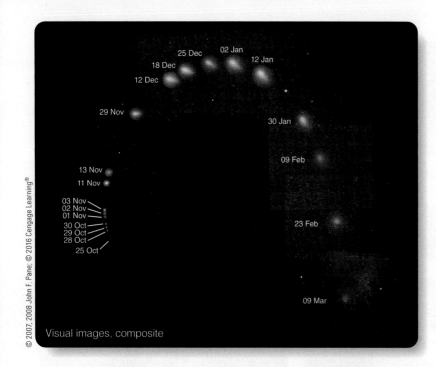

Visual images, composite

▲ **Figure 16-16** Composite of 19 snapshots of Comet Holmes, showing its changing brightness and position spanning the period from October 2007 to March 2008. During its outburst in late October 2007, the comet brightened by a factor of about 500,000 as a large pocket of volatile material exploded through its crust and spread into space.

the impactor at almost 10 km/s (22,000 mph). The probe broke through the crust of the nucleus and blasted vapor and dust out into space where the *Deep Impact* "mother ship," as well as the *Spitzer* and *Hubble* space telescopes and observatories on Earth, could analyze it (see page 369). Those missions produced the surprising discovery that some comet dust is crystalline and must have formed originally in very warm environments close to the Sun but then was incorporated somehow into comet nuclei in the cold outer Solar System.

The *Solar and Heliospheric Observatory (SOHO)* spacecraft was put into space to observe the Sun, but it has also discovered more than a thousand comets, called "Sun grazers," that come very close to the Sun, in some cases 70 times closer to the Sun's surface than the planet Mercury (**Figure 16-17**). As many as three small comets per week plunge into the Sun and are destroyed. Most Sun grazers belong to one of four groups, the comets in each group having similar orbits. Like the Hirayama families of asteroids, these comet groups appear to be made up of fragments of larger comet nuclei. The original comet may have been ripped apart by the violence of gases superheated near the Sun and bursting

can expose ices to sunlight, and vents can occur in those regions. It also seems that some comets have large pockets of volatiles below the crust. When one of those pockets is exposed and begins to vaporize, the comet can suffer a dramatic outburst, as did Comet Holmes in 2007 (**Figure 16-16**).

Astronomers have devised ways to study comet material more directly. The *Stardust* spacecraft passed through the tail of Comet Wild 2 in 2004, collected dust particles that had been ejected from the comet's nucleus, and returned the samples in a sealed capsule to Earth in 2006 for analysis. In 2005, the *Deep Impact* spacecraft released an instrumented impactor probe into the path of comet Tempel 1. As planned, the nucleus of the comet ran into

▶ **Figure 16-17** The *SOHO* observatory can see comets rounding the Sun on very tight orbits. Some Sun-grazing comets, like the two shown here, are destroyed by solar radiation and are not detected emerging on the other side of the Sun.

Sun hidden behind mask

Comets

Visual

through the crust, by solar tidal forces, or both. Comet ISON, which may have been a member of a family of Sun grazers, passed less than two solar radii from the surface of the Sun in November 2013 and disintegrated.

Sun-grazing comets can be destroyed quickly by the Sun, but even normal comets suffer from the effects of solar heating. Each passage around the Sun vaporizes many millions of tons of ices, so the nucleus slowly loses its ices until there is nothing left but dust and rock moving along an orbit around the Sun. The eventual fate of a comet is clear, but a more important question is its origin.

Origin and History of Comets

Family relationships among the comets can give you clues to their origin. Most comets have long, elliptical orbits with periods greater than 200 years and are known as long-period comets. The long-period comet orbits are randomly inclined to the plane of the Solar System, so those comets approach the inner Solar System from all directions. Long-period comets revolve around the Sun in about equal numbers in prograde orbits (the same direction in which the planets move) and retrograde orbits.

In contrast, about 100 or so of the 600 well-studied comets have orbits with periods less than 200 years. These short-period comets usually follow orbits that lie within 30 degrees of the plane of the Solar System, and most revolve around the Sun prograde. Comet Halley, with a period of 76 years, is an unusual short-period comet with a retrograde orbit.

A comet cannot survive long in an orbit that brings it into the inner Solar System. The heat of the Sun vaporizes ices and reduces comets to inactive bodies of rock and dust; such comets can last at most 100 to 1000 orbits around the Sun. Astronomers calculate that even before a comet completely vaporizes from solar heating, it can't survive more than about half a million years crossing the orbits of the planets, especially Jupiter, without having its path rerouted into the Sun, out of the Solar System, or into collision with one of the planets. Therefore, comets visible in our skies now can't have survived in their present orbits for 4.6 billion years since the formation of the Solar System, and that means there must be a continuous supply of new comets. Where do they come from?

Comets from the Oort Cloud and the Kuiper Belt

In the 1950s, astronomer Jan Oort proposed that the long-period comets are objects that fall inward from what has become known as the **Oort Cloud**, a spherical cloud of icy bodies that extends from about 10,000 to 100,000 AU from the Sun (Figure 16-18). Astronomers estimate that the cloud contains several trillion (10^{12}) icy bodies. Far from the Sun, they are very cold, lack comae and tails, and are invisible from Earth. The gravitational influence of occasional passing stars can perturb a

© 2016 Cengage Learning®

▲ Figure 16-18 Long-period comets appear to originate in the spherical Oort Cloud. Objects that fall into the inner Solar System from that cloud arrive from all directions.

few of these objects and causes them to fall into the inner Solar System, where the heat of the Sun warms their ices and transforms them into comets. The fact that long-period comets are observed to fall inward from all directions is explained by their Oort Cloud reservoir being approximately spherically symmetric around the Sun and inner Solar System.

It is not surprising that stars pass close enough to affect the Oort Cloud. For example, data from the *HIPPARCOS* satellite show that the star Gliese 710 will pass within 1 light-year (about 63,000 AU) of the Sun, crossing through the Oort Cloud, in about a million years. The result may be a shower of Oort Cloud objects perturbed into the inner Solar System, where, warmed by the Sun, they will become comets.

Saying that comets come from the Oort Cloud only pushes the mystery back one step. How did those icy bodies get there? In preceding chapters, you have studied the origin and evolution of our Solar System so carefully that the answer may leap out at you. Those Oort Cloud comets are some of the icy planetesimals that formed in the outer solar nebula. The bodies in the Oort Cloud, however, could not have formed at their present location because the solar nebula would not have been dense enough so far from the Sun. And if they had formed from the solar nebula, they would be distributed in a disk and not in a sphere. Astronomers think that the Oort Cloud planetesimals formed in the outer Solar System near the present orbits of the giant planets. As those planets grew, they swept up many of these planetesimals, but they also would have ejected some out of the Solar

System. Most of those ejected objects vanished into interstellar space, but perhaps 10 percent had their orbits modified by the gravity of stars passing nearby and became part of the Oort Cloud. Some of those later became the long-period comets.

Long-period comets originate in the Oort Cloud, and some of the short-period comets do also. A few short-period comets, including Comet Halley, appear to have begun as long-period comets from the Oort Cloud and had their orbits altered by a close encounter with Jupiter. However, that process can't explain all of the short-period comets: Some follow orbits that could not have been reached by Oort Cloud objects interacting with Jupiter or the other planets. There must be another source of icy bodies in our Solar System, which astronomers now understand is the Kuiper Belt.

In 1951, astronomer Gerard Kuiper proposed that the formation of the Solar System should have left behind a belt of small, icy planetesimals beyond the Jovian planets and in the plane of the Solar System. Such objects were first discovered in 1992 and are now known as Kuiper Belt Objects (KBOs). You first learned about the Kuiper Belt in Chapter 10 regarding evidence about the origin of the Solar System and formation of the Jovian planets. In Chapter 15, you learned about the two largest Kuiper Belt Objects, Eris and Pluto, as examples of dwarf planets.

The KBOs are small, icy bodies (Figure 16-19) that orbit in the plane of the Solar System in a zone extending from the orbit of Neptune out to about 50 AU from the Sun. Some objects are known to loop out as far as 1000 AU, but those may have been scattered into those orbits by gravitational interactions with passing stars. The entire Kuiper Belt, containing as many as 100,000 objects larger than 100 km (60 mi) in diameter and hundreds of millions of smaller bodies, would be hidden behind the yellow dot representing the Solar System in Figure 16-18.

How can this belt of ancient, icy worlds generate short-period comets? Because KBOs orbit in the same direction as the planets and in the plane of the Solar System, it is possible for an object perturbed inward by the influence of the giant planets to move into an orbit resembling those of the short-period comets. Rare collisions and interactions among the KBOs could also add to a continuous supply of small, icy bodies from the Kuiper Belt sent into the inner Solar System.

Comets vary in brightness and orbit. Nevertheless, there are two basic types of comets in our Solar System. Some originate in the Oort Cloud far from the Sun. Others come from the Kuiper Belt just beyond Neptune. They all share one characteristic: They are ancient, icy bodies that were born when the Solar System was young.

Kuiper Belt Object 2000 FV53, about 120 km (75 mi) in diameter, moves against stars in the background.

January 26, 2003
19:19 UT

06:50 UT

Visual

NASA/ESA/STScI/AURA/NSF/G. Bernstein and D. Trilling (University of Pennsylvania)

▲ Figure 16-19 (a) Kuiper Belt Objects (KBOs) are small bodies with dark surfaces that are hard to detect from Earth. A KBO originally designated 2000 FV53 was discovered from these two superimposed images of one field of view.

DOING SCIENCE

How do comets help explain the formation of the planets?
This the type of question scientists ask to connect properties of present-day objects with the history of the Solar System.

Recall once more the solar nebula theory. Planetesimals that formed in the inner solar nebula were warm and could not incorporate much ice. The asteroids are understood to be the last remains of such rocky bodies. On the other hand, planetesimals in the outer Solar System contained large amounts of ices. Many of them are gone because they accreted together to make the Jovian planets, but some survived intact. The icy bodies of the Oort Cloud and the Kuiper Belt are understood to be the Solar System's last surviving icy planetesimals. When those icy objects have their orbits perturbed by the gravity of the planets or passing stars, some are redirected into the inner Solar System where you see them as comets. The gases released by comets indicate that they are rich in volatile materials such as water and carbon dioxide. These are the ices you would expect to find in the icy planetesimals. Comets also contain grit with rocklike chemical composition, and the planetesimals must have included large amounts of such dust frozen into the ices when they formed. Thus, the nuclei of comets seem to be frozen samples of the original outer solar nebula.

16-4 Asteroid and Comet Impacts

Meteorite impacts affecting homes and cities, like the one described at the start of this chapter, are not common. Most meteors are small particles ranging from a few centimeters down to microscopic dust. Astronomers estimate that Earth gains about 40,000 tons of mass per year from meteorites of all sizes. (That may seem like a lot, but it is less than a hundred-thousandth of a trillionth of Earth's total mass.) Statistical calculations indicate that a meteorite large enough to cause some damage, like the one that hit Mrs. Hodges and her house in 1954, strikes a building somewhere in the world about once every 16 months.

Objects a few tens of meters or less in diameter such as the 2013 Chelyabinsk meteor are likely to fragment and explode in Earth's atmosphere without reaching the surface, but the shock waves from those explosions obviously can still cause serious damage on Earth's surface. Declassified data from military satellites show that Earth is hit about once a week by meter-size asteroids. An event like the Chelyabinsk explosion is estimated to occur once every 30 years, but these usually happen over an ocean or piece of uninhabited territory rather than directly above a large city. What happens when even larger Solar System objects collide with Earth?

Barringer Crater

Barringer Crater near Flagstaff, Arizona, is 1.2 km (3/4 of a mile) in diameter and 200 m (650 ft) deep. It seems quite large when you stand on the edge, and the hike around it, though beautiful, is long and dry (Figure 16-20). Barringer Crater was the subject of controversy among geologists for years as to whether it was caused by a volcanic event or a large meteorite impact. Finally, in 1963, Eugene Shoemaker proved in his doctoral thesis that the crater must be the result of an impact because quartz crystals in and around it had been subjected to pressures much higher than can be produced by a volcano.

Further studies showed Barringer Crater was created approximately 50,000 years ago by a meteorite estimated to have been about 50 m (160 ft) in diameter, as large as a good-size building, that hit at a speed of 11 km/s, releasing as much energy as a large thermonuclear bomb. An object of that size could be called either a large meteorite or a small asteroid. Debris at the site shows that the impactor was composed of iron.

The Tunguska Event

On a summer morning in 1908, reindeer herders and homesteaders in central Siberia were startled to see a brilliant blue-white fireball brighter than the Sun streak across the sky. Still descending, it exploded with a blinding flash and an intense pulse of heat. One eyewitness account states:

> The whole northern part of the sky appeared to be covered with fire. . . . I felt great heat as if my shirt had caught fire . . . there was a . . . mighty crash. . . . I was thrown on the ground about [7 meters] from the porch. . . . A hot wind, as from a cannon, blew past the huts from the north.

The blast was heard up to 1000 km (600 mi) away, and the resulting pulse of air pressure circled Earth twice. For a number of nights following the blast, European astronomers, who knew nothing of the explosion, observed a glowing reddish haze high in the atmosphere.

When members of a scientific expedition arrived at the site in 1927, they found that the blast had occurred above the Stony Tunguska River valley and had flattened trees in an irregular pattern extending to a radius of about 30 km (20 mi; Figure 16-21). The trees were knocked down pointing away from the center of the blast, and limbs and leaves had been stripped away. The trunks of trees at the very center of the area were still standing, although they had lost all their limbs. No crater has been found, so it seems that the explosion, estimated to have equaled 12 megatons (12 million tons) of TNT, occurred at least a few kilometers above the ground.

In the early 1980s, a detailed analysis of all the Tunguska evidence suggested that the impactor's speed and direction resembled the orbits of Apollo objects. In 1993, astronomers produced computer models of objects entering Earth's atmosphere at various speeds and concluded that the fragile icy head of a comet would have exploded much too high in the atmosphere. On the other hand, a dense, iron-rich meteorite would be so strong that it would have survived to reach the ground and would have formed a large crater. Therefore, the most likely candidate for the Tunguska object seems to be a stony asteroid about 30 m in diameter, perhaps one-tenth the mass of the Barringer impactor. The models indicate that an object of this size with moderate material strength would have fragmented and exploded at just about the right height to produce the observed blast. This conclusion is consistent with modern studies of the Tunguska area showing that thousands of tons of powdered material with a composition resembling carbonaceous chondrites are scattered in the soil.

Planet-Shaking Events

There are some very big craters in the Solar System, for example on the Moon (Chapter 12, pages 247 and 253–254), that show what can happen when a full-size asteroid or comet collides with a planet. Also, Earthlings watched in awe in 1994 as fragments from the nucleus of Comet Shoemaker–Levy 9 (abbreviated SL-9) slammed into Jupiter and produced impacts equaling millions of megatons (that is, trillions of tons) of TNT (Figure 16-22). Note that the Shoemaker in Shoemaker–Levy refers to Carolyn and Eugene Shoemaker who codiscovered the comet with David Levy. Eugene Shoemaker is the person whose analysis showed that Barringer Crater in Arizona is an impact crater.

As you know, Jupiter does not have a solid surface, so SL-9 did not leave any permanent craters, but astronomers have found chains of craters on other Solar System objects that seem to have

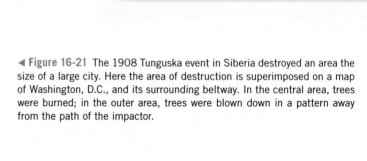

▲ **Figure 16-20** (a) The Barringer Meteorite Crater (near Flagstaff, Arizona) is nearly a mile in diameter and was formed about 50,000 years ago by the impact of an iron meteorite estimated to have been roughly 50 m (160 ft) in diameter. Notice the raised and deformed rock layers all around the crater. The brick museum building visible on the far rim at right provides some idea of scale. (b) Like all larger-impact features, the Barringer Meteorite Crater has a raised rim and scattered ejecta.

◀ **Figure 16-21** The 1908 Tunguska event in Siberia destroyed an area the size of a large city. Here the area of destruction is superimposed on a map of Washington, D.C., and its surrounding beltway. In the central area, trees were burned; in the outer area, trees were blown down in a pattern away from the path of the impactor.

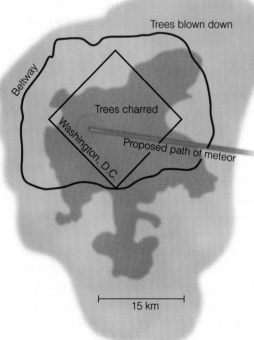

Visual, composite

Impact site just out of sight as seen from Earth

Fragments of comet falling toward Jupiter

At visual wavelengths, impact sites were dark smudges that lasted for many days.

Larger than Earth

b   Visual

a

Panel: a NASA/ESA/STScI/AURA/NSF/JPL-Caltech/H. A. Weaver and T. E. Smith (STScI), J. T. Trauger and R. W. Evans (JPL-Caltech); Panel: b, c, d: NASA; © 2016 Cengage Learning ®

◄ **Figure 16-22** In 1992, Comet Shoemaker–Levy 9 passed within 1.3 planetary radii of Jupiter's center, well within its Roche limit, and tidal forces ripped the nucleus into more than 20 pieces. The fragmented pieces were as large as a few kilometers in diameter and spread into a long string of objects that looped away from Jupiter and then fell back to strike the planet, producing massive impacts over a period of 6 days in July 1994. Note that the impact flash in panel (c) is larger than Earth.

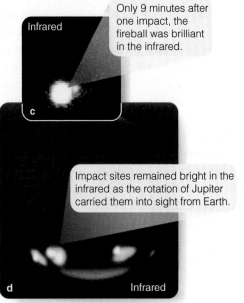

Infrared

Only 9 minutes after one impact, the fireball was brilliant in the infrared.

c

Impact sites remained bright in the infrared as the rotation of Jupiter carried them into sight from Earth.

d   Infrared

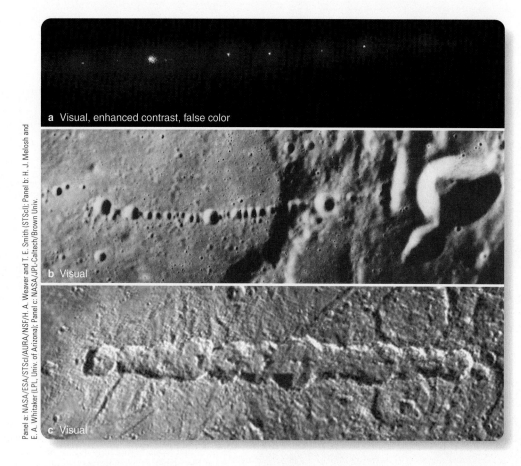

a Visual, enhanced contrast, false color

b Visual

c Visual

Panel a: NASA/ESA/STScI/AURA/NSF/H. A. Weaver and T. E. Smith (STScI); Panel b: H. J. Melosh and E. A. Whitaker (LPL, Univ. of Arizona); Panel c: NASA/JPL-Caltech/Brown Univ.

▲ Figure 16-23 (a) Close-up image of the Comet Shoemaker–Levy 9 fragment train on the way to colliding with Jupiter. (b) A 40-km (25 mi)-long crater chain on Earth's Moon, and (c) a 140-km (90 mi)-long crater chain on Jupiter's moon Ganymede, probably formed by the impact of fragmented comet nuclei similar to Shoemaker–Levy 9.

been formed by fragmented comets (Figure 16-23). Evidently events like the SL-9 collision with Jupiter have occurred many times in the history of the Solar System.

What would happen if an object the size of SL-9, or even larger, were to hit Earth? Sixty-five million years ago, at the end of the Cretaceous period, more than 75 percent of the species on Earth, including the dinosaurs, became extinct. Scientists have found a thin layer of clay all over the world that was laid down at that time, and it is rich in the element iridium—common in meteorites but rare in Earth's crust. This suggests that an impact occurred that was large enough to have altered Earth's climate and caused the worldwide extinction.

Mathematical models combined with lab experiments and observations of craters on other worlds create a plausible scenario of a major impact on Earth. Of course, creatures living near the site of the impact would die in the initial shock, but then things would get bad elsewhere. An impact at sea would create tsunamis many hundreds of meters high that would sweep around the world, devastating regions far inland from coasts. On land or sea, a major impact would eject huge amounts of pulverized rock high above the atmosphere. As this material fell back, Earth's

atmosphere would be turned into a glowing oven of red-hot meteorites streaming through the air, and the heat would trigger massive forest fires around the world. Soot from such fires has been found in the layers of clay laid down at the end of the Cretaceous period. Once the firestorms cooled, the remaining dust in the atmosphere would block sunlight and produce deep darkness for a year or more, killing off most plant life. At the same time, if the impact site were at or near limestone deposits, large amounts of carbon dioxide could be released into the atmosphere and produce intense acid rain.

Geologists have located a crater at least 180 km (110 mi) in diameter centered near the village of **Chicxulub** (pronounced *CHEEK-shoe-lube*) in the northern Yucatán region of Mexico (Figure 16-24). Although the crater is now completely covered by sediments, mineral samples show that it contains shocked quartz typical of impact sites and that it is the right age. The impact of an object 10 to 15 km (6 to 10 mi) in diameter formed the crater about 65 million years ago, just when the dinosaurs and many other species died out. Most scientists now conclude that this is the scar of the impact that ended the Cretaceous period.

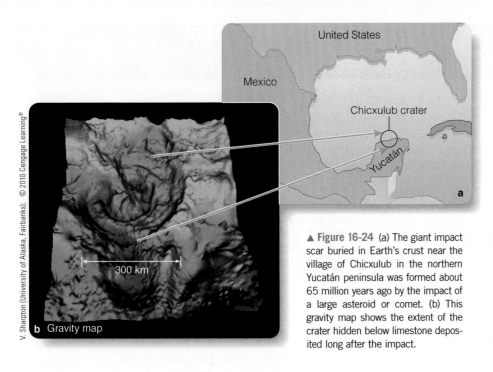

▲ **Figure 16-24** (a) The giant impact scar buried in Earth's crust near the village of Chicxulub in the northern Yucatán peninsula was formed about 65 million years ago by the impact of a large asteroid or comet. (b) This gravity map shows the extent of the crater hidden below limestone deposited long after the impact.

Asteroid 2004 MN4 was initially predicted to have a 2.6 percent chance of striking Earth in 2029. That object is about 400 m (¼ mile) in diameter, large enough to do significant damage over a wide area but not large enough to alter Earth's global climate. Fortunately, further observations and calculations revealed that this object will not hit Earth. There will be no impact by 2004 MN4 in 2029, but there are plenty more asteroids in Earth-crossing orbits to be discovered. For example, a rock designated 2009 DD45, estimated to have been 30 m (100 ft) in diameter, about the size of the Tunguska impactor, passed only 64,000 km (40,000 mi) from Earth in March 2009. That is only twice the distance of humanity's geosynchronous communications and weather satellites, a very near miss indeed; yet DD45 was first spotted only three days before its closest approach.

There are a number of major extinctions in the fossil record, and at least some of these were probably caused by large impacts. Large asteroid impacts on Earth happen very rarely from a human perspective, but they happen often relative to geological and astronomical time scales. For example, astronomers estimate that an Apollo object hits Earth once every 250,000 years on average. A typical Apollo with a diameter of 1 km would strike with the power of a 100,000-megaton bomb and dig a crater more than 10 km in diameter. The good news is that we are certain that no known Apollo object will hit Earth in the foreseeable future; the bad news is that there are about 1000 of them 1 km in size or larger.

Some people have argued that the danger from asteroid and comet impacts is so great that governments should develop massive nuclear-tipped missiles, ready to blast a meteoroid to pieces long before it can reach Earth. Other experts respond that lots of small fragments slamming into Earth may be even worse than one big impact. Astronomers point out that the biggest objects are so rare they can be ignored. The real danger lies in the more common smaller, yet still substantial, meteoroids, and those are difficult to detect with current telescopes. The future of our civilization on Earth may depend on our doing an increasingly careful job of tracking both large and small objects that cross our path with surprising frequency.

What Are We? Targets

Human civilization is spread out over Earth's surface and exposed to anything that falls out of the sky. Meteorites, asteroids, and comets bombard Earth, producing impacts that vary from dust settling gently on rooftops to disasters capable of destroying all life. In this case, the scientific evidence is conclusive and highly unwelcome.

Statistically we are quite safe. The chance that a major impact will occur during your lifetime is so small it is hard to estimate. But the consequences of such an impact are so severe that humanity should be preparing. One way to prepare is to find those objects that could hit us, map their orbits, and identify any that are dangerous.

What we do next isn't clear. Blowing up a dangerous asteroid in space might make a good movie, but converting one big projectile into a thousand small ones might not be very smart. Changing an asteroid's orbit could be difficult without a few decades' advance warning. Unlikely or not, large impacts demand consideration and preparation.

Throughout the Universe, there may be two kinds of inhabited worlds. On one type of world, intelligent creatures have developed ways to prevent asteroid and comet impacts from altering their climates and destroying their civilizations. But on other worlds, including Earth, intelligent races have not yet found ways to protect themselves. Some of those civilizations survive. Some don't.

Study and Review

Summary

▶ Reviewing terms and concepts first presented in Chapter 10: The term *meteoroid* refers to a small, solid particle moving through space, outside Earth's atmosphere. The term *meteor* refers to a visible streak of light from a heated and glowing particle passing through Earth's atmosphere. The term *meteorite* refers to space material that is found on Earth's surface.

▶ Meteorites seen to hit Earth are called **falls (p. 355)**. **Finds (p. 355)** are meteorites discovered on the ground that fell unobserved, perhaps thousands of years ago.

▶ As their name implies, **iron meteorites (p. 355)** are mostly iron, plus some nickel. After being sliced open, polished, and etched, they show metal crystal **Widmanstätten patterns (p. 356)**, which reveal that the metal cooled very slowly from a molten state.

▶ **Stony meteorites (p. 355)** are silicates that resemble Earth rocks. Some include **chondrites (p. 356)**, which contain small, glassy particles called **chondrules (p. 357)**. Chondrules are solidified droplets of once-molten material that formed in the solar nebula by an as-yet-unknown mechanism.

▶ **Stony-iron meteorites (p. 355)** are quite rare and, as the name implies, consist of mixtures of iron and stone.

▶ **Selection effects (p. 355)** cause iron meteorites to be the most common finds, even though stony meteorites are the most common falls.

▶ Stony meteorites that are rich in volatiles and carbon are called **carbonaceous chondrites (p. 355)**, which are among the least modified meteorites. Some carbonaceous chondrites contain calcium–aluminum-rich inclusions **CAIs (p. 357)**, understood to be the very first solid particles to condense in the cooling solar nebula.

▶ An **achondrite (p. 356)** is a stony meteorite that contains no chondrules and no volatiles. Achondrites appear to have been melted after they formed and, in some cases, resemble solidified lavas.

▶ Evidence from the orbits and compositions of meteorites indicates that all meteorites are fragments of asteroids. In contrast, orbital paths shown by the **radiant (p. 358)** points of **meteor showers (p. 358)** plus other evidence indicates that the vast majority of meteors are low-density, fragile bits of cometary material. This applies to meteors in meteor showers as well as isolated **sporadic meteors (p. 358)**.

▶ Many meteorites appear to be from larger bodies that melted, differentiated, and cooled very slowly. After formation, these bodies broke; fragments from the core became iron meteorites, fragments from the outer layers became stony-iron or stony meteorites, and fragments from intermediate layers became the stony-iron meteorites.

▶ Asteroids are irregular in shape and heavily cratered from collisions. Their surfaces are covered by gray, pulverized rock. Some asteroids have such low densities that they must be fragmented rubble piles.

▶ Most asteroids lie in a belt between Mars and Jupiter. **Kirkwood gaps (p. 361)** are ranges of asteroid belt orbit parameters that are nearly unoccupied, caused by orbital resonances with Jupiter.

▶ **Apollo-Amor objects (p. 364)** have orbits that cause them to cross into the inner Solar System. If they pass near Earth, they are called **Near-Earth Objects (NEOs) (p. 364)**. Two groups of asteroids are caught in the Lagrange points along Jupiter's orbit. These **Trojan asteroids (p. 365)** are located 60 degrees ahead of and 60 degrees behind the planet. **Centaurs (p. 365)** are objects with both asteroidal and cometary characteristics that orbit among the planets of the outer Solar System.

▶ Asteroids formed as rocky planetesimals between Mars and Jupiter. Jupiter prevented those planetesimals from accumulating into a planet. Collisions have fragmented all but the largest of the asteroids. Most of the original material in the asteroid belt may have already been gravitationally perturbed and swept up by the planets or tossed out of the Solar System.

▶ Members of asteroid **Hirayama families (p. 366)** each follow similar orbits, and family members have similar spectra. These asteroids appear to be fragments produced in past collisions of asteroids.

▶ **C-type (p. 363)** asteroids are more common in the outer asteroid belt where the solar nebula was cooler. They are darker and may be carbonaceous. **S-type (p. 363)** asteroids are more frequently found in the inner belt where the nebula was warmer. They are the most common asteroid type and may be the source of the most common kind of meteorites, the chondrites. **M-type (p. 363)** asteroids may be the cores of differentiated asteroids shattered by collisions. They appear to have nickel-iron compositions.

▶ A visible comet is produced by a nucleus of ices and rock usually between 1 and 100 km in diameter. A comet nucleus travels in a long, elliptical orbit, and stays frozen unless it comes near the Sun. When nearing the Sun, some of the ices vaporize, releasing dust and gas that are blown away from the Sun to form a prominent head and tails.

▶ The **coma (p. 368)**, or head, of a comet can be up to a million kilometers in diameter. The **gas tail (p. 368)** of a comet is ionized gas that is carried away by the solar wind. The **dust tail (p. 368)** of a comet is composed of solid debris released from the nucleus that is blown outward by the pressure of sunlight. Comet tails always point away from the Sun, regardless of the direction the comet is moving.

▶ Spacecraft flying past comets have revealed that comet nuclei have very dark, rocky crusts. Jets of vapor and dust issue from active regions of the rocky crust on the sunlit side.

▶ The low density of comet nuclei shows that comets are irregular mixtures of ices and silicates, probably containing large voids. At least one comet nucleus has surface features that reveal a surprising amount of material strength.

▶ Comet nuclei are leftover icy planetesimals in the outer Solar System, some of which were ejected to form the **Oort Cloud (p. 373)**. Planetesimals perturbed inward from the Oort Cloud may become long-period comets.

▶ In addition to comet nuclei, other icy bodies also formed in the outer Solar System. These icy bodies now make up the Kuiper Belt, which is located in the plane of the Solar System beyond the orbit of Neptune. Objects from the Kuiper Belt that are perturbed into the inner Solar System can become short-period comets.

▶ A major impact on Earth can trigger extinctions resulting from global forest fires caused by heated material falling back into the atmosphere, tsunamis inundating coastal regions around the world, acid rain resulting from large amounts of carbon dioxide released into the atmosphere, and rapid climate change caused by dust filling the atmosphere that plunges the entire Earth into darkness for years.

▶ An impact 65 million years ago at **Chicxulub (p. 378)** in Mexico's Yucatán region appears to have triggered the extinction of 75 percent of the living species on Earth at that time, including the dinosaurs.

Review Questions

1. I am pebble-sized and rocky, and I am located outside Earth's atmosphere but still in the Solar System. What am I?

2. What do Widmanstätten patterns indicate about the history of iron meteorites?

3. Meteorites from the Moon never land on Earth. True or false?

4. I'm a meteorite that resembles Earth rocks. What kind of meteorite am I?

5. What do chondrules tell you about the history of chondrites?

6. Why are no chondrules seen in achondritic meteorites?

7. Why do astronomers refer to carbonaceous chondrites as unmodified or "primitive" material?

8. Iron meteorites are a primitive type of meteorite, unmodified by heat. True or false?

9. List two differences between achondrite and chondrite meteorites.

10. Of all the meteorites shown in Figure 16-2, which one is the most likely meteorite to be found on the ground? Why?

11. Meteorites were once part of which type of celestial object?

12. Most sporadic meteors were once part of which type of celestial object?

13. Meteors in showers were once part of which type of celestial object?

14. Refer to Appendix Table A-12. Why do meteor showers occur at the same time each year?

15. I am a very dark and gray asteroid. What type of asteroid am I, and where might I be located? (*Hint:* See the plot on the right-hand page of **Observations of Asteroids**.)

16. Why do astronomers conclude that asteroids were never part of a full-sized planet?

17. A fragment from the surface of a differentiated asteroid will yield which kind of meteorite?

18. What evidence indicates that the asteroids are mostly fragments of larger bodies?

19. Kirkwood gaps are ranges of orbital radii in the asteroid belt that contain no asteroids. True or false?

20. What evidence indicates that some asteroids have differentiated? What might have been the source or sources of their internal heat?

21. What evidence indicates that some asteroids once had geologically active surfaces?

22. How is the composition of meteorites related to the formation and evolution of asteroids?

23. Describe four differences between asteroids and comets. Describe four similarities between asteroids and comets.

24. How is the composition of meteoroids related to the formation and evolution of comets and comet nuclei?

25. What is the difference between a centaur and a NEO?

26. What is the difference between a comet's dust tail and a comet's gas tail? What does that tell you about the composition and origin of comets?

27. What evidence indicates that a comet's nucleus is rich in ices?

28. Why do most short-period comets have prograde orbits near the plane of the Solar System?

29. What are possible fates (or end-states) for comets?

30. What are the hypotheses for how the bodies in the Kuiper Belt and the bodies in the Oort Cloud formed?

31. How likely is a major impact on Earth large enough to cause mass extinctions in the next 100 years? In the next 100 thousand years? In the next 100 million years? Cite evidence to support your answer.

32. **How Do We Know?** How would studying the chemical composition of only the largest, brightest, and most easily observed asteroids yield potentially misleading information about asteroids in general? Why is this called a selection effect?

Discussion Questions

1. Many "Trojan" asteroids have been found orbiting the Sun in Jupiter's L_4 and L_5 Lagrange points. A few asteroids have also been found in the Lagrange points of Earth, Mars, Uranus, and Neptune. Does this suggest that careful searches might reveal asteroids in the Lagrange points of Mercury, Venus, and Saturn, also? Do these asteroids in the orbital paths of planets suggest a need to modify the IAU's definitions of *planet* and *dwarf planet*?

2. Humans may someday mine asteroids for materials to build and supply space colonies. What kinds of materials could Earthlings obtain from asteroids? (*Hint:* What materials are in S-, M-, and C-type asteroids, respectively?)

3. If cometary nuclei were heated during the formation of the Solar System by internal radioactive decay rather than by solar radiation, how would comets differ from what is observed?

4. Do you think the government should continue spending money to find NEOs? How serious is the risk of a NEO impact? What do you think is the right amount of spending, given your assessment of the risk?

5. The Moon is heavily cratered, indicating that it took many hits that did not impact Earth. Is it fair to say the Moon provided substantial protection to Earth from impacts?

Problems

1. Assuming a night lasts 12 hours, how many total meteors from Swift Tuttle's comet could you see at the rate listed for its shower in Appendix Table A-12?

2. Large meteoroids are hardly slowed by Earth's atmosphere. Assuming the atmosphere is 100 km thick and that a large meteoroid falls perpendicular to the surface, how long does the meteor take to reach the ground?

3. If a single asteroid 1 km in diameter were to fragment into large meteoroids 1 m in diameter, how many meteoroids would the asteroid yield? Assume the asteroid and meteoroids are spherical. (*Note:* The volume of a sphere $= \frac{4}{3}\pi r^3$.)

4. If a trillion (10^{12}) asteroids, each 1 km in diameter, were assembled into one spherical body, how large would that the spherical body be? Compare that body's size to the size of Earth given in **Celestial Profile 2**. (*Note:* The volume of a sphere $= \frac{4}{3}\pi r^3$.)

5. If each asteroid in Problem 3 has a mass of 1.0×10^{12} kg, how massive would the assembled body be? Compared to the mass of Earth given in **Celestial Profile 2**. What can you conclude?

6. The asteroid Vesta has a mass of 2.6×10^{20} kg and an average radius of about 260 km (2.6×10^2 km). What is its escape velocity? Could you jump off the asteroid? (*Hint:* Use the formula for escape velocity, Chapter 5. The formula is in units of m, kg, and s.)

7. An asteroid orbits the Sun in a 2:1 resonance with Jupiter. What is its orbital period? What is its average distance from the Sun?

How fast is the asteroid moving in its orbit? Assume a circular orbit. Express your answer in units of km/s. (*Hint:* Use Kepler's third law, Chapter 4.) (*Notes:* Necessary data can be found in **Celestial Profile 7**. The circumference of a circle is $2\pi r$.)

8. What is the maximum angular diameter of the dwarf planet, Ceres, when it is closest to Earth? Could Earth-based telescopes detect surface features? Could the *Hubble Space Telescope*? (*Hint:* Use the small-angle formula, Chapter 3.) (*Notes:* Ceres's average distance from the Sun is 2.8 AU and its diameter is 950 km. The best angular resolution of Earth-based telescopes at visual wavelengths is about 1 arc second and of *Hubble* about 0.1 arc second.)

9. At what average distances from the Sun would you expect to find Kirkwood gaps where the orbital period of asteroids are respectively one-third, and one-quarter, of the orbital period of Jupiter? Compare your results with Figure 16-9. (*Hint:* Use Kepler's third law, Chapter 4.)

10. The diameter of Ceres is 950 km and its mass is 9.4×10^{20} kg. What is the density of Ceres in units of g/cm^3? Based on this density, what is its likely composition? (*Notes:* Density = mass divided by volume. The volume of a sphere = $\frac{4}{3}\pi r^3$. The density of water is about 1 g/cm^3, and the densities of various types of rock range from about 2.5 to 5 g/cm^3.)

11. If the velocity of the solar wind is about 4.0×10^2 km/s and the visible tail of a comet is 1.0×10^8 km long, how many days does an atom in the solar wind take to travel from the nucleus to the end of the visible tail? (*Note:* 1 day = 86,400 seconds.)

12. What is the average distance of Comet Halley from the Sun? Approximately when will Comet Halley comet next reach aphelion? Perihelion?

13. If you saw Comet Halley when the comet was 0.7 AU from Earth and you observed a visible tail 5 degrees long, how long was the tail in kilometers? Suppose that the tail was not perpendicular to your line of sight. Is your first answer too large or too small? (*Hint:* Use the small-angle formula, Chapter 3.) (*Note:* 1 AU = 1.5×10^8 km.)

14. What is the orbital period of a comet nucleus in the Oort Cloud? What is its orbital velocity? Assume a circular orbit. (*Hint:* Use Kepler's third law, Chapter 4.) (*Note:* The circumference of a circle is $2\pi r$.)

15. The mass of an average comet's nucleus is about 1.0×10^{14} kg. If the Oort Cloud contains 2.0×10^{11} comet nuclei, what is the mass of the cloud in units of Earth masses? In units of Jupiter's mass? (*Notes:* Earth's mass in kg can be found in **Celestial Profile 2**. Jupiter's mass in kg can be found in **Celestial Profile 7**.)

16. Assume a devastating asteroid impact occurs on average every 10 million years. Furthermore, assume the human race is still around for the next such event, with a world population the same

as in the current year. Finally, assume the impact kills 90 percent of the population. Calculate the average death rate per year from major asteroid impacts. Compare that with the death rate per year from airplane crashes. (*Note:* You will need to find the current world population and the air crash fatality rate via searches on the net.)

Learning to Look

1. Look at Figure 16-2d. Identify the chondrules by color. What is the black material?

2. Look at Figure 16-4. Knowing the size of the typical chondrule as discussed in the text, estimate the dimensions of this meteorite.

3. Compare the meteor shower dates listed in Appendix Table A-12 with the cartoon in Figure 16-6a. Why is only one date range per year listed for each meteor shower?

4. Look at the images of Comet Mrkos on the left page of **Observations of Comets.** Is the comet shown on its way in around the Sun, on its way out, or is it not possible to tell?

5. Compare Figure 16-14a with Figure 16-15. Do you think the artist's conception is an accurate portrayal?

6. What do you see in the image at right that tells you the size of planetesimals when the Solar System was forming?

Russell Kempton, New England Meteoritical Services

7. Discuss the surface of the asteroid Mathilde, shown at right. What do you see that tells you something about the history of the asteroid?

Visual

NASA

8. What do you see in this image of the nucleus of Comet Borrelly that tells you how a comet produces a coma and tails?

Visual

NASA

Astrobiology: Life on Other Worlds

17

Guidepost This chapter is either unnecessary or vital. If you believe that astronomy is the study of the physical Universe above Earth's clouds, then you are done; the previous chapters completed your journey. But if you believe that astronomy is the study not only of the physical Universe but also of your role as a living being in the evolution of the Universe, then everything you have learned so far from this book has been preparation for this final chapter.

As you read this chapter, you will ask four important questions:

► **What is life?**

► **How did life originate on Earth?**

► **Could life begin on other worlds?**

► **Can humans on Earth communicate with intelligent species on other worlds?**

You won't get more than the beginnings of answers to those questions here, but often in science, asking a good question is more important than getting an immediate answer.

You have explored the Universe from the phases of the Moon to the big bang, from the origin of Earth to the death of the Sun. Astronomy is meaningful, not just because it is about the Universe but because it is also about you. Now that you know some astronomy, you can see yourself and your world in a different way. Astronomy has changed you.

Did I solicit thee from darkness to promote me?

ADAM, TO GOD, IN JOHN MILTON'S *PARADISE LOST*

Jim Peaco/National Park Service

Colonies of thermophilic (heat-loving) single-celled organisms thrive around the edge of Yellowstone National Park's Grand Prismatic Spring at temperatures up to 72°C (162°F), producing the green pigments. The blue water in the center of the pool is too hot even for thermophiles.

As a living thing, you have been promoted from darkness. The atoms of carbon, oxygen, and other heavy elements that are necessary components of your body did not exist at the beginning of the Universe but were built up by successive generations of stars.

The elements from which you are made are common everywhere in the observable Universe, and planets with Earth-like conditions are almost certainly also common, so it is possible that life began on other worlds. Future explorers may find alien species completely different from any life on Earth. And, it is possible some of those species have evolved to become intelligent. If so, perhaps those other civilizations will be detectable from Earth.

Your goal in this chapter is to try to understand truly intriguing puzzles—the origin and evolution of life on Earth and what that tells you about whether there is life on other worlds. This new hybrid field of study is called **astrobiology** (see, for example, http://astrobiology.nasa.gov/).

17-1 The Nature of Life

What is life? Philosophers have struggled with that question for thousands of years (How Do We Know? 17-1), and it is not possible to answer it completely in one chapter or even one book. An attempt at a general definition of what living things do, distinguishing them from nonliving things, might be: Life is a *process* by which an organism extracts energy from its surroundings, maintains itself, and modifies the surroundings to foster its own survival and reproduction.

One important observation is that all living things on Earth, no matter how apparently different, share certain characteristics in how they perform the process of life.

The Physical Bases of Life

The physical bases of life on Earth are carbon and water (Figure 17-1). Because of the way carbon atoms bond to each other and to other atoms, they can join into long, complex, stable chains that are capable, among many other feats, of storing and transmitting information. A large amount of information is necessary to control the activities and maintain the forms of living things. And, in all living things on Earth, the chemical reactions making, breaking, and combining carbon chains take place in liquid water within the cells of living organisms.

Is it possible that life on other worlds could, for example, use silicon atoms in the role of carbon? Silicon is right below carbon in the periodic table (Appendix Table A-14), which means that it shares many of carbon's chemical properties. But life based on silicon rather than carbon seems unlikely to astrobiologists because silicon chains are harder to assemble and

disassemble than their carbon counterparts, so they can't be as long and complex nor contain as much information.

Is it possible that the chemistry of life on other worlds could take place in a setting other than water? Some alternatives such as methyl alcohol have been proposed. But carbon compounds dissolve especially easily in water. Also, water has exceptional properties such as a high heat capacity (resistance to temperature change) relative to other cosmically abundant substances that are liquid at the temperatures of planetary surfaces. It's not just because Earth scientists are themselves made of carbon and water that they think carbon and water are crucially important for the existence of life.

Science fiction has proposed even stranger life forms based on, for example, electromagnetic fields and ionized gas, and none of these possibilities can be ruled out. Those hypothetical life-forms make for fascinating speculation, but for now they can't be studied systematically in the way that life on Earth can.

This chapter is concerned first with the origin and evolution of life as it is on Earth, based on carbon and water, not because of lack of imagination but because it is the only form of life about which we know anything. From that basis of knowledge we can then sensibly speculate about carbon-based life elsewhere in the Universe. Even carbon- and water-based life has its mysteries. What makes a lump of carbon-based molecules in little bags of water into a living thing? An important part of the answer lies in the transmission of information from one molecule to another.

Information Storage and Duplication

Most actions performed by living cells are carried out by molecules that are built within the cells. Cells must store recipes for making all those molecules as well as how and when to use them, then somehow pass the recipes on to their offspring.

Study **DNA: The Code of Life** on pages 386–387 and notice three important points and seven new terms:

1 The chemical recipes of life are stored within each cell as information in *DNA (deoxyribonucleic acid)* molecules that resemble a ladder with rungs that are composed of chemical bases. The recipe information is expressed by the sequence of the rungs, providing instructions to guide specific chemical reactions within the cell.

2 The instructions stored in DNA are genetic information passed along to offspring. DNA instructions normally are expressed by being copied into a messenger molecule called *RNA (ribonucleic acid)*. The RNA molecule travels to a location in the cell where its message causes a sequence of molecular units called *amino acids* to be connected into large molecules called *proteins*. Proteins serve as the cell's basic structural molecule or as *enzymes* that control chemical reactions.

How Do We Know? 17-1

The Nature of Scientific Explanation

Must science and religion be in conflict?
Science is a way of understanding the world around you, and at the heart of that understanding are explanations that science gives for natural phenomena. Whether you call these explanations stories, histories, hypotheses, or theories, they are attempts to describe how nature works based on evidence and intellectual honesty. Although you may take these explanations to be factual truth, you can understand that they are not the only explanations offered by humans to describe the Universe.

A separate class of explanations involves religion. For example, scriptural descriptions of creation do not fit well with scientific observations, but they are a way of understanding the Universe nonetheless. Religious explanations are based partly on faith rather than on strict rules of logic and evidence, and it is wrong to demand that they follow the same rules as scientific explanations. In the same way, it is wrong to demand that

scientific explanations take into account religious beliefs. The so-called conflict between science and religion arises when people fail to recognize that science and religion are different ways of knowing about the Universe.

Scientific explanations are compelling because science has been so successful at producing technological innovations that have changed the world you live in. From new vaccines, to digital music players, to telescopes that can observe the most distant galaxies, the products of the scientific process are all around you. Scientific explanations have provided tremendous insights into the workings of nature.

Many people are attracted to the suggestion, made by the late evolutionary biologist Stephen Jay Gould and others, that religious explanations and scientific explanations should be considered as "separate magisteria." In other words, religion and science are devoted to different realms of the mystery of existence.

Science and religion offer differing ways of explaining the Universe, but the two ways follow separate rules and cannot be judged by each other's standards. The trial of Galileo can be understood as a conflict between these two ways of knowing.

Galileo's telescope gave him a new way to know about the Universe.

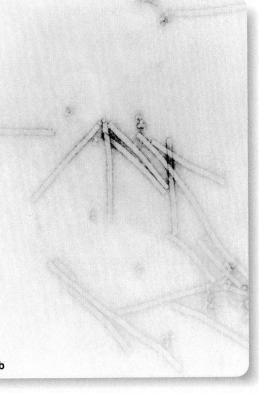

◄ **Figure 17-1** All living things on Earth are based on carbon chemistry. Even the long molecules that carry genetic information, DNA and RNA, have a framework defined by chains of carbon atoms. (a) Jason, a complex mammal, contains more than 200 AU of DNA. (b) Each rodlike tobacco mosaic virus contains a single spiral strand of RNA about 0.01 mm long as its genetic material.

DNA: The Code of Life

1 The key to understanding life is information—the information that guides all of the processes in an organism. In most living things on Earth, that information is stored in a large molecule called **DNA (deoxyribonucleic acid)**.

1a The DNA molecule looks like a spiral ladder with rails made of phosphates and sugars. The rungs of the ladder are made of four chemical bases arranged in pairs. The bases always pair the same way. That is, base A always pairs with base T, and base G always pairs with base C.

1b Information is coded on the DNA molecule by the order in which the base pairs occur. To read that code, molecular biologists have to "sequence" the DNA. That is, they must determine the order in which the base pairs occur along the DNA ladder.

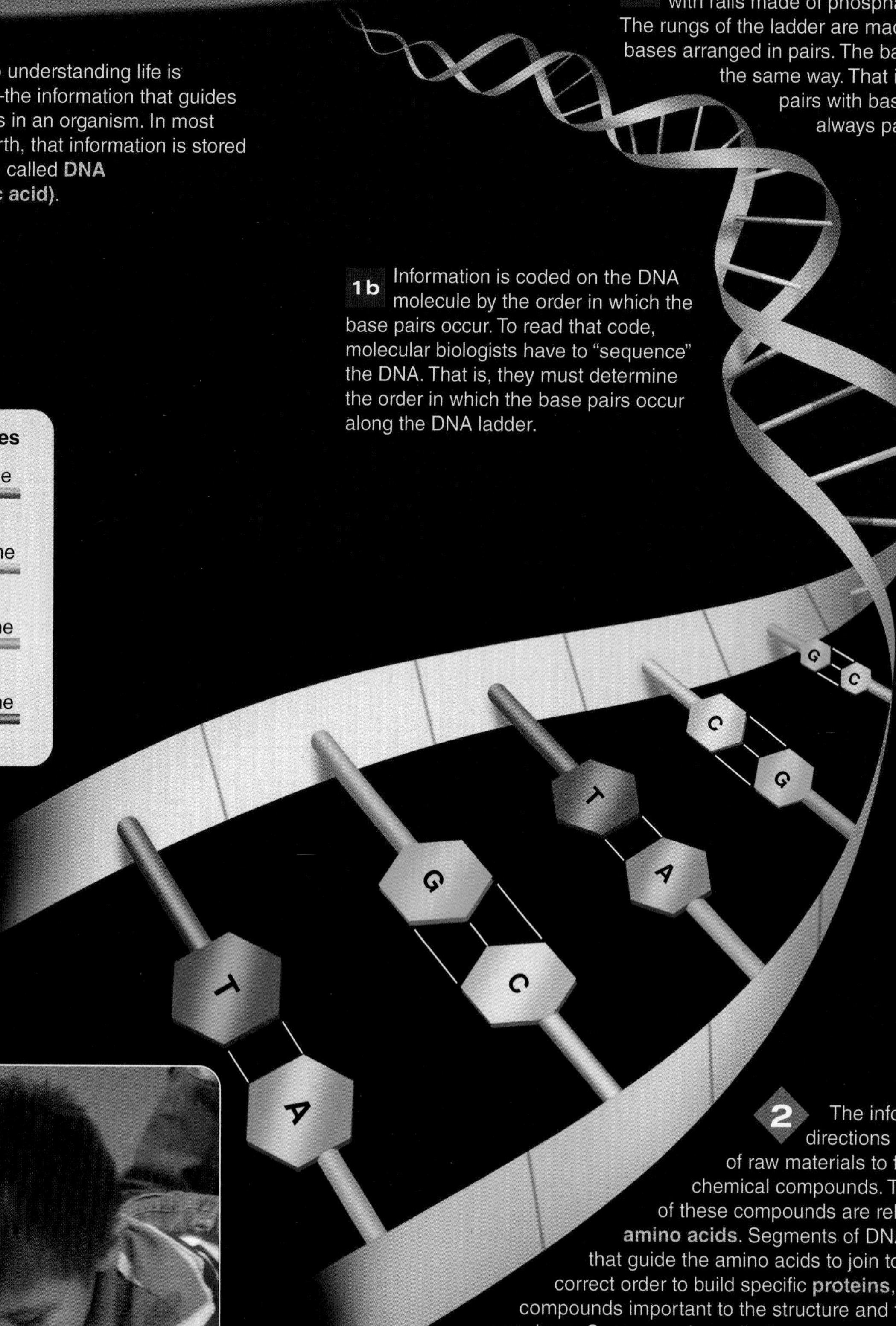

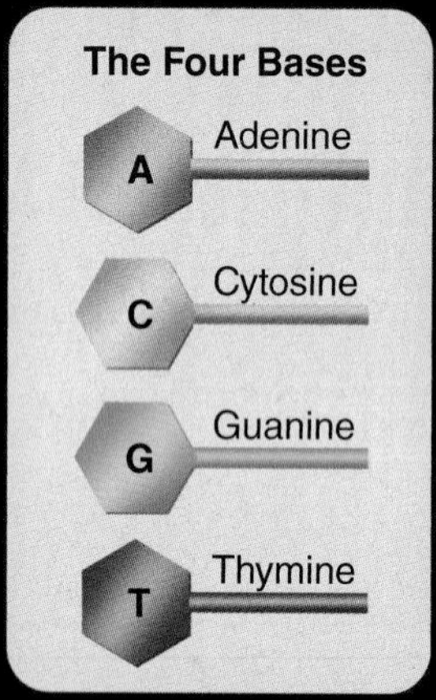

The Four Bases

A — Adenine

C — Cytosine

G — Guanine

T — Thymine

2 The information in DNA is directions for combinations of raw materials to form important chemical compounds. The building blocks of these compounds are relatively simple **amino acids**. Segments of DNA act as templates that guide the amino acids to join together in the correct order to build specific **proteins**, chemical compounds important to the structure and function of organisms. Some proteins called **enzymes** regulate chemical processes. In this way, DNA recipes regulate the production of the compounds of life.

The traits you inherit from your parents, the chemical processes that animate you, and the structure of your body are all encoded in your DNA. When people say, "You have your mother's eyes," they are talking about DNA codes.

Dana E. Backman

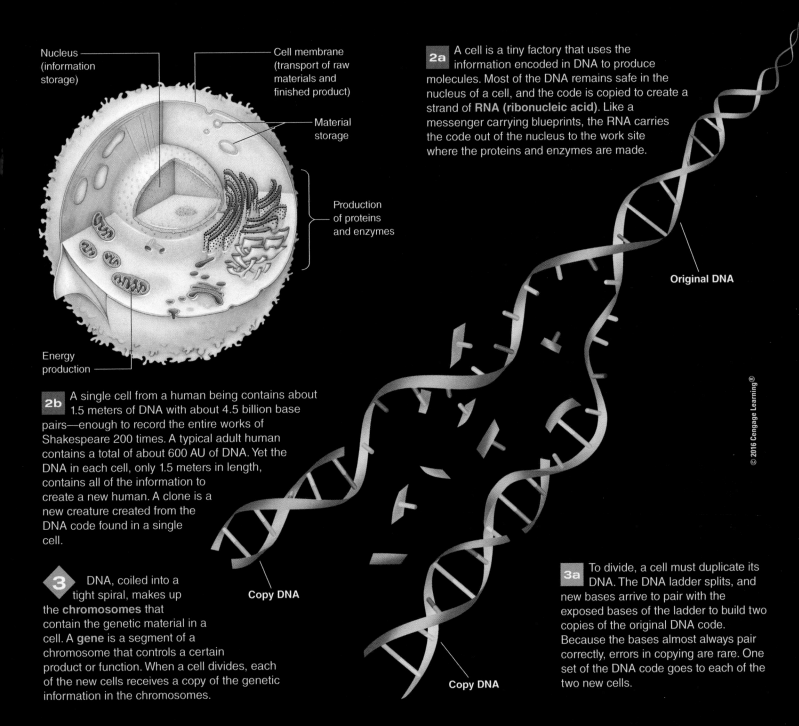

Nucleus
(information
storage)

Cell membrane
(transport of raw
materials and
finished product)

Material
storage

Production
of proteins
and enzymes

Energy
production

2a A cell is a tiny factory that uses the information encoded in DNA to produce molecules. Most of the DNA remains safe in the nucleus of a cell, and the code is copied to create a strand of **RNA (ribonucleic acid)**. Like a messenger carrying blueprints, the RNA carries the code out of the nucleus to the work site where the proteins and enzymes are made.

Original DNA

© 2016 Cengage Learning®

2b A single cell from a human being contains about 1.5 meters of DNA with about 4.5 billion base pairs—enough to record the entire works of Shakespeare 200 times. A typical adult human contains a total of about 600 AU of DNA. Yet the DNA in each cell, only 1.5 meters in length, contains all of the information to create a new human. A clone is a new creature created from the DNA code found in a single cell.

Copy DNA

3 DNA, coiled into a tight spiral, makes up the **chromosomes** that contain the genetic material in a cell. A **gene** is a segment of a chromosome that controls a certain product or function. When a cell divides, each of the new cells receives a copy of the genetic information in the chromosomes.

Copy DNA

3a To divide, a cell must duplicate its DNA. The DNA ladder splits, and new bases arrive to pair with the exposed bases of the ladder to build two copies of the original DNA code. Because the bases almost always pair correctly, errors in copying are rare. One set of the DNA code goes to each of the two new cells.

Cell Reproduction by Division

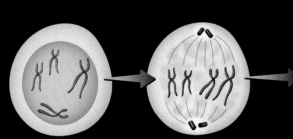
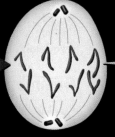

As a cell begins to divide, its DNA duplicates itself.

The duplicated chromosomes move to the middle.

The two sets of chromosomes separate, and . . .

the cell divides to produce . . .

two cells, each containing a full set of the DNA code.

3 The DNA molecule reproduces itself when a cell divides so that each new cell contains a copy of the original information. A sequence of DNA that composes one instruction is called a *gene*. Genes are organized into long coiled chains called *chromosomes*. The genes linked on one chromosome are normally passed on to offspring together.

To produce viable offspring, a cell must be able to make copies of its DNA. Surprisingly, it is important for the continued existence of all life that the copying process includes mistakes.

Modifying the Information

Earth's environment changes continuously. To survive, species must change as their food supply, climate, or environmental conditions change. If the information stored in DNA could not change then life would become extinct. The process by which life adjusts itself to changing environments is called **biological evolution**.

When an organism reproduces, its offspring receive a copy of its DNA. Sometimes external effects such as natural radiation alter the DNA during the parent organism's lifetime, and sometimes mistakes are made in the copying process so that occasionally the copy is slightly different from the original. Such a change is called a **mutation**. Most mutations make no difference, but some mutations are fatal, killing the afflicted organisms before they can reproduce. In rare but vitally important cases, a mutation can actually help an organism survive.

These changes in DNA produce variation among the members of a species, and that variation allows the species to adapt to a changing environment. For example, all of the squirrels in the park may look the same, but they carry a range of genetic variation. Some may have slightly longer tails or faster-growing claws. These variations may make almost no difference until the environment changes. For example, if the environment becomes colder, a squirrel with a heavier coat of fur will, on average, survive longer and produce more offspring than its normal contemporaries. Likewise, the offspring that inherit this beneficial variation will also live longer and have more offspring of their own. In contrast, squirrels containing DNA recipes for thin fur coats will gradually decrease in number.

These differing rates of survival and reproduction are examples of **natural selection**. Over time, the beneficial variation increases in frequency, and a species can evolve until the entire population shares the trait. In this way, natural selection adapts species to their changing environments by selecting, from the huge array of random variations, those that would most benefit the survival of the species.

It is a **Common Misconception** that evolution is random, but that is not true. The underlying mechanisms creating variation within each species may be random, but natural selection is not random because progressive changes in a species are directed by changes in the environment.

DOING SCIENCE

Why is it important that errors occur in copying DNA? Sometimes the most valuable questions for a scientist to ask are those that challenge what would appear to be common sense.

It seems obvious that mistakes shouldn't be made in copying DNA, but in fact variation is necessary for long-term survival of a species. For example, the DNA in a starfish contains all the information the starfish needs to grow, develop, survive, and reproduce. The information must be passed on to the starfish's offspring for them to survive. That information needs to change, however, if the environment changes. A change in the ocean's temperature may kill the specific shellfish that the starfish eat. If the starfish are unable to digest another kind of food—if all the starfish have exactly the same DNA—they all will die. But if a few starfish have slightly different DNA that gives them the ability to make enzymes that can digest a different kind of shellfish, the species may be able to carry on.

Variations in DNA are caused both by external factors such as natural radiation and by occasional mistakes in the copying process. The survival of life depends on this delicate balance between mostly reliable reproduction and the introduction of small variations in DNA.

Now ponder the opposite question: ***Why does the DNA copying process need to be mostly reliable?***

17-2 Life in the Universe

Life as we know it consists of just the single example of life on Earth. It is OK to think of all life on Earth as being just a single type of life because, as you learned in the previous section, all living things on Earth have the same physical basis: the same chemistry and the same genetic code alphabet. How life began on Earth and then developed and evolved into its present variety is the only solid information you have to work with, when considering what might be possible on other worlds.

Everything currently known about life on Earth indicates that the same natural processes should lead to the origin of life on some fraction of other planets with liquid water. If there is life on other worlds, does it use DNA and RNA to carry the information for life processes, different molecules playing the same role, or some radically different scheme? There is no way to know unless another example of life is found on another world. If and when that day comes, even if the non-Earthly life is simple one-celled organisms, the discovery will be one of the most important in the history of science. It will complete the journey of human understanding begun in the Copernican revolution of progressive realizations that Earth is not unique.

Origin of Life on Earth

It is obvious that the 4.5 billion chemical bases that make up human DNA did not come together in the right order just by chance. The key to understanding the origin of life lies in

---

picturing the processes of evolution running "backward." The complex interplay of environmental factors with the DNA of generation after generation of organisms drove some life-forms to become more sophisticated over time, until they became the unique and specialized creatures alive today. Imagining this process in reverse leads to the idea that life on Earth began with simple forms.

Biologists hypothesize that the first living things would have been carbon-chain molecules able to copy themselves. Of course, this is a scientific hypothesis for which you can seek evidence. What evidence exists regarding the origin of life on Earth?

The oldest fossils are the remains of sea creatures, and this indicates that life began in the sea. Identifying the oldest fossils is not easy, however. Ancient rocks from western Australia that are at least 3.4 billion years old contain matlike features that biologists identify as **stromatolites**, fossilized remains of colonies of single-celled organisms that built up layer after layer of trapped sediments (Figure 17-2). Fossils this old are difficult to recognize because the earliest living things did not contain easily preserved hard parts like bones or shells and because the individual organisms were microscopic. Thus, the evidence, although

scarce, indicates that simple organisms lived in Earth's oceans less than 1.2 billion years after Earth formed. Stromatolite colonies of microorganisms are more complex than individual cells, so you can imagine there probably were earlier, simpler organisms. How did those first simplest organisms originate?

An important experiment performed by Stanley Miller and Harold Urey in 1952 sought to recreate the presumed conditions in which life on Earth began. The **Miller experiment** consisted of a sterile, sealed glass container holding water, hydrogen, ammonia, and methane, thought to resemble the young Earth's atmosphere. An electric arc inside the apparatus made sparks to simulate the effects of lightning (Figure 17-3).

Miller and Urey let the experiment run for a week and then analyzed the material inside. They found that the interaction between the electric arc and the simulated atmosphere had produced many organic molecules from the raw material of the experiment, including such important building blocks of life as amino acids. (Recall that an organic molecule is simply a molecule with a carbon-chain structure and need not be derived from a living thing: "Organic" does not necessarily imply "biological.")

▲ Figure 17-2 (a) A fossil stromatolite from western Australia that is more than 3 billion years old, some of the oldest evidence of life on Earth. Stromatolites are formed, layer on layer, by mats of bacteria living in shallow water where they are covered repeatedly by sediments. (b) Artist's conception of a scene on the young Earth, more than 3 billion years ago, with stromatolite bacterial mats growing near the shore of an ocean.

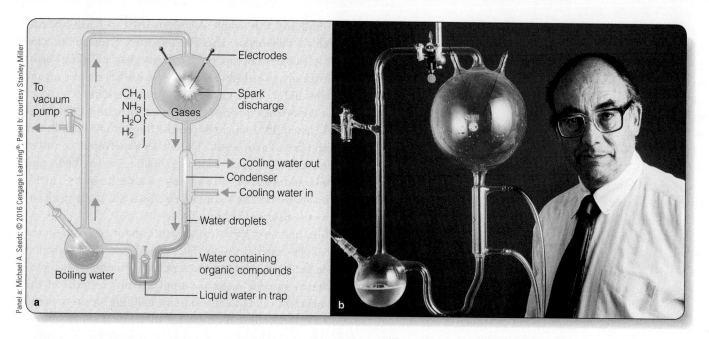

Panel a: Michael A. Seeds; © 2016 Cengage Learning®. Panel b: courtesy Stanley Miller

▲ **Figure 17-3** (a) The Miller experiment circulated gases through water in the presence of an electric arc. This simulation of primitive conditions on Earth produced many complex organic molecules, including amino acids, the building blocks of proteins. (b) Stanley Miller with a Miller apparatus.

When the experiment was run again using different energy sources such as hot silica to represent molten lava spilling into the ocean, similar molecules were produced. Even a light source representing the amount of ultraviolet (UV) radiation in sunlight was sufficient to produce complex organic molecules.

Scientists are professionally skeptical about scientific findings (see "How Do We Know?" 10-2, page 217), and they have reevaluated the Miller–Urey experiment in light of new information. According to updated models of the formation of the Solar System and Earth (look back to Chapters 10 and 11), Earth's early atmosphere probably consisted mostly of carbon dioxide, nitrogen, and water vapor instead of the mix of hydrogen, ammonia, methane, and water vapor assumed by Miller and Urey. When gases corresponding to the newer understanding of the early Earth atmosphere are processed in a Miller apparatus, lesser, but still significant, amounts of organic molecules are produced.

The Miller experiment is important because it shows that complex organic molecules form naturally in a wide variety of circumstances. Lightning, sunlight, and hot lava are just some of the energy sources that can naturally rearrange simple common molecules into the complex molecules that make life possible. If you could travel back in time, you would expect to find Earth's early oceans filled with a rich mixture of organic compounds called the **primordial soup**.

Many of these organic compounds would have been able to link up to form larger molecules. Amino acids, for example, link together naturally to form proteins by joining ends and releasing a water molecule (**Figure 17-4**). That reaction, however, does not proceed easily in a water solution. Scientists hypothesize that this step may have been more likely to happen on shorelines or in Sun-warmed tidal pools where organic molecules from the primordial soup could have been concentrated by water evaporation.

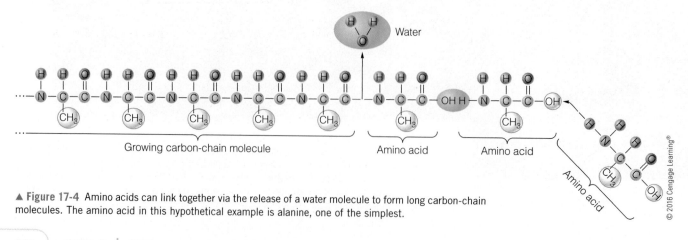

▲ **Figure 17-4** Amino acids can link together via the release of a water molecule to form long carbon-chain molecules. The amino acid in this hypothetical example is alanine, one of the simplest.

The production of large organic molecules may have been aided in such semidry environments by being absorbed by clay crystals that could have acted as templates holding the organic subunits close together.

These complex organic molecules were still not living things. Even though some proteins may have contained hundreds of amino acids, they did not reproduce but rather linked and broke apart at random. Because some molecules are more stable than others, and some bond together more easily than others, scientists hypothesize that a process of **chemical evolution** eventually concentrated the various smaller molecules into the most stable larger forms. Eventually, according to the hypothesis, somewhere in the oceans, after sufficient time, a molecule formed that automatically copied itself, as DNA and RNA are able to do under the right circumstances. At that point, the natural selection and chemical evolution of molecules became the biological evolution of living things.

An alternate theory for the origin of life proposes that reproducing molecules may have arrived here from space. Astronomers have found a wide variety of organic molecules in the interstellar medium, and similar compounds have been found inside meteorites (**Figure 17-5**). The Miller experiment showed how easy it is for complex organic molecules to form naturally from simpler compounds, so it is not surprising to find them in space. Although speculation is fun, the hypothesis that life arrived on Earth from space is presently more difficult to test than the hypothesis that Earth's life originated on Earth.

Whether the first reproducing molecules formed here on Earth or in space, the important thing is that they could have formed by natural processes. Scientists know enough about those processes to feel confident about them, even though some of the steps remain uncertain.

The details of the origin of the first cells are unknown. The structure of cells may have arisen automatically because of the way molecules interact during chemical evolution. If a dry mixture of amino acids is heated, the acids form long, proteinlike molecules that, when poured into water, collect to form microscopic spheres that behave in ways similar to cells (**Figure 17-6**). They have a thin membrane surface, they absorb material from their surroundings, they grow in size, and they divide and bud just as cells do. However, they contain no large molecule that copies itself, so they are not alive. The first reproducing molecule that was surrounded by a protective membrane, resulting in the first cell, would have gained an important survival advantage over other reproducing molecules.

Geologic Time and the Evolution of Life

Biologists infer that the first cells must have been simple, single-celled organisms similar to modern bacteria. As you learned previously, evidence of the presence of these kinds of cells are preserved in stromatolites (see Figure 17-2), mineral formations produced by layers of bacteria and shallow ocean sediments. Stromatolite fossils are found in rocks with radioactive ages of 3.4 billion years, and living stromatolites still form in some places today.

Stromatolites and other photosynthetic organisms would have begun adding oxygen, a product of photosynthesis, to Earth's early atmosphere. Oxygen tends to disappear from the atmosphere almost as soon as it is released because it readily combines with iron in the soil and ocean water. Geological evidence indicates that Earth's surface iron became saturated with oxygen about 2 to 2.5 billion years ago, after which the proportion of oxygen in the atmosphere began steadily increasing. Oxygen metabolism produces much more energy per mass of food than other reactions, and biologists speculate that this greater efficiency allowed for the development of multicelled organisms at about that same time. Also, an oxygen abundance of only 0.1 percent would have created an ozone screen, protecting organisms from the Sun's UV radiation and later allowing life to colonize the land.

Over the course of eons, the natural processes of evolution gave rise to stunningly complex **multicellular** life-forms with their own widely differing ways of life. It is a Common Misconception to imagine that life is too complex to have evolved from such simple beginnings. It is possible because small variations can accumulate, although that accumulation requires great amounts of time.

▲ **Figure 17-5** A piece of the Murchison meteorite, a carbonaceous chondrite (look back to Figures 16-2d and 16-4) that fell near Murchison, Australia, in 1969. Analysis of its interior revealed the presence of amino acids. Whether the first chemical building blocks of life on Earth originated in space is a matter of debate, but the amino acids found in meteorites illustrate how commonly amino acids and other complex organic molecules occur in the Universe, even in the absence of living things.

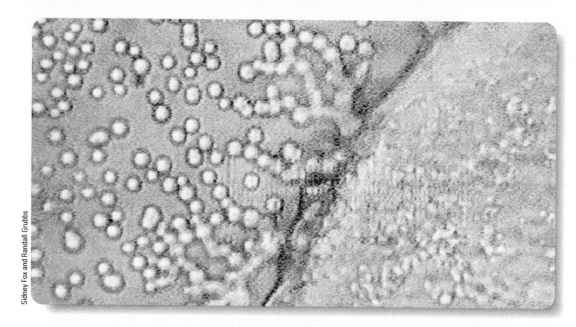

▲ Figure 17-6 Single amino acids can be assembled into long proteinlike molecules. When such material cools in water, it often forms microspheres, tiny globules with double-layered boundaries similar to cell membranes. Microspheres may have been an intermediate stage in the evolution of life between complex but nonliving molecules and living cells holding molecules reproducing genetic information.

There is little evidence of anything more than simple organisms on Earth until about 540 million years ago, almost 3 billion years after the earliest signs of life, at which time fossil evidence indicates that life suddenly developing into a wide variety of complex forms such as the trilobites (Figure 17-7). This sudden increase in biological complexity is known as the **Cambrian explosion**, at the beginning of what geologists refer to as the Cambrian period.

If you represented the entire history of Earth on a scale diagram, the Cambrian explosion would be near the top of the column, as shown at the left of Figure 17-8. The emergence of most animals familiar to you today, including fishes, amphibians, reptiles, birds, and mammals, would be crammed into the topmost part of the chart, above the Cambrian explosion.

If you magnify that portion of the diagram, as shown on the right side of Figure 17-8, you can get a better idea of when these events occurred in the history of life. Humanoid creatures have walked on Earth for about 4 million years. This is a long time by the standard of a human lifetime, but it makes only a narrow red line at the top of the diagram. All of recorded history would be a microscopically thin line at the very top of the column.

▲ Figure 17-7 (a) Trilobites made their first appearance in the Cambrian oceans. The smallest were almost microscopic, and the largest were bigger than dinner plates. This example, about the size of your hand, lived 400 million years ago in an ocean floor that is now a limestone deposit in Pennsylvania. (b) In this artist's conception of a Cambrian sea bottom, Anomalocaris (center-rear and looming at upper right) had specialized organs including eyes, coordinated fins, gripping mandibles, and a powerful, toothed maw. Notice Opabinia at center right with its long snout.

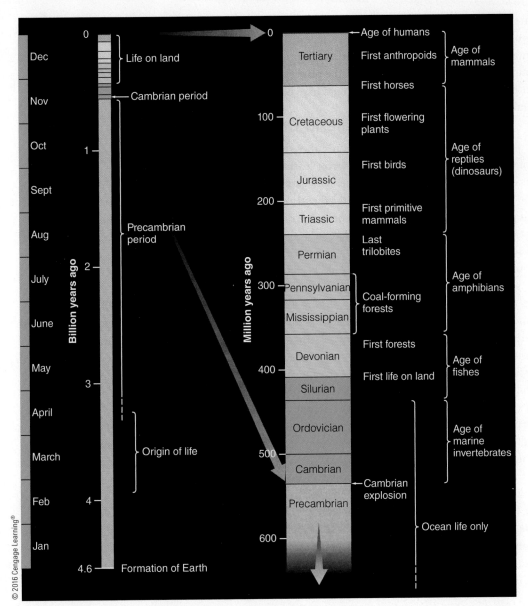

◀ **Figure 17-8** Complex life has developed on Earth only recently. If the entire history of Earth were represented in a time line (*left*), you have to examine the end of the line closely to see details such as life leaving the oceans and dinosaurs appearing. Even on a time line expanded in scale by a factor of about 7 (right column) the age of humans is still only a thin line at the top of the diagram. If the history of Earth were a yearlong video, humans would not appear until the last hours of December 31.

To understand just how thin that line is, imagine that the entire 4.6-billion-year history of the Earth has been compressed onto a yearlong video and that you began watching this video on January 1. You would not see any signs of life until March or early April, and the slow evolution of the first simple forms would take the next six or seven months. Suddenly, in mid-November, you would see the trilobites and other complex organisms of the Cambrian explosion.

You would see no life of any kind on land until the end of November, but once life appeared it would diversify quickly, and by December 12 you would see dinosaurs walking the continents. By the day after Christmas they would be gone, and mammals and birds would be on the rise.

If you watched closely, you might see the first humanoid forms by late afternoon on New Year's Eve, and by late evening you could see humans making the first stone tools. The Stone Age would last until 11:59 pm, after which the first towns, and then cities, would appear. Suddenly things would begin to happen at lightning speed. Babylon would flourish, the pyramids would rise, and Troy would fall. The Christian era would begin 14 seconds before the New Year. Rome would fall, and then the Middle Ages and the Renaissance would flicker past. The American and French revolutions would occur one-and-a-half seconds before the end of the video.

By imagining the history of Earth as a yearlong video, you can gain some perspective on the rise of life. Tremendous amounts of time were needed for the first simple living things to evolve in the oceans. As life became more complex, new forms arose more and more quickly as the hardest problems—how to reproduce, how to take energy efficiently from the environment, how to move around—were "solved" by the process of biological evolution. The easier problems, like what to eat, where to live, and how to raise young, were managed in different ways by different organisms, leading to the diversity that is seen today.

Intelligence—that which appears to set humans apart from other animals—may be a unique solution to an evolutionary problem posed to humanity's ancient ancestors. A smart animal is better able to escape predators, outwit its prey, and feed and shelter itself and its offspring. Under certain conditions it seems plausible that evolution might naturally select for intelligence.

Extremophiles

It is difficult to pin down a range of environments and be sure that life based on carbon and water cannot exist outside those conditions, so long as there is even occasionally some liquid water present. Life has been found in places on Earth previously judged inhospitable, such as the bottoms of ice-covered lakes in Antarctica, far underground inside solid rock, among the cinders at the summits of extinct volcanoes, and in pools of acid (Figure 17-9). An organism that can survive and even thrive in what humans consider an extreme environment is called an **extremophile**. Maybe you have friends like that.

Linguists can figure out the vocabulary of the long-vanished Indo-European language by comparing words in modern languages such as English, Spanish, Russian, Greek, and Hindi that evolved from it. Much the same way, biologists can work out the DNA sequences of ancient species by comparing the sequences of their present-day descendants. This type of analysis indicates that the organism that was ancestor to all life on Earth today most closely resembled present-day simple single-celled organisms known as **archaea**. (The word *archaea* comes from the Greek root *archaios*, meaning ancient.) The most likely common ancestor was an archaea-like extremophile tolerant of high temperatures called a **thermophile** ("heat loving"; look at the photo that opens this chapter, on page 383). Some biologists think this is evidence that life began near volcanic vents on the seafloor or in hot rock deep underground. Others suggest that the heavy bombardment during the end of the Solar System's formation (Chapters 10 and 15) would have repeatedly boiled much of Earth's oceans away. If life had already begun by then, the only organisms that survived this phase of the planet's history to become our ancestors by natural selection would have been heat resistant.

Life in Our Solar System

Could there be carbon-based life elsewhere in our Solar System? As you learned previously, liquid water seems to be a requirement of carbon-based life, necessary both as the medium for vital chemical reactions and to transport nutrients and wastes. It is not surprising that life developed in Earth's oceans and stayed there for billions of years before it was able to colonize the land. On the other hand, scientists searching for life on other worlds should keep in mind Earth's extremophiles and the harsh conditions in which they thrive.

Many worlds in the Solar System can be eliminated immediately as hosts for water-based life because liquid water is not possible there. The Moon and Mercury are airless, and water would boil away into space immediately. Venus has traces of water vapor in its atmosphere, but it is too hot for liquid water to survive on the surface. The Jovian planets have deep atmospheres, and at a certain altitude it is likely that water condenses into liquid droplets. However, it seems unlikely that life could have originated there. The Jovian planets do not have solid surfaces (Chapters 14 and 15), so isolated water droplets cannot mingle as they did in Earth's primordial oceans, where organic molecules were able to grow and interact. In addition, powerful downdraft currents in the atmospheres of the giant planets would quickly carry any reproducing molecules that did form there into inhospitably hot lower regions.

As you learned in Chapter 14, Jupiter's moon Europa appears to have a liquid-water ocean below its icy crust, and

Panel a: Kris Koenig (Coast Learning Systems); Panel b: © catolla/Shutterstock.com

▲ **Figure 17-9** Every life form on Earth has evolved to survive in some ecological niche. (a) Wekiu bugs live with the astronomers at an altitude of 13,800 feet atop the Hawaiian volcano Mauna Kea. The bugs inhabit spaces between the icy cinders and eat insects carried up by ocean breezes. (b) Rio Tinto in Spain hosts a population of eukaryotes and prokaryotes adapted to an extreme environment with high concentrations of acid (pH less than 2) and heavy metals.

▶ **Figure 17-10** About the size of a compact car, a *Viking* lander model sits in a simulated Martian environment on Earth. A claw for grabbing rock and soil samples is in the foreground at the end of the black arm. In 1976, the two *Viking* landers made measurements possibly indicating some sort of microbial activity in the Martian dirt. (b) Meteorite ALH 84001 is one of a dozen meteorites known to have originated on Mars. Its name means this meteorite was the first one found in 1984 near Antarctica's Allan Hills. (c) A group of researchers claimed that ALH 84001 contained chemical and physical traces of ancient life on Mars, including what appear to be fossils of microscopic organisms. That evidence has not been confirmed, and the claim continues to be tested and debated.

minerals dissolved in the water could provide a source of raw material for chemical evolution. Europa's ocean is kept warm and liquid by tidal heating. There also may be liquid-water layers under the surfaces of Ganymede and Callisto. That can change as the moons interact gravitationally and their orbits change; Europa, Ganymede, and Callisto may have been frozen solid at other times in their histories, which would probably have destroyed any living organism that had developed there.

Saturn's moon Titan is rich in organic molecules. You learned in Chapter 14 (pages 315–317) that sunlight converts the methane in Titan's atmosphere into organic smog particles that settle to the surface. The chemistry of life that could have evolved from those molecules and survived in Titan's lakes of methane is unknown. It is fascinating to consider possibilities, but Titan's extremely low temperature of −180°C (−290°F) would make chemical reactions so slow that life processes seem unlikely.

Water containing organic molecules has been observed venting from the south polar region of Saturn's moon Enceladus (Chapter 14, page 318). It is possible that life could exist in that water under Enceladus's crust, but the moon is small, and its tidal heating might operate only occasionally. Enceladus may not have had plentiful liquid water for the extended time necessary for the rise of life.

Aside from Earth, Mars is the most likely place for life to exist in the Solar System because, as you learned in Chapter 13, there is a great deal of evidence that liquid water once flowed on its surface. Even so, results from searches for signs of life on Mars are not encouraging. The robotic spacecraft *Viking 1* and *Viking 2* landed on Mars in 1976 and tested soil samples for living organisms (**Figure 17-10a**). Some of the tests had puzzling semipositive results that scientists now hypothesize were caused by nonbiological chemical reactions in the soil. No evidence clearly indicates the presence of life or even of organic molecules in the Martian soil.

Previously in this chapter you learned that life may have required special circumstances to start on Earth, but once it started, biological evolution allowed life to spread across Earth and adapt to a wide range of conditions. Eventually all niches—even extreme environments—became occupied. Most astrobiologists think this means that, if life begins on a planet, even if

Panel a: NASA/JPL-Caltech/University of Arizona; Panel b and c: NASA JSC/Stanford University

the entire environment of the planet later becomes inhospitable, some life could continue to survive. If life still exists on Mars, it may be hidden below ground where there may be liquid water and where UV radiation from the Sun cannot penetrate.

There was a splash of news stories in the 1990s regarding supposed chemical and physical traces of past life on Mars discovered inside a Martian meteorite found in Antarctica (Figure 17-10b). Scientists were excited by the announcement, but they employed professional skepticism and immediately began testing the evidence. Their results suggest that the unusual chemical signatures in the rock may have formed by processes that did not involve life. Tiny features in the rock that were originally thought to be fossils of ancient Martian microorganisms could possibly be nonbiological mineral formations instead (Figure 17-10c). Although measurements by the *Curiosity* rover have proven conclusively that Mars once had an environment that could have supported Earth-like life, evidence of life on Mars may have to wait until a future rover drills into the soil and discovers signs of metabolizing organisms, or a geologist astronaut scrambles down dry Martian streambeds and cracks rocks open to find fossils.

There is presently no compelling evidence for the existence of life in the Solar System other than on Earth. But that means your search can now take you to distant planetary systems.

Life in Other Planetary Systems

Could life exist in other planetary systems? You already know that there are many different kinds of stars and that many of these stars have planetary systems. As a first step toward answering this question, you can try to identify the kinds of stars that seem most likely to have stable planetary systems where life could evolve.

If a planet is to be a suitable home for living things, it must be in a stable orbit around its sun. That is easy in a planetary system like our own, but planet orbits in binary star systems would be unstable unless the component stars are very close together or very far apart. Astronomers can calculate that, in binary systems with stars separated by intermediate distances of a few AU, the planets should eventually be swallowed up by one of the stars or ejected from the system. Half the stars in the Milky Way Galaxy are members of binary systems, and many of them are unlikely to support life on planets.

Moreover, just because a star is single does not necessarily make it a good candidate for sustaining life. Earth required perhaps as much as 1 billion years to produce the first cells and 4.6 billion years for intelligence to emerge. Massive stars that shine for only a few million years do not meet this criterion. If the history of life on Earth is representative, then stars more massive and luminous than about spectral type F5 last too short a time for complex life to develop. Main-sequence stars of types G and K, and possibly some of the M stars, are the best candidates.

The temperature of a planet is also important, and that depends on the type of star it orbits and its distance from the star. Astronomers have defined a **habitable zone** around a star as a region within which planets that orbit there have temperatures permitting the existence of liquid water. The Sun's habitable zone extends from near the orbit of Venus to the orbit of Mars, with Earth right in the middle. A low-luminosity star has a small and narrow habitable zone, whereas a high-luminosity star has a large and wide one.

Stable planets inside the habitable zones of long-lived stars are the places where life seems most likely, but given the tenacity and resilience of Earth's life forms, there might be other, seemingly inhospitable, places in the Universe where life exists. You should also note that two of the environments considered as possible havens for life—Jupiter's moon Europa and Saturn's moon Enceladus—are in the outer Solar System, far outside the Sun's habitable zone. Those moons have liquid water under their surfaces because of tidal heating due to gravitational interactions with their giant parent planets, a situation that can occur at any distance from a star. Europa and Enceladus show that the conventional definition for habitable zone is probably too limiting.

DOING SCIENCE

What evidence indicates that life is possible on other worlds? Almost all scientists assume that life exists on other worlds. But that is just an assumption. What is the evidence?

Biologists have imagined, and reproduced in the laboratory, likely physical and chemical processes that, over long time intervals, could have changed simple organic compounds into reproducing molecules inside membranes, the first simple life-forms. Fossil evidence, although meager, indicates that life originated in the oceans at least 3.4 billion years ago, soon after the end of the Solar System's heavy bombardment. That indicates to most astrobiologists that, if conditions are right, life begins on a planet relatively quickly. Finally, evidence indicates that Earth-like planets are common in the Universe.

Now carry the question a little further: *What conditions do you expect on worlds that host life?*

17-3 Intelligent Life in the Universe

Could intelligent life arise on other worlds? To try to answer this question, you can estimate the chances of any type of life arising on other worlds and then assess the likelihood of that life developing intelligence. If other civilizations exist, it is possible humans eventually may be able to communicate with them. Nature puts restrictions on the pace of such conversations, but the main problem lies in the unknown life expectancy of civilizations.

Travel Between the Stars

The distances between stars are almost beyond comprehension. The fastest human device ever launched, the *New Horizons* probe currently on its way to Pluto and the Kuiper Belt

How Do We Know? 17-2

UFOs and Space Aliens

Has Earth been visited by aliens? Astronomers, planetary scientists, and astrobiologists get asked this question all the time by members of the public. The reason the question is asked so often is that the public has heard that most scientists believe there is life on other worlds. So then, the logic goes, UFOs are probably alien spacecraft, right? Scientists don't make that connection for two reasons.

First, the reputation of UFO sightings and alien encounters does not inspire confidence that those data are reliable. Most people hear of such events in grocery store tabloids, daytime talk shows, or sensational "specials" on viewer-hungry cable networks. You should take note of the low reputation and motivations of the media that report UFOs and space aliens. Most of these reports, like the reports that Elvis is alive and well, are simply made up for the sake of sensation or to make money, and you cannot use them as reliable evidence.

Second, the few UFO sightings that are not made up do not survive careful examination. Most are mistakes and unintentional misinterpretations, committed by honest people, of natural events or human-made objects. It is important to realize that experts have studied these incidents over many decades and found *not even one report* that is convincing to the professional scientific community.

In short, despite false claims to the contrary on TV shows, there is no dependable evidence that Earth has ever been visited by aliens. Remember, that conclusion does not come from a prejudice on the part of scientists against the idea of extraterrestrial life. Most scientists assume there is life on other worlds but are aware that there is no believable evidence for any such life visiting Earth.

In a way, that's too bad. A confirmed visit by intelligent creatures from beyond our Solar System would answer many questions. It would be exciting, enlightening, and, like any real adventure, a bit scary. Most scientists would love to be part of such a discovery. But scientists must professionally pay attention to what is supported by evidence rather than what might be thrilling. There is not yet any direct evidence of even microbial life on other worlds, never mind intelligent extraterrestrial life visiting Earth.

Michael A. Seeds

Flying saucers from space are fun to think about, but there is no evidence that they are real.

(Chapter 15, page 328), will take about 90,000 years to travel 4 light-years, the distance to the nearest star, Proxima Centauri. The obvious way to overcome these huge distances is with tremendously fast spaceships, but even the closest stars are many light-years away.

Nothing can exceed the speed of light, and accelerating a spaceship close to the speed of light takes huge amounts of energy. Even if you travel slower than light, your rocket would still require massive amounts of fuel. If you wanted to pilot a spaceship with a mass of 100 tons (about the size of a fancy yacht) to Proxima Centauri, and you traveled at half the speed of light so as to arrive in 8 years, the trip would require 400 times as much energy as the entire United States consumes in a year. Don't even think about how much fuel the starship *Enterprise* needs.

These limitations not only make it difficult for humans to leave the Solar System, but they would also make it difficult for aliens to visit Earth. Reputable scientists have studied unidentified flying objects (UFOs) and have never found any evidence that Earth is being visited or has ever been visited by aliens (**How Do We Know? 17-2**). Humans are unlikely ever to meet aliens face to face. However, communication by electromagnetic signals across interstellar distances takes relatively little energy.

Radio Communication

Nature puts restrictions on travel through space, and it also restricts the possibility of communicating with distant civilizations by radio. One restriction is based on simple physics: Radio signals are electromagnetic waves and travel at the speed of light. Because of the distances between the stars, the speed of radio waves would severely limit humanity's ability to carry on normal conversations with distant civilizations. Decades could elapse between asking a question and getting an answer.

So rather than try to begin a conversation, one group of astronomers decided in 1974 to broadcast a message of greeting toward the globular cluster M13, 22,000 light-years away, using the Arecibo radio telescope (look back to Figure 6-17). When the signal arrives 22,000 years in the future, alien astronomers may be able to understand it because the message is **anticoded**, meaning that it is designed to be decoded by beings that know nothing about us or our languages. If they are sophisticated enough to build radio telescopes, the hope is that they should be able to decode the transmission. The message is a string of 1679 pulses and gaps. Pulses represent 1s, and gaps represent 0s. The string can be arranged in two dimensions in only two possible ways: as 23 rows of 73 or as 73 rows of 23. The first way produces

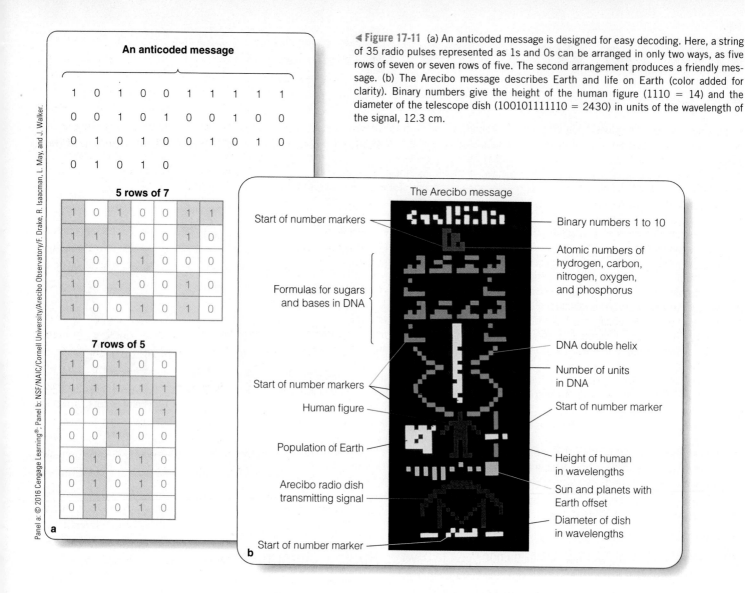

An anticoded message

1	0	1	0	0	1	1	1	1	1
0	0	1	0	1	0	0	1	0	0
0	1	0	1	0	0	1	0	1	0
0	1	0	1	0					

5 rows of 7

1	0	1	0	0	1	1
1	1	1	0	0	1	0
1	0	0	1	0	0	0
1	0	1	0	0	1	0
1	0	0	1	0	1	0

7 rows of 5

1	0	1	0	0
1	1	1	1	1
0	0	1	0	1
0	0	1	0	0
0	1	0	1	0
0	1	0	1	0
0	1	0	1	0

a

The Arecibo message

Start of number markers — Binary numbers 1 to 10

— Atomic numbers of hydrogen, carbon, nitrogen, oxygen, and phosphorus

Formulas for sugars and bases in DNA

— DNA double helix

— Number of units in DNA

Start of number markers

Human figure — Start of number marker

Population of Earth — Height of human in wavelengths

— Sun and planets with Earth offset

Arecibo radio dish transmitting signal — Diameter of dish in wavelengths

Start of number marker

b

◄ **Figure 17-11** (a) An anticoded message is designed for easy decoding. Here, a string of 35 radio pulses represented as 1s and 0s can be arranged in only two ways, as five rows of seven or seven rows of five. The second arrangement produces a friendly message. (b) The Arecibo message describes Earth and life on Earth (color added for clarity). Binary numbers give the height of the human figure (1110 = 14) and the diameter of the telescope dish (100101111110 = 2430) in units of the wavelength of the signal, 12.3 cm.

gibberish, but the second arrangement forms a picture containing information about life on Earth (**Figure 17-11**).

What are the chances that a signal like the Arecibo message would be heard across interstellar distances? Surprisingly, a radio dish the size of the Arecibo telescope, located anywhere in the Milky Way Galaxy, could detect the output from "our" Arecibo. The human race's modest technical capabilities already can put us into cosmic chat rooms.

Although the 1974 Arecibo beacon was the only powerful signal sent purposely from Earth to other star systems, Earth is sending out many other signals more or less accidentally. Shortwave radio signals, including TV and FM, have been leaking into space for the past 60 years or so. Any civilization within 60 light-years could already have detected Earth's civilization. That works both ways: Alien signals, whether intentional messages of friendship or the blather of their equivalent to daytime TV, could be arriving at Earth now. Groups of astronomers from several countries are pointing radio telescopes at the most likely stars and listening for alien civilizations.

Which channels should astronomers monitor? Signals with wavelengths longer than 30 cm would get lost in the background noise of our Milky Way Galaxy, whereas wavelengths shorter than about 1 cm are mostly absorbed in Earth's atmosphere. Between those wavelengths is a radio window that is open for communication. Even this restricted window contains millions of possible radio-frequency bands and is too wide to monitor easily, but astronomers may have thought of a way to narrow the search. Within this broad radio window lie the 21-cm spectral line of neutral hydrogen and the 18-cm line of hydroxyl (OH) (**Figure 17-12**). The interval between those lines has especially low background interference and is named the **water hole** because H plus OH yields water. Any civilizations sophisticated enough to do radio astronomy would know of these lines and might appreciate their significance in the same way as do Earthlings.

A number of searches for extraterrestrial radio signals have been made, and some are now under way. This field of study is known as **Search for Extra-Terrestrial Intelligence (SETI)**, and it has generated heated debate among astronomers, philosophers,

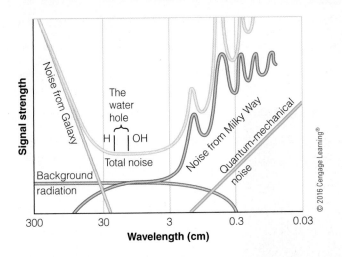

Signal strength

Noise from Galaxy

The water hole

H | OH

Total noise

Noise from Milky Way

Quantum-mechanical noise

Background radiation

300   30   3   0.3   0.03

Wavelength (cm)

© 2016 Cengage Learning®

▲ **Figure 17-12** Radio noise from various astronomical sources and Earth's atmospheric opacity make it difficult to detect distant signals at wavelengths longer than 30 cm or shorter than 1 cm. Within the low-noise range, the wavelengths of radio emission lines from H atoms and from OH molecules mark a small interval named the water hole that could be an agreeable channel for interstellar communication.

theologians, and politicians. You might imagine that the discovery of real alien intelligence would cause a huge change in humanity's worldview, akin to Galileo's discovery that the moons of Jupiter do not go around Earth. Congress funded a NASA SETI search for a short time but ended support in the early 1990s. In fact, the annual cost of a major search is only about as much as a single Air Force helicopter, but much of the reluctance to fund searches stems from issues other than cost. Segments of the population, including some members of Congress, considered the idea of extraterrestrial beings to be so outlandish that continued public funding for the search became impossible.

Despite the controversy, the search continues. The NASA SETI project canceled by Congress was renamed Project Phoenix and completed using private funds. The SETI Institute, founded in 1984, managed Project Phoenix plus several other important searches and is currently building a new radio telescope array in northern California, in collaboration with the University of California–Berkeley and partly funded by Paul Allen, one of the cofounders of Microsoft (Figure 17-13).

There is even a way for you to help with searches. The Berkeley SETI team (which is separate from the SETI Institute), with the support of the Planetary Society, has recruited about 4 million owners of personal computers that are connected to the Internet. You can download a screen saver that searches data files from the Arecibo radio telescope for signals whenever you are not using the computer. For information, locate the SETI@home project at http://setiathome.ssl.berkeley.edu/.

The search continues, but radio astronomers struggle to hear anything against the worsening babble of radio noise from human civilization. Wider and wider sections of the electromagnetic spectrum are being used for Earthly communication, and this, combined with stray electromagnetic radiation from electronic devices including everything from computers to refrigerators, makes hearing faint radio signals difficult. It would be ironic if humans fail to detect faint radio signals from another world because our own world has become too noisy. One alternate search strategy is to look for rapid flashes of laser light at optical or near-infrared wavelengths. Such extraterrestrial signals, if they exist, would have the advantage of being easily distinguished from natural light sources but the disadvantage of being blocked by interstellar dust. Ultimately, the chance of success for any of the searches depends on the number of inhabited worlds in the galaxy.

◄ **Figure 17-13** Part of the Allen Telescope Array (ATA) near Mount Lassen in California, planned eventually to include 350 radio dishes, each 6 m in diameter, in an arrangement designed to maximize their combined angular resolution. Radio astronomers use the telescope to study galaxies and nebulae of astrophysical interest while SETI researchers employ state-of-the-art computer hardware and software to search for signals from distant civilizations.

SETI Institute

The Copernican Principle

Why do astronomers seem so confident that there is life on other worlds? No message has been received from distant worlds and no life has been detected on any of the worlds visited by probes from Earth. Yet astronomers face this absence of evidence with confidence because of the work of an astronomer who lived 500 years ago.

In Chapter 4, you read about Nicolaus Copernicus and the world he lived in. Astronomers before Copernicus accepted that Earth was the unmoving center of the Universe and that the planets and stars were mysterious lights attached to rotating spheres that carried them around Earth. That made Earth a special place unlike any other place. To better explain these motions, Copernicus proposed that the Earth rotates on its axis and revolves around the Sun. That works much better, but as a consequence it made Earth just another one of the planets. Earth was no longer a special place.

Astronomers have adopted the **Copernican Principle**: Earth is not in a special place. That principle can be extended to assuming that Earth also is not special in other ways.

Visual + Infrared

NASA/ESA/STScI/AURA/NSF/HDF-South Team, R. Williams

The Hubble Deep Field South includes many galaxies containing a huge number of stars. There might be more than 1 trillion (10^{12}) planets in this picture.

If Earth is not special, if it is just a planet, then there should be lots of planets like Earth. Of course, many planets are hot, cold, dry, have no atmosphere, or have tremendously deep atmospheres as does Jupiter, but there are a lot of planets. You have learned that a large fraction of stars have planets, that a galaxy contains roughly 100 billion stars, and that there are more than 100 billion galaxies

visible with existing telescopes. There must be many planets like Earth.

Although there is no direct evidence of life on other worlds, you have seen in this chapter how life could have originated by natural chemical processes, and how evolution shapes living things to survive and become better adapted. The evidence for that is strong. For astronomers accepting the Copernican Principle, it seems inevitable that life will arise on planets where conditions permit it, and then evolve to become more complex.

You can see the influence of the Copernican Principle in the Drake equation. The factors in the equation follow the logical steps outlined here, and the final factor represents the likelihood that an alien civilization will survive long enough to communicate with other civilizations. Astronomers tend to think that the lack of direct evidence just means we haven't searched long enough or well enough. As Carl Sagan said: "Absence of evidence is not evidence of absence."

When Copernicus said Earth is just one of the planets, he changed the way humanity sees itself, and that change is still rippling through history.

How Many Inhabited Worlds?

Given enough time, the searches will find other worlds with civilizations, assuming that there are at least a few out there. If intelligence is common, scientists should find signals relatively soon—within the next few decades—but if intelligence is rare, it may take much longer.

Simple arithmetic can give you an estimate of the number of technological civilizations in the Milky Way Galaxy with which you might communicate, N_c. The formula proposed for discussions about N_c is named the **Drake equation** after the radio astronomer Frank Drake, a pioneer in the search for extraterrestrial intelligence. The version of the Drake equation presented here is modified slightly from its original form:

$$N_c = N_* \cdot f_P \cdot n_{HZ} \cdot f_L \cdot f_I \cdot f_S$$

N_* is the number of stars in our galaxy, and f_P represents the fraction of stars that have planets. If all single stars have planets,

f_P is about 0.5. The factor n_{HZ} is the average number of planets in each planetary system suitably located in the habitable zone—meaning, for the sake of the current discussion, the number of planets per planetary system possessing substantial amounts of liquid water. The conventional habitable zone in our system contains Earth's orbit and, arguably, Venus and Mars as well. Europa and Enceladus in our Solar System show that liquid water can exist as a result of tidal heating outside the conventional habitable zone. Thus, n_{HZ} may be larger than had been originally thought. (However, note that, as of this writing, only one Earth-size extrasolar planet, Kepler 186f, has been found orbiting in a habitable zone.) The factor f_L is the fraction of suitable planets on which life begins, and f_I is the fraction of those planets where life evolves to intelligence.

Notice that the Drake equation is, in a sense, based on an extension of the idea put forth by Copernicus that Earth is not unique (**How Do We Know? 17-3**). You can put in numbers for the

TABLE 17-1   The Number of Technological Civilizations per Galaxy

Estimates	Variables	Pessimistic	Optimistic
N_*	Number of stars in a typical large galaxy	2×10^{11}	2×10^{11}
f_P	Fraction of stars with planets	0.1	0.5
n_{HZ}	Number of planets per star that orbit in the habitable zone for longer than 4 billion years	0.01	1
f_L	Fraction of habitable zone planets on which life begins	0.01	1
f_I	Fraction of planets with life on which some species evolves to intelligence	0.01	1
f_S	Fraction of star's existence during which a technological civilization survives	10^{-8}	10^{-3}
N_c	Current number of communicative civilizations per galaxy	2×10^{-4}	1×10^{8}

© 2016 Cengage Learning®

various factors that represent low probabilities or high probabilities, but either way, the implicit assumption of the equation is that Earth, life on Earth, and the human species that desires to make contact with other intelligences in the Universe are single examples of larger sets of such things.

The six factors on the right-hand side of the Drake equation can be roughly estimated, with decreasing certainty as you proceed from left to right. The final factor is extremely uncertain. That factor, f_S, is the fraction of a star's life during which an intelligent species is communicative. If most civilizations last only a short time at a technological level, say for 100 years, there may be none capable of transmitting during the cosmic interval when Earthlings are capable of building radio telescopes to listen for them. On the other hand, a society that stabilizes and remains technologically capable for a long time is much more likely to be detected. For a star with a life span of 10 billion years, f_S might conceivably range from 10^{-8} for extremely short-lived societies to 10^{-3} for societies that survive for ten million years. Table 17-1 summarizes what many scientists consider a reasonable range of values for f_S and the other factors.

If the optimistic estimates are true, there could be a communicative civilization within a few tens of light-years from Earth. On the other hand, if the pessimistic estimates are true,

Earth may be the only planet that is capable of communication within thousands of the nearest galaxies.

DOING SCIENCE

Why does the number of civilizations that could be detected depend on how long civilizations survive at a technological level? Answering this question requires a type of reasoning, used often by scientists, that depends on the timing of events.

Recall from previous chapters that few stars are observed in short-lived life stages because Earthlings' "snapshot" of the Universe catches mostly stars in long-lasting, and therefore more common, parts of their life cycles. Similarly, if you turn a radio telescope to the sky and scan many stars, you would be taking a snapshot of the Universe at a particular time. Broadcasts from other civilizations must be arriving at the time you are observing if you're going to detect them. If most civilizations survive for a long time, there is a much greater chance that you will detect one of them in your snapshot than if civilizations tend to disappear quickly because of, for example, nuclear war or environmental collapse. If most civilizations last only a short time, there may be none transmitting during the cosmically short interval when Earthlings are capable of building radio telescopes to listen for them.

Now consider another aspect of searching for extraterrestrial signals: *Why might the "water hole" be an especially good frequency band in which to listen?*

What Are We?   Matter and Spirit

There are more than 4000 religions around the world, and nearly all hold that humans have a dual nature: We are physical objects made of atoms, but we are also spiritual beings. Science is unable to examine the spiritual side of existence, but it can tell us about our physical nature.

The matter you are made of appeared in the big bang and was cooked into a wide range of elements inside stars. Your

atoms may have been inside at least two or three generations of stars. Eventually, your atoms became part of a nebula that contracted to form our Sun and the planets of the Solar System.

Your atoms have been part of Earth for the past 4.6 billion years. They have been recycled many times through dinosaurs, stromatolites, fish, bacteria, grass, birds, worms, and other living

(continued)

things. You are using your atoms now, but when you are done with them, they will go back to Earth and be used again and again.

When the Sun swells into a red giant star 6 billion years from now, Earth's atmosphere and oceans will be driven away, and at least the outer few kilometers of Earth's crust will be vaporized and blown outward to become part of the nebula expanding into space away from the white-dwarf remains of the Sun. Your atoms are destined to return to the interstellar medium and become part of future generations of stars and planets.

The message of astronomy is that humans are not just observers: We are participants in the Universe. Among all of the

galaxies, stars, planets, planetesimals, and other bits of matter, humans are objects that can think, and that means we can understand what we are.

Is the human race the only thinking species? If so, we bear the sole responsibility to understand and admire the Universe. The detection of signals from another civilization would demonstrate that we are not alone, and such communication would end the self-centered isolation of humanity and stimulate a reevaluation of the meaning of human existence. We may never realize our full potential as humans until we communicate with nonhuman intelligent life.

Study and Review

Summary

▶ The search for life on other worlds and the investigation of possible habitats for such life is the field of study known as **astrobiology (p. 384)**. Some astrobiologists also study life's origin and evolution on Earth. Hence, the field of astrobiology is a hybrid of multiple fields.

▶ Life can be defined as a process by which an organism extracts energy from the surroundings, maintains itself, and modifies the surroundings to promote its survival.

▶ Living things have a physical basis—an arrangement of matter and energy that makes the life process possible. The physical bases of life on Earth are carbon and water: The carbon chemistry of Earth life occurs in bags of water (cells).

▶ Living things have controlling sets of information that can be passed to successive generations. Genetic information for life on Earth is stored in long carbon-chain molecules such as **DNA (deoxyribonucleic acid) (p. 386)**.

▶ The DNA molecule stores information in the form of chemical bases, which are linked together like the rungs of a ladder. Copied by the **RNA (ribonucleic acid) (p. 387)** molecule, the patterns of bases act as recipes for connecting **amino acid (p. 386)** subunits together to construct **proteins (p. 386)**, including **enzymes (p. 386)**. Proteins are the main structural components, and enzymes control components of the life process.

▶ The unit of heredity is a **gene (p. 387)**, a piece or several pieces of DNA that in most cases specifies the construction of one particular protein molecule. Genes are connected together in structures called **chromosomes (p. 387)**, which are essentially single, long DNA molecules.

▶ When a cell divides, the chromosomes split lengthwise and duplicate themselves so that each of the new cells can receive a copy of the genetic information.

▶ **Biological evolution (p. 388)** is the process by which species adjusts to changes in the environment.

▶ Damage or errors in duplicating the DNA molecule can produce **mutations (p. 388)**. Mutations are changes in the DNA information that can result in changes to the properties of an organism. Variation in genetic codes can become widespread among individuals of a species. **Natural selection (p. 388)** determines which of these variations are best suited for survival.

▶ Evolution is not random. Genetic variation is essentially random, but natural selection is controlled by the environment.

▶ The oldest definitely identified fossils on Earth are 3.4 billion-year-old structures called **stromatolites (p. 389)** that are composed of stacks of bacterial mats and sediment layers. Those fossils provide evidence that life began as single-celled organisms in the oceans.

▶ After billions of years, life evolved into more complex, **multicellular (p. 391)** organisms.

▶ The **Miller experiment (p. 389)** shows that the chemical building blocks of life form naturally in a wide range of circumstances.

▶ Scientists hypothesize that **chemical evolution (p. 392)** occurred before biological evolution. Chemical evolution concentrated simple molecules into a diversity of larger, stable organic molecules, which were dissolved in the young Earth's oceans. However, those organic molecules did not reproduce copies of themselves. The hypothetical organic-rich water is sometimes referred to as the **primordial soup (p. 390)**. Biological evolution began when molecules developed the ability to make copies of themselves.

▶ Life-forms did not become large and complex until about 0.5 billion years ago, during a relatively short time known as the **Cambrian explosion (p. 391)**.

▶ Life emerged from the oceans only about 0.4 billion years ago. Human intelligence developed over the past 4 million (0.004 billion) years. Anatomically modern humans appeared only about 200,000 years ago.

▶ Life on Earth requires liquid water and thus a specific range of temperatures and pressures.

▶ Organisms that thrive in extreme environments on Earth are called **extremophiles (p. 394)**. Genetic evidence indicates that the common ancestor of all Earth life was a **thermophile (p. 394)**, a heat-tolerant version of present-day single-celled organisms called **archaea (p. 394)**.

▶ No other celestial object in our Solar System is known to, or appears to, harbor life. Most objects in our Solar System are currently too hot or too cold. Life might have begun on Mars before Mars became too cold and dry. If so, life conceivably could persist today on Mars in restricted environments.

▶ Liquid water exists under the surfaces of Jupiter's moons Europa and Ganymede and Saturn's moon Enceladus, and possibly a few other places in the Solar System. Saturn's moon Titan has abundant organic compounds but does not have long-term presence of liquid surface water.

▶ Because the origin of life and its evolution into intelligent creatures took so long on Earth, scientists do not consider planets that are orbiting middle- and upper-main-sequence stars as likely homes for life. The reason is that massive stars shine for time spans that are too short.

▶ Main-sequence G and K stars are thought to be the likeliest candidates to host exoplanets with life. There is a scientific debate regarding whether planets orbiting main-sequence M stars are also good candidates.

▶ The **habitable zone (p. 396)** around a star, within which planets have surface temperatures allowing liquid water on or near their surfaces, may not be the only possible location for water environments. Tidally heated moons orbiting planets located outside the Solar System's habitable zone may have liquid water near their surfaces and thus could potentially harbor life.

▶ Because of the distances involved and energy required, physical travel between stars seems technically impractical compared with communication between Earth and exoplanetary systems using electromagnetic signals. However, a real conversation would be difficult because of long travel times for such signals.

▶ Broadcasting a pulsed radio (or light) beacon would distinguish the signal from naturally occurring emission and would identify the source as a technological civilization. The signal can be **anticoded (p. 397)** in the hope that another intelligent civilization could understand it.

▶ One good part of the radio spectrum for communication is called the **water hole (p. 398)**, the wavelength range from the 21-cm spectral line of hydrogen to the 18-cm line of hydroxyl (OH). Millions of radio channels fit in this range.

▶ Sophisticated searches are now under way to detect radio transmissions from civilizations on other worlds. However, such **Search for Extra-Terrestrial Intelligence (SETI; p. 398)** programs are hampered by limited computer power and radio noise pollution.

▶ The number of civilizations in our Milky Way Galaxy that are at a technological level and able to communicate while humans are listening can be estimated by the **Drake equation (p. 400)**. This number is limited primarily by the lifetimes of our and other civilizations. Implicit in the Drake equation is the **Copernican Principle (p. 400)**, that Earth should not be assumed to be a special place.

Review Questions

1. Explain how astrobiology is a science and not a pseudoscience. (*Hint:* Refer back to Chapter 2 for the definition of *pseudoscience*.)

2. Carbon and water are the physical bases for all life on Earth. True or false?

3. Earth is the only world known to have life. True or false?

4. How does the DNA molecule produce a copy of itself?

5. What would happen to a life-form if the genetic information handed down to offspring was copied extremely inaccurately?

6. What would happen to a life-form if the genetic information handed down to offspring differs slightly from that in the parents?

7. What would happen to a life-form if the information handed down to offspring was always copied perfectly?

8. Describe an example of natural selection acting on new DNA patterns to select the most advantageous characteristics.

9. Explain how evolution is not random.

10. What evidence do scientists have that life on Earth began in the sea?

11. Organic molecules are all derived from living organisms. True or false?

12. Why is liquid water generally considered necessary for the origin of life?

13. Asteroids could contain amino acids. True or false?

14. What is the difference between chemical evolution and biological evolution?

15. What is the significance of the Miller experiment?

16. In photosynthesis, plants remove carbon dioxide from the atmosphere and produce oxygen. Could the Earth's early atmosphere support photosynthesis?

17. Molecules of which gas were needed in Earth's atmosphere for life to evolve from living in the sea to living on the land?

18. Does intelligence make a creature more likely to survive? Why or why not?

19. Archaea are multicellular organisms. True or false?

20. What is the evidence that the first organisms on Earth most closely resembled today's thermophilic archaea?

21. Name three locations in our Solar System to search for Earth-like life.

22. Why are upper-main-sequence (high-luminosity) host stars unlikely sites for intelligent civilizations?

23. Why is it reasonable to expect that physical travel between stars is highly unlikely?

24. How does the stability of technological civilizations affect the probability that humanity can communicate with them?

25. What is the water hole? Why is the water hole a good "place" to search for extraterrestrial civilizations?

26. Why is anticoding a message difficult? In other words, why is it hard to make a message that potentially can be understood by completely unknown recipients?

27. **How Do We Know?** Do science and religion have complementary explanations of the world? If so, how?

28. **How Do We Know?** Why are scientists confident Earth has never been visited by aliens?

29. **How Do We Know?** Why does the Drake equation implicitly assume the Copernican Principle?

Discussion Questions

1. Would the atmosphere of an exoplanet be a place where life could begin? Why or why not?

2. Could the chemistry of life-forms outside Earth be based on silicon instead of carbon?

3. What environments on Earth do you consider extreme? Do extremophiles live in your idea of an extreme environment?

4. Can you think of anything that is missing in the Arecibo message shown in Figure 17-11? Do you expect that any alien recipient of the Arecibo message will be able to decode it? Why or why not?

5. How well do you think the public would react to an announcement that extraterrestrial intelligence has been detected? Do you think it would tend to upset or confirm human beliefs about ourselves, the world, and the Universe?

6. If decades of careful searches for radio signals from extraterrestrial intelligence turn up nothing, what would that mean?

Problems

1. A single human cell encloses about 1.5 m of DNA. This length of DNA contains about 4.5 billion base pairs. What is the spacing between these base pairs in units of nanometers? That is, how far apart are the rungs on the DNA ladder? How many base pairs are there per millimeter?

2. If you represent Earth's history by a line that is 1 m long, how long a segment would represent the 400 million years since life first moved onto the land? How long a segment would represent the 4-million-year history of humanoid life?

3. Consider Figure 17-8. What is the ratio of the length of time since the origin of fish to the time since the origin of mammals? What does this value indicate?

4. Suppose a human generation is defined as the average time from birth to childbearing, which is about 20 years long. How many generations have passed in the 200,000 years during which anatomically modern humans have existed?

5. If a star must remain on the main sequence for at least 4 billion years for life to evolve to intelligence, what is the most massive a star that can form and still possibly harbor intelligent life on one of its exoplanets? (*Hint:* Use the data given in Appendix Table A-7.)

6. Mathematician Karl Gauss suggested planting forests and fields in gigantic geometric figures as signals to possible Martians that intelligent life exists on Earth. If Martians had telescopes that could resolve details on Earth no smaller than 1 arc second, how large would the smallest element of Gauss's signal have to be for the element to be visible at Mars's closest approach to Earth? (*Hint:* Use the small-angle formula, Chapter 3.) (*Note:* Necessary data are given in Appendix Table A-10.)

7. If you detected radio signals with an average wavelength of 20.000 cm and suspected that they came from a civilization on a distant Earth-like exoplanet, roughly how much of a change in wavelength should you expect to detect as a result of the orbital motion of the distant exoplanet? (*Hint:* Use the Doppler shift formula, Chapter 7.) (*Note:* Earth's orbital velocity is 30 km/s.)

8. What is the minimum length of time humans have to wait for a response to the signal sent from Arecibo in 1974 to M13?

9. The first radio broadcast was made on January 13, 1910. It was a live performance at New York City's Metropolitan Opera House. Since that time, how far has that radio broadcast traveled in light-years? In the solar neighborhood there is, on average, one star system per 400 cubic light-years. How many star systems could have heard those opera singers? (*Note:* The volume of a sphere $= \frac{4}{3}\pi r^3$.)

10. Calculate the number of communicative civilizations per galaxy using your own estimates of the factors in Table 17-1.

Learning to Look

1. Look at Figure 17-11. Since the time we sent the message from Arecibo, what has happened that makes one of the symbols inaccurate?

2. The star cluster shown in the image to the right contains a few red giants as well as main-sequence stars ranging from spectral type B to M. Discuss the likelihood that exoplanets orbiting any of these stars might be home to life. (*Hint:* Estimate the age of the cluster.)

NASA/ESA/STScI/AURA/NSF/E. Olszewski (University of Arizona)

Visual + Infrared

3. If you could search for life in the galaxy shown in the image to the right, would you look among stars in the disk, or in the central bulge, or in the halo, or all of those places? Discuss the factors that influence your decision.

ESO   Visual

Afterword

The aggregate of all our joys and sufferings, thousands of confident religions, ideologies and economic doctrines, every hunter and forager, every hero and coward, every creator and destroyer of civilizations, every king and peasant, every young couple in love, every hopeful child, every mother and father, every inventor and explorer, every teacher of morals, every corrupt politician, every superstar, every supreme leader, every saint and sinner in the history of our species, lived there on a mote of dust, suspended in a sunbeam.

CARL SAGAN (1934–1996)

Earth imaged by *Voyager 1* looking back from the Kuiper Belt, beyond Neptune's orbit.

NASA

Our journey together is over, but before we part company, ponder one final time the primary theme of this book—humanity's place in the physical universe. Astronomy gives us some comprehension of the workings of stars, galaxies, and planets, but its greatest value lies in what it teaches us about ourselves. Now that you have surveyed astronomical knowledge, you can better understand your own position in nature.

To some, the word nature conjures up visions of furry rabbits hopping about in a forest glade. To others, nature is the blue-green ocean depths, and still others think of nature as windswept mountaintops. As diverse as these images are, they are all Earth-bound. Having studied astronomy, you can see nature as a beautiful dance of matter and energy, interacting according to simple rules, forming galaxies, stars, planets, mountaintops, ocean depths, forest glades, and people.

Perhaps the most important astronomical lesson is that humanity is a small but important part of the universe. Most of the universe is probably lifeless. The vast reaches between the galaxies appear to be empty of all but the thinnest gas, and stars are much too hot to preserve the chemical bonds that seem necessary for life to survive and develop. It seems that only on the surfaces of a few planets, where temperatures are moderate, can atoms link together in special ways to form living matter.

If life is special, then intelligence is precious. The universe must contain many planets devoid of life, planets where sunlight has shined unfelt for billions of years. There may also exist planets on which life has developed but has not become complex, planets where the wind stirs wide plains of grass and rustles through dark forests. On some planets, creatures resembling Earth's insects, fish, birds, and animals may watch the passing days only dimly aware of their own existence. It is intelligence, human or otherwise, that gives meaning to the landscape.

Science is the process by which Earth's intelligence has tried to understand the physical universe. Science is not the invention of new devices or processes. It does not create home computers, cure the mumps, or manufacture plastic spoons—those are engineering and technology, the adaptation of scientific understanding for practical purposes. Science is the understanding of nature, and astronomy is that understanding on the grandest scale. Astronomy is the science by which the universe, through its intelligent lumps of matter, tries to understand its own existence.

As the primary intelligent species on this planet, we are the custodians of a priceless gift—a planet filled with living things.

This is especially true if life is rare in the universe. In fact, the rarer life actually is in the Universe, the more overwhelming is our responsibility. We are the only creatures who can take action to preserve the existence of life on Earth; ironically, our own actions are the most serious hazards.

The future of humanity is not secure. We are trapped on a tiny planet with limited resources and a population growing faster than our ability to produce food. We have already driven some creatures to extinction and now threaten others. We are changing the climate of our planet in ways we do not fully understand. Even if we reshape our civilization to preserve our world, the Sun's evolution will eventually destroy Earth.

This may be a sad prospect, but a few factors are comforting. First, everything in the universe is temporary. Stars die, galaxies die; perhaps the entire universe will someday end. As part of a much larger whole, we are reminded that our distant future is limited. Only a few million years ago, our ancestors were starting to walk upright and communicate. A billion years ago, our ancestors were microscopic organisms living in the oceans. To suppose that a billion years hence there will be beings resembling today's humans, or that humans will still be the dominant intelligence on Earth, or that human descendants will even exist, is ultimately a conceit.

Our responsibility is not to save our race for all eternity but to behave as dependable custodians of our planet, preserving it, admiring it, and trying to understand it. That calls for drastic changes in our behavior toward other living things and a revolution in our attitude toward our planet's resources. Whether we can change our ways is debatable—humanity is far from perfect in its understanding, abilities, or intentions. However, you must not imagine that we, and our civilization, are less than precious. We have the gift of intelligence, and that is the finest thing this planet has ever produced.

> We shall not cease from exploration
> And the end of all our exploring
> Will be to arrive where we started
> And know the place for the first time.
> —T. S. Eliot, "Little Gidding"*

---

*Excerpt from "Little Gidding" in *Four Quartets,* copyright 1942 by T. S. Eliot and renewed 1970 by Esme Valerie Eliot, reprinted by permission of Harcourt Inc. and Faber & Faber Ltd.

Appendix A

Units and Astronomical Data

Introduction

A SYSTEM OF UNITS is based on the three fundamental units for length, mass, and time. By international agreement, there is a preferred subset of metric units known as the *Système International d'Unités* (SI units), commonly called the metric system, which is based on the meter, kilogram, and second. Other quantities, such as density and force, are derived from these fundamental units.

Residents of the United States generally use the (British) imperial system of units (officially used only in the United States, Liberia, and Myanmar but, ironically, not in Great Britain). For example, in imperial units the fundamental unit of length is the foot, composed of 12 inches.

SI units employ the decimal system. For example, a meter is composed of 100 centimeters. Because the metric system is a decimal system, it is easy to express quantities in larger or smaller units as is convenient. You can give distances in centimeters, meters, kilometers, and so on. The prefixes specify the relation of the unit to the meter. Just as a cent is 1/1000 of a dollar, so a centimeter is 1/1000 of a meter. A kilometer is 1000 m, and a kilogram is 1000 g. The meanings of the commonly used prefixes are given in Table A-1.

Fundamental and Derived SI Units

THE THREE FUNDAMENTAL SI units define the rest of the units, as given in Table A-2.

The SI unit of force is the newton (N), named after Isaac Newton. It is the force needed to accelerate a 1-kg mass by 1 m/s^2, or the force roughly equivalent to the weight of an apple at Earth's surface. The SI unit of energy is the joule (J), the energy produced by a force of 1 N acting through a distance of 1 m. A joule is roughly the energy in the impact of an apple falling off a table.

Exceptions

Units can help you in two ways. They make it possible to make calculations, and they can help you to conceive of certain quantities. For calculations, the metric system is far superior, and it is used for calculations throughout this book.

In SI units, density should be expressed as kilograms per cubic meter, but no human hand can enclose a cubic meter, so that unit does not help you grasp the significance of a given density. This book refers to density in grams per cubic centimeter. A gram is roughly the mass of a paper clip, and a cubic

TABLE A-1 Metric Prefixes

Prefix	Symbol	Factor
giga	G	10^9
mega	M	10^6
kilo	k	10^3
centi	c	10^{-2}
milli	m	10^{-3}
micro	μ	10^{-6}
nano	n	10^{-9}

TABLE A-2 SI (*Système International*) Metric Units

Quantity	SI Unit
Length	meter (m)
Mass	kilogram (kg)
Time	second (s)
Force	newton (N)
Energy	joule (J)
Power	watt (W)

centimeter is the size of a small sugar cube, so you can easily conceive of a density of 1 g/cm³, roughly the density of water. This is not a bothersome departure from SI units because you will not have to make complex calculations using density.

For conceptual purposes, this book expresses some quantities in both SI and imperial units. Instead of saying the average adult would weigh 111 N on the moon, it might be more helpful to some readers for that weight to be expressed as 25 lb. In such cases, the imperial form is given in parentheses after the SI form. For example, the radius of the moon is 1738 km (1080 mi).

Conversions

To convert from one metric unit to another (from meters to kilometers, for example), you need only look at the prefix. However, converting from metric to English or English to metric is more complicated. The conversion factors are given in Table A-3.

Example: The radius of the moon is 1738 km. What is this in miles? Table A-3 indicates that 1.000 mile equals 1.609 km, so

$$1738 \text{ km} \times \frac{(1.000 \text{ mi})}{(1.609 \text{ km})} = 1080 \text{ mi}$$

Temperature Scales

In astronomy, as in most other sciences, temperatures are expressed on the Kelvin scale, although the centigrade (or Celsius) scale is also used. The Fahrenheit scale commonly used in the United States is not used in scientific work.

The centigrade scale refers temperatures to the freezing point of water (0°C) and to the boiling point of water (100°C). One degree Centigrade is $^1/_{100}$, the temperature difference between the freezing and boiling points of water, thus the prefix *centi*. The centigrade scale is also called the Celsius scale after its inventor, the Swedish astronomer Anders Celsius (1701–1744).

Temperatures on the Kelvin scale are measured in Celsius degrees from absolute zero (−273.15°C), the temperature of an object that contains no extractable heat. In practice, no object can be as cold as absolute zero, although laboratory apparatuses have reached temperatures lower than 10^{-6} K. The Kelvin scale is named after the Scottish mathematical physicist William Thomson, Lord Kelvin (1824–1907).

The Fahrenheit scale fixes the freezing point of water at 32°F and the boiling point at 212°F. Named after the German physicist Gabriel Daniel Fahrenheit (1686–1736), who made the first successful mercury thermometer in 1720, the Fahrenheit scale is used routinely only in the United States.

It is easy to convert temperatures from one scale to another using the information given in Table A-4.

Powers of 10 Notation

Powers of 10 make writing very large numbers much simpler. For example, the nearest star is about 43,000,000,000,000 km from the Sun. Writing this number as 4.3×10^{13} km is much easier.

Very small numbers can also be written with powers of 10. For example, the wavelength of visible light is about 0.0000005 m. In powers of 10, this becomes 5×10^{-7} m.

The powers of 10 used in this notation appear below. The exponent tells you how to move the decimal point. If the exponent is positive, move the decimal point to the right. If the exponent is negative, move the decimal point to the left. For example, 2.0000×10^3 equals 2000, and 2×10^{-3} equals 0.002.

$$10^5 = 100,000$$
$$10^4 = 10,000$$
$$10^3 = 1000$$
$$10^2 = 100$$
$$10^1 = 10$$
$$10^0 = 1$$
$$10^{-1} = 0.1$$
$$10^{-2} = 0.01$$
$$10^{-3} = 0.001$$
$$10^{-4} = 0.0001$$

TABLE A-3 Conversion Factors Between British and Metric Units

1 inch = 2.54 centimeters	1 centimeter = 0.394 inch
1 foot = 0.3048 meter	1 meter = 39.37 inches = 3.28 feet
1 mile = 1.609 kilometers	1 kilometer = 0.6214 mile
1 slug = 14.59 kilograms	1 kilogram = 0.06852 slug
1 pound = 4.448 newtons	1 newton = 0.2248 pound
1 foot-pound = 1.356 joules	1 joule = 0.7375 foot-pound
1 horsepower = 745.7 joules/s	1 joule/s = 0.001341 horsepower
	1 joule/s = 1 watt

TABLE A-4 Temperature Scales and Conversion Formulas

	Kelvin (K)	Centigrade (°C)	Fahrenheit (°F)
Absolute zero	0 K	−273°C	−460°F
Freezing point of water	273 K	0°C	32°F
Boiling point of water	373 K	100°C	212°F

Conversions:

$$K = °C + 273$$
$$°C = \tfrac{5}{9}(°F - 32)$$
$$°F = \tfrac{9}{5}(°C) + 32$$

If you use scientific notation in calculations, be sure you correctly enter the numbers into your calculator. Not all calculators accept scientific notation, but those that can have a key labeled EXP, EEX, or perhaps EE that allows you to enter the exponent of 10. To enter a number such as 3×10^8, press the keys 3 EXP 8. To enter a number with a negative exponent, you must use the change-sign key, usually labeled +/− or CHS. To enter the number 5.2×10^{-3}, press the keys 5.2 EXP +/− 3. Try a few examples.

To read a number in scientific notation from a calculator, you must read the exponent separately. The number 3.1×10^{25} may appear in a calculator display as 3.1 25 or on some calculators as 3.1 10^{25}. Examine your calculator to determine how such numbers are displayed.

Astronomy Units and Constants

Astronomy, and science in general, is a way of learning about nature and understanding the universe. To test hypotheses about how nature works, scientists use observations of nature. The tables that follow contain some of the basic observations that support science's best understanding of the astronomical universe. Of course, these data are expressed in the form of numbers, not because science reduces all understanding to mere numbers, but because the struggle to understand nature is so demanding that science must use every valid means available. Quantitative thinking—reasoning mathematically—is one of the most powerful techniques ever invented by the human brain. Thus, these tables are not nature reduced to mere numbers but numbers supporting humanity's growing understanding of the natural world around us.

TABLE A-5 Astronomical Constants

Velocity of light (c)	$= 3.00 \times 10^8$ m/s
Gravitational constant (G)	$= 6.67 \times 10^{-11}$ m^3/s^{2}kg
Mass of H atom	$= 1.67 \times 10^{-27}$ kg
Mass of Earth $(M_\oplus)$	$= 5.97 \times 10^{24}$ kg
Earth equatorial radius $(R_\oplus)$	$= 6.38 \times 10^3$ km
Mass of Sun $(M_\odot)$	$= 1.99 \times 10^{30}$ kg
Radius of Sun $(R_\odot)$	$= 6.96 \times 10^8$ m
Solar luminosity $(L_\odot)$	$= 3.83 \times 10^{26}$ J/s
Mass of Moon	$= 7.35 \times 10^{22}$ kg
Radius of Moon	$= 1.74 \times 10^3$ km

TABLE A-6 Units Used in Astronomy

1 Angstrom (Å)	$= 10^{-8}$ cm
	$= 10^{-10}$ m
	$= 10$ nm
1 astronomical unit (AU)	$= 1.50 \times 10^{11}$ m
	$= 93.0 \times 10^6$ mi
1 light-year (ly)	$= 6.32 \times 10^4$ AU
	$= 9.46 \times 10^{15}$ m
	$= 5.88 \times 10^{12}$ mi
1 parsec (pc)	$= 2.06 \times 10^5$ AU
	$= 3.09 \times 10^{16}$ m
	$= 3.26$ ly
1 kiloparsec (kpc)	$= 1000$ pc
1 megaparsec (Mpc)	$= 1,000,000$ pc

TABLE A-7   Properties of Main-Sequence Stars

Spectral Type	Absolute Visual Magnitude (M_v)	$L^{1,2}$	Temp. (K)	λ_{max} (nm)	Mass[1]	Radius[1]	Average Density (g/cm³)
O5	−5.7	620,000	42,000	69	60	12	0.03
B0	−4.0	61,000	30,000	97	18	7.4	0.04
B5	−1.2	1,100	15,000	191	5.9	3.9	0.1
A0	0.7	73	9800	296	2.9	2.4	0.2
A5	2.0	18	8200	354	2.0	1.7	0.4
F0	2.7	8.8	7300	397	1.6	1.5	0.5
F5	3.5	4.6	6600	436	1.4	1.3	0.6
G0	4.4	2.1	5900	488	1.05	1.1	0.8
G2	4.7	1.0	5200	558	1.0	1.0	0.1
G5	5.1	0.7	5600	521	0.9	0.9	0.8
K0	5.9	0.6	5200	563	0.8	0.8	1.2
K5	7.4	0.3	4400	657	0.7	0.7	1.8
M0	8.8	0.1	3800	755	0.5	0.6	2.2
M5	16.3	0.01	3200	914	0.2	0.3	10

[1]Luminosity, mass, and radius are given in terms of the Sun's luminosity, mass, and radius. The Sun's luminosity, mass, and radius are given in Table A-5.

[2]Luminosity is computed from radius and temperature.

TABLE A-8   The 15 Brightest Stars

Star	Name	Apparent Visual Magnitude (m_v)	Distance (pc)	Absolute Visual Magnitude (M_v)	Spectral Type
	Sun	−26.74		4.8	G2 V
α CMa	Sirius	−1.47	2.6	1.4	A1 V
α Car	Canopus	−0.72	96	−5.6	F0 II
α Cen	Rigil Kentaurus	−0.29	1.3	4.1	G2 V
α Boo	Arcturus	−0.04	11	−0.3	K2 III
α Lyr	Vega	0.03	7.8	0.6	A0 V
α Aur	Capella	0.08	13	−0.5	G8 III
β Ori	Rigel	0.12	240	−6.8	B8 Iab
α CMi	Procyon	0.34	3.5	2.6	F5 IV-V
α Eri	Achernar	0.50	44	−5.1	B3 V
α Ori	Betelguese	0.58	130	−5.0	M2 Iab
β Cen	Hadar	0.60	160	−5.4	B1 III
α Aql	Altair	0.77	5.1	2.2	A7 V
α Cru	Acrux	0.81	98	−4.2	B0 IV
α Tau	Aldebaran	0.85	20	−0.7	K5 III

TABLE A-9 The 15 Nearest Stars

Name	Distance (ly)	Distance (pc)	Apparent Visual Magnitude (m_v)	Absolute Visual Magnitude (M_v)	Spectral Type
Sun			−26.7	4.8	G2
Proxima Cen	4.2	1.3	11.0	15.5	M6
α Cen A	4.4	1.3	0.0	4.4	G2
α Cen B	4.4	1.3	1.3	5.7	K5
Barnard's Star	5.9	1.8	9.5	13.2	M4
Wolf 359	7.7	2.4	13.5	16.7	M6
Lalande 21185	8.3	2.5	7.5	10.5	M2
α CMa A (Sirius A)	8.6	2.6	−1.5	1.4	A1
α CMa B (Sirius B)	8.6	2.6	8.4	11.3	white dwarf
Luyten 726-8A	8.7	2.7	12.6	15.4	M6
Luyten 726-8B	8.7	2.7	12.0	14.9	M5
Ross 154	9.7	3.0	11.0	13.6	M3
Ross 248	10.4	3.2	12.2	14.8	M6
ε Eri	10.5	3.2	3.7	6.2	K2
Ross 128	10.9	3.3	11.1	13.5	M4

TABLE A-10 Properties of the Planets

ORBITAL PROPERTIES

Planet	Semimajor Axis (a) (AU)	Semimajor Axis (a) (10^6 km)	Orbital Period (P) (y)	Orbital Period (P) (days)	Average Orbital Velocity (km/s)	Orbital Eccentricity	Inclination to Ecliptic
Mercury	0.387	57.9	0.241	88.0	47.9	0.206	7.0°
Venus	0.723	108	0.615	224.7	35.0	0.007	3.4°
Earth	1.00*	150	1.00*	365.3	29.8	0.017	0°*
Mars	1.52	228	1.88	687.0	24.1	0.093	1.8°
Jupiter	5.20	779	11.9	4332	13.1	0.048	1.3°
Saturn	9.58	1433	29.5	10,759	9.7	0.056	2.5°
Uranus	19.23	2877	84.0	30,680	6.8	0.047	0.8°
Neptune	30.10	4503	164.8	60,190	5.4	0.011	1.8°

*By definition.

PHYSICAL PROPERTIES (Earth = ⊕)

Planet	Equatorial Radius (km)	Equatorial Radius (⊕ = 1)	Mass (⊕ = 1)*	Average Density (g/cm³)	Surface Gravity (⊕ = 1)	Escape Velocity (km/s)	Sidereal Period of Rotation	Inclination of Equator to Orbit
Mercury	2440	0.383	0.055	5.43	0.38	4.3	58.6 d	0.0°
Venus	6052	0.945	0.815	5.20	0.90	10.4	243.0 d	177.3°
Earth	6378	1.00	1.000	5.51	1.00	11.2	23.93 h	23.4°
Mars	3396	0.533	0.107	3.93	0.38	5.0	24.62 h	25.2°
Jupiter	71,492	11.2	318	1.33	2.53	59.5	9.92 h	3.1°
Saturn	60,268	9.45	95.2	0.69	1.06	35.5	10.57 h	26.7°
Uranus	25,559	4.01	14.5	1.27	0.89	21.3	17.23 h	97.8°
Neptune	24,764	3.88	17.1	1.64	1.14	23.5	16.11 h	28.3°

*Earth's mass = 5.97 x 10^{24} kg

TABLE A-11 Principal Satellites of the Solar System

Planet	Satellite	Radius (km)	Distance from Planet (10^3 km)	Orbital Period (days)	Orbital Eccentricity	Orbital Inclination**
Earth	Moon	1738	384.4	27.32	0.055	18.3°
Mars	Phobos	14 × 12 × 10	9.4	0.32	0.018	1.0°
	Deimos	8 × 6 × 5	23.5	1.26	0.002	2.8°
Jupiter	Amalthea	135 × 100 × 78	182	0.50	0.003	0.4°
	Io	1820	422	1.77	0.000	0.3°
	Europa	1560	671	3.55	0.000	0.5°
	Ganymede	2630	1071	7.16	0.002	0.2°
	Callisto	2410	1884	16.69	0.008	0.2°
	Himalia	~85*	11,470	250.6	0.158	27.6°
Saturn	Janus	110 × 80 × 100	151.5	0.70	0.007	0.1°
	Mimas	196	185.5	0.94	0.020	1.5°
	Enceladus	260	238.0	1.37	0.004	0.0°
	Tethys	530	294.7	1.89	0.000	1.1°
	Dione	560	377	2.74	0.002	0.0°
	Rhea	765	527	4.52	0.001	0.4°
	Titan	2575	1222	15.94	0.029	0.3°
	Hyperion	205 × 130 × 110	1484	21.28	0.104	~0.5°
	Iapetus	720	3562	79.33	0.028	14.7°
	Phoebe	105	12,930	550.4	0.163	150°
Uranus	Miranda	235	129.9	1.41	0.017	3.4°
	Ariel	580	190.9	2.52	0.003	0°
	Umbriel	585	266.0	4.14	0.003	0°
	Titania	805	436.3	8.71	0.002	0°
	Oberon	790	583.4	13.46	0.001	0°
Neptune	Proteus	205	117.6	1.12	~0	~0°
	Triton	1355	354.59	5.88	0.00	160°
	Nereid	170	5588.6	360.12	0.76	27.7°

**Relative to planet's equator

*The ~ symbol means "approximately."

TABLE A-12 Meteor Showers

Shower	Dates	Rate per Hour	Radiant position		Associated Comet
			Right Ascension	Declination	
Quadrantids	Jan. 2–4	30	15h 24m	+50°	2003 EH1*
Lyrids	April 20–22	8	18h 08m	+33°	Thatcher
η Aquarids	May 2–7	10	22h 32m	−01°	Halley
S. δ Aquarids	July 26–31	15	22h 40m	−10°	?
Perseids	Aug. 10–14	40	03h 12m	+58°	Swift-Tuttle
Orionids	Oct. 18–23	15	06h 20m	+16°	Halley
S. Taurids	Nov. 1–7	8	03h 40m	+15°	Encke
Leonids	Nov. 14–19	6	10h 16m	+22°	Tempel-Tuttle
Geminids	Dec. 10–13	50	07h 24m	+33°	Phaethon*

*Source object appears asteroid-like.

TABLE A-13 The Greek Alphabet

A, α	alpha	H, η	eta	N, ν	nu	T, τ	tau
B, β	beta	Θ, θ	theta	Ξ, ξ	xi	Y, υ	upsilon
Γ, γ	gamma	I, ι	iota	O, o	omicron	Φ, ϕ	phi
Δ, δ	delta	K, κ	kappa	Π, π	pi	X, χ	chi
E, ε	epsilon	Λ, λ	lambda	P, ρ	rho	Ψ, ψ	psi
Z, ζ	zeta	M, μ	mu	Σ, σ	sigma	Ω, ω	omega

Atomic masses are based on carbon-12. Numbers in parentheses are mass numbers of most stable or best-known isotopes of radioactive elements.

Key:
- Atomic number → 11
- Symbol → Na
- Atomic mass → 22.99

Group IA(1)	IIA(2)	IIIB(3)	IVB(4)	VB(5)	VIB(6)	VIIB(7)	VIII (8)	VIII (9)	VIII (10)	IB(11)	IIB(12)	IIIA(13)	IVA(14)	VA(15)	VIA(16)	VIIA(17)	Noble Gases (18)
1 H 1.008																	2 He 4.003
3 Li 6.941	4 Be 9.012											5 B 10.81	6 C 12.01	7 N 14.01	8 O 16.00	9 F 19.00	10 Ne 20.18
11 Na 22.99	12 Mg 24.31											13 Al 26.98	14 Si 28.09	15 P 30.97	16 S 32.06	17 Cl 35.45	18 Ar 39.95
19 K 39.10	20 Ca 40.08	21 Sc 44.96	22 Ti 47.90	23 V 50.94	24 Cr 52.00	25 Mn 54.94	26 Fe 55.85	27 Co 58.93	28 Ni 58.7	29 Cu 63.55	30 Zn 65.38	31 Ga 69.72	32 Ge 72.59	33 As 74.92	34 Se 78.96	35 Br 79.90	36 Kr 83.80
37 Rb 85.47	38 Sr 87.62	39 Y 88.91	40 Zr 91.22	41 Nb 92.91	42 Mo 95.94	43 Tc 98.91	44 Ru 101.1	45 Rh 102.9	46 Pd 106.4	47 Ag 107.9	48 Cd 112.4	49 In 114.8	50 Sn 118.7	51 Sb 121.8	52 Te 127.6	53 I 126.9	54 Xe 131.3
55 Cs 132.9	56 Ba 137.3	57* La 138.9	72 Hf 178.5	73 Ta 180.9	74 W 183.9	75 Re 186.2	76 Os 190.2	77 Ir 192.2	78 Pt 195.1	79 Au 197.0	80 Hg 200.6	81 Tl 204.4	82 Pb 207.2	83 Bi 209.0	84 Po (210)	85 At (210)	86 Rn (222)
87 Fr (223)	88 Ra 226.0	89** Ac (227)	104 Rf (261)	105 Db (262)	106 Sg (263)	107 Bh (262)	108 Hs (265)	109 Mt (266)	110 Ds (269)	111 Uuu (272)	112 Uub (277)	113 Uub (284)	114 Uuq (285)	115 Uub (288)	116 Uuh (289)		

Period 1–7 shown at left. Transition Elements span groups IIIB(3)–IIB(12). VIII spans groups (8), (9), (10).

Inner Transition Elements

Lanthanide Series 6 (*)

58 Ce 140.1	59 Pr 140.9	60 Nd 144.2	61 Pm (145)	62 Sm 150.4	63 Eu 152.0	64 Gd 157.3	65 Tb 158.9	66 Dy 162.5	67 Ho 164.9	68 Er 167.3	69 Tm 168.9	70 Yb 173.0	71 Lu 175.0

Actinide Series 7 (**)

90 Th 232.0	91 Pa 231.0	92 U 238.0	93 Np 237.0	94 Pu (244)	95 Am (243)	96 Cm (247)	97 Bk (247)	98 Cf (251)	99 Es (252)	100 Fm (257)	101 Md (258)	102 No (259)	103 Lr (260)

The Elements and Their Symbols

Element	Symbol	Element	Symbol	Element	Symbol	Element	Symbol	Element	Symbol	Element	Symbol
Actinium	Ac	Cesium	Cs	Hafnium	Hf	Mercury	Hg	Protactinium	Pa	Tellurium	Te
Aluminum	Al	Chlorine	Cl	Hassium	Hs	Molybdenum	Mo	Radium	Ra	Terbium	Tb
Americium	Am	Chromium	Cr	Helium	He	Neodymium	Nd	Radon	Rn	Thallium	Tl
Antimony	Sb	Cobalt	Co	Holmium	Ho	Neon	Ne	Rhenium	Re	Thorium	Th
Argon	Ar	Copper	Cu	Hydrogen	H	Neptunium	Np	Rhodium	Rh	Thulium	Tm
Arsenic	As	Curium	Cm	Indium	In	Nickel	Ni	Rubidium	Rb	Tin	Sn
Astatine	At	Darmstadtium	Ds	Iodine	I	Niobium	Nb	Ruthenium	Ru	Titanium	Ti
Barium	Ba	Dubnium	Db	Iridium	Ir	Nitrogen	N	Rutherfordium	Rf	Tungsten	W
Berkelium	Bk	Dysprosium	Dy	Iron	Fe	Nobelium	No	Samarium	Sm	Uranium	U
Beryllium	Be	Einsteinium	Es	Krypton	Kr	Osmium	Os	Scandium	Sc	Vanadium	V
Bismuth	Bi	Erbium	Er	Lanthanum	La	Oxygen	O	Seaborgium	Sg	Xenon	Xe
Bohrium	Bh	Europium	Eu	Lawrencium	Lr	Palladium	Pd	Selenium	Se	Ytterbium	Yb
Boron	B	Fermium	Fm	Lead	Pb	Phosphorous	P	Silicon	Si	Yttrium	Y
Bromine	Br	Fluorine	F	Lithium	Li	Platinum	Pt	Silver	Ag	Zinc	Zn
Cadmium	Cd	Francium	Fr	Lutetium	Lu	Plutonium	Pu	Sodium	Na	Zirconium	Zr
Calcium	Ca	Gadolinium	Gd	Magnesium	Mg	Polonium	Po	Strontium	Sr		
Californium	Cf	Gallium	Ga	Manganese	Mn	Potassium	K	Sulfur	S		
Carbon	C	Germanium	Ge	Meitnerium	Mt	Praseodymium	Pr	Tantalum	Ta		
Cerium	Ce	Gold	Au	Mendelevium	Md	Promethium	Pm	Technetium	Tc		

Appendix B

Observing the Sky

Observing the sky with the unaided eye is as important to modern astronomy as picking up pretty pebbles is to modern geology. The sky is a natural wonder unimaginably bigger than the Grand Canyon, the Rocky Mountains, or any other site that tourists visit every year. To neglect the beauty of the sky is equivalent to geologists neglecting the beauty of the minerals they study. This supplement is meant to act as a tourist's guide to the sky. You analyzed the universe in the textbook's chapters; here you can admire it.

The brighter stars in the sky are visible even from the centers of cities with their air and light pollution. But in the countryside, only a few miles beyond the cities, the night sky is a velvety blackness strewn with thousands of glittering stars. From a wilderness location, far from the city's glare, and especially from high mountains, the night sky is spectacular.

Using Star Charts

The constellations are a fascinating cultural heritage of our planet, but they are sometimes a bit difficult to learn because of Earth's motion. The constellations above the horizon change with the time of night and the seasons.

Because Earth rotates eastward, the sky appears to rotate westward around Earth. A constellation visible overhead soon after sunset will appear to move westward, and in a few hours it will disappear below the horizon. Other constellations will rise in the east, so the sky changes gradually through the night.

In addition, Earth's orbital motion makes the sun appear to move eastward among the stars. Each day the sun moves about twice its own diameter, about one degree, eastward along the ecliptic. Consequently, each night at sunset, the constellations are about one degree farther toward the west.

Orion, for instance, is visible in the evening sky in January, but, as the days pass, the sun moves closer to Orion. By March, Orion is difficult to see in the western sky soon after sunset. By June, the sun is so close to Orion that the constellation sets with the sun and is invisible. Not until late July is the sun far enough past Orion for the constellation to become visible rising in the eastern sky just before dawn.

Because of the rotation and orbital motion of Earth, you need more than one star chart to map the sky. Which chart you select depends on the month and the time of night. The charts on the following pages show the evening sky for each season, as viewed from the northern hemisphere at a latitude typical of the United States or central Europe.

To use the charts, select the appropriate chart and hold it overhead as shown in Figure B-1. If you face south, turn the chart until the words southern horizon are at the bottom of the chart. If you face other directions, turn the chart appropriately. Note that hours are in standard time; for daylight savings time (spring, summer, and the first half of fall in the United States) add one hour.

© 2016 Cengage Learning®

▲ Figure B-1 To use the star charts in this book, select the appropriate chart for the season and time. Hold it overhead and turn it until the direction at the bottom of the chart is the same as the direction you are facing.

Star Charts

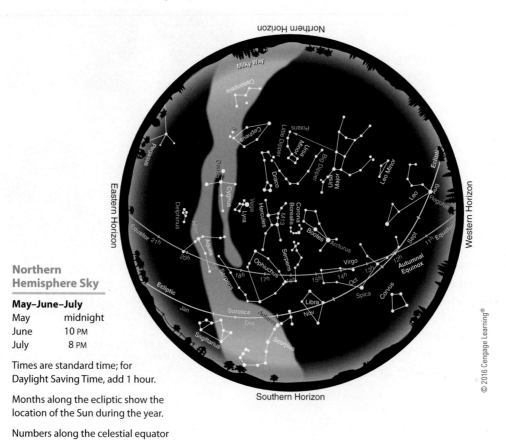

Northern Hemisphere Sky

February–March–April

February	midnight
March	10 PM
April	8 PM

Times are Standard Time; for Daylight Saving Time, add 1 hour.

Months along the ecliptic show the location of the Sun during the year.

Numbers along the celestial equator show Right Ascension (celestial longitude).

Northern Hemisphere Sky

May–June–July

May	midnight
June	10 PM
July	8 PM

Times are standard time; for Daylight Saving Time, add 1 hour.

Months along the ecliptic show the location of the Sun during the year.

Numbers along the celestial equator show Right Ascension (celestial longitude).

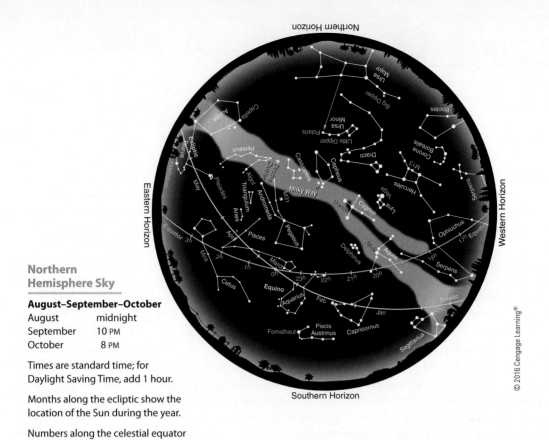

Northern Hemisphere Sky

August–September–October

August	midnight
September	10 PM
October	8 PM

Times are standard time; for Daylight Saving Time, add 1 hour.

Months along the ecliptic show the location of the Sun during the year.

Numbers along the celestial equator show Right Ascension (celestial longitude).

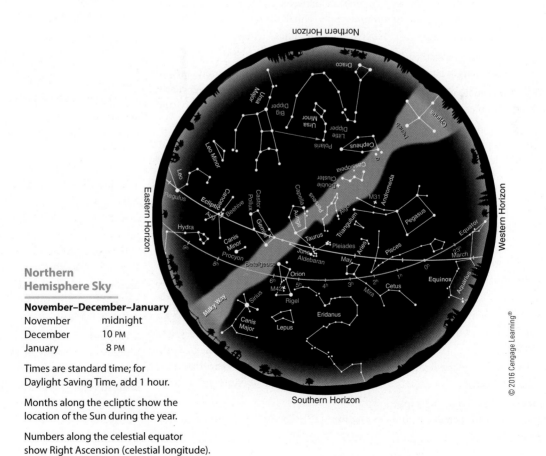

Northern Hemisphere Sky

November–December–January

November	midnight
December	10 PM
January	8 PM

Times are standard time; for Daylight Saving Time, add 1 hour.

Months along the ecliptic show the location of the Sun during the year.

Numbers along the celestial equator show Right Ascension (celestial longitude).

Glossary

Numbers in parentheses refer to the page where the term is first discussed in the text.

absolute age (248) An age determined in years, as from radioactive dating (see also *relative age*).

absolute zero (135) The lowest possible temperature; the temperature at which the particles in a material, atoms or molecules, contain no energy of motion that can be extracted from the body.

absorption (dark-line) spectrum (140) A spectrum that contains absorption lines.

absorption line (140) A dark line in a spectrum; produced by the absence of photons absorbed by atoms or molecules.

acceleration (83) A change in a velocity; a change in either speed or direction. (See *velocity*.)

acceleration of gravity (81) A measure of the strength of gravity at a planet's surface.

accretion (207) The sticking together of solid particles to produce a larger particle.

achondrite (356) Stony meteorite containing no chondrules or volatiles.

achromatic lens (108) A telescope lens composed of two lenses ground from different kinds of glass and designed to bring two selected colors to the same focus and correct for chromatic aberration.

active optics (115) Optical elements whose position or shape is continuously controlled by computers.

active region (159) An area on the Sun where sunspots, prominences, flares, and the like occur.

adaptive optics (124) Computer-controlled telescope mirrors that can at least partially compensate for seeing.

albedo (238) The fraction of the light hitting an object that is reflected.

alt-azimuth mount (115) A telescope mounting capable of motion parallel to and perpendicular to the horizon.

Amazonian period (288) On Mars, the geological era from about 3 billion years ago to the present marked by low-level cratering, wind erosion, and small amounts of water seeping from subsurface ice.

amino acid (386) One of the carbon-chain molecules that are the building blocks of protein.

angstrom (Å) (105) A unit of distance; 1 Å $= 5 \times 10^{-10}$ m; often used to measure the wavelength of light.

angular diameter (19) A measure of the size of an object in the sky; numerically equal to the angle in degrees between two lines extending from the observer's eye to opposite edges of the object.

angular distance (19) A measure of the separation between two objects in the sky; numerically equal to the angle in degrees between two lines extending from the observer's eye to the two objects.

angular momentum (90) The tendency of a rotating body to continue rotating; mathematically, the product of mass, velocity, and radius.

angular momentum problem (204) An objection to Laplace's nebular hypothesis that cited the slow rotation of the Sun.

annular eclipse (42) A solar eclipse in which the solar photosphere appears around the edge of the Moon in a bright ring, or annulus. The corona, chromosphere, and prominences cannot be seen.

anorthosite (251) Rock of aluminum and calcium silicates found in the lunar highlands.

anticoded (203) Describes a message designed to be understood by a recipient about whom the sender knows little or nothing, for example an interstellar broadcast aimed at possible inhabitants of another planet.

aphelion (25) The orbital point of greatest distance from the Sun.

apogee (42) The orbital point of greatest distance from Earth.

Apollo–Amor object (364) Asteroid whose orbit crosses that of Earth (Apollo) and Mars (Amor).

apparent visual magnitude, m_v (16) The brightness of a star as seen by human eyes on Earth.

arc minute (19) An angular measure; each degree is divided into 60 arc minutes.

arc second (19) An angular measure; each arc minute is divided into 60 arc seconds.

archaea (394) A biological kingdom of microorganisms similar to, but distinct from, bacteria, with characteristics most closely resembling those inferred for the common ancestor of all present-day Earth life.

archaeoastronomy (53) The study of the astronomy of ancient cultures.

array detector (121) A grid of photosensitive detectors for the purpose of recording images. Commercial still and video digital cameras contain CCD array detectors. See also charge-coupled device.

asterism (13) A named group of stars not identified as a constellation, for example, the Big Dipper.

asteroid (197) Small, rocky world; most asteroids lie between Mars and Jupiter in the asteroid belt.

astrobiology (384) The field of study involving searches for life on other worlds and investigation of possible habitats for such life. Also known as "exobiology."

astronomical unit (AU) (4) Average distance from Earth to the Sun; 1.5×10^8 km, or 93×10^6 miles.

astrophysics (130) The application of physics to the study of astronomical objects and the entire Universe.

atmospheric window (106) Wavelength region in which Earth's atmosphere is transparent—at visual, infrared, and radio wavelengths.

aurora (163) The glowing light display that results when a planet's magnetic field guides charged particles toward the north and south magnetic poles, where they strike the upper atmosphere and excite atoms to emit photons.

autumnal equinox (24) The point on the celestial sphere where the Sun crosses the celestial equator going southward. Also, the time when the Sun reaches this point and autumn begins in the Northern Hemisphere—about September 22.

Babcock model (157) A model of the Sun's magnetic cycle in which the differential rotation of the Sun winds up and tangles the solar magnetic field in a 22-year cycle. This is thought to be responsible for the 11-year sunspot cycle.

Balmer series (141) Spectral lines in the visible and near-ultraviolet spectrum of hydrogen produced by transitions whose lowest orbit is the second.

barred spiral galaxy (186) A spiral galaxy with an elongated nucleus resembling a bar from which the arms originate.

basalt (234) Dark, igneous rock characteristic of solidified lava.

belt–zone circulation (297) The atmospheric circulation typical of Jovian planets. Dark belts and bright zones encircle the planet parallel to its equator.

big bang (188) The theory that the Universe began with a violent explosion from which the expanding Universe of galaxies eventually formed.

binding energy (132) The energy needed to pull an electron away from its atom.

biological evolution (388) The combined effect of variation and natural selection resulting in new species arising and existing species adapting to the environment or becoming extinct. Also called "Darwinian evolution."

blackbody radiation (135) Radiation emitted by a hypothetical perfect radiator; the spectrum is continuous, and the wavelength of maximum emission depends only on the body's temperature.

blueshift (142) The shortening of the wavelengths of light observed when the source and observer are approaching each other.

Bok globule (178) Small, dark, and dense interstellar cloud only about 1 ly in diameter that contains 10 to 1000 solar masses of gas and dust; thought to be related to star formation.

bow shock (231) The boundary between the undisturbed solar wind and the region being deflected around a planet or comet.

breccia (252) A rock composed of fragments of earlier rocks bonded together.

bright-line spectrum (140) See *emission spectrum*.

butterfly diagram (158) See *Maunder butterfly diagram*.

CAI (357) Calcium–aluminum-rich inclusions found in some meteorites.

Cambrian explosion (392) The sudden appearance of complex life forms at the beginning of the Cambrian period 0.6 to 0.5 billion years ago. Cambrian rocks contain the oldest easily identifiable fossils.

carbonaceous chondrite (355) Stony meteorite that contains both chondrules and volatiles. These may be the least altered remains of the solar nebula still present in the Solar System.

Cassegrain focus (114) The optical design of a reflecting telescope in which the secondary mirror reflects light back down the tube through a hole in the center of the objective mirror.

catastrophic hypothesis (204) Explanation for natural processes that depends on dramatic and unlikely events, such as the collision of two stars to produce our Solar System.

celestial equator (18) The imaginary line around the sky directly above Earth's equator.

celestial sphere (18) An imaginary sphere of very large radius surrounding Earth to which the planets, stars, Sun, and Moon seem to be attached.

centaur (365) An outer Solar System body with an orbit entirely within the region of the Jovian planets, for example Chiron, that orbits between Saturn and Uranus.

center of mass (89) The balance point of a body or system of bodies.

charge-coupled device (CCD) (121) An electronic device consisting of a large array of light-sensitive elements used to record very faint images.

chemical evolution (391) The chemical process that led to the growth of complex molecules on the primitive Earth. This did not involve the reproduction of molecules.

Chicxulub (378) The buried crater associated with the mass extinction event at the end of the age of dinosaurs, named after the town in the coastal region of Mexico's Yucatán peninsula near the center of the crater.

chondrite (356) A stony meteorite that contains chondrules.

chondrule (357) Round, glassy body in some stony meteorites; thought to have solidified very quickly from molten drops of silicate material.

chromatic aberration (107) A distortion found in refracting telescopes because lenses focus different colors at slightly different distances. Images are consequently surrounded by color fringes.

chromosome (387) One of the bodies in a cell that contains the DNA carrying genetic information.

chromosphere (42, 148) Bright gases just above the photosphere of the Sun.

circular velocity (88) The velocity required to remain in a circular orbit about a body.

circumpolar constellation (19) Any of the constellations so close to the celestial pole that they never set (or never rise) as seen from a given latitude.

closed orbit (89) An orbit that returns to its starting point; a circular or elliptical orbit. (See *open orbit*.)

coma (368) The glowing head of a comet.

comet (200) One of the small, icy bodies that orbit the Sun and produce tails of gas and dust when they near the Sun.

comparative planetology (225) The study of planets by comparing the characteristics of different examples.

comparison spectrum (124) A spectrum of known spectral lines used to identify unknown wavelengths in an object's spectrum.

composite volcano (272) A volcano built up of layers of lava flows and ash falls. These are steep sided and typically associated with subduction zones.

condensation (207) The growth of a particle by addition of material from surrounding gas, one atom or molecule at a time.

condensation sequence (207) The sequence in which different materials condense from the solar nebula at increasing distances from the Sun.

constellation (12) One of the stellar patterns identified by name, usually of mythological gods, people, animals, or objects; also, the region of the sky containing that star pattern.

continuous spectrum (140) A spectrum in which there are no absorption or emission lines.

convection (149) Circulation in a fluid driven by heat; hot material rises, and cool material sinks.

convective zone (169) The region inside a star where energy is carried outward as rising hot gas and sinking cool gas.

Copernican Principle (400) The idea that Earth should not be assumed to be in a special location or to be otherwise unique.

corona (42, 148) The faint outer atmosphere of the Sun; composed of low-density, very hot, ionized gas. On Venus, round network of fractures and ridges up to 1000 km in diameter, caused by the intrusion of magma below the crust.

coronagraph (151) A telescope designed to photograph the inner corona of the Sun.

coronal hole (163) An area of the solar surface that is dark at X-ray wavelengths; thought to be associated with divergent magnetic fields and the source of the solar wind.

coronal mass ejection (CME) (163) Gas trapped in the Sun's magnetic field.

cosmic ray (126) A subatomic particle traveling at tremendous velocity that strikes Earth's atmosphere from space.

Coulomb barrier (168) The electrostatic force of repulsion between bodies of like charge; commonly applied to atomic nuclei.

Coulomb force (132) The repulsive force between particles with like electrostatic charge.

critical point (299) The temperature and pressure at which the vapor and liquid phases of a material have the same density.

C-type asteroid (363) A type of asteroid common in the outer asteroid belt, with very low reflectivity and grayish color, probably composed of carbonaceous material.

dark-line spectrum (140) See *absorption spectrum*.

debris disk (213) A disk of dust found by infrared observations around some stars. The dust is debris from collisions among asteroids, comets, and Kuiper Belt objects.

deferent (61) In the Ptolemaic theory, the large circle around Earth along which the center of the epicycle moved.

density (42) The amount of matter per unit volume in a material; measured in grams per cubic centimeter, for example.

deuterium (167) An isotope of hydrogen in which the nucleus contains a proton and a neutron.

diamond ring effect (44) A momentary phenomenon seen during some total solar eclipses when the ring of the corona and a bright spot of photosphere resemble a large diamond set in a silvery ring.

differential rotation (157) The rotation of a body in which different parts of the body have different periods of rotation; this is true of the Sun, the Jovian planets, and the disk of the galaxy.

differentiation (208) The separation of planetary material according to density.

diffraction fringe (109) Blurred fringe surrounding any image caused by the wave properties of light. Because of this, no image detail smaller than the fringe can be seen.

digitize (121) Convert information to numerical form for convenient transfer, storage, and analysis.

direct collapse (210) The hypothetical process by which a Jovian planet might skip the accretion of a solid core, instead forming quickly and directly from the gases of the solar nebula.

DNA (deoxyribonucleic acid) (386) The long carbon-chain molecule that records information to govern the biological activity of the organism. DNA carries the genetic data passed to offspring.

Doppler effect (139) The change in the wavelength of radiation relative radial motion of source and observer.

Drake equation (400) A formula for the number of communicative civilizations in our galaxy.

dust tail (368) The tail of a comet formed of dust blown outward by the pressure of sunlight. (See gas tail.)

dwarf planet (4) An object that orbits the Sun and has pulled itself into a spherical shape but has not cleared its orbital lane of other objects. Pluto is a dwarf planet.

dynamo effect (157) The process by which a rotating, convecting body of conducting matter, such as Earth's core, can generate a magnetic field.

east point (18) The point on the eastern horizon exactly halfway between the north point and the south point; exactly east.

eccentric (58) (noun) An off-center circular path. (Note that the adjective "eccentric" refers instead to an ellipse that is not a perfect circle.)

eccentricity (*e*) (67) A measure of the flattening of an ellipse. An ellipse of $e = 0$ is circular. The closer to 1 that e becomes, the more flattened the ellipse.

eclipse season (46) That period when the Sun is near a node of the Moon's orbit and eclipses are possible.

eclipse year (47) The time the Sun takes to circle the sky and return to a node of the Moon's orbit; 346.62 days.

ecliptic (22) The apparent path of the Sun around the sky.

ejecta (246) Pulverized rock scattered by meteorite impacts on a planetary surface.

electromagnetic radiation (104) Changing electric and magnetic fields that travel through space and transfer energy from one place to another—for example, light, radio waves, and the like.

electron (131) Low-mass atomic particle carrying a negative charge.

ellipse (68) A closed curve enclosing two points (foci) such that the total distance from one focus to any point on the curve back to the other focus equals a constant.

elliptical galaxy (186) A galaxy that is round or elliptical in outline; it contains little gas and dust, no disk or spiral arms, and few hot, bright stars.

emission (bright-line) spectrum (140) A spectrum containing emission lines.

emission line (140) A bright line in a spectrum caused by the emission of photons from atoms.

empirical (69) Description of a phenomenon without explaining why it occurs. Kepler's laws of planetary motion are empirical laws.

energy (91) The capacity of a natural system to perform work—for example, thermal energy.

energy level (134) One of a number of states an electron may occupy in an atom, depending on its binding energy.

enzyme (386) Special protein that controls processes in an organism.

epicycle (61) The small circle followed by a planet in the Ptolemaic theory. The center of the epicycle follows a larger circle (deferent) around Earth.

equant (61) The point off-center in the deferent from which the center of the epicycle appears to move uniformly.

equatorial mount (115) A telescope mounting that allows motion parallel to and perpendicular to the celestial equator.

escape velocity (V_e) (87) The initial velocity an object needs to escape from the surface of a celestial body.

evening star (26) Any planet visible in the sky just after sunset.

evolutionary hypothesis (204) Explanation for natural events that involves gradual changes as opposed to sudden catastrophic changes—for example, the formation of the planets in the gas cloud around the forming Sun.

excited atom (134) An atom in which an electron has moved from a lower to a higher orbit.

extrasolar planet (5, 215) A planet orbiting a star other than the Sun.

extremophile (394) An organism that can survive in an extreme environment, for example, very low or high temperatures, high acidity, extreme dryness, and so on.

eyepiece (107) A short-focal-length lens used to enlarge the image in a telescope; the lens nearest the eye.

fall (355) A meteorite seen to fall. (See *find.*)

false-color image (121) See *representational-color image.*

field (86) A way of explaining action at a distance; a particle produces a field of influence (gravitational, electric, or magnetic) to which another particle in the field responds.

field of view (2) The area visible in an image; usually given as the diameter of the region.

filament (162) (1) On the Sun, a prominence seen silhouetted against the solar surface. (2) A linear region containing many galaxies and galaxy clusters, part of the large-scale structure of the Universe.

filtergram (150) An image (usually of the Sun) taken in the light of a specific region of the spectrum—for example, an H-alpha filtergram.

find (355) A meteorite that is found but was not seen to fall. (See fall.)

first principle (57) A first principle is an idea considered so obviously true that the idea does not need to be questioned. Classical philosophers accepted as a first principle that Earth was the unmoving center of the Universe.

flare (163) A violent eruption on the Sun's surface.

flux (16, 178) A measure of the flow of energy onto or through a surface. Usually applied to light.

focal length (107) The distance from a lens to the point where it focuses parallel rays of light.

folded mountain range (234) A long range of mountains formed by the compression of a planet's crust—for example, the Andes on Earth.

forward scattering (311) The optical property of finely divided particles to preferentially direct light in the original direction of the light's travel.

frequency (ν) (104) The number of times a given event occurs in a given time; for a wave, the number of cycles that pass the observer in 1 second.

frost line (207) In the solar nebula, the boundary beyond which water vapor and other compounds could form ice particles.

galaxy (5) A very large collection of gas, dust, and stars orbiting a common center of mass. The Sun and Earth are located in the Milky Way Galaxy.

Galilean moons or satellites (72) The four largest satellites of Jupiter, named after their discoverer, Galileo.

gamma-ray (106) Electromagnetic wave with extremely short wavelength, high frequency, and large photon energy.

gas tail (368) The tail of a comet produced by gas blown outward by the solar wind. (See dust tail.)

gene (387) A unit of DNA containing genetic information that influences a particular inherited trait.

general theory of relativity (97) Einstein's more sophisticated theory of space and time, which describes gravity as a curvature of space-time.

geocentric Universe (56) A model Universe with Earth at the center, such as the Ptolemaic Universe.

geosynchronous satellite (88) An Earth satellite in an eastward orbit whose period is 24 hours. A satellite in such an orbit remains above the same spot on Earth's surface.

giant (180) Large, cool, highly luminous star in the upper right of the H–R diagram, typically 10 to 100 times the diameter of the Sun.

global warming (238) The gradual increase in the surface temperature of Earth caused by human modifications to Earth's atmosphere.

gossamer ring (312) The dimmest part of Jupiter's ring, produced by dust particles orbiting near small moons.

granulation (149) The fine structure visible on the solar surface caused by rising currents of hot gas and sinking currents of cool gas below the surface.

grating (122) A piece of material in which numerous microscopic parallel lines are scribed; light encountering a grating is dispersed to form a spectrum.

gravitational collapse (208) The stage in the formation of a massive planet when it grows massive enough to begin capturing gas directly from the nebula around it.

greenhouse effect (237) The process by which a carbon dioxide atmosphere traps heat and raises the temperature of a planetary surface.

grooved terrain (305) Region of the surface of Ganymede consisting of bright, parallel grooves.

ground state (134) The lowest permitted electron orbit in an atom.

habitable zone (396) The region around a star within which an orbiting planet can have surface temperatures allowing liquid water.

half-life (200) The time required for half of the atoms in a radioactive sample to decay.

heat (135) Energy flowing from a warm body to a cool body by the agitation of particles such as atoms or molecules.

heat of formation (208) In planetology, the heat released by the infall of matter during the formation of a planetary body.

heavy bombardment (212) The period of intense meteorite impacts early in the formation of the planets, when the Solar System was filled with debris.

heliocentric Universe (56) A model of the Universe with the Sun at the center, such as the Copernican Universe.

heliopause (153) The surface at which the solar wind collides with the interstellar medium. Effectively, the outer boundary of the Solar System.

helioseismology (154) The study of the interior of the Sun by the analysis of its modes of vibration.

Hesperian period (288) On Mars, the geological era from the decline of heavy cratering and lava flows and the melting of subsurface ice to form the outflow channels.

horizon (18) The line that marks the apparent intersection of Earth and the sky.

horoscope (26) A chart showing the positions of the Sun, Moon, planets, and constellations at the time of a person's birth; used in astrology to attempt to read character or foretell the future.

hot Jupiter (217) A massive and presumably Jovian planet that orbits close to its star and consequently has a high temperature.

hypothesis (69) A conjecture, subject to further tests, that accounts for a set of facts.

inertia (82) The property of matter that resists changes in motion.

infrared (IR) radiation (105) Electromagnetic radiation with wavelengths intermediate between visible light and radio waves.

intercrater plain (259) The relatively smooth terrain on Mercury.

interferometry (121) The observing technique in which separated telescopes combine to produce a virtual telescope with the resolution of a much-larger-diameter telescope.

International Astronomical Union (IAU) (13) An international society of astronomers that, among other activities, decides definitions and naming conventions for celestial objects and surface features. The IAU defined the constellation boundaries in 1930 and reclassified Pluto as a dwarf planet in 2006.

interstellar medium (ISM) (205) The gas and dust distributed between the stars.

inverse square law (84) The rule that the strength of an effect (such as gravity) decreases in proportion as the distance squared increases.

Io flux tube (300) A tube of magnetic lines and electric currents connecting Io and Jupiter.

Io plasma torus (300) The doughnut-shaped cloud of ionized gas that encloses the orbit of Jupiter's moon Io.

ion (132) An atom that has lost or gained one or more electrons.

ionization (132) The process in which atoms lose or gain electrons.

iron meteorite (355) A meteorite composed mainly of iron–nickel alloy.

irregular galaxy (187) A galaxy with a chaotic appearance, large clouds of gas and dust, and both population I and population II stars, but without spiral arms.

irregular satellite (297) A moon with an orbit that has large eccentricity or high inclination to the equator of its parent planet or is retrograde. Irregular moons are thought to have been captured.

isotopes (132) Atoms that have the same number of protons but a different number of neutrons.

joule (J) (91) A unit of energy equivalent to a force of 1 newton acting over a distance of 1 meter; 1 joule per second equals 1 watt of power.

Jovian planet (198) Jupiter-like planet with large diameter and low density.

Jovian problem (210) The puzzle that protoplanetary disks around young stars don't seem to survive long enough to form Jovian planets by condensation, accretion, and gravitational collapse, yet Jovian-mass extrasolar planets are common. See also extrasolar planet.

jumbled terrain (254) Disturbed regions of the Moon's surface opposite the locations of the Imbrium Basin and Mare Orientale, possibly focusing of seismic waves from the large impacts that formed those basins.

Kelvin temperature scale (135) The temperature, in Celsius (centigrade) degrees, measured above absolute zero.

kinetic energy (91) Energy of motion. Depends on mass and velocity of a moving body.

Kirchhoff's laws (140) A set of laws that describe the absorption and emission of light by matter.

Kirkwood gaps (361) Regions in the asteroid belt in which there are very few asteroids; caused by orbital resonances with Jupiter.

Kuiper Belt (197) The collection of icy planetesimals that orbit in a region from just beyond Neptune out to about 50.

Kuiper Belt object (KBO) (197) An object in the Kuiper Belt, a region beyond Neptune's orbit containing planetesimals remaining from the formation of the Solar System. Pluto is one of the largest Kuiper Belt objects.

large-impact hypothesis (255) The hypothesis that the Moon formed from debris ejected during a collision between Earth and a large planetesimal.

laser guide star (124) An artificial star image produced by a laser pointing up into Earth's atmosphere, used as a reference in adaptive optics systems. See also *adaptive optics*.

late heavy bombardment (252) The surge in cratering impacts in the Solar System that occurred about 3.8 billion years ago.

light pollution (112) The illumination of the night sky by waste light from cities and outdoor lighting, which prevents the observation of faint objects.

light-gathering power (109) The ability of a telescope to collect light; proportional to the area of the telescope's objective lens or mirror.

light-year (ly) (4) The distance light travels in one year.

limb (150) The edge of the apparent disk of a body, as in "the limb of the Moon."

limb darkening (150) The decrease in brightness of the Sun or other body from its center to its limb.

line of nodes (47) The line across an orbit connecting the nodes; commonly applied to the orbit of the Moon.

liquid metallic hydrogen (299) A form of hydrogen under high pressure that is a good electrical conductor.

lobate scarp (255) A curved cliff such as those found on Mercury.

lunar eclipse (34) The darkening of the Moon when it moves through Earth's shadow.

lunar phase (34) The appearance of the Moon from Earth in terms of which portion is lit by the Sun versus which portion is dark, which changes in a regular monthly cycle.

Lyman series (141) Spectral lines in the ultraviolet spectrum of hydrogen produced by transitions whose lowest orbit is the ground state.

magnetic carpet (152) The widely distributed, low-level magnetic field extending up through the Sun's visible surface.

magnetosphere (232) The volume of space around a planet within which the motion of charged particles is dominated by the planetary magnetic field rather than the solar wind.

magnifying power (112) The ability of a telescope to make an image larger.

magnitude scale (15) The astronomical brightness scale; the larger the number, the fainter the star.

main sequence (177) The region of the H–R diagram running from upper left to lower right, which includes roughly 90 percent of all stars.

mantle (225) The layer of dense rock and metal oxides that lies between the molten core and Earth's surface; also, similar layers in other planets.

mare (244) (plural: maria) One of the lunar lowlands filled by successive flows of dark lava; from the Latin word for "sea."

mass (84) A measure of the amount of matter making up an object.

Maunder butterfly diagram (158) A graph showing the latitude of sunspots versus time; first plotted by W. W. Maunder in 1904.

Maunder minimum (159) A period of less numerous sunspots and other solar activity from 1645 to 1715.

meridional flow (161) Flows of material from the Sun's equator toward the poles. These flows carry portions of the solar magnetic field and are understood to be one mechanism that helps drive the 11- and 22-year solar activity cycles.

meteor (358) A small bit of matter heated by friction to incandescent vapor as it falls into Earth's atmosphere.

meteor shower (358) An event lasting for hours or days in which the number of meteors entering Earth's atmosphere suddenly increases. The meteors in a shower have a common origin and are traveling through space on nearly parallel paths.

meteorite (200) A meteor that has survived its passage through the atmosphere and strikes the ground.

meteoroid (200) A meteor in space before it enters Earth's atmosphere.

microlensing (217) Brightening of a background star due to focusing of its light by the gravity of a foreground extrasolar planet, allowing the planet to be detected and some of its characteristics measured.

micrometeorite (247) Meteorite of microscopic size.

microwave (105) Electromagnetic wave with wavelength, frequency, and photon energy intermediate between infrared and radio waves.

midocean rise (234) One of the undersea mountain ranges that push up from the seafloor in the center of the oceans.

Milankovitch hypothesis (27) The hypothesis that small changes in Earth's orbital and rotational motions cause the ice ages.

Milky Way (6) The hazy band of light that circles the sky, produced by the combined light of billions of stars in our Milky Way Galaxy.

Milky Way Galaxy (6) The spiral galaxy containing the Sun; visible at night as the Milky Way.

Miller experiment (389) An experiment that reproduced the conditions under which life began on Earth and amino acids and other organic compounds were manufactured.

molecule (132) Two or more atoms bonded together.

momentum (83) The tendency of a moving object to continue moving; mathematically, the product of mass and velocity.

morning star (26) Any planet visible in the sky just before sunrise.

M-type asteroid (363) A type of asteroid with relatively high reflectivity and grayish color, probably composed primarily of metal.

multicellular (391) Describes an organism composed of many cells.

multiringed basin (247) Very large impact basin in which there are concentric rings of mountains.

mutation (388) Offspring born with altered DNA.

nadir (18) The point on the bottom of the sky directly under your feet.

nanometer (nm) (105) A unit of length equal to 10^{-9} m.

natural law (69) A conjecture about how nature works in which scientists have overwhelming confidence.

natural motion (81) In Aristotelian physics, the motion of objects toward their natural places—fire and air upward and earth and water downward.

natural selection (388) The process by which the best traits are passed on, allowing the most able to survive.

neap tide (93) Ocean tide of low amplitude occurring at first- and third-quarter Moon.

Near-Earth Object (NEO) (364) An asteroid or comet in an orbit that passes near or intersects Earth's orbit that could potentially collide with Earth.

nebular hypothesis (204) The proposal that the Solar System formed from a rotating cloud of gas.

neutrino (167) A neutral, massless atomic particle that travels at or nearly at the speed of light.

neutron (131) An atomic particle with no charge and about the same mass as a proton.

neutron star (181) A small, highly dense star composed almost entirely of tightly packed neutrons; radius about 10 km.

Newtonian focus (114) The focal arrangement of a reflecting telescope in which a diagonal mirror reflects light out the side of the telescope tube for easier access.

Noachian period (287) On Mars, the era that extends from the formation of the crust to the end of heavy cratering and includes the formation of the valley networks.

node (46) A point where an object's orbit passes through the plane of Earth's orbit.

north celestial pole (18) The point on the celestial sphere directly above Earth's North Pole.

north point (18) The point on the horizon directly below the north celestial pole; exactly north.

nuclear fission (166) Reaction that splits nuclei into less massive fragments.

nuclear fusion (166) Reaction that joins the nuclei of atoms to form more massive nuclei.

nucleus (of an atom) (131) The central core of an atom containing protons and neutrons; carries a net positive charge.

oblateness (299) The flattening of a spherical body, usually caused by rotation.

occultation (338) The passage of a larger body in front of a smaller body.

Oort Cloud (373) The cloud of icy bodies—extending from the outer part of our Solar System out to roughly 100,000 from the Sun—that acts as the source of most comets.

open orbit (89) An orbit that does not return to its starting point; an escape orbit. (See closed orbit.)

optical telescope (108) A telescope that gathers and focuses visible light. See also *radio telescope.*

outflow channel (284) Geological feature on Mars that appears to have been caused by sudden flooding.

outgassing (209) The release of gases from a planet's interior.

ovoid (337) Geological feature on Uranus's moon Miranda thought to be produced by circulation in the solid icy mantle and crust.

ozone layer (236) In Earth's atmosphere, a layer of oxygen ions (O_3) lying 15 to 30 km high that protects the surface by absorbing ultraviolet radiation.

paradigm (64) A commonly accepted set of scientific ideas and assumptions.

parallax (60) The apparent change in the position of an object a change in the location of the observer. Astronomical parallax is measured in seconds of arc.

parsec (pc) (170) The distance to a hypothetical star whose parallax is 1 second of arc; 1 pc = 206,265 AU = 3.26 ly.

partial lunar eclipse (39) A lunar eclipse in which the Moon does not completely enter Earth's shadow.

partial solar eclipse (41) A solar eclipse in which the Moon does not completely cover the Sun.

Paschen series (141) Spectral lines in the infrared spectrum of hydrogen produced by transitions whose lowest orbit is the third.

passing star hypothesis (204) The proposal that our Solar System formed when two stars passed near each other and material was pulled out of one to form the planets.

path of totality (42) The track of the Moon's umbral shadow over Earth's surface. The Sun is totally eclipsed as seen from within this path.

penumbra (38) The portion of a shadow that is only partially shaded.

penumbral lunar eclipse (39) A lunar eclipse in which the Moon enters the penumbra of Earth's shadow but does not reach the umbra.

perigee (42) The orbital point of closest approach to Earth.

perihelion (25) The orbital point of closest approach to the Sun.

permitted orbit (132) One of the energy levels in an atom that an electron may occupy.

photographic plate (121) An old-fashioned means of recording astronomical images and photometric information on a photographic emulsion coating a glass plate. See also *array detector* and *charge-coupled device.*

photometer (121) An instrument attached to a telescope for the purpose of precisely measuring the brightness of stars or other objects at one or more wavelengths.

photon (105) A quantum of electromagnetic energy; carries an amount of energy that depends inversely on its wavelength.

photosphere (42, 148) The bright visible surface of the Sun.

planet (3) A nonluminous object, larger than a comet or asteroid, that orbits a star.

planetary nebula (180) An expanding shell of gas ejected from a star during the latter stages of its evolution.

planetesimal (207) One of the small bodies that formed from the solar nebula and eventually grew into protoplanets.

plastic (230) A material with the properties of a solid but capable of flowing under pressure.

plate tectonics (234) The constant destruction and renewal of Earth's surface by the motion of sections of crust.

plutino (249) One of the icy Kuiper Belt objects that, like Pluto, are caught in a 3:2 orbital resonance with Neptune.

polar axis (115) In an equatorial telescope mounting, the axis that is parallel to Earth's axis of rotation.

polarity (160) Orientation and strength of a magnetic field's manifestation as north and south poles. Also applies to an electrical field's manifestation as positive and negative charges.

positron (167) The antiparticle of the electron.

potential energy (91) The energy a body has by virtue of its position. A weight on a high shelf has more potential energy than a weight on a low shelf.

precession (20) The slow change in the direction of Earth's axis of rotation; one cycle takes nearly 26,000 years.

pressure (P) wave (229) In geophysics, a mechanical wave of compression and rarefaction that travels through Earth's interior.

primary lens (107) The main lens in a refracting telescope.

primary mirror (107) The main optical element in an astronomical telescope. The large lens at the top of the telescope tube or the large mirror at the bottom.

prime focus (114) The point at which the objective mirror forms an image in a reflecting telescope.

primeval atmosphere (233) Earth's first air, composed of gases from the solar nebula.

primordial soup (390) The rich solution of organic molecules in Earth's first oceans.

prograde (297) Rotation or revolution in the direction in common with most such motions in the Solar System. See also retrograde.

prominence (44, 162) Eruption on the solar surface; visible during total solar eclipses.

protein (386) Complex molecule composed of amino acid units.

proton (131) A positively charged atomic particle contained in the nucleus of an atom; the nucleus of a hydrogen atom.

proton–proton chain (167) A series of three nuclear reactions that build a helium atom by adding together protons; the main energy source in the Sun.

protoplanet (208) Massive object resulting from the coalescence of planetesimals in the solar nebula and destined to become a planet.

protostar (176) A collapsing cloud of gas and dust destined to become a star.

pseudoscience (26) A subject that claims to obey the rules of scientific reasoning but does not. Examples include astrology, crystal power, and pyramid power.

quantum mechanics (132) The study of the behavior of atoms and atomic particles.

radial velocity (V_r) (142) That component of an object's velocity directed away from or toward Earth.

radiant (358) The point in the sky from which meteors in a shower seem to come.

radiation pressure (200) The force exerted on the surface of a body by its absorption of light. Small particles floating in the Solar System can be blown outward by the pressure of the sunlight.

radiative zone (169) The region inside a star where energy is carried outward as photons.

radio telescope (108) A telescope that gathers and focuses electromagnetic energy with microwave and radio wavelengths. See also *optical telescope*.

radio wave (105) Electromagnetic wave with extremely long wavelength, low frequency, and small photon energy.

ray (246) Ejecta from a meteorite impact, forming white streamers radiating from some lunar craters.

reconnection event (163) The process in the Sun's atmosphere by which opposing magnetic fields combine and release energy to power solar flares.

red dwarf (177) Cool, low-mass star on the lower main sequence.

redshift (142) The lengthening of the wavelengths of light seen when the source and observer are receding from each other.

reflecting telescope (107) A telescope that uses a concave mirror to focus light into an image.

refracting telescope (107) A telescope that forms images by bending (refracting) light with a lens.

regolith (252) A soil made up of crushed rock fragments.

regular satellite (297) A moon with an orbit that has small eccentricity, low inclination to the equator of its parent planet, and is prograde. Regular moons are thought to have formed with their respective planets rather than having been captured.

relative age (248) The age of a geological feature as referred to other features. For example, relative ages reveal that the lunar maria are younger than the highlands.

representational-color image (121) A representation of graphical data in which the colors are altered or added to reveal details. Also sometimes called a *false-color image*.

resolving power (109) The ability of a telescope to reveal fine detail; depends on the diameter of the telescope objective.

resonance (258) The coincidental agreement between two periodic phenomena; commonly applied to agreements between orbital periods, which can make orbits more stable or less stable.

retrograde motion (60) The apparent backward (westward) motion of planets as seen against the background of stars.

revolution (21) The motion of an object in a closed path about a point outside its volume; Earth revolves around the Sun.

rift valley (234) A long, straight, deep valley produced by the separation of crustal plates.

RNA (ribonucleic acid) (387) A long carbon-chain molecule that uses the information stored in DNA to manufacture complex molecules necessary to the organism.

Roche limit (311) The minimum distance between a planet and a satellite that holds itself together by its own gravity. If a satellite's orbit brings it within its planet's Roche limit, tidal forces will pull the satellite apart.

rotation (21) The turning of a body about an axis that passes through its volume; Earth rotates on its axis.

runaway greenhouse effect (269) A "vicious cycle" (positive feedback) in which a planet's greenhouse effect causes enhanced release or retention of greenhouse gases that increases the greenhouse effect, and so on. There is evidence that Venus suffered a runaway greenhouse effect early in its history with the result that its oceans boiled away and were destroyed by solar .

Saros cycle (47) An 18-year 11 ⅓-day period after which the pattern of lunar and solar eclipses repeats.

Schmidt-Cassegrain focus (114) The optical design of a reflecting telescope in which a thin correcting lens is placed at the top of a Cassegrain telescope.

scientific argument (29) An honest, logical discussion of observations and theories intended to reach a valid conclusion.

scientific method (7) The reasoning style by which scientists test theories against evidence to understand how nature works.

scientific model (17) An intellectual concept designed to help you think about a natural process without necessarily being a conjecture of truth.

scientific notation (3) The system of recording very large or very small numbers by using powers of 10.

secondary atmosphere (236) The gases outgassed from a planet's interior; rich in carbon dioxide.

secondary crater (246) A crater formed by the impact of debris ejected from a larger crater.

secondary mirror (114) In a reflecting telescope, the mirror that reflects the light to a point of easy observation.

seeing (111) Atmospheric conditions on a given night. When the atmosphere is unsteady, producing blurred images, the seeing is said to be poor.

seismic wave (229) A mechanical vibration that travels through Earth; usually caused by an earthquake.

seismograph (229) An instrument that records seismic waves.

selection effect (355) An influence on the probability that certain phenomena will be detected or selected, which can alter the outcome of a survey.

semimajor axis *(a)* (68) Half of the longest axis of an ellipse.

SETI (398) Search for Extra-Terrestrial Intelligence.

shepherd satellite (321) A satellite that, by its gravitational field, confines particles to a planetary ring.

shield volcano (272) Wide, low-profile volcanic cone produced by highly liquid lava.

sidereal drive (115) The motor and gears on a telescope that turn it westward to keep it pointed at a star.

sidereal period (37) The period of rotation or revolution of an astronomical body relative to the stars.

sinuous rille (244) A narrow, winding valley on the Moon caused by ancient lava flows along narrow channels.

small-angle formula (41) The mathematical formula that relates an object's linear diameter and distance to its angular diameter.

smooth plain (260) Apparently young plain on Mercury formed by lava flows at or soon after the formation of the Caloris Basin.

solar constant (165) A measure of the energy output of the Sun; the total solar energy striking 1 m² just above Earth's atmosphere in 1 second.

solar eclipse (41) The event that occurs when the Moon passes directly between Earth and the Sun, blocking your view of the Sun.

solar nebula theory (205) The proposal that the planets formed from the same cloud of gas and dust that formed the Sun.

Solar System (3) The Sun and the nonluminous objects that orbit it, including the planets, comets, and asteroids.

solar wind (153) Rapidly moving atoms and ions that escape from the solar corona and blow outward through the Solar System.

south celestial pole (18) The point of the celestial sphere directly above Earth's South Pole.

south point (18) The point on the horizon directly above the south celestial pole; exactly south.

special theory of relativity (96) The first of Einstein's theories of relativity, which deals with uniform motion.

spectral line (124) A dark or bright line that crosses a spectrum at a specific wavelength.

spectrograph (122) A device that separates light by wavelength to produce a spectrum.

spectrum (105) An arrangement of electromagnetic radiation in order of wavelength or frequency.

spicule (150) Small, flamelike projection in the chromosphere of the Sun.

spiral arm (6) Long, spiral pattern of bright stars, star clusters, gas, and dust that extends from the center to the edge of the disk of spiral galaxies.

sporadic meteor (358) A meteor not part of a meteor shower.

spring tide (93) Ocean tide of high amplitude that occurs at full and new Moon.

star (4) A celestial object composed of gas held together by its own gravity and supported by nuclear fusion occurring in its interior.

Stefan-Boltzmann law (136) The mathematical relation between the temperature of a blackbody (an ideal radiator) and the amount of energy emitted per second from 1 square meter of its surface.

stony meteorite (355) A meteorite composed of silicate (rocky) material.

stony-iron meteorite (355) A meteorite that is a mixture of stone and iron.

stromatolite (389) A layered fossil formation caused by ancient mats of algae or bacteria that build up mineral deposits season after season.

strong nuclear force (166) One of the four forces of nature; the strong force binds protons and neutrons together in atomic nuclei.

S-type asteroid (363) A type of asteroid common in the inner asteroid belt, with relatively high reflectivity and reddish color, probably composed of rocky material.

subduction zone (234) A region of a planetary crust where a tectonic plate slides downward.

subsolar point (268) The point on a planet that is directly below the Sun.

summer solstice (24) The point on the celestial sphere where the Sun is at its most northerly point; also, the time when the Sun passes this point, about June 22, and summer begins in the Northern Hemisphere.

sunspot (156) Relatively dark spot on the Sun that contains intense magnetic fields.

supergiant (181) Exceptionally luminous star, 10 to 1000 times the Sun's diameter.

supergranule (149) A large granule on the Sun's surface including many smaller granules.

supernova (181) A "new" star appearing in Earth's sky and lasting for a year or so before fading. Caused by the violent explosion of a star.

synodic period (37) The period of rotation or revolution of a celestial body with respect to the Sun.

temperature (135) A measure of the velocity of random motions among the atoms or molecules in a material.

terminator (244) The dividing line between daylight and darkness on a planet or moon.

Terrestrial planet (198) Earth-like planet—small, dense, rocky.

theory (69) A system of assumptions and principles applicable to a wide range of phenomena that have been repeatedly verified.

thermal energy (135) The energy stored in an object as agitation among its atoms and molecules.

thermophile (394) A type of microorganism that thrives in high-temperature environments.

tidal coupling (244) The locking of the rotation of a body to its revolution around another body.

tidal heating (244) The heating of a planet or satellite because of friction caused by tides.

total lunar eclipse (38) A lunar eclipse in which the Moon completely enters Earth's dark shadow.

total solar eclipse (41) A solar eclipse in which the Moon completely covers the bright surface of the Sun.

totality (38) The period during a solar eclipse when the Sun's photosphere is completely hidden by the Moon, or the period during a lunar eclipse when the Moon is completely inside the umbra of Earth's shadow.

transit (216) The passage of an extrasolar planet across the disk of its parent star as observed from Earth, partially blocking the light from the star and allowing detection and study of the planet.

transition (141) The movement of an electron from one atomic orbit to another.

Trojan asteroid (365) Small, rocky body caught in Jupiter's orbit at the Lagrange points, 60° ahead of and behind the planet.

ultraviolet (UV) radiation (106) Electromagnetic radiation with wavelengths shorter than visible light but longer than X-rays.

umbra (35) The region of a shadow that is totally shaded.

uncompressed density (206) The density a planet would have if its gravity did not compress it.

uniform circular motion (56) The classical belief that the perfect heavens could move only by the combination of constant motion along circular orbits.

valley networks (284) Dry drainage channels resembling streambeds found on Mars.

Van Allen belt (232) One of the radiation belts of high-energy particles trapped in Earth's magnetosphere.

velocity (83) A rate of travel that specifies both speed and direction.

vernal equinox (24) The place on the celestial sphere where the Sun crosses the celestial equator moving northward; also, the time of year when the Sun crosses this point, about March 21, and spring begins in the Northern Hemisphere.

vesicular (251) A porous basalt rock formed by solidified lava with trapped bubbles.

violent motion (81) In Aristotelian physics, motion other than natural motion. (See natural motion.)

water hole (398) The interval of the radio spectrum between the 21-cm hydrogen radiation and the 18-cm OH radiation, likely wavelengths to use in the search for extraterrestrial life.

wavelength (104) The distance between successive peaks or troughs of a wave; usually represented by λ.

wavelength of maximum intensity (λ_{max}) (137) The wavelength at which a perfect radiator emits the maximum amount of energy; depends only on the object's temperature.

weak nuclear force (166) One of the four forces of nature; the weak nuclear force is responsible for some forms of radioactive decay.

west point (18) The point on the western horizon exactly halfway between the north point and the south point; exactly west.

white dwarf (180) The remains of a dying star that has collapsed to the size of Earth and is slowly cooling off; at the lower left of the H–R diagram.

Widmanstätten pattern (356) Bands in iron meteorites large crystals of nickel–iron alloys.

Wien's law (137) The mathematical relation between the temperature of an ideal radiator (a blackbody) and the wavelength at which it radiates most intensely.

winter solstice (24) The point on the celestial sphere where the Sun is farthest south; also, the time of year when the Sun passes this point, about December 22, and winter begins in the Northern Hemisphere.

X-ray (106) Electromagnetic radiation with short wavelengths, high frequencies, and high photon energies, between gamma-rays and ultraviolet radiation on the electromagnetic spectrum.

Zeeman effect (159) The splitting of spectral lines into multiple components when the atoms are in a magnetic field.

zenith (18) The point directly overhead on the sky.

zodiac (26) The band around the sky centered on the ecliptic within which the planets move.

Answers to Even-Numbered Problems

Chapter 1
2. 2160 miles; **4.** 1×10^{11} stars; **6.** 1.1×10^8 km; **8.** about 1.2 seconds; **10.** 640 ly; approx. beginning of the Renaissance in Europe, beginning of the Ming Dynasty in China; Betelgeuse is nearby compared with most stars in the Universe, but farther than most stars visible with an unaided eye; **12.** about 2.4×10^{22} m

Chapter 2
2. about 2.2 mags; **4.** about 0.75 mag; Star A is brighter; **6.** +3.8; both Star A and Star B can be seen with unaided eyes; **8.** about 170; **10.** a factor of about 1000 (7.5 mags); **12.** 40 degrees; 40 degrees; **14.** about 1.6; about 3.8; the largest climate variations in Figure 2-11 have a period of about 100,000 years, the time scale for variations in Earth's orbit shape

Chapter 3
2. new; full; first quarter; third quarter; **4.** about ½; ½; ½; **6.** about 12 degrees per day; **8.** about 2 to 3 times larger; **10.** 207 degrees; note, in this situation Earth would actually be inside the Sun and the small-angle approximation would not be accurate; **12.** 4/10, or 40 percent; **14.** (a) The Moon won't be full until about October 18; (b) The Moon will no longer be near that node of its orbit; **16.** August 12, 2026 [July 10, 1972 + 3 × (6585 ⅓ days)]. Note that you must take into account the number of leap days in the interval to get the right answer; **18.** S S S S S S N N S S S S N N S S S S S S; an educated guess would be N N

Chapter 4
2. retrograde motion: Mercury, Venus, Earth, Jupiter, Saturn, Uranus, and Neptune; never seen as crescents: Jupiter, Saturn, Uranus, and Neptune; **4.** 2.8 yr; **6.** 19.2 AU; **8.** more; faster; 2.8 times; 30 AU, Saturn and Neptune; **10.** new; full; Venus orbits the Sun, which is the light source

Chapter 5
2. 100 times weaker; 400 times weaker; **4.** increase, by a factor of 4; **6.** 19.6 m/s; 39.2 m/s; **8.** about 3100 m/s (3.1 km/s); about 87,000 s (slightly more than 24 hrs); **10.** about 7900 m/s (7.9 km/s); **12.** The cannonball would move in an elliptical orbit with Earth's center at one focus of the ellipse; **14.** 6320 s (about 1 hr 45 min); **16.** 42 m/s; the fastest pitchers could

Chapter 6
2. 3 m; **4.** less; red and blue, respectively; blue; 1.75 times as much; **6.** Each Keck telescope has an LGP 1.6 million times greater than a human eye; **8.** Telescope A (diameter = 60 in. = 152 cm), 0.076 arc seconds; Telescope B, 2.8 arc seconds; Telescope A has better resolving power; it can resolve smaller angular size; **10.** 0.46 arc seconds; the two stars would be resolved as separate objects; **12.** The Palomar telescope has about 2.1 times better (smaller) resolving power than *HST*, considering only diffraction; *HST* is in space and is not subject to atmospheric seeing; **14.** LGP requires 98 cm diameter; resolving power requires 116 cm diameter, so the larger size is necessary; magnifying power requires the primary focal length be 250 times larger than the eyepiece focal length; no, atmospheric seeing prevents testing 0.1 arc second resolving power from the ground

Chapter 7
2. 5.8×10^4 nm, infrared, object in the outer Solar System; 5.8×10^3 nm, infrared, object in the inner Solar System; 580 nm, visible, medium-temperature star; 58 nm, ultraviolet, hot star; Wien's law: wavelength of blackbody maximum intensity is inversely proportional to temperature; **4.** 1450 K; **6.** about 1.9 times more; **8.** 250 nm; **10.** about 120 km/s; receding; redshift; **12.** about 0.65 nm

Chapter 8
2. core about ¼ to ⅓ of the radius; radiative zone about ⅔; convective zone about ⅓; radiative zone is most of the volume; **4.** about 6×10^{-3} arc seconds; no, that is much smaller than *HST*'s resolution; **6.** 6.4×10^7 watts; increases by a factor of 16; **8.** 656.299 nm; blueshift; rising; **10.** average rotation period is about 28.9 days; **12.** about 6×10^{14} joules; **14.** 2.5×10^9 megatons; **16.** 4.3×10^9 kg

Chapter 10
2. 4.4×10^{10} times fainter, or 26.6 mags; about +22.6; **4.** Saturn, 0.299 M_{Jup}; Uranus, 0.0456 M_{Jup}; 0.0538 M_{Jup}; Earth 0.00314 M_{Jup}; there seem to be three categories of planets: (i) Jupiter and Saturn, (ii) Uranus and Neptune, and (iii) Earth, rather than only 2 categories; **6.** 87.5 percent daughter isotope, 12.5 percent (= ⅛) parent isotope; 13.5 billion years; the sample is material that is older than Earth, almost as old as the Universe; **8.** Mars; it has compressed the least because of its small mass and low original density; **10.** Silicates, metallic iron and nickel, and metal oxides; **12.** about 130 (3.6×10^{-4} impacts per second per m^2)

Chapter 11
2. P waves: about 11 km/s; S waves: about 6 km/s; P waves, about 11 s, S waves, about 16 s; **4.** about 17 percent; **6.** 2.1×10^8 yr (210 million years); **8.** 4.3×10^7 yr (43 million years)

Chapter 12
2. 2.37 km/s; **4.** Heat escapes through a planet's surface; the rate depends on the surface area. Heat is held in a planet's interior; the amount depends on the volume. The ratio of surface area (rate of heat escape) to volume (amount of heat) is inversely proportional to radius. Thus, the larger the planet, the longer it takes for the internal heat to escape; 3–5 days; **6.** about 2.7 km in diameter; about 3–5 km; **8.** about 1.63 km/s; 7080 s (about 1 hr 58 min); **10.** 10.9 arc seconds at $d = 0.613$ AU (average closest approach); average angular diameter of the Moon is 1870 arc seconds, 172 times larger; **12.** for T_{max} of about 700 K, $\lambda_{max} = 4100$ nm (4.1 microns); infrared; **14.** about 730 km; yes, Caloris basin has a diameter of 1550 km

Chapter 13
2. 138 s (2 min 18 s) when $d = 0.277$ AU; 862 s (14 m 22 s) when $d = 1.723$ AU; **4.** 33,400 km (39,500 km from the center of Venus); **6.** 1.9×10^6 yr (1.9 million years); that would be the time since Haleakala was most active; the value of 1 million years given in the figure is the time since the last major eruption; **8.** about 1.7; that means molecular hydrogen can easily escape from Venus; **10.** about 0.5; that means Mars can easily retain CO_2; **12.** 2.3×10^{15} kg

Chapter 14
2. 17.3 km/s; **4.** 2800 m/s; **6.** Io, Europa, and Ganymede are in mutual resonances; **8.** 7.6×10^5 s (about 211 hours); this is much longer than Saturn's 10.57-hour rotation period; **10.** about 0.056 nm

Chapter 15
2. about 42 arc seconds; **4.** about 62,300 km; all of Uranus's rings and the moons Ophelia and Cordelia are within the Roche limit; **6.** about 5.3 m/s; **8.** about 14 arc seconds

Chapter 16
2. between about 1.4 and 10 seconds, for speeds between 70 and 10 km/s; **4.** about 10^4 (10,000) km in diameter; this is almost as large as Earth; **6.** about 370 m/s (830 mph); no; **8.** 0.72 arc seconds; no; yes; **10.** about 2.1 g/cm^3; mostly rock with some ice; **12.** about 18 AU; about the year 2024; 2062; **14.** 1 million years at $r = 10,000$ AU, about 30 million years at $r = 100,000$ AU; 300 m/s at $r = 10,000$ AU, about 100 m/s at $r = 100,000$ AU; **16.** about 650 deaths per year from asteroid impacts; worldwide air crash fatalities averaged about 400 per year during the 10-year span 2004–2013

Chapter 17
2. 8.8 cm; 0.088 cm (0.88 mm); **4.** 10,000 generations; **6.** 380 km (240 mi) at average closest approach, $d = 0.52$ AU; **8.** 44,000 years; **10.** no definitely correct answer; perhaps somewhere between 2×10^{-4} and 1×10^8, the pessimistic and optimistic values in the table

Index

Boldface page numbers indicate definitions of key terms.

temperature, 136
in terminology, 6
theory, 69
Comparative planetology, **225**, 227, 280
Comparison spectrums, **124**
Composite volcanoes, 271, **272**–273
Compton Gamma Ray Observatory (CGRO), 121
Condensation, 206–**207**
Condensation hypothesis, 255
Condensation sequence, **207**
Confidence, scientific, 170
Confirmation, of hypotheses, 161
Consolidation, of hypotheses, 161
Constants, astronomical, A–5
Constellations. *See also specific constellations*
circumpolar, 17, 19
defined, **12**
grouping and naming of, 12–13
Continental drift, 235
Continuous spectrum, 138, **140**
Convection currents, **149**, 157, 160
Convective zones, **169**
Copernican Principle, 400
Copernican Revolution, 59, 62–65
Copernicus, Nicolaus, 59, 62–65, 75, 400
Cordelia, 338
Cores
Earth, 229–230, 233
Jupiter, 299
Mercury, 261
Moon, 252
Neptune, 344
Venus, 267
Corona, **42**, 148, 151–153, 155
Coronae, **275**, 288
Coronagraphs, **151**
Coronal holes, 161, **163**
Coronal mass ejections (CMEs), 161, **163**
Cosmic debris, 197, 200
Cosmic microwave background radiation,
190
Cosmic rays, **126**
Coulomb, Charles-Augustin de, 132
Coulomb barriers, **168**
Coulomb force, **132**, 168
Craters/cratering
age of, 248
Earth, 226–227, 229
heavy bombardment and, 212, 225, 252
Mars, 282
Mercury, 198, 258–261
Moon, 245–248, 252
Venus, 271, 276
Cretaceous period, 378
Crick, Francis, 20
Critical point, **299**

Crust (Earth), 229–230, 233
C-type asteroids, 360, **363**, 366
Curiosity, 126, 143
Curiosity rover, 281, 287, 396
Curvature of space-time, 97, 98–99
CXO (Chandra X-ray Observatory), 121

D

Dactyl, 363
Dark-line spectrum, 138, **140**
Data, manipulation of, 270
Daughter isotopes, 200
Davis, Raymond, Jr., 169
Dawn spacecraft, 363, 367
Death, of stars, 177, 180–184
Debris disks, **213**–215, 220
Deep Impact spacecraft, 369, 372
Deferents, 58, **61**
Deimos, 289, 292
De Laplace, Pierre-Simon, 204
Density
Mercury, 260
uncompressed, 206
Venus, 267
Deoxyribonucleic acid (DNA)
defined, **386**
information storage and duplication functions of,
384–388
modification of information, 388
scientific models of, 20
De Revolutionibus Orbium Coelestium (Copernicus), 62–64
Descartes, René, 203, 204
Deuterium, **167**, 268
Devils Hole (Nevada), 27, 29
Dialogue Concerning the Two Chief World Systems (Galileo Galilei), 73
Diameter
angular, 17, 19, 41
linear, 41
of Moon, 19, 41
of Sun, 19, 41
Diamond ring effect, **44**
Differential rotation, **157**
Differentiation, **208**, 226
Diffraction fringes, **109**
Digitized images, **121**
Dione, 318
Direct collapse, **210**
Disks, planet-forming, 213–215
Disk-shaped galaxies, 185, 187
Distance, angular, 17, 19
DNA. *See* Deoxyribonucleic acid
Doppler effect, **139**, 142–143
Drake, Frank, 400
Drake equation, **400**–401

H

I

J

Lunar landing module, 248–249
Lunar phase cycle, **34**, 36–37
Lutetia, 362
Lyman series, 138, **140**

M

Mab, 337
Magellanic Clouds, 187
Magellan spacecraft, 269–270
Magma, 234, 252, 272
Magnetic carpet, **152**
Magnetic fields
    Earth, 157, 230–233
    Ganymede, 305–306
    Jupiter, 300
    Mercury, 260–261
    Saturn, 314
    Sun, 152, 153, 157–161
    Uranus, 333
Magnetic resonance imaging (MRI), 270
Magnetosphere, **232**
Magnifying power, **112**
Magnitude scale, **15**, 16–17
Main-sequence stars, 177, A–6
Makemake, 349
Mantle (Earth), **225**, 229–230
Mare, **244**
Mare Imbrium, 249, 252, 253–254
Mare Orientale, 247, 252
Maria, **244**–245, 248, 253
Mariner 4 spacecraft, 279
Mariner 10 spacecraft, 257, 259, 260–261
Mars, 278–292
    age of, 202
    atmosphere, 225, 279–281
    canals of, 278–279
    comparison with Earth, 278, 283, 288
    craters, 282
    distance from Sun, 4
    history of, 287–288
    interior of, 279
    moons of, 289, 292
    orbital motion of, 23, 60, 62, 67
    polar caps, 279, 281
    profile, 277
    retrograde motion of, 60, 62
    surface of, 281–284
    volcanoes/volcanism, 273, 274, 282–284, 287, 291
    water on, 284–287, 291
Mars Express spacecraft, 284, 286
Mars Global Surveyor, 292
Mars Odyssey spacecraft, 281, 284, 286
Mars Reconnaissance Orbiter spacecraft, 284, 285

Mass
    center of, 86, 89, 91
    defined, **84**
Massive stars, iron core in, 203
Mathematical models, 180
Mathematike Syntaxis (Ptolemy), 61
Mathilde, 362
Matter
    history of, 190–193
    light, interactions with, 134–137
    misconceptions regarding, 132
Maunder butterfly diagrams, 156, **158**, 161
Maunder minimum, 157, **159**, 165
Maxwell, James Clerk, 319
Mayall Telescope, 114
Mayan civilization, 49, 55
Mechanical determinism, 94
Mendel, Gregor, 8
Menzel 3 Nebula, 183
Mercury, 256–262
    atmosphere, 198, 225, 258
    comparison with Moon, 257, 259–261
    core, 260–261
    craters, 198, 258–261
    density of, 260
    distance from Sun, 4
    greatest elongations of, A–9
    history of, 261–262
    lava flows, 259, 260
    magnetic field of, 260–261
    plains of, 259–261
    profile, 256
    rotation and revolution of, 257–258
    size of, 3–4
    surface of, 258–259
    temperature of, 258
MErcury Surface, Space ENvironment, GEochemistry, and Ranging (MESSENGER) spacecraft, 199, 257, 260–261
Meridional flow, **161**
Meteorite ALH 84001, 395
Meteorites
    composition of, 354–358
    defined, **200**, 354
    impacts affecting Earth, 374–375, 378–379
    micrometeorites, 245, 247, 252
    Moon and, 245, 246–247, 252
    orbits of, 358
    origins of, 358–360
    radioactive dating of, 202
Meteors, **200**, 354–358
Meteor showers, **358**, A–8
Meteoroids, **200**, 354, 358–360
Methane, on Titan, 315–316
Metis, 312
Metric system, 2, A–3, A–4
Microlensing, **217**

Transition, of electrons, 138, **140**
Trapezium cluster, 178, 179
Travel between stars, 396–397
Triton, 344–346
Trojan asteroids, **365**
Tunguska event, 375, 376
Twinkling, of stars, 111
Two New Sciences (Galileo Galilei), 82
Tycho Brahe, 65–66
Tycho's supernova, 65
Type Ia supernovae, 203
Type II supernovae, 203

U

UFOs (unidentified flying objects), 397
Ultraviolet (UV), **106**
Umbra, **35**, 38
Umbriel, 335, 336
Uncompressed densities, **206**
Unidentified flying objects (UFOs), 397
Uniform circular motion, **56**, 60, 62–63
Units, astronomical, A–5
Universal mutual gravitation, 85, 86
Universe
    big bang and, 7, 188–190
    formation of, 7
    geocentric, 56–58, 60, 64
    heliocentric, 56, 59–60, 63–64
    intelligent life in, 396–401
    origins of, 188–190
Upsilon Andromedae, 216
Uranus, 329–341
    atmosphere, 331–332
    discovery of, 329–330
    history of, 337, 340
    interior of, 333–335, 340
    magnetic field of, 333
    moons of, 335–337
    orbital motion of, 330–331
    profile, 341
    rings of, 337, 338
    rotation of, 211, 330–331
    temperature, 334
Urey, Harold, 389
Ursa Major, 13

V

Vaccine, 230
Valles Marineris, 283, 291
Valley networks, **284**, 287
Van Allen belts, **232**
Vega, 15–16, 214–215

Velocity
    calculating, 143
    circular, 86, 88, 90
    defined, **83**
    escape, 86, 87, 89
    measuring, 139, 142
    orbital, 86–87
    radial, 142
Venera spacecraft, 269, 271
Venus, 267–277
    atmosphere, 225, 267–269
    comparison with Earth, 267, 268, 275, 276, 288, 290
    core of, 267
    coronae on, 275, 288
    craters, 271, 276
    density, 267
    distance from Sun, 4
    greatest elongations of, A–9
    greenhouse effect on, 269, 276, 290
    history of, 276
    lava flows, 271, 274–275, 290
    orbital motion of, 23, 26
    profile, 277
    radar maps of, 269–271
    rotation of, 211, 267
    size of, 3
    surface of, 269–271
    temperature of, 258
    volcanoes/volcanism, 271–275
    water on, 268–269, 276
Vernal equinox, 22–24
Very Large Array (VLA) telescope, 125
Very Large Telescope Interferometer (VLTI), 126
Very Large Telescope (VLT), 116, 126
Very Long Baseline Array (VLBA), 125
Vesicular, **251**
Vesta, 363, 366–367
Viking spacecraft, 281, 284, 395
Violent motions, **81**, 82
Viruses, 230
Visual magnitude, 16
VLA (Very Large Array) telescope, 125
VLBA (Very Long Baseline Array), 125
VLTI (Very Large Telescope Interferometer), 126
VLT (Very Large Telescope), 116, 126
Volcanoes/volcanism. *See also* Lava flows
    composite, 271, 272–273
    Earth, 233–236, 271–272
    Io, 308, 309
    Mars, 273, 274, 282–284, 287, 291
    Moon, 244–245, 251
    shield, 271, 272–273, 282–283
    Venus, 271–275
Von Fraunhofer, Joseph, 131